# STUDENT'S SOLUTIONS MANUAL

## DANIEL S. MILLER
*Niagara County Community College*

# COLLEGE ALGEBRA
## SIXTH EDITION

# Robert Blitzer
*Miami Dade College*

Boston  Columbus  Indianapolis  New York  San Francisco  Upper Saddle River
Amsterdam  Cape Town  Dubai  London  Madrid  Milan  Munich  Paris  Montreal  Toronto
Delhi  Mexico City  São Paulo  Sydney  Hong Kong  Seoul  Singapore  Taipei  Tokyo

The author and publisher of this book have used their best efforts in preparing this book. These efforts include the development, research, and testing of the theories and programs to determine their effectiveness. The author and publisher make no warranty of any kind, expressed or implied, with regard to these programs or the documentation contained in this book. The author and publisher shall not be liable in any event for incidental or consequential damages in connection with, or arising out of, the furnishing, performance, or use of these programs.

Reproduced by Pearson from electronic files supplied by the author.

ISBN-13: 978-0-321-85010-2
ISBN-10: 0-321-85010-6

3 4 5 6 7 8 VOZN 16 15 14

www.pearsonhighered.com

# TABLE OF CONTENTS for STUDENT SOLUTIONS

## *COLLEGE ALGEBRA 6E*

# Chapter P
# Fundamental Concepts of Algebra

## Section P.1

### Check Point Exercises

1.  $8 + 6(x-3)^2 = 8 + 6(13-3)^2$

    $\qquad = 8 + 6(10)^2$

    $\qquad = 8 + 6(100)$

    $\qquad = 8 + 600$

    $\qquad = 608$

2.  Since 2015 is 15 years after 2000, substitute 15 for $x$.

    $T = 4x^2 + 341x + 3194$

    $\qquad = 4(15)^2 + 341(15) + 3194$

    $\qquad = 9209$

    If trends continue, the tuition and fees will be $9209.

3.  The elements common to $\{3, 4, 5, 6, 7\}$ and $\{3, 7, 8, 9\}$ are 3 and 7.

    $\{3,4,5,6,7\} \cap \{3,7,8,9\} = \{3,7\}$

4.  The union is the set containing all the elements of either set.

    $\{3,4,5,6,7\} \cup \{3,7,8,9\} = \{3,4,5,6,7,8,9\}$

5.  $\left\{-9,\ -1.3,\ 0,\ 0.\overline{3},\ \dfrac{\pi}{2},\ \sqrt{9},\ \sqrt{10}\right\}$

    a.  Natural numbers: $\sqrt{9}$ because $\sqrt{9} = 3$

    b.  Whole numbers: $0,\ \sqrt{9}$

    c.  Integers: $-9, 0, \sqrt{9}$

    d.  Rational numbers: $-9,\ -1.3,\ 0,\ 0.\overline{3},\ \sqrt{9}$

    e.  Irrational numbers: $\dfrac{\pi}{2},\ \sqrt{10}$

    f.  Real numbers: $-9,\ -1.3,\ 0,\ 0.\overline{3},\ \dfrac{\pi}{2},\ \sqrt{9},\ \sqrt{10}$

6.  a.  $|1 - \sqrt{2}|$

    Because $\sqrt{2} \approx 1.4$, the number inside the absolute value bars is negative. The absolute value of $x$ when $x < 0$ is $-x$. Thus,

    $|1 - \sqrt{2}| = -(1 - \sqrt{2}) = \sqrt{2} - 1$

    b.  $|\pi - 3|$

    Because $\pi \approx 3.14$, the number inside the absolute value bars is positive. The absolute value of a positive number is the number itself. Thus,

    $|\pi - 3| = \pi - 3$.

    c.  $\dfrac{|x|}{x}$

    Because $x > 0$, $|x| = x$.

    Thus, $\dfrac{|x|}{x} = \dfrac{x}{x} = 1$

7.  $|-4 - (5)| = |-9| = 9$

    The distance between $-4$ and 5 is 9.

8.  $7(4x^2 + 3x) + 2(5x^2 + x)$

    $\qquad = 7(4x^2 + 3x) + 2(5x^2 + x)$

    $\qquad = 28x^2 + 21x + 10x^2 + 2x$

    $\qquad = 38x^2 + 23x$

9.  $6 + 4[7 - (x-2)]$

    $\qquad = 6 + 4[7 - x + 2)]$

    $\qquad = 6 + 4[9 - x]$

    $\qquad = 6 + 36 - 4x$

    $\qquad = 42 - 4x$

### Concept and Vocabulary Check P.1

1.  expression

2.  $b$ to the $n$th power;  base;  exponent

3.  formula;  modeling;  models

4.  intersection;  $A \cap B$

5.  union;  $A \cup B$

6.  natural

7.  whole

8.  integers

9.  rational

10. irrational

**11.** rational; irrational

**12.** absolute value; $x,\ -x$

**13.** $b+a$ ; $ba$

**14.** $a+(b+c)$ ; $(ab)c$

**15.** $ab+ac$

**16.** 0; inverse; 0; identity

**17.** inverse; 1; identity

**18.** simplified

**19.** $a$

**Exercise Set P.1**

**1.** $7+5(10)=7+50=57$

**3.** $6(3)-8=18-8=10$

**5.** $8^2+3(8)=64+24=88$

**7.** $7^2-6(7)+3=49-42+3=7+3=10$

**9.** $4+5(9-7)^3=4+5(2)^3$
$\qquad\qquad\ =4+5(8)=4+40=44$

**11.** $8^2-3(8-2)=64-3(6)$
$\qquad\qquad\quad\ =64-18=46$

**13.** $\dfrac{5(x+2)}{2x-14}=\dfrac{5(10+2)}{2(10)-14}$
$\qquad=\dfrac{5(12)}{6}$
$\qquad=5\cdot 2$
$\qquad=10$

**15.** $\dfrac{2x+3y}{x+1}; x=-2, y=4$
$\quad=\dfrac{2(-2)+3(4)}{-2+1}=\dfrac{-4+12}{-1}=\dfrac{8}{-1}=-8$

**17.** $C=\dfrac{5}{9}(50-32)=\dfrac{5}{9}(18)=10$
$50°\text{F}$ is equivalent to $10°\text{C}$.

**19.** $h=4+60t-16t^2=4+60(2)-16(2)^2$
$\qquad=4+120-16(4)=4+120-64$
$\qquad=124-64=60$
Two seconds after it is kicked, the ball's height is 60 feet.

**21.** $\{1,2,3,4\}\cap\{2,4,5\}=\{2,4\}$

**23.** $\{s,e,t\}\cap\{t,e,s\}=\{s,e,t\}$

**25.** $\{1,3,5,7\}\cap\{2,4,6,8,10\}=\{\ \ \}$
The empty set is also denoted by $\varnothing$.

**27.** $\{a,b,c,d\}\cap\varnothing=\varnothing$

**29.** $\{1,2,3,4\}\cup\{2,4,5\}=\{1,2,3,4,5\}$

**31.** $\{1,3,5,7\}\cup\{2,4,6,8,10\}$
$\quad=\{1,2,3,4,5,6,7,8,10\}$

**33.** $\{a,e,i,o,u\}\cup\varnothing=\{a,e,i,o,u\}$

**35. a.** $\sqrt{100}$

  **b.** $0,\sqrt{100}$

  **c.** $-9,0,\sqrt{100}$

  **d.** $-9,-\dfrac{4}{5},0,0.25,9.2,\sqrt{100}$

  **e.** $\sqrt{3}$

  **f.** $-9,-\dfrac{4}{5},0,0.25,\sqrt{3},9.2,\sqrt{100}$

**37. a.** $\sqrt{64}$

  **b.** $0,\sqrt{64}$

  **c.** $-11,0,\sqrt{64}$

  **d.** $-11,-\dfrac{5}{6},0,0.75,\sqrt{64}$

  **e.** $\sqrt{5},\pi$

  **f.** $-11,-\dfrac{5}{6},0,0.75,\sqrt{5},\pi,\sqrt{64}$

**39.** 0

**41.** Answers will vary. An example is 2.

**43.** true; $-13$ is to the left of $-2$ on the number line.

**45.** true; 4 is to the right of $-7$ on the number line.

**47.** true; $-\pi = -\pi$

**49.** true; 0 is to the right of $-6$ on the number line.

**51.** $|300| = 300$

**53.** $|12 - \pi| = 12 - \pi$

**55.** $\left|\sqrt{2} - 5\right| = 5 - \sqrt{2}$

**57.** $\dfrac{-3}{|-3|} = \dfrac{-3}{3} = -1$

**59.** $\left||-3| - |-7|\right| = |3 - 7| = |-4| = 4$

**61.** $|x + y| = |2 + (-5)| = |-3| = 3$

**63.** $|x| + |y| = |2| + |-5| = 2 + 5 = 7$

**65.** $\dfrac{y}{|y|} = \dfrac{-5}{|-5|} = \dfrac{-5}{5} = -1$

**67.** The distance is $|2 - 17| = |-15| = 15$.

**69.** The distance is $|-2 - 5| = |-7| = 7$.

**71.** The distance is $|-19 - (-4)| = |-19 + 4| = |-15| = 15$.

**73.** The distance is
$|-3.6 - (-1.4)| = |-3.6 + 1.4| = |-2.2| = 2.2$.

**75.** $6 + (-4) = (-4) + 6$;
commutative property of addition

**77.** $6 + (2 + 7) = (6 + 2) + 7$;
associative property of addition

**79.** $(2 + 3) + (4 + 5) = (4 + 5) + (2 + 3)$;
commutative property of addition

**81.** $2\,(-8 + 6) = -16 + 12$;
distributive property of multiplication over addition

**83.** $\dfrac{1}{x + 3}(x + 3) = 1; x \neq -3$,
inverse property of multiplication

**85.** $5(3x + 4) - 4 = 5 \cdot 3x + 5 \cdot 4 - 4$
$\phantom{5(3x + 4) - 4} = 15x + 20 - 4$
$\phantom{5(3x + 4) - 4} = 15x + 16$

**87.** $5(3x - 2) + 12x = 5 \cdot 3x - 5 \cdot 2 + 12x$
$\phantom{5(3x - 2) + 12x} = 15x - 10 + 12x$
$\phantom{5(3x - 2) + 12x} = 27x - 10$

**89.** $7(3y - 5) + 2(4y + 3)$
$= 7 \cdot 3y - 7 \cdot 5 + 2 \cdot 4y + 2 \cdot 3$
$= 21y - 35 + 8y + 6$
$= 29y - 29$

**91.** $5(3y - 2) - (7y + 2) = 15y - 10 - 7y - 2$
$\phantom{5(3y - 2) - (7y + 2)} = 8y - 12$

**93.** $7 - 4\left[3 - (4y - 5)\right] = 7 - 4\left[3 - 4y + 5\right]$
$\phantom{7 - 4[3 - (4y - 5)]} = 7 - 4\left[8 - 4y\right]$
$\phantom{7 - 4[3 - (4y - 5)]} = 7 - 32 + 16y$
$\phantom{7 - 4[3 - (4y - 5)]} = 16y - 25$

**95.** $18x^2 + 4 - \left[6\left(x^2 - 2\right) + 5\right]$
$= 18x^2 + 4 - \left[6x^2 - 12 + 5\right]$
$= 18x^2 + 4 - \left[6x^2 - 7\right]$
$= 18x^2 + 4 - 6x^2 + 7$
$= 18x^2 - 6x^2 + 4 + 7$
$= (18 - 6)x^2 + 11 = 12x^2 + 11$

**97.** $-(-14x) = 14x$

**99.** $-(2x - 3y - 6) = -2x + 3y + 6$

**101.** $\dfrac{1}{3}(3x) + \left[(4y) + (-4y)\right] = x + 0 = x$

**103.** $|-6| \,\square\, |-3|$
$\phantom{xx} 6 \,\square\, 3$
$\phantom{xx} 6 > 3$
Since $6 > 3$, $|-6| > |-3|$.

**105.**

$$\left|\frac{3}{5}\right| \square \left|-0.6\right|$$

$$\left|0.6\right| \square \left|-0.6\right|$$

$$0.6 \square 0.6$$

$$0.6 = 0.6$$

Since $0.6 = 0.6$, $\left|\dfrac{3}{5}\right| = \left|-0.6\right|$.

**107.**

$$\frac{30}{40} - \frac{3}{4} \square \frac{14}{15} \cdot \frac{15}{14}$$

$$\frac{30}{40} - \frac{30}{40} \square \frac{\cancel{14}}{\cancel{15}} \cdot \frac{\cancel{15}}{\cancel{14}}$$

$$0 \square 1$$

$$0 < 1$$

Since $0 < 1$, $\dfrac{30}{40} - \dfrac{3}{4} < \dfrac{14}{15} \cdot \dfrac{15}{14}$.

**109.**

$$\frac{8}{13} \div \frac{8}{13} \square \left|-1\right|$$

$$\frac{8}{13} \cdot \frac{13}{8} \square 1$$

$$1 \square 1$$

$$1 = 1$$

Since $1 = 1$, $\dfrac{8}{13} \div \dfrac{8}{13} = \left|-1\right|$.

**111.**
$$8^2 - 16 \div 2^2 \cdot 4 - 3 = 64 - 16 \div 4 \cdot 4 - 3$$
$$= 64 - 4 \cdot 4 - 3$$
$$= 64 - 16 - 3$$
$$= 48 - 3$$
$$= 45$$

**113.**
$$\frac{5 \cdot 2 - 3^2}{[3^2 - (-2)]^2} = \frac{5 \cdot 2 - 9}{[9 - (-2)]^2}$$
$$= \frac{10 - 9}{[9 + 2]^2}$$
$$= \frac{10 - 9}{11^2}$$
$$= \frac{1}{121}$$

**115.** $8 - 3[-2(2 - 5) - 4(8 - 6)] = 8 - 3[-2(-3) - 4(2)]$
$$= 8 - 3[6 - 8]$$
$$= 8 - 3[-2]$$
$$= 8 + 6$$
$$= 14$$

**117.** $\dfrac{2(-2) - 4(-3)}{5 - 8} = \dfrac{-4 + 12}{-3}$
$$= \frac{8}{-3}$$
$$= -\frac{8}{3}$$

**119.** $\dfrac{(5 - 6)^2 - 2|3 - 7|}{89 - 3 \cdot 5^2} = \dfrac{(-1)^2 - 2|-4|}{89 - 3 \cdot 25}$
$$= \frac{1 - 2(4)}{89 - 75}$$
$$= \frac{1 - 8}{14}$$
$$= \frac{-7}{14}$$
$$= -\frac{1}{2}$$

**121.** $x - (x + 4) = x - x - 4 = -4$

**123.** $6(-5x) = -30x$

**125.** $5x - 2x = 3x$

**127.** $8x - (3x + 6) = 8x - 3x - 6 = 5x - 6$

**129. a.** $H = \dfrac{7}{10}(220 - a)$

$$H = \frac{7}{10}(220 - 20)$$
$$= \frac{7}{10}(200)$$
$$= 140$$

The lower limit of the heart rate for a 20-year-old with this exercise goal is 140 beats per minute.

**b.** $H = \dfrac{4}{5}(220 - a)$

$$H = \frac{4}{5}(220 - 20)$$
$$= \frac{4}{5}(200)$$
$$= 160$$

The upper limit of the heart rate for a 20-year-old with this exercise goal is 160 beats per minute.

**131. a.**  $T = 26x^2 + 819x + 15{,}527$

$= 26(10)^2 + 819(10) + 15{,}527$

$= 26{,}317$

The formula estimates the cost to have been $26,317 in 2010.

**b.** This overestimates the value in the graph by $44.

**c.**  $T = 26x^2 + 819x + 15{,}527$

$= 26(13)^2 + 819(13) + 15{,}527$

$= 30{,}568$

The formula projects the cost to be $30,568 in 2013.

**133. a.**  $0.05x + 0.12(10{,}000 - x)$

$= 0.05x + 1200 - 0.12x$

$= 1200 - 0.07x$

**b.**  $1200 - 0.07x = 1200 - 0.07(6000)$

$= \$780$

**135. – 143.** Answers will vary.

**145.** does not make sense; Explanations will vary. Sample explanation: To use the model, substitute 0 for $x$.

**147.** does not make sense; Explanations will vary. Sample explanation: The commutative property changes order and the associative property changes groupings.

**149.** false; Changes to make the statement true will vary. A sample change is: All whole numbers are integers.

**151.** false; Changes to make the statement true will vary. A sample change is: Some irrational numbers are negative.

**153.** false; Changes to make the statement true will vary. A sample change is: $5 + 3(x - 4) = 5 + 3x - 12 = 3x - 7$.

**155.** true

**157.**  $-\pi > -3.5$

**159. a.**  $b^4 \cdot b^3 = (b \cdot b \cdot b \cdot b)(b \cdot b \cdot b) = b^7$

**b.**  $b^5 \cdot b^5 = (b \cdot b \cdot b \cdot b \cdot b)(b \cdot b \cdot b \cdot b \cdot b) = b^{10}$

**c.** add the exponents

**160. a.**  $\dfrac{b^7}{b^3} = \dfrac{b \cdot b \cdot b \cdot b \cdot b \cdot b \cdot b}{b \cdot b \cdot b} = b^4$

**b.**  $\dfrac{b^8}{b^2} = \dfrac{b \cdot b \cdot b \cdot b \cdot b \cdot b \cdot b \cdot b}{b \cdot b} = b^6$

**c.** subtract the exponents

**161.**  $6.2 \times 10^3 = 6.2 \times 10 \times 10 \times 10 = 6200$

It moves the decimal point 3 places to the right.

**Section P.2**

**Check Point Exercises**

**1. a.**  $3^3 3^2 = 3^{3+2} = 3^5$  or  243

**b.**  $\left(4x^3 y^4\right)\left(10x^2 y^6\right) = 4 \cdot 10 \cdot x^3 \cdot x^2 \cdot y^4 \cdot y^6$

$= 40x^{3+2} \cdot y^{4+6}$

$= 40x^5 y^{10}$

**2. a.**  $\dfrac{(-3)^6}{(-3)^3} = (-3)^3 = -27$

**b.**  $\dfrac{27x^{14} y^8}{3x^3 y^5} = \dfrac{27}{3} \cdot \dfrac{x^{14}}{x^3} \cdot \dfrac{y^8}{y^5} = 9x^{14-3} y^{8-5} = 9x^{11} y^3$

**3. a.**  $5^{-2} = \dfrac{1}{5^2} = \dfrac{1}{25}$

**b.**  $(-3)^{-3} = \dfrac{1}{(-3)^3} = \dfrac{1}{-27} = -\dfrac{1}{27}$

**c.**  $\dfrac{1}{4^{-2}} = \dfrac{1}{\dfrac{1}{4^2}} = 1 \cdot \dfrac{4^2}{1} = 4^2 = 16$

**d.**  $3x^{-6} y^4 = 3 \cdot \dfrac{1}{x^6} \cdot y^4 = \dfrac{3y^4}{x^6}$

**4. a.**  $\left(3^3\right)^2 = 3^{3 \cdot 2} = 3^6$  or  729

**b.**  $\left(y^7\right)^{-2} = y^{7(-2)} = y^{-14} = \dfrac{1}{y^{14}}$

**c.**  $\left(b^{-3}\right)^{-4} = b^{-3(-4)} = b^{12}$

**5.** $(-4x)^3 = (-4)^3(x)^3 = -64x^3$

**6.** **a.** $\left(-\dfrac{2}{y}\right)^5 = \dfrac{(-2)^5}{y^5} = \dfrac{-32}{y^5}$

**b.** $\left(\dfrac{x^5}{3}\right)^3 = \dfrac{\left(x^5\right)^3}{3^3} = \dfrac{x^{15}}{27}$

**7.** **a.** $\left(2x^3y^6\right)^4 = (2)^4\left(x^3\right)^4\left(y^6\right)^4 = 16x^{12}y^{24}$

**b.** $\left(-6x^2y^5\right)\left(3xy^3\right) = (-6)\cdot 3\cdot x^2\cdot x\cdot y^5\cdot y^3$
$= -18x^3y^8$

**c.** $\dfrac{100x^{12}y^2}{20x^{16}y^{-4}} = \left(\dfrac{100}{20}\right)\left(\dfrac{x^{12}}{x^{16}}\right)\left(\dfrac{y^2}{y^{-4}}\right)$
$= 5x^{12-16}y^{2-(-4)}$
$= 5x^{-4}y^6$
$= \dfrac{5y^6}{x^4}$

**d.** $\left(\dfrac{5x}{y^4}\right)^{-2} = \dfrac{(5)^{-2}(x)^{-2}}{\left(y^4\right)^{-2}}$
$= \dfrac{(5)^{-2}(x)^{-2}}{\left(y^4\right)^{-2}}$
$= \dfrac{5^{-2}x^{-2}}{y^{-8}}$
$= \dfrac{y^8}{5^2x^2}$
$= \dfrac{y^8}{25x^2}$

**8.** **a.** $-2.6\times10^9 = -2,600,000,000$

**b.** $3.017\times10^{-6} = 0.000003017$

**9.** **a.** $5,210,000,000 = 5.21\times10^9$

**b.** $-0.00000006893 = -6.893\times10^{-8}$

**10.** $410\times10^7 = \left(4.1\times10^2\right)\times10^7$
$= 4.1\times\left(10^2\times10^7\right)$
$= 4.1\times10^9$

**11** **a.** $\left(7.1\times10^5\right)\left(5\times10^{-7}\right)$
$= 7.1\cdot5\times10^5\cdot10^{-7}$
$= 35.5\times10^{-2}$
$= \left(3.55\times10^1\right)\times10^{-2}$
$= 3.55\times\left(10^1\times10^{-2}\right)$
$= 3.55\times10^{-1}$

**b.** $\dfrac{1.2\times10^6}{3\times10^{-3}} = \dfrac{1.2}{3}\cdot\dfrac{10^6}{10^{-3}}$
$= 0.4\times10^{6-(-3)}$
$= 0.4\times10^9$
$= 4\times10^8$

**12.** $\dfrac{2.6\times10^{12}}{3.12\times10^8} = \dfrac{2.6}{3.12}\cdot\dfrac{10^{12}}{10^8}$
$\approx 0.83\cdot10^4$
$\approx 8300$
The cost was \$8300 per citizen.

**Concept and Vocabulary Check P.2**

1. $b^{m+n}$; add

2. $b^{m-n}$; subtract

3. 1

4. $\dfrac{1}{b^n}$

5. false

6. $b^n$

7. true

8. a number greater than or equal to 1 and less than 10; integer

9. true

10. false

**Exercise Set P.2**

**1.** $5^2 \cdot 2 = (5 \cdot 5) \cdot 2 = 25 \cdot 2 = 50$

**3.** $(-2)^6 = (-2)(-2)(-2)(-2)(-2)(-2) = 64$

**5.** $-2^6 = -2 \cdot 2 \cdot 2 \cdot 2 \cdot 2 \cdot 2 = -64$

**7.** $(-3)^0 = 1$

**9.** $-3^0 = -1$

**11.** $4^{-3} = \dfrac{1}{4^3} = \dfrac{1}{4 \cdot 4 \cdot 4} = \dfrac{1}{64}$

**13.** $2^2 \cdot 2^3 = 2^{2+3} = 2^5 = 2 \cdot 2 \cdot 2 \cdot 2 \cdot 2 = 32$

**15.** $(2^2)^3 = 2^{2 \cdot 3} = 2^6 = 2 \cdot 2 \cdot 2 \cdot 2 \cdot 2 \cdot 2 = 64$

**17.** $\dfrac{2^8}{2^4} = 2^{8-4} = 2^4 = 2 \cdot 2 \cdot 2 \cdot 2 = 16$

**19.** $3^{-3} \cdot 3 = 3^{-3+1} = 3^{-2} = \dfrac{1}{3^2} = \dfrac{1}{3 \cdot 3} = \dfrac{1}{9}$

**21.** $\dfrac{2^3}{2^7} = 2^{3-7} = 2^{-4} = \dfrac{1}{2^4} = \dfrac{1}{2 \cdot 2 \cdot 2 \cdot 2} = \dfrac{1}{16}$

**23.** $x^{-2}y = \dfrac{1}{x^2} \cdot y = \dfrac{y}{x^2}$

**25.** $x^0 y^5 = 1 \cdot y^5 = y^5$

**27.** $x^3 \cdot x^7 = x^{3+7} = x^{10}$

**29.** $x^{-5} \cdot x^{10} = x^{-5+10} = x^5$

**31.** $(x^3)^7 = x^{3 \cdot 7} = x^{21}$

**33.** $(x^{-5})^3 = x^{-5 \cdot 3} = x^{-15} = \dfrac{1}{x^{15}}$

**35.** $\dfrac{x^{14}}{x^7} = x^{14-7} = x^7$

**37.** $\dfrac{x^{14}}{x^{-7}} = x^{14-(-7)} = x^{14+7} = x^{21}$

**39.** $(8x^3)^2 = 8^2(x^3)^2 = 8^2 x^{3 \cdot 2} = 64x^6$

**41.** $\left(-\dfrac{4}{x}\right)^3 = \dfrac{(-4)^3}{x^3} = -\dfrac{64}{x^3}$

**43.** $(-3x^2 y^5)^2 = (-3)^2 (x^2)^2 \cdot (y^5)^2$
$$= 9x^{2 \cdot 2} y^{5 \cdot 2}$$
$$= 9x^4 y^{10}$$

**45.** $(3x^4)(2x^7) = 3 \cdot 2x^4 \cdot x^7 = 6x^{4+7} = 6x^{11}$

**47.** $(-9x^3 y)(-2x^6 y^4) = (-9)(-2)x^3 x^6 yy^4$
$$= 18x^{3+6} y^{1+4}$$
$$= 18x^9 y^5$$

**49.** $\dfrac{8x^{20}}{2x^4} = \left(\dfrac{8}{2}\right)\left(\dfrac{x^{20}}{x^4}\right) = 4x^{20-4} = 4x^{16}$

**51.** $\dfrac{25a^{13} \cdot b^4}{-5a^2 \cdot b^3} = \left(\dfrac{25}{-5}\right)\left(\dfrac{a^{13}}{a^2}\right)\left(\dfrac{b^4}{b^3}\right)$
$$= -5a^{13-2} b^{4-3}$$
$$= -5a^{11} b$$

**53.** $\dfrac{14b^7}{7b^{14}} = \left(\dfrac{14}{7}\right)\left(\dfrac{b^7}{b^{14}}\right) = 2 \cdot b^{7-14} = 2b^{-7} = \dfrac{2}{b^7}$

**55.** $(4x^3)^{-2} = (4^{-2})(x^3)^{-2}$
$$= 4^{-2} x^{-6}$$
$$= \dfrac{1}{4^2 x^6}$$
$$= \dfrac{1}{16x^6}$$

**57.** $\dfrac{24x^3 \cdot y^5}{32x^7 y^{-9}} = \dfrac{3}{4} x^{3-7} y^{5-(-9)}$
$$= \dfrac{3}{4} x^{-4} y^{14}$$
$$= \dfrac{3y^{14}}{4x^4}$$

**59.** $\left(\dfrac{5x^3}{y}\right)^{-2} = \dfrac{5^{-2} x^{-6}}{y^{-2}} = \dfrac{y^2}{25x^6}$

**61.** $\left(\dfrac{-15a^4b^2}{5a^{10}b^{-3}}\right)^3 = \left(\dfrac{-3b^{2-(-3)}}{a^{10-4}}\right)^3$

$\qquad = \left(\dfrac{-3b^5}{a^6}\right)^3$

$\qquad = \dfrac{-27b^{15}}{a^{18}}$

**63.** $\left(\dfrac{3a^{-5}b^2}{12a^3b^{-4}}\right)^0 = 1$

**65.** $3.8 \times 10^2 = 380$

**67.** $6 \times 10^{-4} = 0.0006$

**69.** $-7.16 \times 10^6 = -7,160,000$

**71.** $7.9 \times 10^{-1} = 0.79$

**73.** $-4.15 \times 10^{-3} = -0.00415$

**75.** $-6.00001 \times 10^{10} = -60,000,100,000$

**77.** $32,000 = 3.2 \times 10^4$

**79.** $638,000,000,000,000,000$

$\qquad = 6.38 \times 10^{17}$

**81.** $-5716 = -5.716 \times 10^3$

**83.** $0.0027 = 2.7 \times 10^{-3}$

**85.** $-0.00000000504 = -5.04 \times 10^{-9}$

**87.** $\left(3 \times 10^4\right)\left(2.1 \times 10^3\right) = \left(3 \times 2.1\right)\left(10^4 \times 10^3\right)$

$\qquad = 6.3 \times 10^{4+3} = 6.3 \times 10^7$

**89.** $\left(1.6 \times 10^{15}\right)\left(4 \times 10^{-11}\right) = \left(1.6 \times 4\right)\left(10^{15} \times 10^{-11}\right)$

$\qquad = 6.4 \times 10^{15+(-11)}$

$\qquad = 6.4 \times 10^4$

**91.** $\left(6.1 \times 10^{-8}\right)\left(2 \times 10^{-4}\right) = \left(6.1 \times 2\right)\left(10^{-8} \times 10^{-4}\right)$

$\qquad = 12.2 \times 10^{-8+(-4)}$

$\qquad = 12.2 \times 10^{-12}$

$\qquad = 1.22 \times 10^{-11}$

**93.** $\left(4.3 \times 10^8\right)\left(6.2 \times 10^4\right)$

$\qquad = \left(4.3 \times 6.2\right)\left(10^8 \times 10^4\right)$

$\qquad = 26.66 \times 10^{8+4}$

$\qquad = 26.66 \times 10^{12}$

$\qquad = 2.666 \times 10^{13} \approx 2.67 \times 10^{13}$

**95.** $\dfrac{8.4 \times 10^8}{4 \times 10^5} = \dfrac{8.4}{4} \times \dfrac{10^8}{10^5}$

$\qquad = 2.1 \times 10^{8-5} = 2.1 \times 10^3$

**97.** $\dfrac{3.6 \times 10^4}{9 \times 10^{-2}} = \dfrac{3.6}{9} \times \dfrac{10^4}{10^{-2}}$

$\qquad = 0.4 \times 10^{4-(-2)}$

$\qquad = 0.4 \times 10^6 = 4 \times 10^5$

**99.** $\dfrac{4.8 \times 10^{-2}}{2.4 \times 10^6} = \dfrac{4.8}{2.4} \times \dfrac{10^{-2}}{10^6}$

$\qquad = 2 \times 10^{-2-6} = 2 \times 10^{-8}$

**101.** $\dfrac{2.4 \times 10^{-2}}{4.8 \times 10^{-6}} = \dfrac{2.4}{4.8} \times \dfrac{10^{-2}}{10^{-6}}$

$\qquad = 0.5 \times 10^{-2-(-6)}$

$\qquad = 0.5 \times 10^4 = 5 \times 10^3$

**103.** $\dfrac{480,000,000,000}{0.00012} = \dfrac{4.8 \times 10^{11}}{1.2 \times 10^{-4}}$

$\qquad = \dfrac{4.8}{1.2} \times \dfrac{10^{11}}{10^{-4}}$

$\qquad = 4 \times 10^{11-(-4)}$

$\qquad = 4 \times 10^{15}$

**105.** $\dfrac{0.00072 \times 0.003}{0.00024}$

$\qquad = \dfrac{\left(7.2 \times 10^{-4}\right)\left(3 \times 10^{-3}\right)}{2.4 \times 10^{-4}}$

$\qquad = \dfrac{7.2 \times 3}{2.4} \times \dfrac{10^{-4} \cdot 10^{-3}}{10^{-4}} = 9 \times 10^{-3}$

**107.** $\dfrac{\left(x^{-2}y\right)^{-3}}{\left(x^2y^{-1}\right)^3} = \dfrac{x^6y^{-3}}{x^6y^{-3}}$

$\qquad = x^{6-6}y^{-3-(-3)} = x^0y^0 = 1$

**109.** $\left(2x^{-3}yz^{-6}\right)\left(2x\right)^{-5} = 2x^{-3}yz^{-6}\cdot 2^{-5}x^{.-5}$

$$= 2^{-4}x^{-8}yz^{-6} = \frac{y}{2^4 x^8 z^6} = \frac{y}{16x^8 z^6}$$

**111.** $\left(\dfrac{x^3 y^4 z^5}{x^{-3}y^{-4}z^{-5}}\right)^{-2} = \left(x^6 y^8 z^{10}\right)^{-2}$

$$= x^{-12}y^{-16}z^{-20} = \frac{1}{x^{12}y^{16}z^{20}}$$

**113.** $\dfrac{\left(2^{-1}x^{-2}y^{-1}\right)^{-2}\left(2x^{-4}y^3\right)^{-2}\left(16x^{-3}y^3\right)^{0}}{\left(2x^{-3}y^{-5}\right)^{2}}$

$$= \frac{\left(2^2 x^2 y^2\right)\left(2^{-2}x^8 y^{-6}\right)(1)}{\left(2^2 x^{-6} y^{-10}\right)}$$

$$= \frac{x^{18}y^6}{4}$$

**115. a.** $2.17 \times 10^{12}$

**b.** $3.09 \times 10^8$

**c.** $\dfrac{2.17 \times 10^{12}}{3.09 \times 10^8} = \dfrac{2.17}{3.09} \times \dfrac{10^{12}}{10^8}$

$$\approx 0.7023 \times 10^4$$

$$\approx 7023$$

$7023 per American

**117.** $1340 \times 10^6 \cdot 7.90 = 1.34 \times 10^9 \cdot 7.9$

$$= 1.34 \cdot 7.9 \times 10^9$$

$$= 10.586 \times 10^9$$

$$= 1.0586 \times 10^{10}$$

Box-office receipts were $1.0586 \times 10^{10}$ in 2010.

**119.** $5.3 \times 10^{-23} \cdot 20,000 = 5.3 \times 10^{-23} \cdot 2 \times 10^4$

$$= 5.3 \cdot 2 \times 10^{-23} \cdot 10^4$$

$$= 10.6 \times 10^{-19}$$

$$= 1.06 \times 10^1 \cdot 10^{-19}$$

$$= 1.06 \times 10^{-18}$$

The mass is $1.06 \times 10^{-18}$ gram.

**121.** $3.2 \times 10^7 \cdot 127 = 3.2 \times 10^7 \cdot 1.27 \times 10^2$

$$= 3.2 \cdot 1.27 \times 10^7 \cdot 10^2$$

$$= 4.064 \times 10^9$$

Americans eat $4.064 \times 10^9$ chickens per year.

**123. – 131.** Answers will vary.

**133.** makes sense

**135.** makes sense

**137.** true

**139.** false; Changes to make the statement true will vary. A sample change is: $5^2 \cdot 5^{-2} = 2^5 \cdot 2^{-5}$.

**141.** false; Changes to make the statement true will vary. A sample change is:

$$\frac{8 \times 10^{30}}{2 \times 10^{-5}} = 4 \times 10^{30-(-5)} = 4 \times 10^{35}.$$

**143.** true

**145.** $b^A = MN, b^C = M, b^D = N$

$$b^A = b^C b^D$$

$$A = C + D$$

**147.** Answers will vary.

**148. a.** $\sqrt{16} \cdot \sqrt{4} = 4 \cdot 2 = 8$

**b.** $\sqrt{16 \cdot 4} = \sqrt{64} = 8$

**c.** $\sqrt{16} \cdot \sqrt{4} = \sqrt{16 \cdot 4}$

**149. a.** $\sqrt{300} \approx 17.32$

**b.** $10\sqrt{3} \approx 17.32$

**c.** $\sqrt{300} = 10\sqrt{3}$

**150. a.** $21x + 10x = 31x$

**b.** $21\sqrt{2} + 10\sqrt{2} = 31\sqrt{2}$

**Section P.3**

**Check Point Exercises**

1.　a.　$\sqrt{81} = 9$

　　b.　$-\sqrt{9} = -3$

　　c.　$\sqrt{\dfrac{1}{25}} = \dfrac{1}{5}$

　　d.　$\sqrt{36+64} = \sqrt{100} = 10$

　　e.　$\sqrt{36} + \sqrt{64} = 6 + 8 = 14$

2.　a.　$\sqrt{75} = \sqrt{25 \cdot 3} = \sqrt{25}\sqrt{3} = 5\sqrt{3}$

　　b.　$\sqrt{5x} \cdot \sqrt{10x} = \sqrt{5x \cdot 10x}$
　　　　$= \sqrt{50x^2}$
　　　　$= \sqrt{25 \cdot 2x^2}$
　　　　$= \sqrt{25x^2} \cdot \sqrt{2}$
　　　　$= 5x\sqrt{2}$

3.　a.　$\sqrt{\dfrac{25}{16}} = \dfrac{\sqrt{25}}{\sqrt{16}} = \dfrac{5}{4}$

　　b.　$\dfrac{\sqrt{150x^3}}{\sqrt{2x}} = \sqrt{\dfrac{150x^3}{2x}}$
　　　　$= \sqrt{75x^2}$
　　　　$= \sqrt{25x^2} \cdot \sqrt{3}$
　　　　$= 5x\sqrt{3}$

4.　a.　$8\sqrt{13} + 9\sqrt{13} = (8+9)\sqrt{3}$
　　　　　　　　　　$= 17\sqrt{13}$

　　b.　$\sqrt{17x} - 20\sqrt{17x}$
　　　　$= 1\sqrt{17x} - 20\sqrt{17x}$
　　　　$= (1-20)\sqrt{17x}$
　　　　$= -19\sqrt{17x}$

5.　a.　$5\sqrt{27} + \sqrt{12}$
　　　　$= 5\sqrt{9 \cdot 3} + \sqrt{4 \cdot 3}$
　　　　$= 5 \cdot 3\sqrt{3} + 2\sqrt{3}$
　　　　$= 15\sqrt{3} + 2\sqrt{3}$
　　　　$= (15+2)\sqrt{3}$
　　　　$= 17\sqrt{3}$

　　b.　$6\sqrt{18x} - 4\sqrt{8x}$
　　　　$= 6\sqrt{9 \cdot 2x} - 4\sqrt{4 \cdot 2x}$
　　　　$= 6 \cdot 3\sqrt{2x} - 4 \cdot 2\sqrt{2x}$
　　　　$= 18\sqrt{2x} - 8\sqrt{2x}$
　　　　$= (18-8)\sqrt{2x}$
　　　　$= 10\sqrt{2x}$

6.　a.　If we multiply numerator and denominator by $\sqrt{3}$, the denominator becomes $\sqrt{3} \cdot \sqrt{3} = \sqrt{9} = 3$. Therefore, multiply by 1, choosing $\dfrac{\sqrt{3}}{\sqrt{3}}$ for 1.

　　　　$\dfrac{5}{\sqrt{3}} = \dfrac{5}{\sqrt{3}} \cdot \dfrac{\sqrt{3}}{\sqrt{3}} = \dfrac{5\sqrt{3}}{\sqrt{9}} = \dfrac{5\sqrt{3}}{3}$

　　b.　The *smallest* number that will produce a perfect square in the denominator of $\dfrac{6}{\sqrt{12}}$ is $\sqrt{3}$ because $\sqrt{12} \cdot \sqrt{3} = \sqrt{36} = 6$. So multiply by 1, choosing $\dfrac{\sqrt{3}}{\sqrt{3}}$ for 1.

　　　　$\dfrac{6}{\sqrt{12}} = \dfrac{6}{\sqrt{12}} \cdot \dfrac{\sqrt{3}}{\sqrt{3}} = \dfrac{6\sqrt{3}}{\sqrt{36}} = \dfrac{6\sqrt{3}}{6} = \sqrt{3}$

7.　Multiply by $\dfrac{4-\sqrt{5}}{4-\sqrt{5}}$.

　　$\dfrac{8}{4+\sqrt{5}} = \dfrac{8}{4+\sqrt{5}} \cdot \dfrac{4-\sqrt{5}}{4-\sqrt{5}}$
　　　　$= \dfrac{8(4-\sqrt{5})}{4^2 - (\sqrt{5})^2}$
　　　　$= \dfrac{8(4-\sqrt{5})}{16-5}$
　　　　$= \dfrac{8(4-\sqrt{5})}{11}$ or $\dfrac{32-8\sqrt{5}}{11}$

**8.** **a.** $\sqrt[3]{40} = \sqrt[3]{8 \cdot 5} = \sqrt[3]{8} \cdot \sqrt[3]{5} = 2\sqrt[3]{5}$

**b.** $\sqrt[5]{8} \cdot \sqrt[5]{8} = \sqrt[5]{64} = \sqrt[5]{32} \cdot \sqrt[5]{2} = 2\sqrt[5]{2}$

**c.** $\sqrt[3]{\dfrac{125}{27}} = \dfrac{\sqrt[3]{125}}{\sqrt[3]{27}} = \dfrac{5}{3}$

**9.** $3\sqrt[3]{81} - 4\sqrt[3]{3}$
$= 3\sqrt[3]{27 \cdot 3} - 4\sqrt[3]{3}$
$= 3 \cdot 3\sqrt[3]{3} - 4\sqrt[3]{3}$
$= 9\sqrt[3]{3} - 4\sqrt[3]{3}$
$= (9 - 4)\sqrt[3]{3}$
$= 5\sqrt[3]{3}$

**10.** **a.** $25^{\frac{1}{2}} = \sqrt{25} = 5$

**b.** $8^{\frac{1}{3}} = \sqrt[3]{8} = 2$

**c.** $-81^{\frac{1}{4}} = -\sqrt[4]{81} = -3$

**d.** $(-8)^{\frac{1}{3}} = \sqrt[3]{-8} = -2$

**e.** $27^{-\frac{1}{3}} = \dfrac{1}{27^{\frac{1}{3}}} = \dfrac{1}{\sqrt[3]{27}} = \dfrac{1}{3}$

**11.** **a.** $27^{\frac{4}{3}} = \left(\sqrt[3]{27}\right)^4 = (3)^4 = 81$

**b.** $4^{\frac{3}{2}} = \left(\sqrt[2]{4}\right)^3 = (2)^3 = 8$

**c.** $32^{-\frac{2}{5}} = \dfrac{1}{32^{\frac{2}{5}}} = \dfrac{1}{\left(\sqrt[5]{32}\right)^2} = \dfrac{1}{2^2} = \dfrac{1}{4}$

**12.** **a.** $\left(2x^{4/3}\right)\left(5x^{8/3}\right)$
$= 2 \cdot 5x^{4/3} \cdot x^{8/3}$
$= 10x^{(4/3)+(8/3)}$
$= 10x^{12/3}$
$= 10x^4$

**b.** $\dfrac{20x^4}{5x^{3/2}} = \left(\dfrac{20}{5}\right)\left(\dfrac{x^4}{x^{3/2}}\right)$
$= 4x^{4-(3/2)}$
$= 4x^{(8/2)-(3/2)}$
$= 4x^{5/2}$

**13.** $\sqrt[6]{x^3} = x^{3/6} = x^{1/2} = \sqrt{x}$

**Concept and Vocabulary Check P.3**

**1.** principal

**2.** $8^2$

**3.** $|a|$

**4.** $\sqrt{a} \cdot \sqrt{b}$

**5.** $\dfrac{\sqrt{a}}{\sqrt{b}}$

**6.** $18\sqrt{3}$

**7.** $5;\ 6\sqrt{3}$

**8.** $7 - \sqrt{3}$

**9.** $\sqrt{10} + \sqrt{2}$

**10.** index; radicand

**11.** $(-2)^5$

**12.** $a;\ |a|$

**13.** $\sqrt[n]{a}$

**14.** $2;\ 8$

**Exercise Set P.3**

**1.** $\sqrt{36} = \sqrt{6^2} = 6$

**3.** $-\sqrt{36} = -\sqrt{6^2} = -6$

**5.** $\sqrt{-36}$, The square root of a negative number is not real.

**7.** $\sqrt{25-16} = \sqrt{9} = 3$

**9.** $\sqrt{25} - \sqrt{16} = 5 - 4 = 1$

**11.** $\sqrt{(-13)^2} = \sqrt{169} = 13$

**13.** $\sqrt{50} = \sqrt{25 \cdot 2} = \sqrt{25}\sqrt{2} = 5\sqrt{2}$

**15.** $\sqrt{45x^2} = \sqrt{9x^2 \cdot 5}$
$= \sqrt{9x^2}\sqrt{5}$
$= \sqrt{9}\sqrt{x^2}\sqrt{5}$
$= 3|x|\sqrt{5}$

**17.** $\sqrt{2x} \cdot \sqrt{6x} = \sqrt{2x \cdot 6x}$
$= \sqrt{12x^2}$
$= \sqrt{4x^2} \cdot \sqrt{3}$
$= 2x\sqrt{3}$

**19.** $\sqrt{x^3} = \sqrt{x^2} \cdot \sqrt{x} = x\sqrt{x}$

**21.** $\sqrt{2x^2} \cdot \sqrt{6x} = \sqrt{2x^2 \cdot 6x}$
$= \sqrt{12x^3}$
$= \sqrt{4x^2} \cdot \sqrt{3x}$
$= 2x\sqrt{3x}$

**23.** $\sqrt{\dfrac{1}{81}} = \dfrac{\sqrt{1}}{\sqrt{81}} = \dfrac{1}{9}$

**25.** $\sqrt{\dfrac{49}{16}} = \dfrac{\sqrt{49}}{\sqrt{16}} = \dfrac{7}{4}$

**27.** $\dfrac{\sqrt{48x^3}}{\sqrt{3x}} = \sqrt{\dfrac{48x^3}{3x}} = \sqrt{16x^2} = 4x$

**29.** $\dfrac{\sqrt{150x^4}}{\sqrt{3x}} = \sqrt{\dfrac{150x^4}{3x}}$
$= \sqrt{50x^3}$
$= \sqrt{25x^2} \cdot \sqrt{2x}$
$= 5x\sqrt{2x}$

**31.** $\dfrac{\sqrt{200x^3}}{\sqrt{10x^{-1}}} = \sqrt{\dfrac{200x^3}{10x^{-1}}}$
$= \sqrt{20x^{3-(-1)}}$
$= \sqrt{20x^4}$
$= \sqrt{4 \cdot 5x^4}$
$= 2x^2\sqrt{5}$

**33.** $7\sqrt{3} + 6\sqrt{3} = (7+6)\sqrt{3} = 13\sqrt{3}$

**35.** $6\sqrt{17x} - 8\sqrt{17x} = (6-8)\sqrt{17x} = -2\sqrt{17x}$

**37.** $\sqrt{8} + 3\sqrt{2} = \sqrt{4 \cdot 2} + 3\sqrt{2}$
$= 2\sqrt{2} + 3\sqrt{2}$
$= (2+3)\sqrt{2}$
$= 5\sqrt{2}$

**39.** $\sqrt{50x} - \sqrt{8x} = \sqrt{25 \cdot 2x} - \sqrt{4 \cdot 2x}$
$= 5\sqrt{2x} - 2\sqrt{2x}$
$= (5-2)\sqrt{2x}$
$= 3\sqrt{2x}$

**41.** $3\sqrt{18} + 5\sqrt{50} = 3\sqrt{9 \cdot 2} + 5\sqrt{25 \cdot 2}$
$= 3 \cdot 3\sqrt{2} + 5 \cdot 5\sqrt{2}$
$= 9\sqrt{2} + 25\sqrt{2}$
$= (9+25)\sqrt{2}$
$= 34\sqrt{2}$

**43.** $3\sqrt{8} - \sqrt{32} + 3\sqrt{72} - \sqrt{75}$
$= 3\sqrt{4 \cdot 2} - \sqrt{16 \cdot 2} + 3\sqrt{36 \cdot 2} - \sqrt{25 \cdot 3}$
$= 3 \cdot 2\sqrt{2} - 4\sqrt{2} + 3 \cdot 6\sqrt{2} - 5\sqrt{3}$
$= 6\sqrt{2} - 4\sqrt{2} + 18\sqrt{2} - 5\sqrt{3}$
$= 20\sqrt{2} - 5\sqrt{3}$

**45.** $\dfrac{1}{\sqrt{7}} = \dfrac{1}{\sqrt{7}} \cdot \dfrac{\sqrt{7}}{\sqrt{7}} = \dfrac{\sqrt{7}}{7}$

**47.** $\dfrac{\sqrt{2}}{\sqrt{5}} = \dfrac{\sqrt{2}}{\sqrt{5}} \cdot \dfrac{\sqrt{5}}{\sqrt{5}} = \dfrac{\sqrt{10}}{5}$

**49.** $\dfrac{13}{3+\sqrt{11}} = \dfrac{13}{3+\sqrt{11}} \cdot \dfrac{3-\sqrt{11}}{3-\sqrt{11}}$

$= \dfrac{13(3-\sqrt{11})}{3^2 - (\sqrt{11})^2}$

$= \dfrac{13(3-\sqrt{11})}{9-11}$

$= \dfrac{13(3-\sqrt{11})}{-2}$

**51.** $\dfrac{7}{\sqrt{5}-2} = \dfrac{7}{\sqrt{5}-2} \cdot \dfrac{\sqrt{5}+2}{\sqrt{5}+2}$

$= \dfrac{7(\sqrt{5}+2)}{(\sqrt{5})^2 - 2^2}$

$= \dfrac{7(\sqrt{5}+2)}{5-4}$

$= 7(\sqrt{5}+2)$

**53.** $\dfrac{6}{\sqrt{5}+\sqrt{3}} = \dfrac{6}{\sqrt{5}+\sqrt{3}} \cdot \dfrac{\sqrt{5}-\sqrt{3}}{\sqrt{5}-\sqrt{3}}$

$= \dfrac{6(\sqrt{5}-\sqrt{3})}{(\sqrt{5})^2 - (\sqrt{3})^2}$

$= \dfrac{6(\sqrt{5}-\sqrt{3})}{5-3}$

$= \dfrac{6(\sqrt{5}-\sqrt{3})}{2}$

$= 3(\sqrt{5}-\sqrt{3})$

**55.** $\sqrt[3]{125} = \sqrt[3]{5^3} = 5$

**57.** $\sqrt[3]{-8} = \sqrt[3]{(-2)^3} = -2$

**59.** $\sqrt[4]{-16}$ is not a real number.

**61.** $\sqrt[4]{(-3)^4} = |-3| = 3$

**63.** $\sqrt[5]{(-3)^5} = -3$

**65.** $\sqrt[5]{-\dfrac{1}{32}} = \sqrt[5]{-\dfrac{1}{2^5}} = -\dfrac{1}{2}$

**67.** $\sqrt[3]{32} = \sqrt[3]{8 \cdot 4} = \sqrt[3]{8}\sqrt[3]{4} = 2 \cdot \sqrt[3]{4}$

**69.** $\sqrt[3]{x^4} = \sqrt[3]{x^3 \cdot x} = x \cdot \sqrt[3]{x}$

**71.** $\sqrt[3]{9} \cdot \sqrt[3]{6} = \sqrt[3]{54} = \sqrt[3]{27 \cdot 2} = \sqrt[3]{27}\sqrt[3]{2} = 3\sqrt[3]{2}$

**73.** $\dfrac{\sqrt[5]{64x^6}}{\sqrt[5]{2x}} = \sqrt[5]{\dfrac{64x^6}{2x}} = \sqrt[5]{32x^5} = 2x$

**75.** $4\sqrt[5]{2} + 3\sqrt[5]{2} = 7\sqrt[5]{2}$

**77.** $5\sqrt[3]{16} + \sqrt[3]{54} = 5\sqrt[3]{8 \cdot 2} + \sqrt[3]{27 \cdot 2}$

$= 5 \cdot 2\sqrt[3]{2} + 3\sqrt[3]{2}$

$= 10\sqrt[3]{2} + 3\sqrt[3]{2}$

$= 13\sqrt[3]{2}$

**79.** $\sqrt[3]{54xy^3} - y\sqrt[3]{128x}$

$= \sqrt[3]{27 \cdot 2xy^3} - y\sqrt[3]{64 \cdot 2x}$

$= 3y\sqrt[3]{2x} - 4y\sqrt[3]{2x}$

$= -y\sqrt[3]{2x}$

**81.** $\sqrt{2} + \sqrt[3]{8} = \sqrt{2} + 2$

**83.** $36^{1/2} = \sqrt{36} = 6$

**85.** $8^{1/3} = \sqrt[3]{8} = 2$

**87.** $125^{2/3} = \left(\sqrt[3]{125}\right)^2 = 5^2 = 25$

**89.** $32^{-4/5} = \dfrac{1}{32^{4/5}} = \dfrac{1}{2^4} = \dfrac{1}{16}$

**91.** $\left(7x^{1/3}\right)\left(2x^{1/4}\right) = 7 \cdot 2x^{1/3} \cdot x^{1/4}$

$= 14 \cdot x^{1/3+1/4}$

$= 14x^{7/12}$

**93.** $\dfrac{20x^{1/2}}{5x^{1/4}} = \left(\dfrac{20}{5}\right)\left(\dfrac{x^{1/2}}{x^{1/4}}\right)$

$= 4 \cdot x^{1/2-1/4}$

$= 4x^{1/4}$

**95.** $\left(x^{2/3}\right)^3 = x^{2/3 \cdot 3} = x^2$

**97.** $(25x^4y^6)^{1/2} = 25^{1/2}x^{4 \cdot 1/2}y^{6 \cdot 1/2} = 5x^2|y|^3$

**99.** $\dfrac{\left(3y^{\frac{1}{4}}\right)^3}{y^{\frac{1}{12}}} = \dfrac{27y^{\frac{3}{4}}}{y^{\frac{1}{12}}} = 27y^{\frac{3}{4}-\frac{1}{12}}$

$= 27y^{\frac{8}{12}} = 27y^{\frac{2}{3}}$

**101.** $\sqrt[4]{5^2} = 5^{2/4} = 5^{1/2} = \sqrt{5}$

**103.** $\sqrt[3]{x^6} = x^{6/3} = x^2$

**105.** $\sqrt[6]{x^4} = \sqrt[6/2]{x^{4/2}} = \sqrt[3]{x^2}$

**107.** $\sqrt[9]{x^6 y^3} = x^{\frac{6}{9}} y^{\frac{3}{9}} = x^{\frac{2}{3}} y^{\frac{1}{3}} = \sqrt[3]{x^2 y}$

**109.** $\sqrt[3]{\sqrt[4]{16} + \sqrt{625}} = \sqrt[3]{2+25} = \sqrt[3]{27} = 3$

**111.** $\left(49x^{-2}y^4\right)^{-1/2}\left(xy^{1/2}\right)$

$= (49)^{-1/2}\left(x^{-2}\right)^{-1/2}\left(y^4\right)^{-1/2}\left(xy^{1/2}\right)$

$= \dfrac{1}{49^{1/2}} x^{(-2)(-1/2)} y^{(4)(-1/2)}\left(xy^{1/2}\right)$

$= \dfrac{1}{7} x^1 y^{-2} \cdot xy^{1/2} = \dfrac{1}{7} x^{1+1} y^{-2+(1/2)}$

$= \dfrac{1}{7} x^2 y^{-3/2} = \dfrac{x^2}{7y^{3/2}}$

**113.** $\left(\dfrac{x^{-5/4}y^{1/3}}{x^{-3/4}}\right)^{-6} = \left(x^{(-5/4)-(-3/4)}y^{1/3}\right)^{-6}$

$= \left(x^{-2/4}y^{1/3}\right)^{-6} = x^{(-2/4)(-6)}y^{(1/3)(-6)}$

$= x^3 y^{-2} = \dfrac{x^3}{y^2}$

**115.** The message is "Paige Fox is bad at math."

**117.** $\dfrac{2}{\sqrt{5}-1} \cdot \dfrac{\sqrt{5}+1}{\sqrt{5}+1} = \dfrac{2(\sqrt{5}+1)}{5-1}$

$= \dfrac{2(\sqrt{5}+1)}{4}$

$= \dfrac{\sqrt{5}+1}{2}$

$\approx 1.62$

About 1.62 to 1.

**119.** Perimeter:
$P = 2l + 2w$

$= 2\cdot\sqrt{125} + 2\cdot 2\sqrt{20}$

$= 2\cdot\sqrt{25\cdot 5} + 4\sqrt{4\cdot 5}$

$= 2\cdot 5\sqrt{5} + 4\cdot 2\sqrt{5}$

$= 10\sqrt{5} + 8\sqrt{5}$

$= 18\sqrt{5}$ feet

Area:
$A = lw$

$= \sqrt{125}\cdot 2\sqrt{20}$

$= 2\sqrt{125\cdot 20}$

$= 2\sqrt{2500}$

$= 2\cdot 50$

$= 100$ square feet

**121. – 127.** Answers will vary.

**129.** does not make sense; Explanations will vary. Sample explanation: The denominator is rationalized correctly.

**131.** does not make sense; Explanations will vary. Sample explanation: $2\sqrt{20} + 4\sqrt{75}$ simplifies to $4\sqrt{5} + 20\sqrt{3}$ and thus the radical terms are not common.

**133.** false; Changes to make the statement true will vary. A sample change is: $7^{1/2}\cdot 7^{1/2} = 7^1 = 7$.

**135.** false; Changes to make the statement true will vary. The cube root of –8 is the real number –2.

**137.** $\left(5+\sqrt{\boxed{3}}\right)\left(5-\sqrt{\boxed{3}}\right) = 22$

$25 - \boxed{3} = 22$

$\boxed{3} = 3$

**139.** $\sqrt{13+\sqrt{2}+\dfrac{7}{3+\sqrt{2}}}$

$= \sqrt{13+\sqrt{2}+\dfrac{7}{3+\sqrt{2}}\cdot\dfrac{3-\sqrt{2}}{3-\sqrt{2}}}$

$= \sqrt{13+\sqrt{2}+\dfrac{21-7\sqrt{2}}{9-2}}$

$= \sqrt{13+\sqrt{2}+\dfrac{21-7\sqrt{2}}{7}}$

$= \sqrt{13+\sqrt{2}+3-\sqrt{2}}$

$= \sqrt{16}$

$= 4$

**141. a.**

$$\frac{ab}{a^2+ab+b^2}+\left(\frac{ac-ad-bc+bd}{ac-ad+bc-bd}\div\frac{a^3-b^3}{a^3+b^3}\right)=\frac{ab}{a^2+ab+b^2}+\left(\frac{a(c-d)-b(c-d)}{a(c-d)+b(c-d)}\cdot\frac{a^3+b^3}{a^3-b^3}\right)$$

$$=\frac{ab}{a^2+ab+b^2}+\left(\frac{(c-d)(a-b)}{(c-d)(a+b)}\cdot\frac{(a+b)(a^2-ab+b^2)}{(a-b)(a^2+ab+b^2)}\right)=\frac{ab}{a^2+ab+b^2}+\frac{a^2-ab+b^2}{a^2+ab+b^2}$$

$$=\frac{ab+a^2-ab+b^2}{a^2+ab+b^2}=\frac{a^2+b^2}{a^2+ab+b^2}$$

Her son is 8 years old.

**b.**   Son's portion:

$$\frac{8^{-\frac{4}{3}}+2^{-2}}{16^{-\frac{3}{4}}+2^{-1}}=\frac{\dfrac{1}{\left(\sqrt[3]{8}\right)^4}+\dfrac{1}{2^2}}{\dfrac{1}{\left(\sqrt[4]{16}\right)^3}+\dfrac{1}{2}}$$

$$=\frac{\dfrac{1}{2^4}+\dfrac{1}{4}}{\dfrac{1}{2^3}+\dfrac{1}{2}}$$

$$=\frac{\dfrac{1}{16}+\dfrac{1}{4}}{\dfrac{1}{8}+\dfrac{1}{2}}$$

$$=\frac{\dfrac{5}{16}}{\dfrac{5}{8}}$$

$$=\frac{8}{16}$$

$$=\frac{1}{2}$$

Mom's portion:

$$\frac{1}{2}\left(1-\frac{1}{2}\right)=\frac{1}{2}\left(\frac{1}{2}\right)=\frac{1}{4}$$

**142.**  $(2x^3y^2)(5x^4y^7)=10x^7y^9$

**143.**  $2x^4(8x^4+3x)=2x^4(8x^4)+2x^4(3x)$
$$=16x^8+6x^5$$

**144.**  $2x(x^2+4x+5)+3(x^2+4x+5)$
$$=2x^3+8x^2+10x+3x^2+12x+15$$
$$=2x^3+8x^2+3x^2+10x+12x+15$$
$$=2x^3+11x^2+22x+15$$

## Section P.4

### Check Point Exercises

1.  **a.**  $(-17x^3 + 4x^2 - 11x - 5) + (16x^3 - 3x^2 + 3x - 15)$

    $= (-17x^3 + 16x^3) + (4x^2 - 3x^2) + (-11x + 3x) + (-5 - 15)$

    $= -x^3 + x^2 - 8x - 20$

    **b.**  $(13x^2 - 9x^2 - 7x + 1) - (-7x^3 + 2x^2 - 5x + 9)$

    $= (13x^3 - 9x^2 - 7x + 1) + (7x^3 - 2x^2 + 5x - 9)$

    $= (13x^3 + 7x^3) + (-9x^2 - 2x^2) + (-7x + 5x) + (1 - 9)$

    $= 20x^3 - 11x^2 - 2x - 8$

2.  $(5x - 2)(3x^2 - 5x + 4)$

    $= 5x(3x^2 - 5x + 4) - 2(3x^2 - 5x + 4)$

    $= 5x \cdot 3x^2 - 5x \cdot 5x + 5x \cdot 4 - 2 \cdot 3x^2 + 2 \cdot 5x - 2 \cdot 4$

    $= 15x^3 - 25x^2 + 20x - 6x^2 + 10x - 8$

    $= 15x^3 - 31x^2 + 30x - 8$

3.  $(7x - 5)(4x - 3) = 7x \cdot 4x + 7x(-3) + (-5)4x + (-5)(-3)$

    $= 28x^2 - 21x - 20x + 15$

    $= 28x^2 - 41x + 15$

4.  **a.**  Use the special-product formula shown.

    $(A + B)(A - B) = A^2 - B^2$

    $(7x + 8)(7x - 8) = (7x)^2 - (8)^2$

    $= 49x^2 - 64$

    **b.**  Use the special-product formula shown.

    $(A + B)(A - B) = A^2 - B^2$

    $(2y^3 - 5)(2y^3 + 5) = (2y^3 + 5)(2y^3 - 5)$

    $= (2y^3)^2 - (5)^2$

    $= 4y^6 - 25$

5.  **a.**  Use the special-product formula shown.

    $(A + B)^2 = A^2 + 2AB + B^2$

    $(x + 10)^2 = x^2 + 2(x)(10) + 10^2$

    $= x^2 + 20x + 100$

    **b.**  Use the special-product formula shown.

    $(A + B)^2 = A^2 + 2AB + B^2$

    $(5x + 4)^2 = (5x)^2 + 2(5x)(4) + 4^2$

    $= 25x^2 + 40x + 16$

**6.** **a.** Use the special-product formula shown.

$$(A - B)^2 = A^2 - 2AB + B^2$$
$$(x - 9)^2 = x^2 - 2(x)(9) + 9^2$$
$$= x^2 - 18x + 81$$

**b.** Use the special-product formula shown.

$$(A - B)^2 = A^2 - 2AB + B^2$$
$$(7x - 3)^2 = (7x)^2 - 2(7x)(3) + 3^2$$
$$= 49x^2 - 42x + 9$$

**7.** $(x^3 - 4x^2y + 5xy^2 - y^3) - (x^3 - 6x^2y + y^3)$

$$= (x^3 - 4x^2y + 5xy^2 - y^3) + (-x^3 + 6x^2y - y^3)$$
$$= (x^3 - x^3) + (-4x^2y + 6x^2y) + (5xy^2) + (-y^3 - y^3)$$
$$= 2x^2y + 5xy^2 - 2y^3$$

**8.** **a.** $(7x - 6y)(3x - y) = (7x)(3x) + (7x)(-y) + (-6y)(3x) + (-6y)(-y)$

$$= 21x^2 - 7xy - 18xy + 6y^2$$
$$= 21x^2 - 25xy + 6y^2$$

**b.** $(2x + 4y)^2 = (2x)^2 + 2(2x)(4y) + (4y)^2$

$$= 4x^2 + 16xy + 16y^2$$

## Concept and Vocabulary Check P.4

**1.** whole

**2.** standard

**3.** monomial

**4.** binomial

**5.** trinomial

**6.** $n$

**7.** like;

**8.** distributive; $4x^3 - 8x^2 + 6$; $7x^3$

**9.** $5x$; 3; like

**10.** $3x^2$; $5x$; $21x$; 35

**11.** $A^2 - B^2$; minus

**12.** $A^2 + 2AB + B^2$; squared; product of the terms; squared

**13.** $A^2 - 2AB + B^2$; minus; product of the terms; plus

**14.** $n + m$

### Exercise Set P.4

**1.** yes; $2x + 3x^2 - 5 = 3x^2 + 2x - 5$

**3.** no; The form of a polynomial involves addition and subtraction, not division.

**5.** $3x^2$ has degree 2
$-5x$ has degree 1
4 has degree 0
$3x^2 - 5x + 4$ has degree 2.

**7.** $x^2$ has degree 2
$-4x^3$ has degree 3
$9x$ has degree 1
$-12x^4$ has degree 4
63 has degree 0
$x^2 - 4x^3 + 9x - 12x^4 + 63$ has degree 4.

**9.** $(-6x^3 + 5x^2 - 8x + 9) + (17x^3 + 2x^2 - 4x - 13) = (-6x^3 + 17x^3) + (5x^2 + 2x^2) + (-8x - 4x) + (9 - 13)$
$$= 11x^3 + 7x^2 - 12x - 4$$
The degree is 3.

**11.** $(17x^3 - 5x^2 + 4x - 3) - (5x^3 - 9x^2 - 8x + 11) = (17x^3 - 5x^2 + 4x - 3) + (-5x^3 + 9x^2 + 8x - 11)$
$$= (17x^3 - 5x^3) + (-5x^2 + 9x^2) + (4x + 8x) + (-3 - 11)$$
$$= 12x^3 + 4x^2 + 12x - 14$$
The degree is 3.

**13.** $(5x^2 - 7x - 8) + (2x^2 - 3x + 7) - (x^2 - 4x - 3) = (5x^2 - 7x - 8) + (2x^2 - 3x + 7) + (-x^2 + 4x + 3)$
$$= (5x^2 + 2x^2 - x^2) + (-7x - 3x + 4x) + (-8 + 7 + 3)$$
$$= 6x^2 - 6x + 2$$
The degree is 2.

**15.** $(x + 1)(x^2 - x + 1) = x(x^2) - x \cdot x + x \cdot 1 + 1(x^2) - 1 \cdot x + 1 \cdot 1$
$$= x^3 - x^2 + x + x^2 - x + 1$$
$$= x^3 + 1$$

**17.** $(2x - 3)(x^2 - 3x + 5) = (2x)(x^2) + (2x)(-3x) + (2x)(5) + (-3)(x^2) + (-3)(-3x) + (-3)(5)$
$$= 2x^3 - 6x^2 + 10x - 3x^2 + 9x - 15$$
$$= 2x^3 - 9x^2 + 19x - 15$$

**19.** $(x + 7)(x + 3) = x^2 + 3x + 7x + 21 = x^2 + 10x + 21$

**21.** $(x-5)(x+3) = x^2 + 3x - 5x - 15 = x^2 - 2x - 15$

**23.** $(3x+5)(2x+1) = (3x)(2x) + 3x(1) + 5(2x) + 5 = 6x^2 + 3x + 10x + 5 = 6x^2 + 13x + 5$

**25.** $(2x-3)(5x+3) = (2x)(5x) + (2x)(3) + (-3)(5x) + (-3)(3) = 10x^2 + 6x - 15x - 9 = 10x^2 - 9x - 9$

**27.** $(5x^2-4)(3x^2-7) = (5x^2)(3x^2) + (5x^2)(-7) + (-4)(3x^2) + (-4)(-7) = 15x^4 - 35x^2 - 12x^2 + 28 = 15x^4 - 47x^2 + 28$

**29.** $\left(8x^3+3\right)\left(x^2-5\right) = \left(8x^3\right)\left(x^2\right) + \left(8x^3\right)(-5) + (3)\left(x^2\right) + (3)(-5) = 8x^5 - 40x^3 + 3x^2 - 15$

**31.** $(x+3)(x-3) = x^2 - 3^2 = x^2 - 9$

**33.** $(3x+2)(3x-2) = (3x)^2 - 2^2 = 9x^2 - 4$

**35.** $(5-7x)(5+7x) = 5^2 - (7x)^2 = 25 - 49x^2$

**37.** $(4x^2+5x)(4x^2-5x) = (4x^2)^2 - (5x)^2 = 16x^4 - 25x^2$

**39.** $\left(1-y^5\right)\left(1+y^5\right) = (1)^2 - \left(y^5\right)^2 = 1 - y^{10}$

**41.** $(x+2)^2 = x^2 + 2 \cdot x \cdot 2 + 2^2 = x^2 + 4x + 4$

**43.** $(2x+3)^2 = (2x)^2 + 2(2x)(3) + 3^2 = 4x^2 + 12x + 9$

**45.** $(x-3)^2 = x^2 - 2 \cdot x \cdot 3 + 3^2 = x^2 - 6x + 9$

**47.** $(4x^2-1)^2 = (4x^2)^2 - 2(4x^2)(1) + 1^2 = 16x^4 - 8x^2 + 1$

**49.** $(7-2x)^2 = 7^2 - 2(7)(2x) + (2x)^2 = 49 - 28x + 4x^2 = 4x^2 - 28x + 49$

**51.** $(x+1)^3 = x^3 + 3 \cdot x^2 \cdot 1 + 3x \cdot 1^2 + 1^3 = x^3 + 3x^2 + 3x + 1$

**53.** $(2x+3)^3 = (2x)^3 + 3 \cdot (2x)^2 \cdot 3 + 3(2x) \cdot 3^2 + 3^3 = 8x^3 + 36x^2 + 54x + 27$

**55.** $(x-3)^3 = x^3 - 3 \cdot x^3 \cdot 3 + 3 \cdot x \cdot 3^2 - 3^3 = x^3 - 9x^2 + 27x - 27$

**57.** $(3x-4)^3 = (3x)^3 - 3(3x)^2 \cdot 4 + 3(3x) \cdot 4^2 - 4^3 = 27x^3 - 108x^2 + 144x - 64$

**59.** $(5x^2y - 3xy) + (2x^2y - xy) = (5x^2y + 2x^2y) + (-3xy - xy)$
$$= (5+2)x^2y + (-3-1)xy$$
$$= 7x^2y - 4xy \text{ is of degree 3.}$$

**61.** $(4x^2y + 8xy + 11) + (-2x^2y + 5xy + 2) = (4x^2y - 2x^2y) + (8xy + 5xy) + (11+2)$
$$= (4-2)x^2y + (8+5)xy + 13$$
$$= 2x^2y + 13xy + 13 \text{ is of degree 3.}$$

**63.** $(x^3 + 7xy - 5y^2) - (6x^3 - xy + 4y^2) = (x^3 + 7xy - 5y^2)$
$$= (x^3 - 6x^3) + (7xy + xy) + (-5y^2 - 4y^2)$$
$$= (1-6)x^3 + (7+1)xy + (-5-4)y^2$$
$$= -5x^3 + 8xy - 9y^2 \text{ is of degree 3.}$$

**65.** $(3x^4y^2 + 5x^3y - 3y) - (2x^4y^2 - 3x^3y - 4y + 6x) = (3x^4y^2 + 5x^3y - 3y) + (-2x^4y^2 + 3x^3y + 4y - 6x)$
$$= (3x^4y^2 - 2x^4y^2) + (5x^3y + 3x^3y) + (-3y + 4y) - 6x$$
$$= (3-2)x^4y^2 + (5+3)x^3y + (-3+4)y - 6x$$
$$= x^4y^2 + 8x^3y + y - 6x \text{ is of degree 6.}$$

**67.** $(x + 5y)(7x + 3y) = x(7x) + x(3y) + (5y)(7x) + (5y)(3y)$
$$= 7x^2 + 3xy + 35xy + 15y^2$$
$$= 7x^2 + 38xy + 15y^2$$

**69.** $(x - 3y)(2x + 7y) = x(2x) + x(7y) + (-3y)(2x) + (-3y)(7y)$
$$= 2x^2 + 7xy - 6xy - 21y^2$$
$$= 2x^2 + xy - 21y^2$$

**71.** $(3xy - 1)(5xy + 2) = (3xy)(5xy) + (3xy)(2) + (-1)(5xy) + (-1)(2)$
$$= 15x^2y^2 + 6xy - 5xy - 2$$
$$= 15x^2y^2 + xy - 2$$

**73.** $(7x + 5y)^2 = (7x)^2 + 2(7x)(5y) + (5y)^2 = 49x^2 + 70xy + 25y^2$

**75.** $(x^2y^2 - 3)^2 = (x^2y^2)^2 - 2(x^2y^2)(3) + 3^2 = x^4y^4 - 6x^2y^2 + 9$

**77.** $(x - y)(x^2 + xy + y^2) = x(x^2) + x(xy) + x(y^2) + (-y)(x^2) + (-y)(xy) + (-y)(y^2)$
$$= x^3 + x^2y + xy^2 - x^2y - xy^2 - y^3$$
$$= x^3 - y^3$$

**79.** $(3x + 5y)(3x - 5y) = (3x)^2 - (5y)^2 = 9x^2 - 25y^2$

**81.** $\left(7xy^2 - 10y\right)\left(7xy^2 + 10y\right) = \left(7xy^2\right)^2 - \left(10y\right)^2 = 49x^2y^4 - 100y^2$

**83.** $(3x + 4y)^2 - (3x - 4y)^2 = \left[(3x)^2 + 2(3x)(4y) + (4y)^2\right] - \left[(3x)^2 - 2(3x)(4y) + (4y)^2\right]$
$$= \left(9x^2 + 24xy + 16y^2\right) - \left(9x^2 - 24xy + 16y^2\right)$$
$$= 9x^2 + 24xy + 16y^2 - 9x^2 + 24xy - 16y^2$$
$$= 48xy$$

**85.** $(5x-7)(3x-2)-(4x-5)(6x-1)$

$=\left[15x^2-10x-21x+14\right]-\left[24x^2-4x-30x+5\right]$

$=\left(15x^2-31x+14\right)-\left(24x^2-34x+5\right)$

$=15x^2-31x+14-24x^2+34x-5$

$=-9x^2+3x+9$

**87.** $(2x+5)(2x-5)\left(4x^2+25\right)$

$=\left[(2x)^2-5^2\right]\left(4x^2+25\right)$

$=\left(4x^2-25\right)\left(4x^2+25\right)$

$=\left(4x^2\right)^2-(25)^2$

$=16x^4-625$

**89.** $\dfrac{(2x-7)^5}{(2x-7)^3}=(2x-7)^{5-3}$

$=(2x-7)^2$

$=(2x)^2-2(2x)(7)+(7)^2$

$=4x^2-28x+49$

**91.** **a.** $M=312x^2-2615x+16,615$

$M=312(16)^2-2615(16)+16,615=54,647$

The model estimates the median annual income for a man with 16 years of education to be $54,647.
The model overestimates the actual value of $54,091 shown in the bar graph by $556.

**b.** $M-W=(0.6x^3+285x^2-2256x+15,112)-(-1.2x^3+367x^2-4900x+26,561)$

$M-W=0.6x^3+285x^2-2256x+15,112+1.2x^3-367x^2+4900x-26,561$

$M-W=0.6x^3+1.2x^3+285x^2-367x^2-2256x+4900x+15,112-26,561$

$M-W=1.8x^3-82x^2+2644x-11,449$

**c.** $M-W=1.8x^3-82x^2+2644x-11,449$

$M-W=1.8(14)^3-82(14)^2+2644(14)-11,449=14,434$

The difference in the median income between men and women with 14 years experience is $14,434.

**d.** $42,163-27,027=15,136$

The actual difference displayed in the graph in the median income between men and women with 14 years experience is $15,136.
The model underestimates this difference by $15,136-\$14,434=\$702$.

**93.** $x(8-2x)(10-2x)=x\left(80-36x+4x^2\right)$

$=80x-36x^2+4x^3$

$=4x^3-36x^2+80x$

**95.** $(x+9)(x+3)-(x+5)(x+1)$

$\quad = x^2+12x+27-\left(x^2+6x+5\right)$

$\quad = x^2+12x+27-x^2-6x-5$

$\quad = 6x+22$

**97. – 101.** Answers will vary.

**103.** makes sense

**105.** makes sense

**107.** false; Changes to make the statement true will vary. A sample change is: $(3x^3+2)(3x^3-2)=9x^6-4$

**109.** false; Changes to make the statement true will vary. A sample change is: $(x+1)^2=x^2+2x+1$

**111.** $[(7x+5)+4y][(7x+5)-4y]=(7x+5)^2-4y^2$

$\quad\quad\quad\quad\quad\quad\quad\quad\quad = (7x)^2+2(7x)(5)+5^2-16y^2$

$\quad\quad\quad\quad\quad\quad\quad\quad\quad = 49x^2+70x+25-16y^2$

**113.** $(x^n+2)(x^n-2)-(x^n-3)^2$

$\quad (x^n+2)(x^n-2)-(x^n-3)^2$

$\quad = (x^{2n}-4)-(x^{2n}-6x^n+9)$

$\quad = x^{2n}-4-x^{2n}+6x^n-9$

$\quad = 6x^n-13$

**115.** $(x+3)(x+\boxed{4})=x^2+7x+12$

**116.** $(x-\boxed{2})(x-12)=x^2-14x+24$

**117.** $(4x+1)(2x-\boxed{3})=8x^2-10x-3$

**Mid-Chapter P Check Point**

**1.** $(3x+5)(4x-7)=(3x)(4x)+(3x)(-7)+(5)(4x)+(5)(-7)$

$\quad\quad\quad\quad\quad\quad\quad = 12x^2-21x+20x-35$

$\quad\quad\quad\quad\quad\quad\quad = 12x^2-x-35$

**2.** $(3x+5)-(4x-7)=3x+5-4x+7$

$\quad\quad\quad\quad\quad\quad\quad = 3x-4x+5+7$

$\quad\quad\quad\quad\quad\quad\quad = -x+12$

**3.** $\sqrt{6}+9\sqrt{6}=10\sqrt{6}$

**4.** $3\sqrt{12}-\sqrt{27}=3\cdot2\sqrt{3}-3\sqrt{3}=6\sqrt{3}-3\sqrt{3}=3\sqrt{3}$

**5.** $7x+3[9-(2x-6)]=7x+3[9-2x+6]=7x+3[15-2x]=7x+45-6x=x+45$

**6.**  $(8x-3)^2 = (8x)^2 - 2(8x)(3) + (3)^2 = 64x^2 - 48x + 9$

**7.**  $\left( x^{\frac{1}{3}} y^{-\frac{1}{2}} \right)^6 = x^{\frac{1}{3} \cdot 6} y^{-\frac{1}{2} \cdot 6} = x^2 y^{-3} = \dfrac{x^2}{y^3}$

**8.**  $\left( \dfrac{2}{7} \right)^0 - 32^{-\frac{2}{5}} = 1 - \dfrac{1}{\left( \sqrt[5]{32} \right)^2} = 1 - \dfrac{1}{(2)^2} = 1 - \dfrac{1}{4} = \dfrac{3}{4}$

**9.**  $(2x-5) - (x^2 - 3x + 1) = 2x - 5 - x^2 + 3x - 1 = -x^2 + 5x - 6$

**10.**  $(2x-5)(x^2 - 3x + 1) = 2x(x^2 - 3x + 1) - 5(x^2 - 3x + 1)$
$$\begin{aligned} &= 2x(x^2 - 3x + 1) - 5(x^2 - 3x + 1) \\ &= 2x^3 - 6x^2 + 2x - 5x^2 + 15x - 5 \\ &= 2x^3 - 6x^2 - 5x^2 + 2x + 15x - 5 \\ &= 2x^3 - 11x^2 + 17x - 5 \end{aligned}$$

**11.**  $x^3 + x^3 - x^3 \cdot x^3 = 2x^3 - x^6 = -x^6 + 2x^3$

**12.**  $(9a - 10b)(2a + b) = (9a)(2a) + (9a)(b) + (-10b)(2a) + (-10b)(b)$
$$\begin{aligned} &= (9a)(2a) + (9a)(b) + (-10b)(2a) + (-10b)(b) \\ &= 18a^2 + 9ab - 20ab - 10b^2 \\ &= 18a^2 - 11ab - 10b^2 \end{aligned}$$

**13.**  $\{a,c,d,e\} \cup \{c,d,f,h\} = \{a,c,d,e,f,h\}$

**14.**  $\{a,c,d,e\} \cap \{c,d,f,h\} = \{c,d\}$

**15.**  $\left( 3x^2 y^3 - xy + 4y^2 \right) - \left( -2x^2 y^3 - 3xy + 5y^2 \right) = 3x^2 y^3 - xy + 4y^2 + 2x^2 y^3 + 3xy - 5y^2$
$$\begin{aligned} &= 3x^2 y^3 - xy + 4y^2 + 2x^2 y^3 + 3xy - 5y^2 \\ &= 3x^2 y^3 + 2x^2 y^3 - xy + 3xy + 4y^2 - 5y^2 \\ &= 5x^2 y^3 + 2xy - y^2 \end{aligned}$$

**16.**  $\dfrac{24x^2 y^{13}}{-2x^5 y^{-2}} = -12x^{2-5} y^{13-(-2)} = -12x^{-3} y^{15} = -\dfrac{12y^{15}}{x^3}$

**17.**  $\left( \dfrac{1}{3} x^{-5} y^4 \right) \left( 18x^{-2} y^{-1} \right) = 6x^{-5-2} y^{4-1} = \dfrac{6y^3}{x^7}$

**18.**  $\sqrt[12]{x^4} = x^{\frac{4}{12}} = \left| x^{\frac{1}{3}} \right| = \left| \sqrt[3]{x} \right|$

**19.**  $\dfrac{24 \times 10^3}{2 \times 10^6} = \dfrac{24}{2} \cdot \dfrac{10^3}{10^6} = 12 \times 10^{-3} = \left( 1.2 \times 10^1 \right) \times 10^{-3} = 1.2 \times \left( 10^1 \times 10^{-3} \right) = 1.2 \times 10^{-2}$

**20.** $\dfrac{\sqrt[3]{32}}{\sqrt[3]{2}} = \sqrt[3]{\dfrac{32}{2}} = \sqrt[3]{16} = \sqrt[3]{2^4} = 2\sqrt[3]{2}$

**21.** $(x^3 + 2)(x^3 - 2) = x^6 - 4$

**22.** $(x^2 + 2)^2 = (x^2)^2 + 2(x^2)(2) + (2)^2 = x^4 + 4x^2 + 4$

**23.** $\sqrt{50} \cdot \sqrt{6} = 5\sqrt{2} \cdot \sqrt{6} = 5\sqrt{2 \cdot 6} = 5\sqrt{12} = 5 \cdot 2\sqrt{3} = 10\sqrt{3}$

**24.** $\dfrac{11}{7 - \sqrt{3}} = \dfrac{11}{7 - \sqrt{3}} \cdot \dfrac{7 + \sqrt{3}}{7 + \sqrt{3}} = \dfrac{77 + 11\sqrt{3}}{49 - 3} = \dfrac{77 + 11\sqrt{3}}{46}$

**25.** $\dfrac{11}{\sqrt{3}} = \dfrac{11}{\sqrt{3}} \cdot \dfrac{\sqrt{3}}{\sqrt{3}} = \dfrac{11\sqrt{3}}{3}$

**26.** $\left\{ -11, \ -\dfrac{3}{7}, \ 0, \ 0.45, \ \sqrt{25} \right\}$

**27.** Since $2 - \sqrt{13} < 0$ then $\left| 2 - \sqrt{13} \right| = \sqrt{13} - 2$

**28.** Since $x < 0$ then $|x| = -x$. Thus $x^2 |x| = -x^2 x = -x^3$

**29.** $4.6 \cdot 3.0 \times 10^8 = 4.6 \times 10^8 = 13.8 \times 10^8 = 1.38 \times 10^9$

The U.S. produces $1.38 \times 10^9$ pounds of garbage per day.

**30.** $\dfrac{3 \times 10^{10}}{7.5 \times 10^9} = \dfrac{3}{7.5} \cdot \dfrac{10^{10}}{10^9} = 0.4 \times 10 = 4$

A human brain has 4 times as many neurons as a gorilla brain.

**31. a.** Model 1:
$P = -86x + 890$
$P = -86(0) + 890$
$P = 890$
Model 2:
$P = 18x^2 - 175x + 950$
$P = 18(0)^2 - 175(0) + 950$
$P = 950$

Model 2 best describes the data in 2007.

**b.** $P = 18x^2 - 175x + 950$
$P = 18(5)^2 - 175(5) + 950$
$P = 525$
Model 2 overestimates the average price in 2012 by $5.

**Section P.5**

**Check Point Exercises**

1.  **a.**   $10x^3 - 4x^2$
$$= 2x^2(5x) - 2x^2(2)$$
$$= 2x^2(5x - 2)$$

    **b.**   $2x(x-7) + 3(x-7)$
$$= (x-7)(2x+3)$$

2.  $x^3 + 5x^2 - 2x - 10$
$$= (x^3 + 5x^2) - (2x + 10)$$
$$= x^2(x+5) - 2(x+5)$$
$$= (x+5)(x^2 - 2)$$

3.  Find two numbers whose product is 40 and whose sum is 13. The required integers are 8 and 5. Thus, $x^2 + 13x + 40 = (x+8)(x+5)$ or $(x+5)(x+8)$.

4.  Find two numbers whose product is –14 and whose sum is –5. The required integers are –7 and 2. Thus, $x^2 - 5x - 14 = (x-7)(x+2)$ or $(x+2)(x-7)$.

5.  Find two First terms whose product is $6x^2$.
$$6x^2 + 19x - 7 = (6x\quad)(x\quad)$$
$$6x^2 + 19x - 7 = (3x\quad)(2x\quad)$$

    Find two Last terms whose product is –7.
    The possible factors are 1(–7) and –1(7).

    Try various combinations of these factors to find the factorization in which the sum of the Outside and Inside products is 19x.

| Possible Factors of $6x^2 + 19x - 7$ | Sum of Outside and Inside Products (Should Equal 19x) |
|---|---|
| $(6x+1)(x-7)$ | $-42x + x = -41x$ |
| $(6x-7)(x+1)$ | $6x - 7x = -x$ |
| $(6x-1)(x+7)$ | $42x - x = 41x$ |
| $(6x+7)(x-1)$ | $-6x + 7x = x$ |
| $(3x+1)(2x-7)$ | $-21x + 2x = -19x$ |
| $(3x-7)(2x+1)$ | $3x - 14x = -11x$ |
| $(3x-1)(2x+7)$ | $21x - 2x = 19x$ |
| $(3x+7)(2x-1)$ | $-3x + 14x = 11x$ |

Thus, $6x^2 + 19x - 7 = (3x-1)(2x+7)$ or $(2x+7)(3x-1)$.

**6.** Find two First terms whose product is $3x^2$.

$3x^2 - 13xy + 4y^2 = (3x \quad )(x \quad )$

Find two Last terms whose product is $4y^2$.
The possible factors are $(2y)(2y)$, $(-2y)(-2y)$, $(4y)(y)$, and $(-4y)(-y)$.

Try various combinations of these factors to find the factorization in which the sum of the Outside and Inside products is $-13xy$.

$3x^2 - 13xy + y^2 = (3x - y)(x - 4y)$ or $(x - 4y)(3x - y)$.

**7.** Express each term as the square of some monomial. Then use the formula for factoring $A^2 - B^2$.

**a.** $x^2 - 81 = x^2 - 9^2 = (x+9)(x-9)$

**b.** $36x^2 - 25 = (6x)^2 - 5^2 = (6x+5)(6x-5)$

**8.** Express $81x^4 - 16$ as the difference of two squares and use the formula for factoring $A^2 - B^2$.
$81x^4 - 16 = (9x^2)^2 - 4^2 = (9x^2 + 4)(9x^2 - 4)$

The factor $9x^2 - 4$ is the difference of two squares and can be factored. Express $9x^2 - 4$ as the difference of two squares and again use the formula for factoring $A^2 - B^2$.
$(9x^2 + 4)(9x^2 - 4) = (9x^2 + 4)\left[(3x)^2 - 2^2\right] = (9x^2 + 4)(3x+2)(3x-2)$

Thus, factored completely,
$81x^4 - 16 = (9x^2 + 4)(3x+2)(3x-2)$.

**9. a.** $x^2 + 14x + 49 = x^2 + 2 \cdot x \cdot 7 + 7^2 = (x+7)^2$

**b.** Since $16x^2 = (4x)^2$ and $49 = 7^2$, check to see if the middle term can be expressed as twice the product of $4x$ and 7. Since $2 \cdot 4x \cdot 7 = 56x$, $16x^2 - 56x + 49$ is a perfect square trinomial. Thus, $16x^2 - 56x + 49 = (4x)^2 - 2 \cdot 4x \cdot 7 + 7^2$
$$= (4x-7)^2$$

**10. a.** $x^3 + 1 = x^3 + 1^3$
$$= (x+1)(x^2 - x \cdot 1 + 1^2)$$
$$= (x+1)(x^2 - x + 1)$$

**b.** $125x^3 - 8 = (5x)^3 - 2^3$
$$= (5x-2)\left[(5x)^2 + (5x)(2) + 2^2\right]$$
$$= (5x-2)(25x^2 + 10x + 4)$$

**11.** Factor out the greatest common factor.
$3x^3 - 30x^2 + 75x = 3x\left(x^2 - 10x + 25\right)$
Factor the perfect square trinomial.
$3x\left(x^2 - 10x + 25\right) = 3x\left(x - 5\right)^2$

**12.** Reorder to write as a difference of squares.

$x^2 - 36a^2 + 20x + 100$

$= x^2 + 20x + 100 - 36a^2$

$= \left(x^2 + 20x + 100\right) - 36a^2$

$= (x+10)^2 - 36a^2$

$= (x+10+6a)(x+10-6a)$

**13.** $x(x-1)^{-\frac{1}{2}} + (x-1)^{\frac{1}{2}}$

$= (x-1)^{-\frac{1}{2}} \left[ x + (x-1)^{\frac{1}{2} - \left(-\frac{1}{2}\right)} \right]$

$= (x-1)^{-\frac{1}{2}} \left[ x + (x-1) \right]$

$= (x-1)^{-\frac{1}{2}} (2x-1)$

$= \dfrac{2x-1}{(x-1)^{\frac{1}{2}}}$

**Concept and Vocabulary Check P.5**

**1.** d

**2.** g

**3.** b

**4.** c

**5.** c

**6.** a

**7.** f

**8.** $(x+1)^{\frac{1}{2}}$

**Exercise Set P.5**

**1.** $18x + 27 = 9 \cdot 2x + 9 \cdot 3 = 9(2x+3)$

**3.** $3x^2 + 6x = 3x \cdot x + 3x \cdot 2 = 3x(x+2)$

**5.** $9x^4 - 18x^3 + 27x^2$

$= 9x^2(x^2) + 9x^2(-2x) + 9x^2(3)$

$= 9x^2(x^2 - 2x + 3)$

**7.** $x(x+5) + 3(x+5) = (x+5)(x+3)$

**9.** $x^2(x-3) + 12(x-3) = (x-3)(x^2+12)$

**11.** $x^3 - 2x^2 + 5x - 10 = x^2(x-2) + 5(x-2)$
$$= (x^2 + 5)(x-2)$$

**13.** $x^3 - x^2 + 2x - 2 = x^2(x-1) + 2(x-1)$
$$= (x-1)(x^2 + 2)$$

**15.** $3x^3 - 2x^2 - 6x + 4 = x^2(3x-2) - 2(3x-2)$
$$= (3x-2)(x^2 - 2)$$

**17.** $x^2 + 5x + 6 = (x+2)(x+3)$

**19.** $x^2 - 2x - 15 = (x-5)(x+3)$

**21.** $x^2 - 8x + 15 = (x-5)(x-3)$

**23.** $3x^2 - x - 2 = (3x+2)(x-1)$

**25.** $3x^2 - 25x - 28 = (3x-28)(x+1)$

**27.** $6x^2 - 11x + 4 = (2x-1)(3x-4)$

**29.** $4x^2 + 16x + 15 = (2x+3)(2x+5)$

**31.** $9x^2 - 9x + 2 = (3x-1)(3x-2)$

**33.** $20x^2 + 27x - 8 = (5x+8)(4x-1)$

**35.** $2x^2 + 3xy + y^2 = (2x+y)(x+y)$

**37.** $6x^2 - 5xy - 6y^2 = (3x+2y)(2x-3y)$

**39.** $x^2 - 100 = x^2 - 10^2 = (x+10)(x-10)$

**41.** $36x^2 - 49 = (6x)^2 - 7^2 = (6x+7)(6x-7)$

**43.** $9x^2 - 25y^2 = (3x)^2 - (5y)^2$
$$= (3x+5y)(3x-5y)$$

**45.** $x^4 - 16 = (x^2)^2 - 4^2$
$$= (x^2 + 4)(x^2 - 4)$$
$$= (x^2 + 4)(x+2)(x-2)$$

**47.** $16x^4 - 81 = (4x^2)^2 - 9^2$
$$= (4x^2 + 9)(4x^2 - 9)$$
$$= (4x^2 + 9)[(2x)^2 - 3^2]$$
$$= (4x^2 + 9)(2x+3)(2x-3)$$

**49.** $x^2 + 2x + 1 = x^2 + 2 \cdot x \cdot 1 + 1^2 = (x+1)^2$

**51.** $x^2 - 14x + 49 = x^2 - 2 \cdot x \cdot 7 + 7^2$
$\qquad\qquad\qquad = (x-7)^2$

**53.** $4x^2 + 4x + 1 = (2x)^2 + 2 \cdot 2x \cdot 1 + 1^2$
$\qquad\qquad\qquad = (2x+1)^2$

**55.** $9x^2 - 6x + 1 = (3x)^2 - 2 \cdot 3x \cdot 1 + 1^2$
$\qquad\qquad\qquad = (3x-1)^2$

**57.** $x^3 + 27 = x^3 + 3^3$
$\qquad\qquad = (x+3)(x^2 - x \cdot 3 + 3^2)$
$\qquad\qquad = (x+3)(x^2 - 3x + 9)$

**59.** $x^3 - 64 = x^3 - 4^3$
$\qquad\qquad = (x-4)(x^2 + x \cdot 4 + 4^2)$
$\qquad\qquad = (x-4)(x^2 + 4x + 16)$

**61.** $8x^3 - 1 = (2x)^3 - 1^3$
$\qquad\qquad = (2x-1)[(2x)^2 + (2x)(1) + 1^2]$
$\qquad\qquad = (2x-1)(4x^2 + 2x + 1)$

**63.** $64x^3 + 27 = (4x)^3 + 3^3$
$\qquad\qquad\quad = (4x+3)[(4x)^2 - (4x)(3) + 3^2]$
$\qquad\qquad\quad = (4x+3)(16x^2 - 12x + 9)$

**65.** $3x^3 - 3x = 3x(x^2 - 1) = 3x(x+1)(x-1)$

**67.** $4x^2 - 4x - 24 = 4(x^2 - x - 6)$
$\qquad\qquad\qquad\quad = 4(x+2)(x-3)$

**69.** $2x^4 - 162 = 2(x^4 - 81)$
$\qquad\qquad\quad = 2[(x^2)^2 - 9^2]$
$\qquad\qquad\quad = 2(x^2 + 9)(x^2 - 9)$
$\qquad\qquad\quad = 2(x^2 + 9)(x^2 - 3^2)$
$\qquad\qquad\quad = 2(x^2 + 9)(x+3)(x-3)$

**71.** $x^3 + 2x^2 - 9x - 18 = (x^3 + 2x^2) - (9x + 18)$
$$= x^2(x+2) - 9(x+2)$$
$$= (x^2 - 9)(x+2)$$
$$= (x^2 - 3^2)(x+2)$$
$$= (x-3)(x+3)(x+2)$$

**73.** $2x^2 - 2x - 112 = 2(x^2 - x - 56) = 2(x-8)(x+7)$

**75.** $x^3 - 4x = x(x^2 - 4)$
$$= x(x^2 - 2^2)$$
$$= x(x-2)(x+2)$$

**77.** $x^2 + 64$ is prime.

**79.** $x^3 + 2x^2 - 4x - 8 = (x^3 + 2x^2) + (-4x - 8)$
$$= x^2(x+2) - 4(x+2) = (x^2 - 4)(x+2) = (x^2 - 2^2)(x+2) = (x-2)(x+2)(x+2) = (x-2)(x+2)^2$$

**81.** $y^5 - 81y$
$$= y(y^4 - 81) = y[(y^2)^2 - 9^2] = y(y^2 + 9)(y^2 - 9) = y(y^2 + 9)(y^2 - 3^2) = y(y^2 + 9)(y+3)(y-3)$$

**83.** $20y^4 - 45y^2 = 5y^2(4y^2 - 9) = 5y^2[(2y)^2 - 3^2] = 5y^2(2y+3)(2y-3)$

**85.** $x^2 - 12x + 36 - 49y^2 = (x^2 - 12x + 36) - 49y^2 = (x-6)^2 - 49y^2 = (x-6+7y)(x-6-7y)$

**87.** $9b^2x - 16y - 16x + 9b^2y$
$$= (9b^2x + 9b^2y) + (-16x - 16y) = 9b^2(x+y) - 16(x+y) = (x+y)(9b^2 - 16) = (x+y)(3b+4)(3b-4)$$

**89.** $x^2y - 16y + 32 - 2x^2$
$$= (x^2y - 16y) + (-2x^2 + 32) = y(x^2 - 16) - 2(x^2 - 16) = (x^2 - 16)(y-2) = (x+4)(x-4)(y-2)$$

**91.** $2x^3 - 8a^2x + 24x^2 + 72x$
$$= 2x(x^2 - 4a^2 + 12x + 36) = 2x[(x^2 + 12x + 36) - 4a^2] = 2x[(x+6)^2 - 4a^2] = 2x(x+6-2a)(x+6+2a)$$

**93.** $x^{\frac{3}{2}} - x^{\frac{1}{2}} = x^{\frac{1}{2}}\left(x^{\frac{3}{2} - \frac{1}{2}}\right) - 1 = x^{\frac{1}{2}}(x-1)$

**95.** $4x^{-\frac{2}{3}} + 8x^{\frac{1}{3}} = 4x^{-\frac{2}{3}}\left(1 + 2x^{\frac{1}{3} - \left(-\frac{2}{3}\right)}\right) = 4x^{-\frac{2}{3}}(1+2x) = \dfrac{4(1+2x)}{x^{\frac{2}{3}}}$

**97.** $(x+3)^{\frac{1}{2}} - (x+3)^{\frac{3}{2}} = (x+3)^{\frac{1}{2}}\left[1 - (x+3)^{\frac{3}{2}-\frac{1}{2}}\right] = (x+3)^{\frac{1}{2}}\left[1 - (x+3)\right] = (x+3)^{\frac{1}{2}}(-x-2) = -(x+3)^{\frac{1}{2}}(x+2)$

**99.** $(x+5)^{-\frac{1}{2}} - (x+5)^{-\frac{3}{2}} = (x+5)^{-\frac{3}{2}}\left[(x+5)^{-\frac{1}{2}-\left(-\frac{3}{2}\right)} - 1\right] = (x+5)^{-\frac{3}{2}}\left[(x+5) - 1\right] = (x+5)^{-\frac{3}{2}}(x+4) = \dfrac{x+4}{(x+5)^{\frac{3}{2}}}$

**101.** $(4x-1)^{\frac{1}{2}} - \dfrac{1}{3}(4x-1)^{\frac{3}{2}}$

$= (4x-1)^{\frac{1}{2}}\left[1 - \dfrac{1}{3}(4x-1)^{\frac{3}{2}-\frac{1}{2}}\right] = (4x-1)^{\frac{1}{2}}\left[1 - \dfrac{1}{3}(4x-1)\right] = (4x-1)^{\frac{1}{2}}\left[1 - \dfrac{4}{3}x + \dfrac{1}{3}\right]$

$= (4x-1)^{\frac{1}{2}}\left(\dfrac{4}{3} - \dfrac{4}{3}x\right) = (4x-1)^{\frac{1}{2}} \dfrac{4}{3}(1-x) = \dfrac{-4(4x-1)^{\frac{1}{2}}(x-1)}{3}$

**103.** $10x^2(x+1) - 7x(x+1) - 6(x+1) = (x+1)\left(10x^2 - 7x - 6\right) = (x+1)(5x-6)(2x+1)$

**105.** $6x^4 + 35x^2 - 6 = \left(x^2 + 6\right)\left(6x^2 - 1\right)$

**107.** $y^7 + y = y\left(y^6 + 1\right) = y\left[\left(y^2\right)^3 + 1^3\right] = y\left(y^2 + 1\right)\left(y^4 - y^2 + 1\right)$

**109.** $x^4 - 5x^2y^2 + 4y^4 = \left(x^2 - 4y^2\right)\left(x^2 - y^2\right) = (x+2y)(x-2y)(x+y)(x-y)$

**111.** $(x-y)^4 - 4(x-y)^2$

$= (x-y)^2\left((x-y)^2 - 4\right) = (x-y)^2\left((x-y)+2\right)\left((x-y)-2\right) = (x-y)^2(x-y+2)(x-y-2)$

**113.** $2x^2 - 7xy^2 + 3y^4 = \left(2x - y^2\right)\left(x - 3y^2\right)$

**115. a.** $(x - 0.4x) - 0.4(x - 0.4x) = (x - 0.4x)(1 - 0.4) = (0.6x)(0.6) = 0.36x$

    **b.** No, the computer is selling at 36% of its original price.

**117. a.** $(3x)^2 - 4 \cdot 2^2 = 9x^2 - 16$

    **b.** $9x^2 - 16 = (3x+4)(3x-4)$

**119. a.** $x(x+y) - y(x+y)$

    **b.** $x(x+y) - y(x+y) = (x+y)(x-y)$

**121.** $V_{\text{shaded}} = V_{\text{outside}} - V_{\text{inside}}$

$\phantom{V_{\text{shaded}}} = a \cdot a \cdot 4a - b \cdot b \cdot 4a$

$\phantom{V_{\text{shaded}}} = 4a^3 - 4ab^2$

$\phantom{V_{\text{shaded}}} = 4a\left(a^2 - b^2\right)$

$\phantom{V_{\text{shaded}}} = 4a(a+b)(a-b)$

**123. – 129.** Answers will vary.

**131.** makes sense

**133.** makes sense

**135.** true

**137.** false; Changes to make the statement true will vary. A sample change is: $x^3 - 64 = (x-4)(x+4x+16)$

**139.** $-x^2 - 4x + 5 = -1\left(x^2 + 4x - 5\right) = -1(x+5)(x-1) = -(x+5)(x-1)$

**141.** $(x-5)^{-\frac{1}{2}}(x+5)^{-\frac{1}{2}} - (x+5)^{\frac{1}{2}}(x-5)^{-\frac{3}{2}} = (x-5)^{-\frac{3}{2}}(x+5)^{-\frac{1}{2}}\left[(x-5)^{\frac{1}{2}-\left(-\frac{3}{2}\right)} - (x+5)^{\frac{1}{2}-\left(-\frac{1}{2}\right)}\right]$

$\phantom{xxxxxxxxxxxxxxxxxxxxxxxxxxxxxxxxxxx} = (x-5)^{-\frac{3}{2}}(x+5)^{-\frac{1}{2}}\left[(x-5) - (x+5)\right]$

$\phantom{xxxxxxxxxxxxxxxxxxxxxxxxxxxxxxxxxxx} = (x-5)^{-\frac{3}{2}}(x+5)^{-\frac{1}{2}}(-10) = \dfrac{-10}{(x-5)^{3/2}(x+5)^{1/2}}$

**143.** $b = 0, 3, 4,$ or $-c(c+4)$, where $c > 0$ is an integer.

**144.** $\dfrac{x^2 + 6x + 5}{x^2 - 25} = \dfrac{(x+5)(x+1)}{(x+5)(x-5)} = \dfrac{x+1}{x-5}$

**145.** $\dfrac{5}{4} \cdot \dfrac{8}{15} = \dfrac{5}{4} \cdot \dfrac{4 \cdot 2}{5 \cdot 3} = \dfrac{1}{1} \cdot \dfrac{2}{3} = \dfrac{2}{3}$

**146.** $\dfrac{1}{2} + \dfrac{2}{3} = \dfrac{3}{6} + \dfrac{4}{6} = \dfrac{7}{6}$

**Section P.6**

**Check Point Exercises**

1.  **a.**  The denominator would equal zero if $x = -5$, so $-5$ must be excluded from the domain.

    **b.**  $x^2 - 36 = (x+6)(x-6)$

    The denominator would equal zero if $x = -6$ or $x = 6$, so $-6$ and $6$ must both must be excluded from the domain.

    **c.**  $x^2 - 5x - 14 = (x+2)(x-7)$

    The denominator would equal zero if $x = -2$ or $x = 7$, so $-2$ and $7$ must both must be excluded from the domain.

2.  **a.**  $\dfrac{x^3 + 3x^2}{x+3} = \dfrac{x^2(x+3)}{x+3}$

    $= \dfrac{x^2(x+3)}{x+3}$

    $= x^2, \ x \neq -3$

    Because the denominator is $x + 3$, $x \neq -3$

    **b.**  $\dfrac{x^2 - 1}{x^2 + 2x + 1} = \dfrac{(x-1)(x+1)}{(x+1)(x+1)} = \dfrac{x-1}{x+1}, x \neq -1$

    Because the denominator is
    $(x+1)(x+1), x \neq -1$

3.  $\dfrac{x+3}{x^2 - 4} \cdot \dfrac{x^2 - x - 6}{x^2 + 6x + 9}$

    $= \dfrac{x+3}{(x+2)(x-2)} \cdot \dfrac{(x-3)(x+2)}{(x+3)(x+3)}$

    $= \dfrac{x+3}{(x+2)(x-2)} \cdot \dfrac{(x-3)(x+2)}{(x+3)(x+3)}$

    $= \dfrac{x-3}{(x-2)(x+3)}, \ x \neq -2, \ x \neq 2, \ x \neq -3$

    Because the denominator has factors of
    $x+2$, $x-2$, and $x+3$, $x \neq -2$, $x \neq 2$, and $x \neq -3$.

4.  $\dfrac{x^2 - 2x + 1}{x^3 + x} \div \dfrac{x^2 + x - 2}{3x^2 + 3}$

    $= \dfrac{x^2 - 2x + 1}{x^3 + x} \cdot \dfrac{3x^2 + 3}{x^2 + x - 2}$

    $= \dfrac{(x-1)(x-1)}{x(x^2+1)} \cdot \dfrac{3(x^2+1)}{(x+2)(x-1)}$

    $= \dfrac{3(x-1)}{x(x+2)}, \ x \neq 1, \ x \neq 0, \ x \neq -2$

5.  $\dfrac{x}{x+1} - \dfrac{3x+2}{x+1} = \dfrac{x - 3x - 2}{x+1}$

    $= \dfrac{-2x - 2}{x+1}$

    $= \dfrac{-2(x+1)}{x+1}$

    $= -2, x \neq -1$

6.  $\dfrac{3}{x+1} + \dfrac{5}{x-1}$

    $= \dfrac{3x(x-1) + 5(x+1)}{(x+1)(x-1)}$

    $= \dfrac{3x - 3 + 5x + 5}{(x+1)(x-1)}$

    $= \dfrac{8x + 2}{(x+1)(x-1)}$

    $= \dfrac{2(4x+1)}{(x+1)(x-1)}$

    $= \dfrac{2(4x+1)}{(x+1)(x-1)}, \ x \neq -1 \text{ and } x \neq 1.$

7.  Factor each denominator completely.
    $x^2 - 6x + 9 = (x-3)^2$

    $x^2 - 9 = (x+3)(x-3)$

    List the factors of the first denominator.
    $x - 3, \ x - 3$

    Add any unlisted factors from the second denominator.
    $x - 3, \ x - 3, \ x + 3$

    The least common denominator is the product of all factors in the final list.
    $(x-3)(x-3)(x+3)$ or $(x-3)^2(x+3)$ is the least common denominator.

8.  Find the least common denominator.
    $x^2 - 10x + 25 = (x-5)^2$
    $2x - 10 = 2(x-5)$

    The least common denominator is $2(x-5)(x-5)$.
    Write all rational expressions in terms of the least common denominator.

    $$\dfrac{x}{x^2 - 10x + 25} - \dfrac{x-4}{2x-10}$$

    $$= \dfrac{x}{(x-5)(x-5)} - \dfrac{x-4}{2(x-5)}$$

    $$= \dfrac{2x}{2(x-5)(x-5)} - \dfrac{(x-4)(x-5)}{2(x-5)(x-5)}$$

Add numerators, putting this sum over the least common denominator.

$$= \frac{2x-(x-4)(x-5)}{2(x-5)(x-5)}$$

$$= \frac{2x-(x^2-5x-4x+20)}{2(x-5)(x-5)}$$

$$= \frac{2x-x^2+5x+4x-20}{2(x-5)(x-5)}$$

$$= \frac{2x-x^2+5x+4x-20}{2(x-5)(x-5)}$$

$$= \frac{-x^2+11x-20}{2(x-5)(x-5)}$$

$$= \frac{-x^2+11x-20}{2(x-5)^2}, \ x \neq 5$$

9. $\dfrac{\dfrac{1}{x}-\dfrac{3}{2}}{\dfrac{1}{x}+\dfrac{3}{4}} = \dfrac{\dfrac{2}{2x}-\dfrac{3x}{2x}}{\dfrac{4}{4x}+\dfrac{3x}{4x}}, \ x \neq 0$

$$= \frac{\dfrac{2-3x}{2x}}{\dfrac{4+3x}{4x}}, \ x \neq \frac{-4}{3}$$

$$= \frac{2-3x}{2x} \div \frac{4+3x}{4x}$$

$$= \frac{2-3x}{2x} \cdot \frac{4x}{4+3x}$$

$$= \frac{2-3x}{4+3x} \cdot \frac{4}{2}$$

$$= \frac{2-3x}{4+3x} \cdot \frac{2}{1}$$

$$= \frac{2(2-3x)}{4+3x}, \ x \neq 0, \ x \neq \frac{-4}{3}$$

10. Multiply each of the three terms, $\dfrac{1}{x+7}$, $\dfrac{1}{x}$, and 7 by the least common denominator of $x(x+7)$.

$$\frac{\dfrac{1}{x+7}-\dfrac{1}{x}}{7} = \frac{x(x+7)\left(\dfrac{1}{x+7}\right)-x(x+7)\left(\dfrac{1}{x}\right)}{7x(x+7)}$$

$$= \frac{x-(x+7)}{7x(x+7)}$$

$$= \frac{-7}{7x(x+7)}$$

$$= -\frac{1}{x(x+7)}, \ x \neq 0, \ x \neq -7$$

**Concept and Vocabulary Check P.6**

1. polynomials

2. domain; 0

3. factoring; common factors

4. $\dfrac{x^2}{15}$

5. $\dfrac{3}{5}$

6. $\dfrac{x^2-x+4}{3}$

7. $x+3$ and $x-2$; $x+3$ and $x+1$; $(x+3)(x-2)(x+1)$

8. $3x+4$

9. complex; complex

10. $x$; $x+3$; $-3$; $\dfrac{1}{x(x+3)}$

**Exercise Set P.6**

1. $\dfrac{7}{x-3}, \ x \neq 3$

3. $\dfrac{x+5}{x^2-25} = \dfrac{x+5}{(x+5)(x-5)}, \ x \neq 5, -5$

5. $\dfrac{x-1}{x^2+11x+10} = \dfrac{x-1}{(x+1)(x+10)}, \ x \neq -1, -10$

7. $\dfrac{3x-9}{x^2-6x+9} = \dfrac{3(x-3)}{(x-3)(x-3)}$
   $= \dfrac{3}{x-3}, \ x \neq 3$

9. $\dfrac{x^2-12x+36}{4x-24} = \dfrac{(x-6)(x-6)}{4(x-6)} = \dfrac{x-6}{4}.$
   $x \neq 6$

11. $\dfrac{y^2+7y-18}{y^2-3y+2} = \dfrac{(y+9)(y-2)}{(y-2)(y-1)} = \dfrac{y+9}{y-1},$
    $y \neq 1, 2$

**13.** $\dfrac{x^2+12x+36}{x^2-36}=\dfrac{(x+6)^2}{(x+6)(x-6)}=\dfrac{x+6}{x-6},$

$x \neq 6, -6$

**15.** $\dfrac{x-2}{3x+9}\cdot\dfrac{2x+6}{2x-4}=\dfrac{x-2}{3(x+3)}\cdot\dfrac{2(x+3)}{2(x-2)}$

$\qquad\qquad = \dfrac{2}{6}=\dfrac{1}{3}, x \neq 2, -3$

**17.** $\dfrac{x^2-9}{x^2}\cdot\dfrac{x^2-3x}{x^2+x-12}$

$= \dfrac{(x-3)(x+3)}{x^2}\cdot\dfrac{x(x-3)}{(x+4)(x-3)}$

$= \dfrac{(x-3)(x+3)}{x(x+4)}, x \neq 0, -4, 3$

**19.** $\dfrac{x^2-5x+6}{x^2-2x-3}\cdot\dfrac{x^2-1}{x^2-4}$

$= \dfrac{(x-3)(x-2)}{(x-3)(x+1)}\cdot\dfrac{(x+1)(x-1)}{(x-2)(x+2)}$

$= \dfrac{x-1}{x+2}, \ x \neq -2, -1, 2, 3$

**21.** $\dfrac{x^3-8}{x^2-4}\cdot\dfrac{x+2}{3x}=\dfrac{(x-2)\left(x^2+2x+4\right)}{(x-2)(x+2)}\cdot\dfrac{x+2}{3x}$

$= \dfrac{x^2+2x+4}{3x}, x \neq -2, 0, 2$

**23.** $\dfrac{x+1}{3}\div\dfrac{3x+3}{7}=\dfrac{x+1}{3}\div\dfrac{3(x+1)}{7}$

$\qquad\qquad = \dfrac{x+1}{3}\cdot\dfrac{7}{3(x+1)}$

$\qquad\qquad = \dfrac{7}{9}, x \neq -1$

**25.** $\dfrac{x^2-4}{x}\div\dfrac{x+2}{x-2}=\dfrac{(x-2)(x+2)}{x}\cdot\dfrac{x-2}{x+2}$

$\qquad\qquad = \dfrac{(x-2)^2}{x}; x \neq 0, -2, 2$

**27.** $\dfrac{4x^2+10}{x-3}\div\dfrac{6x^2+15}{x^2-9}$

$= \dfrac{2(2x^2+5)}{x-3}\div\dfrac{3(2x^2+5)}{(x-3)(x+3)}$

$= \dfrac{2(2x^2+5)}{x-3}\cdot\dfrac{(x-3)(x+3)}{3(2x^2+5)}$

$= \dfrac{2(x+3)}{3}, x \neq 3, -3$

**29.** $\dfrac{x^2-25}{2x-2}\div\dfrac{x^2+10x+25}{x^2+4x-5}$

$= \dfrac{(x-5)(x+5)}{2(x-1)}\div\dfrac{(x+5)^2}{(x+5)(x-1)}$

$= \dfrac{(x-5)(x+5)}{2(x-1)}\cdot\dfrac{(x+5)(x-1)}{(x+5)^2}$

$= \dfrac{x-5}{2}, x \neq 1, -5$

**31.** $\dfrac{x^2+x-12}{x^2+x-30}\cdot\dfrac{x^2+5x+6}{x^2-2x-3}\div\dfrac{x+3}{x^2+7x+6}$

$= \dfrac{(x+4)(x-3)}{(x+6)(x-5)}\cdot\dfrac{(x+2)(x+3)}{(x+1)(x-3)}\cdot\dfrac{(x+6)(x+1)}{x+3}$

$= \dfrac{(x+4)(x+2)}{x-5}$

$x \neq -6, -3, -1, 3, 5$

**33.** $\dfrac{4x+1}{6x+5}+\dfrac{8x+9}{6x+5}=\dfrac{4x+1+8x+9}{6x+5}$

$\qquad\qquad = \dfrac{12x+10}{6x+5}$

$\qquad\qquad = \dfrac{2(6x+5)}{6x+5}=2, x \neq -\dfrac{5}{6}$

**35.** $\dfrac{x^2-2x}{x^2+3x}+\dfrac{x^2+x}{x^2+3x}=\dfrac{x^2-2x+x^2+x}{x^2+3x}$

$\qquad\qquad = \dfrac{2x^2-x}{x^2+3x}$

$\qquad\qquad = \dfrac{x(2x-1)}{x(x+3)}$

$\qquad\qquad = \dfrac{2x-1}{x+3}, x \neq 0, -3$

**37.** $\dfrac{4x-10}{x-2}-\dfrac{x-4}{x-2}=\dfrac{4x-10-(x-4)}{x-2}$

$=\dfrac{4x-10-x+4}{x-2}$

$=\dfrac{3x-6}{x-2}$

$=\dfrac{3(x-2)}{x-2}$

$=3,\ x\neq 2$

**39.** $\dfrac{x^2+3x}{x^2+x-12}-\dfrac{x^2-12}{x^2+x-12}$

$=\dfrac{x^2+3x-(x^2-12)}{x^2+x-12}$

$=\dfrac{x^2+3x-x^2+12}{x^2+x-12}$

$=\dfrac{3x+12}{x^2+x-12}$

$=\dfrac{3(x+4)}{(x+4)(x-3)}$

$=\dfrac{3}{x-3},\ x\neq 3,-4$

**41.** $\dfrac{3}{x+4}+\dfrac{6}{x+5}=\dfrac{3(x+5)+6(x+4)}{(x+4)(x+5)}$

$=\dfrac{3x+15+6x+24}{(x+4)(x+5)}$

$=\dfrac{9x+39}{(x+4)(x+5)},\ x\neq -4,-5$

**43.** $\dfrac{3}{x+1}-\dfrac{3}{x}=\dfrac{3x-3(x+1)}{x(x+1)}$

$=\dfrac{3x-3x-3}{x(x+1)}=-\dfrac{3}{x(x+1)},\ x\neq -1,0$

**45.** $\dfrac{2x}{x+2}+\dfrac{x+2}{x-2}=\dfrac{2x(x-2)+(x+2)(x+2)}{(x+2)(x-2)}$

$=\dfrac{2x^2-4x+x^2+4x+4}{(x+2)(x-2)}$

$=\dfrac{3x^2+4}{(x+2)(x-2)},\ x\neq -2,2$

**47.** $\dfrac{x+5}{x-5}+\dfrac{x-5}{x+5}$

$=\dfrac{(x+5)(x+5)+(x-5)(x-5)}{(x-5)(x+5)}$

$=\dfrac{x^2+10x+25+x^2-10x+25}{(x-5)(x+5)}$

$=\dfrac{2x^2+50}{(x-5)(x+5)},\ x\neq -5,5$

**49.** $\dfrac{3}{2x+4}+\dfrac{2}{3x+6}=\dfrac{3}{2(x+2)}+\dfrac{2}{3(x+2)}$

$=\dfrac{9}{6(x+2)}+\dfrac{4}{6(x+2)}$

$=\dfrac{9+4}{6(x+2)}$

$=\dfrac{13}{6(x+2)}$

$x\neq -2$

**51.** $\dfrac{4}{x^2+6x+9}+\dfrac{4}{x+3}=\dfrac{4}{(x+3)^2}+\dfrac{4}{x+3}$

$=\dfrac{4+4(x+3)}{(x+3)^2}=\dfrac{4+4x+12}{(x+3)^2}=\dfrac{4x+16}{(x+3)^2},$

$x\neq -3$

**53.** $\dfrac{3x}{x^2+3x-10}-\dfrac{2x}{x^2+x-6}$

$=\dfrac{3x}{(x+5)(x-2)}-\dfrac{2x}{(x+3)(x-2)}$

$=\dfrac{3x(x+3)-2x(x+5)}{(x+5)(x-2)(x+3)}$

$=\dfrac{3x^2+9x-2x^2-10x}{(x+5)(x-2)(x+3)}$

$=\dfrac{x^2-x}{(x+5)(x-2)(x+3)},\ x\neq -5,2,-3$

**55.** $\dfrac{x+3}{x^2-1}-\dfrac{x+2}{x-1}$

$=\dfrac{x+3}{(x+1)(x-1)}-\dfrac{x+2}{x-1}$

$=\dfrac{x+3}{(x+1)(x-1)}-\dfrac{(x+1)(x+2)}{(x+1)(x-1)}$

$=\dfrac{x+3}{(x+1)(x-1)}-\dfrac{x^2+3x+2}{(x+1)(x-1)}$

$=\dfrac{x+3-x^2-3x-2}{(x+1)(x-1)}$

$=\dfrac{-x^2-2x+1}{(x+1)(x-1)}$

$x\neq1,-1$

**57.** $\dfrac{4x^2+x-6}{x^2+3x+2}-\dfrac{3x}{x+1}+\dfrac{5}{x+2}$

$=\dfrac{4x^2+x-6}{(x+1)(x+2)}+\dfrac{-3x}{x+1}+\dfrac{5}{x+2}$

$=\dfrac{4x^2+x-5}{(x+1)(x+2)}+\dfrac{-3x(x+2)}{(x+1)(x+2)}+\dfrac{5(x+1)}{(x+1)(x+2)}$

$=\dfrac{4x^2+x-6-3x^2-6x+5x+5}{(x+1)(x+2)}$

$=\dfrac{x^2-1}{(x+1)(x+2)}$

$=\dfrac{(x-1)(x+1)}{(x+1)(x+2)}$

$=\dfrac{x-1}{x+2};x\neq-2,-1$

**59.** $\dfrac{\frac{x}{3}-1}{x-3}=\dfrac{3\left[\frac{x}{3}-1\right]}{3\left[x-3\right]}=\dfrac{x-3}{3(x-3)}=\dfrac{1}{3},\ x\neq3$

**61.** $\dfrac{1+\frac{1}{x}}{3-\frac{1}{x}}=\dfrac{x\left[1+\frac{1}{x}\right]}{x\left[3-\frac{1}{x}\right]}=\dfrac{x+1}{3x-1},x\neq0,\dfrac{1}{3}$

**63.** $\dfrac{\frac{1}{x}+\frac{1}{y}}{x+y}=\dfrac{xy\left[\frac{1}{x}+\frac{1}{y}\right]}{xy\left[x+y\right]}=\dfrac{y+x}{xy(x+y)}=\dfrac{1}{xy},$

$x\neq0,\ y\neq0,\ x\neq-y$

**65.** $\dfrac{x-\frac{x}{x+3}}{x+2}=\dfrac{(x+3)\left[x-\frac{x}{x+3}\right]}{(x+3)(x+2)}=\dfrac{x(x+3)-x}{(x+3)(x+2)}$

$=\dfrac{x^2+3x-x}{(x+3)(x+2)}=\dfrac{x^2+2x}{(x+3)(x+2)}$

$=\dfrac{x(x+2)}{(x+3)(x+2)}=\dfrac{x}{x+3},x\neq-2,-3$

**67.** $\dfrac{\frac{3}{x-2}-\frac{4}{x+2}}{x^2-4}=\dfrac{\frac{3}{x-2}-\frac{4}{x+2}}{\frac{7}{(x-2)(x+2)}}$

$=\dfrac{\left[\frac{3}{x-2}-\frac{4}{x+2}\right](x-2)(x+2)}{\left[\frac{7}{(x-2)(x+2)}\right](x-2)(x+2)}$

$=\dfrac{3(x+2)-4(x-2)}{7}$

$=\dfrac{3x+6-4x+8}{7}=\dfrac{-x+14}{7}$

$=-\dfrac{x-14}{7}\ \ x\neq-2,2$

**69.** $\dfrac{\frac{1}{x+1}}{\frac{1}{x^2-2x-3}+\frac{1}{x-3}}=\dfrac{\frac{1}{x+1}}{\frac{1}{(x+1)(x-3)}+\frac{1}{x-3}}$

$=\dfrac{\frac{(x+1)(x-3)}{x+1}}{\frac{(x+1)(x-3)}{(x+1)(x-3)}+\frac{(x+1)(x-3)}{x-3}}$

$=\dfrac{x-3}{1+x+1}$

$=\dfrac{x-3}{x+2}\ \ x\neq-2,-1,3$

**71.**
$$\frac{\dfrac{1}{(x+h)^2}-\dfrac{1}{x^2}}{h}=\frac{\dfrac{x^2(x+h)^2}{(x+h)^2}-\dfrac{x^2(x+h)^2}{x^2}}{hx^2(x+h)^2}$$

$$=\frac{x^2-(x+h)^2}{hx^2(x+h)^2}$$

$$=\frac{x^2-(x^2+2hx+h^2)}{hx^2(x+h)^2}$$

$$=\frac{x^2-x^2-2hx-h^2}{hx^2(x+h)^2}$$

$$=\frac{-2hx-h^2}{hx^2(x+h)^2}$$

$$=\frac{-h(2x+h)}{hx^2(x+h)^2}$$

$$=-\frac{(2x+h)}{x^2(x+h)^2}$$

**73.**
$$\left(\frac{2x+3}{x+1}\cdot\frac{x^2+4x-5}{2x^2+x-3}\right)-\frac{2}{x+2}=\left(\frac{\cancel{(2x+3)}}{x+1}\cdot\frac{(x+5)\cancel{(x-1)}}{\cancel{(2x+3)}\cancel{(x-1)}}\right)-\frac{2}{x+2}=\frac{x+5}{x+1}-\frac{2}{x+2}$$

$$=\frac{(x+5)(x+2)}{(x+1)(x+2)}-\frac{2(x+1)}{(x+1)(x+2)}=\frac{(x+5)(x+2)-2(x+1)}{(x+1)(x+2)}=\frac{x^2+2x+5x+10-2x-2}{(x+1)(x+2)}=\frac{x^2+5x+8}{(x+1)(x+2)}$$

**75.**
$$\left(2-\frac{6}{x+1}\right)\left(1+\frac{3}{x-2}\right)=\left(\frac{2(x+1)}{(x+1)}-\frac{6}{(x+1)}\right)\left(\frac{(x-2)}{(x-2)}+\frac{3}{(x-2)}\right)$$

$$=\left(\frac{2x+2-6}{x+1}\right)\left(\frac{x-2+3}{x-2}\right)=\left(\frac{2x-4}{x+1}\right)\left(\frac{x+1}{x-2}\right)=\frac{2\cancel{(x-2)}\cancel{(x+1)}}{\cancel{(x+1)}\cancel{(x-2)}}=2$$

**77.**
$$\frac{y^{-1}-(y+5)^{-1}}{5}=\frac{\dfrac{1}{y}-\dfrac{1}{y+5}}{5}$$

LCD $=y(y+5)$

$$\frac{\dfrac{1}{y}-\dfrac{1}{y+5}}{5}=\frac{y(y+5)\left(\dfrac{1}{y}-\dfrac{1}{y+5}\right)}{y(y+5)(5)}=\frac{y+5-y}{5y(y+5)}=\frac{5}{5y(y+5)}=\frac{1}{y(y+5)}$$

**79.**
$$\left(\frac{1}{a^3-b^3}\cdot\frac{ac+ad-bc-bd}{1}\right)-\frac{c-d}{a^2+ab+b^2}=\left(\frac{1}{(a-b)(a^2+ab+b^2)}\cdot\frac{a(c+d)-b(c+d)}{1}\right)-\frac{c-d}{a^2+ab+b^2}$$

$$=\left(\frac{1}{\cancel{(a-b)}(a^2+ab+b^2)}\cdot\frac{(c+d)\cancel{(a-b)}}{1}\right)-\frac{c-d}{a^2+ab+b^2}=\frac{c+d}{a^2+ab+b^2}-\frac{c-d}{a^2+bd+b^2}$$

$$=\frac{c+d-c+d}{a^2+ab+b^2}=\frac{2d}{a^2+ab+b^2}$$

**81. a.**  $\dfrac{130x}{100-x}$ is equal to

1. $\dfrac{130\cdot 40}{100-40}=\dfrac{130\cdot 40}{60}=86.67$,
 when $x=40$

2. $\dfrac{130\cdot 80}{100-80}=\dfrac{130\cdot 80}{20}=520$,
 when $x=80$

3. $\dfrac{130\cdot 90}{100-90}=\dfrac{130\cdot 90}{10}=1170$,
 when $x=90$

It costs \$86,670,000 to inoculate 40% of the population against this strain of flu, and \$520,000,000 to inoculate 80% of the population, and \$1,170,000,000 to inoculate 90% of the population.

**b.** For $x=100$, the function is not defined.

**c.** As $x$ approaches 100, the value of the function increases rapidly. So it costs an astronomical amount of money to inoculate almost all of the people, and it is impossible to inoculate 100% of the population.

**83. a.** Substitute 4 for $x$ in the model.

$W=-66x^2+526x+1030$

$W=-66(4)^2+526(4)+1030$

$W=2078$

According to the model, women between the ages of 19 and 30 with this lifestyle need 2078 calories per day. This underestimates the actual value shown in the bar graph by 22 calories.

**b.** Substitute 4 for $x$ in the model.

$M=-120x^2+998x+590$

$M=-120(4)^2+998(4)+590$

$M=2662$

According to the model, men between the ages of 19 and 30 with this lifestyle need 2662 calories per day. This underestimates the actual value shown in the bar graph by 38 calories.

**c.** $\dfrac{W}{M}=\dfrac{-66x^2+526x+1030}{-120x^2+998x+590}$

$=\dfrac{2(-33x^2+263x+515)}{2(-60x^2+499x+295)}$

$=\dfrac{-33x^2+263x+515}{-60x^2+499x+295}$

**85.**   $P = 2L + 2W$

$$= 2\left(\frac{x}{x+5}\right) + 2\left(\frac{x}{x+6}\right)$$

$$= \frac{2x}{x+5} + \frac{2x}{x+6}$$

$$= \frac{2x(x+6)}{(x+5)(x+6)} + \frac{2x(x+5)}{(x+5)(x+6)}$$

$$= \frac{2x^2 + 12x + 2x^2 + 10x}{(x+5)(x+6)}$$

$$= \frac{4x^2 + 22x}{(x+5)(x+6)}$$

**87. – 97.**   Answers will vary.

**99.**   does not make sense; Explanations will vary. Sample explanation: The numerator and denominator of $\dfrac{7}{14+x}$ do not share a common factor.

**101.**   makes sense

**103.**   true

**105.**   false; Changes to make the statement true will vary. A sample change is: $6 + \dfrac{1}{x} = \dfrac{6x}{x} + \dfrac{1}{x} = \dfrac{6x+1}{x}$

**107.**   $\left(1-\dfrac{1}{x}\right)\left(1-\dfrac{1}{x+1}\right)\left(1-\dfrac{1}{x+2}\right)\left(1-\dfrac{1}{x+3}\right) = \left(\dfrac{x}{x} - \dfrac{1}{x}\right)\left(\dfrac{x+1}{x+1} - \dfrac{1}{x+1}\right)\left(\dfrac{x+2}{x+2} - \dfrac{1}{x+2}\right)\left(\dfrac{x+3}{x+3} - \dfrac{1}{x+3}\right)$

$$= \left(\frac{x-1}{x}\right)\left(\frac{(x+1)-1}{x+1}\right)\left(\frac{(x+2)-1}{x+2}\right)\left(\frac{(x+3)-1}{x+3}\right)$$

$$= \frac{x-1}{\cancel{x}} \cdot \frac{\cancel{x}}{\cancel{x+1}} \cdot \frac{\cancel{x+1}}{\cancel{x+2}} \cdot \frac{\cancel{x+2}}{x+3} = \frac{x-1}{x+3}$$

**109.**   It cubes $x$.

$$\frac{\dfrac{1}{x} + \dfrac{1}{x^2} + \dfrac{1}{x^3}}{\dfrac{1}{x^4} + \dfrac{1}{x^5} + \dfrac{1}{x^6}} = \frac{\dfrac{x^6}{x} + \dfrac{x^6}{x^2} + \dfrac{x^6}{x^3}}{\dfrac{x^6}{x^4} + \dfrac{x^6}{x^5} + \dfrac{x^6}{x^6}} = \frac{x^5 + x^4 + x^3}{x^2 + x + 1} = \frac{x^3(x^2 + x + 1)}{x^2 + x + 1} = x^3$$

**110.**   $y = 4 - x^2$

**111.**   $y = 1 - x^2$

**112.** $y = |x+1|$

| $x$ | $y = |x+1|$ |
|---|---|
| $-4$ | $|-4+1| = 3$ |
| $-3$ | $|-3+1| = 2$ |
| $-2$ | $|-2+1| = 1$ |
| $-1$ | $|-1+1| = 0$ |
| $0$ | $|0+1| = 1$ |
| $1$ | $|1+1| = 2$ |
| $2$ | $|2+1| = 3$ |

## Chapter P Review Exercises

**1.** $3 + 6(x-2)^3 = 3 + 6(4-2)^3$
$$= 3 + 6(2)^3$$
$$= 3 + 6(8)$$
$$= 3 + 48$$
$$= 51$$

**2.** $x^2 - 5(x-y) = 6^2 - 5(6-2)$
$$= 36 - 5(4)$$
$$= 36 - 20$$
$$= 16$$

**3.** $S = 0.015x^2 + x + 10$
$S = 0.015(60)^2 + (60) + 10$
$$= 0.015(3600) + 60 + 10$$
$$= 54 + 60 + 10$$
$$= 124$$

**4.** $A = \{a,b,c\}$  $B = \{a,c,d,e\}$
$\{a,b,c\} \cap \{a,c,d,e\} = \{a,c\}$

**5.** $A = \{a,b,c\}$  $B = \{a,c,d,e\}$
$\{a,b,c\} \cup \{a,c,d,e\} = \{a,b,c,d,e\}$

**6.** $A = \{a,b,c\}$  $C = \{a,d,f,g\}$
$\{a,b,c\} \cup \{a,d,f,g\} = \{a,b,c,d,f,g\}$

**7.** $A = \{a,b,c\}$  $C = \{a,d,f,g\}$
$\{a,d,f,g\} \cap \{a,b,c\} = \{a\}$

**8.** **a.** $\sqrt{81}$

　　**b.** $0, \sqrt{81}$

　　**c.** $-17, 0, \sqrt{81}$

　　**d.** $-17, -\dfrac{9}{13}, 0, 0.75, \sqrt{81}$

　　**e.** $\sqrt{2}, \pi$

　　**f.** $-17, -\dfrac{9}{13}, 0, 0.75, \sqrt{2}, \pi, \sqrt{81}$

**9.** $|-103| = 103$

**10.** $|\sqrt{2}-1| = \sqrt{2}-1$

**11.** $|3 - \sqrt{17}| = \sqrt{17} - 3$ since $\sqrt{17}$ is greater than 3.

**12.** $|4-(-17)| = |4+17| = |21| = 21$

**13.** $3 + 17 = 17 + 3$;
commutative property of addition.

**14.** $(6 \cdot 3) \cdot 9 = 6 \cdot (3 \cdot 9)$;
associative property of multiplication.

**15.** $\sqrt{3}(\sqrt{5}+\sqrt{3}) = \sqrt{15}+3$ ;

distributive property of multiplication over addition.

**16.** $(6\cdot9)\cdot2 = 2\cdot(6\cdot9)$ ;

commutative property of multiplication.

**17.** $\sqrt{3}(\sqrt{5}+\sqrt{3}) = (\sqrt{5}+\sqrt{3})\sqrt{3}$ ;

commutative property of multiplication.

**18.** $(3\cdot7)+(4\cdot7) = (4\cdot7)+(3\cdot7)$ ;

commutative property of addition.

**19.** $5(2x-3)+7x = 10x-15+7x = 17x-15$

**20.** $\dfrac{1}{5}(5x)+\left[(3y)+(-3y)\right]-(-x) = x+\left[0\right]+x = 2x$

**21.** $3(4y-5)-(7y+2) = 12y-15-7y-2 = 5y-17$

**22.** $8-2[3-(5x-1)] = 8-2[3-5x+1]$
$= 8-2[4-5x]$
$= 8-8+10x$
$= 10x$

**23.** $D = 0.005x^2+0.55x+34$

$D = 0.005(30)^2+0.55(30)+34$

$\quad = 55$

The U.S. diversity index was 55% in 2010.
This is the same as the value displayed in the bar graph.

**24.** $(-3)^3(-2)^2 = (-27)\cdot(4) = -108$

**25.** $2^{-4}+4^{-1} = \dfrac{1}{2^4}+\dfrac{1}{4}$

$= \dfrac{1}{16}+\dfrac{1}{4}$

$= \dfrac{1}{16}+\dfrac{4}{16}$

$= \dfrac{5}{16}$

**26.** $5^{-3}\cdot5 = 5^{-3}5^1 = 5^{-3+1} = 5^{-2} = \dfrac{1}{5^2} = \dfrac{1}{25}$

**27.** $\dfrac{3^3}{3^6} = 3^{3-6} = 3^{-3} = \dfrac{1}{3^3} = \dfrac{1}{27}$

**28.** $(-2x^4y^3)^3 = (-2)^3(x^4)^3(y^3)^3$
$= (-2)^3x^{4\cdot3}y^{3\cdot3}$
$= -8x^{12}y^9$

**29.** $(-5x^3y^2)(-2x^{-11}y^{-2})$
$= (-5)(-2)x^3x^{-11}y^2y^{-2}$
$= 10\cdot x^{3-11}y^{2-2}$
$= 10x^{-8}y^0$
$= \dfrac{10}{x^8}$

**30.** $(2x^3)^{-4} = (2)^{-4}(x^3)^{-4}$
$= 2^{-4}x^{-12}$
$= \dfrac{1}{2^4x^{12}}$
$= \dfrac{1}{16x^{12}}$

**31.** $\dfrac{7x^5y^6}{28x^{15}y^{-2}} = \left(\dfrac{7}{28}\right)(x^{5-15})(y^{6-(-2)})$
$= \dfrac{1}{4}x^{-10}y^8$
$= \dfrac{y^8}{4x^{10}}$

**32.** $3.74\times10^4 = 37,400$

**33.** $7.45\times10^{-5} = 0.0000745$

**34.** $3,590,000 = 3.59\times10^6$

**35.** $0.00725 = 7.25\times10^{-3}$

**36.** $(3\times10^3)(1.3\times10^2) = (3\times1.3)\times(10^3\times10^2)$
$= 3.9\times10^5$
$= 390,000$

**37.** $\dfrac{6.9\times10^3}{3\times10^5} = \left(\dfrac{6.9}{3}\right)\times10^{3-5}$
$= 2.3\times10^{-2}$
$= 0.023$

**38.** $1.35\times10^{12}$

**39.** $32,000,000 = 3.2\times10^7$

**40.** $\dfrac{1.35 \times 10^{12}}{3.2 \times 10^{7}} = \dfrac{1.35}{3.2} \cdot \dfrac{10^{12}}{10^{7}} \approx 0.42188 \times 10^{5} = 42{,}188$

$1.35 \times 10^{12}$ seconds is approximately 42,188 years.

**41.** $\sqrt{300} = \sqrt{100 \cdot 3} = \sqrt{100} \cdot \sqrt{3} = 10\sqrt{3}$

**42.** $\sqrt{12x^2} = \sqrt{4x^2 \cdot 3} = \sqrt{4x^2} \cdot \sqrt{3} = 2|x|\sqrt{3}$

**43.** $\sqrt{10x} \cdot \sqrt{2x} = \sqrt{20x^2}$
$\phantom{\sqrt{10x} \cdot \sqrt{2x}} = \sqrt{4x^2} \cdot \sqrt{5}$
$\phantom{\sqrt{10x} \cdot \sqrt{2x}} = 2x\sqrt{5}$

**44.** $\sqrt{r^3} = \sqrt{r^2} \cdot \sqrt{r} = r\sqrt{r}$

**45.** $\sqrt{\dfrac{121}{4}} = \dfrac{\sqrt{121}}{\sqrt{4}} = \dfrac{11}{2}$

**46.** $\dfrac{\sqrt{96x^3}}{\sqrt{2x}} = \sqrt{\dfrac{96x^3}{2x}}$
$\phantom{\dfrac{\sqrt{96x^3}}{\sqrt{2x}}} = \sqrt{48x^2}$
$\phantom{\dfrac{\sqrt{96x^3}}{\sqrt{2x}}} = \sqrt{16x^2} \cdot \sqrt{3}$
$\phantom{\dfrac{\sqrt{96x^3}}{\sqrt{2x}}} = 4x\sqrt{3}$

**47.** $7\sqrt{5} + 13\sqrt{5} = (7+13)\sqrt{5} = 20\sqrt{5}$

**48.** $2\sqrt{50} + 3\sqrt{8} = 2\sqrt{25 \cdot 2} + 3\sqrt{4 \cdot 2}$
$\phantom{2\sqrt{50} + 3\sqrt{8}} = 2 \cdot 5\sqrt{2} + 3 \cdot 2\sqrt{2}$
$\phantom{2\sqrt{50} + 3\sqrt{8}} = 10\sqrt{2} + 6\sqrt{2}$
$\phantom{2\sqrt{50} + 3\sqrt{8}} = 16\sqrt{2}$

**49.** $4\sqrt{72} - 2\sqrt{48} = 4\sqrt{36 \cdot 2} - 2\sqrt{16 \cdot 3}$
$\phantom{4\sqrt{72} - 2\sqrt{48}} = 4 \cdot 6\sqrt{2} - 2 \cdot 4\sqrt{3}$
$\phantom{4\sqrt{72} - 2\sqrt{48}} = 24\sqrt{2} - 8\sqrt{3}$

**50.** $\dfrac{30}{\sqrt{5}} = \dfrac{30}{\sqrt{5}} \cdot \dfrac{\sqrt{5}}{\sqrt{5}} = \dfrac{30\sqrt{5}}{5} = 6\sqrt{5}$

**51.** $\dfrac{\sqrt{2}}{\sqrt{3}} = \dfrac{\sqrt{2}}{\sqrt{3}} \cdot \dfrac{\sqrt{3}}{\sqrt{3}} = \dfrac{\sqrt{6}}{3}$

**52.** $\dfrac{5}{6+\sqrt{3}} = \dfrac{5}{6+\sqrt{3}} \cdot \dfrac{6-\sqrt{3}}{6-\sqrt{3}}$
$\phantom{\dfrac{5}{6+\sqrt{3}}} = \dfrac{5(6-\sqrt{3})}{36-3}$
$\phantom{\dfrac{5}{6+\sqrt{3}}} = \dfrac{5(6-\sqrt{3})}{33}$

**53.**
$\dfrac{14}{\sqrt{7}-\sqrt{5}} = \dfrac{14}{\sqrt{7}-\sqrt{5}} \cdot \dfrac{\sqrt{7}+\sqrt{5}}{\sqrt{7}+\sqrt{5}}$
$\phantom{\dfrac{14}{\sqrt{7}-\sqrt{5}}} = \dfrac{14(\sqrt{7}+\sqrt{5})}{7-5}$
$\phantom{\dfrac{14}{\sqrt{7}-\sqrt{5}}} = \dfrac{14(\sqrt{7}+\sqrt{5})}{2}$
$\phantom{\dfrac{14}{\sqrt{7}-\sqrt{5}}} = 7(\sqrt{7}+\sqrt{5})$

**54.** $\sqrt[3]{125} = 5$

**55.** $\sqrt[5]{-32} = -2$

**56.** $\sqrt[4]{-125}$ is not a real number.

**57.** $\sqrt[4]{(-5)^4} = \sqrt[4]{625} = \sqrt[4]{5^4} = 5$

**58.** $\sqrt[3]{81} = \sqrt[3]{27 \cdot 3} = \sqrt[3]{27} \cdot \sqrt[3]{3} = 3\sqrt[3]{3}$

**59.** $\sqrt[3]{y^5} = \sqrt[3]{y^3 y^2} = y\sqrt[3]{y^2}$

**60.** $\sqrt[4]{8} \cdot \sqrt[4]{10} = \sqrt[4]{80} = \sqrt[4]{16 \cdot 5} = \sqrt[4]{16} \cdot \sqrt[4]{5} = 2\sqrt[4]{5}$

**61.** $4\sqrt[3]{16} + 5\sqrt[3]{2} = 4\sqrt[3]{8 \cdot 2} + 5\sqrt[3]{2}$
$\phantom{4\sqrt[3]{16} + 5\sqrt[3]{2}} = 4 \cdot 2\sqrt[3]{2} + 5\sqrt[3]{2}$
$\phantom{4\sqrt[3]{16} + 5\sqrt[3]{2}} = 8\sqrt[3]{2} + 5\sqrt[3]{2}$
$\phantom{4\sqrt[3]{16} + 5\sqrt[3]{2}} = 13\sqrt[3]{2}$

**62.** $\dfrac{\sqrt[4]{32x^5}}{\sqrt[4]{16x}} = \sqrt[4]{\dfrac{32x^5}{16x}} = \sqrt[4]{2x^4} = x\sqrt[4]{2}$

**63.** $16^{1/2} = \sqrt{16} = 4$

**64.** $25^{-1/2} = \dfrac{1}{25^{1/2}} = \dfrac{1}{\sqrt{25}} = \dfrac{1}{5}$

**65.** $125^{1/3} = \sqrt[3]{125} = 5$

**66.** $27^{-1/3} = \dfrac{1}{27^{1/3}} = \dfrac{1}{\sqrt[3]{27}} = \dfrac{1}{3}$

**67.** $64^{2/3} = (\sqrt[3]{64})^2 = 4^2 = 16$

**68.** $27^{-4/3} = \dfrac{1}{27^{4/3}} = \dfrac{1}{(\sqrt[3]{27})^4} = \dfrac{1}{3^4} = \dfrac{1}{81}$

**69.** $(5x^{2/3})(4x^{1/4}) = 5 \cdot 4 x^{2/3+1/4} = 20x^{11/12}$

**70.** $\dfrac{15x^{3/4}}{5x^{1/2}} = \left(\dfrac{15}{5}\right) x^{3/4-1/2} = 3x^{1/4}$

**71.** $(125 \cdot x^6)^{2/3} = (\sqrt[3]{125x^6})^2$
$\qquad\qquad\quad = (5x^2)^2$
$\qquad\qquad\quad = 25x^4$

**72.** $\sqrt[6]{y^3} = (y^3)^{1/6} = y^{3 \cdot 1/6} = y^{1/2} = \sqrt{y}$

**73.** $(-6x^3 + 7x^2 - 9x + 3) + (14x^3 + 3x^2 - 11x - 7) = (-6x^3 + 14x^3) + (7x^2 + 3x^2) + (-9x - 11x) + (3 - 7)$
$\qquad\qquad\qquad\qquad\qquad\qquad\qquad\qquad\qquad = 8x^3 + 10x^2 - 20x - 4$

The degree is 3.

**74.** $(13x^4 - 8x^3 + 2x^2) - (5x^4 - 3x^3 + 2x^2 - 6) = (13x^4 - 8x^3 + 2x^2) + (-5x^4 + 3x^3 - 2x^2 + 6)$
$\qquad\qquad\qquad\qquad\qquad\qquad\qquad\qquad\qquad = (13x^4 - 5x^4) + (-8x^3 + 3x^3) + (2x^2 - 2x^2) + 6$
$\qquad\qquad\qquad\qquad\qquad\qquad\qquad\qquad\qquad = 8x^4 - 5x^3 + 6$

The degree is 4.

**75.** $(3x - 2)(4x^2 + 3x - 5) = (3x)(4x^2) + (3x)(3x) + (3x)(-5) + (-2)(4x^2) + (-2)(3x) + (-2)(-5)$
$\qquad\qquad\qquad\qquad\qquad = 12x^3 + 9x^2 - 15x - 8x^2 - 6x + 10$
$\qquad\qquad\qquad\qquad\qquad = 12x^3 + x^2 - 21x + 10$

**76.** $(3x - 5)(2x + 1) = (3x)(2x) + (3x)(1) + (-5)(2x) + (-5)(1)$
$\qquad\qquad\qquad\qquad = 6x^2 + 3x - 10x - 5$
$\qquad\qquad\qquad\qquad = 6x^2 - 7x - 5$

**77.** $(4x + 5)(4x - 5) = (4x^2) - 5^2 = 16x^2 - 25$

**78.** $(2x + 5)^2 = (2x)^2 + 2(2x) \cdot 5 + 5^2 = 4x^2 + 20x + 25$

**79.** $(3x-4)^2 = (3x)^2 - 2(3x)\cdot 4 + (-4)^2 = 9x^2 - 24x + 16$

**80.** $(2x+1)^3 = (2x)^3 + 3(2x)^2(1) + 3(2x)(1)^2 + 1^3 = 8x^3 + 12x^2 + 6x + 1$

**81.** $(5x-2)^3 = (5x)^3 - 3(5x)^2(2) + 3(5x)(2)^2 - 2^3 = 125x^3 - 150x^2 + 60x - 8$

**82.** $(7x^2 - 8xy + y^2) + (-8x^2 - 9xy - 4y^2) = (7x^2 - 8x^2) + (-8xy - 9xy) + (y^2 - 4y^2)$
$$= -x^2 - 17xy - 3y^2$$

The degree is 2.

**83.** $(13x^3 y^2 - 5x^2 y - 9x^2) - (-11x^3 y^2 - 6x^2 y + 3x^2 - 4)$
$= (13x^3 y^2 - 5x^2 y - 9x^2) + (11x^3 y^2 + 6x^2 y - 3x^2 + 4)$
$= (13x^3 y^2 + 11x^3 y^2) + (-5x^2 y + 6x^2 y) + (-9x^2 - 3x^2) + 4$
$= 24x^3 y^2 + x^2 y - 12x^2 + 4$

The degree is 5.

**84.** $(x + 7y)(3x - 5y) = x(3x) + (x)(-5y) + (7y)(3x) + (7y)(-5y)$
$$= 3x^2 - 5xy + 21xy - 35y^2$$
$$= 3x^2 + 16xy - 35y^2$$

**85.** $(3x - 5y)^2 = (3x)^2 - 2(3x)(5y) + (-5y)^2$
$$= 9x^2 - 30xy + 25y^2$$

**86.** $(3x^2 + 2y)^2 = (3x^2)^2 + 2(3x^2)(2y) + (2y)^2$
$$= 9x^4 + 12x^2 y + 4y^2$$

**87.** $(7x + 4y)(7x - 4y) = (7x)^2 - (4y)^2$
$$= 49x^2 - 16y^2$$

**88.** $(a - b)(a^2 + ab + b^2)$
$= a(a^2) + a(ab) + a(b^2) + (-b)(a^2) + (-b)(ab) + (-b)(b^2)$
$= a^3 + a^2 b + ab^2 - a^2 b - ab^2 - b^3$
$= a^3 - b^3$

**89.** $15x^3 + 3x^2 = 3x^2 \cdot 5x + 3x^2 \cdot 1$
$$= 3x^2(5x + 1)$$

**90.** $x^2 - 11x + 28 = (x-4)(x-7)$

**91.** $15x^2 - x - 2 = (3x+1)(5x-2)$

**92.** $64 - x^2 = 8^2 - x^2 = (8-x)(8+x)$

**93.** $x^2 + 16$ is prime.

**94.** $3x^4 - 9x^3 - 30x^2 = 3x^2(x^2 - 3x - 10)$
$$= 3x^2(x-5)(x+2)$$

**95.** $20x^7 - 36x^3 = 4x^3(5x^4 - 9)$

**96.** $x^3 - 3x^2 - 9x + 27 = x^2(x-3) - 9(x-3)$
$$= (x^2 - 9)(x-3)$$
$$= (x+3)(x-3)(x-3)$$
$$= (x+3)(x-3)^2$$

**97.** $16x^2 - 40x + 25 = (4x-5)(4x-5)$
$$= (4x-5)^2$$

**98.** $x^4 - 16 = (x^2)^2 - 4^2$
$$= (x^2 + 4)(x^2 - 4)$$
$$= (x^2 + 4)(x+2)(x-2)$$

**99.** $y^3 - 8 = y^3 - 2^3 = (y-2)(y^2 + 2y + 4)$

**100.** $x^3 + 64 = x^3 + 4^3 = (x+4)(x^2 - 4x + 16)$

**101.** $3x^4 - 12x^2 = 3x^2(x^2 - 4)$
$$= 3x^2(x-2)(x+2)$$

**102.** $27x^3 - 125 = (3x)^3 - 5^3$
$$= (3x-5)[(3x)^2 + (3x)(5) + 5^2]$$
$$= (3x-5)(9x^2 + 15x + 25)$$

**103.** $x^5 - x = x(x^4 - 1)$
$$= x(x^2 - 1)(x^2 + 1)$$
$$= x(x-1)(x+1)(x^2 + 1)$$

**104.** $x^3 + 5x^2 - 2x - 10 = x^2(x+5) - 2(x+5)$
$$= (x^2 - 2)(x+5)$$

**105.** $x^2 + 18x + 81 - y^2 = (x^2 + 18x + 81) - y^2$
$$= (x+9)^2 - y^2$$
$$= (x+9-y)(x+9+y)$$

**106.** $16x^{-\frac{3}{4}} + 32x^{\frac{1}{4}} = 16x^{-\frac{3}{4}}\left(1 + 2x^{\frac{1}{4}-\left(-\frac{3}{4}\right)}\right)$
$$= 16x^{-\frac{3}{4}}(1+2x)$$
$$= \frac{16(1+2x)}{x^{\frac{3}{4}}}$$

**107.** $\left(x^2 - 4\right)\left(x^2 + 3\right)^{\frac{1}{2}} - \left(x^2 - 4\right)^2\left(x^2 + 3\right)^{\frac{3}{2}}$
$$= \left(x^2 - 4\right)\left(x^2 + 3\right)^{\frac{1}{2}}\left[1 - \left(x^2 - 4\right)\left(x^2 + 3\right)\right]$$
$$= (x-2)(x+2)\left(x^2 + 3\right)^{\frac{1}{2}}\left[1 - (x-2)(x+2)\left(x^2 + 3\right)\right]$$
$$= (x-2)(x+2)(x^2 + 3)^{\frac{1}{2}}(-x^4 + x^2 + 13)$$

**108.** $12x^{-\frac{1}{2}} + 6x^{-\frac{3}{2}} = 6x^{-\frac{3}{2}}(2x+1) = \frac{6(2x+1)}{x^{\frac{3}{2}}}$

**109.** $\dfrac{x^3 + 2x^2}{x+2} = \dfrac{x^2(x+2)}{x+2} = x^2, x \neq -2$

**110.** $\dfrac{x^2 + 3x - 18}{x^2 - 36} = \dfrac{(x+6)(x-3)}{(x+6)(x-6)} = \dfrac{x-3}{x-6}$,
$x \neq -6, 6$

**111.** $\dfrac{x^2 + 2x}{x^2 + 4x + 4} = \dfrac{x(x+2)}{(x+2)^2} = \dfrac{x}{x+2}$,
$x \neq -2$

**112.** $\dfrac{x^2 + 6x + 9}{x^2 - 4} \cdot \dfrac{x+3}{x-2} = \dfrac{(x+3)^2}{(x-2)(x+2)} \cdot \dfrac{x+3}{x-2}$
$$= \dfrac{(x+3)^3}{(x-2)^2(x+2)},$$
$x \neq 2, -2$

**113.** $\dfrac{6x+2}{x^2-1} \div \dfrac{3x^2+x}{x-1}$

$= \dfrac{2(3x+1)}{(x-1)(x+1)} \div \dfrac{x(3x+1)}{x-1}$

$= \dfrac{2(3x+1)}{(x-1)(x+1)} \cdot \dfrac{x-1}{x(3x+1)}$

$= \dfrac{2}{x(x+1)}$,

$x \neq 0, 1, -1, -\dfrac{1}{3}$

**114.** $\dfrac{x^2-5x-24}{x^2-x-12} \div \dfrac{x^2-10x+16}{x^2+x-6}$

$= \dfrac{(x-8)(x+3)}{(x-4)(x+3)} \div \dfrac{(x-2)(x-8)}{(x+3)(x-2)}$

$= \dfrac{x-8}{x-4} \cdot \dfrac{x+3}{x-8}$

$= \dfrac{x+3}{x-4}$,

$x \neq -3, 4, 2, 8$

**115.** $\dfrac{2x-7}{x^2-9} - \dfrac{x-10}{x^2-9} = \dfrac{2x-7-(x-10)}{x^2-9}$

$= \dfrac{x+3}{(x+3)(x-3)}$

$= \dfrac{1}{x-3}$,

$x \neq 3, -3$

**116.** $\dfrac{3x}{x+2} + \dfrac{x}{x-2} = \dfrac{3x}{x+2} \cdot \dfrac{x-2}{x-2} + \dfrac{x}{x-2} \cdot \dfrac{x+2}{x+2}$

$= \dfrac{3x^2-6x+x^2+2x}{(x+2)(x-2)}$

$= \dfrac{4x^2-4x}{(x+2)(x-2)}$

$= \dfrac{4x(x-1)}{(x+2)(x-2)}$,

$x \neq 2, -2$

**117.** $\dfrac{x}{x^2-9} + \dfrac{x-1}{x^2-5x+6}$

$= \dfrac{x}{(x-3)(x+3)} + \dfrac{x-1}{(x-2)(x-3)}$

$= \dfrac{x}{(x-3)(x+3)} \cdot \dfrac{x-2}{x-2} + \dfrac{x-1}{(x-2)(x-3)} \cdot \dfrac{x+3}{x+3}$

$= \dfrac{x(x-2)+(x-1)(x+3)}{(x-3)(x+3)(x-2)}$

$= \dfrac{x^2-2x+x^2+2x-3}{(x-3)(x+3)(x-2)}$

$= \dfrac{2x^2-3}{(x-3)(x+3)(x-2)}$

$x \neq 3, -3, 2$

**118.** $\dfrac{4x-1}{2x^2+5x-3} - \dfrac{x+3}{6x^2+x-2}$

$= \dfrac{4x-1}{(2x-1)(x+3)} - \dfrac{x+3}{(2x-1)(3x+2)}$

$= \dfrac{4x-1}{(2x-1)(x+3)} \cdot \dfrac{3x+2}{3x+2}$

$- \dfrac{x+3}{(2x-1)(3x+2)} \cdot \dfrac{x+3}{x+3}$

$= \dfrac{12x^2+8x-3x-2-x^2-6x-9}{(2x-1)(x+3)(3x+2)}$

$= \dfrac{11x^2-x-11}{(2x-1)(x+3)(3x+2)}$,

$x \neq \dfrac{1}{2}, -3, -\dfrac{2}{3}$

**119.** $\dfrac{\dfrac{1}{x}-\dfrac{1}{2}}{\dfrac{1}{3}-\dfrac{x}{6}}=\dfrac{\dfrac{1}{x}-\dfrac{1}{2}}{\dfrac{1}{3}-\dfrac{x}{6}}\cdot\dfrac{6x}{6x}$

$=\dfrac{6-3x}{2x-x^2}$

$=\dfrac{-3(x-2)}{-x(x-2)}$

$=\dfrac{3}{x},$

$x\neq 0,\,2$

**120.** $\dfrac{3+\dfrac{12}{x}}{1-\dfrac{16}{x^2}}=\dfrac{3+\dfrac{12}{x}}{1-\dfrac{16}{x^2}}\cdot\dfrac{x^2}{x^2}$

$=\dfrac{3x^2+12x}{x^2-16}$

$=\dfrac{3x(x+4)}{(x+4)(x-4)}$

$=\dfrac{3x}{x-4},$

$x\neq 0,\,4,\,-4$

**121.** $\dfrac{3-\dfrac{1}{x+3}}{3+\dfrac{1}{x+3}}=\dfrac{3-\dfrac{1}{x+3}}{3+\dfrac{1}{x+3}}\cdot\dfrac{x+3}{x+3}$

$=\dfrac{3(x+3)-1}{3(x+3)+1}$

$=\dfrac{3x+9-1}{3x+9+1}$

$=\dfrac{3x+8}{3x+10},$

$x\neq -3,\,-\dfrac{10}{3}$

**Chapter P Test**

**1.** $5(2x^2-6x)-(4x^2-3x)=10x^2-30x-4x^2+3x$
$$=6x^2-27x$$

**2.** $7+2[3(x+1)-2(3x-1)]$
$=7+2[3x+3-6x+2]$
$=7+2[-3x+5]$
$=7-6x+10$
$=-6x+17$

**3.** $\{1,2,5\}\cap\{5,a\}=\{5\}$

**4.** $\{1,2,5\}\cup\{5,a\}=\{1,2,5,a\}$

**5.** $(2x^2y^3-xy+y^2)-(-4x^2y^3-5xy-y^2)$
$=2x^2y^3-xy+y^2+4x^2y^3+5xy+y^2$
$=2x^2y^3+4x^2y^3-xy+5xy+y^2+y^2$
$=6x^2y^3+4xy+2y^2$

**6.** $\dfrac{30x^3y^4}{6x^9y^{-4}}=5x^{3-9}y^{4-(-4)}=5x^{-6}y^8=\dfrac{5y^8}{x^6}$

**7.** $\sqrt{6r}\cdot\sqrt{3r}=\sqrt{18r^2}=\sqrt{9r^2}\cdot\sqrt{2}=3r\sqrt{2}$

**8.** $4\sqrt{50}-3\sqrt{18}=4\sqrt{25\cdot2}-3\sqrt{9\cdot2}$
$=4\cdot5\sqrt{2}-3\cdot3\sqrt{2}$
$=20\sqrt{2}-9\sqrt{2}$
$=11\sqrt{2}$

**9.** $\dfrac{3}{5+\sqrt{2}}=\dfrac{3}{5+\sqrt{2}}\cdot\dfrac{5-\sqrt{2}}{5-\sqrt{2}}$
$=\dfrac{3(5-\sqrt{2})}{25-2}$
$=\dfrac{3(5-\sqrt{2})}{23}$

**10.** $\sqrt[3]{16x^4} = \sqrt[3]{8x^3 \cdot 2x}$

$\qquad = \sqrt[3]{8x^3} \cdot \sqrt[3]{2x}$

$\qquad = 2x\sqrt[3]{2x}$

**11.** $\dfrac{x^2 + 2x - 3}{x^2 - 3x + 2} = \dfrac{(x+3)(x-1)}{(x-2)(x-1)} = \dfrac{x+3}{x-2}$,

$\qquad x \neq 2, 1$

**12.** $\dfrac{5 \times 10^{-6}}{20 \times 10^{-8}} = \dfrac{5}{20} \cdot \dfrac{10^{-6}}{10^{-8}} = 0.25 \times 10^2 = 2.5 \times 10^1$

**13.** $(2x - 5)(x^2 - 4x + 3)$

$\qquad = 2x^3 - 8x^2 + 6x - 5x^2 + 20x - 15$

$\qquad = 2x^3 - 13x^2 + 26x - 15$

**14.** $(5x + 3y)^2 = (5x)^2 + 2(5x)(3y) + (3y)^2$

$\qquad = 25x^2 + 30xy + 9y^2$

**15.** $\dfrac{2x + 8}{x - 3} \div \dfrac{x^2 + 5x + 4}{x^2 - 9}$

$\qquad = \dfrac{2(x+4)}{x-3} \div \dfrac{(x+1)(x+4)}{(x-3)(x+3)}$

$\qquad = \dfrac{2(x+4)}{x-3} \cdot \dfrac{(x-3)(x+3)}{(x+1)(x+4)}$

$\qquad = \dfrac{2(x+3)}{x+1}$,

$\qquad x \neq 3, -1, -4, -3$

**16.** $\dfrac{x}{x+3} + \dfrac{5}{x-3}$

$\qquad = \dfrac{x}{x+3} \cdot \dfrac{x-3}{x-3} + \dfrac{5}{x-3} \cdot \dfrac{x+3}{x+3}$

$\qquad = \dfrac{x(x-3) + 5(x+3)}{(x+3)(x-3)}$

$\qquad = \dfrac{x^2 - 3x + 5x + 15}{(x+3)(x-3)}$

$\qquad = \dfrac{x^2 + 2x + 15}{(x+3)(x-3)}, \; x \neq 3, -3$

**17.** $\dfrac{2x + 3}{x^2 - 7x + 12} - \dfrac{2}{x - 3}$

$\qquad = \dfrac{2x + 3}{(x-3)(x-4)} - \dfrac{2}{x-3}$

$\qquad = \dfrac{2x + 3}{(x-3)(x-4)} - \dfrac{2}{x-3} \cdot \dfrac{x-4}{x-4}$

$\qquad = \dfrac{2x + 3 - 2(x-4)}{(x-3)(x-4)}$

$\qquad = \dfrac{2x + 3 - 2(x-4)}{(x-3)(x-4)}$

$\qquad = \dfrac{2x + 3 - 2x + 8}{(x-3)(x-4)}$

$\qquad = \dfrac{11}{(x-3)(x-4)}$,

$\qquad x \neq 3, 4$

**18.** $\dfrac{\frac{1}{x} - \frac{1}{3}}{\frac{1}{x}} = \dfrac{\frac{1}{x} - \frac{1}{3}}{\frac{1}{x}} \cdot \dfrac{3x}{3x} = \dfrac{3 - x}{3}$,

$\qquad x \neq 0$

**19.** $x^2 - 9x + 18 = (x - 3)(x - 6)$

**20.** $x^3 + 2x^2 + 3x + 6 = x^2(x+2) + 3(x+2)$

$\qquad = (x^2 + 3)(x + 2)$

**21.** $25x^2 - 9 = (5x)^2 - 3^2 = (5x - 3)(5x + 3)$

**22.** $36x^2 - 84x + 49 = (6x)^2 - 2(6x) \cdot 7 + 7^2$

$\qquad = (6x - 7)^2$

**23.** $y^3 - 125 = y^3 - 5^3 = (y - 5)(y^2 + 5y + 25)$

**24.** $(x^2 + 10x + 25) - 9y^2$

$\qquad = (x + 5)^2 - 9y^2$

$\qquad = (x + 5 - 3y)(x + 5 + 3y)$

**25.** $x(x+3)^{-\frac{3}{5}} + (x+3)^{\frac{2}{5}}$

$\qquad = (x+3)^{-\frac{3}{5}} \left[ x + (x+3) \right]$

$\qquad = (x+3)^{-\frac{3}{5}} (2x + 3) = \dfrac{2x + 3}{(x+3)^{\frac{3}{5}}}$

**26.** $-7, -\dfrac{4}{5}, 0, 0.25, \sqrt{4}, \dfrac{22}{7}$ are rational numbers.

**27.** $3(2 + 5) = 3(5 + 2)$;
commutative property of addition

**28.** $6(7+4) = 6 \cdot 7 + 6 \cdot 4$
distributive property of multiplication over addition

**29.** $0.00076 = 7.6 \times 10^{-4}$

**30.** $27^{-\frac{5}{3}} = \dfrac{1}{27^{\frac{5}{3}}} = \dfrac{1}{\left(\sqrt[3]{27}\right)^5} = \dfrac{1}{(3)^5} = \dfrac{1}{243}$

**31.** $2\left(6.6 \times 10^9\right) = 13.2 \times 10^9 = 1.32 \times 10^{10}$

**32.** **a.** 2003 is 14 years after 1989.
$M = -0.28n + 47$
$M = -0.28(14) + 47$
$\quad = 43.08$
In 2003, 43.08% of bachelor's degrees were awarded to men. This overestimates the actual percent shown by the bar graph by 0.08%.

    **b.** $R = \dfrac{M}{W} = \dfrac{-0.28n + 47}{0.28n + 53}$

    **c.** $R = \dfrac{-0.28n + 47}{0.28n + 53}$
$R = \dfrac{-0.28(25) + 47}{0.28(25) + 53}$
$\quad = \dfrac{2}{3}$
Three women will receive bachelor's degrees for every two men. This describes the projections exactly.

# Chapter 1
# Equations and Inequalities

## Section 1.1

### Check Point Exercises

**1.**

**2.**

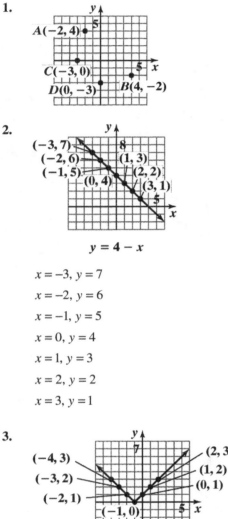

$$y = 4 - x$$

$x = -3, y = 7$

$x = -2, y = 6$

$x = -1, y = 5$

$x = 0, y = 4$

$x = 1, y = 3$

$x = 2, y = 2$

$x = 3, y = 1$

**3.**

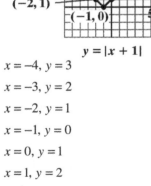

$$y = |x + 1|$$

$x = -4, y = 3$

$x = -3, y = 2$

$x = -2, y = 1$

$x = -1, y = 0$

$x = 0, y = 1$

$x = 1, y = 2$

$x = 2, y = 3$

**4.** The meaning of a
$[-100, 100, 50]$ by $[-100, 100, 10]$
viewing rectangle is as follows:

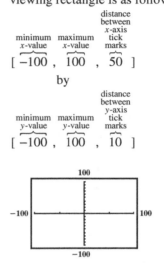

**5. a.** The graph crosses the $x$-axis at (–3, 0).
Thus, the $x$-intercept is –3.
The graph crosses the $y$-axis at (0, 5).
Thus, the $y$-intercept is 5.

**b.** The graph does not cross the $x$-axis.
Thus, there is no $x$-intercept.
The graph crosses the $y$-axis at (0, 4).
Thus, the $y$-intercept is 4.

**c.** The graph crosses the $x$- and $y$-axes at
the origin (0, 0).
Thus, the $x$-intercept is 0 and the
$y$-intercept is 0.

**6. a.** $d = 4n + 5$

$d = 4(15) + 5 = 65$

65% of marriages end in divorce after 15 years
when the wife is under 18 at the time of
marriage.

**b.** According to the line graph, 60% of marriages
end in divorce after 15 years when the wife is
under 18 at the time of marriage.

**c.** The mathematical model overestimates the
actual percentage shown in the graph by 5%.

## Concept and Vocabulary Check 1.1

1. *x*-axis

2. *y*-axis

3. origin

4. quadrants; four

5. *x*-coordinate; *y*-coordinate

6. solution; satisfies

7. *x*-intercept; zero

8. *y*-intercept; zero

## Exercise Set 1.1

1.

3.

5.

7.

9.

11.

13.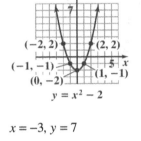

$x = -3, y = 7$

$x = -2, y = 2$

$x = -1, y = -1$

$x = 0, y = -2$

$x = 1, y = -1$

$x = 2, y = 2$

$x = 3, y = 7$

15.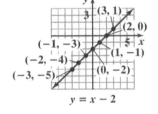

$x = -3, y = -5$

$x = -2, y = -4$

$x = -1, y = -3$

$x = 0, y = -2$

$x = 1, y = -1$

$x = 2, y = 0$

$x = 3, y = 1$

**17.**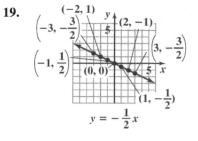

$x = -3, y = -5$

$x = -2, y = -3$

$x = -1, y = -1$

$x = 0, y = 1$

$x = 1, y = 3$

$x = 2, y = 5$

$x = 3, y = 7$

**19.**

$x = -3, y = \dfrac{3}{2}$

$x = -2, y = 1$

$x = -1, y = \dfrac{1}{2}$

$x = 0, y = 0$

$x = 1, y = -\dfrac{1}{2}$

$x = 2, y = -1$

$x = 3, y = -\dfrac{3}{2}$

**21.**

$x = -3, y = 6$

$x = -2, y = 4$

$x = -1, y = 2$

$x = 0, y = 0$

$x = 1, y = 2$

$x = 2, y = 4$

$x = 3, y = 6$

**23.**

$x = -3, y = 4$

$x = -2, y = 3$

$x = -1, y = 2$

$x = 0, y = 1$

$x = 1, y = 2$

$x = 2, y = 3$

$x = 3, y = 4$

**25.**

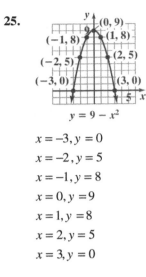

$x = -3, y = 0$

$x = -2, y = 5$

$x = -1, y = 8$

$x = 0, y = 9$

$x = 1, y = 8$

$x = 2, y = 5$

$x = 3, y = 0$

**27.**

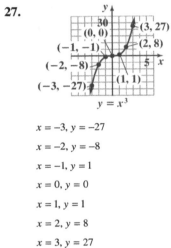

$x = -3, y = -27$

$x = -2, y = -8$

$x = -1, y = 1$

$x = 0, y = 0$

$x = 1, y = 1$

$x = 2, y = 8$

$x = 3, y = 27$

**29.** (c) *x*-axis tick marks –5, –4, –3, –2, –1, 0, 1, 2, 3, 4, 5; *y*-axis tick marks are the same.

**31.** (b); *x*-axis tick marks –20, –10, 0, 10, 20, 30, 40, 50, 60, 70, 80; *y*-axis tick marks –30, –20, –10, 0, 10, 20, 30, 40, 50, 60, 70

**33.** The equation that corresponds to $Y_2$ in the table is (c), $y_2 = 2 - x$. We can tell because all of the points $(-3, 5)$, $(-2, 4)$, $(-1, 3)$, $(0, 2)$, $(1, 1)$, $(2, 0)$, and $(3, -1)$ are on the line $y = 2 - x$, but all are not on any of the others.

**35.** No. It passes through the point $(0, 2)$.

**37.** $(2, 0)$

**39.** The graphs of $Y_1$ and $Y_2$ intersect at the points $(-2, 4)$ and $(1, 1)$.

**41. a.** 2; The graph intersects the *x*-axis at (2, 0).

**b.** –4; The graph intersects the *y*-axis at (0,–4).

**43. a.** 1, –2; The graph intersects the *x*-axis at (1, 0) and (–2, 0).

**b.** 2; The graph intersects the *y*-axis at (0, 2).

**45. a.** –1; The graph intersects the *x*-axis at (–1, 0).

**b.** none; The graph does not intersect the *y*-axis.

**47.**

$y = 2x + 4$

**49.**

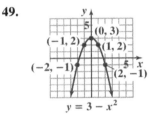

$y = 3 - x^2$

**51.**

| $x$ | $(x, y)$ |
|---|---|
| –3 | $(-3, 5)$ |
| –2 | $(-2, 5)$ |
| –1 | $(-1, 5)$ |
| 0 | $(0, 5)$ |
| 1 | $(1, 5)$ |
| 2 | $(2, 5)$ |
| 3 | $(3, 5)$ |

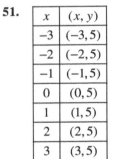

$y = 5$

**53.**

| $x$ | $(x, y)$ |
|---|---|
| $-2$ | $\left(-2, -\dfrac{1}{2}\right)$ |
| $-1$ | $(-1, -1)$ |
| $-\dfrac{1}{2}$ | $\left(-\dfrac{1}{2}, -2\right)$ |
| $-\dfrac{1}{3}$ | $\left(-\dfrac{1}{3}, -3\right)$ |
| $\dfrac{1}{3}$ | $\left(\dfrac{1}{3}, 3\right)$ |
| $\dfrac{1}{2}$ | $\left(\dfrac{1}{2}, 2\right)$ |
| $1$ | $(1, 1)$ |
| $2$ | $\left(2, \dfrac{1}{2}\right)$ |

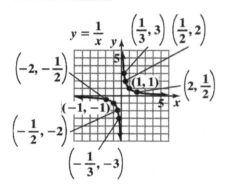

**55.**

   **a.** According to the line graph, 20% of seniors used marijuana in 2005.

   **b.** 2005 is 25 years after 1980.
$M = -0.3n + 27$
$M = -0.3(25) + 27 = 19.5$
According to formula, 19.5% of seniors used marijuana in 2005. This underestimates the value in the graph by 0.5%.

   **c.** According to the line graph, about 47% of seniors used alcohol in 2005.

   **d.** 2005 is 25 years after 1980.
$A = -0.9n + 69$
$A = -0.9(25) + 69 = 46.5$
According to formula, 46.5% of seniors used alcohol in 2005. It is less than the estimate, although answers may vary.

   **e.** The minimum for marijuana was reached in 1990. According to the line graph, about 14% of seniors used marijuana in 1990.

**57.** At age 8, women have the least number of awakenings, averaging about 1 awakening per night.

**59.** The difference between the number of awakenings for 25-year-old men and women is about 1.9.

**61. – 65.** Answers will vary.

**67.** makes sense

**69.** does not make sense; Explanations will vary. Sample explanation: These three points are not collinear.

**71.** false; Changes to make the statement true will vary. A sample change is: The product of the coordinates of a point in quadrant III is also positive.

**73.** true

**75.** I, III

**77.** IV

**79.** (a)

**81.** (b)

**83.** (b)

**85.** (c)

**87.** $2(x-3) - 17 = 13 - 3(x+2)$
$2(6-3) - 17 = 13 - 3(6+2)$
$2(3) - 17 = 13 - 3(8)$
$6 - 17 = 13 - 24$
$-11 = -11, \text{ true}$

**88.** $12\left(\dfrac{x+2}{4} - \dfrac{x-1}{3}\right) = 12\left(\dfrac{x+2}{4}\right) - 12\left(\dfrac{x-1}{3}\right)$
$$= 3(x+2) - 4(x-1)$$
$$= 3x + 6 - 4x + 4$$
$$= -x + 10$$

**89.** $(x-3)\ \dfrac{3}{x-3} + 9 = (x-3)\ \dfrac{3}{x-3} + (x-3)(9)$
$$= 3 + 9x - 27$$
$$= 9x - 24$$

## Section 1.2

## Check Point Exercises

**1.**
$$4x + 5 = 29$$
$$4x + 5 - 5 = 29 - 5$$
$$4x = 24$$
$$\frac{4x}{4} = \frac{24}{4}$$
$$x = 6$$
Check:
$$4x + 5 = 29$$
$$4(6) + 5 = 29$$
$$24 + 5 = 29$$
$$29 = 29 \text{ true}$$
The solution set is $\{6\}$.

**2.**
$$4(2x + 1) - 29 = 3(2x - 5)$$
$$8x + 4 - 29 = 6x - 15$$
$$8x - 25 = 6x - 15$$
$$8x - 25 - 6x = 6x - 15 - 6x$$
$$2x - 25 = -15$$
$$2x - 25 + 25 = -15 + 25$$
$$2x = 10$$
$$\frac{2x}{2} = \frac{10}{2}$$
$$x = 5$$
Check:
$$4(2x + 1) - 29 = 3(2x - 5)$$
$$4[2(5) + 1] - 29 = 3[2(5) - 5]$$
$$4[10 + 1] - 29 = 3[10 - 5]$$
$$4[11] - 29 = 3[5]$$
$$44 - 29 = 15$$
$$15 = 15 \text{ true}$$
The solution set is $\{5\}$.

**3.**
$$\frac{x-3}{4} = \frac{5}{14} - \frac{x+5}{7}$$
$$28 \cdot \frac{x-3}{4} = 28\left(\frac{5}{14} - \frac{x+5}{7}\right)$$
$$7(x-3) = 2(5) - 4(x+5)$$
$$7x - 21 = 10 - 4x - 20$$
$$7x - 21 = -4x - 10$$
$$7x + 4x = -10 + 21$$
$$11x = 11$$
$$\frac{11x}{11} = \frac{11}{11}$$
$$x = 1$$
Check:
$$\frac{x-3}{4} = \frac{5}{14} - \frac{x+5}{7}$$
$$\frac{1-3}{4} = \frac{5}{14} - \frac{1+5}{7}$$
$$\frac{-2}{4} = \frac{5}{14} - \frac{6}{7}$$
$$-\frac{1}{2} = -\frac{1}{2}$$
The solution set is $\{1\}$.

**4.**
$$\frac{5}{2x} = \frac{17}{18} - \frac{1}{3x}, \ x \neq 0$$
$$18x \cdot \frac{5}{2x} = 18x\left(\frac{17}{18} - \frac{1}{3x}\right)$$
$$18 \cdot \frac{5}{2x} = 18x \cdot \frac{17}{18} - 18x \cdot \frac{1}{3x}$$
$$45 = 17x - 6$$
$$45 + 6 = 17x - 6 + 6$$
$$51 = 17x$$
$$\frac{51}{17} = \frac{17x}{17}$$
$$3 = x$$
The solution set is $\{3\}$.

**5.**
$$\frac{x}{x-2} = \frac{2}{x-2} - \frac{2}{3}, \quad x \neq 2$$

$$3(x-2) \cdot \frac{x}{x-2} = 3(x-2)\left[\frac{2}{x-2} - \frac{2}{3}\right]$$

$$3(x-2) \cdot \frac{x}{x-2} = (3x-2) \cdot \frac{2}{x-2} - 3(x-2) \cdot \frac{2}{3}$$

$$3x = 6 - (x-2) \cdot 2$$

$$3x = 6 - 2(x-2)$$

$$3x = 6 - 2x + 4$$

$$3x = 10 - 2x$$

$$3x + 2x = 10 - 2x + 2x$$

$$5x = 10$$

$$\frac{5x}{5} = \frac{10}{5}$$

$$x = 2$$

The solution set is the empty set, $\varnothing$.

**6.**   Set $y_1 = y_2$.

$$\frac{1}{x+4} + \frac{1}{x-4} = \frac{22}{x^2-16}$$

$$\frac{1}{x+4} + \frac{1}{x-4} = \frac{22}{(x+4)(x-4)}$$

$$\frac{(x+4)(x-4)}{x+4} + \frac{(x+4)(x-4)}{x-4} = \frac{22(x+4)(x-4)}{(x+4)(x-4)}$$

$$(x-4) + (x+4) = 22$$

$$x - 4 + x + 4 = 22$$

$$2x = 22$$

$$x = 11$$

Check:

$$\frac{1}{x+4} + \frac{1}{x-4} = \frac{22}{x^2-16}$$

$$\frac{1}{11+4} + \frac{1}{11-4} = \frac{22}{11^2-16}$$

$$\frac{1}{15} + \frac{1}{7} = \frac{22}{105}$$

$$\frac{22}{105} = \frac{22}{105} \quad \text{true}$$

**7.**   $4x - 7 = 4(x-1) + 3$

$4x - 7 = 4(x-1) + 3$

$4x - 7 = 4x - 4 + 3$

$4x - 7 = 4x - 1$

$\quad -7 = -1$

The original equation is equivalent to the statement $-7 = -1$, which is false for every value of $x$.
The solution set is the empty set, $\varnothing$.
The equation is an inconsistent equation.

**8.**   $4x - 7 = 4(x-1) + 3$

$7x + 9 = 9(x+1) - 2x$

$7x + 9 = 9x + 9 - 2x$

$7x + 9 = 7x + 9$

$\quad 9 = 9$

The original equation is equivalent to the statement $9 = 9$, which is true for every value of $x$.
The equation is an identity, and all real numbers are solutions. The solution set $\{x \mid x \text{ is a real number}\}$.

**9.**
$$D = \frac{10}{9}x + \frac{53}{9}$$

$$10 = \frac{10}{9}x + \frac{53}{9}$$

$$9 \cdot 10 = 9\left(\frac{10}{9}x + \frac{53}{9}\right)$$

$$90 = 10x + 53$$

$$90 - 53 = 10x + 53 - 53$$

$$37 = 10x$$

$$\frac{37}{10} = \frac{10x}{10}$$

$$3.7 = x$$

$$x = 3.7$$

The formula indicates that if the low-humor group averages a level of depression of 10 in response to a negative life event, the intensity of that event is 3.7. This is shown as the point whose corresponding value on the vertical axis is 10 and whose value on the horizontal axis is 3.7.

**Concept and Vocabulary Check 1.2**

**1.**   linear

**2.**   equivalent

**3.**   apply the distributive property

**4.**   least common denominator; 12

**5.**   0

**6.**   $2x$

**7.**   $(x+5)(x+1)$

**8.**   $x \neq 2$ ;  $x \neq 4$

**9.**   $5(x+3) + 3(x+4) = 12x + 9$

**10.** identity

**11.** inconsistent

**Exercise Set 1.2**

**1.** $7x - 5 = 72$
$\qquad 7x = 77$
$\qquad\quad x = 11$
Check:
$\qquad 7x - 5 = 72$
$\qquad 7(11) - 5 = 72$
$\qquad\quad 77 - 5 = 72$
$\qquad\qquad 72 = 72$
The solution set is $\{11\}$.

**3.** $11x - (6x - 5) = 40$
$\qquad 11x - 6x + 5 = 40$
$\qquad\quad 5x + 5 = 40$
$\qquad\qquad 5x = 35$
$\qquad\qquad\; x = 7$
The solution set is $\{7\}$.

Check:
$\qquad 11x - (6x - 5) = 40$
$\qquad 11(7) - [6(7) - 5] = 40$
$\qquad\quad 77 - (42 - 5) = 40$
$\qquad\qquad 77 - (37) = 40$
$\qquad\qquad\qquad\quad 40 = 40$

**5.** $2x - 7 = 6 + x$
$\quad x - 7 = 6$
$\qquad\; x = 13$
The solution set is $\{13\}$.

Check:
$\quad 2(13) - 7 = 6 + 13$
$\qquad\; 26 - 7 = 19$
$\qquad\qquad 19 = 19$

**7.** $7x + 4 = x + 16$
$\quad 6x + 4 = 16$
$\qquad 6x = 12$
$\qquad\; x = 2$
The solution set is $\{2\}$.

Check:
$\quad 7(2) + 4 = 2 + 16$
$\qquad 14 + 4 = 18$
$\qquad\qquad 18 = 18$

**9.** $3(x - 2) + 7 = 2(x + 5)$
$\quad 3x - 6 + 7 = 2x + 10$
$\qquad 3x + 1 = 2x + 10$
$\qquad\; x + 1 = 10$
$\qquad\qquad x = 9$
The solution set is $\{9\}$.

Check:
$\quad 3(9 - 2) + 7 = 2(9 + 5)$
$\qquad 3(7) + 7 = 2(14)$
$\qquad 21 + 7 = 28$
$\qquad\qquad 28 = 28$

**11.** $3(x - 4) - 4(x - 3) = x + 3 - (x - 2)$
$\quad 3x - 12 - 4x + 12 = x + 3 - x + 2$
$\qquad\qquad\qquad\; -x = 5$
$\qquad\qquad\qquad\quad x = -5$
The solution set is $\{-5\}$.

Check:
$\; 3(-5 - 4) - 4(-5 - 3) = -5 + 3 - (-5 - 2)$
$\qquad 3(-9) - 4(-8) = -2 - (-7)$
$\qquad\qquad -27 + 32 = -2 + 7$
$\qquad\qquad\qquad\quad 5 = 5$

**13.** $16 = 3(x - 1) - (x - 7)$
$\quad 16 = 3x - 3 - x + 7$
$\quad 16 = 2x + 4$
$\quad 12 = 2x$
$\qquad 6 = x$
The solution set is $\{6\}$.

Check:
$\quad 16 = 3(6 - 1) - (6 - 7)$
$\quad 16 = 3(5) - (-1)$
$\quad 16 = 15 + 1$
$\quad 16 = 16$

**15.** $25-[2+5y-3(y+2)]=-3(2y-5)-[5(y-1)-3y+3]$

$25-[2+5y-3y-6]=-6y+15-[5y-5-3y+3]$

$25-[2y-4]=-6y+15-[2y-2]$

$25-2y+4=-6y+15-2y+2$

$-2y+29=-8y+17$

$6y=-12$

$y=-2$

The solution set is $\{-2\}$.

Check:

$25-[2+5y-3(y+2)]=-3(2y-5)-[5(y-1)-3y+3]$

$25-[2+5(-2)-3(-2+2)]=-3[2(-2)-5]-[5(-2-1)-3(-2)+3]$

$25-[2-10-3(0)]=-3[-4-5]-[5(-3)+6+3]$

$25-[-8]=-3(-9)-[-15+9]$

$25+8=27-(-6)$

$33=27+6$

$33=33$

**17.** $\dfrac{x}{3}=\dfrac{x}{2}-2$

$6\left[\dfrac{x}{3}=\dfrac{x}{2}-2\right]$

$2x=3x-12$

$12=3x-2x$

$x=12$

The solution set is $\{12\}$.

**19.** $20-\dfrac{x}{3}=\dfrac{x}{2}$

$6\left[20-\dfrac{x}{3}=\dfrac{x}{2}\right]$

$120-2x=3x$

$120=3x+2x$

$120=5x$

$x=\dfrac{120}{5}$

$x=24$

The solution set is $\{24\}$.

**21.** $\dfrac{3x}{5}=\dfrac{2x}{3}+1$

$15\left[\dfrac{3x}{5}=\dfrac{2x}{3}+1\right]$

$9x=10x+15$

$9x-10x=15$

$-x=15$

$x=-15$

The solution set is $\{-15\}$.

**23.**
$$\frac{3x}{5} - x = \frac{x}{10} - \frac{5}{2}$$
$$10\left[\frac{3x}{5} - x = \frac{x}{10} - \frac{5}{2}\right]$$
$$6x - 10x = x - 25$$
$$-4x - x = -25$$
$$-5x = -25$$
$$x = 5$$
The solution set is {5}.

**25.**
$$\frac{x+3}{6} = \frac{3}{8} + \frac{x-5}{4}$$
$$24\left[\frac{x+3}{6} = \frac{3}{8} + \frac{x-5}{4}\right]$$
$$4x + 12 = 9 + 6x - 30$$
$$4x - 6x = -21 - 12$$
$$-2x = -33$$
$$x = \frac{33}{2}$$
The solution set is $\left\{\frac{33}{2}\right\}$.

**27.**
$$\frac{x}{4} = 2 + \frac{x-3}{3}$$
$$12\left[\frac{x}{4} = 2 + \frac{x-3}{3}\right]$$
$$3x = 24 + 4x - 12$$
$$3x - 4x = 12$$
$$-x = 12$$
$$x = -12$$
The solution set is {−12}.

**29.**
$$\frac{x+1}{3} = 5 - \frac{x+2}{7}$$
$$21\left[\frac{x+1}{3} = 5 - \frac{x+2}{7}\right]$$
$$7x + 7 = 105 - 3x - 6$$
$$7x + 3x = 99 - 7$$
$$10x = 92$$
$$x = \frac{92}{10}$$
$$x = \frac{46}{5}$$
The solution set is $\left\{\frac{46}{5}\right\}$.

**31.  a.**  $\dfrac{4}{x} = \dfrac{5}{2x} + 3 \,(x \neq 0)$

**b.**
$$\frac{4}{x} = \frac{5}{2x} + 3$$
$$8 = 5 + 6x$$
$$3 = 6x$$
$$\frac{1}{2} = x$$
The solution set is $\left\{\dfrac{1}{2}\right\}$.

**33.  a.**  $\dfrac{2}{x} + 3 = \dfrac{5}{2x} + \dfrac{13}{4} \,(x \neq 0)$

**b.**
$$\frac{2}{x} + 3 = \frac{5}{2x} + \frac{13}{4}$$
$$8 + 12x = 10 + 13x$$
$$-x = 2$$
$$x = -2$$
The solution set is {−2}.

**35.  a.**  $\dfrac{2}{3x} + \dfrac{1}{4} = \dfrac{11}{6x} - \dfrac{1}{3} \,(x \neq 0)$

**b.**
$$\frac{2}{3x} + \frac{1}{4} = \frac{11}{6x} - \frac{1}{3}$$
$$8 + 3x = 22 - 4x$$
$$7x = 14$$
$$x = 2$$
The solution set is {2}.

**37.  a.**  $\dfrac{x-2}{2x} + 1 = \dfrac{x+1}{x} \quad (x \neq 0)$

**b.**
$$\frac{x-2}{2x} + 1 = \frac{x+1}{x}$$
$$x - 2 + 2x = 2x + 2$$
$$x - 2 = 2$$
$$x = 4$$
The solution set is {4}.

**39.  a.**  $\dfrac{1}{x-1} + 5 = \dfrac{11}{x-1} \,(x \neq 1)$

**b.**
$$\frac{1}{x-1} + 5 = \frac{11}{x-1}$$
$$1 + 5(x-1) = 11$$
$$1 + 5x - 5 = 11$$
$$5x - 4 = 11$$
$$5x = 15$$
$$x = 3$$
The solution set is {3}.

**41. a.** $\dfrac{8x}{x+1} = 4 - \dfrac{8}{x+1} \ (x \ne -1)$

**b.** $\dfrac{8x}{x+1} = 4 - \dfrac{8}{x+1}$

$8x = 4(x+1) - 8$

$8x = 4x + 4 - 8$

$4x = -4$

$x = -1 \Rightarrow$ no solution

The solution set is the empty set, $\varnothing$.

**43. a.** $\dfrac{3}{2x-2} + \dfrac{1}{2} = \dfrac{2}{x-1} \ (x \ne 1)$

**b.** $\dfrac{3}{2x-2} + \dfrac{1}{2} = \dfrac{2}{x-1}$

$\dfrac{3}{2(x-1)} + \dfrac{1}{2} = \dfrac{2}{x-1}$

$3 + 1(x-1) = 4$

$3 + x - 1 = 4$

$x = 2$

The solution set is $\{2\}$.

**45. a.** $\dfrac{3}{x+2} + \dfrac{2}{x-2} = \dfrac{8}{(x+2)(x-2)}; (x \ne -2, 2)$

**b.** $\dfrac{3}{x+2} + \dfrac{2}{x-2} = \dfrac{8}{(x+2)(x-2)}$

$(x \ne 2, x \ne -2)$

$3(x-2) + 2(x+2) = 8$

$3x - 6 + 2x + 4 = 8$

$5x = 10$

$x = 2 \Rightarrow$ no solution

The solution set is the empty set, $\varnothing$.

**47. a.** $\dfrac{2}{x+1} - \dfrac{1}{x-1} = \dfrac{2x}{x^2-1} \ (x \ne 1, x \ne -1)$

**b.**

$\dfrac{2}{x+1} - \dfrac{1}{x-1} = \dfrac{2x}{x^2-1}$

$\dfrac{2}{x+1} - \dfrac{1}{x-1} = \dfrac{2x}{(x+1)(x-1)}$

$2(x-1) - 1(x+1) = 2x$

$2x - 2 - x - 1 = 2x$

$-x = 3$

$x = -3$

The solution set is $\{-3\}$.

**49. a.** $\dfrac{1}{x-4} - \dfrac{5}{x+2} = \dfrac{6}{(x-4)(x+2)}; (x \ne -2, 4)$

**b.** $\dfrac{1}{x-4} - \dfrac{5}{x+2} = \dfrac{6}{x^2-2x-8}$

$\dfrac{1}{x-4} - \dfrac{5}{x+2} = \dfrac{6}{(x-4)(x+2)}$

$(x \ne 4, x \ne -2)$

$1(x+2) - 5(x-4) = 6$

$x + 2 - 5x + 20 = 6$

$-4x = -16$

$x = 4 \Rightarrow$ no solution

The solution set is the empty set, $\varnothing$.

**51.** Set $y_1 = y_2$.

$5(2x-8) - 2 = 5(x-3) + 3$

$10x - 40 - 2 = 5x - 15 + 3$

$10x - 42 = 5x - 12$

$10x - 5x = -12 + 42$

$5x = 30$

$x = 6$

The solution set is $\{6\}$.

**53.** Set $y_1 - y_2 = 1$.

$\dfrac{x-3}{5} - \dfrac{x-5}{4} = 1$

$20 \cdot \dfrac{x-3}{5} - 20 \cdot \dfrac{x-5}{4} = 20 \cdot 1$

$4(x-3) - 5(x-5) = 20$

$4x - 12 - 5x + 25 = 20$

$-x + 13 = 20$

$-x = 7$

$x = -7$

The solution set is $\{-7\}$.

**55.** Set $y_1 + y_2 = y_3$.

$$\frac{5}{x+4} + \frac{3}{x+3} = \frac{12x+19}{x^2+7x+12}$$

$$\frac{5}{x+4} + \frac{3}{x+3} = \frac{12x+19}{(x+4)(x+3)}$$

$$(x+4)(x+3)\left(\frac{5}{x+4} + \frac{3}{x+3}\right) = (x+4)(x+3)\frac{12x+19}{(x+4)(x+3)}$$

$$5(x+3) + 3(x+4) = 12x+19$$

$$5x+15+3x+12 = 12x+19$$

$$8x+27 = 12x+19$$

$$-4x = -8$$

$$x = 2$$

The solution set is $\{2\}$.

**57.**  $0 = 4[x-(3-x)] - 7(x+1)$

$0 = 4[x-3+x] - 7x - 7$

$0 = 4[2x-3] - 7x - 7$

$0 = 8x-12-7x-7$

$0 = x-19$

$-x = -19$

$x = 19$

The solution set is $\{19\}$.

**59.**  
$$0 = \frac{x+6}{3x-12} - \frac{5}{x-4} - \frac{2}{3}$$

$$0 = \frac{x+6}{3(x-4)} - \frac{5}{x-4} - \frac{2}{3}$$

$$3(x-4)\cdot 0 = 3(x-4)\left(\frac{x+6}{3(x-4)} - \frac{5}{x-4} - \frac{2}{3}\right)$$

$$0 = \frac{3(x-4)(x+6)}{3(x-4)} - \frac{5\cdot 3(x-4)}{x-4} - \frac{2\cdot 3(x-4)}{3}$$

$$0 = (x+6) - 15 - 2(x-4)$$

$$0 = x+6-15-2x+8$$

$$0 = -x-1$$

$$x = -1$$

The solution set is $\{-1\}$.

**61.** $5x+9=9(x+1)-4x$

$5x+9=9x+9-4x$

$5x+9=5x+9$

$9=9$

The solution set $\{x|x \text{ is a real number}\}$.

The given equation is an identity.

**63.** $3(x+2)=7+3x$

$3x+6=7+3x$

$6=7$

The solution set $\varnothing$.

The given equation is an inconsistent equation.

**65.** $10x+3=8x+3$

$2x+3=3$

$2x=0$

$x=0$

The solution set $\{0\}$.

The given equation is a conditional equation.

**67.** $\dfrac{2x}{x-3}=\dfrac{6}{x-3}+4$

$2x=6+4(x-3)$

$2x=6+4x-12$

$-2x=-6$

$x=3 \Rightarrow$ no solution

The given equation is an inconsistent equation.

**69.** $\dfrac{x+5}{2}-4=\dfrac{2x-1}{3}$

$3(x+5)-24=2(2x-1)$

$3x+15-24=4x-2$

$-x=7$

$x=-7$

The solution set is $\{-7\}$.

The given equation is a conditional equation.

**71.** $\dfrac{2}{x-2}=3+\dfrac{x}{x-2}$

$2=3(x-2)+x$

$2=3x-6+x$

$-4x=-8$

$x=2 \Rightarrow$ no solution

The solution set is the empty set, $\varnothing$.

The given equation is an inconsistent equation.

**73.** $8x-(3x+2)+10=3x$

$8x-3x-2+10=3x$

$2x=-8$

$x=-4$

The solution set is $\{-4\}$.

The given equation is a conditional equation.

**75.** $\dfrac{2}{x}+\dfrac{1}{2}=\dfrac{3}{4}$

$8+2x=3x$

$-x=-8$

$x=8$

The solution set is $\{8\}$.

The given equation is a conditional equation.

**77.** $\dfrac{4}{x-2}+\dfrac{3}{x+5}=\dfrac{7}{(x+5)(x-2)}$

$4(x+5)+3(x-2)=7$

$4x+20+3x-6=7$

$7x=-7$

$x=-1$

The solution set is $\{-1\}$.

The given equation is a conditional equation.

**79.** $\dfrac{4x}{x+3}-\dfrac{12}{x-3}=\dfrac{4x^2+36}{x^2-9}; x\neq 3, -3$

$4x(x-3)-12(x+3)=4x^2+36$

$4x^2-12x-12x-36=4x^2+36$

$4x^2-24x-36=4x^2+36$

$-24x-36=36$

$-24x=72$

$x=-3 \quad$ No solution

The solution set is $\{\ \}$.

The given equation is an inconsistent equation.

**81.** The equation is $3(x-4)=3(2-2x)$, and the solution is $x=2$.

**83.** The equation is $-3(x-3)=5(2-x)$, and the solution is $x=0.5$.

**85.** Solve: $4(x-2)+2=4x-2(2-x)$
$$4x-8+2=4x-4+2x$$
$$4x-6=6x-4$$
$$-2x-6=-4$$
$$-2x=2$$
$$x=-1$$
Now, evaluate $x^2-x$ for $x=-1$:
$$x^2-x=(-1)^2-(-1)$$
$$=1-(-1)=1+1=2$$

**87.** Solve for $x$: $\dfrac{3(x+3)}{5}=2x+6$
$$3(x+3)=5(2x+6)$$
$$3x+9=10x+30$$
$$-7x+9=30$$
$$-7x=21$$
$$x=-3$$
Solve for $y$: $-2y-10=5y+18$
$$-7y-10=18$$
$$-7y=28$$
$$y=-4$$
Now, evaluate $x^2-(xy-y)$ for $x=-3$ and $y=-4$:
$$x^2-(xy-y)$$
$$=(-3)^2-[-3(-4)-(-4)]$$
$$=(-3)^2-[12-(-4)]$$
$$=9-(12+4)=9-16=-7$$

**89.** $\left[(3+6)^2\div3\right]\cdot4=-54x$
$$\left(9^2\div3\right)\cdot4=-54x$$
$$(81\div3)\cdot4=-54x$$
$$27\cdot4=-54x$$
$$108=-54x$$
$$-2=x$$
The solution set is $\{-2\}$.

**91.** $5-12x=8-7x-\left[6\div3\left(2+5^3\right)+5x\right]$
$$5-12x=8-7x-\left[6\div3(2+125)+5x\right]$$
$$5-12x=8-7x-[6\div3\cdot127+5x]$$
$$5-12x=8-7x-[2\cdot127+5x]$$
$$5-12x=8-7x-[254+5x]$$
$$5-12x=8-7x-254-5x$$
$$5-12x=-12x-246$$
$$5=-246$$
The final statement is a contradiction, so the equation has no solution. The solution set is $\varnothing$.

**93.** $0.7x+0.4(20)=0.5(x+20)$
$$0.7x+8=0.5x+10$$
$$0.2x+8=10$$
$$0.2x=2$$
$$x=10$$
The solution set is $\{10\}$.

**95.** $4x+13-\left\{2x-\left[4(x-3)-5\right]\right\}=2(x-6)$
$$4x+13-\left\{2x-[4x-12-5]\right\}=2x-12$$
$$4x+13-\left\{2x-[4x-17]\right\}=2x-12$$
$$4x+13-\{2x-4x+17\}=2x-12$$
$$4x+13-\{-2x+17\}=2x-12$$
$$4x+13+2x-17=2x-12$$
$$6x-4=2x-12$$
$$4x-4=-12$$
$$4x=-8$$
$$x=-2$$
The solution set is $\{-2\}$.

**97. a.** $p+\dfrac{x}{2}=37$
$$p+\dfrac{40}{2}=37$$
$$p+20=37$$
$$p=17$$
According to the model, 17% of American adults smoked cigarettes in 2010. This underestimates the value shown in the bar graph by 2%.

**b.** $p + \dfrac{x}{2} = 37$

$7 + \dfrac{x}{2} = 37$

$\dfrac{x}{2} = 30$

$x = 60$

According to the model, only 7% of American adults will smoke cigarettes 60 years after 1970, or 2030.

**99. a.** What cost $10,000 in 1967 would cost about $52,000 in 2000.

**b.** $C = 1388x + 24{,}963$

$= 1388(20) + 24{,}963$

$= 52{,}723$

According to Model 1, what cost $10,000 in 1967 would cost about $52,723 in 2000. This describes the estimate from part (a) reasonably well.

**c.** $C = 3x^2 + 1308x + 25{,}268$

$= 3(20)^2 + 1308(20) + 25{,}268$

$= 52{,}628$

According to Model 2, what cost $10,000 in 1967 would cost about $52,628 in 2000. This describes the estimate from part (a) reasonably well.

**101.** $C = 1388x + 24{,}963$

$77{,}707 = 1388x + 24{,}963$

$52{,}744 = 1388x$

$38 = x$

Model 1 predicts the cost will be $77,707 38 years after 1980, or 2018.

**103.** 11 learning trials; represented by the point $(11, 0.95)$ on the graph.

**105.** $C = \dfrac{x + 0.1(500)}{x + 500}$

$0.28 = \dfrac{x + 0.1(500)}{x + 500}$

$0.28(x + 500) = x + 0.1(500)$

$0.28x + 140 = x + 50$

$-0.72x = -90$

$\dfrac{-0.72x}{-0.72} = \dfrac{-90}{-0.72}$

$x = 125$

125 liters of pure peroxide must be added.

**107. – 115.** Answers will vary.

**117.** $\{5\}$

**119.** $\{-5\}$

**121.** makes sense

**123.** makes sense

**125.** false; Changes to make the statement true will vary. A sample change is: In the first equation, $x \neq 4$.

**127.** false; Changes to make the statement true will vary. A sample change is: If $a = 0$, then $ax + b = 0$ is equivalent to $b = 0$, which either has no solution ($b \neq 0$) or infinitely many solutions ($b = 0$).

**129.**
$$\frac{7x+4}{b}+13=x$$
$$\frac{7(-6)+4}{b}+13=-6$$
$$\frac{-42+4}{b}+13=-6$$
$$\frac{-38}{b}+13=-6$$
$$\frac{-38}{b}=-19$$
$$-38=-19b$$
$$b=2$$

**131.** $x+150$

**132.** $20+0.05x$

**133.** $4x+400$

## Section 1.3

### Check Point Exercises

**1.** Let $x=$ the median starting salary, in thousands of dollars, of education majors.
Let $x+21=$ the median starting salary, in thousands of dollars, of computer science majors.
Let $x+14=$ the median starting salary, in thousands of dollars, of economics majors.
$$x+(x+21)+(x+14)=140$$
$$x+x+21+x+14=140$$
$$3x+35=140$$
$$3x=105$$
$$x=35$$
$$x+21=56$$
$$x+14=49$$
The median starting salary of education majors is \$35 thousand, of computer science majors is \$56 thousand, and of economics majors is \$49 thousand.

**2.** Let $x=$ the number of years after 1969.
$$85-0.9x=25$$
$$-0.9x=25-85$$
$$-0.9x=-60$$
$$x=\frac{-60}{-0.9}$$
$$x\approx67$$
25% of freshmen will respond this way 67 years after 1969, or 2036.

**3.** Let $x=$ the number of text messages at which the costs of the two plans are the same.
$$\underset{\text{Plan A}}{\underbrace{15+0.08x}}=\underset{\text{Plan B}}{\underbrace{3+0.12x}}$$
$$15+0.08x-15=3+0.12x-15$$
$$0.08x=0.12x-12$$
$$0.08x-0.12x=0.12x-12-0.12x$$
$$-0.04x=-12$$
$$\frac{-0.04x}{-0.04}=\frac{-12}{-0.04}$$
$$x=300$$
The two plans are the same at 300 text messages.

**4.** Let $x=$ the computer's price before the reduction.
$$x-0.30x=840$$
$$0.70x=840$$
$$x=\frac{840}{0.70}$$
$$x=1200$$
Before the reduction the computer's price was \$1200.

**5.** Let $x=$ the amount invested at 9%.
Let $5000-x=$ the amount invested at 11%.
$$0.09x+0.11(5000-x)=487$$
$$0.09x+550-0.11x=487$$
$$-0.02x+550=487$$
$$-0.02x=-63$$
$$x=\frac{-63}{-0.02}$$
$$x=3150$$
$$5000-x=1850$$
\$3150 was invested at 9% and \$1850 was invested at 11%.

**6.** Let $x=$ the width of the court.
Let $x+44=$ the length of the court.
$$2l+2w=P$$
$$2(x+44)+2x=288$$
$$2x+88+2x=288$$
$$4x+88=288$$
$$4x=200$$
$$x=\frac{200}{4}$$
$$x=50$$
$$x+44=94$$
The dimensions of the court are 50 feet by 94 feet.

**7.**
$$2l + 2w = P$$
$$2l + 2w - 2l = P - 2l$$
$$2w = P - 2l$$
$$\frac{2w}{2} = \frac{P - 2l}{2}$$
$$w = \frac{P - 2l}{2}$$

**8.**
$$P = C + MC$$
$$P = C(1 + M)$$
$$\frac{P}{1 + M} = \frac{C(1 + M)}{1 + M}$$
$$\frac{P}{1 + M} = C$$
$$C = \frac{P}{1 + M}$$

## Concept and Vocabulary Check 1.3

**1.** $x + 658.6$

**2.** $31 + 2.4x$

**3.** $4 + 0.15x$

**4.** $x - 0.15x$ or $0.85x$

**5.** $0.12x + 0.09(30,000 - x)$

**6.** isolated on one side

**7.** factoring

## Exercise Set 1.3

**1.** Let $x$ = the number
$$5x - 4 = 26$$
$$5x = 30$$
$$x = 6$$
The number is 6.

**3.** Let $x$ = the number
$$x - 0.20x = 20$$
$$0.80x = 20$$
$$x = 25$$
The number is 25.

**5.** Let $x$ = the number
$$0.60x + x = 192$$
$$1.6x = 192$$
$$x = 120$$
The number is 120.

**7.** Let $x$ = the number
$$0.70x = 224$$
$$x = 320$$
The number is 320.

**9.** Let $x$ = the number
$x + 26$ = the other number
$$x + (x + 26) = 64$$
$$x + x + 26 = 64$$
$$2x + 26 = 64$$
$$2x = 38$$
$$x = 19$$
If $x = 19$, then $x + 26 = 45$.
The numbers are 19 and 45.

**11.**
$$y_1 - y_2 = 2$$
$$(13x - 4) - (5x + 10) = 2$$
$$13x - 4 - 5x - 10 = 2$$
$$8x - 14 = 2$$
$$8x = 16$$
$$\frac{8x}{8} = \frac{16}{8}$$
$$x = 2$$

**13.**
$$y_1 = 8y_2 + 14$$
$$10(2x - 1) = 8(2x + 1) + 14$$
$$20x - 10 = 16x + 8 + 14$$
$$20x - 10 = 16x + 22$$
$$4x = 32$$
$$\frac{4x}{4} = \frac{32}{4}$$
$$x = 8$$

**15.**
$$3y_1 - 5y_2 = y_3 - 22$$
$$3(2x + 6) - 5(x + 8) = (x) - 22$$
$$6x + 18 - 5x - 40 = x - 22$$
$$x - 22 = x - 22$$
$$x - x = -22 + 22$$
$$0 = 0$$
The solution set is the set of all real numbers.

**17.**
$$3y_1 + 4y_2 = 4y_3$$
$$3\left(\frac{1}{x}\right) + 4\left(\frac{1}{2x}\right) = 4\left(\frac{1}{x-1}\right)$$
$$\frac{3}{x} + \frac{2}{x} = \frac{4}{x-1}$$
$$\frac{5}{x} = \frac{4}{x-1}$$
$$\frac{5x(x-1)}{x} = \frac{4x(x-1)}{x-1}$$
$$5(x-1) = 4x$$
$$5x - 5 = 4x$$
$$x = 5$$

**19.** Let $x$ = the number of years spent watching TV.
Let $x + 19$ = the number of years spent sleeping.
$$x + (x + 19) = 37$$
$$x + x + 19 = 37$$
$$2x + 19 = 37$$
$$2x = 18$$
$$x = 9$$
$$x + 19 = 28$$
Americans will spend 9 years watching TV and 28 years sleeping.

**21.** Let $x$ = the average salary, in thousands, for an American whose final degree is a bachelor's.
Let $2x - 49$ = the average salary, in thousands, for an American whose final degree is a master's.
$$x + (2x - 49) = 116$$
$$x + 2x - 49 = 116$$
$$3x - 49 = 116$$
$$3x = 165$$
$$x = 55$$
$$2x - 49 = 61$$
The average salary for an American whose final degree is a bachelor's is \$55 thousand and for an American whose final degree is a master's is \$61 thousand.

**23.** Let $x$ = the number of years since 2000.
$$31 + 2.4x = 67$$
$$2.4x = 67 - 31$$
$$2.4x = 36$$
$$x = \frac{36}{2.4}$$
$$x = 15$$
67% of American adults will view college education as essential 15 years after 2000, or 2015.

**25. a.** $y = 24,000 - 3000x$

**b.**
$$y = 24,000 - 3000x$$
$$9000 = 24,000 - 3000x$$
$$9000 - 24,000 = -3000x$$
$$-15,000 = -3000x$$
$$x = \frac{-15,000}{-3000}$$
$$x = 5$$
The car's value will drop to \$9000 after 5 years.

**c.** Graph:

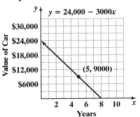

**27.** Let $x$ = the number of months.
The cost for Club A: $25x + 40$
The cost for Club B: $30x + 15$
$$25x + 40 = 30x + 15$$
$$-5x + 40 = 15$$
$$-5x = -25$$
$$x = 5$$
The total cost for the clubs will be the same at 5 months. The cost will be
$$25(5) + 40 = 30(5) + 15 = \$165$$

**29.** Let $x$ = the number of uses.
Cost without discount pass: $1.25x$
Cost with discount pass: $15 + 0.75x$
$$1.25x = 15 + 0.75x$$
$$0.50x = 15$$
$$x = 30$$
The bus must be used 30 times in a month for the costs to be equal.

**31. a.** Let $x$ = the number of years (after 2010).
College A's enrollment: $13,300 + 1000x$

College B's enrollment: $26,800 - 500x$
$$13,300 + 1000x = 26,800 - 500x$$
$$13,300 + 1500x = 26,800$$
$$1500x = 13,500$$
$$x = 9$$
The two colleges will have the same enrollment 9 years after 2010, or 2019. That year the enrollment will be
$$13,300 + 1000(9)$$
$$= 26,800 - 500(9)$$
$$= 22,300 \text{ students}$$

**b.** Check points to determine that
$y_1 = 13,300 + 1000x$ and
$y_2 = 26,800 - 500x$ .

**33.** Let $x$ = the cost of the television set.
$$x - 0.20x = 336$$
$$0.80x = 336$$
$$x = 420$$
The television set's price is $420.

**35.** Let $x$ = the nightly cost
$$x + 0.08x = 162$$
$$1.08x = 162$$
$$x = 150$$
The nightly cost is $150.

**37.** Let c = the dealer's cost
$$584 = c + 0.25c$$
$$584 = 1.25c$$
$$467.20 = c$$
The dealer's cost is $467.20.

**39.** Let $x$ = the amount invested at 6%.
Let $7000 - x$ = the amount invested at 8%.
$$0.06x + 0.08(7000 - x) = 520$$
$$0.06x + 560 - 0.08x = 520$$
$$-0.02x + 560 = 520$$
$$-0.02x = -40$$
$$x = \frac{-40}{-0.02}$$
$$x = 2000$$
$$7000 - x = 5000$$
$2000 was invested at 6% and $5000 was invested at 8%.

**41.** Let $x$ = amount invested at 12%
$8000 - x$ = amount invested at 5% loss
$$.12x - .05(8000 - x) = 620$$
$$.12x - 400 + .05x = 620$$
$$.17x = 1020$$
$$x = 6000$$
$$8000 - x = 2000$$
$6000 at 12%, $2000 at 5% loss

**43.** Let $w$ = the width of the field
Let $2w$ = the length of the field
$$P = 2(\text{length}) + 2(\text{width})$$
$$300 = 2(2w) + 2(w)$$
$$300 = 4w + 2w$$
$$300 = 6w$$
$$50 = w$$
If $w = 50$ , then $2w = 100$ . Thus, the dimensions are 50 yards by 100 yards.

**45.** Let $w$ = the width of the field
Let $2w + 6$ = the length of the field
$$228 = 6w + 12$$
$$216 = 6w$$
$$36 = w$$
If $w = 36$ , then $2w + 6 = 2(36) + 6 = 78$ . Thus, the dimensions are 36 feet by 78 feet.

**47.** Let $x$ = the width of the frame.
Total length: $16 + 2x$
Total width: $12 + 2x$
$$P = 2(\text{length}) + 2(\text{width})$$
$$72 = 2(16 + 2x) + 2(12 + 2x)$$
$$72 = 32 + 4x + 24 + 4x$$
$$72 = 8x + 56$$
$$16 = 8x$$
$$2 = x$$
The width of the frame is 2 inches.

**49.** Let $x$ = number of hours
$35x$ = labor cost
$$35x + 63 = 448$$
$$35x = 385$$
$$x = 11$$
It took 11 hours.

**51.** Let $x$ = inches over 5 feet
$$100 + 5x = 135$$
$$5x = 35$$
$$x = 7$$
A height of 5 feet 7 inches corresponds to 135 pounds.

**53.** Let $x$ = the weight of unpeeled bananas.
$$\frac{7}{8}x = \text{ weight of peeled bananas}$$
$$x = \frac{7}{8}x + \frac{7}{8}$$
$$\frac{1}{8}x = \frac{7}{8}$$
$$x = 7$$
The banana with peel weighs 7 ounces.

**55.** $A = lw$
$$w = \frac{A}{l}$$
area of rectangle

**57.** $A = \frac{1}{2}bh$
$$2A = bh$$
$$b = \frac{2A}{h};$$
area of triangle

**59.** $I = Prt$
$$P = \frac{I}{rt};$$
interest

**61.** $E = mc^2$
$$m = \frac{E}{c^2};$$
Einstein's equation

**63.** $T = D + pm$
$$T - D = pm$$
$$\frac{T - D}{m} = \frac{pm}{m}$$
$$\frac{T - D}{m} = p$$
total of payment

**65.** $A = \frac{1}{2}h(a + b)$
$$2A = h(a + b)$$
$$\frac{2A}{h} = a + b$$
$$\frac{2A}{h} - b = a$$
area of trapezoid

**67.** $S = P + Prt$
$$S - P = Prt$$
$$\frac{S - P}{Pt} = r;$$
interest

**69.** $B = \frac{F}{S - V}$
$$B(S - V) = F$$
$$S - V = \frac{F}{B}$$
$$S = \frac{F}{B} + V$$

**71.** $IR + Ir = E$
$$I(R + r) = E$$
$$I = \frac{E}{R + r}$$
electric current

**73.** $\frac{1}{p} + \frac{1}{q} = \frac{1}{f}$
$$qf + pf = pq$$
$$f(q + p) = pq$$
$$f = \frac{pq}{p + q}$$
thin lens equation

**75. – 79.** Answers will vary.

**81.** does not make sense; Explanations will vary. Sample explanation: Though mathematical models can often provide excellent estimates about future attitudes, they cannot guaranty perfect precision.

**83.** does not make sense; Explanations will vary. Sample explanation: Solving a formula for one of its variables does not produce a numerical value for the variable.

**85.**  $0.1x + .9(1000 - x) = 420$
$0.1 + 900 - 0.9x = 420$
$-0.8x = -480$
$x = 600$

600 students at the north campus, 400 students at south campus.

**87.**  Let $x$ = woman's age
$3x$ = Coburn's age
$3x + 20 = 2(x + 20)$
$3x + 20 = 2x + 40$
$x + 20 = 40$
$x = 20$

Coburn is 60 years old the woman is 20 years old.

**89.**  Let $x$ = mother's amount
$2x$ = boy's amount
$\dfrac{x}{2}$ = girl's amount

$x + 2x + \dfrac{x}{2} = 14,000$

$\dfrac{7}{2}x = 14,000$

$x = \$4,000$

The mother received $4000, the boy received $8000, and the girl received $2000.

**91.**  $V = C - \dfrac{C - S}{L}N$
$VL = CL - CN + SN$
$VL - SN = CL - CN$
$VL - SN = C(L - N)$
$\dfrac{VL - SN}{L - N} = C$
$C = \dfrac{VL - SN}{L - N}$

**93.**  $(7 - 3x)(-2 - 5x) = -14 - 35x + 6x + 15x^2$
$= -14 - 29x + 15x^2$
or
$= 15x^2 - 29x - 14$

**94.**  $\sqrt{18} - \sqrt{8} = \sqrt{9 \cdot 2} - \sqrt{4 \cdot 2}$
$= 3\sqrt{2} - 2\sqrt{2}$
$= \sqrt{2}$

**95.**  $\dfrac{7 + 4\sqrt{2}}{2 - 5\sqrt{2}} \cdot \dfrac{2 + 5\sqrt{2}}{2 + 5\sqrt{2}} = \dfrac{14 + 35\sqrt{2} + 8\sqrt{2} + 40}{4 + 10\sqrt{2} - 10\sqrt{2} - 50}$

$= \dfrac{54 + 43\sqrt{2}}{-46}$

$= -\dfrac{54 + 43\sqrt{2}}{46}$

## Section 1.4

### Check Point Exercises

**1.**  **a.**  $(5 - 2i) + (3 + 3i)$
$= 5 - 2i + 3 + 3i$
$= (5 + 3) + (-2 + 3)i$
$= 8 + i$

  **b.**  $(2 + 6i) - (12 - i)$
$= 2 + 6i - 12 + i$
$= (2 - 12) + (6 + 1)i$
$= -10 + 7i$

**2.**  **a.**  $7i(2 - 9i) = 7i(2) - 7i(9i$
$= 14i - 63i^2$
$= 14i - 63(-1)$
$= 63 + 14i$

  **b.**  $(5 + 4i)(6 - 7i) = 30 - 35i + 24i - 28i^2$
$= 30 - 35i + 24i - 28(-1)$
$= 30 + 28 - 35i + 24i$
$= 58 - 11i$

**3.**  $\dfrac{5 + 4i}{4 - i} = \dfrac{5 + 4i}{4 - i} \cdot \dfrac{4 + i}{4 + i}$

$= \dfrac{20 + 5i + 16i + 4i^2}{16 + 4i - 4i - i^2}$

$= \dfrac{20 + 21i - 4}{16 + 1}$

$= \dfrac{16 + 21i}{17}$

$= \dfrac{16}{17} + \dfrac{21}{17}i$

**4.** **a.** $\sqrt{-27} + \sqrt{-48} = i\sqrt{27} + i\sqrt{48}$
$= i\sqrt{9 \cdot 3} + i\sqrt{16 \cdot 3}$
$= 3i\sqrt{3} + 4i\sqrt{3}$
$= 7i\sqrt{3}$

**b.** $(-2 + \sqrt{-3})^2 = (-2 + i\sqrt{3})^2$
$= (-2)^2 + 2(-2)(i\sqrt{3}) + (i\sqrt{3})^2$
$= 4 - 4i\sqrt{3} + 3i^2$
$= 4 - 4i\sqrt{3} + 3(-1)$
$= 1 - 4i\sqrt{3}$

**c.** $\dfrac{-14 + \sqrt{-12}}{2} = \dfrac{-14 + i\sqrt{12}}{2}$
$= \dfrac{-14 + 2i\sqrt{3}}{2}$
$= \dfrac{-14}{2} + \dfrac{2i\sqrt{3}}{2}$
$= -7 + i\sqrt{3}$

**Concept and Vocabulary Check 1.4**

**1.** $\sqrt{-1}$; $-1$

**2.** complex; imaginary; real

**3.** $-6i$

**4.** $14i$

**5.** $18$; $-15i$; $12i$; $-10i^2$; $10$

**6.** $2 + 9i$

**7.** $2 + 5i$

**8.** $i$; $2i\sqrt{5}$

**Exercise Set 1.4**

**1.** $(7 + 2i) + (1 - 4i) = 7 + 2i + 1 - 4i$
$= 7 + 1 + 2i - 4i$
$= 8 - 2i$

**3.** $(3 + 2i) - (5 - 7i) = 3 - 5 + 2i + 7i$
$= 3 + 2i - 5 + 7i$
$= -2 + 9i$

**5.** $6 - (-5 + 4i) - (-13 - i) = 6 + 5 - 4i + 13 + i$
$= 24 - 3i$

**7.** $8i - (14 - 9i) = 8i - 14 + 9i$
$= -14 + 8i + 9i$
$= -14 + 17i$

**9.** $-3i(7i - 5) = -21i^2 + 15i$
$= -21(-1) + 15i$
$= 21 + 15i$

**11.** $(-5 + 4i)(3 + i) = -15 - 5i + 12i + 4i^2$
$= -15 + 7i - 4$
$= -19 + 7i$

**13.** $(7 - 5i)(-2 - 3i) = -14 - 21i + 10i + 15i^2$
$= -14 - 15 - 11i$
$= -29 - 11i$

**15.** $(3 + 5i)(3 - 5i) = 9 - 15i + 15i - 25i^2$
$= 9 + 25$
$= 34$

**17.** $(-5 + i)(-5 - i) = 25 + 5i - 5i - i^2$
$= 25 + 1$
$= 26$

**19.** $(2 + 3i)^2 = 4 + 12i + 9i^2$
$= 4 + 12i - 9$
$= -5 + 12i$

**21.** $\dfrac{2}{3 - i} = \dfrac{2}{3 - i} \cdot \dfrac{3 + i}{3 + i}$
$= \dfrac{2(3 + i)}{9 + 1}$
$= \dfrac{2(3 + i)}{10}$
$= \dfrac{3 + i}{5}$
$= \dfrac{3}{5} + \dfrac{1}{5}i$

**23.** $\dfrac{2i}{1 + i} = \dfrac{2i}{1 + i} \cdot \dfrac{1 - i}{1 - i} = \dfrac{2i - 2i^2}{1 + 1} = \dfrac{2 + 2i}{2} = 1 + i$

25. $\dfrac{8i}{4-3i} = \dfrac{8i}{4-3i} \cdot \dfrac{4+3i}{4+3i}$

$= \dfrac{32i + 24i^2}{16+9}$

$= \dfrac{-24+32i}{25}$

$= -\dfrac{24}{25} + \dfrac{32}{25}i$

27. $\dfrac{2+3i}{2+i} = \dfrac{2+3i}{2+i} \cdot \dfrac{2-i}{2-i}$

$= \dfrac{4+4i-3i^2}{4+1}$

$= \dfrac{7+4i}{5}$

$= \dfrac{7}{5} + \dfrac{4}{5}i$

29. $\sqrt{-64} - \sqrt{-25} = i\sqrt{64} - i\sqrt{25}$

$= 8i - 5i = 3i$

31. $5\sqrt{-16} + 3\sqrt{-81} = 5(4i) + 3(9i)$

$= 20i + 27i = 47i$

33. $\left(-2+\sqrt{-4}\right)^2 = (-2+2i)^2$

$= 4 - 8i + 4i^2$

$= 4 - 8i - 4$

$= -8i$

35. $\left(-3-\sqrt{-7}\right)^2 = \left(-3 - i\sqrt{7}\right)^2$

$= 9 + 6i\sqrt{7} + i^2(7)$

$= 9 - 7 + 6i\sqrt{7}$

$= 2 + 6i\sqrt{7}$

37. $\dfrac{-8+\sqrt{-32}}{24} = \dfrac{-8 + i\sqrt{32}}{24}$

$= \dfrac{-8 + i\sqrt{16 \cdot 2}}{24}$

$= \dfrac{-8 + 4i\sqrt{2}}{24}$

$= -\dfrac{1}{3} + \dfrac{\sqrt{2}}{6}i$

39. $\dfrac{-6-\sqrt{-12}}{48} = \dfrac{-6 - i\sqrt{12}}{48}$

$= \dfrac{-6 - i\sqrt{4 \cdot 3}}{48}$

$= \dfrac{-6 - 2i\sqrt{3}}{48}$

$= -\dfrac{1}{8} - \dfrac{\sqrt{3}}{24}i$

41. $\sqrt{-8}\left(\sqrt{-3}-\sqrt{5}\right) = i\sqrt{8}(i\sqrt{3}-\sqrt{5})$

$= 2i\sqrt{2}\left(i\sqrt{3}-\sqrt{5}\right)$

$= -2\sqrt{6} - 2i\sqrt{10}$

43. $\left(3\sqrt{-5}\right)\left(-4\sqrt{-12}\right) = \left(3i\sqrt{5}\right)\left(-8i\sqrt{3}\right)$

$= -24i^2\sqrt{15}$

$= 24\sqrt{15}$

45. $(2-3i)(1-i) - (3-i)(3+i)$

$= \left(2-2i-3i+3i^2\right) - \left(3^2 - i^2\right)$

$= 2 - 5i + 3i^2 - 9 + i^2$

$= -7 - 5i + 4i^2$

$= -7 - 5i + 4(-1)$

$= -11 - 5i$

47. $(2+i)^2 - (3-i)^2$

$= \left(4+4i+i^2\right) - \left(9-6i+i^2\right)$

$= 4 + 4i + i^2 - 9 + 6i - i^2$

$= -5 + 10i$

49. $5\sqrt{-16} + 3\sqrt{-81}$

$= 5\sqrt{16}\sqrt{-1} + 3\sqrt{81}\sqrt{-1}$

$= 5 \cdot 4i + 3 \cdot 9i$

$= 20i + 27i$

$= 47i \text{ or } 0 + 47i$

51. $f(x) = x^2 - 2x + 2$

$f(1+i) = (1+i)^2 - 2(1+i) + 2$

$= 1 + 2i + i^2 - 2 - 2i + 2$

$= 1 + i^2$

$= 1 - 1$

$= 0$

**53.** $f(x) = \dfrac{x^2 + 19}{2 - x}$

$f(3i) = \dfrac{(3i)^2 + 19}{2 - 3i}$

$= \dfrac{9i^2 + 19}{2 - 3i}$

$= \dfrac{-9 + 19}{2 - 3i}$

$= \dfrac{10}{2 - 3i}$

$= \dfrac{10}{2 - 3i} \cdot \dfrac{2 + 3i}{2 + 3i}$

$= \dfrac{20 + 30i}{4 - 9i^2}$

$= \dfrac{20 + 30i}{4 + 9}$

$= \dfrac{20 + 30i}{13}$

$= \dfrac{20}{13} + \dfrac{30}{13}i$

**55.** $E = IR = (4 - 5i)(3 + 7i)$

$= 12 + 28i - 15i - 35i^2$

$= 12 + 13i - 35(-1)$

$= 12 + 35 + 13i = 47 + 13i$

The voltage of the circuit is
$(47 + 13i)$ volts.

**57.** Sum:

$\left(5 + i\sqrt{15}\right) + \left(5 - i\sqrt{15}\right)$

$= 5 + i\sqrt{15} + 5 - i\sqrt{15}$

$= 5 + 5$

$= 10$

Product:

$\left(5 + i\sqrt{15}\right)\left(5 - i\sqrt{15}\right)$

$= 25 - 5i\sqrt{15} + 5i\sqrt{15} - 15i^2$

$= 25 + 15$

$= 40$

**59. – 65.** Answers will vary.

**67.** makes sense

**69.** does not make sense; Explanations will vary.
Sample explanation: $i = \sqrt{-1}$ ; It is not a variable in this context.

**71.** false; Changes to make the statement true will vary.
A sample change is: All irrational numbers are complex numbers.

**73.** false; Changes to make the statement true will vary.
A sample change is:

$\dfrac{7 + 3i}{5 + 3i} = \dfrac{7 + 3i}{5 + 3i} \cdot \dfrac{5 - 3i}{5 - 3i} = \dfrac{44 - 6i}{34} = \dfrac{22}{17} - \dfrac{3}{17}i$

**75.** $\dfrac{4}{(2 + i)(3 - i)} = \dfrac{4}{6 - 2i + 3i - i^2}$

$= \dfrac{4}{6 + i + 1}$

$= \dfrac{4}{7 + i}$

$= \dfrac{4}{7 + i} \cdot \dfrac{7 - i}{7 - i}$

$= \dfrac{28 - 4i}{49 - i^2}$

$= \dfrac{28 - 4i}{49 + 1}$

$= \dfrac{28 - 4i}{50}$

$= \dfrac{28}{50} - \dfrac{4}{50}i$

$= \dfrac{14}{25} - \dfrac{2}{25}i$

**77.**
$$\dfrac{8}{1+\dfrac{2}{i}} = \dfrac{8}{\dfrac{i}{i}+\dfrac{2}{i}}$$

$$= \dfrac{8}{\dfrac{2+i}{i}}$$

$$= \dfrac{8i}{2+i}$$

$$= \dfrac{8i}{2+i}\cdot\dfrac{2-i}{2-i}$$

$$= \dfrac{16i-8i^2}{4-i^2}$$

$$= \dfrac{16i+8}{4+1}$$

$$= \dfrac{8+16i}{5}$$

$$= \dfrac{8}{5}+\dfrac{16}{5}i$$

**78.** $2x^2 + 7x - 4 = (2x-1)(x+4)$

**79.** $x^2 - 6x + 9 = (x-3)(x-3) = (x-3)^2$

**80.**
$$\dfrac{-b-\sqrt{b^2-4ac}}{2a} = \dfrac{-(9)-\sqrt{(9)^2-4(2)(-5)}}{2(2)}$$

$$= \dfrac{-9-\sqrt{81+40}}{4}$$

$$= \dfrac{-9-\sqrt{121}}{4}$$

$$= \dfrac{-9-11}{4}$$

$$= -5$$

## Section 1.5

### Check Point Exercises

**1.**  **a.**  $3x^2 - 9x = 0$
$3x(x-3) = 0$
$3x = 0$  or  $x-3 = 0$
$x = 0$ $\qquad$ $x = 3$
The solution set is $\{0, 3\}$.

**b.**  $2x^2 + x = 1$
$2x^2 + x - 1 = 0$
$(2x-1)(x+1) = 0$
$2x-1 = 0$  or  $x+1 = 0$
$2x = 1$ $\qquad$ $x = -1$
$x = \dfrac{1}{2}$
The solution set is $\left\{-1, \dfrac{1}{2}\right\}$.

**2.**  **a.**  $3x^2 = 21$
$\dfrac{3x^2}{3} = \dfrac{21}{3}$
$x^2 = 7$
$x = \pm\sqrt{7}$
The solution set is $\left\{-\sqrt{7}, \sqrt{7}\right\}$.

**b.**  $5x^2 + 45 = 0$
$5x^2 = -45$
$x^2 = -9$
$x = \pm\sqrt{-9}$
$x = \pm 3i$
The solution set is $\{-3i, 3i\}$.

**c.**  $(x+5)^2 = 11$
$x+5 = \pm\sqrt{11}$
$x = -5 \pm\sqrt{11}$
The solution set is $\left\{-5+\sqrt{11}, -5-\sqrt{11}\right\}$.

**3.**  **a.**  The coefficient of the *x*-term is 6.  Half of 6 is 3, and $3^2$ is 9.
9 should be added to the binomial.
$x^2 + 6x + 9 = (x+3)^2$

**b.**  The coefficient of the *x*-term is –5.
Half of –5 is $-\dfrac{5}{2}$, and $\left(-\dfrac{5}{2}\right)^2$ is $\dfrac{25}{4}$.
$\dfrac{25}{4}$ should be added to the binomial.
$x^2 - 5x + \dfrac{25}{4} = \left(x-\dfrac{5}{2}\right)^2$

**c.** The coefficient of the $x$-term is $\frac{2}{3}$.

Half of $\frac{2}{3}$ is $\frac{1}{3}$, and $\left(\frac{1}{3}\right)^2$ is $\frac{1}{9}$.

$\frac{1}{9}$ should be added to the binomial.

$$x^2 + \frac{2}{3}x + \frac{1}{9} = \left(x + \frac{1}{3}\right)^2$$

**4.**
$$x^2 + 4x - 1 = 0$$
$$x^2 + 4x = 1$$
$$x^2 + 4x + 4 = 1 + 4$$
$$(x+2)^2 = 5$$
$$x + 2 = \pm\sqrt{5}$$
$$x = -2 \pm \sqrt{5}$$
The solution set is $\left\{-2 \pm \sqrt{5}\right\}$.

**5.**
$$2x^2 + 3x - 4 = 0$$
$$x^2 + \frac{3}{2}x - 2 = 0$$
$$x^2 + \frac{3}{2}x = 2$$
$$x^2 + \frac{3}{2}x + \frac{9}{16} = 2 + \frac{9}{16}$$
$$\left(x + \frac{3}{4}\right)^2 = \frac{41}{16}$$
$$x + \frac{3}{4} = \pm\sqrt{\frac{41}{16}}$$
$$x + \frac{3}{4} = \pm\frac{\sqrt{41}}{4}$$
$$x = -\frac{3}{4} \pm \frac{\sqrt{41}}{4}$$
$$x = \frac{-3 \pm \sqrt{41}}{4}$$
The solution set is $\left\{\frac{-3 \pm \sqrt{41}}{4}\right\}$.

**6.**
$$2x^2 + 2x - 1 = 0$$
$$a = 2, b = 2, c = -1$$
$$x = \frac{-b \pm \sqrt{b^2 - 4ac}}{2a}$$
$$= \frac{-2 \pm \sqrt{2^2 - 4(2)(-1)}}{2(2)}$$
$$= \frac{-2 \pm \sqrt{4+8}}{4}$$
$$= \frac{-2 \pm \sqrt{12}}{4}$$
$$= \frac{-2 \pm 2\sqrt{3}}{4}$$
$$= \frac{2(-1 \pm \sqrt{3})}{4}$$
$$= \frac{-1 \pm \sqrt{3}}{2}$$
The solution set is $\left\{\frac{-1 \pm \sqrt{3}}{2}\right\}$.

**7.**
$$x^2 - 2x + 2 = 0$$
$$a = 1, b = -2, c = 2$$
$$x = \frac{-b \pm \sqrt{b^2 - 4ac}}{2a}$$
$$x = \frac{-(-2) \pm \sqrt{(-2)^2 - 4(1)(2)}}{2(1)}$$
$$x = \frac{2 \pm \sqrt{4-8}}{2}$$
$$x = \frac{2 \pm \sqrt{-4}}{2}$$
$$x = \frac{2 \pm 2i}{2}$$
$$x = 1 \pm i$$
The solution set is $\{1+i, 1-i\}$.

**8. a.** $a = 1, \quad b = 6, \quad c = 9$
$$b^2 - 4ac = (6)^2 - 4(1)(9)$$
$$= 36 - 36$$
$$= 0$$
Since $b^2 - 4ac = 0$, the equation has one real solution that is rational.

**b.**  $a = 2, \quad b = -7, \quad c = -4$

$$b^2 - 4ac = (-7)^2 - 4(2)(-4)$$
$$= 49 + 32$$
$$= 81$$

Since $b^2 - 4ac > 0$, the equation has two real solutions. Since 81 is a perfect square, the two solutions are rational.

**c.**  $a = 3, \quad b = -2, \quad c = 4$

$$b^2 - 4ac = (-2)^2 - 4(3)(4)$$
$$= 4 - 48$$
$$= -44$$

Since $b^2 - 4ac < 0$, the equation has two imaginary solutions that are complex conjugates.

**9.**  $P = 0.01A^2 + 0.05A + 107$

$$115 = 0.01A^2 + 0.05A + 107$$
$$0 = 0.01A^2 + 0.05A - 8$$
$$a = 0.01, \quad b = 0.05, \quad c = -8$$

$$A = \frac{-b \pm \sqrt{b^2 - 4ac}}{2a}$$

$$A = \frac{-(0.05) \pm \sqrt{(0.05)^2 - 4(0.01)(-8)}}{2(0.01)}$$

$$A = \frac{-0.05 \pm \sqrt{0.3225}}{0.02}$$

$$A \approx \frac{-0.05 + \sqrt{0.3225}}{0.02} \qquad A \approx \frac{-0.05 - \sqrt{0.3225}}{0.02}$$

$$A \approx 26 \qquad\qquad A \approx -31$$

Age cannot be negative, reject the negative answer. Thus, a woman whose normal systolic blood pressure is 115 mm Hg is approximately 26 years old.

**10.** Let w = the screen's width.

$$w^2 + l^2 = d^2$$
$$w^2 + 15^2 = 25^2$$
$$w^2 + 225 = 625$$
$$w^2 = 400$$
$$w = \pm\sqrt{400}$$
$$w = \pm 20$$

Reject the negative value.
The width of the television is 20 inches.

**Concept and Vocabulary Check 1.5**

1. quadratic

2. $A = 0$ or $B = 0$

3. $x$-intercepts

4. $\pm\sqrt{d}$

5. $\pm\sqrt{7}$

6. $\dfrac{9}{4}$

7. $\dfrac{4}{25}$

8. 9

9. $\dfrac{1}{9}$

10. $\dfrac{-b \pm \sqrt{b^2 - 4ac}}{2a}$

11. 2; 9; −5

12. 1; −4; −1

13. $2 \pm \sqrt{2}$

14. $-1 \pm i\dfrac{\sqrt{6}}{2}$

15. $b^2 - 4ac$

16. no

17. two

18. the square root property

19. the quadratic formula

20. factoring and the zero-product principle

21. right; hypotenuse; legs

22. right; legs; the square of the length of the hypotenuse

**Exercise Set 1.5**

**1.** $x^2 - 3x - 10 = 0$
$(x+2)(x-5) = 0$
$x + 2 = 0$   or   $x - 5 = 0$
$x = -2$   or   $x = 5$
The solution set is $\{-2, 5\}$.

**3.** $x^2 = 8x - 15$
$x^2 - 8x + 15 = 0$
$(x-3)(x-5) = 0$
$x - 3 = 0$   or   $x - 5 = 0$
$x = 3$   or   $x = 5$
The solution set is $\{3, 5\}$.

**5.** $6x^2 + 11x - 10 = 0$
$(2x+5)(3x-2) = 0$
$2x + 5 = 0$   or   $3x - 2 = 0$
$2x = -5$          $3x = 2$
$x = -\dfrac{5}{2}$   or   $x = \dfrac{2}{3}$
The solution set is $\left\{-\dfrac{5}{2}, \dfrac{2}{3}\right\}$.

**7.** $3x^2 - 2x = 8$
$3x^2 - 2x - 8 = 0$
$(3x+4)(x-2) = 0$
$3x + 4 = 0$   or   $x - 2 = 0$
$3x = -4$
$x = -\dfrac{4}{3}$   or   $x = 2$
The solution set is $\left\{-\dfrac{4}{3}, 2\right\}$.

**9.** $3x^2 + 12x = 0$
$3x(x+4) = 0$
$3x = 0$   or   $x + 4 = 0$
$x = 0$   or   $x = -4$
The solution set is $\{-4, 0\}$.

**11.** $2x(x-3) = 5x^2 - 7x$
$2x^2 - 6x - 5x^2 + 7x = 0$
$-3x^2 + x = 0$
$x(-3x + 1) = 0$
$x = 0$   or   $-3x + 1 = 0$
$-3x = -1$
$x = \dfrac{1}{3}$
The solution set is $\left\{0, \dfrac{1}{3}\right\}$.

**13.** $7 - 7x = (3x+2)(x-1)$
$7 - 7x = 3x^2 - x - 2$
$7 - 7x - 3x^2 + x + 2 = 0$
$-3x^2 - 6x + 9 = 0$
$-3(x+3)(x-1) = 0$
$x + 3 = 0$   or   $x - 1 = 0$
$x = -3$   or   $x = 1$
The solution set is $\{-3, 1\}$.

**15.** $3x^2 = 27$
$x^2 = 9$
$x = \pm\sqrt{9} = \pm 3$
The solution set is $\{-3, 3\}$.

**17.** $5x^2 + 1 = 51$
$5x^2 = 50$
$x^2 = 10$
$x = \pm\sqrt{10}$
The solution set is $\left\{-\sqrt{10}, \sqrt{10}\right\}$.

**19.** $2x^2 - 5 = -55$
$2x^2 = -50$
$x^2 = -25$
$x = \pm\sqrt{-25} = \pm 5i$
The solution set is $\{5i, -5i\}$.

**21.** $(x+2)^2 = 25$
$x + 2 = \pm\sqrt{25}$
$x + 2 = \pm 5$
$x = -2 \pm 5$
$x = -2 + 5$   or   $x = -2 - 5$
$x = 3$          $x = -7$
The solution set is $\{-7, 3\}$.

**23.** $3(x-4)^2 = 15$

$\qquad (x-4)^2 = 5$

$\qquad\quad x-4 = \pm\sqrt{5}$

$\qquad\qquad x = 4\pm\sqrt{5}$

The solution set is $\left\{4+\sqrt{5}, 4-\sqrt{5}\right\}$.

**25.** $(x+3)^2 = -16$

$\qquad x+3 = \pm\sqrt{-16}$

$\qquad x+3 = \pm 4i$

$\qquad\quad x = -3\pm 4i$

The solution set is $\{-3+4i, -3-4i\}$.

**27.** $(x-3)^2 = -5$

$\qquad x-3 = \pm\sqrt{-5}$

$\qquad x-3 = \pm i\sqrt{5}$

$\qquad\quad x = 3\pm i\sqrt{5}$

The solution set is $\left\{3+i\sqrt{5}, 3-i\sqrt{5}\right\}$.

**29.** $(3x+2)^2 = 9$

$\quad 3x+2 = \pm\sqrt{9} = \pm 3$

$\quad 3x+2 = -3 \quad \text{or} \quad 3x+2 = 3$

$\quad 3x = -5 \qquad\qquad 3x = 1$

$\quad x = -\dfrac{5}{3} \qquad \text{or} \quad x = \dfrac{1}{3}$

The solution set is $\left\{-\dfrac{5}{3}, \dfrac{1}{3}\right\}$.

**31.** $(5x-1)^2 = 7$

$\quad 5x-1 = \pm\sqrt{7}$

$\quad 5x = 1\pm\sqrt{7}$

$\quad x = \dfrac{1\pm\sqrt{7}}{5}$

The solution set is $\left\{\dfrac{1-\sqrt{7}}{5}, \dfrac{1+\sqrt{7}}{5}\right\}$.

**33.** $(3x-4)^2 = 8$

$\quad 3x-4 = \pm\sqrt{8} = \pm 2\sqrt{2}$

$\quad 3x = 4\pm 2\sqrt{2}$

$\quad x = \dfrac{4\pm 2\sqrt{2}}{3}$

The solution set is $\left\{\dfrac{4-2\sqrt{2}}{3}, \dfrac{4+2\sqrt{2}}{3}\right\}$.

**35.** $x^2 + 12x$

$\left(\dfrac{12}{2}\right)^2 = 6^2 = 36$

$x^2 + 12x + 36 = (x+6)^2$

**37.** $x^2 - 10x$

$\left(\dfrac{10}{2}\right)^2 = 5^2 = 25$

$x^2 - 10x + 25 = (x-5)^2$

**39.** $x^2 + 3x$

$\left(\dfrac{3}{2}\right)^2 = \dfrac{9}{4}$

$x^2 + 3x + \dfrac{9}{4} = \left(x+\dfrac{3}{2}\right)^2$

**41.** $x^2 - 7x$

$\left(\dfrac{7}{2}\right)^2 = \dfrac{49}{4}$

$x^2 - 7x + \dfrac{49}{4} = \left(x-\dfrac{7}{2}\right)^2$

**43.** $x^2 - \dfrac{2}{3}x$

$\left(\dfrac{\frac{2}{3}}{2}\right)^2 = \left(\dfrac{1}{3}\right)^2 = \dfrac{1}{9}$

$x^2 - \dfrac{2}{3}x + \dfrac{1}{9} = \left(x-\dfrac{1}{3}\right)^2$

**45.** $x^2 - \dfrac{1}{3}x$

$\left(\dfrac{\frac{1}{3}}{2}\right)^2 = \left(\dfrac{1}{6}\right)^2 = \dfrac{1}{36}$

$x^2 - \dfrac{1}{3}x + \dfrac{1}{36} = \left(x-\dfrac{1}{6}\right)^2$

**47.** $\qquad x^2 + 6x = 7$

$\qquad x^2 + 6x + 9 = 7 + 9$

$\qquad\quad (x+3)^2 = 16$

$\qquad\qquad x+3 = \pm 4$

$\qquad\qquad\quad x = -3\pm 4$

The solution set is $\{-7, 1\}$.

**49.** $x^2 - 2x = 2$

$x^2 - 2x + 1 = 2 + 1$

$(x-1)^2 = 3$

$x - 1 = \pm\sqrt{3}$

$x = 1 \pm \sqrt{3}$

The solution set is $\left\{1+\sqrt{3}, 1-\sqrt{3}\right\}$.

**51.** $x^2 - 6x - 11 = 0$

$x^2 - 6x = 11$

$x^2 - 6x + 9 = 11 + 9$

$(x-3)^2 = 20$

$x - 3 = \pm\sqrt{20}$

$x = 3 \pm 2\sqrt{5}$

The solution set is $\left\{3+2\sqrt{5}, 3-2\sqrt{5}\right\}$.

**53.** $x^2 + 4x + 1 = 0$

$x^2 + 4x = -1$

$x^2 + 4x + 4 = -1 + 4$

$(x+2)^2 = 3$

$x + 2 = \pm\sqrt{3}$

$x = -2 \pm \sqrt{3}$

The solution set is $\left\{-2+\sqrt{3}, -2-\sqrt{3}\right\}$.

**55.** $x^2 - 5x + 6 = 0$

$x^2 - 5x = -6$

$x^2 - 5x + \dfrac{25}{4} = -6 + \dfrac{25}{4}$

$\left(x - \dfrac{5}{2}\right)^2 = \dfrac{1}{4}$

$x - \dfrac{5}{2} = \pm\sqrt{\dfrac{1}{4}}$

$x - \dfrac{5}{2} = \pm\dfrac{1}{2}$

$x = \dfrac{5}{2} \pm \dfrac{1}{2}$

$x = \dfrac{5}{2} + \dfrac{1}{2}$ or $x = \dfrac{5}{2} - \dfrac{1}{2}$

$x = 3$ $\qquad\qquad x = 2$

The solution set is $\{2, 3\}$.

**57.** $x^2 + 3x - 1 = 0$

$x^2 + 3x = 1$

$x^2 + 3x + \dfrac{9}{4} = 1 + \dfrac{9}{4}$

$\left(x + \dfrac{3}{2}\right)^2 = \dfrac{13}{4}$

$x + \dfrac{3}{2} = \pm\dfrac{\sqrt{13}}{2}$

$x = \dfrac{-3 \pm \sqrt{13}}{2}$

The solution set is $\left\{\dfrac{-3+\sqrt{13}}{2}, \dfrac{-3-\sqrt{13}}{2}\right\}$.

**59.** $2x^2 - 7x + 3 = 0$

$x^2 - \dfrac{7}{2}x + \dfrac{3}{2} = 0$

$x^2 - \dfrac{7}{2}x = \dfrac{-3}{2}$

$x^2 - \dfrac{7}{2}x + \dfrac{49}{16} = -\dfrac{3}{2} + \dfrac{49}{16}$

$\left(x - \dfrac{7}{4}\right)^2 = \dfrac{25}{16}$

$x - \dfrac{7}{4} = \pm\dfrac{5}{4}$

$x = \dfrac{7}{4} \pm \dfrac{5}{4}$

The solution set is $\left\{\dfrac{1}{2}, 3\right\}$.

**61.** $4x^2 - 4x - 1 = 0$

$4x^2 - 4x - 1 = 0$

$x^2 - x - \dfrac{1}{4} = 0$

$x^2 - x = \dfrac{1}{4}$

$x^2 - x + \dfrac{1}{4} = \dfrac{1}{4} + \dfrac{1}{4}$

$\left(x - \dfrac{1}{2}\right)^2 = \dfrac{2}{4}$

$x - \dfrac{1}{2} = \dfrac{\pm\sqrt{2}}{2}$

$x = \dfrac{1 \pm \sqrt{2}}{2}$

The solution set is $\left\{\dfrac{1+\sqrt{2}}{2}, \dfrac{1-\sqrt{2}}{2}\right\}$.

**63.** $3x^2 - 2x - 2 = 0$

$x^2 - \dfrac{2}{3}x - \dfrac{2}{3} = 0$

$x^2 - \dfrac{2}{3}x = \dfrac{2}{3}$

$x^2 - \dfrac{2}{3}x + \dfrac{1}{9} = \dfrac{2}{3} + \dfrac{1}{9}$

$\left(x - \dfrac{1}{3}\right)^2 = \dfrac{7}{9}$

$x - \dfrac{1}{3} = \dfrac{\pm\sqrt{7}}{3}$

$x = \dfrac{1 \pm \sqrt{7}}{3}$

The solution set is $\left\{\dfrac{1+\sqrt{7}}{3}, \dfrac{1-\sqrt{7}}{3}\right\}$.

**65.** $x^2 + 8x + 15 = 0$

$x = \dfrac{-8 \pm \sqrt{8^2 - 4(1)(15)}}{2(1)}$

$x = \dfrac{-8 \pm \sqrt{64 - 60}}{2}$

$x = \dfrac{-8 \pm \sqrt{4}}{2}$

$x = \dfrac{-8 \pm 2}{2}$

The solution set is $\{-5, -3\}$.

**67.** $x^2 + 5x + 3 = 0$

$x = \dfrac{-5 \pm \sqrt{5^2 - 4(1)(3)}}{2(1)}$

$x = \dfrac{-5 \pm \sqrt{25 - 12}}{2}$

$x = \dfrac{-5 \pm \sqrt{13}}{2}$

The solution set is $\left\{\dfrac{-5+\sqrt{13}}{2}, \dfrac{-5-\sqrt{13}}{2}\right\}$.

**69.** $3x^2 - 3x - 4 = 0$

$x = \dfrac{3 \pm \sqrt{(-3)^2 - 4(3)(-4)}}{2(3)}$

$x = \dfrac{3 \pm \sqrt{9 + 48}}{6}$

$x = \dfrac{3 \pm \sqrt{57}}{6}$

The solution set is $\left\{\dfrac{3+\sqrt{57}}{6}, \dfrac{3-\sqrt{57}}{6}\right\}$.

**71.** $\qquad 4x^2 = 2x + 7$

$4x^2 - 2x - 7 = 0$

$x = \dfrac{2 \pm \sqrt{(-2)^2 - 4(4)(-7)}}{2(4)}$

$x = \dfrac{2 \pm \sqrt{4 + 112}}{8}$

$x = \dfrac{2 \pm \sqrt{116}}{8}$

$x = \dfrac{2 \pm 2\sqrt{29}}{8}$

$x = \dfrac{1 \pm \sqrt{29}}{4}$

The solution set is $\left\{\dfrac{1+\sqrt{29}}{4}, \dfrac{1-\sqrt{29}}{4}\right\}$.

**73.** $x^2 - 6x + 10 = 0$

$$x = \frac{6 \pm \sqrt{(-6)^2 - 4(1)(10)}}{2(1)}$$

$$x = \frac{6 \pm \sqrt{36 - 40}}{2}$$

$$x = \frac{6 \pm \sqrt{-4}}{2}$$

$$x = \frac{6 \pm 2i}{2}$$

$$x = 3 \pm i$$

The solution set is $\{3 + i, 3 - i\}$.

**75.** $x^2 - 4x - 5 = 0$

$(-4)^2 - 4(1)(-5)$

$= 16 + 20$

$= 36$; 2 unequal real solutions

**77.** $2x^2 - 11x + 3 = 0$

$(-11)^2 - 4(2)(3)$

$= 121 - 24$

$= 97$; 2 unequal real solutions

**79.** $x^2 - 2x + 1 = 0$

$(-2)^2 - 4(1)(1)$

$= 4 - 4$

$= 0$; 1 real solution

**81.** $x^2 - 3x - 7 = 0$

$(-3)^2 - 4(1)(-7)$

$= 9 + 28$

$= 37$; 2 unequal real solutions

**83.** $\qquad 2x^2 - x = 1$

$\qquad 2x^2 - x - 1 = 0$

$\qquad (2x + 1)(x - 1) = 0$

$\qquad\qquad 2x + 1 = 0 \text{ or } x - 1 = 0$

$\qquad\qquad\qquad 2x = -1$

$$x = -\frac{1}{2} \text{ or } x = 1$$

The solution set is $\left\{ -\frac{1}{2}, 1 \right\}$.

**85.** $\qquad 5x^2 + 2 = 11x$

$\qquad 5x^2 - 11x + 2 = 0$

$\qquad (5x - 1)(x - 2) = 0$

$\qquad\qquad 5x - 1 = 0 \text{ or } x - 2 = 0$

$\qquad\qquad\qquad 5x = 1$

$$x = \frac{1}{5} \text{ or } x = 2$$

The solution set is $\left\{ \frac{1}{5}, 2 \right\}$.

**87.** $3x^2 = 60$

$x^2 = 20$

$x = \pm\sqrt{20}$

$x = \pm 2\sqrt{5}$

The solution set is $\left\{ -2\sqrt{5}, 2\sqrt{5} \right\}$.

**89.** $\qquad x^2 - 2x = 1$

$\qquad x^2 - 2x + 1 = 1 + 1$

$\qquad (x - 1)^2 = 2$

$\qquad x - 1 = \pm\sqrt{2}$

$\qquad x = 1 \pm \sqrt{2}$

The solution set is $\left\{ 1 + \sqrt{2}, 1 - \sqrt{2} \right\}$.

**91.** $\qquad (2x + 3)(x + 4) = 1$

$\qquad 2x^2 + 8x + 3x + 12 = 1$

$\qquad\qquad 2x^2 + 11x + 11 = 0$

$$x = \frac{-11 \pm \sqrt{11^2 - 4(2)(11)}}{2(2)}$$

$$x = \frac{-11 \pm \sqrt{121 - 88}}{4}$$

$$x = \frac{-11 \pm \sqrt{33}}{4}$$

The solution set is $\left\{ \dfrac{-11 + \sqrt{33}}{4}, \dfrac{-11 - \sqrt{33}}{4} \right\}$.

**93.** $(3x-4)^2 = 16$

$$3x - 4 = \pm\sqrt{16}$$
$$3x - 4 = \pm 4$$
$$3x = 4 \pm 4$$
$$3x = 8 \text{ or } 3x = 0$$
$$x = \frac{8}{3} \text{ or } x = 0$$

The solution set is $\left\{0, \frac{8}{3}\right\}$.

**95.** $3x^2 - 12x + 12 = 0$

$$x^2 - 4x + 4 = 0$$
$$(x-2)(x-2) = 0$$
$$x - 2 = 0$$
$$x = 2$$

The solution set is $\{2\}$.

**97.** $4x^2 - 16 = 0$

$$4x^2 = 16$$
$$x^2 = 4$$
$$x = \pm 2$$

The solution set is $\{-2, 2\}$.

**99.** $x^2 - 6x + 13 = 0$

$$x^2 - 6x = -13$$
$$x^2 - 6x + 9 = -13 + 9$$
$$(x-3)^2 = -4$$
$$x - 3 = \pm 2i$$
$$x = 3 \pm 2i$$

The solution set is $\{3 + 2i, 3 - 2i\}$.

**101.**
$$x^2 = 4x - 7$$
$$x^2 - 4x = -7$$
$$x^2 - 4x + 4 = -7 + 4$$
$$(x-2)^2 = -3$$
$$x - 2 = \pm i\sqrt{3}$$
$$x = 2 \pm i\sqrt{3}$$

The solution set is $\left\{2 + i\sqrt{3}, 2 - i\sqrt{3}\right\}$.

**103.** $2x^2 - 7x = 0$

$$x(2x - 7) = 0$$
$$x = 0 \text{ or } 2x - 7 = 0$$
$$2x = 7$$
$$x = 0 \text{ or } x = \frac{7}{2}$$

The solution set is $\left\{0, \frac{7}{2}\right\}$.

**105.** $\dfrac{1}{x} + \dfrac{1}{x+2} = \dfrac{1}{3}; x \neq 0, -2$

$$3x + 6 + 3x = x^2 + 2x$$
$$0 = x^2 - 4x - 6$$
$$x = \frac{-(-4) \pm \sqrt{(-4)^2 - 4(1)(-6)}}{2(1)}$$
$$x = \frac{4 \pm \sqrt{16 + 24}}{2}$$
$$x = \frac{4 \pm \sqrt{40}}{2}$$
$$x = \frac{4 \pm 2\sqrt{10}}{2}$$
$$x = 2 \pm \sqrt{10}$$

The solution set is $\{2 + \sqrt{10}, \ 2 - \sqrt{10}\}$.

**107.**
$$\frac{2x}{x-3} + \frac{6}{x+3} = \frac{-28}{x^2-9}; x \neq 3, -3$$
$$2x(x+3) + 6(x-3) = -28$$
$$2x^2 + 6x + 6x - 18 = -28$$
$$2x^2 + 12x + 10 = 0$$
$$x^2 + 6x + 5 = 0$$
$$(x+1)(x+5) = 0$$

The solution set is $\{-5, \ -1\}$.

**109.** $x^2 - 4x - 5 = 0$

$$(x+1)(x-5) = 0$$
$$x + 1 = 0 \quad\text{or}\quad x - 5 = 0$$
$$x = -1 \qquad\qquad x = 5$$

This equation matches graph (d).

**111.** $\quad 0 = -(x+1)^2 + 4$

$\qquad (x+1)^2 = 4$

$\qquad\quad x+1 = \pm 2$

$\qquad\qquad x = -1 \pm 2$

$\qquad\qquad x = -3, \quad x = 1$

This equation matches graph (f).

**113.** $x^2 - 2x + 2 = 0$

$\qquad a = 1, \quad b = -2, \quad c = 2$

$\qquad x = \dfrac{-b \pm \sqrt{b^2 - 4ac}}{2a}$

$\qquad x = \dfrac{-(-2) \pm \sqrt{(-2)^2 - 4(1)(2)}}{2(1)}$

$\qquad x = \dfrac{2 \pm \sqrt{-4}}{2}$

$\qquad x = \dfrac{2 \pm 2i}{2}$

$\qquad x = 1 \pm i$

This equation has no real roots. Thus, its equation has no x-intercepts. This equation matches graph (b).

**115.** $y = 2x^2 - 3x$

$\qquad 2 = 2x^2 - 3x$

$\qquad 0 = 2x^2 - 3x - 2$

$\qquad 0 = (2x+1)(x-2)$

$\qquad x = -\dfrac{1}{2}, \quad x = 2$

**117.** $\quad y_1 y_2 = 14$

$\qquad (x-1)(x+4) = 14$

$\qquad\quad x^2 + 3x - 4 = 14$

$\qquad\quad x^2 + 3x - 18 = 0$

$\qquad (x+6)(x-3) = 0$

$\qquad x = -6, \quad x = 3$

**119.**

$$y_1 + y_2 = 1$$

$$\frac{2x}{x+2} + \frac{3}{x+4} = 1$$

$$(x+2)(x+4)\left(\frac{2x}{x+2} + \frac{3}{x+4}\right) = 1(x+2)(x+4)$$

$$\frac{2x(x+2)(x+4)}{x+2} + \frac{3(x+2)(x+4)}{x+4} = (x+2)(x+4)$$

$$2x(x+4) + 3(x+2) = (x+2)(x+4)$$

$$2x^2 + 8x + 3x + 6 = x^2 + 6x + 8$$

$$x^2 + 5x - 2 = 0$$

$$x = \frac{-b \pm \sqrt{b^2 - 4ac}}{2a}$$

$$x = \frac{-(5) \pm \sqrt{(5)^2 - 4(1)(-2)}}{2(1)}$$

$$x = \frac{-5 \pm \sqrt{33}}{2}$$

The solution set is $\left\{ \dfrac{-5 + \sqrt{33}}{2}, \dfrac{-5 - \sqrt{33}}{2} \right\}$.

**121.**

$$y_1 - y_2 = 0$$

$$(2x^2 + 5x - 4) - (-x^2 + 15x - 10) = 0$$

$$2x^2 + 5x - 4 + x^2 - 15x + 10 = 0$$

$$3x^2 - 10x + 6 = 0$$

$$x = \frac{-b \pm \sqrt{b^2 - 4ac}}{2a}$$

$$x = \frac{-(-10) \pm \sqrt{(-10)^2 - 4(3)(6)}}{2(3)}$$

$$x = \frac{10 \pm \sqrt{28}}{6}$$

$$x = \frac{10 \pm 2\sqrt{7}}{6}$$

$$x = \frac{5 \pm \sqrt{7}}{3}$$

The solution set is $\left\{ \dfrac{5 + \sqrt{7}}{3}, \dfrac{5 - \sqrt{7}}{3} \right\}$.

**123.** Values that make the denominator zero must be excluded.

$$2x^2 + 4x - 9 = 0$$

$$x = \frac{-b \pm \sqrt{b^2 - 4ac}}{2a}$$

$$x = \frac{-(4) \pm \sqrt{(4)^2 - 4(2)(-9)}}{2(2)}$$

$$x = \frac{-4 \pm \sqrt{88}}{4}$$

$$x = \frac{-4 \pm 2\sqrt{22}}{4}$$

$$x = \frac{-2 \pm \sqrt{22}}{2}$$

**125.** $x^2 - (6 + 2x) = 0$

$$x^2 - 2x - 6 = 0$$

Apply the quadratic formula.

$$a = 1 \quad b = -2 \quad c = -6$$

$$x = \frac{-(-2) \pm \sqrt{(-2)^2 - 4(1)(-6)}}{2(1)}$$

$$= \frac{2 \pm \sqrt{4 - (-24)}}{2}$$

$$= \frac{2 \pm \sqrt{28}}{2}$$

$$= \frac{2 \pm \sqrt{4 \cdot 7}}{2} = \frac{2 \pm 2\sqrt{7}}{2} = 1 \pm \sqrt{7}$$

We disregard $1 - \sqrt{7}$ because it is negative, and we are looking for a positive number.

Thus, the number is $1 + \sqrt{7}$.

**127.**

$$\frac{1}{x^2 - 3x + 2} = \frac{1}{x + 2} + \frac{5}{x^2 - 4}$$

$$\frac{1}{(x-1)(x-2)} = \frac{1}{x+2} + \frac{5}{(x+2)(x-2)}$$

Multiply both sides of the equation by the least common denominator, $(x-1)(x-2)(x+2)$. This results in the following:

$$x + 2 = (x-1)(x-2) + 5(x-1)$$

$$x + 2 = x^2 - 2x - x + 2 + 5x - 5$$

$$x + 2 = x^2 + 2x - 3$$

$$0 = x^2 + x - 5$$

Apply the quadratic formula:

$$a = 1 \quad b = 1 \quad c = -5.$$

$$x = \frac{-1 \pm \sqrt{1^2 - 4(1)(-5)}}{2(1)} = \frac{-1 \pm \sqrt{1 - (-20)}}{2}$$

$$= \frac{-1 \pm \sqrt{21}}{2}$$

The solutions are $\dfrac{-1 \pm \sqrt{21}}{2}$, and the solution set is

$$\left\{ \frac{-1 \pm \sqrt{21}}{2} \right\}.$$

**129.** $\sqrt{2}x^2 + 3x - 2\sqrt{2} = 0$

Apply the quadratic formula:

$$a = \sqrt{2} \quad b = 3 \quad c = -2\sqrt{2}$$

$$x = \frac{-3 \pm \sqrt{3^2 - 4(\sqrt{2})(-2\sqrt{2})}}{2(\sqrt{2})}$$

$$= \frac{-3 \pm \sqrt{9 - (-16)}}{2\sqrt{2}}$$

$$= \frac{-3 \pm \sqrt{25}}{2\sqrt{2}} = \frac{-3 \pm 5}{2\sqrt{2}}$$

Evaluate the expression to obtain two solutions.

$$x = \frac{-3 - 5}{2\sqrt{2}} \quad \text{or} \quad x = \frac{-3 + 5}{2\sqrt{2}}$$

$$= \frac{-8}{2\sqrt{2}} \cdot \frac{\sqrt{2}}{\sqrt{2}} \qquad = \frac{2}{2\sqrt{2}} \cdot \frac{\sqrt{2}}{\sqrt{2}}$$

$$= \frac{-8\sqrt{2}}{4} \qquad = \frac{2\sqrt{2}}{4}$$

$$= -2\sqrt{2} \qquad = \frac{\sqrt{2}}{2}$$

The solutions are $-2\sqrt{2}$ and $\dfrac{\sqrt{2}}{2}$, and the solution

set is $\left\{ -2\sqrt{2}, \dfrac{\sqrt{2}}{2} \right\}$.

**131.** $N = \dfrac{x^2 - x}{2}$

$21 = \dfrac{x^2 - x}{2}$

$42 = x^2 - x$

$0 = x^2 - x - 42$

$0 = (x+6)(x-7)$

$x + 6 = 0 \quad$ or $\quad x - 7 = 0$

$\quad x = -6 \qquad\qquad x = 7$

Reject the negative value.
There were 7 players.

**133.** This is represented on the graph as point $(7, 21)$.

**135.** $f(x) = 0.013x^2 - 1.19x + 28.24$

$3 = 0.013x^2 - 1.19x + 28.24$

$0 = 0.013x^2 - 1.19x + 25.24$

Apply the quadratic formula.
$a = 0.013 \quad b = -1.19 \quad c = 25.24$

$x = \dfrac{-(-1.19) \pm \sqrt{(-1.19)^2 - 4(0.013)(25.24)}}{2(0.013)}$

$= \dfrac{1.19 \pm \sqrt{1.4161 - 1.31248}}{0.026}$

$= \dfrac{1.19 \pm \sqrt{0.10362}}{0.026}$

$\approx \dfrac{1.19 \pm 0.32190}{0.026}$

$\approx 58.15$ or $33.39$

The solutions are approximately 33.39 and 58.15.
Thus, 33 year olds and 58 year olds are expected to be in 3 fatal crashes per 100 million miles driven. The function models the actual data well.

**137.** Let $y_1 = -0.01x^2 + 0.7x + 6.1$

Using the TRACE feature, we find that the height of the shot put is approximately 0 feet when the distance is 77.8 feet. Graph (b) shows the shot' path.

**139.** **a.** $\dfrac{1}{\Phi - 1}$

**b.** $\dfrac{\Phi}{1} = \dfrac{1}{\Phi - 1}$

$(\Phi - 1)\dfrac{\Phi}{1} = (\Phi - 1)\dfrac{1}{\Phi - 1}$

$\Phi^2 - \Phi = 1$

$\Phi^2 - \Phi - 1 = 0$

$\Phi = \dfrac{-b \pm \sqrt{b^2 - 4ac}}{2a}$

$\Phi = \dfrac{-(-1) \pm \sqrt{(-1)^2 - 4(1)(-1)}}{2(1)}$

$\Phi = \dfrac{1 \pm \sqrt{1 + 4}}{2}$

$\Phi = \dfrac{1 \pm \sqrt{5}}{2}$, reject negative

$\Phi = \dfrac{1 + \sqrt{5}}{2}$

**c.** The golden ratio is $\dfrac{1 + \sqrt{5}}{2}$ to 1.

**141.** $x^2 = 4^2 + 2^2$

$x^2 = 16 + 4$

$x^2 = 20$

$x = \pm\sqrt{20}$

$x = \pm 2\sqrt{5}$

We disregard $-2\sqrt{5}$ because we can't have a negative measurement. The path is $2\sqrt{5}$ miles, or approximately 4.5 miles.

**143.** $90^2 + 90^2 = x^2$

$8100 + 8100 = x^2$

$16200 = x^2$

$x \approx \pm 127.28$

The distance is 127.28 feet.

**145.** Let $w$ = the width
Let $w + 3$ = the length

Area = $lw$

$$54 = (w+3)w$$

$$54 = w^2 + 3w$$

$$0 = w^2 + 3w - 54$$

$$0 = (w+9)(w-6)$$

Apply the zero product principle.

$$w+9 = 0 \qquad w-6 = 0$$

$$w = -9 \qquad w = 6$$

The solution set is $\{-9, 6\}$. Disregard –9

because we can't have a negative length measurement. The width is 6 feet and the length is $6+3 = 9$ feet.

**147.** Let $x$ = the length of the side of the original square
Let $x + 3$ = the length of the side of the new, larger square

$$(x+3)^2 = 64$$

$$x^2 + 6x + 9 = 64$$

$$x^2 + 6x - 55 = 0$$

$$(x+11)(x-5) = 0$$

Apply the zero product principle.

$$x+11 = 0 \qquad x-5 = 0$$

$$x = -11 \qquad x = 5$$

The solution set is $\{-11, 5\}$. Disregard –11

because we can't have a negative length measurement. This means that $x$, the length of the side of the original square, is 5 inches.

**149.** Let $x$ = the width of the path

$$(20+2x)(10+2x) = 600$$

$$200 + 40x + 20x + 4x^2 = 600$$

$$200 + 60x + 4x^2 = 600$$

$$4x^2 + 60x + 200 = 600$$

$$4x^2 + 60x - 400 = 0$$

$$4(x^2 + 15x - 100) = 0$$

$$4(x+20)(x-5) = 0$$

Apply the zero product principle.

$$4(x+20) = 0 \qquad x-5 = 0$$

$$x+20 = 0 \qquad x = 5$$

$$x = -20$$

The solution set is $\{-20, 5\}$. Disregard –20

because we can't have a negative width measurement. The width of the path is 5 meters.

**151.** $x(x)(2) = 200$

$$2x^2 = 200$$

$$x^2 = 100$$

$$x = \pm 10$$

The length and width are 10 inches.

**153.** $x(20-2x) = 13$

$$20x - 2x^2 = 13$$

$$0 = 2x^2 - 20x + 13$$

$$x = \frac{-(-20) \pm \sqrt{(-20)^2 - 4(2)(13)}}{2(2)}$$

$$x = \frac{20 \pm \sqrt{296}}{4}$$

$$x = \frac{10 \pm 17.2}{4}$$

$$x = 9.3, 0.7$$

9.3 in and 0.7 in

**155. – 165.** Answers will vary.

**167.** does not make sense; Explanations will vary. Sample explanation: Higher degree polynomial equations can have only one $x$-intercept.

**169.** makes sense

**171.** true

**173.** false; Changes to make the statement true will vary. A sample change is: The first step is to collect all the terms on one side and have 0 on the other.

**175.** $s = -16t^2 + v_0 t$

$$0 = -16t^2 + v_0 t - s$$

$$a = -16, \; b = v_0, \; c = -s$$

$$t = \frac{-v_0 \pm \sqrt{(v_0)^2 - 4(-16)(-s)}}{2(-16)}$$

$$t = \frac{-v_0 \pm \sqrt{(v_0)^2 - 64s}}{-32}$$

$$t = \frac{v_0 \pm \sqrt{v_0^2 - 64s}}{32}$$

**177.** $x^3 + x^2 - 4x - 4 = x^2(x+1) - 4(x+1)$
$$= (x+1)(x^2 - 4)$$
$$= (x+1)(x+2)(x-2)$$

**178.** $(\sqrt{x+4} + 1)^2 = \sqrt{x+4}^2 + 2(\sqrt{x+4})(1) + 1^2$
$$= x + 4 + 2\sqrt{x+4} + 1$$
$$= x + 5 + 2\sqrt{x+4}$$

**179.** $\qquad 5x^{2/3} + 11x^{1/3} + 2 = 0$
$$5(-8)^{2/3} + 11(-8)^{1/3} + 2 = 0$$
$$5(-2)^2 + 11(-2)^1 + 2 = 0$$
$$5(4) + 11(-2) + 2 = 0$$
$$20 - 22 + 2 = 0$$
$$0 = 0, \text{ true}$$

The statement is true.

## Mid-Chapter 1 Check Point

**1.** $\quad -5 + 3(x+5) = 2(3x-4)$
$$-5 + 3x + 15 = 6x - 8$$
$$3x + 10 = 6x - 8$$
$$-3x = -18$$
$$\frac{-3x}{-3} = \frac{-18}{-3}$$
$$x = 6$$

The solution set is $\{6\}$.

**2.** $\quad 5x^2 - 2x = 7$
$$5x^2 - 2x - 7 = 0$$
$$(5x-7)(x+1) = 0$$
$$5x - 7 = 0 \quad \text{or} \quad x + 1 = 0$$
$$5x = 7 \qquad\qquad x = -1$$
$$x = \frac{7}{5}$$

The solution set is $\left\{-1, \dfrac{7}{5}\right\}$.

**3.** $\qquad\qquad \dfrac{x-3}{5} - 1 = \dfrac{x-5}{4}$
$$20\left(\frac{x-3}{5} - 1\right) = 20\left(\frac{x-5}{4}\right)$$
$$\frac{20(x-3)}{5} - 20(1) = \frac{20(x-5)}{4}$$
$$4(x-3) - 20 = 5(x-5)$$
$$4x - 12 - 20 = 5x - 25$$
$$4x - 32 = 5x - 25$$
$$-x = 7$$
$$x = -7$$

The solution set is $\{-7\}$.

**4.** $\quad 3x^2 - 6x - 2 = 0$
$$x = \frac{-b \pm \sqrt{b^2 - 4ac}}{2a}$$
$$x = \frac{-(-6) \pm \sqrt{(-6)^2 - 4(3)(-2)}}{2(3)}$$
$$x = \frac{6 \pm \sqrt{60}}{6}$$
$$x = \frac{6 \pm 2\sqrt{15}}{6}$$
$$x = \frac{3 \pm \sqrt{15}}{3}$$

The solution set is $\left\{\dfrac{3+\sqrt{15}}{3}, \dfrac{3-\sqrt{15}}{3}\right\}$.

**5.** $\quad 4x - 2(1-x) = 3(2x+1) - 5$
$$4x - 2(1-x) = 3(2x+1) - 5$$
$$4x - 2 + 2x = 6x + 3 - 5$$
$$6x - 2 = 6x - 2$$
$$0 = 0$$

The equation is an identity.
The solution set is $\{x | x \text{ is a real number}\}$.

**6.** $5x^2 + 1 = 37$

$5x^2 = 36$

$\dfrac{5x^2}{5} = \dfrac{36}{5}$

$x^2 = \dfrac{36}{5}$

$x = \pm\sqrt{\dfrac{36}{5}}$

$x = \pm\dfrac{6}{\sqrt{5}}$

$x = \pm\dfrac{6}{\sqrt{5}} \cdot \dfrac{\sqrt{5}}{\sqrt{5}}$

$x = \pm\dfrac{6\sqrt{5}}{5}$

The solution set is $\left\{-\dfrac{6\sqrt{5}}{5}, \dfrac{6\sqrt{5}}{5}\right\}$.

**7.** $x(2x - 3) = -4$

$2x^2 - 3x = -4$

$2x^2 - 3x + 4 = 0$

$x = \dfrac{-b \pm \sqrt{b^2 - 4ac}}{2a}$

$x = \dfrac{-(-3) \pm \sqrt{(-3)^2 - 4(2)(4)}}{2(2)}$

$x = \dfrac{3 \pm \sqrt{-23}}{4}$

$x = \dfrac{3 \pm i\sqrt{23}}{4}$

The solution set is $\left\{\dfrac{3 + i\sqrt{23}}{4}, \dfrac{3 - i\sqrt{23}}{4}\right\}$.

**8.** $\dfrac{3x}{4} - \dfrac{x}{3} + 1 = \dfrac{4x}{5} - \dfrac{3}{20}$

$\dfrac{3x}{4} - \dfrac{x}{3} + 1 = \dfrac{4x}{5} - \dfrac{3}{20}$

$60\left(\dfrac{3x}{4} - \dfrac{x}{3} + 1\right) = 60\left(\dfrac{4x}{5} - \dfrac{3}{20}\right)$

$\dfrac{60(3x)}{4} - \dfrac{60x}{3} + 60(1) = \dfrac{60(4x)}{5} - \dfrac{60(3)}{20}$

$45x - 20x + 60 = 48x - 9$

$25x + 60 = 48x - 9$

$-23x = -69$

$\dfrac{-23x}{-23} = \dfrac{-69}{-23}$

$x = 3$

The solution set is $\{3\}$.

**9.** $(x + 3)^2 = 24$

$x + 3 = \pm\sqrt{24}$

$x = -3 \pm 2\sqrt{6}$

The solution set is $\left\{-3 + 2\sqrt{6}, -3 - 2\sqrt{6}\right\}$.

**10.** $\dfrac{1}{x^2} - \dfrac{4}{x} + 1 = 0$

$x^2\left(\dfrac{1}{x^2} - \dfrac{4}{x} + 1\right) = x^2(0)$

$\dfrac{x^2}{x^2} - \dfrac{4x^2}{x} + x^2 = 0$

$1 - 4x + x^2 = 0$

$x^2 - 4x + 1 = 0$

$x = \dfrac{-b \pm \sqrt{b^2 - 4ac}}{2a}$

$x = \dfrac{-(-4) \pm \sqrt{(-4)^2 - 4(1)(1)}}{2(1)}$

$x = \dfrac{4 \pm \sqrt{12}}{2}$

$x = \dfrac{4 \pm 2\sqrt{3}}{2}$

$x = 2 \pm \sqrt{3}$

The solution set is $\left\{2 + \sqrt{3}, 2 - \sqrt{3}\right\}$.

**11.**

$$3x+1-(x-5)=2x-4$$
$$2x+6=2x-4$$
$$6=-4$$

The solution set is $\varnothing$ .

**12.**

$$\frac{2x}{x^2+6x+8}=\frac{x}{x+4}-\frac{2}{x+2}, \quad x\neq-2, x\neq-4$$

$$\frac{2x}{(x+4)(x+2)}=\frac{x}{x+4}-\frac{2}{x+2}$$

$$\frac{2x(x+4)(x+2)}{(x+4)(x+2)}=(x+4)(x+2)\left(\frac{x}{x+4}-\frac{2}{x+2}\right)$$

$$2x=\frac{x(x+4)(x+2)}{x+4}-\frac{2(x+4)(x+2)}{x+2}$$

$$2x=x(x+2)-2(x+4)$$

$$2x=x^2+2x-2x-8$$

$$0=x^2-2x-8$$

$$0=(x+2)(x-4)$$

$$x+2=0 \quad \text{or} \quad x-4=0$$
$$x=-2 \qquad\qquad x=4$$

$-2$ must be rejected.

The solution set is $\{4\}$.

**13.** Let $y=0$.

$$0=x^2+6x+2$$

$$x=\frac{-b\pm\sqrt{b^2-4ac}}{2a}$$

$$x=\frac{-(6)\pm\sqrt{(6)^2-4(1)(2)}}{2(1)}$$

$$x=\frac{-6\pm\sqrt{28}}{2}$$

$$x=\frac{-6\pm2\sqrt{7}}{2}$$

$$x=-3\pm\sqrt{7}$$

$x$-intercepts: $-3+\sqrt{7}$ and $-3-\sqrt{7}$.

**14.** Let $y=0$.

$$0=4(x+1)-3x-(6-x)$$

$$0=4x+4-3x-6+x$$

$$0=2x-2$$

$$-2x=-2$$

$$x=1$$

$x$-intercept: 1.

**15.** Let $y=0$.

$$0=2x^2+26$$

$$-2x^2=26$$

$$x^2=-13$$

$$x=\pm\sqrt{-13}$$

$$x=\pm i\sqrt{13}$$

There are no $x$-intercepts.

**16.** Let $y=0$.

$$0=\frac{x^2}{3}+\frac{x}{2}-\frac{2}{3}$$

$$6(0)=6\left(\frac{x^2}{3}+\frac{x}{2}-\frac{2}{3}\right)$$

$$0=\frac{6\cdot x^2}{3}+\frac{6\cdot x}{2}-\frac{6\cdot 2}{3}$$

$$0=2x^2+3x-4$$

$$x=\frac{-b\pm\sqrt{b^2-4ac}}{2a}$$

$$x=\frac{-(3)\pm\sqrt{(3)^2-4(2)(-4)}}{2(2)}$$

$$x=\frac{-3\pm\sqrt{41}}{4}$$

$x$-intercepts: $\dfrac{-3+\sqrt{41}}{4}$ and $\dfrac{-3-\sqrt{41}}{4}$.

**17.** Let $y=0$.

$$0=x^2-5x+8$$

$$x=\frac{-b\pm\sqrt{b^2-4ac}}{2a}$$

$$x=\frac{-(-5)\pm\sqrt{(-5)^2-4(1)(8)}}{2(1)}$$

$$x=\frac{5\pm\sqrt{-7}}{2}$$

$$x=\frac{5\pm i\sqrt{7}}{2}$$

There are no $x$-intercepts.

**18.**

$$y_1=y_2$$

$$3(2x-5)-2(4x+1)=-5(x+3)-2$$

$$6x-15-8x-2=-5x-15-2$$

$$-2x-17=-5x-17$$

$$3x=0$$

$$x=0$$

The solution set is $\{0\}$.

**19.**
$$y_1y_2 = 10$$
$$(2x+3)(x+2) = 10$$
$$2x^2 + 7x + 6 = 10$$
$$2x^2 + 7x - 4 = 0$$
$$(2x-1)(x+4) = 0$$
$$2x - 1 = 0 \quad \text{or} \quad x + 4 = 0$$
$$x = \frac{1}{2} \qquad\qquad x = -4$$

The solution set is $\left\{-4, \frac{1}{2}\right\}$.

**20.** $x^2 + 10x - 3 = 0$
$$x^2 + 10x \quad = 3$$

Since $b = 10$, we add $\left(\dfrac{10}{2}\right)^2 = 5^2 = 25$.

$$x^2 + 10x + 25 = 3 + 25$$
$$(x+5)^2 = 28$$

Apply the square root principle:
$$x + 5 = \pm\sqrt{28}$$
$$x + 5 = \pm\sqrt{4 \cdot 7} = \pm 2\sqrt{7}$$
$$x = -5 \pm 2\sqrt{7}$$

The solutions are $-5 \pm 2\sqrt{7}$, and the solution set is $\left\{-5 \pm 2\sqrt{7}\right\}$.

**21.** $2x^2 + 5x + 4 = 0$
$$a = 2 \quad b = 5 \quad c = 4$$
$$b^2 - 4ac = 5^2 - 4(2)(4)$$
$$= 25 - 32 = -7$$

Since the discriminant is negative, there are no real solutions. There are two imaginary solutions that are complex conjugates.

**22.**
$$10x(x+4) = 15x - 15$$
$$10x^2 + 40x = 15x - 15$$
$$10x^2 - 25x + 15 = 0$$
$$a = 10 \quad b = -25 \quad c = 15$$
$$b^2 - 4ac = (-25)^2 - 4(10)(15)$$
$$= 625 - 600 = 25$$

Since the discriminant is positive and a perfect square, there are two rational solutions.

**23.**

| $x$ | $(x, y)$ |
|----|----|
| $-2$ | $-5$ |
| $-1$ | $-3$ |
| $0$ | $-1$ |
| $1$ | $1$ |
| $2$ | $3$ |

$y = 2x - 1$

**24.**

| $x$ | $(x, y)$ |
|----|----|
| $-3$ | $-2$ |
| $-2$ | $-1$ |
| $-1$ | $0$ |
| $0$ | $1$ |
| $1$ | $0$ |
| $2$ | $-1$ |
| $3$ | $-2$ |

$y = 1 - |x|$

**25.**

| $x$ | $(x, y)$ |
|----|----|
| $-2$ | $6$ |
| $-1$ | $3$ |
| $0$ | $2$ |
| $1$ | $3$ |
| $2$ | $6$ |

$y = x^2 + 2$

**26.**
$$L = a + (n-1)d$$
$$L = a + dn - d$$
$$-dn = a - d - L$$
$$\frac{-dn}{-d} = \frac{a}{-d} - \frac{d}{-d} - \frac{L}{-d}$$
$$n = -\frac{a}{d} + 1 + \frac{L}{d}$$
$$n = \frac{L}{d} - \frac{a}{d} + 1$$
$$n = \frac{L-a}{d} + 1$$

**27.**
$$A = 2lw + 2lh + 2wh$$
$$-2lw - 2lh = 2wh - A$$
$$l(-2w - 2h) = 2wh - A$$
$$l = \frac{2wh - A}{-2w - 2h}$$
$$l = \frac{A - 2wh}{2w + 2h}$$

**28.**
$$f = \frac{f_1 f_2}{f_1 + f_2}$$
$$(f_1 + f_2)(f) = (f_1 + f_2)\frac{f_1 f_2}{f_1 + f_2}$$
$$f_1 f + f_2 f = f_1 f_2$$
$$f_1 f - f_1 f_2 = -f_2 f$$
$$f_1(f - f_2) = -f_2 f$$
$$f_1 = \frac{-f_2 f}{f - f_2}$$
$$f_1 = -\frac{ff_2}{f - f_2} \text{ or } f_1 = \frac{ff_2}{f_2 - f}$$

**29.** Let $x =$ the number of times "sorry" was used.
Let $x + 419 =$ the number of times "love" was used.
Let $x + 32 =$ the number of times "thanks" was used.
$$x + (x + 419) + (x + 32) = 1084$$
$$x + x + 419 + x + 32 = 1084$$
$$3x + 451 = 1084$$
$$3x = 633$$
$$x = 211$$
$$x + 419 = 630$$
$$x + 32 = 243$$
The word "sorry" was used 211 times, the word "love" was used 630 times, and the word "thanks" was used 243 times.

**30.** Let $x =$ the number of years since 1960.
$$23 - 0.28x = 0$$
$$-0.28x = -23$$
$$\frac{-0.28x}{-0.28} = \frac{-23}{-0.28}$$
$$x \approx 82$$
If this trend continues, corporations will pay zero taxes 82 years after 1960, or 2042.

**31.** Let $x =$ the amount invested at 8%.
Let $25,000 - x =$ the amount invested at 9%.
$$0.08x + 0.09(25,000 - x) = 2135$$
$$0.08x + 2250 - 0.09x = 2135$$
$$-0.01x + 2250 = 2135$$
$$-0.01x = -115$$
$$x = \frac{-115}{-0.01}$$
$$x = 11,500$$
$$25,000 - x = 13,500$$
$11,500 was invested at 8% and $13,500 was invested at 9%.

**32.** Let $x =$ the number of text messages.
Plan A:  $C = 15 + 0.05x$
Plan B:  $C = 10 + 0.075x$
Set the costs equal to each other.
$$15 + 0.05x = 10 + 0.075x$$
$$15 = 10 + 0.025x$$
$$5 = 0.025x$$
$$200 = x$$
The cost will be the same for 200 text messages.

$$C = 15 + 0.05x$$
$$C = 15 + 0.05(200)$$
$$= 25$$
The cost for 200 text messages will be $25.

**33.** Let $x =$ the price before the reduction.
$$x - 0.40x = 468$$
$$0.60x = 468$$
$$\frac{0.60x}{0.60} = \frac{468}{0.60}$$
$$x = 780$$
The price before the reduction was $780.

**34.** Let $x$ = the amount invested at 4%.
Let $4000 - x$ = the amount invested that lost 3%.
$$0.04x - 0.03(4000 - x) = 55$$
$$0.04x - 120 + 0.03x = 55$$
$$0.07x - 120 = 55$$
$$0.07x = 175$$
$$x = \frac{175}{0.07}$$
$$x = 2500$$
$$4000 - x = 1500$$
$2500 was invested at 4% and $1500 lost 3%.

**35.** Let $x$ = the width of the rectangle
Let $2x + 5$ = the length of the rectangle
$$2l + 2w = P$$
$$2(2x + 5) + 2x = 46$$
$$4x + 10 + 2x = 46$$
$$6x + 10 = 46$$
$$6x = 36$$
$$\frac{6x}{6} = \frac{36}{6}$$
$$x = 6$$
$$2x + 5 = 17$$
The dimensions of the rectangle are 6 ft by 17 ft.

**36.** Let $x$ = the width of the rectangle
Let $2x - 1$ = the length of the rectangle
$$lw = A$$
$$(2x - 1)x = 28$$
$$2x^2 - x = 28$$
$$2x^2 - x - 28 = 0$$
$$(2x + 7)(x - 4) = 0$$
$$2x + 7 = 0 \quad \text{or} \quad x - 4 = 0$$
$$2x = -7 \qquad\qquad x = 4$$
$$x = -\frac{7}{2}$$
$-\frac{7}{2}$ must be rejected.
If $x = 4$, then $2x - 1 = 7$
The dimensions of the rectangle are 4 ft by 7 ft.

**37.** Let $x$ = the height up the pole at which the wires are attached.
$$x^2 + 5^2 = 13^2$$
$$x^2 + 25 = 169$$
$$x^2 = 144$$
$$x = \pm 12$$
$-12$ must be rejected.
The wires are attached 12 yards up the pole.

**38. a.** 
$$P = -10x^2 + 475x + 3500$$
$$5990 = -10x^2 + 475x + 3500$$
$$0 = -10x^2 + 475x - 2490$$
$$0 = 2x^2 - 95x + 498$$
$$0 = (x - 6)(2x - 83)$$
$$x - 6 = 0 \quad \text{or} \quad 2x - 83 = 0$$
$$x = 6 \qquad\qquad 2x = 83$$
$$x = 41.5$$
The population reached 5990 after 6 years.

**b.** This is represented by the point (6, 5990).

**39.** 
$$P = 0.004x^2 - 0.37x + 14.1$$
$$25 = 0.004x^2 - 0.37x + 14.1$$
$$0 = 0.004x^2 - 0.37x - 10.9$$
$$x = \frac{-b \pm \sqrt{b^2 - 4ac}}{2a}$$
$$x = \frac{-(-0.37) \pm \sqrt{(-0.37)^2 - 4(0.004)(-10.9)}}{2(0.004)}$$
$$x = \frac{0.37 \pm \sqrt{0.1369 + 0.1744}}{0.008}$$
$$x \approx 116, \quad x \approx -23 \text{ (rejected)}$$
The percentage of foreign born Americans will be 25% about 116 years after 1920, or 2036.

**40.** $(6 - 2i) - (7 - i) = 6 - 2i - 7 + i = -1 - i$

**41.** $3i(2 + i) = 6i + 3i^2 = -3 + 6i$

**42.** $(1 + i)(4 - 3i) = 4 - 3i + 4i - 3i^2$
$$= 4 + i + 3 = 7 + i$$

**43.** $\dfrac{1+i}{1-i} = \dfrac{1+i}{1-i} \cdot \dfrac{1+i}{1+i} = \dfrac{1+i+i+i^2}{1-i^2}$
$$= \frac{1 + 2i - 1}{1 + 1}$$
$$= \frac{2i}{2}$$
$$= i$$

**44.** $\sqrt{-75} - \sqrt{-12} = 5i\sqrt{3} - 2i\sqrt{3} = 3i\sqrt{3}$

**45.** $\left(2 - \sqrt{-3}\right)^2 = \left(2 - i\sqrt{3}\right)^2$
$$= 4 - 4i\sqrt{3} + 3i^2$$
$$= 4 - 4i\sqrt{3} - 3$$
$$= 1 - 4i\sqrt{3}$$

**Section 1.6**

**Check Point Exercises**

1.
$$4x^4 = 12x^2$$
$$4x^4 - 12x^2 = 0$$
$$4x^2(x^2 - 3) = 0$$
$$4x^2 = 0 \quad \text{or} \quad x^2 - 3 = 0$$
$$x^2 = 0 \qquad\qquad x^2 = 3$$
$$x = \pm\sqrt{0} \qquad\quad x = \pm\sqrt{3}$$
$$x = 0 \qquad\qquad x = \pm\sqrt{3}$$
The solution set is $\left\{-\sqrt{3}, 0, \sqrt{3}\right\}$.

2.
$$2x^3 + 3x^2 = 8x + 12$$
$$x^2(2x+3) - 4(2x+3) = 10$$
$$(2x+3)(x^2 - 4) = 0$$
$$2x + 3 = 0 \quad \text{or} \quad x^2 - 4 = 0$$
$$2x = -3 \qquad\qquad x^2 = 4$$
$$x = -\frac{3}{2} \qquad\qquad x = \pm 2$$
The solution set is $\left\{-2, -\dfrac{3}{2}, 2\right\}$.

3.
$$\sqrt{x+3} + 3 = x$$
$$\sqrt{x+3} = x - 3$$
$$\left(\sqrt{x+3}\right)^2 = (x-3)^2$$
$$x + 3 = x^2 - 6x + 9$$
$$0 = x^2 - 7x + 6$$
$$0 = (x-6)(x-1)$$
$$x - 6 = 0 \quad \text{or} \quad x - 1 = 0$$
$$x = 6 \qquad\qquad x = 1$$
1 does not check and must be rejected.
The solution set is $\{6\}$.

4.
$$\sqrt{x+5} - \sqrt{x-3} = 2$$
$$\sqrt{x+5} = 2 + \sqrt{x-3}$$
$$\left(\sqrt{x+5}\right)^2 = \left(2 + \sqrt{x-3}\right)^2$$
$$x + 5 = (2)^2 + 2(2)\left(\sqrt{x-3}\right) + \left(\sqrt{x-3}\right)^2$$
$$x + 5 = 4 + 4\sqrt{x-3} + x - 3$$
$$4 = 4\sqrt{x-3}$$
$$\frac{4}{4} = \frac{4\sqrt{x-3}}{4}$$
$$1 = \sqrt{x-3}$$
$$(1)^2 = \left(\sqrt{x-3}\right)^2$$
$$1 = x - 3$$
$$4 = x$$
The check indicates that 4 is a solution.
The solution set is $\{4\}$.

5.  **a.**
$$5x^{3/2} - 25 = 0$$
$$5x^{3/2} = 25$$
$$x^{3/2} = 5$$
$$\left(x^{3/2}\right)^{2/3} = (5)^{2/3}$$
$$x = 5^{2/3} \quad \text{or} \quad \sqrt[3]{25}$$

Check:
$$5\left(5^{2/3}\right)^{3/2} - 25 = 0$$
$$5(5) - 25 = 0$$
$$25 - 25 = 0$$
$$0 = 0$$
The solution set is $\left\{5^{2/3}\right\}$ or $\left\{\sqrt[3]{25}\right\}$.

   **b.**
$$x^{\frac{2}{3}} - 8 = -4$$
$$x^{2/3} = 4$$
$$\left(x^{2/3}\right)^{3/2} = 4^{3/2} \qquad \text{or}$$
$$x = \left(2^2\right)^{3/2}$$
$$x = 2^3 \qquad\qquad x = (-2)^3$$
$$x = 8 \qquad\qquad x = -8$$
The solution set is $\{-8, 8\}$.

**6.**
$$x^4 - 5x^2 + 6 = 0$$
$$\left(x^2\right)^2 - 5x^2 + 6 = 0$$
Let $t = x^2$.
$$t^2 - 5t + 6 = 0$$
$$(t-3)(t-2) = 0$$

| $t - 3 = 0$ | or | $t - 2 = 0$ |
|---|---|---|
| $t = 3$ | or | $t = 2$ |
| $x^2 = 3$ | or | $x^2 = 2$ |
| $x = \pm\sqrt{3}$ | or | $x = \pm\sqrt{2}$ |

The solution set is $\left\{-\sqrt{3}, -\sqrt{2}, \sqrt{2}, \sqrt{3}\right\}$.

**7.**
$$3x^{2/3} - 11x^{1/3} - 4 = 0$$
Let $t = x^{1/3}$.
$$3t^2 - 11t - 4 = 0$$
$$(3t+1)(t-4) = 0$$
$$3t + 1 = 0 \quad \text{or} \quad t - 4 = 0$$
$$3t = -1$$

$$t = -\frac{1}{3} \qquad t = 4$$

$$x^{1/3} = -\frac{1}{3} \qquad x^{1/3} = 4$$

$$x = \left(-\frac{1}{3}\right)^3 \qquad x = 4^3$$

$$x = -\frac{1}{27} \qquad x = 64$$

The solution set is $\left\{-\frac{1}{27}, 64\right\}$.

**8.**
$$|2x - 1| = 5$$

| $2x - 1 = 5$ | or | $2x - 1 = -5$ |
|---|---|---|
| $2x = 6$ | | $2x = -4$ |
| $x = 3$ | | $x = -2$ |

The solution set is $\{-2, 3\}$.

**9.**
$$4|1 - 2x| - 20 = 0$$
$$4|1 - 2x| = 20$$
$$|1 - 2x| = 5$$

| $1 - 2x = 5$ | or | $1 - 2x = -5$ |
|---|---|---|
| $-2x = 4$ | | $-2x = -6$ |
| $x = -2$ | | $x = 3$ |

The solution set is $\{-2, 3\}$.

**10.**
$$H = -2.3\sqrt{I} + 67.6$$
$$33.1 = -2.3\sqrt{I} + 67.6$$
$$-34.5 = -2.3\sqrt{I}$$
$$\frac{-34.5}{-2.3} = \frac{-2.3\sqrt{I}}{-2.3}$$
$$15 = \sqrt{I}$$
$$15^2 = \left(\sqrt{I}\right)^2$$
$$225 = I$$

The model indicates that an annual income of 225 thousand dollars, or $225,000, corresponds to 33.1 hours per week watching TV.

**Concept and Vocabulary Check 1.6**

**1.** subtract $8x$ and subtract 12 from both sides

**2.** radical

**3.** extraneous

**4.** $2x + 1$; $x^2 + 14x + 49$

**5.** $x + 2$; $x + 8 - 6\sqrt{x-1}$

**6.** $5^{\frac{4}{3}}$

**7.** $\pm 5^{\frac{3}{2}}$

**8.** $x^2$; $u^2 - 13u + 36 = 0$

**9.** $x^{\frac{1}{3}}$; $u^2 + 2u - 3 = 0$

**10.** $c$; $-c$

**11.** $3x - 1 = 7$; $3x - 1 = -7$

**Exercise Set 1.6**

**1.**
$$3x^4 - 48x^2 = 0$$
$$3x^2(x^2 - 16) = 0$$
$$3x^2(x+4)(x-4) = 0$$
$$3x^2 = 0 \quad x+4=0 \quad x-4=0$$
$$x^2 = 0 \qquad x=-4 \qquad x=4$$
$$x = 0$$
The solution set is {–4, 0, 4}.

**3.**
$$3x^3 + 2x^2 = 12x + 8$$
$$3x^3 + 2x^2 - 12x - 8 = 0$$
$$x^2(3x+2) - 4(3x+2) = 0$$
$$(3x+2)(x^2-4) = 0$$
$$3x+2=0 \qquad x^2-4=0$$
$$3x=-2 \qquad x^2=4$$
$$x=-\frac{2}{3} \qquad x=\pm 2$$
The solution set is $\left\{-2, -\frac{2}{3}, 2\right\}$.

**5.**
$$2x-3 = 8x^3 - 12x^2$$
$$8x^3 - 12x^2 - 2x + 3 = 0$$
$$4x^2(2x-3) - (2x-3) = 0$$
$$(2x-3)(4x^2-1) = 0$$
$$2x-3=0 \quad 4x^2-1=0$$
$$2x=3 \qquad 4x^2=1$$
$$x^2 = \frac{1}{4}$$
$$x=\frac{3}{2} \qquad x=\pm\frac{1}{2}$$
The solution set is $\left\{\frac{3}{2}, \frac{1}{2}, -\frac{1}{2}\right\}$.

**7.**
$$4y^3 - 2 = y - 8y^2$$
$$4y^3 + 8y^2 - y - 2 = 0$$
$$4y^2(y+2) - (y+2) = 0$$
$$(y+2)(4y^2-1) = 0$$
$$y+2=0 \qquad 4y^2-1=0$$
$$4y^2 = 1$$
$$y^2 = \frac{1}{4}$$
$$y=-2 \qquad y=\pm\frac{1}{2}$$
The solution set is $\left\{-2, \frac{1}{2}, -\frac{1}{2}\right\}$.

**9.**
$$2x^4 = 16x$$
$$2x^4 - 16x = 0$$
$$2x(x^3 - 8) = 0$$
$$2x=0 \qquad\qquad x^3-8=0$$
$$x=0 \qquad (x-2)(x^2+2x+2)=0$$
$$x-2=0 \quad x^2+2x+4=0$$
$$x=2 \quad x=\frac{-2\pm\sqrt{2^2-4(1)(4)}}{2(1)}$$
$$x=\frac{-2\pm\sqrt{-12}}{2}$$
$$x=\frac{-2\pm 2i\sqrt{3}}{2}$$
$$x=-1\pm i\sqrt{3}$$
The solution set is $\left\{0, 2, -1\pm i\sqrt{3}\right\}$.

**11.**
$$\sqrt{3x+18} = x$$
$$3x+18 = x^2$$
$$x^2 - 3x - 18 = 0$$
$$(x+3)(x-6) = 0$$
$$x+3=0 \quad x-6=0$$
$$x=-3 \qquad x=6$$
$$\sqrt{3(-3)+18} = -3 \quad \sqrt{3(6)+18} = 6$$
$$\sqrt{-9+18} = -3 \qquad \sqrt{18+18} = 6$$
$$\sqrt{9} = -3 \text{ False} \quad \sqrt{36} = 6$$
The solution set is {6}.

**13.**
$$\sqrt{x+3} = x-3$$
$$x+3 = x^2 - 6x + 9$$
$$x^2 - 7x + 6 = 0$$
$$(x-1)(x-6) = 0$$
$$x - 1 = 0 \quad x - 6 = 0$$
$$x = 1 \qquad x = 6$$
$$\sqrt{1+3} = 1-3 \qquad \sqrt{6+3} = 6-3$$
$$\sqrt{4} = -2 \quad \text{False} \ \sqrt{9} = 3$$
The solution set is $\{6\}$.

**15.**
$$\sqrt{2x+13} = x+7$$
$$2x+13 = (x+7)^2$$
$$2x+13 = x^2 + 14x + 49$$
$$x^2 + 12x + 36 = 0$$
$$(x+6)^2 = 0$$
$$x+6 = 0$$
$$x = -6$$
$$\sqrt{2(-6)+13} = -6+7$$
$$\sqrt{-12+13} = 1$$
$$\sqrt{1} = 1$$
The solution set is $\{-6\}$.

**17.**
$$x - \sqrt{2x+5} = 5$$
$$x - 5 = \sqrt{2x+5}$$
$$(x-5)^2 = 2x+5$$
$$x^2 - 10x + 25 = 2x + 5$$
$$x^2 - 12x + 20 = 0$$
$$(x-2)(x-10) = 0$$
$$x - 2 = 0 \quad x - 10 = 0$$
$$x = 2 \qquad x = 10$$
$$2 - \sqrt{2(2)+5} = 5 \quad 10 - \sqrt{2(10)+5} = 5$$
$$2 - \sqrt{9} = 5 \qquad 10 - \sqrt{25} = 5$$
$$2 - 3 = 5 \quad \text{False} \quad 10 - 5 = 5$$
The solution set is $\{10\}$.

**19.**
$$\sqrt{2x+19} - 8 = x$$
$$\sqrt{2x+19} = x+8$$
$$\left(\sqrt{2x+19}\right)^2 = (x+8)^2$$
$$2x+19 = x^2 + 16x + 64$$
$$0 = x^2 + 14x + 45$$
$$0 = (x+9)(x+5)$$
$$x+9 = 0 \quad \text{or} \quad x+5 = 0$$
$$x = -9 \qquad x = -5$$
$-9$ does not check and must be rejected.
The solution set is $\{-5\}$.

**21.**
$$\sqrt{3x} + 10 = x+4$$
$$\sqrt{3x} = x-6$$
$$3x = (x-6)^2$$
$$3x = x^2 - 12x + 36$$
$$x^2 - 15x + 36 = 0$$
$$(x-12)(x-3) = 0$$
$$x - 12 = 0 \quad x - 3 = 0$$
$$x = 12 \qquad x = 3$$
$$\sqrt{3(12)} + 10 = 12 + 4 \quad \sqrt{3(3)} + 10 = 3+4$$
$$\sqrt{36} + 10 = 16 \qquad \sqrt{9} + 10 = 7$$
$$6 + 10 = 16 \qquad 3 + 10 = 7 \ \text{False}$$
The solution set is $\{12\}$.

**23.**
$$\sqrt{x+8} - \sqrt{x-4} = 2$$
$$\sqrt{x+8} = \sqrt{x-4} + 2$$
$$x+8 = (\sqrt{x-4} + 2)^2$$
$$x+8 = x-4 + 4\sqrt{x-4} + 4$$
$$x+8 = x + 4\sqrt{x-4}$$
$$8 = 4\sqrt{x-4}$$
$$2 = \sqrt{x-4}$$
$$4 = x-4$$
$$x = 8$$
$$\sqrt{8+8} - \sqrt{8-4} = 2$$
$$\sqrt{16} - \sqrt{4} = 2$$
$$4 - 2 = 2$$
The solution set is $\{8\}$.

**25.** $\sqrt{x-5} - \sqrt{x-8} = 3$

$\qquad \sqrt{x-5} = \sqrt{x-8} + 3$

$\qquad x-5 = (\sqrt{x-8} + 3)^2$

$\qquad x-5 = x-8 + 6\sqrt{x-8} + 9$

$\qquad x-5 = x+1 + 6\sqrt{x-8}$

$\qquad -6 = 6\sqrt{x-8}$

$\qquad -1 = \sqrt{x-8}$

$\qquad 1 = x-8$

$\qquad x = 9$

$\sqrt{9-5} - \sqrt{9-8} = 3$

$\qquad \sqrt{4} - \sqrt{1} = 3$

$\qquad 2-1 = 3$ False

The solution set is the empty set, $\varnothing$.

**27.** $\sqrt{2x+3} + \sqrt{x-2} = 2$

$\qquad \sqrt{2x+3} = 2 - \sqrt{x-2}$

$\qquad 2x+3 = (2 - \sqrt{x-2})^2$

$\qquad 2x+3 = 4 - 4\sqrt{x-2} + x-2$

$\qquad x+1 = -4\sqrt{x-2}$

$\qquad (x+1)^2 = 16(x-2)$

$\qquad x^2 + 2x+1 = 16x - 32$

$\qquad x^2 - 14x + 33 = 0$

$\qquad (x-11)(x-3) = 0$

$\qquad x-11 = 0 \quad x-3 = 0$

$\qquad x = 11 \qquad x = 3$

$\sqrt{2(11)+3} + \sqrt{11-2} = 2$

$\qquad \sqrt{22+3} + \sqrt{9} = 2$

$\qquad\qquad\qquad 5+3 = 2$ False

$\sqrt{2(3)+3} + \sqrt{3-2} = 2$

$\qquad \sqrt{6+3} + \sqrt{1} = 2$

$\qquad\qquad 3+1 = 2$ False

The solution set is the empty set, $\varnothing$.

**29.** $\sqrt{3\sqrt{x+1}} = \sqrt{3x-5}$

$\qquad 3\sqrt{x+1} = 3x-5$

$\qquad 9(x+1) = 9x^2 - 30x + 25$

$9x^2 - 39x + 16 = 0$

$\qquad x = \dfrac{39 \pm \sqrt{945}}{18} = \dfrac{13 \pm \sqrt{105}}{6}$

Check proposed solutions.

The solution set is $\left\{ \dfrac{13+\sqrt{105}}{6} \right\}$.

**31.** $\qquad x^{3/2} = 8$

$\qquad (x^{3/2})^{2/3} = 8^{2/3}$

$\qquad\qquad x = \sqrt[3]{8}^{\,2}$

$\qquad\qquad x = 2^2$

$\qquad\qquad x = 4$

$\qquad 4^{3/2} = 8$

$\qquad \sqrt{4}^{\,3} = 8$

$\qquad 2^3 = 8$

The solution set is $\{4\}$.

**33.** $\qquad (x-4)^{3/2} = 27$

$\qquad ((x-4)^{3/2})^{2/3} = 27^{2/3}$

$\qquad\qquad x-4 = \sqrt[3]{27}^{\,2}$

$\qquad\qquad x-4 = 3^2$

$\qquad\qquad x-4 = 9$

$\qquad\qquad\quad x = 13$

$\qquad (13-4)^{3/2} = 27$

$\qquad\qquad 9^{3/2} = 27$

$\qquad\qquad \sqrt{9}^{\,3} = 27$

$\qquad\qquad 3^3 = 27$

The solution set is $\{13\}$.

**35.** $6x^{5/2} - 12 = 0$

$\qquad 6x^{5/2} = 12$

$\qquad x^{5/2} = 2$

$\qquad (x^{5/2})^{2/5} = 2^{2/5}$

$\qquad\qquad x = \sqrt[5]{2^2}$

$\qquad\qquad x = \sqrt[5]{4}$

$6(\sqrt[5]{4})^{5/2} - 12 = 0$

$6(4^{1/5})^{5/2} - 12 = 0$

$\quad 6(4^{1/2}) - 12 = 0$

$\qquad 6(2) - 12 = 0$

The solution set is $\left\{\sqrt[5]{4}\right\}$.

**37.**

$\qquad (x-4)^{2/3} = 16$

$\left[(x-4)^{2/3}\right]^{3/2} = (16)^{3/2}$

$\qquad x - 4 = \left(2^4\right)^{3/2}$

$\quad x - 4 = 4^3 \qquad x - 4 = (-4)^3$

$\quad x - 4 = 64 \qquad x - 4 = -64$

$\qquad x = 68 \qquad\qquad x = -60$

The solution set is $\{-60, 68\}$.

**39.** $(x^2 - x - 4)^{3/4} - 2 = 6$

$\qquad (x^2 - x - 4)^{3/4} = 8$

$\quad ((x^2 - x - 4)^{3/4})^{4/3} = 8^{4/3}$

$\qquad x^2 - x - 4 = \sqrt[3]{8}^4$

$\qquad x^2 - x - 4 = 2^4$

$\qquad x^2 - x - 4 = 16$

$\qquad x^2 - x - 20 = 0$

$\qquad (x-5)(x+4) = 0$

$\quad x - 5 = 0 \quad x + 4 = 0$

$\qquad x = 5 \qquad x = -4$

$(5^2 - 5 - 4)^{3/4} - 2 = 6$

$\qquad (25 - 9)^{3/4} - 2 = 6$

$\qquad\qquad 16^{3/4} - 2 = 6$

$\qquad\qquad \sqrt[4]{16}^3 - 2 = 6$

$\qquad\qquad\qquad 2^3 - 2 = 6$

$\qquad\qquad\qquad 8 - 2 = 6$

$((-4)^2 - (-4) - 4)^{3/4} - 2 = 6$

$\qquad (16 + 4 - 4)^{3/4} - 2 = 6$

$\qquad\qquad 16^{3/4} - 2 = 6$

$\qquad\qquad \sqrt[4]{16}^3 - 2 = 6$

$\qquad\qquad\qquad 2^3 - 2 = 6$

$\qquad\qquad\qquad 8 - 2 = 6$

The solution set is $\{5, -4\}$.

**41.** $x^4 - 5x^2 + 4 = 0$ let $t = x^2$

$\qquad t^2 - 5t + 4 = 0$

$\qquad (t-1)(t-4) = 0$

$\quad t - 1 = 0 \quad t - 4 = 0$

$\qquad t = 1 \qquad t = 4$

$\qquad x^2 = 1 \qquad x^2 = 4$

$\qquad x = \pm 1 \qquad x = \pm 2$

The solution set is $\{1, -1, 2, -2\}$.

**43.** $9x^4 = 25x^2 - 16$

$\quad 9x^4 - 25x^2 + 16 = 0$ let $t = x^2$

$\qquad 9t^2 - 25t + 16 = 0$

$\qquad (9t - 16)(t - 1) = 0$

$\quad 9t - 16 = 0 \qquad t - 1 = 0$

$\qquad 9t = 16 \qquad\qquad t = 1$

$\qquad t = \dfrac{16}{9} \qquad\qquad x^2 = 1$

$\qquad\qquad\qquad\qquad x = \pm 1$

$\qquad x^2 = \dfrac{16}{9}$

$\qquad x = \pm \dfrac{4}{3}$

The solution set is $\left\{1, -1, \dfrac{4}{3}, -\dfrac{4}{3}\right\}$.

**45.** $x - 13\sqrt{x} + 40 = 0$   Let $t = \sqrt{x}$.

$t^2 - 13t + 40 = 0$

$(t-8)(t-5) = 0$

$t - 8 = 0 \quad t - 5 = 0$

$\quad t = 8 \qquad t = 5$

$\quad \sqrt{x} = 8 \qquad \sqrt{x} = 5$

$\quad\quad x = 64 \qquad x = 25$

The solution set is $\{25, 64\}$.

**47.** $x^{-2} - x^{-1} - 20 = 0$   Let $t = x^{-1}$

$\quad t^2 - t - 20 = 0$

$(t-5)(t+4) = 0$

$t - 5 = 0 \quad t + 4 = 0$

$t = 5 \qquad t = -4$

$x^{-1} = 5 \quad x^{-1} = -4$

$\dfrac{1}{x} = 5 \qquad \dfrac{1}{x} = -4$

$1 = 5x \qquad 1 = -4x$

$\dfrac{1}{5} = x \qquad -\dfrac{1}{4} = x$

The solution set is $\left\{ -\dfrac{1}{4}, \dfrac{1}{5} \right\}$.

**49.** $x^{2/3} - x^{1/3} - 6 = 0$ let $t = x^{1/3}$

$\quad t^2 - t - 6 = 0$

$(t-3)(t+2) = 0$

$t - 3 = 0 \quad t + 2 = 0$

$\quad t = 3 \qquad t = -2$

$\quad x^{1/3} = 3 \qquad x^{1/3} = -2$

$\quad\quad x = 3^3 \qquad x = (-2)^3$

$\quad\quad x = 27 \qquad x = -8$

The solution set is $\{27, -8\}$.

**51.** $x^{3/2} - 2x^{3/4} + 1 = 0$ let $t = x^{3/4}$

$\quad t^2 - 2t + 1 = 0$

$(t-1)(t-1) = 0$

$\quad t - 1 = 0$

$\quad\quad t = 1$

$\quad x^{3/4} = 1$

$\quad\quad x = 1^{4/3}$

$\quad\quad x = 1$

The solution set is $\{1\}$.

**53.** $2x - 3x^{1/2} + 1 = 0$ let $t = x^{1/2}$

$\quad 2t^2 - 3t + 1 = 0$

$(2t-1)(t-1) = 0$

$2t - 1 = 0 \qquad t - 1 = 0$

$2t = 1$

$\quad t = \dfrac{1}{2} \qquad t = 1$

$x^{1/2} = \dfrac{1}{2} \qquad x^{1/2} = 1$

$\quad x = \left(\dfrac{1}{2}\right)^2 \qquad x = 1^2$

$\quad x = \dfrac{1}{4} \qquad x = 1$

The solution set is $\left\{ \dfrac{1}{4}, 1 \right\}$.

**55.** $(x-5)^2 - 4(x-5) - 21 = 0$ let $t = x - 5$

$\quad t^2 - 4t - 21 = 0$

$\quad (t+3)(t-7) = 0$

$t + 3 = 0 \qquad t - 7 = 0$

$\quad t = -3 \qquad t = 7$

$x - 5 = -3 \quad x - 5 = 7$

$\quad x = 2 \qquad x = 12$

The solution set is $\{2, 12\}$.

**57.** $\left(x^2 - x\right)^2 - 14\left(x^2 - x\right) + 24 = 0$

Let $t = x^2 - x$.

$t^2 - 14t + 24 = 0$

$(t-2)(t-12) = 0$

$t = 2$ or $t = 12$

$x^2 - x = 2 \qquad\quad$ or $\quad x^2 - x = 12$

$x^2 - x - 2 = 0 \qquad\qquad x^2 - x - 12 = 0$

$(x-2)(x+1) = 0 \qquad (x-4)(x+3) = 0$

The solution set is $\{-3, -1, 2, 4\}$.

**59.** $\left(y-\dfrac{8}{y}\right)^2+5\left(y-\dfrac{8}{y}\right)-14=0$

Let $t=y-\dfrac{8}{y}$.

$t^2+5t-14=0$
$(t+7)(t-2)=0$
$t=-7$ or $t=2$

$\begin{array}{ll}y-\dfrac{8}{y}=-7 & \text{or}\quad y-\dfrac{8}{y}=2 \\ y^2+7y-8=0 & y^2-2y-8=0 \\ (y+8)(y-1)=0 & (y-4)(y+2)=0\end{array}$

The solution set is $\{-8,-2,1,4\}$.

**61.** $|x|=8$
$x=8,\ x=-8$
The solution set is $\{8,-8\}$.

**63.** $|x-2|=7$
$\begin{array}{ll}x-2=7 & x-2=-7 \\ x=9 & x=-5\end{array}$
The solution set is $\{9,-5\}$.

**65.** $|2x-1|=5$
$\begin{array}{ll}2x-1=5 & 2x-1=-5 \\ 2x=6 & 2x=-4 \\ x=3 & x=-2\end{array}$
The solution set is $\{3,-2\}$.

**67.** $2|3x-2|=14$
$|3x-2|=7$
$\begin{array}{ll}3x-2=7 & 3x-2=-7 \\ 3x=9 & 3x=-5 \\ x=3 & x=-5/3\end{array}$
The solution set is $\{3,-5/3\}$

**69.** $7|5x|+2=16$
$7|5x|=14$
$|5x|=2$
$\begin{array}{ll}5x=2 & 5x=-2 \\ x=2/5 & x=-2/5\end{array}$
The solution set is $\left\{\dfrac{2}{5},-\dfrac{2}{5}\right\}$.

**71.** $2\left|4-\dfrac{5}{2}x\right|+6=18$

$2\left|4-\dfrac{5}{2}x\right|=12$

$\left|4-\dfrac{5}{2}x\right|=6$

$\begin{array}{lll}4-\dfrac{5}{2}x=6 & \text{or} & 4-\dfrac{5}{2}x=-6 \\ -\dfrac{5}{2}x=2 & & -\dfrac{5}{2}x=-10 \\ -\dfrac{2}{5}\left(-\dfrac{5}{2}\right)x=-\dfrac{2}{5}(2) & & -\dfrac{2}{5}\left(-\dfrac{5}{2}\right)x=-\dfrac{2}{5}(-10) \\ x=-\dfrac{4}{5} & & x=4\end{array}$

The solution set is $\left\{-\dfrac{4}{5},4\right\}$.

**73.** $|x+1|+5=3$
$|x+1|=-2$
No solution
The solution set is $\{\ \}$.

**75.** $|2x-1|+3=3$

$|2x-1|=0$

$2x-1=0$

$2x=1$

$x=\dfrac{1}{2}$

The solution set is $\left\{\dfrac{1}{2}\right\}$.

**77.** $|3x-1|=|x+5|$
$\begin{array}{ll}3x-1=x+5 & 3x-1=-x-5 \\ 2x-1=5 & 4x-1=-5 \\ 2x=6 & 4x=-4 \\ x=3 & x=-1\end{array}$
The solution set is $\{3,-1\}$.

**79.** Set $y = 0$ to find the $x$–intercept(s).

$$0 = \sqrt{x+2} + \sqrt{x-1} - 3$$
$$-\sqrt{x+2} = \sqrt{x-1} - 3$$
$$\left(-\sqrt{x+2}\right)^2 = \left(\sqrt{x-1} - 3\right)^2$$
$$x+2 = \left(\sqrt{x-1}\right)^2 - 2\left(\sqrt{x-1}\right)(3) + (3)^2$$
$$x+2 = x-1 - 6\sqrt{x-1} + 9$$
$$x+2 = x-1 - 6\sqrt{x-1} + 9$$
$$2 = 8 - 6\sqrt{x-1}$$
$$-6 = -6\sqrt{x-1}$$
$$\frac{-6}{-6} = \frac{-6\sqrt{x-1}}{-6}$$
$$1 = \sqrt{x-1}$$
$$(1)^2 = \left(\sqrt{x-1}\right)^2$$
$$1 = x-1$$
$$2 = x$$

The $x$-intercept is 2.
The corresponding graph is graph (c).

**81.** Set $y = 0$ to find the $x$–intercept(s).

$$0 = x^{\frac{1}{3}} + 2x^{\frac{1}{6}} - 3$$

Let $t = x^{\frac{1}{6}}$.

$$x^{\frac{1}{3}} + 2x^{\frac{1}{6}} - 3 = 0$$
$$\left(x^{\frac{1}{6}}\right)^2 + 2x^{\frac{1}{6}} - 3 = 0$$
$$t^2 + 2t - 3 = 0$$
$$(t+3)(t-1) = 0$$
$$t+3 = 0 \quad \text{or} \quad t-1 = 0$$
$$t = -3 \qquad\qquad t = 1$$

Substitute $x^{\frac{1}{6}}$ for $t$.

$$x^{\frac{1}{6}} = -3 \quad \text{or} \quad x^{\frac{1}{6}} = 1$$
$$\left(x^{\frac{1}{6}}\right)^6 = (-3)^6 \qquad \left(x^{\frac{1}{6}}\right)^6 = (1)^6$$
$$x = 729 \qquad\qquad x = 1$$

729 does not check and must be rejected.
The $x$-intercept is 1.
The corresponding graph is graph (e).

**83.** Set $y = 0$ to find the $x$–intercept(s).

$$(x+2)^2 - 9(x+2) + 20 = 0$$

Let $t = x+2$.

$$(x+2)^2 - 9(x+2) + 20 = 0$$
$$t^2 - 9t + 20 = 0$$
$$(t-5)(t-4) = 0$$
$$t-5 = 0 \quad \text{or} \quad t-4 = 0$$
$$t = 5 \qquad\qquad t = 4$$

Substitute $x+2$ for $t$.

$$x+2 = 5 \quad \text{or} \quad x+2 = 4$$
$$x = 3 \qquad\qquad x = 2$$

The $x$-intercepts are 2 and 3.
The corresponding graph is graph (f).

**85.** $|5 - 4x| = 11$

$$5 - 4x = 11 \qquad\qquad 5 - 4x = -11$$
$$-4x = 6 \quad \text{or} \quad -4x = -16$$
$$x = -\frac{3}{2} \qquad\qquad x = 4$$

The solution set is $\left\{ -\dfrac{3}{2}, 4 \right\}$.

**87.** $x + \sqrt{x+5} = 7$

$$\sqrt{x+5} = 7 - x$$
$$\left(\sqrt{x+5}\right)^2 = (7-x)^2$$
$$x + 5 = 49 - 14x + x^2$$
$$0 = x^2 - 15x + 44$$
$$0 = (x-4)(x-11)$$
$$x - 4 = 0 \quad \text{or} \quad x - 11 = 0$$
$$x = 4 \qquad\qquad x = 11$$

11 does not check and must be rejected.
The solution set is $\{4\}$.

**89.**
$$2x^3 + x^2 - 8x + 2 = 6$$
$$2x^3 + x^2 - 8x - 4 = 0$$
$$x^2(2x+1) - 4(2x+1) = 0$$
$$(2x+1)(x^2 - 4) = 0$$
$$(2x+1)(x+2)(x-2) = 0$$
$$2x+1 = 0 \quad \text{or} \quad x+2 = 0 \quad \text{or} \quad x-2 = 0$$
$$x = -\frac{1}{2} \qquad\qquad x = -2 \qquad\qquad x = 2$$

The solution set is $\left\{ -\dfrac{1}{2}, -2, 2 \right\}$.

**91.**  $(x+4)^{\frac{3}{2}} = 8$

$$\left((x+4)^{\frac{3}{2}}\right)^{\frac{2}{3}} = (8)^{\frac{2}{3}}$$

$$x+4 = \left(\sqrt[3]{8}\right)^2$$

$$x+4 = (2)^2$$

$$x+4 = 4$$

$$x = 0$$

The solution set is $\{0\}$.

**93.**
$$y_1 = y_2 + 3$$
$$\left(x^2-1\right)^2 = 2\left(x^2-1\right)+3$$
$$\left(x^2-1\right)^2 - 2\left(x^2-1\right)-3 = 0$$

Let $t = x^2 - 1$ and substitute.

$$t^2 - 2t - 3 = 0$$
$$(t+1)(t-3) = 0$$
$$t+1 = 0 \quad \text{or} \quad t-3 = 0$$
$$t = -1 \qquad\qquad t = 3$$

Substitute $x^2 - 1$ for $t$.

$$x^2 - 1 = -1 \quad \text{or} \quad x^2 - 1 = 3$$
$$x^2 = 0 \qquad\qquad x^2 = 4$$
$$x = 0 \qquad\qquad x = \pm 2$$

The solution set is $\{-2, 0, 2\}$.

**95.**  $\left|x^2 + 2x - 36\right| = 12$

$$x^2 + 2x - 36 = 12 \qquad x^2 + 2x - 36 = -12$$
$$x^2 + 2x - 48 = 0 \quad \text{or} \quad x^2 + 2x - 24 = 0$$
$$(x+8)(x-6) = 0 \qquad (x+6)(x-4) = 0$$

Setting each of the factors above equal to zero gives $x = -8$, $x = 6$, $x = -6$, and $x = 4$.
The solution set is $\{-8, -6, 4, 6\}$.

**97.**  $x(x+1)^3 - 42(x+1)^2 = 0$

$$(x+1)^2\left(x(x+1)-42\right) = 0$$
$$(x+1)^2\left(x^2+x-42\right) = 0$$
$$(x+1)^2(x+7)(x-6) = 0$$

Setting each of the factors above equal to zero gives $x = -7$, $x = -1$, and $x = 6$.
The solution set is $\{-7, -1, 6\}$.

**99.**  Let $x$ = the number.
$$\sqrt{5x-4} = x-2$$
$$\left(\sqrt{5x-4}\right)^2 = (x-2)^2$$
$$5x - 4 = x^2 - 4x + 4$$
$$0 = x^2 - 9x + 8$$
$$0 = (x-8)(x-1)$$
$$x - 8 = 0 \quad \text{or} \quad x - 1 = 0$$
$$x = 8 \qquad\qquad x = 1$$

Check $x = 8$: $\sqrt{5(8)-4} = 8-2$
$$\sqrt{40-4} = 6$$
$$\sqrt{36} = 6$$
$$6 = 6$$

Check $x = 1$: $\sqrt{5(1)-4} = 1-2$
$$\sqrt{5-4} = -1$$
$$\sqrt{-1} \ne -1$$

Discard $x = 1$. The number is 8.

**101.**
$$r = \sqrt{\frac{3V}{\pi h}}$$
$$r^2 = \left(\sqrt{\frac{3V}{\pi h}}\right)^2$$
$$r^2 = \frac{3V}{\pi h}$$
$$\pi r^2 h = 3V$$
$$\frac{\pi r^2 h}{3} = V$$
$$V = \frac{\pi r^2 h}{3} \quad \text{or} \quad V = \frac{1}{3}\pi r^2 h$$

**103.** Exclude any value that causes the denominator to equal zero.

$$|x+2|-14=0$$
$$|x+2|=14$$
$$x+2=14 \quad \text{or} \quad x+2=-14$$
$$x=12 \qquad\qquad x=-16$$

$-16$ and $12$ must be excluded from the domain.

**105.** $t=\dfrac{\sqrt{d}}{2}$

$$1.16=\frac{\sqrt{d}}{2}$$
$$2.32=\sqrt{d}$$
$$2.32^2=\left(\sqrt{d}\right)^2$$
$$d\approx 5.4$$

The vertical distance was about 5.4 feet.

**107.** It is represented by the point $(5.4, 1.16)$.

**109. a.** According to the line graph, about 48% $\pm1\%$ of U.S. women participated in the labor force in 2000.

**b.** $p=2.2\sqrt{t}+36.2$

$$p=2.2\sqrt{28}+36.2\approx 47.8$$

According to the formula, 47.8% of U.S. women participated in the labor force in 2000.

**c.** $p=2.2\sqrt{t}+36.2$

$$52=2.2\sqrt{t}+36.2$$
$$15.8=2.2\sqrt{t}$$
$$\frac{15.8}{2.2}=\frac{2.2\sqrt{t}}{2.2}$$
$$\frac{15.8}{2.2}=\sqrt{t}$$
$$\left(\frac{15.8}{2.2}\right)^2=\left(\sqrt{t}\right)^2$$
$$52\approx t$$

According to the formula, 52% of U.S. women will participate in the labor force 52 years after 1972, or 2024.

**111.**

$$365=0.2x^{3/2}$$
$$\frac{365}{0.2}=\frac{0.2x^{3/2}}{0.2}$$
$$1825=x^{3/2}$$
$$1825^2=\left(x^{3/2}\right)^2$$
$$3,330,625=x^3$$
$$\sqrt[3]{3,330,625}=\sqrt[3]{x^3}$$
$$149.34\approx x$$

The average distance of the Earth from the sun is approximately 149 million kilometers.

**113.** $\sqrt{6^2+x^2}+\sqrt{8^2+(10-x)^2}=18$

$$\sqrt{36+x^2}=18-\sqrt{64+100-20x+x^2}$$
$$36+x^2=324-36\sqrt{x^2-20x+164}+x^2-20x+164$$
$$36\sqrt{x^2-20x+164}=-20x+452$$
$$9\sqrt{x^2-20x+164}=-5x+113$$
$$81(x^2-20x+164)=25x^2-1130x+12769$$
$$81x^2-1620x+13284=25x^2-1130x+12769$$
$$56x^2-490x+515=0$$
$$x=\frac{490\pm\sqrt{(-490)^2-4(56)(515)}}{2(56)}$$
$$x=\frac{490\pm353.19}{112}$$
$$x\approx 1.2 \qquad x\approx 7.5$$

The point should be located approximately either 1.2 feet or 7.5 feet from the base of the 6-foot pole.

**115. – 121.** Answers will vary.

**123.** $-x^4+4x^3-4x^2=0$

The solution set is $\{0,2\}$.
$$-(0)^4+4(0)^3-4(0)^2=0$$
$$0=0$$
$$-(2)^4+4(2)^3-4(2)^2=0$$
$$-16+32-16=0$$
$$0=0$$

**125.** does not make sense; Explanations will vary. Sample explanation: You should substitute into the original equation.

**127.** does not make sense; Explanations will vary. Sample explanation: Changing the order of the terms does not change the fact that this equation is quadratic in form.

**129.** false; Changes to make the statement true will vary.
A sample change is: Squaring $x+2$ results in $x^2 + 4x + 4$.

**131.** false; Changes to make the statement true will vary.
A sample change is: To solve the equation, let $u^2 = x$.

**133.** $\sqrt{6x-2} = \sqrt{2x+3} - \sqrt{4x-1}$

$6x-2 = 2x+3 - 2\sqrt{(2x+3)(4x-1)} + 4x - 1$

$-4 = -2\sqrt{(2x+3)(4x-1)}$

$2 = \sqrt{8x^2 + 10x - 3}$

$4 = 8x^2 + 10x - 3$

$8x^2 + 10x - 7 = 0$

$x = \dfrac{-10 \pm \sqrt{10^2 - 4(8)(-7)}}{2(8)}$

$x = \dfrac{-10 \pm \sqrt{100 + 224}}{16}$

$x = \dfrac{-10 \pm \sqrt{324}}{16}$

$x = \dfrac{-10 \pm 18}{16}$

$x = \dfrac{-28}{26}, \dfrac{8}{16}$

$x = \dfrac{1}{2}$

The solution set is $\left\{ \dfrac{1}{2} \right\}$.

**135.** $\sqrt[3]{x\sqrt{x}} = 9$

$\sqrt[3]{x\sqrt{x}} = 9$

$\sqrt[3]{x^1 x^{\frac{1}{2}}} = 9$

$\left( x^1 x^{\frac{1}{2}} \right)^{\frac{1}{3}} = 9$

$\left( x^{\frac{3}{2}} \right)^{\frac{1}{3}} = 9$

$x^{\frac{1}{2}} = 9$

$\left( x^{\frac{1}{2}} \right)^2 = (9)^2$

$x = 81$

The solution set is $\{81\}$.

**137.**  $3 - 2x \le 11$

$3 - 2(-1) \le 11$

$3 + 2 \le 11$

$5 \le 11,$ true

Yes, $-1$ is a solution.

**138.** $-2x - 4 = x + 5$

$-2x - x = 5 + 4$

$-3x = 9$

$x = \dfrac{9}{-3}$

$x = -3$

The solution set is $\{-3\}$.

**139.**     $\dfrac{x+3}{4} = \dfrac{x-2}{3} + \dfrac{1}{4}$

$12\left( \dfrac{x+3}{4} \right) = 12\left( \dfrac{x-2}{3} + \dfrac{1}{4} \right)$

$3(x+3) = 4(x-2) + 3$

$3x + 9 = 4x - 8 + 3$

$3x + 9 = 4x - 5$

$3x - 4x = -5 - 9$

$-x = -14$

$x = 14$

The solution set is $\{14\}$.

**Section 1.7**

**Check Point Exercises**

**1.**  **a.**   $[-2, 5) = \left\{ x \middle| -2 \le x < 5 \right\}$

**b.**   $[1, 3.5] = \left\{ x \middle| 1 \le x \le 3.5 \right\}$

**c.**   $[-\infty, -1) = \left\{ x \middle| x < -1 \right\}$

**2.   a.**     Graph $[1,3]$:

Graph $(2,6)$:

To find the intersection, take the portion of the number line that the two graphs have in common.

Numbers in both
$[1,3]$ and $(2,6)$:

Thus, $[1,3] \cap (2,6) = (2,3]$.

**b.**     Graph $[1,3]$:

Graph $(2,6)$:

To find the union, take the portion of the number line representing the total collection of numbers in the two graphs.

Numbers in either $[1,3]$
or $(2,6)$ or both:

Thus, $[1,3] \cup (2,6) = [1,6)$.

**3.**     $2 - 3x \le 5$

$-3x \le 3$

$x \ge -1$

The solution set is $\{x | x \ge -1\}$ or $[-1, \infty)$.

**4.**     $3x + 1 > 7x - 15$

$-4x > -16$

$\dfrac{-4x}{-4} < \dfrac{-16}{-4}$

$x < 4$

The solution set is $\{x | x < 4\}$ or $(-\infty, 4)$.

**5.**     $\dfrac{x-4}{2} \ge \dfrac{x-2}{3} + \dfrac{5}{6}$

$6\left(\dfrac{x-4}{2}\right) \ge 6\left(\dfrac{x-2}{3} + \dfrac{5}{6}\right)$

$3(x-4) \ge 2(x-2) + 5$

$3x - 12 \ge 2x - 4 + 5$

$3x - 12 \ge 2x + 1$

$3x - 2x \ge 1 + 12$

$x \ge 13$

The solution set is $\{x | x \ge 13\}$ or $[13, \infty)$.

**6.**    **a.**    $3(x+1) > 3x+2$

$3x+3 > 3x+2$

$3 > 2$

$3 > 2$ is true for all values of $x$.

The solution set is $\{x | x \text{ is a real number}\}$ or $\Re$

or $(-\infty, \infty)$.

     **b.**    $x+1 \le x-1$

$1 \le -1$

$1 \le -1$ is false for all values of $x$.

The solution set is $\varnothing$.

**7.**    $1 \le 2x+3 < 11$

$-2 \le 2x < 8$

$-1 \le x < 4$

The solution set is $\{x | -1 \le x < 4\}$ or $[-1, 4)$.

**8.**    $|x-2| < 5$

$-5 < x-2 < 5$

$-3 < x < 7$

The solution set is $\{x | -3 < x < 7\}$ or $(-3, 7)$.

**9.**    $-3|5x-2| + 20 \ge -19$

$-3|5x-2| \ge -39$

$\dfrac{-3|5x-2|}{-3} \le \dfrac{-39}{-3}$

$|5x-2| \le 13$

$-13 \le 5x-2 \le 13$

$-11 \le 5x \le 15$

$\dfrac{-11}{5} \le \dfrac{5x}{5} \le \dfrac{15}{5}$

$-\dfrac{11}{5} \le x \le 3$

The solution set is $\left\{x \left| -\dfrac{11}{5} \le x \le 3\right.\right\}$ or $\left[-\dfrac{11}{5}, 3\right]$.

**10.**    $18 < |6-3x|$

$\begin{array}{ccc} 6-3x < -18 & & 6-3x > 18 \\ -3x < -24 & \text{or} & -3x > 12 \\ \dfrac{-3x}{-3} > \dfrac{-24}{-3} & & \dfrac{-3x}{-3} < \dfrac{12}{-3} \\ x > 8 & & x < -4 \end{array}$

The solution set is $\{x | x < -4 \text{ or } x > 8\}$

or $(-\infty, -4) \cup (8, \infty)$.

**11.**    Let $x$ = the number of miles driven in a week.

$260 < 80 + 0.25x$

$180 < 0.25x$

$720 < x$

Driving more than 720 miles in a week makes Basic the better deal.

**Concept and Vocabulary Check 1.7**

**1.**    2; 5; 2; 5

**2.**    greater than

**3.**    less than or equal to

**4.**    $(-\infty, 9)$; intersection

**5.**    $(-\infty, 12)$; union

**6.**    adding 4; dividing; $-3$; direction; $>$; $<$

**7.**    $\varnothing$

**8.**    $(-\infty, \infty)$

**9.**    middle

**10.**    $-c$; $c$

**11.**    $-c$; $c$

**12.**    $-2 < x-7 < 2$

**13.**    $x-7 < -2$ or $-7 > 2$

**Exercise Set 1.7**

1.   $1 < x \le 6$

3.   $-5 \le x < 2$

5.   $-3 \le x \le 1$

7.   $x > 2$

9.   $x \ge -3$

11.  $x < 3$

13.  $x < 5.5$

15.      Graph $(-3,0)$:

      Graph $[-1,2]$:

      To find the intersection, take the portion of the number line that the two graphs have in common.
      Numbers in both $(-3,0)$ and $[-1,2]$:

      Thus, $(-3,0) \cap [-1,2] = [-1,0)$.

17.       Graph $(-3,0)$:

       Graph $[-1,2]$:

      To find the union, take the portion of the number line representing the total collection of numbers in the two graphs.

      Numbers in either $(-3,0)$ or $[-1,2]$ or both:

      Thus, $(-3,0) \cup [-1,2] = (-3,2]$.

19.      Graph $(-\infty,5)$:

      Graph $[1,8]$:

      To find the intersection, take the portion of the number line that the two graphs have in common.
      Numbers in both $(-\infty,5)$ and $[1,8]$:

      Thus, $(-\infty,5) \cap [1,8] = [1,5)$.

**21.**

Graph $(-\infty, 5)$:

Graph $[1, 8]$:

To find the union, take the portion of the number line representing the total collection of numbers in the two graphs.

Numbers in either $(-\infty, 5)$ or $[1, 8]$ or both:

Thus, $(-\infty, 5) \cup [1, 8] = (-\infty, 8]$.

**23.**

Graph $[3, \infty)$:

Graph $(6, \infty)$:

To find the intersection, take the portion of the number line that the two graphs have in common.

Numbers in both $[3, \infty)$ and $(6, \infty)$:

Thus, $[3, \infty) \cap (6, \infty) = (6, \infty)$.

**25.**

Graph $[3, \infty)$:

Graph $(6, \infty)$:

To find the union, take the portion of the number line representing the total collection of numbers in the two graphs.

Numbers in either $[3, \infty)$ or $(6, \infty)$ or both:

Thus, $[3, \infty) \cup (6, \infty) = [3, \infty)$.

**27.** $5x + 11 < 26$
$5x < 15$
$x < 3$
The solution set is $\{x \mid x < 3\}$, or $(-\infty, 3)$.

**29.** $3x - 7 \geq 13$
$3x \geq 20$
$x \geq \dfrac{20}{3}$

The solution set is $\left\{ x \mid x > \dfrac{20}{3} \right\}$, or $\left[ \dfrac{20}{3}, \infty \right)$.

**31.** $-9x \geq 36$
$x \leq -4$
The solution set is $\{x \mid x \leq -4\}$, or $(-\infty, -4]$.

**33.** $8x - 11 \leq 3x - 13$
$8x - 3x \leq -13 + 11$
$5x \leq -2$
$x \leq -\dfrac{2}{5}$

The solution set is $\left\{x \,\middle|\, x \leq -\dfrac{2}{5}\right\}$, or $\left(-\infty, -\dfrac{2}{5}\right]$.

**35.** $4(x + 1) + 2 \geq 3x + 6$
$4x + 4 + 2 \geq 3x + 6$
$4x + 6 \geq 3x + 6$
$4x - 3x \geq 6 - 6$
$x \geq 0$

The solution set is $\left\{x \,\middle|\, x > 0\right\}$, or $[0, \infty)$.

**37.** $2x - 11 < -3(x + 2)$
$2x - 11 < -3x - 6$
$5x < 5$
$x < 1$

The solution set is $\left\{x \,\middle|\, x < 1\right\}$, or $(-\infty, 1)$.

**39.** $1 - (x + 3) \geq 4 - 2x$
$1 - x - 3 \geq 4 - 2x$
$-x - 2 \geq 4 - 2x$
$x \geq 6$

The solution set is $\left\{x \,\middle|\, x \geq 6\right\}$, or $[6, \infty)$.

**41.** $\dfrac{x}{4} - \dfrac{3}{2} \leq \dfrac{x}{2} + 1$
$\dfrac{4x}{4} - \dfrac{4 \cdot 3}{2} \leq \dfrac{4 \cdot x}{2} + 4 \cdot 1$
$x - 6 \leq 2x + 4$
$-x \leq 10$
$x \geq -10$

The solution set is $\left\{x \,\middle|\, x \geq -10\right\}$, or $[-10, \infty)$.

**43.** $1 - \dfrac{x}{2} > 4$
$-\dfrac{x}{2} > 3$
$x < -6$

The solution set is $\left\{x \,\middle|\, x, -6\right\}$, or $(-\infty, -6)$.

**45.** $\dfrac{x - 4}{6} \geq \dfrac{x - 2}{9} + \dfrac{5}{18}$
$3(x - 4) \geq 2(x - 2) + 5$
$3x - 12 \geq 2x - 4 + 5$
$x \geq 13$

The solution set is $\left\{x \,\middle|\, x \geq 13\right\}$, or $[13, \infty)$.

**47.** $4(3x - 2) - 3x < 3(1 + 3x) - 7$
$12x - 8 - 3x < 3 + 9x - 7$
$9x - 8 < -4 + 9x$
$-8 < -4$
True for all $x$
The solution set is $\left\{x \,\middle|\, x \text{ is any real number}\right\}$, or $(-\infty, \infty)$.

**49.** $5(x - 2) - 3(x + 4) \geq 2x - 20$
$5x - 10 - 3x - 12 \geq 2x - 20$
$2x - 22 \geq 2x - 20$
$-22 \geq -20$
Not true for any $x$.
The solution set is the empty set, $\varnothing$.

**51.** $6 < x + 3 < 8$
$6 - 3 < x + 3 - 3 < 8 - 3$
$3 < x < 5$
The solution set is $\left\{x \,\middle|\, 3 < x < 5\right\}$, or $(3, 5)$.

**53.** $-3 \leq x - 2 < 1$
$-1 \leq x < 3$
The solution set is $\left\{x \,\middle|\, -1 \leq x < 3\right\}$, or $[-1, 3)$.

**55.** $-11 < 2x - 1 \leq -5$
$-10 < 2x \leq -4$
$-5 < x \leq -2$
The solution set is $\left\{x \,\middle|\, -5 < x \leq -2\right\}$, or $(-5, -2]$.

**57.** $-3 \le \dfrac{2}{3}x - 5 < -1$

$2 \le \dfrac{2}{3}x < 4$

$3 \le x < 6$

The solution set is $\left\{x \mid 3 \le x < 6\right\}$, or [3, 6).

**59.** $|x| < 3$

$-3 < x < 3$

The solution set is $\left\{x \mid -3 < x < 3\right\}$, or (−3, 3).

**61.** $|x - 1| \le 2$

$-2 \le x - 1 \le 2$

$-1 \le x \le 3$

The solution set is $\left\{x \mid -1 \le x \le 3\right\}$, or [−1, 3].

**63.** $|2x - 6| < 8$

$-8 < 2x - 6 < 8$

$-2 < 2x < 14$

$-1 < x < 7$

The solution set is $\left\{x \mid -1 < x < 7\right\}$, or (−1, 7).

**65.** $|2(x - 1) + 4| \le 8$

$-8 \le 2(x - 1) + 4 \le 8$

$-8 \le 2x - 2 + 4 \le 8$

$-8 \le 2x + 2 \le 8$

$-10 \le 2x \le 6$

$-5 \le x \le 3$

The solution set is $\left\{x \mid -5 \le x \le 3\right\}$, or [−5, 3].

**67.** $\left|\dfrac{2y + 6}{3}\right| < 2$

$-2 < \dfrac{2y + 6}{3} < 2$

$-6 < 2y + 6 < 6$

$-12 < 2y < 0$

$-6 < y < 0$

The solution set is $\left\{x \mid -6 < y < 0\right\}$, or (−6, 0).

**69.** $|x| > 3$

$x > 3$ or $x < -3$

The solution set is $\left\{x \mid x > 3 \text{ or } x < -3\right\}$, that is,

$(-\infty, -3)$ or $(3, \infty)$.

**71.** $|x - 1| \ge 2$

$x - 1 \ge 2$ or $x - 1 \le -2$

$x \ge 3$ $\qquad$ $x \le -1$

The solution set is $\left\{x \mid x \le -1 \text{ or } x \ge 3\right\}$, that is,

$(-\infty, -1]$ or $[3, \infty)$.

**73.** $|3x - 8| > 7$

$3x - 8 > 7$ or $3x - 8 < -7$

$3x > 15$ $\qquad$ $3x < 1$

$x > 5$ $\qquad$ $x < \dfrac{1}{3}$

The solution set is $\left\{x \mid x < \dfrac{1}{3} \text{ or } x > 5\right\}$, that is,

$\left(-\infty, \dfrac{1}{3}\right)$ or $(5, \infty)$ .

**75.** $\left|\dfrac{2x + 2}{4}\right| \ge 2$

$\dfrac{2x + 2}{4} \ge 2$ or $\dfrac{2x + 2}{4} \le -2$

$2x + 2 \ge 8$ $\qquad$ $2x + 2 \le -8$

$2x \ge 6$ $\qquad$ $2x \le -10$

$x \ge 3$ $\qquad$ $x \le -5$

The solution set is $\left\{x \mid x \le -5 \text{ or } x \ge 3\right\}$, that is,

$(-\infty, -5]$ or $[3, \infty)$ .

**77.** $\left|3 - \dfrac{2}{3}x\right| > 5$

$3 - \dfrac{2}{3}x > 5$ or $3 - \dfrac{2}{3}x < -5$

$-\dfrac{2}{3}x > 2$ $\qquad$ $-\dfrac{2}{3}x < -8$

$x < -3$ $\qquad$ $x > 12$

The solution set is $\left\{x \mid x < -3 \text{ or } x > 12\right\}$, that is,

$(-\infty, -3)$ or $(12, \infty)$ .

**79.** $3|x - 1| + 2 \ge 8$

$3|x - 1| \ge 6$

$|x - 1| \ge 2$

$x - 1 \ge 2$ or $x - 1 \le -2$

$x \ge 3$ $\qquad$ $x \le -1$

The solution set is $\left\{x \mid x \le 1 \text{ or } x \ge 3\right\}$, that is,

$(-\infty, -1]$ or $[3, \infty)$ .

**81.** $-2|x-4| \geq -4$

$$\frac{-2|x-4|}{-2} \leq \frac{-4}{-2}$$

$$|x-4| \leq 2$$

$$-2 \leq x-4 \leq 2$$

$$2 \leq x \leq 6$$

The solution set is $\{x | 2 \leq x \leq 6\}$.

**83.** $-4|1-x| < -16$

$$\frac{-4|1-x|}{-4} > \frac{-16}{-4}$$

$$|1-x| > 4$$

$$1-x > 4 \qquad 1-x < -4$$

$$-x > 3 \quad \text{or} \quad -x < -5$$

$$x < -3 \qquad x > 5$$

The solution set is $\{x | x < -3 \text{ or } x > 5\}$.

**85.** $3 \leq |2x-1|$

$$2x-1 \geq 3 \qquad 2x-1 \leq -3$$

$$2x \geq 4 \quad \text{or} \quad 2x \leq -2$$

$$x \geq 2 \qquad x \leq -1$$

The solution set is $\{x | x \leq -1 \text{ or } x \geq 2\}$.

**87.** $5 > |4-x|$ is equivalent to $|4-x| < 5$.

$$-5 < 4-x < 5$$

$$-9 < -x < 1$$

$$\frac{-9}{-1} > \frac{-x}{-1} > \frac{1}{-1}$$

$$9 > x > -1$$

$$-1 < x < 9$$

The solution set is $\{x | -1 < x < 9\}$.

**89.** $1 < |2-3x|$ is equivalent to $|2-3x| > 1$.

$$\begin{array}{ll}
2-3x > 1 & 2-3x < -1 \\
-3x > -1 & -3x < -3 \\
\frac{-3x}{-3} < \frac{-1}{-3} \quad \text{or} & \frac{-3x}{-3} > \frac{-3}{-3} \\
x < \frac{1}{3} & x > 1
\end{array}$$

The solution set is $\left\{ x \middle| x < \frac{1}{3} \text{ or } x > 1 \right\}$.

**91.** $12 < \left| -2x + \frac{6}{7} \right| + \frac{3}{7}$

$$\frac{81}{7} < \left| -2x + \frac{6}{7} \right|$$

$$-2x + \frac{6}{7} > \frac{81}{7} \quad \text{or} \quad -2x + \frac{6}{7} < -\frac{81}{7}$$

$$-2x > \frac{75}{7} \qquad\qquad -2x < -\frac{87}{7}$$

$$x < -\frac{75}{14} \qquad\qquad x > \frac{87}{14}$$

The solution set is $\left\{ x \middle| x < -\frac{75}{14} \text{ or } x > \frac{87}{14} \right\}$, that is,

$$\left( -\infty, -\frac{75}{14} \right) \text{ or } \left( \frac{87}{14}, \infty \right).$$

**93.** $4 + \left| 3 - \frac{x}{3} \right| \geq 9$

$$\left| 3 - \frac{x}{3} \right| \geq 5$$

$$3 - \frac{x}{3} \geq 5 \quad \text{or} \quad 3 - \frac{x}{3} \leq -5$$

$$-\frac{x}{3} \geq 2 \qquad\qquad -\frac{x}{3} \leq -8$$

$$x \leq -6 \qquad\qquad x \geq 24$$

The solution set is $\{x | x \leq -6 \text{ or } x \geq 24\}$, that is,

$$(-\infty, -6] \text{ or } [24, \infty).$$

**95.**

$$y_1 \leq y_2$$

$$\frac{x}{2} + 3 \leq \frac{x}{3} + \frac{5}{2}$$

$$6\left( \frac{x}{2} + 3 \right) \leq 6\left( \frac{x}{3} + \frac{5}{2} \right)$$

$$\frac{6x}{2} + 6(3) \leq \frac{6x}{3} + \frac{6(5)}{2}$$

$$3x + 18 \leq 2x + 15$$

$$x \leq -3$$

The solution set is $(-\infty, -3]$.

**97.**

$$y \geq 4$$

$$1 - (x+3) + 2x \geq 4$$

$$1 - x - 3 + 2x \geq 4$$

$$x - 2 \geq 4$$

$$x \geq 6$$

The solution set is $[6, \infty)$.

**99.**
$$y < 8$$
$$|3x - 4| + 2 < 8$$
$$|3x - 4| < 6$$
$$-6 < 3x - 4 < 6$$
$$-2 < 3x < 10$$
$$\frac{-2}{3} < \frac{3x}{3} < \frac{10}{3}$$
$$\frac{-2}{3} < x < \frac{10}{3}$$

The solution set is $\left(\dfrac{-2}{3}, \dfrac{10}{3}\right)$.

**101.**
$$y \le 4$$
$$7 - \left|\frac{x}{2} + 2\right| \le 4$$
$$-\left|\frac{x}{2} + 2\right| \le -3$$
$$\left|\frac{x}{2} + 2\right| \ge 3$$
$$\frac{x}{2} + 2 \ge 3 \quad \text{or} \quad \frac{x}{2} + 2 \le -3$$
$$x + 4 \ge 6 \qquad\qquad x + 4 \le -6$$
$$x \ge 2 \qquad\qquad\quad x \le -10$$

The solution set is $(-\infty, -10] \cup [2, \infty)$.

**103.** The graph's height is below 5 on the interval $(-1, 9)$.

**105.** The solution set is $\{x \mid -1 \le x < 2\}$ or $[-1, 2)$.

**107.** Let $x$ be the number.
$$|4 - 3x| \ge 5 \quad \text{or} \quad |3x - 4| \ge 5$$
$$3x - 4 \ge 5 \qquad\qquad 3x - 4 \le -5$$
$$3x \ge 9 \quad \text{or} \quad 3x \le -1$$
$$x \ge 3 \qquad\qquad x \le -\frac{1}{3}$$

The solution set is $\left\{x \mid x \le -\dfrac{1}{3} \text{ or } x \ge 3\right\}$ or

**109.** $(0, 4)$

**111.** passion $\le$ intimacy or intimacy $\ge$ passion

**113.** passion<commitment or commitment > passion

**115.** 9, after 3 years

**117. a.** $I = \frac{1}{4}x + 26$
$$\frac{1}{4}x + 26 > 33$$
$$\frac{1}{4}x > 7$$
$$x > 28$$
More than 33% of U.S. households will have an interfaith marriage in years after 2016 (i.e. $1988 + 28$).

**b.** $N = \frac{1}{4}x + 6$
$$\frac{1}{4}x + 6 > 14$$
$$\frac{1}{4}x > 8$$
$$x > 32$$
More than 14% of U.S. households will have a person of faith married to someone with no religion in years after 2020 (i.e. $1988 + 32$).

**c.** More than 33% of U.S. households will have an interfaith marriage *and* more than 14% of U.S. households will have a person of faith married to someone with no religion in years after 2020.

**d.** More than 33% of U.S. households will have an interfaith marriage *or* more than 14% of U.S. households will have a person of faith married to someone with no religion in years after 2016.

**119.** $28 \le 20 + 0.40(x - 60) \le 40$
$$28 \le 20 + 0.40x - 24 \le 40$$
$$28 \le 0.40x - 4 \le 40$$
$$32 \le 0.40x \le 44$$
$$80 \le x \le 110$$
Between 80 and 110 ten minutes, inclusive.

**121.**
$$\left|\frac{h - 50}{5}\right| \ge 1.645$$
$$\frac{h - 50}{5} \ge 1.645 \quad \text{or} \quad \frac{h - 50}{5} \le -1.645$$
$$h - 50 \ge 8.225 \qquad\qquad h - 50 \le -8.225$$
$$h \ge 58.225 \qquad\qquad\quad h \le 41.775$$
The number of outcomes would be 59 or more, or 41 or less.

**123.** $15 + 0.08x < 3 + .12x$
$$12 < 0.04x$$
$$300 < x$$
Plan A is a better deal when texting more than 300 times per month.

**125.** $2 + 0.08x < 8 + 0.05x$

$0.03x < 6$

$x < 200$

The credit union is a better deal when writing less than 200 checks.

**127.** $3000 + 3x < 5.5x$

$3000 < 2.5x$

$1200 < x$

More then 1200 packets of stationary need to be sold each week to make a profit.

**129.** $245 + 95x \le 3000$

$95x \le 2755$

$x \le 29$

29 bags or less can be lifted safely.

**131. a.** $\dfrac{86 + 88 + x}{3} \ge 90$

$\dfrac{174 + x}{3} \ge 90$

$174 + x \ge 270$

$x \ge 96$

You must get at least a 96.

**b.** $\dfrac{86 + 88 + x}{3} < 80$

$\dfrac{174 + x}{3} < 80$

$174 + x < 240$

$x < 66$

This will happen if you get a grade less than 66.

**133.** Let $x$ = the number of times the bridge is crossed per three month period

The cost with the 3-month pass is $C_3 = 7.50 + 0.50x$.

The cost with the 6-month pass is $C_6 = 30$.

Because we need to buy two 3-month passes per 6-month pass, we multiply the cost with the 3-month pass by 2.

$2(7.50 + 0.50x) < 30$

$15 + x < 30$

$x < 15$

We also must consider the cost without purchasing a pass. We need this cost to be less than the cost with a 3-month pass.

$3x > 7.50 + 0.50x$

$2.50x > 7.50$

$x > 3$

The 3-month pass is the best deal when making more than 3 but less than 15 crossings per 3-month period.

**135. – 141.** Answers will vary.

**143.** $x < -3$

**145. a.** The cost of Plan A is $4 + 0.10x$; The cost of Plan B is $2 + 0.15x$.

**b.** Graph:

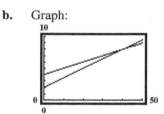

**c.** 41 or more checks make Plan A better.

**d.** $4 + 0.10x < 2 + 0.15x$

$2 < 0.05x$

$x > 40$

The solution set is $\{x \mid x > 40\}$ or $(40, \infty)$.

**147.** makes sense

**149.** makes sense

**151.** false; Changes to make the statement true will vary. A sample change is: $(-\infty, 3) \cup (-\infty, -2) = (-\infty, 3)$

**153.** true

**155. a.** $|x - 4| < 3$

**b.** $|x - 4| \ge 3$

**157.** Set 1 has each $x$-coordinate paired with only one $y$-coordinate.

**158.**

| $x$ | $y = 2x$ | $(x, y)$ |
|---|---|---|
| $-2$ | $y = 2(-2) = -4$ | $(-2, -4)$ |
| $-1$ | $y = 2(-1) + 4 = 2$ | $(-1, -2)$ |
| $0$ | $y = 2(0) = 0$ | $(0, 0)$ |
| $1$ | $y = 2(1) = 2$ | $(1, 2)$ |
| $2$ | $y = 2(2) = 4$ | $(2, 4)$ |

| $x$ | $y = 2x + 4$ | $(x, y)$ |
|---|---|---|
| $-2$ | $y = 2(-2) + 4 = 0$ | $(-2, 0)$ |
| $-1$ | $y = 2(-1) + 4 = 2$ | $(-1, 2)$ |
| $0$ | $y = 2(0) + 4 = 4$ | $(0, 4)$ |
| $1$ | $y = 2(1) + 4 = 6$ | $(1, 6)$ |
| $2$ | $y = 2(2) + 4 = 8$ | $(2, 8)$ |

$$y = 2x$$
$$y = 2x + 4$$

**159. a.** When the $x$-coordinate is 2, the $y$-coordinate is 3.

**b.** When the $y$-coordinate is 4, the $x$-coordinates are $-3$ and 3.

**c.** The $x$-coordinates are all real numbers.

**d.** The $y$-coordinates are all real numbers greater than or equal to 1.

## Chapter 1 Review Exercises

**1.**

$$y = 2x - 2$$

$x = -3, y = -8$
$x = -2, y = -6$
$x = -1, y = -4$
$x = 0, y = -2$
$x = 1, y = 0$
$x = 2, y = 2$
$x = 3, y = 4$

**2.**

$$y = x^2 - 3$$

$x = -3, y = 6$
$x = -2, y = 1$
$x = -1, y = -2$
$x = 0, y = -3$
$x = 1, y = -2$
$x = 2, y = 1$
$x = 3, y = 6$

**3.**

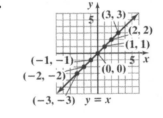

$$y = x$$

$x = -3, y = -3$
$x = -2, y = -2$
$x = -1, y = -1$
$x = 0, y = 0$
$x = 1, y = 1$
$x = 2, y = 2$
$x = 3, y = 3$

**4.**

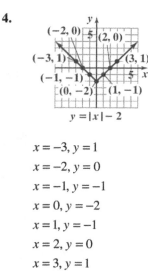

$x = -3, y = 1$

$x = -2, y = 0$

$x = -1, y = -1$

$x = 0, y = -2$

$x = 1, y = -1$

$x = 2, y = 0$

$x = 3, y = 1$

**5.** A portion of Cartesian coordinate plane with minimum $x$-value equal to $-20$, maximum $x$-value equal to 40, $x$-scale equal to 10 and with minimum $y$-value equal to $-5$, maximum $y$-value equal to 5, and $y$-scale equal to 1.

**6.** $x$-intercept: $-2$; The graph intersects the $x$-axis at $(-2, 0)$.

$y$-intercept: 2; The graph intersects the $y$-axis at $(0, 2)$.

**7.** $x$-intercepts: 2, $-2$; The graph intersects the $x$-axis at $(-2, 0)$ and $(2, 0)$.

$y$-intercept: $-4$; The graph intercepts the $y$-axis at $(0, -4)$.

**8.** $x$-intercept: 5; The graph intersects the $x$-axis at $(5, 0)$.

$y$-intercept: None; The graph does not intersect the $y$-axis.

**9.** The coordinates are $(20, 8)$. This means that 8% of college students anticipated a starting salary of $20 thousand.

**10.** The starting salary that was anticipated by the greatest percentage of college students was $30 thousand. 22% of students anticipated this salary.

**11.** The starting salary that was anticipated by the least percentage of college students was $70 thousand. 2% of students anticipated this salary.

**12.** Starting salaries of $25 thousand and $30 thousand were anticipated by more than 20% of college students

**13.** 14% of students anticipated a starting salary of $40 thousand.

**14.** $p = -0.01s^2 + 0.8s + 3.7$

$p = -0.01(40)^2 + 0.8(40) + 3.7$

$p = 19.7$

This is greater than the estimate of the previous question.

**15.** $2x - 5 = 7$

$\quad 2x = 12$

$\quad\quad x = 6$

The solution set is $\{6\}$.

This is a conditional equation.

**16.** $5x + 20 = 3x$

$\quad\quad 2x = -20$

$\quad\quad\; x = -10$

The solution set is $\{-10\}$.

This is a conditional equation.

**17.** $7(x - 4) = x + 2$

$7x - 28 = x + 2$

$\quad\quad 6x = 30$

$\quad\quad\; x = 5$

The solution set is $\{5\}$.

This is a conditional equation.

**18.** $1 - 2(6 - x) = 3x + 2$

$1 - 12 + 2x = 3x + 2$

$\quad -11 - x = 2$

$\quad\quad\; -x = 13$

$\quad\quad\quad x = -13$

The solution set is $\{-13\}$.

This is a conditional equation.

**19.** $2(x - 4) + 3(x + 5) = 2x - 2$

$2x - 8 + 3x + 15 = 2x - 2$

$\quad\quad 5x + 7 = 2x - 2$

$\quad\quad\quad 3x = -9$

$\quad\quad\quad\; x = -3$

The solution set is $\{-3\}$.

This is a conditional equation.

**20.** $2x - 4(5x + 1) = 3x + 17$
$2x - 20x - 4 = 3x + 17$
$-18x - 4 = 3x + 17$
$-21x = 21$
$x = -1$
The solution set is $\{-1\}$.
This is a conditional equation.

**21.** $7x + 5 = 5(x + 3) + 2x$
$7x + 5 = 5x + 15 + 2x$
$7x + 5 = 7x + 15$
$5 = 15$
The solution set is $\varnothing$.
This is an inconsistent equation.

**22.** $7x + 13 = 2(2x - 5) + 3x + 23$
$7x + 13 = 2(2x - 5) + 3x + 23$
$7x + 13 = 4x - 10 + 3x + 23$
$7x + 13 = 7x + 13$
$13 = 13$
The solution set is all real numbers.
This is an identity.

**23.** $\dfrac{2x}{3} = \dfrac{x}{6} + 1$
$2(2x) = x + 6$
$4x = x + 6$
$3x = 6$
$x = 2$
The solution set is $\{2\}$.
This is a conditional equation.

**24.** $\dfrac{x}{2} - \dfrac{1}{10} = \dfrac{x}{5} + \dfrac{1}{2}$
$5x - 1 = 2x + 5$
$3x = 6$
$x = 2$
The solution set is $\{2\}$.
This is a conditional equation.

**25.** $\dfrac{2x}{3} = 6 - \dfrac{x}{4}$
$4(2x) = 12(6) - 3x$
$8x = 72 - 3x$
$11x = 72$
$x = \dfrac{72}{11}$
The solution set is $\left\{\dfrac{72}{11}\right\}$.
This is a conditional equation.

**26.** $\dfrac{x}{4} = 2 - \dfrac{x-3}{3}$
$\dfrac{12 \cdot x}{4} = 12(2) - \dfrac{12(x-3)}{3}$
$3x = 24 - 4x + 12$
$7x = 36$
$x = \dfrac{36}{7}$
The solution set is $\left\{\dfrac{36}{7}\right\}$.
This is a conditional equation.

**27.** $\dfrac{3x+1}{3} - \dfrac{13}{2} = \dfrac{1-x}{4}$
$4(3x + 1) - 6(13) = 3(1 - x)$
$12x + 4 - 78 = 3 - 3x$
$12x - 74 = 3 - 3x$
$15x = 77$
$x = \dfrac{77}{15}$
The solution set is $\left\{\dfrac{77}{15}\right\}$.
This is a conditional equation.

**28.** $\dfrac{9}{4} - \dfrac{1}{2x} = \dfrac{4}{x}$
$9x - 2 = 16$
$9x = 18$
$x = 2$
The solution set is $\{2\}$.
This is a conditional equation.

**29.** $\dfrac{7}{x-5} + 2 = \dfrac{x+2}{x-5}$
$7 + 2(x - 5) = x + 2$
$7 + 2x - 10 = x + 2$
$2x - 3 = x + 2$
$x = 5$
5 does not check and must be rejected.
The solution set is the empty set, $\varnothing$.
This is an inconsistent equation.

**30.**
$$\frac{1}{x-1} - \frac{1}{x+1} = \frac{2}{x^2-1}$$
$$\frac{1}{x-1} - \frac{1}{x+1} = \frac{2}{(x+1)(x-1)}$$
$$x+1-(x-1) = 2$$
$$x+1-x+1 = 2$$
$$2 = 2$$

The solution set is all real numbers except $-1$ and $1$.
This is a conditional equation.

**31.**
$$\frac{5}{x+3} + \frac{1}{x-2} = \frac{8}{x^2+x-6}$$
$$\frac{5}{x+3} + \frac{1}{x-2} = \frac{8}{(x+3)(x-2)}$$
$$\frac{5(x+3)(x-2)}{x+3} + \frac{(x+3)(x-2)}{x-2} = \frac{8(x+3)(x-2)}{(x+3)(x-2)}$$
$$5(x-2)+1(x+3) = 8$$
$$5x-10+x+3 = 8$$
$$6x-7 = 8$$
$$6x = 15$$
$$x = \frac{15}{6}$$
$$x = \frac{5}{2}$$

The solution set is $\left\{\frac{5}{2}\right\}$.
This is a conditional equation.

**32.**
$$\frac{1}{x+5} = 0$$
$$(x+5)\frac{1}{x+5} = (x+5)(0)$$
$$1 = 0$$

The solution set is the empty set, $\varnothing$.
This is an inconsistent equation.

**33.**
$$\frac{4}{x+2} + \frac{3}{x} = \frac{10}{x^2+2x}$$
$$\frac{4}{x+2} + \frac{3}{x} = \frac{10}{x(x+2)}$$
$$\frac{4 \cdot x(x+2)}{x+2} + \frac{3 \cdot x(x+2)}{x} = \frac{10 \cdot x(x+2)}{x(x+2)}$$
$$4x+3(x+2) = 10$$
$$4x+3x+6 = 10$$
$$7x+6 = 10$$
$$7x = 4$$
$$x = \frac{4}{7}$$

The solution set is $\left\{\frac{4}{7}\right\}$.
This is a conditional equation.

**34.**
$$3-5(2x+1)-2(x-4) = 0$$
$$3-5(2x+1)-2(x-4) = 0$$
$$3-10x-5-2x+8 = 0$$
$$-12x+6 = 0$$
$$-12x = -6$$
$$x = \frac{-6}{-12}$$
$$x = \frac{1}{2}$$

The solution set is $\left\{\frac{1}{2}\right\}$.
This is a conditional equation.

**35.**
$$\frac{x+2}{x+3} + \frac{1}{x^2+2x-3} - 1 = 0$$
$$\frac{x+2}{x+3} + \frac{1}{(x+3)(x-1)} - 1 = 0$$
$$\frac{x+2}{x+3} + \frac{1}{(x+3)(x-1)} = 1$$
$$\frac{(x+2)(x+3)(x-1)}{x+3} + 1 = (x+3)(x-1)$$
$$(x+2)(x-1)+1 = (x+3)(x-1)$$
$$x^2+x-2+1 = x^2+2x-3$$
$$x-1 = 2x-3$$
$$-x = -2$$
$$x = 2$$

The solution set is $\{2\}$.
This is a conditional equation.

**36.** Let $x =$ the body count in *Scream*.
Let $x + 2 =$ the body count in *Scream 2*.
Let $x + 3 =$ the body count in *Scream 3*.
$$x + (x + 2) + (x + 3) + 10 = 33$$
$$x + x + 2 + x + 3 + 10 = 33$$
$$3x + 15 = 33$$
$$3x = 18$$
$$x = 6$$
$$x + 2 = 8$$
$$x + 3 = 9$$
The body count in *Scream*, *Scream 2*, and *Scream 3*, respectively, is 6, 8, and 9.

**37.** Let $x =$ the number of years after 1980.
$$2.69 + 0.15x = 8.69$$
$$0.15x = 6$$
$$x = 40$$
The average price of a movie ticket will be $8.69 40 years after 1980, or 2020.

**38.**
$$15 + .05x = 5 + .07x$$
$$10 = .02x$$
$$500 = x$$
Both plans cost the same at 500 text messages.

**39.** Let $x =$ the original price of the phone
$$48 = x - 0.20x$$
$$48 = 0.80x$$
$$60 = x$$
The original price is $60.

**40.** Let $x =$ the amount sold to earn $800 in one week
$$800 = 300 + 0.05x$$
$$500 = 0.05x$$
$$10,000 = x$$
Sales must be $10,000 in one week to earn $800.

**41.** Let $x =$ the amount invested at 4%
Let $y =$ the amount invested at 7%
$$x + \quad y = 9000$$
$$0.04x + 0.07y = 555$$
Multiply the first equation by –0.04 and add.
$$-0.04x - 0.04y = -360$$
$$\underline{0.04x + 0.07y = 555}$$
$$0.03y = 195$$
$$y = 6500$$

Back-substitute 6500 for $y$ in one of the original equations to find $x$.
$$x + y = 9000$$
$$x + 6500 = 9000$$
$$x = 2500$$
There was $2500 invested at 4% and $6500 invested at 7%.

**42.** Let $x =$ the amount invested at 2%
Let $8000 - x =$ the amount invested at 5%.
$$0.05(8000 - x) = 0.02x + 85$$
$$400 - 0.05x = 0.02x + 85$$
$$-0.05x - 0.02x = 85 - 400$$
$$-0.07x = -315$$
$$\frac{-0.07x}{-0.07} = \frac{-315}{-0.07}$$
$$x = 4500$$
$$8000 - x = 3500$$
$4500 was invested at 2% and $3500 was invested at 5%.

**43.** Let $w =$ the width of the playing field,
Let $3w - 6 =$ the length of the playing field
$$P = 2(\text{length}) + 2(\text{width})$$
$$340 = 2(3w - 6) + 2w$$
$$340 = 6w - 12 + 2w$$
$$340 = 8w - 12$$
$$352 = 8w$$
$$44 = w$$
The dimensions are 44 yards by 126 yards.

**44. a.** Let $x =$ the number of years after 2010.
College A's enrollment: $14,100 + 1500x$
College B's enrollment: $41,700 - 800x$
$$14,100 + 1500x = 41,700 - 800x$$

**b.** Check some points to determine that
$y_1 = 14,100 + 1500x$ and
$y_2 = 41,700 - 800x$. Since
$y_1 = y_2 = 32,100$ when $x = 12$, the two colleges will have the same enrollment in the year $2010 + 12 = 2022$. That year the enrollments will be 32,100 students.

**45.**
$$vt + gt^2 = s$$
$$gt^2 = s - vt$$
$$\frac{gt^2}{t^2} = \frac{s - vt}{t^2}$$
$$g = \frac{s - vt}{t^2}$$

**46.**
$$T = gr + gvt$$
$$T = g(r + vt)$$
$$\frac{T}{r + vt} = \frac{g(r + vt)}{r + vt}$$
$$\frac{T}{r + vt} = g$$
$$g = \frac{T}{r + vt}$$

**47.**
$$T = \frac{A - P}{\text{Pr}}$$
$$Pr(T) = Pr\frac{A - P}{\text{Pr}}$$
$$PrT = A - P$$
$$PrT + P = A$$
$$P(rT + 1) = A$$
$$P = \frac{A}{1 + rT}$$

**48.** $(8 - 3i) - (17 - 7i) = 8 - 3i - 17 + 7i$
$$= -9 + 4i$$

**49.** $4i(3i - 2) = (4i)(3i) + (4i)(-2)$
$$= 12i^2 - 8i$$
$$= -12 - 8i$$

**50.** $(7 - i)(2 + 3i)$
$$= 7 \cdot 2 + 7(3i) + (-i)(2) + (-i)(3i)$$
$$= 14 + 21i - 2i + 3$$
$$= 17 + 19i$$

**51.** $(3 - 4i)^2 = 3^2 + 2 \cdot 3(-4i) + (-4i)^2$
$$= 9 - 24i - 16$$
$$= -7 - 24i$$

**52.** $(7 + 8i)(7 - 8i) = 7^2 + 8^2 = 49 + 64 = 113$

**53.** $\dfrac{6}{5 + i} = \dfrac{6}{5 + i} \cdot \dfrac{5 - i}{5 - i}$
$$= \frac{30 - 6i}{25 + 1}$$
$$= \frac{30 - 6i}{26}$$
$$= \frac{15 - 3i}{13}$$
$$= \frac{15}{13} - \frac{3}{13}i$$

**54.** $\dfrac{3 + 4i}{4 - 2i} = \dfrac{3 + 4i}{4 - 2i} \cdot \dfrac{4 + 2i}{4 + 2i}$
$$= \frac{12 + 6i + 16i + 8i^2}{16 - 4i^2}$$
$$= \frac{12 + 22i - 8}{16 + 4}$$
$$= \frac{4 + 22i}{20}$$
$$= \frac{1}{5} + \frac{11}{10}i$$

**55.** $\sqrt{-32} - \sqrt{-18} = i\sqrt{32} - i\sqrt{18}$
$$= i\sqrt{16 \cdot 2} - i\sqrt{9 \cdot 2}$$
$$= 4i\sqrt{2} - 3i\sqrt{2}$$
$$= (4i - 3i)\sqrt{2}$$
$$= i\sqrt{2}$$

**56.** $(-2 + \sqrt{-100})^2 = (-2 + i\sqrt{100})^2$
$$= (-2 + 10i)^2$$
$$= 4 - 40i + (10i)^2$$
$$= 4 - 40i - 100$$
$$= -96 - 40i$$

**57.** $\dfrac{4 + \sqrt{-8}}{2} = \dfrac{4 + i\sqrt{8}}{2} = \dfrac{4 + 2i\sqrt{2}}{2} = 2 + i\sqrt{2}$

**58.**
$$2x^2 + 15x = 8$$
$$2x^2 + 15x - 8 = 0$$
$$(2x - 1)(x + 8) = 0$$
$$2x - 1 = 0 \quad x + 8 = 0$$
$$x = \frac{1}{2} \text{ or } x = -8$$
The solution set is $\left\{\dfrac{1}{2}, -8\right\}$.

**59.** $5x^2 + 20x = 0$

$5x(x+4) = 0$

$5x = 0 \quad x + 4 = 0$

$x = 0$ or $x = -4$

The solution set is $\{0, -4\}$.

**60.** $2x^2 - 3 = 125$

$2x^2 = 128$

$x^2 = 64$

$x = \pm 8$

The solution set is $\{8, -8\}$.

**61.** $\dfrac{x^2}{2} + 5 = -3$

$\dfrac{x^2}{2} = -8$

$x^2 = -16$

$\sqrt{x^2} = \pm\sqrt{-16}$

$x = \pm 4i$

**62.** $(x+3)^2 = -10$

$\sqrt{(x+3)^2} = \pm\sqrt{-10}$

$x + 3 = \pm i\sqrt{10}$

$x = -3 \pm i\sqrt{10}$

**63.** $(3x-4)^2 = 18$

$\sqrt{(3x-4)^2} = \pm\sqrt{18}$

$3x - 4 = \pm 3\sqrt{2}$

$3x = 4 \pm 3\sqrt{2}$

$\dfrac{3x}{3} = \dfrac{4 \pm 3\sqrt{2}}{3}$

$x = \dfrac{4 \pm 3\sqrt{2}}{3}$

**64.** $x^2 + 20x$

$\left(\dfrac{20}{2}\right)^2 = 10^2 = 100$

$x^2 + 20x + 100 = (x+10)^2$

**65.** $x^2 - 3x$

$\left(\dfrac{3}{2}\right)^2 = \dfrac{9}{4}$

$x^2 - 3x + \dfrac{9}{4} = \left(x - \dfrac{3}{2}\right)^2$

**66.** $x^2 - 12x = -27$

$x^2 - 12x + 36 = -27 + 36$

$(x-6)^2 = 9$

$x - 6 = \pm 3$

$x = 6 \pm 3$

$x = 9, 3$

The solution set is $\{9, 3\}$.

**67.** $3x^2 - 12x + 11 = 0$

$x^2 - 4x = -\dfrac{11}{3}$

$x^2 - 4x + 4 = -\dfrac{11}{3} + 4$

$(x-2)^2 = \dfrac{1}{3}$

$x - 2 = \pm\sqrt{\dfrac{1}{3}}$

$x = 2 \pm \dfrac{\sqrt{3}}{3}$

The solution set is $\left\{2 + \dfrac{\sqrt{3}}{3}, 2 - \dfrac{\sqrt{3}}{3}\right\}$.

**68.** $x^2 = 2x + 4$

$x^2 - 2x - 4 = 0$

$x = \dfrac{2 \pm \sqrt{(-2)^2 - 4(1)(-4)}}{2(1)}$

$x = \dfrac{2 \pm \sqrt{4 + 16}}{2}$

$x = \dfrac{2 \pm \sqrt{20}}{2}$

$x = \dfrac{2 \pm 2\sqrt{5}}{2}$

$x = 1 \pm \sqrt{5}$

The solution set is $\left\{1 + \sqrt{5}, 1 - \sqrt{5}\right\}$.

**69.** $x^2 - 2x + 19 = 0$

$$x = \frac{2 \pm \sqrt{(-2)^2 - 4(1)(19)}}{2(1)}$$

$$x = \frac{2 \pm \sqrt{4 - 76}}{2}$$

$$x = \frac{2 \pm \sqrt{-72}}{2}$$

$$x = \frac{2 \pm 6i\sqrt{2}}{2}$$

$$x = 1 \pm 3i\sqrt{2}$$

The solution set is $\left\{ 1 + 3i\sqrt{2}, 1 - 3i\sqrt{2} \right\}$.

**70.** $2x^2 = 3 - 4x$

$2x^2 + 4x - 3 = 0$

$$x = \frac{-4 \pm \sqrt{4^2 - 4(2)(-3)}}{2(2)}$$

$$x = \frac{-4 \pm \sqrt{16 + 24}}{4}$$

$$x = \frac{-4 \pm \sqrt{40}}{4}$$

$$x = \frac{-4 \pm 2\sqrt{10}}{4}$$

$$x = \frac{-2 \pm \sqrt{10}}{2}$$

The solution set is $\left\{ \dfrac{-2 + \sqrt{10}}{2}, \dfrac{-2 - \sqrt{10}}{2} \right\}$.

**71.** $x^2 - 4x + 13 = 0$

$(-4)^2 - 4(1)(13)$

$= 16 - 52$

$= -36$; 2 complex imaginary solutions

**72.** $9x^2 = 2 - 3x$

$9x^2 + 3x - 2 = 0$

$3^2 - 4(9)(-2)$

$= 9 + 72$

$= 81$; 2 unequal real solutions

**73.** $2x^2 - 11x + 5 = 0$

$(2x - 1)(x - 5) = 0$

$2x - 1 = 0 \quad x - 5 = 0$

$x = \dfrac{1}{2}$ or $x = 5$

The solution set is $\left\{ 5, \dfrac{1}{2} \right\}$.

**74.** $\quad (3x + 5)(x - 3) = 5$

$3x^2 + 5x - 9x - 15 = 5$

$\quad 3x^2 - 4x - 20 = 0$

$$x = \frac{4 \pm \sqrt{(-4)^2 - 4(3)(-20)}}{2(3)}$$

$$x = \frac{4 \pm \sqrt{16 + 240}}{6}$$

$$x = \frac{4 \pm \sqrt{256}}{6}$$

$$x = \frac{4 \pm 16}{6}$$

$$x = \frac{20}{6}, \frac{-12}{6}$$

$$x = \frac{10}{3}, -2$$

The solution set is $\left\{ -2, \dfrac{10}{3} \right\}$.

**75.** $3x^2 - 7x + 1 = 0$

$$x = \frac{7 \pm \sqrt{(-7)^2 - 4(3)(1)}}{2(3)}$$

$$x = \frac{7 \pm \sqrt{49 - 12}}{6}$$

$$x = \frac{7 \pm \sqrt{37}}{6}$$

The solution set is $\left\{ \dfrac{7 + \sqrt{37}}{6}, \dfrac{7 - \sqrt{37}}{6} \right\}$.

**76.** $x^2 - 9 = 0$

$x^2 = 9$

$x = \pm 3$

The solution set is $\{-3, 3\}$.

**77.** $(x-3)^2 - 25 = 0$

$$(x-3)^2 = 25$$

$$x-3 = \pm 5$$

$$x = 3 \pm 5$$

$$x = 8, -2$$

The solution set is $\{8, -2\}$.

**78.** $3x^2 - x + 2 = 0$

$$x = \frac{1 \pm \sqrt{(-1)^2 - 4(3)(2)}}{2(3)}$$

$$x = \frac{1 \pm \sqrt{1 - 24}}{6}$$

$$x = \frac{1 \pm \sqrt{-23}}{6}$$

$$x = \frac{1 \pm i\sqrt{23}}{6}$$

The solution set is $\left\{ \dfrac{1+i\sqrt{23}}{6}, \dfrac{1-i\sqrt{23}}{6} \right\}$.

**79.** $3x^2 - 10x = 8$

$$3x^2 - 10x - 8 = 0$$

$$(3x+2)(x-4) = 0$$

$$3x+2 = 0 \quad \text{or} \quad x-4 = 0$$

$$3x = -2 \qquad\qquad x = 4$$

$$x = -\frac{2}{3}$$

The solution set is $\left\{ -\dfrac{2}{3}, 4 \right\}$.

**80.** $(x+2)^2 + 4 = 0$

$$(x+2)^2 = -4$$

$$\sqrt{(x+2)^2} = \pm\sqrt{-4}$$

$$x+2 = \pm 2i$$

$$x = -2 \pm 2i$$

The solution set is $\{-2+2i, -2-2i\}$.

**81.**
$$\frac{5}{x+1} + \frac{x-1}{4} = 2$$

$$\frac{5 \cdot 4(x+1)}{x+1} + \frac{(x-1) \cdot 4(x+1)}{4} = 2 \cdot 4(x+1)$$

$$20 + (x-1)(x+1) = 8(x+1)$$

$$20 + x^2 - 1 = 8x + 8$$

$$x^2 - 8x - 11 = 0$$

$$x = \frac{-b \pm \sqrt{b^2 - 4ac}}{2a}$$

$$x = \frac{-(-8) \pm \sqrt{(-8)^2 - 4(1)(11)}}{2(1)}$$

$$x = \frac{8 \pm \sqrt{20}}{2}$$

$$x = \frac{8 \pm 2\sqrt{5}}{2}$$

$$x = 4 \pm \sqrt{5}$$

The solution set is $\left\{ 4+\sqrt{5}, 4-\sqrt{5} \right\}$.

**82.** $W(t) = 3t^2$

$$588 = 3t^2$$

$$196 = t^2$$

Apply the square root property.

$$t^2 = 196$$

$$t = \pm\sqrt{196}$$

$$t = \pm 14$$

The solutions are $-14$ and $14$. We disregard $-14$, because we cannot have a negative time measurement. The fetus will weigh 588 grams after 14 weeks.

**83.**  **a.**  2011 is 8 years after 2003.

$$B = 1.7x^2 + 6x + 26$$

$$B = 1.7(8)^2 + 6(8) + 26$$

$$= 182.8$$

$$\approx 183$$

According to the function, in 2011 there were 183 "Bicycle Friendly" communities. This overestimates the number shown in the graph by 3.

**b.**
$$B = 1.7x^2 + 6x + 26$$
$$826 = 1.7x^2 + 6x + 26$$
$$0 = 1.7x^2 + 6x - 800$$
$$x = \frac{-b \pm \sqrt{b^2 - 4ac}}{2a}$$
$$x = \frac{-(6) \pm \sqrt{(6)^2 - 4(1.7)(-800)}}{2(1.7)}$$
$$= \frac{-6 \pm \sqrt{36 + 5440}}{2(1.7)}$$
$$= \frac{-6 \pm \sqrt{5476}}{3.4}$$
$$= \frac{-6 \pm 74}{3.4}$$
$$x = 20 \ \text{ or } \ -23\frac{9}{17}$$

According to the function, there will be 826 "Bicycle Friendly" communities 20 years after 2003, or 2023.

**84.**
$$A = lw$$
$$15 = l(2l - 7)$$
$$15 = 2l^2 - 7l$$
$$0 = 2l^2 - 7l - 15$$
$$0 = (2l + 3)(l - 5)$$
$$l = 5$$
$$2l - 7 = 3$$
The length is 5 yards, the width is 3 yards.

**85.** Let $x$ = height of building
$2x$ = shadow height
$$x^2 + (2x)^2 = 300^2$$
$$x^2 + 4x^2 = 90,000$$
$$5x^2 = 90,000$$
$$x^2 = 18,000$$
$$x \approx \pm 134.164$$
Discard negative height.
The building is approximately 134 meters high.

**86.**
$$2x^4 = 50x^2$$
$$2x^4 - 50x^2 = 0$$
$$2x^2(x^2 - 25) = 0$$
$$x = 0$$
$$x = \pm 5$$
The solution set is $\{-5, 0, 5\}$.

**87.**
$$2x^3 - x^2 - 18x + 9 = 0$$
$$x^2(2x - 1) - 9(2x - 1) = 0$$
$$(x^2 - 9)(2x - 1) = 0$$
$$x = \pm 3, \ x = \frac{1}{2}$$
The solution set is $\left\{-3, \frac{1}{2}, 3\right\}$.

**88.**
$$\sqrt{2x - 3} + x = 3$$
$$\sqrt{2x - 3} = 3 - x$$
$$2x - 3 = 9 - 6x + x^2$$
$$x^2 - 8x + 12 = 0$$
$$x^2 - 8x = -12$$
$$x^2 - 8x + 16 = -12 + 16$$
$$(x - 4)^2 = 4$$
$$x - 4 = \pm 2$$
$$x = 4 + 2$$
$$x = 6, 2$$
The solution set is $\{2\}$.

**89.**
$$\sqrt{x - 4} + \sqrt{x + 1} = 5$$
$$\sqrt{x - 4} = 5 - \sqrt{x + 1}$$
$$x - 4 = 25 - 10\sqrt{x + 1} + (x + 1)$$
$$x - 4 = 26 + x - 10\sqrt{x + 1}$$
$$-30 = -10\sqrt{x + 1}$$
$$3 = \sqrt{x + 1}$$
$$9 = x + 1$$
$$x = 8$$
The solution set is $\{8\}$.

**90.**
$$3x^{\frac{3}{4}} - 24 = 0$$
$$3x^{\frac{3}{4}} = 24$$
$$x^{\frac{3}{4}} = 8$$
$$\left(x^{\frac{3}{4}}\right)^{\frac{4}{3}} = (8)^{\frac{4}{3}}$$
$$x = 16$$
The solution set is $\{16\}$.

**91.** $(x-7)^{\frac{2}{3}} = 25$

$$\left[(x-7)^{\frac{2}{3}}\right]^{\frac{3}{2}} = 25^{\frac{3}{2}}$$

$$x - 7 = \left(5^2\right)^{\frac{3}{2}}$$

$$x - 7 = 5^3$$

$$x - 7 = 125$$

$$x = 132$$

The solution set is $\{132\}$.

**92.** $x^4 - 5x^2 + 4 = 0$

Let $t = x^2$

$t^2 - 5t + 4 = 0$

$t = 4$    or    $t = 1$

$x^2 = 4$        $x^2 = 1$

$x = \pm 2$       $x = \pm 1$

The solution set is $\{-2, -1, 1, 2\}$.

**93.** $x^{1/2} + 3x^{1/4} - 10 = 0$

Let $t = x^{1/4}$

$t^2 + 3t - 10 = 0$

$(t+5)(t-2) = 0$

$t = -5$    or    $t = 2$

$x^{\frac{1}{4}} = -5$       $x^{\frac{1}{4}} = 2$

$\left(x^{\frac{1}{4}}\right)^4 = (-5)^4$    $\left(x^{\frac{1}{4}}\right)^4 = (2)^4$

$x = 625$        $x = 16$

625 does not check and must be rejected.
The solution set is $\{16\}$.

**94.** $|2x+1| = 7$

$2x + 1 = 7$   or   $2x + 1 = -7$

$2x = 6$         $2x = -8$

$x = 3$          $x = -8$

The solution set is $\{-4, 3\}$.

**95.** $2|x-3| - 6 = 10$

$2|x-3| = 16$

$|x-3| = 8$

$x - 3 = 8$   or   $x - 3 = -8$

$x = 11$        $x = -5$

The solution set is $\{-5, 11\}$.

**96.** $3x^{4/3} - 5x^{2/3} + 2 = 0$

Let $t = x^{\frac{2}{3}}$.

$3t^2 - 5t + 2 = 0$

$(3t-2)(t-1) = 0$

$3t - 2 = 0$     or     $t - 1 = 0$

$3t = 2$              $t = 1$

$t = \dfrac{2}{3}$          $x^{\frac{2}{3}} = 1$

$x^{\frac{2}{3}} = \dfrac{2}{3}$      $\left(x^{\frac{2}{3}}\right)^{\frac{3}{2}} = \pm(1)^{\frac{3}{2}}$

$\left(x^{\frac{2}{3}}\right)^{\frac{3}{2}} = \pm\left(\dfrac{2}{3}\right)^{\frac{3}{2}}$     $x = \pm 1$

$x = \pm\sqrt[2]{\left(\dfrac{2}{3}\right)^3}$

$x = \pm\dfrac{2}{3}\sqrt{\dfrac{2}{3}}$

$x = \pm\dfrac{2}{3} \cdot \dfrac{\sqrt{2}}{\sqrt{3}} \cdot \dfrac{\sqrt{3}}{\sqrt{3}}$

$x = \pm\dfrac{2\sqrt{6}}{9}$

The solution set is $\left\{-\dfrac{2\sqrt{6}}{9}, \dfrac{2\sqrt{6}}{9}, -1, 1\right\}$.

**97.** $2\sqrt{x-1} = x$

$4(x-1) = x^2$

$4x - 4 = x^2$

$x^2 - 4x + 4 = 0$

$(x-2)^2 = 0$

$x = 2$

The solution set is $\{2\}$.

**98.** $|2x-5| - 3 = 0$

$2x - 5 = 3$   or   $2x - 5 = -3$

$2x = 8$         $2x = 2$

$x = 4$          $x = 1$

The solution set is $\{4, 1\}$.

**99.** $x^3 + 2x^2 - 9x - 18 = 0$

$x^2(x+2) - 9(x+2) = 0$

$(x+2)(x^2-9) = 0$

$(x+2)(x+3)(x-3) = 0$

The solution set is $\{-3, -2, 3\}$.

**100.** $\sqrt{8-2x} - x = 0$

$\sqrt{8-2x} = x$

$\left(\sqrt{8-2x}\right)^2 = (x)^2$

$8 - 2x = x^2$

$0 = x^2 + 2x - 8$

$0 = (x+4)(x-2)$

$x+4 = 0 \quad$ or $\quad x-2 = 0$

$x = -4 \qquad\qquad x = 2$

–4 does not check.

The solution set is $\{2\}$.

**101.** $x^3 + 3x^2 - 2x - 6 = 0$

$x^2(x+3) - 2(x+3) = 0$

$(x+3)(x^2-2) = 0$

$x+3 = 0 \quad$ or $\quad x^2 - 2 = 0$

$x = -3 \qquad\qquad x^2 = 2$

$\qquad\qquad\qquad\qquad x = \pm\sqrt{2}$

The solution set is $\left\{-3, -\sqrt{2}, \sqrt{2}\right\}$.

**102.** $-4|x+1| + 12 = 0$

$-4|x+1| = -12$

$|x+1| = 3$

$x+1 = 3 \quad$ or $\quad x+1 = -3$

$x = 2 \qquad\qquad x = -4$

The solution set is $\{-4, 2\}$.

**103.** $\quad p = -2.5\sqrt{t} + 17$

$\quad 7 = -2.5\sqrt{t} + 17$

$-10 = -2.5\sqrt{t}$

$\quad 4 = \sqrt{t}$

$\quad 16 = t$

The percentage will drop to 7% 16 years after 1993, or 2009.

**104.** $\left\{x \mid -3 \le x < 5\right\}$

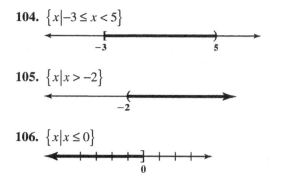

**105.** $\left\{x \mid x > -2\right\}$

**106.** $\left\{x \mid x \le 0\right\}$

**107.** Graph $(-2,1]$:

Graph $[-1,3)$:

To find the intersection, take the portion of the number line that the two graphs have in common.

Numbers in both
$(-2,1]$ and $[-1,3)$:

Thus, $(-2,1] \cap [-1,3) = [-1,1]$.

**108.** Graph $(-2,1]$:

Graph $[-1,3)$:

To find the union, take the portion of the number line representing the total collection of numbers in the two graphs.

Numbers in either $(-2,1]$
or $[-1,3)$ or both:

Thus, $(-2,1] \cup [-1,3) = (-2,3)$.

**109.** Graph $[1,3)$:

Graph $(0,4)$:

To find the intersection, take the portion of the number line that the two graphs have in common.

Numbers in both
$[1,3)$ and $(0,4)$:

Thus, $[1,3) \cap (0,4) = [1,3)$.

**110.** Graph $[1,3)$:

Graph $(0,4)$:

To find the union, take the portion of the number line representing the total collection of numbers in the two graphs.

Numbers in either $[1,3)$
or $(0,4)$ or both:

Thus, $[1,3) \cup (0,4) = (0,4)$.

**111.** $-6x + 3 \le 15$
$-6x \le 12$
$x \ge 2$

The solution set is $[-2, \infty)$.

**112.** $6x - 9 \geq -4x - 3$

$\qquad 10x \geq 6$

$\qquad x \geq \dfrac{3}{5}$

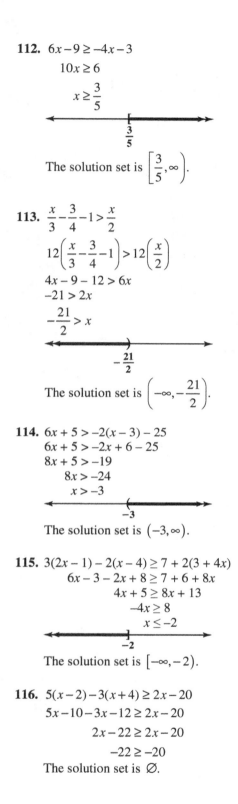

The solution set is $\left[\dfrac{3}{5}, \infty\right)$.

**113.** $\dfrac{x}{3} - \dfrac{3}{4} - 1 > \dfrac{x}{2}$

$12\left(\dfrac{x}{3} - \dfrac{3}{4} - 1\right) > 12\left(\dfrac{x}{2}\right)$

$4x - 9 - 12 > 6x$

$-21 > 2x$

$-\dfrac{21}{2} > x$

The solution set is $\left(-\infty, -\dfrac{21}{2}\right)$.

**114.** $6x + 5 > -2(x - 3) - 25$

$\qquad 6x + 5 > -2x + 6 - 25$

$\qquad 8x + 5 > -19$

$\qquad 8x > -24$

$\qquad x > -3$

The solution set is $(-3, \infty)$.

**115.** $3(2x - 1) - 2(x - 4) \geq 7 + 2(3 + 4x)$

$\qquad 6x - 3 - 2x + 8 \geq 7 + 6 + 8x$

$\qquad 4x + 5 \geq 8x + 13$

$\qquad -4x \geq 8$

$\qquad x \leq -2$

The solution set is $[-\infty, -2)$.

**116.** $5(x - 2) - 3(x + 4) \geq 2x - 20$

$\qquad 5x - 10 - 3x - 12 \geq 2x - 20$

$\qquad 2x - 22 \geq 2x - 20$

$\qquad -22 \geq -20$

The solution set is $\varnothing$.

**117.** $7 < 2x + 3 \leq 9$

$\qquad 4 < 2x \leq 6$

$\qquad 2 < x \leq 3$

$\qquad (2, 3]$

The solution set is $[2, 3)$.

**118.** $|2x + 3| \leq 15$

$\qquad -15 \leq 2x + 3 \leq 15$

$\qquad -18 \leq 2x \leq 12$

$\qquad -9 \leq x \leq 6$

The solution set is $[-9, 6]$.

**119.** $\left|\dfrac{2x + 6}{3}\right| > 2$

$\dfrac{2x + 6}{3} > 2 \qquad \dfrac{2x + 6}{3} < -2$

$2x + 6 > 6 \qquad 2x + 6 < -6$

$\quad 2x > 0 \qquad\qquad 2x < -12$

$\quad\ x > 0 \qquad\qquad\ x < -6$

The solution set is $(-\infty, -6)$ or $(0, \infty)$.

**120.** $|2x + 5| - 7 \geq -6$

$\qquad |2x + 5| \geq 1$

$2x + 5 \geq 1$ or $2x + 5 \leq -1$

$\quad 2x \geq -4 \qquad\quad 2x \leq -6$

$\quad\ x \geq -2 \ $ or $\ x \leq -3$

The solution set is $(-\infty, -3]$ or $[-2, \infty)$.

**121.** $-4|x + 2| + 5 \leq -7$

$\qquad -4|x + 2| \leq -12$

$\qquad |x + 2| \geq 3$

$x + 2 \geq 3 \qquad x + 2 \leq -3$

$\qquad\qquad$ or

$\ \ x \geq 1 \qquad\qquad x \leq -5$

The solution set is $(-\infty, -5] \cup [1, \infty)$.

**122.**
$$y_1 > y_2$$
$$-10-3(2x+1) > 8x+1$$
$$-10-6x-3 > 8x+1$$
$$-6x-13 > 8x+1$$
$$-14x > 14$$
$$\frac{-14x}{-14} < \frac{14}{-14}$$
$$x < -1$$
The solution set is $(-\infty, -1)$.

**123.** $3-|2x-5| \geq -6$
$$-|2x-5| \geq -9$$
$$\frac{-|2x-5|}{-1} \leq \frac{-9}{-1}$$
$$|2x-5| \leq 9$$
$$-9 \leq 2x-5 \leq 9$$
$$-4 \leq 2x \leq 14$$
$$-2 \leq x \leq 7$$
The solution set is $[-2, 7]$.

**124.** $0.20x + 24 \leq 40$
$$0.20x \leq 16$$
$$\frac{0.20x}{0.20} \leq \frac{16}{0.20}$$
$$x \leq 80$$
A customer can drive no more than 80 miles.

**125.** $80 \leq \dfrac{95+79+91+86+x}{5} < 90$
$$400 \leq 95+79+91+86+x < 450$$
$$400 \leq 351+x < 450$$
$$49 \leq x < 99$$
A grade of at least 49% but less than 99% will result in a B.

**126.** $0.075x \geq 9000$
$$\frac{0.075x}{0.075} \geq \frac{9000}{0.075}$$
$$x \geq 120,000$$
The investment must be at least $120,000.

## Chapter 1 Test

**1.** $7(x-2) = 4(x+1)-21$
$$7x-14 = 4x+4-21$$
$$7x-14 = 4x-17$$
$$3x = -3$$
$$x = -1$$
The solution set is $\{-1\}$.

**2.** $-10-3(2x+1)-8x-1 = 0$
$$-10-6x-3-8x-1 = 0$$
$$-14x-14 = 0$$
$$-14x = 14$$
$$x = -1$$
The solution set is $\{-1\}$.

**3.** $\dfrac{2x-3}{4} = \dfrac{x-4}{2} - \dfrac{x+1}{4}$
$$2x-3 = 2(x-4)-(x+1)$$
$$2x-3 = 2x-8-x-1$$
$$2x-3 = x-9$$
$$x = -6$$
The solution set is $\{-6\}$.

**4.** $\dfrac{2}{x-3} - \dfrac{4}{x+3} = \dfrac{8}{(x-3)(x+3)}$
$$2(x+3)-4(x-3) = 8$$
$$2x+6-4x+12 = 8$$
$$-2x+18 = 8$$
$$-2x = -10$$
$$x = 5$$
The solution set is $\{5\}$.

**5.** $2x^2 - 3x - 2 = 0$
$(2x+1)(x-2) = 0$
$2x+1 = 0$ or $x-2 = 0$
$$x = -\frac{1}{2} \quad \text{or} \quad x = 2$$
The solution set is $\left\{-\dfrac{1}{2}, 2\right\}$.

**6.** $(3x-1)^2 = 75$
$$3x-1 = \pm\sqrt{75}$$
$$3x = 1 \pm 5\sqrt{3}$$
$$x = \frac{1 \pm 5\sqrt{3}}{3}$$
The solution set is $\left\{\dfrac{1-5\sqrt{3}}{3}, \dfrac{1+5\sqrt{3}}{3}\right\}$.

**7.** $(x+3)^2 + 25 = 0$

$(x+3)^2 = -25$

$x+3 = \pm\sqrt{-25}$

$x = -3 \pm 5i$

The solution set is $\{-3+5i, -3-5i\}$.

**8.** $x(x-2) = 4$

$x^2 - 2x - 4 = 0$

$x = \dfrac{-b \pm \sqrt{b^2 - 4ac}}{2a}$

$x = \dfrac{2 \pm \sqrt{(-2)^2 - 4(1)(-4)}}{2}$

$x = \dfrac{2 \pm 2\sqrt{5}}{2}$

$x = 1 \pm \sqrt{5}$

The solution set is $\{1-\sqrt{5}, 1+\sqrt{5}\}$.

**9.** $4x^2 = 8x - 5$

$4x^2 - 8x + 5 = 0$

$x = \dfrac{-b \pm \sqrt{b^2 - 4ac}}{2a}$

$x = \dfrac{8 \pm \sqrt{(-8)^2 - 4(4)(5)}}{2(4)}$

$x = \dfrac{8 \pm \sqrt{-16}}{8}$

$x = \dfrac{8 \pm 4i}{8}$

$x = 1 \pm \dfrac{1}{2}i$

The solution set is $\left\{1+\dfrac{1}{2}i, 1-\dfrac{1}{2}i\right\}$.

**10.** $x^3 - 4x^2 - x + 4 = 0$

$x^2(x-4) - 1(x-4) = 0$

$(x^2 - 1)(x-4) = 0$

$(x-1)(x+1)(x-4) = 0$

$x = 1$ or $x = -1$ or $x = 4$

The solution set is $\{-1, 1, 4\}$.

**11.** $\sqrt{x-3} + 5 = x$

$\sqrt{x-3} = x - 5$

$x - 3 = x^2 - 10x + 25$

$x^2 - 11x + 28 = 0$

$x = \dfrac{11 \pm \sqrt{11^2 - 4(1)(28)}}{2(1)}$

$x = \dfrac{11 \pm \sqrt{121 - 112}}{2}$

$x = \dfrac{11 \pm \sqrt{9}}{2}$

$x = \dfrac{11 \pm 3}{2}$

$x = 7$ or $x = 4$

4 does not check and must be rejected.
The solution set is $\{7\}$.

**12.** $\sqrt{8-2x} - x = 0$

$\sqrt{8-2x} = x$

$\left(\sqrt{8-2x}\right)^2 = (x)^2$

$8 - 2x = x^2$

$0 = x^2 + 2x - 8$

$0 = (x+4)(x-2)$

$x+4 = 0$ or $x-2 = 0$

$x = -4$ $\quad$ $x = 2$

−4 does not check and must be rejected.
The solution set is $\{2\}$.

**13.** $\sqrt{x+4} + \sqrt{x-1} = 5$

$\sqrt{x+4} = 5 - \sqrt{x-1}$

$x+4 = 25 - 10\sqrt{x-1} + (x-1)$

$x+4 = 25 - 10\sqrt{x-1} + x - 1$

$-20 = -10\sqrt{x-1}$

$2 = \sqrt{x-1}$

$4 = x - 1$

$x = 5$

The solution set is $\{5\}$.

**14.** $5x^{3/2} - 10 = 0$

$5x^{3/2} = 10$

$x^{3/2} = 2$

$x = 2^{2/3}$

$x = \sqrt[3]{4}$

The solution set is $\left\{\sqrt[3]{4}\right\}$.

**15.** $x^{2/3} - 9x^{1/3} + 8 = 0$ let $t = x^{1/3}$

$\quad\quad t^2 - 9t + 8 = 0$

$\quad\quad (t-1)(t-8) = 0$

$\quad\quad\quad t = 1 \quad\quad t = 8$

$\quad\quad x^{1/3} = 1 \quad x^{1/3} = 8$

$\quad\quad\quad x = 1 \quad\quad x = 512$

The solution set is $\{1, 512\}$.

**16.** $\left|\dfrac{2}{3}x - 6\right| = 2$

$\quad \dfrac{2}{3}x - 6 = 2 \quad\quad \dfrac{2}{3}x - 6 = -2$

$\quad\quad \dfrac{2}{3}x = 8 \quad\quad\quad \dfrac{2}{3}x = 4$

$\quad\quad\quad x = 12 \quad\quad\quad\quad x = 6$

The solution set is $\{6, 12\}$.

**17.** $-3|4x - 7| + 15 = 0$

$\quad -3|4x - 7| = -15$

$\quad\quad |4x - 7| = 5$

$\quad 4x - 7 = 5 \quad\quad\quad 4x - 7 = -5$

$\quad\quad\quad\quad\quad\quad \text{or}$

$\quad\quad 4x = 12 \quad\quad\quad\quad 4x = 2$

$\quad\quad\quad x = 3 \quad\quad\quad\quad\quad x = \dfrac{1}{2}$

The solution set is $\left\{\dfrac{1}{2}, 3\right\}$.

**18.** $\dfrac{1}{x^2} - \dfrac{4}{x} + 1 = 0$

$\quad \dfrac{x^2}{x^2} - \dfrac{4x^2}{x} + x^2 = 0$

$\quad\quad 1 - 4x + x^2 = 0$

$\quad\quad\quad x^2 - 4x + 1 = 0$

$\quad x = \dfrac{-b \pm \sqrt{b^2 - 4ac}}{2a}$

$\quad x = \dfrac{-(-4) \pm \sqrt{(-4)^2 - 4(1)(1)}}{2(1)}$

$\quad x = \dfrac{4 \pm \sqrt{12}}{2}$

$\quad x = \dfrac{4 \pm 2\sqrt{3}}{2}$

$\quad x = 2 \pm \sqrt{3}$

The solution set is $\left\{2 + \sqrt{3}, 2 - \sqrt{3}\right\}$.

**19.** $\dfrac{2x}{x^2 + 6x + 8} + \dfrac{2}{x + 2} = \dfrac{x}{x + 4}$

$\quad \dfrac{2x}{(x+4)(x+2)} + \dfrac{2}{x+2} = \dfrac{x}{x+4}$

$\dfrac{2x(x+4)(x+2)}{(x+4)(x+2)} + \dfrac{2(x+4)(x+2)}{x+2} = \dfrac{x(x+4)(x+2)}{x+4}$

$\quad\quad 2x + 2(x+4) = x(x+2)$

$\quad\quad 2x + 2x + 8 = x^2 + 2x$

$\quad\quad\quad 2x + 8 = x^2$

$\quad\quad\quad\quad 0 = x^2 - 2x - 8$

$\quad\quad\quad\quad 0 = (x-4)(x+2)$

$\quad x - 4 = 0 \quad \text{or} \quad x + 2 = 0$

$\quad\quad x = 4 \quad\quad\quad x = -2 \text{ (rejected)}$

The solution set is $\{4\}$.

**20.** $3(x + 4) \geq 5x - 12$

$\quad 3x + 12 \geq 5x - 12$

$\quad\quad -2x \geq -24$

$\quad\quad\quad x \leq 12$

The solution set is $(-\infty, 12]$.

**21.** $\dfrac{x}{6} + \dfrac{1}{8} \leq \dfrac{x}{2} - \dfrac{3}{4}$

$\quad 4x + 3 \leq 12x - 18$

$\quad\quad -8x \leq -21$

$\quad\quad\quad x \geq \dfrac{21}{8}$

The solution set is $\left[\dfrac{21}{8}, \infty\right)$.

**22.** $-3 \leq \dfrac{2x + 5}{3} < 6$

$\quad -9 \leq 2x + 5 < 18$

$\quad -14 \leq 2x < 13$

$\quad -7 \leq x < \dfrac{13}{2}$

The solution set is $\left[-7, \dfrac{13}{2}\right)$.

**23.** $|3x + 2| \geq 3$

$3x + 2 \geq 3$  or  $3x + 2 \leq -3$

$3x \geq 1$      $3x \leq -5$

$x \geq \dfrac{1}{3}$     $x \leq -\dfrac{5}{3}$

The solution set is $\left(-\infty, -\dfrac{5}{3}\right] \cup \left[\dfrac{1}{3}, \infty\right)$.

**24.** $-3 \leq y \leq 7$

$-3 \leq 2x - 5 \leq 7$

$2 \leq 2x \leq 12$

$1 \leq x \leq 6$

The solution set is $[1, 6]$.

**25.**      $y \geq 1$

$\left|\dfrac{2-x}{4}\right| \geq 1$

$\dfrac{2-x}{4} \geq 1$   or   $\dfrac{2-x}{4} \leq -1$

$2 - x \geq 4$         $2 - x \leq -4$

$-x \geq 2$           $-x \leq -6$

$x \leq -2$           $x \geq 6$

The solution set is $(-\infty, -2] \cup [6, \infty)$.

**26.**     Graph $[-1, 2)$:

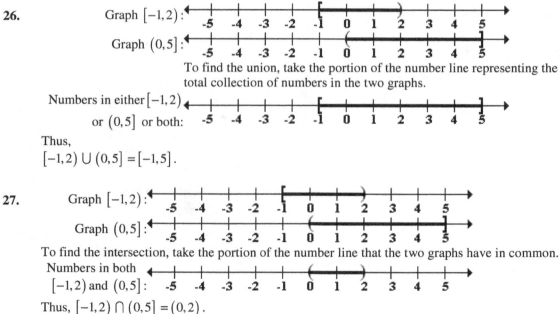

Graph $(0, 5]$:

To find the union, take the portion of the number line representing the total collection of numbers in the two graphs.

Numbers in either $[-1, 2)$ or $(0, 5]$ or both:

Thus,

$[-1, 2) \cup (0, 5] = [-1, 5]$.

**27.**     Graph $[-1, 2)$:

Graph $(0, 5]$:

To find the intersection, take the portion of the number line that the two graphs have in common.

Numbers in both $[-1, 2)$ and $(0, 5]$:

Thus, $[-1, 2) \cap (0, 5] = (0, 2)$.

**28.** $V = \frac{1}{3}lwh$

$3V = lwh$

$\frac{3V}{lw} = \frac{lwh}{lw}$

$\frac{3V}{lw} = h$

$h = \frac{3V}{lw}$

**29.** $y - y_1 = m(x - x_1)$

$y - y_1 = mx - mx_1$

$-mx = y_1 - mx_1 - y$

$\frac{-mx}{-m} = \frac{y_1 - mx_1 - y}{-m}$

$x = \frac{y - y_1}{m} + x_1$

**30.**

$y = 2 - |x|$

**31.**

$y = x^2 - 4$

**32.** $(6 - 7i)(2 + 5i) = 12 + 30i - 14i - 35i^2$

$= 12 + 16i + 35$

$= 47 + 16i$

**33.** $\frac{5}{2-i} = \frac{5}{2-i} \cdot \frac{2+i}{2+i}$

$= \frac{5(2+i)}{4+1}$

$= \frac{5(2+i)}{5}$

$= 2 + i$

**34.** $2\sqrt{-49} + 3\sqrt{-64} = 2(7i) + 3(8i)$

$= 14i + 24i$

$= 38i$

**35.** $43x + 575 = 1177$

$43x = 602$

$x = 14$

The system's income will be \$1177 billion 14 years after 2004, or 2018.

**36.** $B = 0.07x^2 + 47.4x + 500$

$1177 = 0.07x^2 + 47.4x + 500$

$0 = 0.07x^2 + 47.4x - 677$

$0 = 0.07x^2 + 47.4x - 677$

$x = \frac{-b \pm \sqrt{b^2 - 4ac}}{2a}$

$x = \frac{-(47.4) \pm \sqrt{(47.4)^2 - 4(0.07)(-677)}}{2(0.07)}$

$x \approx 14, \quad x \approx -691 \text{ (rejected)}$

The system's income will be \$1177 billion 14 years after 2004, or 2018.

**37.** The formulas model the data quite well.

**38.** Let $x =$ the percentage of strikingly-attractive men.
Let $x + 57 =$ the percentage of average-looking men.
Let $x + 25 =$ the percentage of good-looking men.

$(x) + (x + 57) + (x + 25) = 88$

$x + x + 57 + x + 25 = 88$

$3x + 82 = 88$

$3x = 6$

$x = 2$

$x + 57 = 59$

$x + 25 = 27$

2% of men are strikingly-attractive.
59% of men are average-looking.
27% of men are good-looking.

**39.** $29700 + 150x = 5000 + 1100x$

$24700 = 950x$

$26 = x$

In 26 years, the cost will be \$33,600.

**40.** Let $x$ = amount invested at 8%
$10000 - x$ = amount invested at 10%
$$0.08x + 0.1(10000 - x) = 940$$
$$0.08x + 1000 - 0.1x = 940$$
$$-0.02x = -60$$
$$x = 3000$$
$$10000 - x = 7000$$
$3000 at 8%, $7000 at 10%

**41.** $l = 2w + 4$
$A = lw$
$48 = (2w + 4)w$
$48 = 2w^2 + 4w$
$0 = 2w^2 + 4w - 48$
$0 = w^2 + 2w - 24$
$0 = (w + 6)(w - 4)$

$w + 6 = 0 \qquad w - 4 = 0$
$\quad w = -6 \qquad \quad w = 4$
$2w + 4 = 2(4) + 4 = 12$
width is 4 feet, length is 12 feet

**42.** $24^2 + x^2 = 26^2$
$576 + x^2 = 676$
$\quad x^2 = 100$
$\quad x = \pm 10$
The wire should be attached 10 feet up the pole.

**43.** Let $x$ = the original selling price
$20 = x - 0.60x$
$20 = 0.40x$
$50 = x$
The original price is $50.

**44.** Let $x$ = the number text messages.
The monthly cost using Plan $A$ is $C_A = 25$.
The monthly cost using Plan $B$ is $C_B = 13 + 0.06x$.
For Plan $A$ to be better deal, it must cost less than Plan $B$.
$C_A < C_B$
$25 < 13 + 0.06x$
$12 < 0.06x$
$200 < x$
$\quad x > 200$
Plan $A$ is a better deal when more than 200 text messages per month are sent/received.

# Chapter 2
# Functions and Graphs

**Section 2.1**

**Check Point Exercises**

1.  The domain is the set of all first components: {0, 10, 20, 30, 40}. The range is the set of all second components: {9.1, 6.7, 10.7, 13.2, 21.2}.

2.  a.  The relation is not a function since the two ordered pairs (5, 6) and (5, 8) have the same first component but different second components.

    b.  The relation is a function since no two ordered pairs have the same first component and different second components.

3.  a.  $2x + y = 6$
    $$y = 6 - 2x$$
    For each value of $x$, there is one and only one value for $y$, so the equation defines $y$ as a function of $x$.

    b.  $x^2 + y^2 = 1$
    $$y^2 = 1 - x^2$$
    $$y = \pm\sqrt{1 - x^2}$$
    Since there are values of $x$ (all values between –1 and 1 exclusive) that give more than one value for $y$ (for example, if $x = 0$, then $y = \pm\sqrt{1 - 0^2} = \pm 1$), the equation does not define $y$ as a function of $x$.

4.  a.  $f(-5) = (-5)^2 - 2(-5) + 7$
    $$= 25 - (-10) + 7$$
    $$= 42$$

    b.  $f(x+4) = (x+4)^2 - 2(x+4) + 7$
    $$= x^2 + 8x + 16 - 2x - 8 + 7$$
    $$= x^2 + 6x + 15$$

    c.  $f(-x) = (-x)^2 - 2(-x) + 7$
    $$= x^2 - (-2x) + 7$$
    $$= x^2 + 2x + 7$$

5.

| $x$ | $f(x) = 2x$ | $(x, y)$ |
|---|---|---|
| -2 | –4 | $(-2, -4)$ |
| -1 | –2 | $(-1, -2)$ |
| 0 | 0 | $(0, 0)$ |
| 1 | 2 | $(1, 2)$ |
| 2 | 4 | $(2, 4)$ |

| $x$ | $g(x) = 2x - 3$ | $(x, y)$ |
|---|---|---|
| -2 | $g(-2) = 2(-2) - 3 = -7$ | $(-2, -7)$ |
| -1 | $g(-1) = 2(-1) - 3 = -5$ | $(-1, -5)$ |
| 0 | $g(0) = 2(0) - 3 = -3$ | $(0, -3)$ |
| 1 | $g(1) = 2(1) - 3 = -1$ | $(1, -1)$ |
| 2 | $g(2) = 2(2) - 3 = 1$ | $(2, 1)$ |

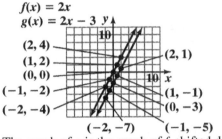

The graph of $g$ is the graph of $f$ shifted down 3 units.

6.  The graph (a) passes the vertical line test and is therefore is a <u>function</u>.
    The graph (b) fails the vertical line test and is therefore <u>not a function</u>.
    The graph (c) passes the vertical line test and is therefore is a <u>function</u>.
    The graph (d) fails the vertical line test and is therefore <u>not a function</u>.

7.  a.  $f(5) = 400$

    b.  $x = 9$, $f(9) = 100$

    c.  The minimum T cell count in the asymptomatic stage is approximately 425.

**8.** **a.** domain: $\{x|-2 \le x \le 1\}$ or $[-2,1]$.

range: $\{y|0 \le y \le 3\}$ or $[0,3]$.

**b.** domain: $\{x|-2 < x \le 1\}$ or $(-2,1]$.

range: $\{y|-1 \le y < 2\}$ or $[-1,2)$.

**c.** domain: $\{x|-3 \le x < 0\}$ or $[-3,0)$.

range: $\{y|y = -3,-2,-1\}$.

**Concept and Vocabulary Check 2.1**

**1.** relation; domain; range

**2.** function

**3.** $f$; $x$

**4.** true

**5.** false

**6.** $x$; $x+6$

**7.** ordered pairs

**8.** more than once; function

**9.** $[0,3)$; domain

**10.** $[1,\infty)$; range

**11.** 0; 0; zeros

**12.** false

**Exercise Set 2.1**

**1.** The relation is a function since no two ordered pairs have the same first component and different second components. The domain is {1, 3, 5} and the range is {2, 4, 5}.

**3.** The relation is not a function since the two ordered pairs (3, 4) and (3, 5) have the same first component but different second components (the same could be said for the ordered pairs (4, 4) and (4, 5)). The domain is {3, 4} and the range is {4, 5}.

**5.** The relation is a function because no two ordered pairs have the same first component and different second components The domain is {3, 4, 5, 7} and the range is {-2, 1, 9}.

**7.** The relation is a function since there are no same first components with different second components. The domain is {-3, -2, -1, 0} and the range is {-3, -2, -1, 0}.

**9.** The relation is not a function since there are ordered pairs with the same first component and different second components. The domain is {1} and the range is {4, 5, 6}.

**11.** $x + y = 16$

$y = 16 - x$

Since only one value of $y$ can be obtained for each value of $x$, $y$ is a function of $x$.

**13.** $x^2 + y = 16$

$y = 16 - x^2$

Since only one value of $y$ can be obtained for each value of $x$, $y$ is a function of $x$.

**15.** $x^2 + y^2 = 16$

$y^2 = 16 - x^2$

$y = \pm\sqrt{16 - x^2}$

If $x = 0$, $y = \pm 4$.

Since two values, $y = 4$ and $y = -4$, can be obtained for one value of $x$, $y$ is not a function of $x$.

**17.** $x = y^2$

$y = \pm\sqrt{x}$

If $x = 1$, $y = \pm 1$.

Since two values, $y = 1$ and $y = -1$, can be obtained for $x = 1$, $y$ is not a function of $x$.

**19.** $y = \sqrt{x+4}$

Since only one value of $y$ can be obtained for each value of $x$, $y$ is a function of $x$.

**21.** $x + y^3 = 8$

$y^3 = 8 - x$

$y = \sqrt[3]{8-x}$

Since only one value of $y$ can be obtained for each value of $x$, $y$ is a function of $x$.

**23.** $xy + 2y = 1$

$y(x+2) = 1$

$y = \dfrac{1}{x+2}$

Since only one value of $y$ can be obtained for each value of $x$, $y$ is a function of $x$.

**25.** $|x| - y = 2$
$$-y = -|x| + 2$$
$$y = |x| - 2$$
Since only one value of $y$ can be obtained for each value of $x$, $y$ is a function of $x$.

**27. a.** $f(6) = 4(6) + 5 = 29$

**b.** $f(x + 1) = 4(x + 1) + 5 = 4x + 9$

**c.** $f(-x) = 4(-x) + 5 = -4x + 5$

**29. a.** $g(-1) = (-1)^2 + 2(-1) + 3$
$$= 1 - 2 + 3$$
$$= 2$$

**b.** $g(x + 5) = (x + 5)^2 + 2(x + 5) + 3$
$$= x^2 + 10x + 25 + 2x + 10 + 3$$
$$= x^2 + 12x + 38$$

**c.** $g(-x) = (-x)^2 + 2(-x) + 3$
$$= x^2 - 2x + 3$$

**31. a.** $h(2) = 2^4 - 2^2 + 1$
$$= 16 - 4 + 1$$
$$= 13$$

**b.** $h(-1) = (-1)^4 - (-1)^2 + 1$
$$= 1 - 1 + 1$$
$$= 1$$

**c.** $h(-x) = (-x)^4 - (-x)^2 + 1 = x^4 - x^2 + 1$

**d.** $h(3a) = (3a)^4 - (3a)^2 + 1$
$$= 81a^4 - 9a^2 + 1$$

**33. a.** $f(-6) = \sqrt{-6+6} + 3 = \sqrt{0} + 3 = 3$

**b.** $f(10) = \sqrt{10+6} + 3$
$$= \sqrt{16} + 3$$
$$= 4 + 3$$
$$= 7$$

**c.** $f(x - 6) = \sqrt{x-6+6} + 3 = \sqrt{x} + 3$

**35. a.** $f(2) = \dfrac{4(2)^2 - 1}{2^2} = \dfrac{15}{4}$

**b.** $f(-2) = \dfrac{4(-2)^2 - 1}{(-2)^2} = \dfrac{15}{4}$

**c.** $f(-x) = \dfrac{4(-x)^2 - 1}{(-x)^2} = \dfrac{4x^2 - 1}{x^2}$

**37. a.** $f(6) = \dfrac{6}{|6|} = 1$

**b.** $f(-6) = \dfrac{-6}{|-6|} = \dfrac{-6}{6} = -1$

**c.** $f(r^2) = \dfrac{r^2}{|r^2|} = \dfrac{r^2}{r^2} = 1$

**39.**

| $x$ | $f(x) = x$ | $(x, y)$ |
|---|---|---|
| $-2$ | $f(-2) = -2$ | $(-2, -2)$ |
| $-1$ | $f(-1) = -1$ | $(-1, -1)$ |
| $0$ | $f(0) = 0$ | $(0, 0)$ |
| $1$ | $f(1) = 1$ | $(1, 1)$ |
| $2$ | $f(2) = 2$ | $(2, 2)$ |

| $x$ | $g(x) = x + 3$ | $(x, y)$ |
|---|---|---|
| $-2$ | $g(-2) = -2 + 3 = 1$ | $(-2, 1)$ |
| $-1$ | $g(-1) = -1 + 3 = 2$ | $(-1, 2)$ |
| $0$ | $g(0) = 0 + 3 = 3$ | $(0, 3)$ |
| $1$ | $g(1) = 1 + 3 = 4$ | $(1, 4)$ |
| $2$ | $g(2) = 2 + 3 = 5$ | $(2, 5)$ |

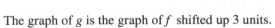

The graph of $g$ is the graph of $f$ shifted up 3 units.

**41.**

| $x$ | $f(x) = -2x$ | $(x, y)$ |
|---|---|---|
| –2 | $f(-2) = -2(-2) = 4$ | $(-2, 4)$ |
| –1 | $f(-1) = -2(-1) = 2$ | $(-1, 2)$ |
| 0 | $f(0) = -2(0) = 0$ | $(0, 0)$ |
| 1 | $f(1) = -2(1) = -2$ | $(1, -2)$ |
| 2 | $f(2) = -2(2) = -4$ | $(2, -4)$ |

| $x$ | $g(x) = -2x - 1$ | $(x, y)$ |
|---|---|---|
| –2 | $g(-2) = -2(-2) - 1 = 3$ | $(-2, 3)$ |
| –1 | $g(-1) = -2(-1) - 1 = 1$ | $(-1, 1)$ |
| 0 | $g(0) = -2(0) - 1 = -1$ | $(0, -1)$ |
| 1 | $g(1) = -2(1) - 1 = -3$ | $(1, -3)$ |
| 2 | $g(2) = -2(2) - 1 = -5$ | $(2, -5)$ |

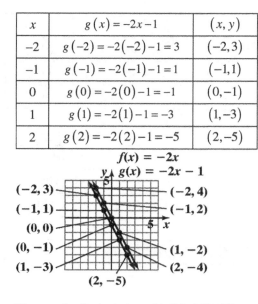

The graph of $g$ is the graph of $f$ shifted down 1 unit.

**43.**

| $x$ | $f(x) = x^2$ | $(x, y)$ |
|---|---|---|
| –2 | $f(-2) = (-2)^2 = 4$ | $(-2, 4)$ |
| –1 | $f(-1) = (-1)^2 = 1$ | $(-1, 1)$ |
| 0 | $f(0) = (0)^2 = 0$ | $(0, 0)$ |
| 1 | $f(1) = (1)^2 = 1$ | $(1, 1)$ |
| 2 | $f(2) = (2)^2 = 4$ | $(2, 4)$ |

| $x$ | $g(x) = x^2 + 1$ | $(x, y)$ |
|---|---|---|
| –2 | $g(-2) = (-2)^2 + 1 = 5$ | $(-2, 5)$ |
| –1 | $g(-1) = (-1)^2 + 1 = 2$ | $(-1, 2)$ |
| 0 | $g(0) = (0)^2 + 1 = 1$ | $(0, 1)$ |
| 1 | $g(1) = (1)^2 + 1 = 2$ | $(1, 2)$ |
| 2 | $g(2) = (2)^2 + 1 = 5$ | $(2, 5)$ |

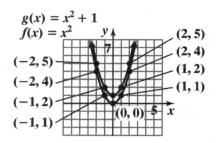

The graph of $g$ is the graph of $f$ shifted up 1 unit.

**45.**

| $x$ | $f(x) = |x|$ | $(x, y)$ |
|---|---|---|
| –2 | $f(-2) = |-2| = 2$ | $(-2, 2)$ |
| –1 | $f(-1) = |-1| = 1$ | $(-1, 1)$ |
| 0 | $f(0) = |0| = 0$ | $(0, 0)$ |
| 1 | $f(1) = |1| = 1$ | $(1, 1)$ |
| 2 | $f(2) = |2| = 2$ | $(2, 2)$ |

| $x$ | $g(x) = |x| - 2$ | $(x, y)$ |
|---|---|---|
| –2 | $g(-2) = |-2| - 2 = 0$ | $(-2, 0)$ |
| –1 | $g(-1) = |-1| - 2 = -1$ | $(-1, -1)$ |
| 0 | $g(0) = |0| - 2 = -2$ | $(0, -2)$ |
| 1 | $g(1) = |1| - 2 = -1$ | $(1, -1)$ |
| 2 | $g(2) = |2| - 2 = 0$ | $(2, 0)$ |

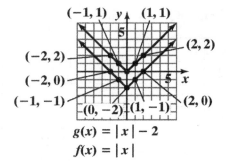

The graph of $g$ is the graph of $f$ shifted down 2 units.

**47.**

| $x$ | $f(x) = x^3$ | $(x, y)$ |
|---|---|---|
| $-2$ | $f(-2) = (-2)^3 = -8$ | $(-2, -8)$ |
| $-1$ | $f(-1) = (-1)^3 = -1$ | $(-1, -1)$ |
| $0$ | $f(0) = (0)^3 = 0$ | $(0, 0)$ |
| $1$ | $f(1) = (1)^3 = 1$ | $(1, 1)$ |
| $2$ | $f(2) = (2)^3 = 8$ | $(2, 8)$ |

| $x$ | $g(x) = x^3 + 2$ | $(x, y)$ |
|---|---|---|
| $-2$ | $g(-2) = (-2)^3 + 2 = -6$ | $(-2, -6)$ |
| $-1$ | $g(-1) = (-1)^3 + 2 = 1$ | $(-1, 1)$ |
| $0$ | $g(0) = (0)^3 + 2 = 2$ | $(0, 2)$ |
| $1$ | $g(1) = (1)^3 + 2 = 3$ | $(1, 3)$ |
| $2$ | $g(2) = (2)^3 + 2 = 10$ | $(2, 10)$ |

The graph of g is the graph of f shifted up 2 units.

**49.**

| $x$ | $f(x) = 3$ | $(x, y)$ |
|---|---|---|
| $-2$ | $f(-2) = 3$ | $(-2, 3)$ |
| $-1$ | $f(-1) = 3$ | $(-1, 3)$ |
| $0$ | $f(0) = 3$ | $(0, 3)$ |
| $1$ | $f(1) = 3$ | $(1, 3)$ |
| $2$ | $f(2) = 3$ | $(2, 3)$ |

| $x$ | $g(x) = 5$ | $(x, y)$ |
|---|---|---|
| $-2$ | $g(-2) = 5$ | $(-2, 5)$ |
| $-1$ | $g(-1) = 5$ | $(-1, 5)$ |
| $0$ | $g(0) = 5$ | $(0, 5)$ |
| $1$ | $g(1) = 5$ | $(1, 5)$ |
| $2$ | $g(2) = 5$ | $(2, 5)$ |

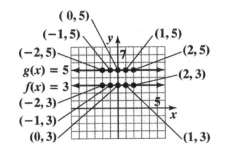

The graph of g is the graph of f shifted up 2 units.

**51.**

| $x$ | $f(x) = \sqrt{x}$ | $(x, y)$ |
|---|---|---|
| $0$ | $f(0) = \sqrt{0} = 0$ | $(0, 0)$ |
| $1$ | $f(1) = \sqrt{1} = 1$ | $(1, 1)$ |
| $4$ | $f(4) = \sqrt{4} = 2$ | $(4, 2)$ |
| $9$ | $f(9) = \sqrt{9} = 3$ | $(9, 3)$ |

| $x$ | $g(x) = \sqrt{x} - 1$ | $(x, y)$ |
|---|---|---|
| $0$ | $g(0) = \sqrt{0} - 1 = -1$ | $(0, -1)$ |
| $1$ | $g(1) = \sqrt{1} - 1 = 0$ | $(1, 0)$ |
| $4$ | $g(4) = \sqrt{4} - 1 = 1$ | $(4, 1)$ |
| $9$ | $g(9) = \sqrt{9} - 1 = 2$ | $(9, 2)$ |

The graph of g is the graph of f shifted down 1 unit.

**53.**

| $x$ | $f(x)=\sqrt{x}$ | $(x,y)$ |
|---|---|---|
| 0 | $f(0)=\sqrt{0}=0$ | $(0,0)$ |
| 1 | $f(1)=\sqrt{1}=1$ | $(1,1)$ |
| 4 | $f(4)=\sqrt{4}=2$ | $(4,2)$ |
| 9 | $f(9)=\sqrt{9}=3$ | $(9,3)$ |

| $x$ | $g(x)=\sqrt{x-1}$ | $(x,y)$ |
|---|---|---|
| 1 | $g(1)=\sqrt{1-1}=0$ | $(1,0)$ |
| 2 | $g(2)=\sqrt{2-1}=1$ | $(2,1)$ |
| 5 | $g(5)=\sqrt{5-1}=2$ | $(5,2)$ |
| 10 | $g(10)=\sqrt{10-1}=3$ | $(10,3)$ |

The graph of g is the graph of f shifted right 1 unit.

**55.** function

**57.** function

**59.** not a function

**61.** function

**63.** function

**65.** $f(-2)=-4$

**67.** $f(4)=4$

**69.** $f(-3)=0$

**71.** $g(-4)=2$

**73.** $g(-10)=2$

**75.** When $x=-2$, $g(x)=1$.

**77. a.** domain: $(-\infty,\infty)$

**b.** range: $[-4,\infty)$

**c.** $x$-intercepts: $-3$ and $1$

**d.** $y$-intercept: $-3$

**e.** $f(-2)=-3$ and $f(2)=5$

**79. a.** domain: $(-\infty,\infty)$

**b.** range: $[1,\infty)$

**c.** $x$-intercept: none

**d.** $y$-intercept: 1

**e.** $f(-1)=2$ and $f(3)=4$

**81. a.** domain: $[0,5)$

**b.** range: $[-1,5)$

**c.** $x$-intercept: 2

**d.** $y$-intercept: $-1$

**e.** $f(3)=1$

**83. a.** domain: $[0,\infty)$

**b.** range: $[1,\infty)$

**c.** $x$-intercept: none

**d.** $y$-intercept: 1

**e.** $f(4)=3$

**85. a.** domain: $[-2,6]$

**b.** range: $[-2,6]$

**c.** $x$-intercept: 4

**d.** $y$-intercept: 4

**e.** $f(-1)=5$

**87. a.** domain: $(-\infty,\infty)$

**b.** range: $(-\infty,-2]$

**c.** $x$-intercept: none

**d.** $y$-intercept: $-2$

**e.** $f(-4)=-5$ and $f(4)=-2$

**89. a.** domain: $(-\infty, \infty)$

**b.** range: $(0, \infty)$

**c.** *x*-intercept: none

**d.** *y*-intercept: 1.5

**e.** $f(4) = 6$

**91. a.** domain: $\{-5, -2, 0, 1, 3\}$

**b.** range: $\{2\}$

**c.** *x*-intercept: none

**d.** *y*-intercept: 2

**e.** $f(-5) + f(3) = 2 + 2 = 4$

**93.** $g(1) = 3(1) - 5 = 3 - 5 = -2$

$f(g(1)) = f(-2) = (-2)^2 - (-2) + 4$
$= 4 + 2 + 4 = 10$

**95.** $\sqrt{3 - (-1)} - (-6)^2 + 6 \div (-6) \cdot 4$
$= \sqrt{3+1} - 36 + 6 \div (-6) \cdot 4$
$= \sqrt{4} - 36 + -1 \cdot 4$
$= 2 - 36 + -4$
$= -34 + -4$
$= -38$

**97.** $f(-x) - f(x)$
$= (-x)^3 + (-x) - 5 - (x^3 + x - 5)$
$= -x^3 - x - 5 - x^3 - x + 5 = -2x^3 - 2x$

**99. a.** $\{(\text{Philippines, 12}), (\text{Spain, 13}),$
$(\text{Italy, 14}), (\text{Germany, 14}),$
$(\text{Russia, 16})\}$

**b.** Yes, the relation is a function. Each country (element in the domain) corresponds to only one age (element in the range).

**c.** $\{(12, \text{Philippines}), (13, \text{Spain}),$
$(14, \text{Italy}), (14, \text{Germany}),$
$(16, \text{Russia})\}$

**d.** No, the relation is not a function. 14 in the domain corresponds to two members in the range, Italy and Germany.

**101. a.** $f(70) = 83$ which means the chance that a 60-year old will survive to age 70 is 83%.

**b.** $g(70) = 76$ which means the chance that a 60-year old will survive to age 70 is 76%.

**c.** Function *f* is the better model.

**103. a.** $G(30) = -0.01(30)^2 + (30) + 60 = 81$
In 2010, the wage gap was 81%. This is represented as $(30, 81)$ on the graph.

**b.** $G(30)$ underestimates the actual data shown by the bar graph by 2%.

**105.** $C(x) = 100,000 + 100x$
$C(90) = 100,000 + 100(90) = \$109,000$
It will cost \$109,000 to produce 90 bicycles.

**107.** $T(x) = \dfrac{40}{x} + \dfrac{40}{x+30}$
$T(30) = \dfrac{40}{30} + \dfrac{40}{30+30}$
$= \dfrac{80}{60} + \dfrac{40}{60}$
$= \dfrac{120}{60}$
$= 2$
If you travel 30 mph going and 60 mph returning, your total trip will take 2 hours.

**109. – 117.** Answers will vary.

**119.** does not make sense; Explanations will vary. Sample explanation: The parentheses used in function notation, such as $f(x)$, do not imply multiplication.

**121.** does not make sense; Explanations will vary. Sample explanation: This would not be a function because some elements in the domain would correspond to more than one age in the range.

**123.** false; Changes to make the statement true will vary. A sample change is: The range is $[-2, 2]$.

**125.** false; Changes to make the statement true will vary. A sample change is: $f(0) = 0.8$

**127.** Answers will vary.
An example is $\{(1,1),(2,1)\}$

**129.** $C(t) = 20 + 0.40(t-60)$

$C(100) = 20 + 0.40(100-60)$

$= 20 + 0.40(40)$

$= 20 + 16$

$= 36$

For 100 calling minutes, the monthly cost is $36.

**130.** $f(x) = x + 2, \ x \le 1$

**131.** $2(x+h)^2 + 3(x+h) + 5 - (2x^2 + 3x + 5)$

$= 2(x^2 + 2xh + h^2) + 3x + 3h + 5 - 2x^2 - 3x - 5$

$= 2x^2 + 4xh + 2h^2 + 3x + 3h + 5 - 2x^2 - 3x - 5$

$= 2x^2 - 2x^2 + 4xh + 2h^2 + 3x - 3x + 3h + 5 - 5$

$= 4xh + 2h^2 + 3h$

## Section 2.2

### Check Point Exercises

**1.** The function is increasing on the interval $(-\infty, -1)$, decreasing on the interval $(-1, 1)$, and increasing on the interval $(1, \infty)$.

**2. a.** $f(-x) = (-x)^2 + 6 = x^2 + 6 = f(x)$
The function is even.

**b.** $g(-x) = 7(-x)^3 - (-x) = -7x^3 + x = -f(x)$
The function is odd.

**c.** $h(-x) = (-x)^5 + 1 = -x^5 + 1$
The function is neither even nor odd.

**3.** $C(t) = \begin{cases} 20 & \text{if } 0 \le t \le 60 \\ 20 + 0.40(t-60) & \text{if } t > 60 \end{cases}$

**b.** Since $0 \le 40 \le 60$, $C(40) = 20$
With 40 calling minutes, the cost is $20.
This is represented by $(40, 20)$.

**c.** Since $80 > 60$,
$C(80) = 20 + 0.40(80-60) = 28$
With 80 calling minutes, the cost is $28.
This is represented by $(80, 28)$.

**4.**

$f(x) = \begin{cases} 3 & \text{if } x \le -1 \\ x - 2 & \text{if } x > -1 \end{cases}$

**5. a.** $f(x) = -2x^2 + x + 5$

$f(x+h) = -2(x+h)^2 + (x+h) + 5$

$= -2(x^2 + 2xh + h^2) + x + h + 5$

$= -2x^2 - 4xh - 2h^2 + x + h + 5$

**b.** $\dfrac{f(x+h) - f(x)}{h}$

$= \dfrac{-2x^2 - 4xh - 2h^2 + x + h + 5 - \left(-2x^2 + x + 5\right)}{h}$

$= \dfrac{-2x^2 - 4xh - 2h^2 + x + h + 5 + 2x^2 - x - 5}{h}$

$= \dfrac{-4xh - 2h^2 + h}{h}$

$= \dfrac{h\left(-4x - 2h + 1\right)}{h}$

$= -4x - 2h + 1, \quad h \ne 0$

### Concept and Vocabulary Check 2.2

**1.** $< f(x_2)$; $> f(x_2)$; $= f(x_2)$

**2.** maximum; minimum

**3.** $f(x)$; $y$-axis

**4.** $-f(x)$; origin

**5.** piecewise

**6.** less than or equal to $x$; 2; $-3$; 0

**7.** difference quotient; $x + h$; $f(x)$; $h$; $h$

**8.** false

**9.** false

**Exercise Set 2.2**

1.  **a.** increasing: $(-1, \infty)$

    **b.** decreasing: $(-\infty, -1)$

    **c.** constant: none

3.  **a.** increasing: $(0, \infty)$

    **b.** decreasing: none

    **c.** constant: none

5.  **a.** increasing: none

    **b.** decreasing: $(-2, 6)$

    **c.** constant: none

7.  **a.** increasing: $(-\infty, -1)$

    **b.** decreasing: none

    **c.** constant: $(-1, \infty)$

9.  **a.** increasing: $(-\infty, 0)$ or $(1.5, 3)$

    **b.** decreasing: $(0, 1.5)$ or $(3, \infty)$

    **c.** constant: none

11. **a.** increasing: $(-2, 4)$

    **b.** decreasing: none

    **c.** constant: $(-\infty, -2)$ or $(4, \infty)$

13. **a.** $x = 0$, relative maximum $= 4$

    **b.** $x = -3, 3$, relative minimum $= 0$

15. **a.** $x = -2$, relative maximum $= 21$

    **b.** $x = 1$, relative minimum $= -6$

17. $f(x) = x^3 + x$
    $f(-x) = (-x)^3 + (-x)$
    $f(-x) = -x^3 - x = -(x^3 + x)$
    $f(-x) = -f(x)$, odd function

19. $g(x) = x^2 + x$
    $g(-x) = (-x)^2 + (-x)$
    $g(-x) = x^2 - x$, neither

21. $h(x) = x^2 - x^4$
    $h(-x) = (-x)^2 - (-x)^4$
    $h(-x) = x^2 - x^4$
    $h(-x) = h(x)$, even function

23. $f(x) = x^2 - x^4 + 1$
    $f(-x) = (-x)^2 - (-x)^4 + 1$
    $f(-x) = x^2 - x^4 + 1$
    $f(-x) = f(x)$, even function

25. $f(x) = \frac{1}{5}x^6 - 3x^2$
    $f(-x) = \frac{1}{5}(-x)^6 - 3(-x)^2$
    $f(-x) = \frac{1}{5}x^6 - 3x^2$
    $f(-x) = f(x)$, even function

27. $f(x) = x\sqrt{1-x^2}$
    $f(-x) = -x\sqrt{1-(-x)^2}$
    $f(-x) = -x\sqrt{1-x^2}$
    $= -\left(x\sqrt{1-x^2}\right)$
    $f(-x) = -f(x)$, odd function

29. The graph is symmetric with respect to the $y$-axis. The function is even.

31. The graph is symmetric with respect to the origin. The function is odd.

33. **a.** domain: $(-\infty, \infty)$

    **b.** range: $[-4, \infty)$

    **c.** $x$-intercepts: 1, 7

    **d.** $y$-intercept: 4

    **e.** $(4, \infty)$

    **f.** $(0, 4)$

    **g.** $(-\infty, 0)$

    **h.** $x = 4$

    **i.** $y = -4$

**j.**   $f(-3) = 4$

**k.**   $f(2) = -2$ and $f(6) = -2$

**l.**   neither ; $f(-x) \neq x$ , $f(-x) \neq -x$

**35.** **a.**   domain: $(-\infty, 3]$

**b.**   range: $(-\infty, 4]$

**c.**   $x$-intercepts: $-3$, $3$

**d.**   $f(0) = 3$

**e.**   $(-\infty, 1)$

**f.**   $(1, 3)$

**g.**   $(-\infty, -3]$

**h.**   $f(1) = 4$

**i.**   $x = 1$

**j.**   positive; $f(-1) = +2$

**37.** **a.**   $f(-2) = 3(-2) + 5 = -1$

**b.**   $f(0) = 4(0) + 7 = 7$

**c.**   $f(3) = 4(3) + 7 = 19$

**39.** **a.**   $g(0) = 0 + 3 = 3$

**b.**   $g(-6) = -(-6 + 3) = -(-3) = 3$

**c.**   $g(-3) = -3 + 3 = 0$

**41.** **a.**   $h(5) = \dfrac{5^2 - 9}{5 - 3} = \dfrac{25 - 9}{2} = \dfrac{16}{2} = 8$

**b.**   $h(0) = \dfrac{0^2 - 9}{0 - 3} = \dfrac{-9}{-3} = 3$

**c.**   $h(3) = 6$

**43.** **a.**

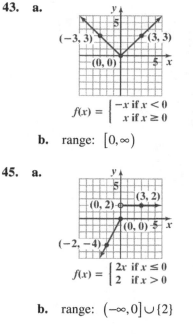

$f(x) = \begin{cases} -x \text{ if } x < 0 \\ x \text{ if } x \geq 0 \end{cases}$

**b.**   range: $[0, \infty)$

**45.** **a.**

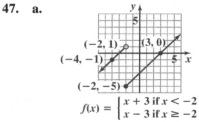

$f(x) = \begin{cases} 2x \text{ if } x \leq 0 \\ 2 \text{ if } x > 0 \end{cases}$

**b.**   range: $(-\infty, 0] \cup \{2\}$

**47.** **a.**

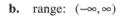

$f(x) = \begin{cases} x + 3 \text{ if } x < -2 \\ x - 3 \text{ if } x \geq -2 \end{cases}$

**b.**   range: $(-\infty, \infty)$

**49.** **a.**

$f(x) = \begin{cases} 3 \text{ if } x \leq -1 \\ -3 \text{ if } x > -1 \end{cases}$

**b.**   range: $\{-3, 3\}$

**51.** **a.**

$f(x) = \begin{cases} \dfrac{1}{2}x^2 \quad \text{if } x < 1 \\ 2x - 1 \text{ if } x \geq 1 \end{cases}$

**b.**   range: $[0, \infty)$

**53.** **a.**

$$f(x) = \begin{cases} 0 & \text{if } x < -4 \\ -x & \text{if } -4 \le x < 0 \\ x^2 & \text{if } x \ge 0 \end{cases}$$

**b.** range: $[0, \infty)$

**55.**
$$\frac{f(x+h) - f(x)}{h}$$
$$= \frac{4(x+h) - 4x}{h}$$
$$= \frac{4x + 4h - 4x}{h}$$
$$= \frac{4h}{h}$$
$$= 4$$

**57.**
$$\frac{f(x+h) - f(x)}{h}$$
$$= \frac{3(x+h) + 7 - (3x+7)}{h}$$
$$= \frac{3x + 3h + 7 - 3x - 7}{h}$$
$$= \frac{3h}{h}$$
$$= 3$$

**59.**
$$\frac{f(x+h) - f(x)}{h}$$
$$= \frac{(x+h)^2 - x^2}{h}$$
$$= \frac{x^2 + 2xh + h^2 - x^2}{h}$$
$$= \frac{2xh + h^2}{h}$$
$$= \frac{h(2x+h)}{h}$$
$$= 2x + h$$

**61.**
$$\frac{f(x+h) - f(x)}{h}$$
$$= \frac{(x+h)^2 - 4(x+h) + 3 - (x^2 - 4x + 3)}{h}$$
$$= \frac{x^2 + 2xh + h^2 - 4x - 4h + 3 - x^2 + 4x - 3}{h}$$
$$= \frac{2xh + h^2 - 4h}{h}$$
$$= \frac{h(2x + h - 4)}{h}$$
$$= 2x + h - 4$$

**63.**
$$\frac{f(x+h) - f(x)}{h}$$
$$= \frac{2(x+h)^2 + (x+h) - 1 - (2x^2 + x - 1)}{h}$$
$$= \frac{2x^2 + 4xh + 2h^2 + x + h - 1 - 2x^2 - x + 1}{h}$$
$$= \frac{4xh + 2h^2 + h}{h}$$
$$= \frac{h(4x + 2h + 1)}{h}$$
$$= 4x + 2h + 1$$

**65.**
$$\frac{f(x+h) - f(x)}{h}$$
$$= \frac{-(x+h)^2 + 2(x+h) + 4 - (-x^2 + 2x + 4)}{h}$$
$$= \frac{-x^2 - 2xh - h^2 + 2x + 2h + 4 + x^2 - 2x - 4}{h}$$
$$= \frac{-2xh - h^2 + 2h}{h}$$
$$= \frac{h(-2x - h + 2)}{h}$$
$$= -2x - h + 2$$

**67.** $\dfrac{f(x+h)-f(x)}{h}$

$=\dfrac{-2(x+h)^2+5(x+h)+7-(-2x^2+5x+7)}{h}$

$=\dfrac{-2x^2-4xh-2h^2+5x+5h+7+2x^2-5x-7}{h}$

$=\dfrac{-4xh-2h^2+5h}{h}$

$=\dfrac{h(-4x-2h+5)}{h}$

$=-4x-2h+5$

**69.** $\dfrac{f(x+h)-f(x)}{h}$

$=\dfrac{-2(x+h)^2-(x+h)+3-(-2x^2-x+3)}{h}$

$=\dfrac{-2x^2-4xh-2h^2-x-h+3+2x^2+x-3}{h}$

$=\dfrac{-4xh-2h^2-h}{h}$

$=\dfrac{h(-4x-2h-1)}{h}$

$=-4x-2h-1$

**71.** $\dfrac{f(x+h)-f(x)}{h}=\dfrac{6-6}{h}=\dfrac{0}{h}=0$

**73.** $\dfrac{f(x+h)-f(x)}{h}$

$=\dfrac{\dfrac{1}{x+h}-\dfrac{1}{x}}{h}$

$=\dfrac{\dfrac{x}{x(x+h)}+\dfrac{-(x+h)}{x(x+h)}}{h}$

$=\dfrac{\dfrac{x-x-h}{x(x+h)}}{h}$

$=\dfrac{\dfrac{-h}{x(x+h)}}{h}$

$=\dfrac{-h}{x(x+h)}\cdot\dfrac{1}{h}$

$=\dfrac{-1}{x(x+h)}$

**75.** $\dfrac{f(x+h)-f(x)}{h}$

$=\dfrac{\sqrt{x+h}-\sqrt{x}}{h}$

$=\dfrac{\sqrt{x+h}-\sqrt{x}}{h}\cdot\dfrac{\sqrt{x+h}+\sqrt{x}}{\sqrt{x+h}+\sqrt{x}}$

$=\dfrac{x+h-x}{h\left(\sqrt{x+h}+\sqrt{x}\right)}$

$=\dfrac{h}{h\left(\sqrt{x+h}+\sqrt{x}\right)}$

$=\dfrac{1}{\sqrt{x+h}+\sqrt{x}}$

**77.** $\sqrt{f(-1.5)+f(-0.9)}-\left[f(\pi)\right]^2+f(-3)\div f(1)\cdot f(-\pi)$

$=\sqrt{1+0}-\left[-4\right]^2+2\div(-2)\cdot3$

$=\sqrt{1}-16+(-1)\cdot3$

$=1-16-3$

$=-18$

**79.** $30+0.30(t-120)=30+0.3t-36=0.3t-6$

**81.** $C(t)=\begin{cases}50 & \text{if } 0\le t\le 400\\ 50+0.30(t-400) & \text{if } t>400\end{cases}$

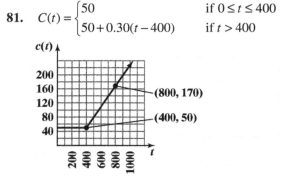

**83.** increasing: (25, 55); decreasing: (55, 75)

**85.** The percent body fat in women reaches a maximum at age 55. This maximum is 38%.

**87.** domain: [25, 75]; range: [34, 38]

**89.** This model describes percent body fat in men.

**91.** $T(20,000) = 850 + 0.15(20,000 - 8500)$
$$= 2575$$
A single taxpayer with taxable income of $20,000 owes $2575.

**93.** $42,449 + 0.33(x - 174,400)$

**95.** $f(3) = 0.85$
The cost of mailing a first-class letter weighing 3 ounces is $0.85.

**97.** The cost to mail a letter weighing 1.5 ounces is $0.65.

**99.**

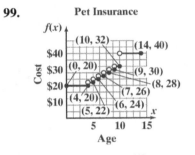

**101. – 105.** Answers will vary.

**107.**

Increasing: $(-\infty, 1)$ or $(3, \infty)$
Decreasing: $(1, 3)$

**109.**

Increasing: $(2, \infty)$
Decreasing: $(-\infty, -2)$
Constant: $(-2, 2)$

**111.**

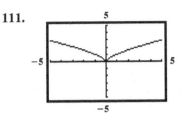

Increasing: $(0, \infty)$
Decreasing: $(-\infty, 0)$

**113. a.**

**b.**

**c.** Increasing: $(0, \infty)$
Decreasing: $(-\infty, 0)$

**d.** $f(x) = x^n$ is increasing from $(-\infty, \infty)$ when $n$ is odd.

**e.**

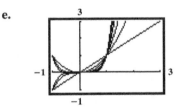

**115.** makes sense

**117.** makes sense

**119.** answers will vary

**121.** answers will vary

**122.** $\dfrac{y_2 - y_1}{x_2 - x_1} = \dfrac{4 - 1}{-2 - (-3)} = \dfrac{3}{1} = 3$

**123.** When $y = 0$:
$$4x - 3y - 6 = 0$$
$$4x - 3(0) - 6 = 0$$
$$4x - 6 = 0$$
$$4x = 6$$
$$x = \frac{3}{2}$$
The point is $\left(\frac{3}{2}, 0\right)$.

When $x = 0$:
$$4x - 3y - 6 = 0$$
$$4(0) - 3y - 6 = 0$$
$$-3y - 6 = 0$$
$$-3y = 6$$
$$x = -2$$
The point is $(0, -2)$.

**124.** $3x + 2y - 4 = 0$
$$2y = -3x + 4$$
$$y = \frac{-3x + 4}{2}$$
or
$$y = -\frac{3}{2}x + 2$$

## Section 2.3

### Check Point Exercises

**1.** **a.** $m = \dfrac{-2 - 4}{-4 - (-3)} = \dfrac{-6}{-1} = 6$

   **b.** $m = \dfrac{5 - (-2)}{-1 - 4} = \dfrac{7}{-5} = -\dfrac{7}{5}$

**2.** Point-slope form:
$$y - y_1 = m(x - x_1)$$
$$y - (-5) = 6(x - 2)$$
$$y + 5 = 6(x - 2)$$

Slope-intercept form:
$$y + 5 = 6(x - 2)$$
$$y + 5 = 6x - 12$$
$$y = 6x - 17$$

**3.** $m = \dfrac{-6 - (-1)}{-1 - (-2)} = \dfrac{-5}{1} = -5$,

so the slope is –5.
Using the point $(-2, -1)$, we get the following point-slope equation:
$$y - y_1 = m(x - x_1)$$
$$y - (-1) = -5[x - (-2)]$$
$$y + 1 = -5(x + 2)$$
Using the point $(-1, -6)$, we get the following point-slope equation:
$$y - y_1 = m(x - x_1)$$
$$y - (-6) = -5[x - (-1)]$$
$$y + 6 = -5(x + 1)$$
Solve the equation for $y$:
$$y + 1 = -5(x + 2)$$
$$y + 1 = -5x - 10$$
$$y = -5x - 11.$$

**4.** The slope $m$ is $\frac{3}{5}$ and the $y$-intercept is 1, so one point on the line is $(1, 0)$. We can find a second point on the line by using the slope $m = \frac{3}{5} = \frac{\text{Rise}}{\text{Run}}$: starting at the point $(0, 1)$, move 3 units up and 5 units to the right, to obtain the point $(5, 4)$.

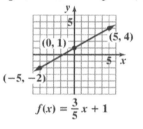

$$f(x) = \frac{3}{5}x + 1$$

**5.** $y = 3$ is a horizontal line.

**6.** All ordered pairs that are solutions of $x = -3$ have a value of $x$ that is always $-3$. Any value can be used for $y$.

**7.** $3x + 6y - 12 = 0$

$$6y = -3x + 12$$

$$y = \frac{-3}{6}x + \frac{12}{6}$$

$$y = -\frac{1}{2}x + 2$$

$$3x + 6y - 12 = 0$$

The slope is $-\dfrac{1}{2}$ and the *y*-intercept is 2.

**8.** Find the *x*-intercept:

$$3x - 2y - 6 = 0$$

$$3x - 2(0) - 6 = 0$$

$$3x - 6 = 0$$

$$3x = 6$$

$$x = 2$$

Find the *y*-intercept:

$$3x - 2y - 6 = 0$$

$$3(0) - 2y - 6 = 0$$

$$-2y - 6 = 0$$

$$-2y = 6$$

$$y = -3$$

$$3x - 2y = 6$$

**9.** First find the slope.

$$m = \frac{\text{Change in } y}{\text{Change in } x} = \frac{57.64 - 57.04}{354 - 317} = \frac{0.6}{37} \approx 0.016$$

Use the point-slope form and then find slope-intercept form.

$$y - y_1 = m(x - x_1)$$

$$y - 57.04 = 0.016(x - 317)$$

$$y - 57.04 = 0.016x - 5.072$$

$$y = 0.016x + 51.968$$

$$f(x) = 0.016x + 52.0$$

Find the temperature at a concentration of 600 parts per million.

$$f(x) = 0.016x + 52.0$$

$$f(600) = 0.016(600) + 52.0$$

$$= 61.6$$

The temperature at a concentration of 600 parts per million would be $61.6°F$.

**Concept and Vocabulary Check 2.3**

**1.** scatter plot; regression

**2.** $\dfrac{y_2 - y_1}{x_2 - x_1}$

**3.** positive

**4.** negative

**5.** zero

**6.** undefined

**7.** $y - y_1 = m(x - x_1)$

**8.** $y = mx + b$; slope; *y*-intercept

**9.** $(0,3)$; 2; 5

**10.** horizontal

**11.** vertical

**12.** general

**Exercise Set 2.3**

**1.** $m = \dfrac{10 - 7}{8 - 4} = \dfrac{3}{4}$; rises

**3.** $m = \dfrac{2 - 1}{2 - (-2)} = \dfrac{1}{4}$; rises

**5.** $m = \dfrac{2 - (-2)}{3 - 4} = \dfrac{0}{-1} = 0$; horizontal

**7.** $m = \dfrac{-1 - 4}{-1 + (-2)} = \dfrac{-5}{1} = -5$; falls

**9.** $m = \dfrac{-2 - 3}{5 - 5} = \dfrac{-5}{0}$ undefined; vertical

**11.** $m = 2$, $x_1 = 3$, $y_1 = 5$;
point-slope form: $y - 5 = 2(x - 3)$;
slope-intercept form: $y - 5 = 2x - 6$
$$y = 2x - 1$$

**13.** $m = 6$, $x_1 = -2$, $y_1 = 5$;
point-slope form: $y - 5 = 6(x + 2)$;
slope-intercept form: $y - 5 = 6x + 12$
$$y = 6x + 17$$

**15.** $m = -3$, $x_1 = -2$, $y_1 = -3$;
point-slope form: $y + 3 = -3(x + 2)$;
slope-intercept form: $y + 3 = -3x - 6$
$$y = -3x - 9$$

**17.** $m = -4$, $x_1 = -4$, $y_1 = 0$;
point-slope form: $y - 0 = -4(x + 4)$;
slope-intercept form: $y = -4(x + 4)$
$$y = -4x - 16$$

**19.** $m = -1$, $x_1 = \dfrac{-1}{2}$, $y_1 = -2$;

point-slope form: $y + 2 = -1\left(x + \dfrac{1}{2}\right)$;

slope-intercept form: $y + 2 = -x - \dfrac{1}{2}$
$$y = -x - \dfrac{5}{2}$$

**21.** $m = \dfrac{1}{2}$, $x_1 = 0$, $y_1 = 0$;

point-slope form: $y - 0 = \dfrac{1}{2}(x - 0)$;

slope-intercept form: $y = \dfrac{1}{2}x$

**23.** $m = -\dfrac{2}{3}$, $x_1 = 6$, $y_1 = -2$;

point-slope form: $y + 2 = -\dfrac{2}{3}(x - 6)$;

slope-intercept form: $y + 2 = -\dfrac{2}{3}x + 4$
$$y = -\dfrac{2}{3}x + 2$$

**25.** $m = \dfrac{10 - 2}{5 - 1} = \dfrac{8}{4} = 2$;
point-slope form: $y - 2 = 2(x - 1)$ using
$(x_1, y_1) = (1, 2)$, or $y - 10 = 2(x - 5)$ using
$(x_1, y_1) = (5, 10)$;
slope-intercept form:   $y - 2 = 2x - 2$ or
$$y - 10 = 2x - 10,$$
$$y = 2x$$

**27.** $m = \dfrac{3 - 0}{0 - (-3)} = \dfrac{3}{3} = 1$;
point-slope form: $y - 0 = 1(x + 3)$ using
$(x_1, y_1) = (-3, 0)$, or $y - 3 = 1(x - 0)$ using
$(x_1, y_1) = (0, 3)$; slope-intercept form: $y = x + 3$

**29.** $m = \dfrac{4 - (-1)}{2 - (-3)} = \dfrac{5}{5} = 1$;
point-slope form: $y + 1 = 1(x + 3)$ using
$(x_1, y_1) = (-3, -1)$, or $y - 4 = 1(x - 2)$ using
$(x_1, y_1) = (2, 4)$; slope-intercept form:
$y + 1 = x + 3$ or
$y - 4 = x - 2$
$$y = x + 2$$

**31.** $m = \dfrac{6 - (-2)}{3 - (-3)} = \dfrac{8}{6} = \dfrac{4}{3}$;
point-slope form: $y + 2 = \dfrac{4}{3}(x + 3)$ using

$(x_1, y_1) = (-3, -2)$, or $y - 6 = \dfrac{4}{3}(x - 3)$ using

$(x_1, y_1) = (3, 6)$;

slope-intercept form: $y + 2 = \dfrac{4}{3x} + 4$ or

$$y - 6 = \dfrac{4}{3}x - 4,$$
$$y = \dfrac{4}{3}x + 2$$

**33.** $m = \dfrac{-1 - (-1)}{4 - (-3)} = \dfrac{0}{7} = 0$;
point-slope form: $y + 1 = 0(x + 3)$ using
$(x_1, y_1) = (-3, -1)$, or $y + 1 = 0(x - 4)$ using
$(x_1, y_1) = (4, -1)$;
slope-intercept form: $y + 1 = 0$, so
$$y = -1$$

**35.** $m = \dfrac{0-4}{-2-2} = \dfrac{-4}{-4} = 1$;

point-slope form: $y - 4 = 1(x - 2)$ using
$(x_1, y_1) = (2, 4)$, or $y - 0 = 1(x + 2)$ using
$(x_1, y_1) = (-2, 0)$;

slope-intercept form: $y - 9 = x - 2$, or
$$y = x + 2$$

**37.** $m = \dfrac{4-0}{0-\left(-\frac{1}{2}\right)} = \dfrac{4}{\frac{1}{2}} = 8$;

point-slope form: $y - 4 = 8(x - 0)$ using
$(x_1, y_1) = (0, 4)$, or $y - 0 = 8\left(x + \frac{1}{2}\right)$ using
$(x_1, y_1) = \left(-\frac{1}{2}, 0\right)$; or $y - 0 = 8\left(x + \frac{1}{2}\right)$

slope-intercept form: $y = 8x + 4$

**39.** $m = 2; b = 1$

$y = 2x + 1$

**41.** $m = -2; b = 1$

$f(x) = -2x + 1$

**43.** $m = \dfrac{3}{4}; b = -2$

$f(x) = \dfrac{3}{4}x - 2$

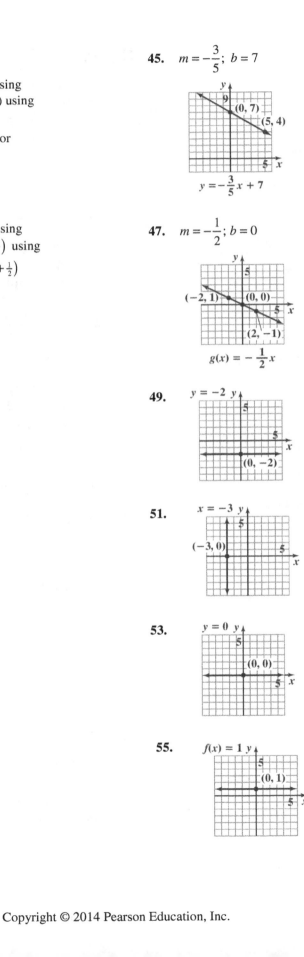

**45.** $m = -\dfrac{3}{5}; b = 7$

$y = -\dfrac{3}{5}x + 7$

**47.** $m = -\dfrac{1}{2}; b = 0$

$g(x) = -\dfrac{1}{2}x$

**49.** $y = -2$

**51.** $x = -3$

**53.** $y = 0$

**55.** $f(x) = 1$

**57.** $3x - 18 = 0$

$\qquad 3x = 18$

$\qquad x = 6$

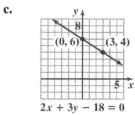

**59. a.** $3x + y - 5 = 0$

$\qquad y - 5 = -3x$

$\qquad y = -3x + 5$

**b.** $m = -3; b = 5$

**c.**

**61. a.** $2x + 3y - 18 = 0$

$\qquad 2x - 18 = -3y$

$\qquad -3y = 2x - 18$

$\qquad y = \dfrac{2}{-3}x - \dfrac{18}{-3}$

$\qquad y = -\dfrac{2}{3}x + 6$

**b.** $m = -\dfrac{2}{3}; b = 6$

**c.**

**63. a.** $8x - 4y - 12 = 0$

$\qquad 8x - 12 = 4y$

$\qquad 4y = 8x - 12$

$\qquad y = \dfrac{8}{4}x - \dfrac{12}{4}$

$\qquad y = 2x - 3$

**b.** $m = 2; b = -3$

**c.**

**65. a.** $3y - 9 = 0$

$\qquad 3y = 9$

$\qquad y = 3$

**b.** $m = 0; b = 3$

**c.**

**67.** Find the *x*-intercept:
$$6x - 2y - 12 = 0$$
$$6x - 2(0) - 12 = 0$$
$$6x - 12 = 0$$
$$6x = 12$$
$$x = 2$$
Find the *y*-intercept:
$$6x - 2y - 12 = 0$$
$$6(0) - 2y - 12 = 0$$
$$-2y - 12 = 0$$
$$-2y = 12$$
$$y = -6$$

$$6x - 2y - 12 = 0$$

**69.** Find the *x*-intercept:
$$2x + 3y + 6 = 0$$
$$2x + 3(0) + 6 = 0$$
$$2x + 6 = 0$$
$$2x = -6$$
$$x = -3$$
Find the *y*-intercept:
$$2x + 3y + 6 = 0$$
$$2(0) + 3y + 6 = 0$$
$$3y + 6 = 0$$
$$3y = -6$$
$$y = -2$$

**71.** Find the *x*-intercept:
$$8x - 2y + 12 = 0$$
$$8x - 2(0) + 12 = 0$$
$$8x + 12 = 0$$
$$8x = -12$$
$$\frac{8x}{8} = \frac{-12}{8}$$
$$x = \frac{-3}{2}$$
Find the *y*-intercept:
$$8x - 2y + 12 = 0$$
$$8(0) - 2y + 12 = 0$$
$$-2y + 12 = 0$$
$$-2y = -12$$
$$y = -6$$

$$8x - 2y + 12 = 0$$

**73.**
$$m = \frac{0 - a}{b - 0} = \frac{-a}{b} = -\frac{a}{b}$$

Since *a* and *b* are both positive, $-\frac{a}{b}$ is negative. Therefore, the line falls.

**75.**
$$m = \frac{(b + c) - b}{a - a} = \frac{c}{0}$$
The slope is undefined.
The line is vertical.

**77.**
$$Ax + By = C$$
$$By = -Ax + C$$
$$y = -\frac{A}{B}x + \frac{C}{B}$$

The slope is $-\frac{A}{B}$ and the *y* – intercept is $\frac{C}{B}$.

**79.**
$$-3 = \frac{4-y}{1-3}$$
$$-3 = \frac{4-y}{-2}$$
$$6 = 4-y$$
$$2 = -y$$
$$-2 = y$$

**81.**
$$3x - 4f(x) = 6$$
$$-4f(x) = -3x + 6$$
$$f(x) = \frac{3}{4}x - \frac{3}{2}$$

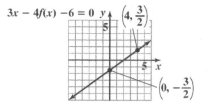

**83.** Using the slope-intercept form for the equation of a line:
$$-1 = -2(3) + b$$
$$-1 = -6 + b$$
$$5 = b$$

**85.** $m_1, m_3, m_2, m_4$

**87.**

**a.** First, find the slope using $(20, 38.9)$ and $(10, 31.1)$.
$$m = \frac{38.9 - 31.1}{20 - 10} = \frac{7.8}{10} = 0.78$$
Then use the slope and one of the points to write the equation in point-slope form.
$$y - y_1 = m(x - x_1)$$
$$y - 31.1 = 0.78(x - 10)$$
or
$$y - 38.9 = 0.78(x - 20)$$

**b.**
$$y - 31.1 = 0.78(x - 10)$$
$$y - 31.1 = 0.78x - 7.8$$
$$y = 0.78x + 23.3$$
$$f(x) = 0.78x + 23.3$$

**c.** $f(40) = 0.78(40) + 23.3 = 54.5$
The linear function predicts the percentage of never married American females, ages 25 – 29, to be 54.5% in 2020.

**89. a.** **Life Expectancy for United States Males, by Year of Birth**

**b.** $m = \dfrac{\text{Change in } y}{\text{Change in } x} = \dfrac{74.3 - 70.0}{40 - 20} = 0.215$
$$y - y_1 = m(x - x_1)$$
$$y - 70.0 = 0.215(x - 20)$$
$$y - 70.0 = 0.215x - 4.3$$
$$y = 0.215x + 65.7$$
$$E(x) = 0.215x + 65.7$$

**c.** $E(x) = 0.215x + 65.7$
$$E(60) = 0.215(60) + 65.7$$
$$= 78.6$$
The life expectancy of American men born in 2020 is expected to be 78.6.

**91.** $(10, 230)$ $(60, 110)$ Points may vary.
$$m = \frac{110 - 230}{60 - 10} = -\frac{120}{50} = -2.4$$
$$y - 230 = -2.4(x - 10)$$
$$y - 230 = -2.4x + 24$$
$$y = -2.4x + 254$$
Answers will vary for predictions.

**93. – 99.** Answers will vary.

**101.** Two points are $(0, 6)$ and $(10, -24)$.
$$m = \frac{-24 - 6}{10 - 0} = \frac{-30}{10} = -3.$$
Check: $y = mx + b$: $y = -3x + 6$.

**103.** Two points are $(0, -2)$ and $(10, 5.5)$.

$$m = \frac{5.5 - (-2)}{10 - 0} = \frac{7.5}{10} = 0.75 \text{ or } \frac{3}{4}.$$

Check: $y = mx + b$: $y = \frac{3}{4}x - 2$.

**105.** does not make sense; Explanations will vary. Sample explanation: Linear functions never change from increasing to decreasing.

**107.** does not make sense; Explanations will vary. Sample explanation: The slope of line's whose equations are in this form can be determined in several ways. One such way is to rewrite the equation in slope-intercept form.

**109.** false; Changes to make the statement true will vary. A sample change is: It is possible for $m$ to equal $b$.

**111.** true

**113.** We are given that the $x$-intercept is $-2$ and the $y$-intercept is $4$. We can use the points $(-2, 0)$ and $(0, 4)$ to find the slope.

$$m = \frac{4 - 0}{0 - (-2)} = \frac{4}{0 + 2} = \frac{4}{2} = 2$$

Using the slope and one of the intercepts, we can write the line in point-slope form.

$$y - y_1 = m(x - x_1)$$
$$y - 0 = 2(x - (-2))$$
$$y = 2(x + 2)$$
$$y = 2x + 4$$
$$-2x + y = 4$$

Find the $x$– and $y$–coefficients for the equation of the line with right-hand-side equal to 12. Multiply both sides of $-2x + y = 4$ by 3 to obtain 12 on the right-hand-side.

$$-2x + y = 4$$
$$3(-2x + y) = 3(4)$$
$$-6x + 3y = 12$$

Therefore, the coefficient of $x$ is $-6$ and the coefficient of $y$ is 3.

**115.** Answers will vary.

**117.** Answers will vary.

**118.** Since the slope is the same as the slope of $y = 2x + 1$, then $m = 2$.

$$y - y_1 = m(x - x_1)$$
$$y - 1 = 2(x - (-3))$$
$$y - 1 = 2(x + 3)$$
$$y - 1 = 2x + 6$$
$$y = 2x + 7$$

**119.** Since the slope is the negative reciprocal of $-\frac{1}{4}$, then $m = 4$.

$$y - y_1 = m(x - x_1)$$
$$y - (-5) = 4(x - 3)$$
$$y + 5 = 4x - 12$$
$$-4x + y + 17 = 0$$
$$4x - y - 17 = 0$$

**120.**
$$\frac{f(x_2) - f(x_1)}{x_2 - x_1} = \frac{f(4) - f(1)}{4 - 1}$$
$$= \frac{4^2 - 1^2}{4 - 1}$$
$$= \frac{15}{3}$$
$$= 5$$

**Section 2.4**

**Check Point Exercises**

**1.** The slope of the line $y = 3x + 1$ is 3.

$$y - y_1 = m(x - x_1)$$
$$y - 5 = 3(x - (-2))$$
$$y - 5 = 3(x + 2) \quad \text{point-slope}$$
$$y - 5 = 3x + 6$$
$$y = 3x + 11 \quad \text{slope-intercept}$$

**2. a.** Write the equation in slope-intercept form:
$$x + 3y - 12 = 0$$
$$3y = -x + 12$$
$$y = -\frac{1}{3}x + 4$$

The slope of this line is $-\frac{1}{3}$ thus the slope of any line perpendicular to this line is 3.

**b.** Use $m = 3$ and the point $(-2, -6)$ to write the equation.
$$y - y_1 = m(x - x_1)$$
$$y - (-6) = 3(x - (-2))$$
$$y + 6 = 3(x + 2)$$
$$y + 6 = 3x + 6$$
$$-3x + y = 0$$
$$3x - y = 0 \text{ general form}$$

**3.** $m = \dfrac{\text{Change in } y}{\text{Change in } x} = \dfrac{14.7 - 9.0}{2008 - 1990} = \dfrac{5.7}{18} \approx 0.32$

The slope indicates that the number of U.S. men living alone increased at a rate of 0.32 million each year.
The rate of change is 0.32 million men per year.

**4.** **a.** $\dfrac{f(x_2) - f(x_1)}{x_2 - x_1} = \dfrac{1^3 - 0^3}{1 - 0} = 1$

  **b.** $\dfrac{f(x_2) - f(x_1)}{x_2 - x_1} = \dfrac{2^3 - 1^3}{2 - 1} = \dfrac{8 - 1}{1} = 7$

  **c.** $\dfrac{f(x_2) - f(x_1)}{x_2 - x_1} = \dfrac{0^3 - (-2)^3}{0 - (-2)} = \dfrac{8}{2} = 4$

**5.** $\dfrac{f(x_2) - f(x_1)}{x_2 - x_1} = \dfrac{f(3) - f(1)}{3 - 1}$
$$= \dfrac{0.05 - 0.03}{3 - 1}$$
$$= 0.01$$
The average rate of change in the drug's concentration between 1 hour and 3 hours is 0.01 mg per 100 mL per hour.

## Concept and Vocabulary Check 2.4

**1.** the same

**2.** $-1$

**3.** $-\dfrac{1}{3}$ ; 3

**4.** $-2$ ; $\dfrac{1}{2}$

**5.** $y$; $x$

**6.** $\dfrac{f(x_2) - f(x_1)}{x_2 - x_1}$

## Exercise Set 2.4

**1.** Since $L$ is parallel to $y = 2x$, we know it will have slope $m = 2$. We are given that it passes through $(4, 2)$. We use the slope and point to write the equation in point-slope form.
$$y - y_1 = m(x - x_1)$$
$$y - 2 = 2(x - 4)$$
Solve for $y$ to obtain slope-intercept form.
$$y - 2 = 2(x - 4)$$
$$y - 2 = 2x - 8$$
$$y = 2x - 6$$
In function notation, the equation of the line is $f(x) = 2x - 6$.

**3.** Since $L$ is perpendicular to $y = 2x$, we know it will have slope $m = -\dfrac{1}{2}$. We are given that it passes through $(2, 4)$. We use the slope and point to write the equation in point-slope form.
$$y - y_1 = m(x - x_1)$$
$$y - 4 = -\dfrac{1}{2}(x - 2)$$
Solve for $y$ to obtain slope-intercept form.
$$y - 4 = -\dfrac{1}{2}(x - 2)$$
$$y - 4 = -\dfrac{1}{2}x + 1$$
$$y = -\dfrac{1}{2}x + 5$$
In function notation, the equation of the line is $f(x) = -\dfrac{1}{2}x + 5$.

**5.** $m = -4$ since the line is parallel to $y = -4x + 3$; $x_1 = -8$, $y_1 = -10$; point-slope form:
$$y + 10 = -4(x + 8)$$
slope-intercept form: $y + 10 = -4x - 32$
$$y = -4x - 42$$

**7.** $m = -5$ since the line is perpendicular to $y = \dfrac{1}{5}x + 6$; $x_1 = 2$, $y_1 = -3$;
point-slope form: $y + 3 = -5(x - 2)$
slope-intercept form: $y + 3 = -5x + 10$
$$y = -5x + 7$$

**9.** $2x - 3y - 7 = 0$

$$-3y = -2x + 7$$

$$y = \frac{2}{3}x - \frac{7}{3}$$

The slope of the given line is $\frac{2}{3}$, so $m = \frac{2}{3}$ since the lines are parallel.

point-slope form: $y - 2 = \frac{2}{3}(x + 2)$

general form: $2x - 3y + 10 = 0$

**11.** $x - 2y - 3 = 0$

$$-2y = -x + 3$$

$$y = \frac{1}{2}x - \frac{3}{2}$$

The slope of the given line is $\frac{1}{2}$, so $m = -2$ since the lines are perpendicular.

point-slope form: $y + 7 = -2(x - 4)$

general form: $2x + y - 1 = 0$

**13.** $\frac{15 - 0}{5 - 0} = \frac{15}{5} = 3$

**15.** $\frac{5^2 + 2 \cdot 5 - (3^2 + 2 \cdot 3)}{5 - 3} = \frac{25 + 10 - (9 + 6)}{2}$

$$= \frac{20}{2}$$

$$= 10$$

**17.** $\frac{\sqrt{9} - \sqrt{4}}{9 - 4} = \frac{3 - 2}{5} = \frac{1}{5}$

**19.** Since the line is perpendicular to $x = 6$ which is a vertical line, we know the graph of $f$ is a horizontal line with 0 slope. The graph of $f$ passes through $(-1, 5)$, so the equation of $f$ is $f(x) = 5$.

**21.** First we need to find the equation of the line with $x$-intercept of 2 and $y$-intercept of $-4$. This line will pass through $(2, 0)$ and $(0, -4)$. We use these points to find the slope.

$$m = \frac{-4 - 0}{0 - 2} = \frac{-4}{-2} = 2$$

Since the graph of $f$ is perpendicular to this line, it will have slope $m = -\frac{1}{2}$.

Use the point $(-6, 4)$ and the slope $-\frac{1}{2}$ to find the equation of the line.

$$y - y_1 = m(x - x_1)$$

$$y - 4 = -\frac{1}{2}(x - (-6))$$

$$y - 4 = -\frac{1}{2}(x + 6)$$

$$y - 4 = -\frac{1}{2}x - 3$$

$$y = -\frac{1}{2}x + 1$$

$$f(x) = -\frac{1}{2}x + 1$$

**23.** First put the equation $3x - 2y - 4 = 0$ in slope-intercept form.

$$3x - 2y - 4 = 0$$

$$-2y = -3x + 4$$

$$y = \frac{3}{2}x - 2$$

The equation of $f$ will have slope $-\frac{2}{3}$ since it is perpendicular to the line above and the same $y$-intercept $-2$.

So the equation of $f$ is $f(x) = -\frac{2}{3}x - 2$.

**25.** $p(x) = -0.25x + 22$

**27.** $m = \frac{1163 - 617}{1998 - 1994} = \frac{546}{4} \approx 137$

There was an average increase of approximately 137 discharges per year.

**29.** **a.** $f(x) = 1.1x^3 - 35x^2 + 264x + 557$

$$f(0) = 1.1(0)^3 - 35(0)^2 + 264(0) + 557 = 557$$

$$f(4) = 1.1(4)^3 - 35(4)^2 + 264(4) + 557 = 1123.4$$

$$m = \frac{1123.4 - 557}{4 - 0} \approx 142$$

**b.** This overestimates by 5 discharges per year.

**31. – 35.** Answers will vary.

**37.** $y = \dfrac{1}{3}x + 1$

$y = -3x - 2$

  **a.** The lines are perpendicular because their slopes are negative reciprocals of each other. This is verified because product of their slopes is $-1$.

  **b.** The lines do not appear to be perpendicular.

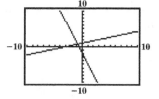

  **c.** The lines appear to be perpendicular. The calculator screen is rectangular and does not have the same width and height. This causes the scale of the $x$–axis to differ from the scale on the $y$–axis despite using the same scale in the window settings. In part (b), this causes the lines not to appear perpendicular when indeed they are. The zoom square feature compensates for this and in part (c), the lines appear to be perpendicular.

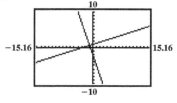

**39.** makes sense

**41.** makes sense

**43.** The slope of the line containing $(1, -3)$ and $(-2, 4)$

has slope $m = \dfrac{4 - (-3)}{-2 - 1} = \dfrac{4 + 3}{-3} = \dfrac{7}{-3} = -\dfrac{7}{3}$

Solve $Ax + y - 2 = 0$ for $y$ to obtain slope-intercept form.

$Ax + y - 2 = 0$

$y = -Ax + 2$

So the slope of this line is $-A$.

This line is perpendicular to the line above so its

slope is $\dfrac{3}{7}$. Therefore, $-A = \dfrac{3}{7}$ so $A = -\dfrac{3}{7}$.

**44.  a.**

  **b.**

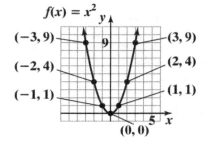

  **c.** The graph in part (b) is the graph in part (a) shifted down 4 units.

**45.  a.**

  **b.**

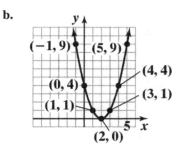

  **c.** The graph in part (b) is the graph in part (a) shifted to the right 2 units.

**46. a.**

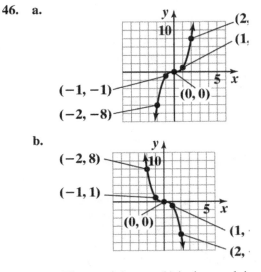

(2,
(1,
(−1, −1)
(0, 0)
(−2, −8)

**b.**

(−2, 8)
(−1, 1)
(0, 0)
(1,
(2,

**c.** The graph in part (b) is the graph in part (a) reflected across the *y*-axis.

## Mid-Chapter 2 Check Point

**1.** The relation is not a function.
The domain is $\{1, 2\}$.
The range is $\{-6, 4, 6\}$.

**2.** The relation is a function.
The domain is $\{0, 2, 3\}$.
The range is $\{1, 4\}$.

**3.** The relation is a function.
The domain is $\{x \mid -2 \le x < 2\}$.
The range is $\{y \mid 0 \le y \le 3\}$.

**4.** The relation is not a function.
The domain is $\{x \mid -3 < x \le 4\}$.
The range is $\{y \mid -1 \le y \le 2\}$.

**5.** The relation is not a function.
The domain is $\{-2, -1, 0, 1, 2\}$.
The range is $\{-2, -1, 1, 3\}$.

**6.** The relation is a function.
The domain is $\{x \mid x \le 1\}$.
The range is $\{y \mid y \ge -1\}$.

**7.** $x^2 + y = 5$
$$y = -x^2 + 5$$
For each value of *x*, there is one and only one value for *y*, so the equation defines *y* as a function of *x*.

**8.** $x + y^2 = 5$
$$y^2 = 5 - x$$
$$y = \pm\sqrt{5 - x}$$
Since there are values of *x* that give more than one value for *y* (for example, if $x = 4$, then $y = \pm\sqrt{5 - 4} = \pm 1$), the equation does not define *y* as a function of *x*.

**9.** No vertical line intersects the graph in more than one point. Each value of *x* corresponds to exactly one value of *y*.

**10.** Domain: $(-\infty, \infty)$

**11.** Range: $(-\infty, 4]$

**12.** *x*-intercepts: −6 and 2

**13.** *y*-intercept: 3

**14.** increasing: $(-\infty, -2)$

**15.** decreasing: $(-2, \infty)$

**16.** $x = -2$

**17.** $f(-2) = 4$

**18.** $f(-4) = 3$

**19.** $f(-7) = -2$ and $f(3) = -2$

**20.** $f(-6) = 0$ and $f(2) = 0$

**21.** $(-6, 2)$

**22.** $f(100)$ is negative.

**23.** neither; $f(-x) \ne x$ and $f(-x) \ne -x$

**24.** $\dfrac{f(x_2) - f(x_1)}{x_2 - x_1} = \dfrac{f(4) - f(-4)}{4 - (-4)} = \dfrac{-5 - 3}{4 + 4} = -1$

**25.** $y = -2x$

(−1, 2)
(0, 0)
(1, −2)

Copyright © 2014 Pearson Education, Inc.

**26.** $y = -2$

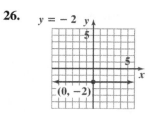

$(0, -2)$

**27.** $x + y = -2$

$(-2, 0)$  $(0, -2)$

**28.** $y = \frac{1}{3}x - 2$

$(0, -2)$  $(3, -1)$

**29.** $x = 3.5$

$(3.5, 0)$

**30.**

$(2, 0)$  $(0, -4)$  $4x - 2y = 8$

**31.**

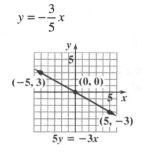

$(-2, 0)$  $(2, 0)$  $(-1, -3)$  $(1, -3)$  $(0, -4)$  $f(x) = x^2 - 4$

**32.**

$(4, 0)$  $(0, -4)$  $f(x) = x - 4$

**33.**

$(-4, 0)$  $(4, 0)$  $(-2, -2)$  $(2, -2)$  $(0, -4)$  $f(x) = |x| - 4$

**34.** $5y = -3x$

$y = -\frac{3}{5}x$

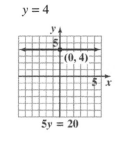

$(-5, 3)$  $(0, 0)$  $(5, -3)$  $5y = -3x$

**35.** $5y = 20$

$y = 4$

$(0, 4)$  $5y = 20$

**36.**

$(0, 1)$  $(1, 3)$  $(-4, -1)$  $(0, -1)$

$f(x) = \begin{cases} -1 & \text{if } x \le 0 \\ 2x + 1 & \text{if } x > 0 \end{cases}$

**37. a.** $f(-x) = -2(-x)^2 - x - 5$

$= -2x^2 - x - 5$

neither; $f(-x) \neq x$ and $f(-x) \neq -x$

**b.** $\dfrac{f(x+h) - f(x)}{h}$

$= \dfrac{-2(x+h)^2 + (x+h) - 5 - (-2x^2 + x - 5)}{h}$

$= \dfrac{-2x^2 - 4xh - 2h^2 + x + h - 5 + 2x^2 - x + 5}{h}$

$= \dfrac{-4xh - 2h^2 + h}{h}$

$= \dfrac{h(-4x - 2h + 1)}{h}$

$= -4x - 2h + 1$

**38.** $C(x) = \begin{cases} 30 & \text{if } 0 \le t \le 200 \\ 30 + 0.40(t-200) & \text{if } t > 200 \end{cases}$

**a.** $C(150) = 30$

**b.** $C(250) = 30 + 0.40(250 - 200) = 50$

**39.** $y - y_1 = m(x - x_1)$

$y - 3 = -2(x - (-4))$

$y - 3 = -2(x + 4)$

$y - 3 = -2x - 8$

$y = -2x - 5$

$f(x) = -2x - 5$

**40.** $m = \dfrac{\text{Change in } y}{\text{Change in } x} = \dfrac{1 - (-5)}{2 - (-1)} = \dfrac{6}{3} = 2$

$y - y_1 = m(x - x_1)$

$y - 1 = 2(x - 2)$

$y - 1 = 2x - 4$

$y = 2x - 3$

$f(x) = 2x - 3$

**41.** $3x - y - 5 = 0$

$-y = -3x + 5$

$y = 3x - 5$

The slope of the given line is 3, and the lines are parallel, so $m = 3$.

$y - y_1 = m(x - x_1)$

$y - (-4) = 3(x - 3)$

$y + 4 = 3x - 9$

$y = 3x - 13$

$f(x) = 3x - 13$

**42.** $2x - 5y - 10 = 0$

$-5y = -2x + 10$

$\dfrac{-5y}{-5} = \dfrac{-2x}{-5} + \dfrac{10}{-5}$

$y = \dfrac{2}{5}x - 2$

The slope of the given line is $\dfrac{2}{5}$, and the lines are perpendicular, so $m = -\dfrac{5}{2}$.

$y - y_1 = m(x - x_1)$

$y - (-3) = -\dfrac{5}{2}(x - (-4))$

$y + 3 = -\dfrac{5}{2}x - 10$

$y = -\dfrac{5}{2}x - 13$

$f(x) = -\dfrac{5}{2}x - 13$

**43.** $m_1 = \dfrac{\text{Change in } y}{\text{Change in } x} = \dfrac{0 - (-4)}{7 - 2} = \dfrac{4}{5}$

$m_2 = \dfrac{\text{Change in } y}{\text{Change in } x} = \dfrac{6 - 2}{1 - (-4)} = \dfrac{4}{5}$

The slope of the lines are equal thus the lines are parallel.

**44. a.** $m = \dfrac{\text{Change in } y}{\text{Change in } x} = \dfrac{42 - 26}{180 - 80} = \dfrac{16}{100} = 0.16$

**b.** For each minute of brisk walking, the percentage of patients with depression in remission increased by 0.16%. The rate of change is 0.16% per minute of brisk walking.

**45.** $\dfrac{f(x_2) - f(x_1)}{x_2 - x_1} = \dfrac{f(2) - f(-1)}{2 - (-1)}$

$= \dfrac{(3(2)^2 - 2) - (3(-1)^2 - (-1))}{2 + 1}$

$= 2$

**Section 2.5**

**Check Point Exercises**

1.  Shift up vertically 3 units.

2.  Shift to the right 4 units.

3.  Shift to the right 1 unit and down 2 units.

    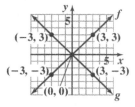

4.  Reflect about the *x*-axis.

5.  Reflect about the *y*-axis.

    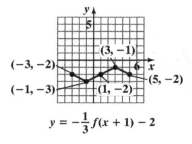

6.  Vertically stretch the graph of $f(x) = |x|$.

7.  **a.**  Horizontally shrink the graph of $y = f(x)$.

    $$g(x) = f(2x)$$

    **b.**  Horizontally stretch the graph of $y = f(x)$.

    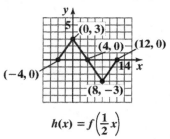

    $$h(x) = f\left(\frac{1}{2}x\right)$$

8.  The graph of $y = f(x)$ is shifted 1 unit left, shrunk by a factor of $\frac{1}{3}$, reflected about the x-axis, then shifted down 2 units.

    $$y = -\frac{1}{3}f(x+1) - 2$$

**9.** The graph of $f(x) = x^2$ is shifted 1 unit right, stretched by a factor of 2, then shifted up 3 units.

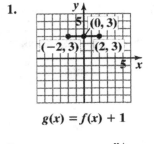

$$g(x) = 2(x-1)^2 + 3$$

**Concept and Vocabulary Check 2.5**

**1.** vertical; down

**2.** horizontal; to the right

**3.** *x*-axis

**4.** *y*-axis

**5.** vertical; *y*

**6.** horizontal; *x*

**7.** false

**Exercise Set 2.5**

**1.**

$$g(x) = f(x) + 1$$

**3.**

$$g(x) = f(x+1)$$

**5.**

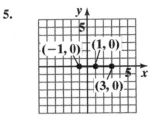

$$g(x) = f(x-1) - 2$$

**7.**

$$g(x) = -f(x)$$

**9.**

$$g(x) = -f(x) + 3$$

**11.**

$$g(x) = \frac{1}{2}f(x)$$

**13.** $g(x) = f\left(\frac{1}{2}x\right)$

**15.**

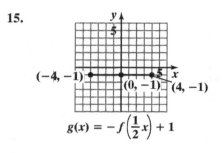

(−4, −1)    (0, −1)    (4, −1)

$$g(x) = -f\left(\frac{1}{2}x\right) + 1$$

**17.**

(2, 1)
(−4, −1)    (4, −1)
(−2, −3)  (0, −1)

$$g(x) = f(x) - 1$$

**19.**

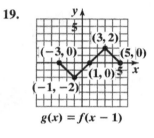

(3, 2)
(−3, 0)    (5, 0)
(1, 0)
(−1, −2)

$$g(x) = f(x - 1)$$

**21.**

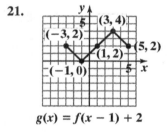

(−3, 2)    (3, 4)
(1, 2)    (5, 2)
(−1, 0)

$$g(x) = f(x - 1) + 2$$

**23.**

(−2, 2)
(4, 0)
(−4, 0)
(0, 0)  (2, −2)

$$g(x) = -f(x)$$

**25.**

(−2, 3)
(−4, 1)    (4, 1)
(0, 1)  (2, −1)

$$g(x) = f(-x) + 1$$

**27.**

(2, 4)
(−4, 0)    (0, 0)    5
(4, 0)
(−2, −4)

$$g(x) = 2f(x)$$

**29.**

(1, 2)
(−2, 0)    (2, 0)
(0, 0)
(−1, −2)

$$g(x) = f(2x)$$

**31.**

(0, 5)
(−2, 1)
(−6, 1)    (2, 1)
(−4, −3)

$$g(x) = 2f(x + 2) + 1$$

**33.**

(−4, 2)    (0, 2)
(0, 0)  (4, 0)

$$g(x) = f(x) + 2$$

**35.**

(−6, 0)    (−2, 0)
(−2, −2)  (2, −2)

$$g(x) = f(x + 2)$$

**37.**

(−2, 2)    (2, 2)
(−6, 0)
(−2, 0)

$$g(x) = -f(x + 2)$$

**39.**

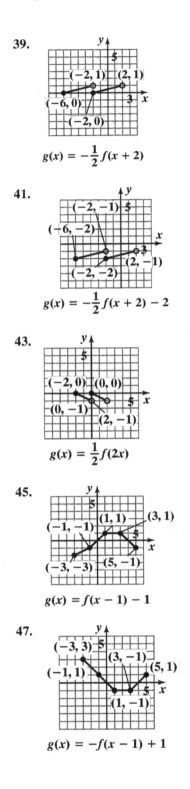

$$g(x) = -\frac{1}{2}f(x+2)$$

**41.**

$$g(x) = -\frac{1}{2}f(x+2) - 2$$

**43.**

$$g(x) = \frac{1}{2}f(2x)$$

**45.**

$$g(x) = f(x-1) - 1$$

**47.**

$$g(x) = -f(x-1) + 1$$

**49.**

$$g(x) = 2f\left(\frac{1}{2}x\right)$$

**51.**

$$g(x) = \frac{1}{2}f(x+1)$$

**53.**

**55.**

**57.**

**59.**

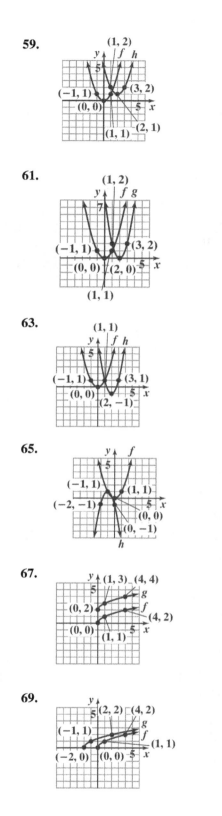

**61.**

**63.**

**65.**

**67.**

**69.**

**71.**

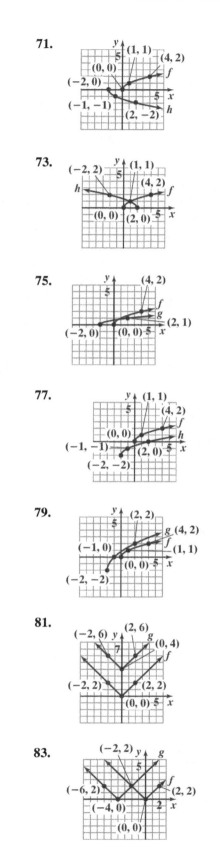

**73.**

**75.**

**77.**

**79.**

**81.**

**83.**

**85.**

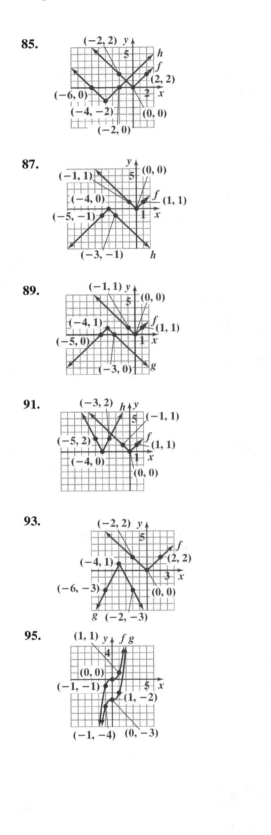

**87.**

**89.**

**91.**

**93.**

**95.**

**97.**

**99.**

**101.**

**103.**

**105.**

**107.**

**109.**

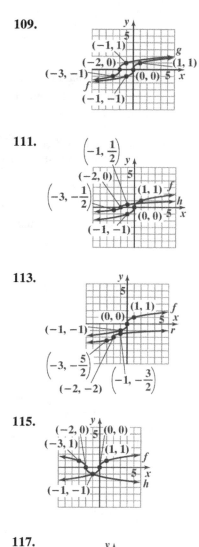

**111.**

**113.**

**115.**

**117.**

**119.**

$g(x) = 2 \operatorname{int}(x+1)$

**121.**

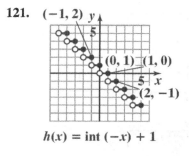

$h(x) = \operatorname{int}(-x) + 1$

**123.**  $y = \sqrt{x-2}$

**125.**  $y = (x+1)^2 - 4$

**127. a.**  First, vertically stretch the graph of $f(x) = \sqrt{x}$ by the factor 2.9; then shift the result up 20.1 units.

**b.**  $f(x) = 2.9\sqrt{x} + 20.1$

$f(48) = 2.9\sqrt{48} + 20.1 \approx 40.2$

The model describes the actual data very well.

**c.**  $\dfrac{f(x_2) - f(x_1)}{x_2 - x_1}$

$= \dfrac{f(10) - f(0)}{10 - 0}$

$= \dfrac{\left(2.9\sqrt{10} + 20.1\right) - \left(2.9\sqrt{0} + 20.1\right)}{10 - 0}$

$= \dfrac{29.27 - 20.1}{10}$

$\approx 0.9$

0.9 inches per month

**d.**  $\dfrac{f(x_2) - f(x_1)}{x_2 - x_1}$

$= \dfrac{f(60) - f(50)}{60 - 50}$

$= \dfrac{\left(2.9\sqrt{60} + 20.1\right) - \left(2.9\sqrt{50} + 20.1\right)}{60 - 50}$

$= \dfrac{42.5633 - 40.6061}{10}$

$\approx 0.2$

This rate of change is lower than the rate of change in part (c). The relative leveling off of the curve shows this difference.

**129. – 133.**  Answers will vary.

**135. a.**

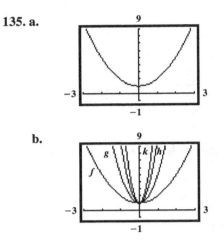

**b.**

**137.** makes sense

**139.** does not make sense; Explanations will vary. Sample explanation: The reprogram should be $y = f(t+1)$.

**141.** false; Changes to make the statement true will vary. A sample change is: The graph of *g* is a translation of *f* three units to the <u>left</u> and three units upward.

**143.** false; Changes to make the statement true will vary. A sample change is: The stretch will be 5 units and the downward shift will be 10 units.

**145.** $g(x) = -(x+4)^2$

**147.** $g(x) = -\sqrt{x-2} + 2$

**149.** $(-a, b)$

**151.** $(a + 3, b)$

**153.** $(2x-1)(x^2 + x - 2) = 2x(x^2 + x - 2) - 1(x^2 + x - 2)$
$$= 2x^3 + 2x^2 - 4x - x^2 - x + 2$$
$$= 2x^3 + 2x^2 - x^2 - 4x - x + 2$$
$$= 2x^3 + x^2 - 5x + 2$$

**154.** $(f(x))^2 - 2f(x) + 6 = (3x-4)^2 - 2(3x-4) + 6$
$$= 9x^2 - 24x + 16 - 6x + 8 + 6$$
$$= 9x^2 - 24x - 6x + 16 + 8 + 6$$
$$= 9x^2 - 30x + 30$$

**155.** $\dfrac{2}{\dfrac{3}{x} - 1} = \dfrac{2x}{\dfrac{3x}{x} - x} = \dfrac{2x}{3 - x}$

**Section 2.6**

**Check Point Exercises**

**1.  a.** The function $f(x) = x^2 + 3x - 17$ contains neither division nor an even root. The domain of *f* is the set of all real numbers or $(-\infty, \infty)$.

  **b.** The denominator equals zero when $x = 7$ or $x = -7$. These values must be excluded from the domain.
  domain of $g = (-\infty, -7) \cup (-7, 7) \cup (7, \infty)$.

  **c.** Since $h(x) = \sqrt{9x - 27}$ contains an even root; the quantity under the radical must be greater than or equal to 0.
  $$9x - 27 \ge 0$$
  $$9x \ge 27$$
  $$x \ge 3$$
  Thus, the domain of *h* is $\{x \mid x \ge 3\}$, or the interval $[3, \infty)$.

  **d.** Since the denominator of $j(x)$ contains an even root; the quantity under the radical must be greater than or equal to 0. But that quantity must also not be 0 (because we cannot have division by 0). Thus, $24 - 3x$ must be strictly greater than 0.
  $$24 - 3x > 0$$
  $$-3x > -24$$
  $$x < 8$$
  Thus, the domain of *j* is $\{x \mid x < 8\}$, or the interval $(-\infty, 8)$.

**2.  a.** $(f + g)(x) = f(x) + g(x)$
  $$= x - 5 + (x^2 - 1)$$
  $$= x - 5 + x^2 - 1$$
  $$= -x^2 + x - 6$$
  domain: $(-\infty, \infty)$

  **b.** $(f - g)(x) = f(x) - g(x)$
  $$= x - 5 - (x^2 - 1)$$
  $$= x - 5 - x^2 + 1$$
  $$= -x^2 + x - 4$$
  domain: $(-\infty, \infty)$

**c.** $(fg)(x) = (x-5)(x^2-1)$

$\qquad = x(x^2-1)-5(x^2-1)$

$\qquad = x^3-x-5x^2+5$

$\qquad = x^3-5x^2-x+5$

domain: $(-\infty,\infty)$

**d.** $\left(\dfrac{f}{g}\right)(x) = \dfrac{f(x)}{g(x)}$

$\qquad = \dfrac{x-5}{x^2-1}, \; x \neq \pm 1$

domain: $(-\infty,-1)\cup(-1,1)\cup(1,\infty)$

**3. a.** $(f+g)(x) = f(x)+g(x)$

$\qquad = \sqrt{x-3}+\sqrt{x+1}$

**b.** domain of *f*: $\quad x-3 \geq 0$

$\qquad\qquad\qquad\quad x \geq 3$

$\qquad\qquad\qquad [3,\infty)$

domain of g: $\quad x+1 \geq 0$

$\qquad\qquad\qquad\quad x \geq -1$

$\qquad\qquad\qquad [-1,\infty)$

The domain of *f* + *g* is the set of all real numbers that are common to the domain of *f* and the domain of *g*. Thus, the domain of *f* + *g* is [3, ∞).

**4. a.** $(B+D)(x)$

$= B(x)+D(x)$

$= (-2.6x^2+49x+3994)+(-0.6x^2+7x+2412)$

$= -2.6x^2+49x+3994-0.6x^2+7x+2412$

$= -3.2x^2+56x+6406$

**b.** $(B+D)(x) = -3.2x^2+56x+6406$

$(B+D)(5) = -3.2(3)^2+56(3)+6406$

$\qquad\qquad = 6545.2$

The number of births and deaths in the U.S. in 2003 was 6545.2 thousand.

**c.** $(B+D)(x)$ overestimates the actual number of births and deaths in 2003 by 7.2 thousand.

**5. a.** $(f \circ g)(x) = f(g(x))$

$\qquad = 5(2x^2-x-1)+6$

$\qquad = 10x^2-5x-5+6$

$\qquad = 10x^2-5x+1$

**b.** $(g \circ f)(x) = g(f(x))$

$\qquad = 2(5x+6)^2-(5x+6)-1$

$\qquad = 2(25x^2+60x+36)-5x-6-1$

$\qquad = 50x^2+120x+72-5x-6-1$

$\qquad = 50x^2+115x+65$

**c.** $(f \circ g)(x) = 10x^2-5x+1$

$(f \circ g)(-1) = 10(-1)^2-5(-1)+1$

$\qquad\qquad = 10+5+1$

$\qquad\qquad = 16$

**6. a.** $(f \circ g)(x) = \dfrac{4}{\dfrac{1}{x}+2} = \dfrac{4x}{1+2x}$

**b.** domain: $\left\{x \mid x \neq 0, x \neq -\dfrac{1}{2}\right\}$

or $\left(-\infty,-\dfrac{1}{2}\right)\cup\left(-\dfrac{1}{2},0\right)\cup(0,\infty)$

**7.** $h(x) = f \circ g$ where $f(x) = \sqrt{x}; \; g(x) = x^2+5$

**Concept and Vocabulary Check 2.6**

**1.** zero

**2.** negative

**3.** $f(x)+g(x)$

**4.** $f(x)-g(x)$

**5.** $f(x) \cdot g(x)$

**6.** $\dfrac{f(x)}{g(x)}; \; g(x)$

**7.** $(-\infty,\infty)$

**8.** $(2,\infty)$

**9.** $(0,3); \; (3,\infty)$

**10.** composition; $f(g(x))$

**11.** *f*; $g(x)$

**12.** composition; $g(f(x))$

**13.** $g; \ f(x)$

**14.** false

**15.** false

**16.** 2

**Exercise Set 2.6**

**1.** The function contains neither division nor an even root. The domain $= (-\infty, \infty)$

**3.** The denominator equals zero when $x = 4$. This value must be excluded from the domain.
domain: $(-\infty, 4) \cup (4, \infty)$.

**5.** The function contains neither division nor an even root. The domain $= (-\infty, \infty)$

**7.** The values that make the denominator equal zero must be excluded from the domain.
domain: $(-\infty, -3) \cup (-3, 5) \cup (5, \infty)$

**9.** The values that make the denominators equal zero must be excluded from the domain.
domain: $(-\infty, -7) \cup (-7, 9) \cup (9, \infty)$

**11.** The first denominator cannot equal zero. The values that make the second denominator equal zero must be excluded from the domain.
domain: $(-\infty, -1) \cup (-1, 1) \cup (1, \infty)$

**13.** Exclude $x$ for $x = 0$.
Exclude $x$ for $\dfrac{3}{x} - 1 = 0$.
$$\frac{3}{x} - 1 = 0$$
$$x\left(\frac{3}{x} - 1\right) = x(0)$$
$$3 - x = 0$$
$$-x = -3$$
$$x = 3$$
domain: $(-\infty, 0) \cup (0, 3) \cup (3, \infty)$

**15.** Exclude $x$ for $x - 1 = 0$.
$$x - 1 = 0$$
$$x = 1$$

Exclude $x$ for $\dfrac{4}{x-1} - 2 = 0$.
$$\frac{4}{x-1} - 2 = 0$$
$$(x-1)\left(\frac{4}{x-1} - 2\right) = (x-1)(0)$$
$$4 - 2(x-1) = 0$$
$$4 - 2x + 2 = 0$$
$$-2x + 6 = 0$$
$$-2x = -6$$
$$x = 3$$
domain: $(-\infty, 1) \cup (1, 3) \cup (3, \infty)$

**17.** The expression under the radical must not be negative.
$$x - 3 \geq 0$$
$$x \geq 3$$
domain: $[3, \infty)$

**19.** The expression under the radical must be positive.
$$x - 3 > 0$$
$$x > 3$$
domain: $(3, \infty)$

**21.** The expression under the radical must not be negative.
$$5x + 35 \geq 0$$
$$5x \geq -35$$
$$x \geq -7$$
domain: $[-7, \infty)$

**23.** The expression under the radical must not be negative.
$$24 - 2x \geq 0$$
$$-2x \geq -24$$
$$\frac{-2x}{-2} \leq \frac{-24}{-2}$$
$$x \leq 12$$
domain: $(-\infty, 12]$

**25.** The expressions under the radicals must not be negative.
$$x - 2 \geq 0 \quad \text{and} \quad x + 3 \geq 0$$
$$x \geq 2 \qquad\qquad x \geq -3$$
To make both inequalities true, $x \geq 2$.
domain: $[2, \infty)$

**27.** The expression under the radical must not be negative.

$x - 2 \geq 0$

$x \geq 2$

The denominator equals zero when $x = 5$.

domain: $[2,5) \cup (5,\infty)$.

**29.** Find the values that make the denominator equal zero and must be excluded from the domain.

$x^3 - 5x^2 - 4x + 20$

$= x^2(x-5) - 4(x-5)$

$= (x-5)(x^2-4)$

$= (x-5)(x+2)(x-2)$

−2, 2, and 5 must be excluded.

domain: $(-\infty,-2) \cup (-2,2) \cup (2,5) \cup (5,\infty)$

**31.** $(f+g)(x) = 3x + 2$

domain: $(-\infty, \infty)$

$(f-g)(x) = f(x) - g(x)$

$= (2x+3) - (x-1)$

$= x + 4$

domain: $(-\infty, \infty)$

$(fg)(x) = f(x) \cdot g(x)$

$= (2x+3) \cdot (x-1)$

$= 2x^2 + x - 3$

domain: $(-\infty, \infty)$

$\left(\dfrac{f}{g}\right)(x) = \dfrac{f(x)}{g(x)} = \dfrac{2x+3}{x-1}$

domain: $(-\infty,1) \cup (1,\infty)$

**33.** $(f+g)(x) = 3x^2 + x - 5$

domain: $(-\infty, \infty)$

$(f-g)(x) = -3x^2 + x - 5$

domain: $(-\infty, \infty)$

$(fg)(x) = (x-5)(3x^2) = 3x^3 - 15x^2$

domain: $(-\infty, \infty)$

$\left(\dfrac{f}{g}\right)(x) = \dfrac{x-5}{3x^2}$

domain: $(-\infty,0) \cup (0,\infty)$

**35.** $(f+g)(x) = 2x^2 - 2$

domain: $(-\infty, \infty)$

$(f-g)(x) = 2x^2 - 2x - 4$

domain: $(-\infty, \infty)$

$(fg)(x) = (2x^2 - x - 3)(x+1)$

$= 2x^3 + x^2 - 4x - 3$

domain: $(-\infty, \infty)$

$\left(\dfrac{f}{g}\right)(x) = \dfrac{2x^2 - x - 3}{x+1}$

$= \dfrac{(2x-3)(x+1)}{(x+1)} = 2x - 3$

domain: $(-\infty,-1) \cup (-1,\infty)$

**37.** $(f+g)(x) = (3-x^2) + (x^2 + 2x - 15)$

$= 2x - 12$

domain: $(-\infty, \infty)$

$(f-g)(x) = (3-x^2) - (x^2 + 2x - 15)$

$= -2x^2 - 2x + 18$

domain: $(-\infty, \infty)$

$(fg)(x) = (3-x^2)(x^2 + 2x - 15)$

$= -x^4 - 2x^3 + 18x^2 + 6x - 45$

domain: $(-\infty, \infty)$

$\left(\dfrac{f}{g}\right)(x) = \dfrac{3-x^2}{x^2 + 2x - 15}$

domain: $(-\infty,-5) \cup (-5,3) \cup (3,\infty)$

**39.** $(f+g)(x) = \sqrt{x} + x - 4$

domain: $[0, \infty)$

$(f-g)(x) = \sqrt{x} - x + 4$

domain: $[0, \infty)$

$(fg)(x) = \sqrt{x}(x-4)$

domain: $[0, \infty)$

$\left(\dfrac{f}{g}\right)(x) = \dfrac{\sqrt{x}}{x-4}$

domain: $[0,4) \cup (4,\infty)$

**41.** $(f+g)(x) = 2 + \dfrac{1}{x} + \dfrac{1}{x} = 2 + \dfrac{2}{x} = \dfrac{2x+2}{x}$

domain: $(-\infty,0) \cup (0,\infty)$

$(f-g)(x) = 2 + \dfrac{1}{x} - \dfrac{1}{x} = 2$

domain: $(-\infty,0) \cup (0,\infty)$

$(fg)(x) = \left(2+\dfrac{1}{x}\right) \cdot \dfrac{1}{x} = \dfrac{2}{x} + \dfrac{1}{x^2} = \dfrac{2x+1}{x^2}$

domain: $(-\infty,0) \cup (0,\infty)$

$\left(\dfrac{f}{g}\right)(x) = \dfrac{2+\frac{1}{x}}{\frac{1}{x}} = \left(2+\dfrac{1}{x}\right) \cdot x = 2x+1$

domain: $(-\infty,0) \cup (0,\infty)$

**43.** $(f+g)(x) = f(x)+g(x)$

$$= \frac{5x+1}{x^2-9} + \frac{4x-2}{x^2-9}$$

$$= \frac{9x-1}{x^2-9}$$

domain: $(-\infty,-3)\cup(-3,3)\cup(3,\infty)$

$(f-g)(x) = f(x)-g(x)$

$$= \frac{5x+1}{x^2-9} - \frac{4x-2}{x^2-9}$$

$$= \frac{x+3}{x^2-9}$$

$$= \frac{1}{x-3}$$

domain: $(-\infty,-3)\cup(-3,3)\cup(3,\infty)$

$(fg)(x) = f(x)\cdot g(x)$

$$= \frac{5x+1}{x^2-9} \cdot \frac{4x-2}{x^2-9}$$

$$= \frac{(5x+1)(4x-2)}{\left(x^2-9\right)^2}$$

domain: $(-\infty,-3)\cup(-3,3)\cup(3,\infty)$

$$\left(\frac{f}{g}\right)(x) = \frac{\frac{5x+1}{x^2-9}}{\frac{4x-2}{x^2-9}}$$

$$= \frac{5x+1}{x^2-9} \cdot \frac{x^2-9}{4x-2}$$

$$= \frac{5x+1}{4x-2}$$

The domain must exclude –3, 3, and any values that make $4x-2=0$.

$4x-2=0$

$4x=2$

$x=\frac{1}{2}$

domain: $(-\infty,-3)\cup\left(-3,\frac{1}{2}\right)\cup\left(\frac{1}{2},3\right)\cup(3,\infty)$

**45.** $(f+g)(x) = f(x)+g(x)$

$$= \frac{8x}{x-2} + \frac{6}{x+3}$$

$$= \frac{8x(x+3)}{(x-2)(x+3)} + \frac{6(x-2)}{(x-2)(x+3)}$$

$$= \frac{8x^2+24x}{(x-2)(x+3)} + \frac{6x-12}{(x-2)(x+3)}$$

$$= \frac{8x^2+30x-12}{(x-2)(x+3)}$$

domain: $(-\infty,-3)\cup(-3,2)\cup(2,\infty)$

$(f+g)(x) = f(x)-g(x)$

$$= \frac{8x}{x-2} - \frac{6}{x+3}$$

$$= \frac{8x(x+3)}{(x-2)(x+3)} - \frac{6(x-2)}{(x-2)(x+3)}$$

$$= \frac{8x^2+24x}{(x-2)(x+3)} - \frac{6x-12}{(x-2)(x+3)}$$

$$= \frac{8x^2+18x+12}{(x-2)(x+3)}$$

domain: $(-\infty,-3)\cup(-3,2)\cup(2,\infty)$

$(fg)(x) = f(x)\cdot g(x)$

$$= \frac{8x}{x-2} \cdot \frac{6}{x+3}$$

$$= \frac{48x}{(x-2)(x+3)}$$

domain: $(-\infty,-3)\cup(-3,2)\cup(2,\infty)$

$$\left(\frac{f}{g}\right)(x) = \frac{\frac{8x}{x-2}}{\frac{6}{x+3}}$$

$$= \frac{8x}{x-2} \cdot \frac{x+3}{6}$$

$$= \frac{4x(x+3)}{3(x-2)}$$

The domain must exclude –3, 2, and any values that make $3(x-2)=0$.

$3(x-2)=0$

$3x-6=0$

$3x=6$

$x=2$

domain: $(-\infty,-3)\cup(-3,2)\cup(2,\infty)$

**47.** $(f+g)(x) = \sqrt{x+4} + \sqrt{x-1}$
domain: $[1, \infty)$
$(f-g)(x) = \sqrt{x+4} - \sqrt{x-1}$
domain: $[1, \infty)$
$(fg)(x) = \sqrt{x+4} \cdot \sqrt{x-1} = \sqrt{x^2+3x-4}$
domain: $[1, \infty)$
$\left(\dfrac{f}{g}\right)(x) = \dfrac{\sqrt{x+4}}{\sqrt{x-1}}$
domain: $(1, \infty)$

**49.** $(f+g)(x) = \sqrt{x-2} + \sqrt{2-x}$
domain: $\{2\}$
$(f-g)(x) = \sqrt{x-2} - \sqrt{2-x}$
domain: $\{2\}$
$(fg)(x) = \sqrt{x-2} \cdot \sqrt{2-x} = \sqrt{-x^2+4x-4}$
domain: $\{2\}$
$\left(\dfrac{f}{g}\right)(x) = \dfrac{\sqrt{x-2}}{\sqrt{2-x}}$
domain: $\varnothing$

**51.** $f(x) = 2x;\ g(x) = x+7$

   **a.**  $(f \circ g)(x) = 2(x+7) = 2x+14$

   **b.**  $(g \circ f)(x) = 2x+7$

   **c.**  $(f \circ g)(2) = 2(2)+14 = 18$

**53.** $f(x) = x+4;\ g(x) = 2x+1$

   **a.**  $(f \circ g)(x) = (2x+1)+4 = 2x+5$

   **b.**  $(g \circ f)(x) = 2(x+4)+1 = 2x+9$

   **c.**  $(f \circ g)(2) = 2(2)+5 = 9$

**55.** $f(x) = 4x-3;\ g(x) = 5x^2-2$

   **a.**  $(f \circ g)(x) = 4(5x^2-2)-3$
             $= 20x^2-11$

   **b.**  $(g \circ f)(x) = 5(4x-3)^2-2$
             $= 5(16x^2-24x+9)-2$
             $= 80x^2-120x+43$

   **c.**  $(f \circ g)(2) = 20(2)^2-11 = 69$

**57.** $f(x) = x^2+2;\ g(x) = x^2-2$

   **a.**  $(f \circ g)(x) = (x^2-2)^2+2$
             $= x^4-4x^2+4+2$
             $= x^4-4x^2+6$

   **b.**  $(g \circ f)(x) = (x^2+2)^2-2$
             $= x^4+4x^2+4-2$
             $= x^4+4x^2+2$

   **c.**  $(f \circ g)(2) = 2^4-4(2)^2+6 = 6$

**59.** $f(x) = 4-x;\ \ g(x) = 2x^2+x+5$

   **a.**  $(f \circ g)(x) = 4-\left(2x^2+x+5\right)$
             $= 4-2x^2-x-5$
             $= -2x^2-x-1$

   **b.**  $(g \circ f)(x) = 2(4-x)^2+(4-x)+5$
             $= 2(16-8x+x^2)+4-x+5$
             $= 32-16x+2x^2+4-x+5$
             $= 2x^2-17x+41$

   **c.**  $(f \circ g)(2) = -2(2)^2-2-1 = -11$

**61.** $f(x) = \sqrt{x};\ g(x) = x-1$

   **a.**  $(f \circ g)(x) = \sqrt{x-1}$

   **b.**  $(g \circ f)(x) = \sqrt{x}-1$

   **c.**  $(f \circ g)(2) = \sqrt{2-1} = \sqrt{1} = 1$

**63.** $f(x) = 2x-3;\ g(x) = \dfrac{x+3}{2}$

   **a.**  $(f \circ g)(x) = 2\left(\dfrac{x+3}{2}\right)-3$
             $= x+3-3$
             $= x$

   **b.**  $(g \circ f)(x) = \dfrac{(2x-3)+3}{2} = \dfrac{2x}{2} = x$

   **c.**  $(f \circ g)(2) = 2$

**65.** $f(x) = \dfrac{1}{x}$; $g(x) = \dfrac{1}{x}$

    **a.** $(f \circ g)(x) = \dfrac{1}{\frac{1}{x}} = x$

    **b.** $(g \circ f)(x) = \dfrac{1}{\frac{1}{x}} = x$

    **c.** $(f \circ g)(2) = 2$

**67.** **a.** $(f \circ g)(x) = f\left(\dfrac{1}{x}\right) = \dfrac{2}{\frac{1}{x} + 3}, x \neq 0$

       $= \dfrac{2(x)}{\left(\frac{1}{x} + 3\right)(x)}$

       $= \dfrac{2x}{1 + 3x}$

    **b.** We must exclude 0 because it is excluded from $g$.

       We must exclude $-\dfrac{1}{3}$ because it causes the denominator of $f \circ g$ to be 0.

       domain: $\left(-\infty, -\dfrac{1}{3}\right) \cup \left(-\dfrac{1}{3}, 0\right) \cup (0, \infty)$.

**69.** **a.** $(f \circ g)(x) = f\left(\dfrac{4}{x}\right) = \dfrac{\frac{4}{x}}{\frac{4}{x} + 1}$

       $= \dfrac{\left(\frac{4}{x}\right)(x)}{\left(\frac{4}{x} + 1\right)(x)}$

       $= \dfrac{4}{4 + x}, x \neq -4$

    **b.** We must exclude 0 because it is excluded from $g$.
       We must exclude $-4$ because it causes the denominator of $f \circ g$ to be 0.

       domain: $(-\infty, -4) \cup (-4, 0) \cup (0, \infty)$.

**71.** **a.** $f \circ g(x) = f(x - 2) = \sqrt{x - 2}$

    **b.** The expression under the radical in $f \circ g$ must not be negative.

$x - 2 \geq 0$

$x \geq 2$

domain: $[2, \infty)$.

**73.** **a.** $(f \circ g)(x) = f(\sqrt{1 - x})$

       $= \left(\sqrt{1 - x}\right)^2 + 4$

       $= 1 - x + 4$

       $= 5 - x$

    **b.** The domain of $f \circ g$ must exclude any values that are excluded from $g$.

       $1 - x \geq 0$

       $-x \geq -1$

       $x \leq 1$

       domain: $(-\bullet, 1]$.

**75.** $f(x) = x^4$    $g(x) = 3x - 1$

**77.** $f(x) = \sqrt[3]{x}$    $g(x) = x^2 - 9$

**79.** $f(x) = |x|$    $g(x) = 2x - 5$

**81.** $f(x) = \dfrac{1}{x}$    $g(x) = 2x - 3$

**83.** $(f + g)(-3) = f(-3) + g(-3) = 4 + 1 = 5$

**85.** $(fg)(2) = f(2)g(2) = (-1)(1) = -1$

**87.** The domain of $f + g$ is $[-4, 3]$.

**89.** The graph of $f + g$

**91.** $(f \circ g)(-1) = f(g(-1)) = f(-3) = 1$

**93.** $(g \circ f)(0) = g(f(0)) = g(2) = -6$

**95.**
$$(f \circ g)(x) = 7$$
$$2\left(x^2 - 3x + 8\right) - 5 = 7$$
$$2x^2 - 6x + 16 - 5 = 7$$
$$2x^2 - 6x + 11 = 7$$
$$2x^2 - 6x + 4 = 0$$
$$x^2 - 3x + 2 = 0$$
$$(x-1)(x-2) = 0$$
$$x - 1 = 0 \quad \text{or} \quad x - 2 = 0$$
$$x = 1 \qquad \qquad x = 2$$

**97.** **a.** $(M+F)(x) = M(x) + F(x)$
$$= (1.53x + 114.8) + (1.46x + 120.7)$$
$$= 2.99x + 235.5$$

**b.** $(M+F)(x) = 2.99x + 235.5$
$(M+F)(20) = 2.99(20) + 235.5$
$$= 295.3$$
The total U.S. population in 2005 was 295.3 million.

**c.** The result in part (b) underestimates the actual total by 2.7 million.

**99.** $(R - C)(20,000)$
$= 65(20,000) - (600,000 + 45(20,000))$
$= -200,000$
The company lost $200,000 since costs exceeded revenues.
$(R - C)(30,000)$
$= 65(30,000) - (600,000 + 45(30,000))$
$= 0$
The company broke even.

**101. a.** *f* gives the price of the computer after a $400 discount. *g* gives the price of the computer after a 25% discount.

**b.** $(f \circ g)(x) = 0.75x - 400$
This models the price of a computer after first a 25% discount and then a $400 discount.

**c.** $(g \circ f)(x) = 0.75(x - 400)$
This models the price of a computer after first a $400 discount and then a 25% discount.

**d.** The function $f \circ g$ models the greater discount, since the 25% discount is taken on the regular price first.

**103. – 107.** Answers will vary.

**109.**

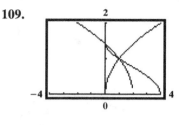

$$(f \circ g)(x) = \sqrt{2 - \sqrt{x}}$$
The domain of g is $[0, \infty)$.
The expression under the radical in $f \circ g$ must not be negative.
$$2 - \sqrt{x} \geq 0$$
$$-\sqrt{x} \geq -2$$
$$\sqrt{x} \leq 2$$
$$x \leq 4$$
domain: $[0, 4]$

**111.** makes sense

**113.** does not make sense; Explanations will vary. Sample explanation: The diagram illustrates
$$g\left(f(x)\right) = x^2 + 4.$$

**115.** false; Changes to make the statement true will vary. A sample change is:
$$f(x) = 2x; g(x) = 3x$$
$$(f \circ g)(x) = f\left(g(x)\right) = f(3x) = 2(3x) = 6x$$
$$(g \circ f)(x) = g\left(f(x)\right) = g(f(x)) = 3(2x) = 6x$$

**117.** true

**119.** Answers will vary.

**120.** $\{(4, -2), (1, -1), (1, 1), (4, 2)\}$
The element 1 in the domain corresponds to two elements in the range.
Thus, the relation is not a function.

**121.** $x = \dfrac{5}{y} + 4$

$$y(x) = y\left(\dfrac{5}{y} + 4\right)$$

$$xy = 5 + 4y$$

$$xy - 4y = 5$$

$$y(x - 4) = 5$$

$$y = \dfrac{5}{x - 4}$$

**122.** $x = y^2 - 1$

$$x + 1 = y^2$$

$$\sqrt{x + 1} = \sqrt{y^2}$$

$$\sqrt{x + 1} = y$$

$$y = \sqrt{x + 1}$$

## Section 2.7

## Check Point Exercises

**1.** $f\left(g(x)\right) = 4\left(\dfrac{x + 7}{4}\right) - 7$

$$= x + 7 - 7$$

$$= x$$

$$g\left(f(x)\right) = \dfrac{(4x - 7) + 7}{4}$$

$$= \dfrac{4x - 7 + 7}{4}$$

$$= \dfrac{4x}{4}$$

$$= x$$

$$f\left(g(x)\right) = g\left(f(x)\right) = x$$

**2.** $f(x) = 2x + 7$

Replace $f(x)$ with $y$:

$y = 2x + 7$

Interchange $x$ and $y$:

$x = 2y + 7$

Solve for $y$:

$$x = 2y + 7$$

$$x - 7 = 2y$$

$$\dfrac{x - 7}{2} = y$$

Replace y with $f^{-1}(x)$:

$$f^{-1}(x) = \dfrac{x - 7}{2}$$

**3.** $f(x) = 4x^3 - 1$

Replace $f(x)$ with $y$:

$y = 4x^3 - 1$

Interchange $x$ and $y$:

$x = 4y^3 - 1$

Solve for $y$:

$$x = 4y^3 - 1$$

$$x + 1 = 4y^3$$

$$\dfrac{x + 1}{4} = y^3$$

$$\sqrt[3]{\dfrac{x + 1}{4}} = y$$

Replace $y$ with $f^{-1}(x)$:

$$f^{-1}(x) = \sqrt[3]{\dfrac{x + 1}{4}}$$

Alternative form for answer:

$$f(x)^{-1} = \sqrt[3]{\dfrac{x + 1}{4}} = \dfrac{\sqrt[3]{x + 1}}{\sqrt[3]{4}}$$

$$= \dfrac{\sqrt[3]{x + 1}}{\sqrt[3]{4}} \cdot \dfrac{\sqrt[3]{2}}{\sqrt[3]{2}} = \dfrac{\sqrt[3]{2x + 2}}{\sqrt[3]{8}}$$

$$= \dfrac{\sqrt[3]{2x + 2}}{2}$$

**4.** $f(x) = \dfrac{3}{x} - 1$

Replace $f(x)$ with $y$:

$$y = \dfrac{3}{x} - 1$$

Interchange $x$ and $y$:

$$x = \dfrac{3}{y} - 1$$

Solve for $y$:

$$x = \dfrac{3}{y} - 1$$

$$xy = 3 - y$$

$$xy + y = 3$$

$$y(x + 1) = 3$$

$$y = \dfrac{3}{x + 1}$$

Replace $y$ with $f^{-1}(x)$:

$$f^{-1}(x) = \dfrac{3}{x + 1}$$

**5.** The graphs of (b) and (c) pass the horizontal line test and thus have an inverse.

**6.** Find points of $f^{-1}$.

| $f(x)$ | $f^{-1}(x)$ |
|---|---|
| $(-2,-2)$ | $(-2,-2)$ |
| $(-1,0)$ | $(0,-1)$ |
| $(1,2)$ | $(2,1)$ |

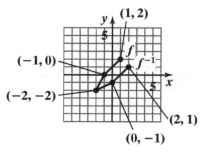

**7.** $f(x) = x^2 + 1$

Replace $f(x)$ with $y$:

$y = x^2 + 1$

Interchange $x$ and $y$:

$x = y^2 + 1$

Solve for $y$:

$x = y^2 + 1$

$x - 1 = y^2$

$\sqrt{x-1} = y$

Replace $y$ with $f^{-1}(x)$:

$f^{-1}(x) = \sqrt{x-1}$

**Concept and Vocabulary Check 2.7**

**1.** inverse

**2.** $x$; $x$

**3.** horizontal; one-to-one

**4.** $y = x$

**Exercise Set 2.7**

**1.** $f(x) = 4x; g(x) = \dfrac{x}{4}$

$f(g(x)) = 4\left(\dfrac{x}{4}\right) = x$

$g(f(x)) = \dfrac{4x}{4} = x$

$f$ and $g$ are inverses.

**3.** $f(x) = 3x + 8;\ g(x) = \dfrac{x-8}{3}$

$f(g(x)) = 3\left(\dfrac{x-8}{3}\right) + 8 = x - 8 + 8 = x$

$g(f(x)) = \dfrac{(3x+8)-8}{3} = \dfrac{3x}{3} = x$

$f$ and $g$ are inverses.

**5.** $f(x) = 5x - 9;\ g(x) = \dfrac{x+5}{9}$

$f(g(x)) = 5\left(\dfrac{x+5}{9}\right) - 9$

$= \dfrac{5x+25}{9} - 9$

$= \dfrac{5x-56}{9}$

$g(f(x)) = \dfrac{5x-9+5}{9} = \dfrac{5x-4}{9}$

$f$ and $g$ are not inverses.

**7.** $f(x) = \dfrac{3}{x-4};\ g(x) = \dfrac{3}{x} + 4$

$f(g(x)) = \dfrac{3}{\frac{3}{x}+4-4} = \dfrac{3}{\frac{3}{x}} = x$

$g(f(x)) = \dfrac{3}{\frac{3}{x-4}} + 4$

$= 3 \cdot \left(\dfrac{x-4}{3}\right) + 4$

$= x - 4 + 4$

$= x$

$f$ and $g$ are inverses.

**9.** $f(x) = -x; g(x) = -x$

$f(g(x)) = -(-x) = x$

$g(f(x)) = -(-x) = x$

$f$ and $g$ are inverses.

**11. a.**
$$f(x) = x + 3$$
$$y = x + 3$$
$$x = y + 3$$
$$y = x - 3$$
$$f^{-1}(x) = x - 3$$

**b.**
$$f(f^{-1}(x)) = x - 3 + 3 = x$$
$$f^{-1}(f(x)) = x + 3 - 3 = x$$

**13. a.**
$$f(x) = 2x$$
$$y = 2x$$
$$x = 2y$$
$$y = \frac{x}{2}$$
$$f^{-1}(x) = \frac{x}{2}$$

**b.**
$$f(f^{-1}(x)) = 2\left(\frac{x}{2}\right) = x$$
$$f^{-1}(f(x)) = \frac{2x}{2} = x$$

**15. a.**
$$f(x) = 2x + 3$$
$$y = 2x + 3$$
$$x = 2y + 3$$
$$x - 3 = 2y$$
$$y = \frac{x-3}{2}$$
$$f^{-1}(x) = \frac{x-3}{2}$$

**b.**
$$f(f^{-1}(x)) = 2\left(\frac{x-3}{2}\right) + 3$$
$$= x - 3 + 3$$
$$= x$$
$$f^{-1}(f(x)) = \frac{2x+3-3}{2} = \frac{2x}{2} = x$$

**17. a.**
$$f(x) = x^3 + 2$$
$$y = x^3 + 2$$
$$x = y^3 + 2$$
$$x - 2 = y^3$$
$$y = \sqrt[3]{x-2}$$
$$f^{-1}(x) = \sqrt[3]{x-2}$$

**b.**
$$f(f^{-1}(x)) = \left(\sqrt[3]{x-2}\right)^3 + 2$$
$$= x - 2 + 2$$
$$= x$$
$$f^{-1}(f(x)) = \sqrt[3]{x^3 + 2 - 2} = \sqrt[3]{x^3} = x$$

**19. a.**
$$f(x) = (x+2)^3$$
$$y = (x+2)^3$$
$$x = (y+2)^3$$
$$\sqrt[3]{x} = y + 2$$
$$y = \sqrt[3]{x} - 2$$
$$f^{-1}(x) = \sqrt[3]{x} - 2$$

**b.**
$$f(f^{-1}(x)) = \left(\sqrt[3]{x} - 2 + 2\right)^3 = \left(\sqrt[3]{x}\right)^3 = x$$
$$f^{-1}(f(x)) = \sqrt[3]{(x+2)^3} - 2$$
$$= x + 2 - 2$$
$$= x$$

**21. a.**
$$f(x) = \frac{1}{x}$$
$$y = \frac{1}{x}$$
$$x = \frac{1}{y}$$
$$xy = 1$$
$$y = \frac{1}{x}$$
$$f^{-1}(x) = \frac{1}{x}$$

**b.**
$$f(f^{-1}(x)) = \frac{1}{\frac{1}{x}} = x$$
$$f^{-1}(f(x)) = \frac{1}{\frac{1}{x}} = x$$

**23. a.**
$$f(x) = \sqrt{x}$$
$$y = \sqrt{x}$$
$$x = \sqrt{y}$$
$$y = x^2$$
$$f^{-1}(x) = x^2, x \geq 0$$

**b.**
$$f(f^{-1}(x)) = \sqrt{x^2} = |x| = x \text{ for } x \geq 0.$$
$$f^{-1}(f(x)) = (\sqrt{x})^2 = x$$

**25. a.**
$$f(x) = \frac{7}{x} - 3$$
$$y = \frac{7}{x} - 3$$
$$x = \frac{7}{y} - 3$$
$$xy = 7 - 3y$$
$$xy + 3y = 7$$
$$y(x+3) = 7$$
$$y = \frac{7}{x+3}$$
$$f^{-1}(x) = \frac{7}{x+3}$$

**b.**
$$f\left(f^{-1}(x)\right) = \frac{7}{\frac{7}{x+3}} - 3 = x$$
$$f^{-1}\left(f(x)\right) = \frac{7}{\frac{7}{x} - 3 + 3} = x$$

**27. a.**
$$f(x) = \frac{2x+1}{x-3}$$
$$y = \frac{2x+1}{x-3}$$
$$x = \frac{2y+1}{y-3}$$
$$x(y-3) = 2y+1$$
$$xy - 3x = 2y+1$$
$$xy - 2y = 3x+1$$
$$y(x-2) = 3x+1$$
$$y = \frac{3x+1}{x-2}$$
$$f^{-1}(x) = \frac{3x+1}{x-2}$$

**b.**
$$f(f^{-1}(x)) = \frac{2\left(\frac{3x+1}{x-2}\right)+1}{\frac{3x+1}{x-2}-3}$$
$$= \frac{2(3x+1)+x-2}{3x+1-3(x-2)} = \frac{6x+2+x-2}{3x+1-3x+6}$$
$$= \frac{7x}{7} = x$$

$$f^{-1}(f(x)) = \frac{3\left(\frac{2x+1}{x-3}\right)+1}{\frac{2x+1}{x-3}-2}$$
$$= \frac{3(2x+1)+x-3}{2x+1-2(x-3)}$$
$$= \frac{6x+3+x-3}{2x+1-2x+6} = \frac{7x}{7} = x$$

**29.** The function fails the horizontal line test, so it does not have an inverse function.

**31.** The function fails the horizontal line test, so it does not have an inverse function.

**33.** The function passes the horizontal line test, so it does have an inverse function.

**35.**

**37.**

**39. a.**
$$f(x) = 2x-1$$
$$y = 2x-1$$
$$x = 2y-1$$
$$x+1 = 2y$$
$$\frac{x+1}{2} = y$$
$$f^{-1}(x) = \frac{x+1}{2}$$

**b.**

**c.**    domain of $f$ : $(-\infty,\infty)$

range of $f$ : $(-\infty,\infty)$

domain of $f^{-1}$ : $(-\infty,\infty)$

range of $f^{-1}$ : $(-\infty,\infty)$

**41. a.**    $f(x) = x^2 - 4$

$$y = x^2 - 4$$
$$x = y^2 - 4$$
$$x + 4 = y^2$$
$$\sqrt{x+4} = y$$
$$f^{-1}(x) = \sqrt{x+4}$$

**b.**

**c.**    domain of $f$ : $[0,\infty)$

range of $f$ : $[-4,\infty)$

domain of $f^{-1}$ : $[-4,\infty)$

range of $f^{-1}$ : $[0,\infty)$

**43. a.**    $f(x) = (x-1)^2$

$$y = (x-1)^2$$
$$x = (y-1)^2$$
$$-\sqrt{x} = y - 1$$
$$-\sqrt{x} + 1 = y$$
$$f^{-1}(x) = 1 - \sqrt{x}$$

**b.**

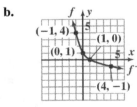

**c.**    domain of $f$ : $(-\infty,1]$

range of $f$ : $[0,\infty)$

domain of $f^{-1}$ : $[0,\infty)$

range of $f^{-1}$ : $(-\infty,1]$

**45. a.**    $f(x) = x^3 - 1$

$$y = x^3 - 1$$
$$x = y^3 - 1$$
$$x + 1 = y^3$$
$$\sqrt[3]{x+1} = y$$
$$f^{-1}(x) = \sqrt[3]{x+1}$$

**b.**

**c.**    domain of $f$ : $(-\infty,\infty)$

range of $f$ : $(-\infty,\infty)$

domain of $f^{-1}$ : $(-\infty,\infty)$

range of $f^{-1}$ : $(-\infty,\infty)$

**47. a.**    $f(x) = (x+2)^3$

$$y = (x+2)^3$$
$$x = (y+2)^3$$
$$\sqrt[3]{x} = y + 2$$
$$\sqrt[3]{x} - 2 = y$$
$$f^{-1}(x) = \sqrt[3]{x} - 2$$

**b.**

**c.**    domain of $f$ : $(-\infty,\infty)$

range of $f$ : $(-\infty,\infty)$

domain of $f^{-1}$ : $(-\infty,\infty)$

range of $f^{-1}$ : $(-\infty,\infty)$

**49. a.**
$$f(x) = \sqrt{x-1}$$
$$y = \sqrt{x-1}$$
$$x = \sqrt{y-1}$$
$$x^2 = y - 1$$
$$x^2 + 1 = y$$
$$f^{-1}(x) = x^2 + 1$$

**b.**

**c.** domain of $f$ : $[1, \infty)$
range of $f$ : $[0, \infty)$
domain of $f^{-1}$ : $[0, \infty)$
range of $f^{-1}$ : $[1, \infty)$

**51. a.**
$$f(x) = \sqrt[3]{x} + 1$$
$$y = \sqrt[3]{x} + 1$$
$$x = \sqrt[3]{y} + 1$$
$$x - 1 = \sqrt[3]{y}$$
$$(x-1)^3 = y$$
$$f^{-1}(x) = (x-1)^3$$

**b.**

**c.** domain of $f$ : $(-\infty, \infty)$
range of $f$ : $(-\infty, \infty)$
domain of $f^{-1}$ : $(-\infty, \infty)$
range of $f^{-1}$ : $(-\infty, \infty)$

**53.** $f(g(1)) = f(1) = 5$

**55.** $(g \circ f)(-1) = g(f(-1)) = g(1) = 1$

**57.** $f^{-1}(g(10)) = f^{-1}(-1) = 2$, since $f(2) = -1$.

**59.**
$$(f \circ g)(0) = f(g(0))$$
$$= f(4 \cdot 0 - 1)$$
$$= f(-1) = 2(-1) - 5 = -7$$

**61.** Let $f^{-1}(1) = x$. Then
$$f(x) = 1$$
$$2x - 5 = 1$$
$$2x = 6$$
$$x = 3$$
Thus, $f^{-1}(1) = 3$

**63.**
$$g(f[h(1)]) = g(f[1^2 + 1 + 2])$$
$$= g(f(4))$$
$$= g(2 \cdot 4 - 5)$$
$$= g(3)$$
$$= 4 \cdot 3 - 1 = 11$$

**65. a.** {(Zambia, 4.2), (Colombia, 4.5),
(Poland, 3.3), (Italy, 3.3),
(United States, 2.5)}

**b.** {(4.2, Zambia), (4.5 , Colombia),
(3.3 , Poland), (3.3, Italy),
(2.5, United States)}
$f$ is not a one-to-one function because the
inverse of $f$ is not a function.

**67. a.** It passes the horizontal line test and is one-to-one.

**b.** $f^{-1}(0.25) = 15$ If there are 15 people in the
room, the probability that 2 of them have the
same birthday is 0.25.
$f^{-1}(0.5) = 21$ If there are 21 people in the room,
the probability that 2 of them have the same
birthday is 0.5.
$f^{-1}(0.7) = 30$ If there are 30 people in the room,
the probability that 2 of them have the same
birthday is 0.7.

**69.**
$$f(g(x)) = \frac{9}{5}\left[\frac{5}{9}(x - 32)\right] + 32$$
$$= x - 32 + 32$$
$$= x$$
$$g(f(x)) = \frac{5}{9}\left[\left(\frac{9}{5}x + 32\right) - 32\right]$$
$$= x + 32 - 32$$
$$= x$$
$f$ and $g$ are inverses.

**71. – 75.** Answers will vary.

**77.**

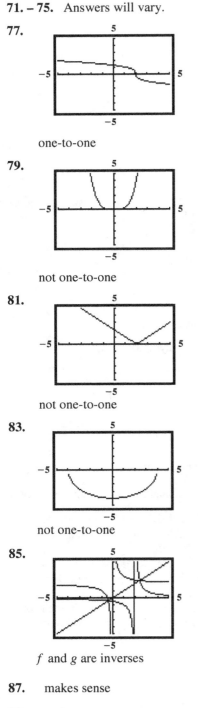

one-to-one

**79.**

not one-to-one

**81.**

not one-to-one

**83.**

not one-to-one

**85.**

$f$ and $g$ are inverses

**87.** makes sense

**89.** makes sense

**91.** false; Changes to make the statement true will vary. A sample change is: The inverse is $\{(4,1), (7,2)\}$.

**93.** false; Changes to make the statement true will vary.

A sample change is: $f^{-1}(x) = \dfrac{x}{3}$.

**95.** $(f \circ g)(x) = 3(x+5) = 3x+15$.

$$y = 3x+15$$
$$x = 3y+15$$
$$y = \frac{x-15}{3}$$
$$(f \circ g)^{-1}(x) = \frac{x-15}{3}$$
$$g(x) = x+5$$
$$y = x+5$$
$$x = y+5$$
$$y = x-5$$
$$g^{-1}(x) = x-5$$
$$f(x) = 3x$$
$$y = 3x$$
$$x = 3y$$
$$y = \frac{x}{3}$$
$$f^{-1}(x) = \frac{x}{3}$$
$$\left(g^{-1} \circ f^{-1}\right)(x) = \frac{x}{3} - 5 = \frac{x-15}{3}$$

**97.** No, there will be 2 times when the spacecraft is at the same height, when it is going up and when it is coming down.

**99.** Answers will vary.

**100.** $\sqrt{(x_2 - x_1)^2 + (y_2 - y_1)^2} = \sqrt{(1-7)^2 + (-1-2)^2}$

$$= \sqrt{(-6)^2 + (-3)^2}$$
$$= \sqrt{36+9}$$
$$= \sqrt{45}$$
$$= 3\sqrt{5}$$

**101.**

**102.** $y^2 - 6y - 4 = 0$

$$y^2 - 6y = 4$$
$$y^2 - 6y + 9 = 4+9$$
$$(y-3)^2 = 13$$
$$y-3 = \pm\sqrt{13}$$
$$y = 3 \pm \sqrt{13}$$

Solution set: $\left\{3 \pm \sqrt{13}\right\}$

## Section 2.8

## Check Point Exercises

1. $d = \sqrt{(x_2 - x_1)^2 + (y_2 - y_1)^2}$

   $d = \sqrt{(2 - (-1))^2 + (3 - (-3))^2}$

   $= \sqrt{3^2 + 6^2}$

   $= \sqrt{9 + 36}$

   $= \sqrt{45}$

   $= 3\sqrt{5}$

   $\approx 6.71$

2. $\left( \dfrac{1+7}{2}, \dfrac{2+(-3)}{2} \right) = \left( \dfrac{8}{2}, \dfrac{-1}{2} \right) = \left( 4, -\dfrac{1}{2} \right)$

3. $h = 0,\ k = 0,\ r = 4;$

   $(x - 0)^2 + (y - 0)^2 = 4^2$

   $x^2 + y^2 = 16$

4. $h = 0,\ k = -6,\ r = 10;$

   $(x - 0)^2 + [y - (-6)]^2 = 10^2$

   $(x - 0)^2 + (y + 6)^2 = 100$

   $x^2 + (y + 6)^2 = 100$

5. **a.** $(x + 3)^2 + (y - 1)^2 = 4$

   $[x - (-3)]^2 + (y - 1)^2 = 2^2$

   So in the standard form of the circle's equation
   $(x - h)^2 + (y - k)^2 = r^2$,

   we have $h = -3,\ k = 1,\ r = 2.$

   center: $(h, k) = (-3, 1)$

   radius: $r = 2$

   **b.**

   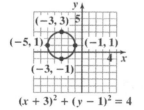

   $(x + 3)^2 + (y - 1)^2 = 4$

   **c.** domain: $[-5, -1]$

   range: $[-1, 3]$

6. $x^2 + y^2 + 4x - 4y - 1 = 0$

   $x^2 + y^2 + 4x - 4y - 1 = 0$

   $(x^2 + 4x\ \ \ ) + (y^2 - 4y\ \ \ ) = 0$

   $(x^2 + 4x + 4) + (y^2 + 4y + 4) = 1 + 4 + 4$

   $(x + 2)^2 + (y - 2)^2 = 9$

   $[x - (-x)]^2 + (y - 2)^2 = 3^2$

   So in the standard form of the circle's equation
   $(x - h)^2 + (y - k)^2 = r^2$, we have
   $h = -2,\ k = 2,\ r = 3$.

   $x^2 + y^2 + 4x - 4y - 1 = 0$

## Concept and Vocabulary Check 2.8

1. $\sqrt{(x_2 - x_1)^2 + (y_2 - y_1)^2}$

2. $\dfrac{x_1 + x_2}{2}$ ; $\dfrac{y_1 + y_2}{2}$

3. circle; center; radius

4. $(x - h)^2 + (y - k)^2 = r^2$

5. general

6. 4; 16

## Exercise Set 2.8

1. $d = \sqrt{(14 - 2)^2 + (8 - 3)^2}$

   $= \sqrt{12^2 + 5^2}$

   $= \sqrt{144 + 25}$

   $= \sqrt{169}$

   $= 13$

**3.** $d = \sqrt{(-6-4)^2 + (3-(-1))^2}$
$= \sqrt{(-10)^2 + (4)^2}$
$= \sqrt{100+16}$
$= \sqrt{116}$
$= 2\sqrt{29}$
$\approx 10.77$

**5.** $d = \sqrt{(-3-0)^2 + (4-0)^2}$
$= \sqrt{3^2 + 4^2}$
$= \sqrt{9+16}$
$= \sqrt{25}$
$= 5$

**7.** $d = \sqrt{[3-(-2)]^2 + [-4-(-6)]^2}$
$= \sqrt{5^2 + 2^2}$
$= \sqrt{25+4}$
$= \sqrt{29}$
$\approx 5.39$

**9.** $d = \sqrt{(4-0)^2 + [1-(-3)]^2}$
$= \sqrt{4^2 + 4^2}$
$= \sqrt{16+16}$
$= \sqrt{32}$
$= 4\sqrt{2}$
$\approx 5.66$

**11.** $d = \sqrt{(-.5-3.5)^2 + (6.2-8.2)^2}$
$= \sqrt{(-4)^2 + (-2)^2}$
$= \sqrt{16+4}$
$= \sqrt{20}$
$= 2\sqrt{5}$
$\approx 4.47$

**13.** $d = \sqrt{(\sqrt{5}-0)^2 + [0-(-\sqrt{3})]^2}$
$= \sqrt{(\sqrt{5})^2 + (\sqrt{3})^2}$
$= \sqrt{5+3}$
$= \sqrt{8}$
$= 2\sqrt{2}$
$\approx 2.83$

**15.** $d = \sqrt{(-\sqrt{3}-3\sqrt{3})^2 + (4\sqrt{5}-\sqrt{5})^2}$
$= \sqrt{(-4\sqrt{3})^2 + (3\sqrt{5})^2}$
$= \sqrt{16(3)+9(5)}$
$= \sqrt{48+45}$
$= \sqrt{93}$
$\approx 9.64$

**17.** $d = \sqrt{\left(\dfrac{1}{3}-\dfrac{7}{3}\right)^2 + \left(\dfrac{6}{5}-\dfrac{1}{5}\right)^2}$
$= \sqrt{(-2)^2 + 1^2}$
$= \sqrt{4+1}$
$= \sqrt{5}$
$\approx 2.24$

**19.** $\left(\dfrac{6+2}{2}, \dfrac{8+4}{2}\right) = \left(\dfrac{8}{2}, \dfrac{12}{2}\right) = (4,6)$

**21.** $\left(\dfrac{-2+(-6)}{2}, \dfrac{-8+(-2)}{2}\right)$
$= \left(\dfrac{-8}{2}, \dfrac{-10}{2}\right) = (-4,-5)$

**23.** $\left(\dfrac{-3+6}{2}, \dfrac{-4+(-8)}{2}\right)$
$= \left(\dfrac{3}{2}, \dfrac{-12}{2}\right) = \left(\dfrac{3}{2}, -6\right)$

**25.** $\left(\dfrac{\dfrac{-7}{2}+\left(-\dfrac{5}{2}\right)}{2}, \dfrac{\dfrac{3}{2}+\left(-\dfrac{11}{2}\right)}{2}\right)$
$= \left(\dfrac{\dfrac{-12}{2}}{2}, \dfrac{\dfrac{-8}{2}}{2}\right) = \left(-\dfrac{6}{2}, \dfrac{-4}{2}\right) = (-3,-2)$

**27.** $\left(\dfrac{8+(-6)}{2}, \dfrac{3\sqrt{5}+7\sqrt{5}}{2}\right)$
$= \left(\dfrac{2}{2}, \dfrac{10\sqrt{5}}{2}\right) = \left(1, 5\sqrt{5}\right)$

**29.** $\left(\dfrac{\sqrt{18}+\sqrt{2}}{2}, \dfrac{-4+4}{2}\right)$

$=\left(\dfrac{3\sqrt{2}+\sqrt{2}}{2}, \dfrac{0}{2}\right)=\left(\dfrac{4\sqrt{2}}{2},0\right)=(2\sqrt{2},0)$

**31.** $(x-0)^2+(y-0)^2=7^2$

$\qquad\qquad x^2+y^2=49$

**33.** $(x-3)^2+(y-2)^2=5^2$

$\quad(x-3)^2+(y-2)^2=25$

**35.** $[x-(-1)]^2+(y-4)^2=2^2$

$\qquad(x+1)^2+(y-4)^2=4$

**37.** $[x-(-3)]^2+[y-(-1)]^2=\left(\sqrt{3}\right)^2$

$\qquad(x+3)^2+(y+1)^2=3$

**39.** $[x-(-4)]^2+(y-0)^2=10^2$

$\qquad(x+4)^2+(y-0)^2=100$

**41.** $\qquad\qquad x^2+y^2=16$

$(x-0)^2+(y-0)^2=y^2$

$h=0,\ k=0,\ r=4;$

center $=(0,0);$ radius $=4$

$x^2+y^2=16$

domain: $[-4,4]$

range: $[-4,4]$

**43.** $(x-3)^2+(y-1)^2=36$

$\quad(x-3)^2+(y-1)^2=6^2$

$h=3,\ k=1,\ r=6;$

center $=(3,1);$ radius $=6$

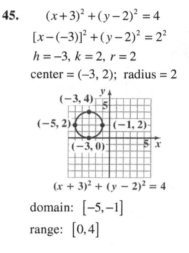

$(x-3)^2+(y-1)^2=36$

domain: $[-3,9]$

range: $[-5,7]$

**45.** $\qquad(x+3)^2+(y-2)^2=4$

$[x-(-3)]^2+(y-2)^2=2^2$

$h=-3,\ k=2,\ r=2$

center $=(-3,2);$ radius $=2$

$(x+3)^2+(y-2)^2=4$

domain: $[-5,-1]$

range: $[0,4]$

**47.** $\qquad(x+2)^2+(y+2)^2=4$

$[x-(-2)]^2+[y-(-2)]^2=2^2$

$h=-2,\ k=-2,\ r=2$

center $=(-2,-2);$ radius $=2$

$(x+2)^2+(y+2)^2=4$

domain: $[-4,0]$

range: $[-4,0]$

**49.** $x^2+(y-1)^2=1$

$h=0, k=1, r=1;$

center $=(0,1);$ radius $=1$

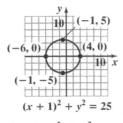

$x^2+(y-1)^2=1$

domain: $[-1,1]$

range: $[0,2]$

**51.** $(x+1)^2+y^2=25$

$h=-1, k=0, r=5;$

center $=(-1,0);$ radius $=5$

$(x+1)^2+y^2=25$

domain: $[-6,4]$

range: $[-5,5]$

**53.** 
$$x^2+y^2+6x+2y+6=0$$
$$(x^2+6x)+(y^2+2y)=-6$$
$$(x^2+6x+9)+(y^2+2y+1)=9+1-6$$
$$(x+3)^2+(y+1)^2=4$$
$$[x-(-3)]^2+[9-(-1)]^2=2^2$$
center $=(-3,-1);$ radius $=2$

$x^2+y^2+6x+2y+6=0$

**55.** 
$$x^2+y^2-10x-6y-30=0$$
$$(x^2-10x)+(y^2-6y)=30$$
$$(x^2-10x+25)+(y^2-6y+9)=25+9+30$$
$$(x-5)^2+(y-3)^2=64$$
$$(x-5)^2+(y-3)^2=8^2$$
center $=(5,3);$ radius $=8$

$x^2+y^2-10x-6y-30=0$

**57.** 
$$x^2+y^2+8x-2y-8=0$$
$$(x^2+8x)+(y^2-2y)=8$$
$$(x^2+8x+16)+(y^2-2y+1)=16+1+8$$
$$(x+4)^2+(y-1)^2=25$$
$$[x-(-4)]^2+(y-1)^2=5^2$$
center $=(-4,1);$ radius $=5$

$x^2+y^2+8x-2y-8=0$

**59.** 
$$x^2-2x+y^2-15=0$$
$$(x^2-2x)+y^2=15$$
$$(x^2-2x+1)+(y-0)^2=1+0+15$$
$$(x-1)^2+(y-0)^2=16$$
$$(x-1)^2+(y-0)^2=4^2$$
center $=(1,0);$ radius $=4$

$x^2-2x+y^2-15=0$

**61.** $x^2 + y^2 - x + 2y + 1 = 0$

$x^2 - x + y^2 + 2y = -1$

$x^2 - x + \dfrac{1}{4} + y^2 + 2y + 1 = -1 + \dfrac{1}{4} + 1$

$\left(x - \dfrac{1}{2}\right)^2 + (y + 1)^2 = \dfrac{1}{4}$

center $= \left(\dfrac{1}{2}, -1\right)$; radius $= \dfrac{1}{2}$

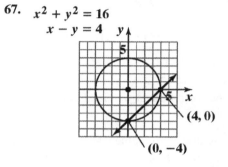

**63.** $x^2 + y^2 + 3x - 2y - 1 = 0$

$x^2 + 3x + y^2 - 2y = 1$

$x^2 + 3x + \dfrac{9}{4} + y^2 - 2y + 1 = 1 + \dfrac{9}{4} + 1$

$\left(x + \dfrac{3}{2}\right)^2 + (y - 1)^2 = \dfrac{17}{4}$

center $= \left(-\dfrac{3}{2}, 1\right)$; radius $= \dfrac{\sqrt{17}}{2}$

**65. a.** Since the line segment passes through the center, the center is the midpoint of the segment.

$M = \left(\dfrac{x_1 + x_2}{2}, \dfrac{y_1 + y_2}{2}\right)$

$= \left(\dfrac{3 + 7}{2}, \dfrac{9 + 11}{2}\right) = \left(\dfrac{10}{2}, \dfrac{20}{2}\right)$

$= (5, 10)$

The center is $(5, 10)$.

**b.** The radius is the distance from the center to one of the points on the circle. Using the point $(3, 9)$, we get:

$d = \sqrt{(5 - 3)^2 + (10 - 9)^2}$

$= \sqrt{2^2 + 1^2} = \sqrt{4 + 1}$

$= \sqrt{5}$

The radius is $\sqrt{5}$ units.

**c.** $(x - 5)^2 + (y - 10)^2 = \left(\sqrt{5}\right)^2$

$(x - 5)^2 + (y - 10)^2 = 5$

**67.** $x^2 + y^2 = 16$

$x - y = 4$

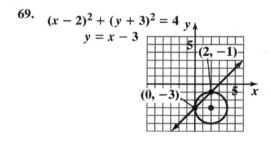

Intersection points: $(0, -4)$ and $(4, 0)$

Check $(0, -4)$:

$0^2 + (-4)^2 = 16 \qquad 0 - (-4) = 4$

$16 = 16$ true $\qquad 4 = 4$ true

Check $(4, 0)$:

$4^2 + 0^2 = 16 \qquad 4 - 0 = 4$

$16 = 16$ true $\qquad 4 = 4$ true

The solution set is $\{(0, -4), (4, 0)\}$.

**69.** $(x - 2)^2 + (y + 3)^2 = 4$

$y = x - 3$

Intersection points: $(0, -3)$ and $(2, -1)$

Check $(0,-3)$:

$(0-2)^2 + (-3+3)^2 = 9 \qquad -3 = 0-3$

$\qquad (-2)^2 + 0^2 = 4 \qquad\qquad -3 = -3 \text{ true}$

$\qquad\qquad\qquad 4 = 4$

$\qquad\qquad\qquad\qquad \text{true}$

Check $(2,-1)$:

$(2-2)^2 + (-1+3)^2 = 4 \qquad -1 = 2-3$

$\qquad\qquad 0^2 + 2^2 = 4 \qquad\quad -1 = -1 \text{ true}$

$\qquad\qquad\qquad 4 = 4$

$\qquad\qquad\qquad\qquad \text{true}$

The solution set is $\{(0,-3),(2,-1)\}$.

**71.** $d = \sqrt{(8495-4422)^2 + (8720-1241)^2} \cdot \sqrt{0.1}$

$d = \sqrt{72,524,770} \cdot \sqrt{0.1}$

$d \approx 2693$

The distance between Boston and San Francisco is about 2693 miles.

**73.** If we place L.A. at the origin, then we want the equation of a circle with center at $(-2.4,-2.7)$ and radius 30.

$(x-(-2.4))^2 + (y-(-2.7))^2 = 30^2$

$\qquad (x+2.4)^2 + (y+2.7)^2 = 900$

**75. – 81.** Answers will vary.

**83.**

**85.**

**87.** makes sense

**89.** makes sense

**91.** false; Changes to make the statement true will vary.
A sample change is: The center is at (3, –5).

**93.** false; Changes to make the statement true will vary.
A sample change is: Since $r^2 = -36$ this is not the equation of a circle.

**95. a.** $d_1$ is distance from $(x_1, x_2)$ to midpoint

$d_1 = \sqrt{\left(\dfrac{x_1+x_2}{2}-x_1\right)^2 + \left(\dfrac{y_1+y_2}{2}-y_1\right)^2}$

$d_1 = \sqrt{\left(\dfrac{x_1+x_2-2x_1}{2}\right)^2 + \left(\dfrac{y_1+y_2-2y_1}{2}\right)^2}$

$d_1 = \sqrt{\left(\dfrac{x_2-x_1}{2}\right)^2 + \left(\dfrac{y_2-y_1}{2}\right)^2}$

$d_1 = \sqrt{\dfrac{x_2-2x_1x_2+x_1^2}{4} + \dfrac{y_2^2-2y_2y_1+y_1^2}{4}}$

$d_1 = \sqrt{\dfrac{1}{4}\left(x_2-2x_1x_2+x_1+y_2^2-2y_2y_1+y_1^2\right)}$

$d_1 = \dfrac{1}{2}\sqrt{x_2-2x_1x_2+x_1+y_2^2-2y_2y_1+y_1^2}$

$d_2$ is distance from midpoint to $(x_2, y_2)$

$d_2 = \sqrt{\left(\dfrac{x_1+x_2}{2}-x_2\right)^2 + \left(\dfrac{y_1+y_2}{2}-y_2\right)^2}$

$d_2 = \sqrt{\left(\dfrac{x_1+x_2-2x_2}{2}\right)^2 + \left(\dfrac{y_1+y_2-2y_2}{2}\right)^2}$

$d_2 = \sqrt{\left(\dfrac{x_1-x_2}{2}\right)^2 + \left(\dfrac{y_1-y_2}{2}\right)^2}$

$d_2 = \sqrt{\dfrac{x_1^2-2x_1x_2+x_2^2}{4} + \dfrac{y_1^2-2y_2y_1+y_2^2}{4}}$

$d_2 = \sqrt{\dfrac{1}{4}\left(x_1^2-2x_1x_2+x_2^2+y_1^2-2y_2y_1+y_2^2\right)}$

$d_2 = \dfrac{1}{2}\sqrt{x_1^2-2x_1x_2+x_2^2+y_1^2-2y_2y_1+y_2^2}$

$d_1 = d_2$

**b.** $d_3$ is the distance from $(x_1, y_1)$ to $(x_2, y_2)$

$d_3 = \sqrt{(x_2-x_1)^2 + (y_2-y_1)^2}$

$d_3 = \sqrt{x_2^2-2x_1x_2+x_1^2+y_2^2-2y_2y_1+y_1^2}$

$d_1 + d_2 = d_3$ because $\dfrac{1}{2}\sqrt{a}+\dfrac{1}{2}\sqrt{a} = \sqrt{a}$

**97.** The circle is centered at (0,0). The slope of the radius with endpoints (0,0) and (3,–4) is

$$m = -\frac{-4-0}{3-0} = -\frac{4}{3}.$$ The line perpendicular to the

radius has slope $\frac{3}{4}$. The tangent line has slope $\frac{3}{4}$ and

passes through (3,–4), so its equation is:

$$y + 4 = \frac{3}{4}(x-3).$$

**99.** $-x^2 - 2x + 1 = 0$

$$x^2 + 2x - 1 = 0$$

$$x = \frac{-b \pm \sqrt{b^2 - 4ac}}{2a}$$

$$x = \frac{-(-2) \pm \sqrt{(-2)^2 - 4(1)(-1)}}{2(1)}$$

$$= \frac{2 \pm \sqrt{8}}{2}$$

$$= \frac{2 \pm 2\sqrt{2}}{2}$$

$$= 1 \pm \sqrt{2}$$

The solution set is $\{1 \pm \sqrt{2}\}$.

**100.** The graph of $g$ is the graph of $f$ shifted 1 unit up and 3 units to the left.

$$f(x) = x^2$$
$$g(x) = (x + 3)^2 + 1$$

## Chapter 2 Review Exercises

**1.** function
domain: {2, 3, 5}
range: {7}

**2.** function
domain: {1, 2, 13}
range: {10, 500, $\pi$}

**3.** not a function
domain: {12, 14}
range: {13, 15, 19}

**4.** $2x + y = 8$

$$y = -2x + 8$$

Since only one value of $y$ can be obtained for each value of $x$, $y$ is a function of $x$.

**5.** $3x^2 + y = 14$

$$y = -3x^2 + 14$$

Since only one value of $y$ can be obtained for each value of $x$, $y$ is a function of $x$.

**6.** $2x + y^2 = 6$

$$y^2 = -2x + 6$$

$$y = \pm\sqrt{-2x+6}$$

Since more than one value of $y$ can be obtained from some values of $x$, $y$ is not a function of $x$.

**7.** $f(x) = 5 - 7x$

 **a.** $f(4) = 5 - 7(4) = -23$

 **b.** $f(x+3) = 5 - 7(x+3)$
$$= 5 - 7x - 21$$
$$= -7x - 16$$

 **c.** $f(-x) = 5 - 7(-x) = 5 + 7x$

**8.** $g(x) = 3x^2 - 5x + 2$

 **a.** $g(0) = 3(0)^2 - 5(0) + 2 = 2$

 **b.** $g(-2) = 3(-2)^2 - 5(-2) + 2$
$$= 12 + 10 + 2$$
$$= 24$$

 **c.** $g(x-1) = 3(x-1)^2 - 5(x-1) + 2$
$$= 3(x^2 - 2x + 1) - 5x + 5 + 2$$
$$= 3x^2 - 11x + 10$$

 **d.** $g(-x) = 3(-x)^2 - 5(-x) + 2$
$$= 3x^2 + 5x + 2$$

**9. a.** $g(13) = \sqrt{13-4} = \sqrt{9} = 3$

 **b.** $g(0) = 4 - 0 = 4$

 **c.** $g(-3) = 4 - (-3) = 7$

**10. a.**  $f(-2) = \dfrac{(-2)^2 - 1}{-2 - 1} = \dfrac{3}{-3} = -1$

**b.**  $f(1) = 12$

**c.**  $f(2) = \dfrac{2^2 - 1}{2 - 1} = \dfrac{3}{1} = 3$

**11.**  The vertical line test shows that this is not the graph of a function.

**12.**  The vertical line test shows that this is the graph of a function.

**13.**  The vertical line test shows that this is the graph of a function.

**14.**  The vertical line test shows that this is not the graph of a function.

**15.**  The vertical line test shows that this is not the graph of a function.

**16.**  The vertical line test shows that this is the graph of a function.

**17. a.**  domain: $[-3, 5)$

**b.**  range: $[-5, 0]$

**c.**  $x$-intercept: $-3$

**d.**  $y$-intercept: $-2$

**e.**  increasing: $(-2, 0)$ or $(3, 5)$
decreasing: $(-3, -2)$ or $(0, 3)$

**f.**  $f(-2) = -3$ and $f(3) = -5$

**18. a.**  domain: $(-\infty, \infty)$

**b.**  range: $(-\infty, 3]$

**c.**  $x$-intercepts: $-2$ and $3$

**d.**  $y$-intercept: $3$

**e.**  increasing: $(-\infty, 0)$
decreasing: $(0, \infty)$

**f.**  $f(-2) = 0$ and $f(6) = -3$

**19. a.**  domain: $(-\infty, \infty)$

**b.**  range: $[-2, 2]$

**c.**  $x$-intercept: $0$

**d.**  $y$-intercept: $0$

**e.**  increasing: $(-2, 2)$
constant: $(-\infty, -2)$ or $(2, \infty)$

**f.**  $f(-9) = -2$ and $f(14) = 2$

**20. a.**  $0$, relative maximum $-2$

**b.**  $-2, 3$, relative minimum $-3, -5$

**21. a.**  $0$, relative maximum $3$

**b.**  none

**22.**  $f(x) = x^3 - 5x$
$f(-x) = (-x)^3 - 5(-x)$
$\qquad = -x^3 + 5x$
$\qquad = -f(x)$

The function is odd. The function is symmetric with respect to the origin.

**23.**  $f(x) = x^4 - 2x^2 + 1$
$f(-x) = (-x)^4 - 2(-x)^2 + 1$
$\qquad = x^4 - 2x^2 + 1$
$\qquad = f(x)$

The function is even. The function is symmetric with respect to the $y$-axis.

**24.**    $f(x) = 2x\sqrt{1-x^2}$

$f(-x) = 2(-x)\sqrt{1-(-x)^2}$

$= -2x\sqrt{1-x^2}$

$= -f(x)$

The function is odd. The function is symmetric with respect to the origin.

**25.   a.**

$f(x) = \begin{cases} 5 \text{ if } x \le -1 \\ -3 \text{ if } x > -1 \end{cases}$

  **b.**   range: $\{-3, 5\}$

**26.   a.**

$f(x) = \begin{cases} 2x \text{ if } x < 0 \\ -x \text{ if } x \ge 0 \end{cases}$

  **b.**   range: $\left\{ y \mid y \le 0 \right\}$

**27.**    $\dfrac{8(x+h) - 11 - (8x - 11)}{h}$

$= \dfrac{8x + 8h - 11 - 8x + 11}{h}$

$= \dfrac{8h}{8}$

$= 8$

**28.**    $\dfrac{-2(x+h)^2 + (x+h) + 10 - \left(-2x^2 + x + 10\right)}{h}$

$= \dfrac{-2\left(x^2 + 2xh + h^2\right) + x + h + 10 + 2x^2 - x - 10}{h}$

$= \dfrac{-2x^2 - 4xh - 2h^2 + x + h + 10 + 2x^2 - x - 10}{h}$

$= \dfrac{-4xh - 2h^2 + h}{h}$

$= \dfrac{h\left(-4x - 2h + 1\right)}{h}$

$-4x - 2h + 1$

**29.   a.**   Yes, the eagle's height is a function of time since the graph passes the vertical line test.

  **b.**   Decreasing: (3, 12)
      The eagle descended.

  **c.**   Constant: (0, 3) or (12, 17)
      The eagle's height held steady during the first 3 seconds and the eagle was on the ground for 5 seconds.

  **d.**   Increasing: (17, 30)
      The eagle was ascending.

**30.**

**31.**    $m = \dfrac{1-2}{5-3} = \dfrac{-1}{2} = -\dfrac{1}{2}$; falls

**32.**    $m = \dfrac{-4-(-2)}{-3-(-1)} = \dfrac{-2}{-2} = 1$; rises

**33.**    $m = \dfrac{\frac{1}{4} - \frac{1}{4}}{6-(-3)} = \dfrac{0}{9} = 0$; horizontal

**34.**    $m = \dfrac{10-5}{-2-(-2)} = \dfrac{5}{0}$ undefined; vertical

**35.**    point-slope form: $y - 2 = -6(x + 3)$
      slope-intercept form: $y = -6x - 16$

**36.** $m = \dfrac{2-6}{-1-1} = \dfrac{-4}{-2} = 2$

point-slope form: $y - 6 = 2(x - 1)$
or $y - 2 = 2(x + 1)$
slope-intercept form: $y = 2x + 4$

**37.** $3x + y - 9 = 0$
$y = -3x + 9$
$m = -3$
point-slope form:
$y + 7 = -3(x - 4)$
slope-intercept form:
$y = -3x + 12 - 7$
$y = -3x + 5$

**38.** perpendicular to $y = \dfrac{1}{3}x + 4$

$m = -3$
point-slope form:
$y - 6 = -3(x + 3)$
slope-intercept form:
$y = -3x - 9 + 6$
$y = -3x - 3$

**39.** Write $6x - y - 4 = 0$ in slope intercept form.
$6x - y - 4 = 0$
$\qquad -y = -6x + 4$
$\qquad y = 6x - 4$
The slope of the perpendicular line is 6, thus the
slope of the desired line is $m = -\dfrac{1}{6}$.
$$y - y_1 = m(x - x_1)$$
$$y - (-1) = -\tfrac{1}{6}\left(x - (-12)\right)$$
$$y + 1 = -\tfrac{1}{6}(x + 12)$$
$$y + 1 = -\tfrac{1}{6}x - 2$$
$$6y + 6 = -x - 12$$
$$x + 6y + 18 = 0$$

**40.** slope: $\dfrac{2}{5}$; $y$-intercept: $-1$

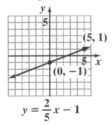

$y = \dfrac{2}{5}x - 1$

**41.** slope: $-4$; $y$-intercept: 5

$f(x) = -4x + 5$

**42.** $2x + 3y + 6 = 0$
$$3y = -2x - 6$$
$$y = -\dfrac{2}{3}x - 2$$

slope: $-\dfrac{2}{3}$; $y$-intercept: $-2$

$2x + 3y + 6 = 0$

**43.** $2y - 8 = 0$
$$2y = 8$$
$$y = 4$$
slope: 0; $y$-intercept: 4

$2y - 8 = 0$

**44.** $2x - 5y - 10 = 0$

Find $x$-intercept:
$2x - 5(0) - 10 = 0$
$2x - 10 = 0$
$2x = 10$
$x = 5$

Find $y$-intercept:
$2(0) - 5y - 10 = 0$
$-5y - 10 = 0$
$-5y = 10$
$y = -2$

**45.** $2x - 10 = 0$
$2x = 10$
$x = 5$

**46. a.** $m = \dfrac{11 - 2.3}{90 - 15} = \dfrac{8.7}{75} = 0.116$

$y - y_1 = m(x - x_1)$
$y - 11 = 0.116(x - 90)$

or

$y - 2.3 = 0.116(x - 15)$

**b.** $y - 11 = 0.116(x - 90)$
$y - 11 = 0.116x - 10.44$
$y = 0.116x + 0.56$
$f(x) = 0.116x + 0.56$

**c.** According to the graph, France has about 5 deaths per 100,000 persons.

**d.** $f(x) = 0.116x + 0.56$
$f(32) = 0.116(32) + 0.56$
$= 4.272$
$\approx 4.3$

According to the function, France has about 4.3 deaths per 100,000 persons.
This underestimates the value in the graph by 0.7 deaths per 100,000 persons.
The line passes below the point for France.

**47. a.** $m = \dfrac{52 - 64}{2010 - 1985} = \dfrac{-12}{25} = -0.48$

**b.** For each year from 1985 through 2010, the percentage of U.S. college freshmen rating their emotional health high or above average decreased by 0.48. The rate of change was $-0.48\%$ per year.

**48.** $\dfrac{f(x_2) - f(x_1)}{x_2 - x_1} = \dfrac{[9^2 - 4(9)] - [4^2 - 4 \cdot 5]}{9 - 5} = 10$

**49.**

**50.**

**51.**

**52.**

**53.**

**54.**

**55.**

**56.**

**57.**

**58.**

**59.**

**60.**

**61.**

**62.**

**63.**

**64.**

**65.**

**66.**

**67.**

**68.**

**69.**

**70.** domain: $(-\infty, \infty)$

**71.** The denominator is zero when $x = 7$. The domain is $(-\infty, 7) \cup (7, \infty)$.

**72.** The expressions under each radical must not be negative.
$8 - 2x \geq 0$
$-2x \geq -8$
$x \leq 4$
domain: $(-\infty, 4]$.

**73.** The denominator is zero when $x = -7$ or $x = 3$.
domain: $(-\infty, -7) \cup (-7, 3) \cup (3, \infty)$

**74.** The expressions under each radical must not be negative. The denominator is zero when $x = 5$.
$x - 2 \geq 0$
$x \geq 2$
domain: $[2, 5) \cup (5, \infty)$

**75.** The expressions under each radical must not be negative.
$x - 1 \geq 0$  and  $x + 5 \geq 0$
$x \geq 1$           $x \geq -5$
domain: $[1, \infty)$

**76.** $f(x) = 3x - 1; g(x) = x - 5$
$(f + g)(x) = 4x - 6$
domain: $(-\infty, \infty)$
$(f - g)(x) = (3x - 1) - (x - 5) = 2x + 4$
domain: $(-\infty, \infty)$
$(fg)(x) = (3x-1)(x-5) = 3x^2 - 16x + 5$
domain: $(-\infty, \infty)$
$\left(\dfrac{f}{g}\right)(x) = \dfrac{3x-1}{x-5}$
domain: $(-\infty, 5) \cup (5, \infty)$

**77.** $f(x) = x^2 + x + 1; g(x) = x^2 - 1$

$(f+g)(x) = 2x^2 + x$

domain: $(-\infty, \infty)$

$(f-g)(x) = (x^2 + x + 1) - (x^2 - 1) = x + 2$

domain: $(-\infty, \infty)$

$(fg)(x) = (x^2 + x + 1)(x^2 - 1)$

$\qquad = x^4 + x^3 - x - 1$

$\left(\dfrac{f}{g}\right)(x) = \dfrac{x^2 + x + 1}{x^2 - 1}$

domain: $(-\infty, -1) \cup (-1, 1) \cup (1, \infty)$

**78.** $f(x) = \sqrt{x+7}; g(x) = \sqrt{x-2}$

$(f+g)(x) = \sqrt{x+7} + \sqrt{x-2}$

domain: $[2, \infty)$

$(f-g)(x) = \sqrt{x+7} - \sqrt{x-2}$

domain: $[2, \infty)$

$(fg)(x) = \sqrt{x+7} \cdot \sqrt{x-2}$

$\qquad = \sqrt{x^2 + 5x - 14}$

domain: $[2, \infty)$

$\left(\dfrac{f}{g}\right)(x) = \dfrac{\sqrt{x+7}}{\sqrt{x-2}}$

domain: $(2, \infty)$

**79.** $f(x) = x^2 + 3; g(x) = 4x - 1$

**a.** $(f \circ g)(x) = (4x - 1)^2 + 3$

$\qquad = 16x^2 - 8x + 4$

**b.** $(g \circ f)(x) = 4(x^2 + 3) - 1$

$\qquad = 4x^2 + 11$

**c.** $(f \circ g)(3) = 16(3)^2 - 8(3) + 4 = 124$

**80.** $f(x) = \sqrt{x}; \quad g(x) = x + 1$

**a.** $(f \circ g)(x) = \sqrt{x+1}$

**b.** $(g \circ f)(x) = \sqrt{x} + 1$

**c.** $(f \circ g)(3) = \sqrt{3+1} = \sqrt{4} = 2$

**81. a.** $(f \circ g)(x) = f\left(\dfrac{1}{x}\right)$

$\qquad = \dfrac{\dfrac{1}{x} + 1}{\dfrac{1}{x} - 2} = \dfrac{\left(\dfrac{1}{x} + 1\right)x}{\left(\dfrac{1}{x} - 2\right)x} = \dfrac{1+x}{1-2x}$

**b.** $x \neq 0 \qquad\qquad 1 - 2x \neq 0$

$\qquad\qquad\qquad\qquad x \neq \dfrac{1}{2}$

$(-\infty, 0) \cup \left(0, \dfrac{1}{2}\right) \cup \left(\dfrac{1}{2}, \infty\right)$

**82. a.** $(f \circ g)(x) = f(x+3) = \sqrt{x+3-1} = \sqrt{x+2}$

**b.** $x + 2 \geq 0 \qquad [-2, \infty)$

$\qquad x \geq -2$

**83.** $f(x) = x^4 \qquad g(x) = x^2 + 2x - 1$

**84.** $f(x) = \sqrt[3]{x} \qquad g(x) = 7x + 4$

**85.** $f(x) = \dfrac{3}{5}x + \dfrac{1}{2}; g(x) = \dfrac{5}{3}x - 2$

$f(g(x)) = \dfrac{3}{5}\left(\dfrac{5}{3}x - 2\right) + \dfrac{1}{2}$

$\qquad = x - \dfrac{6}{5} + \dfrac{1}{2}$

$\qquad = x - \dfrac{7}{10}$

$g(f(x)) = \dfrac{5}{3}\left(\dfrac{3}{5}x + \dfrac{1}{2}\right) - 2$

$\qquad = x + \dfrac{5}{6} - 2$

$\qquad = x - \dfrac{7}{6}$

*f* and *g* are not inverses of each other.

**86.** $f(x) = 2 - 5x; g(x) = \dfrac{2-x}{5}$

$f(g(x)) = 2 - 5\left(\dfrac{2-x}{5}\right)$

$\qquad = 2 - (2 - x)$

$\qquad = x$

$g(f(x)) = \dfrac{2 - (2 - 5x)}{5} = \dfrac{5x}{5} = x$

*f* and *g* are inverses of each other.

**87. a.** $f(x) = 4x - 3$

$\qquad y = 4x - 3$

$\qquad x = 4y - 3$

$\qquad y = \dfrac{x+3}{4}$

$\qquad f^{-1}(x) = \dfrac{x+3}{4}$

**b.** $f(f^{-1}(x)) = 4\left(\dfrac{x+3}{4}\right) - 3$

$\qquad = x + 3 - 3$

$\qquad = x$

$f^{-1}(f(x)) = \dfrac{(4x-3)+3}{4} = \dfrac{4x}{4} = x$

**88. a.** $f(x) = 8x^3 + 1$

$\qquad y = 8x^3 + 1$

$\qquad x = 8y^3 + 1$

$\qquad x - 1 = 8y^3$

$\qquad \dfrac{x-1}{8} = y^3$

$\qquad \sqrt[3]{\dfrac{x-1}{8}} = y$

$\qquad \dfrac{\sqrt[3]{x-1}}{2} = y$

$\qquad f^{-1}(x) = \dfrac{\sqrt[3]{x-1}}{2}$

**b.** $f(f^{-1}(x)) = 8\left(\dfrac{\sqrt[3]{x-1}}{2}\right)^3 + 1$

$\qquad = 8\left(\dfrac{x-1}{8}\right) + 1$

$\qquad = x - 1 + 1$

$\qquad = x$

$f^{-1}(f(x)) = \dfrac{\sqrt[3]{(8x^3+1)-1}}{2}$

$\qquad = \dfrac{\sqrt[3]{8x^3}}{2}$

$\qquad = \dfrac{2x}{2}$

$\qquad = x$

**89. a.** $f(x) = \dfrac{2}{x} + 5$

$\qquad y = \dfrac{2}{x} + 5$

$\qquad x = \dfrac{2}{y} + 5$

$\qquad xy = 2 + 5y$

$\qquad xy - 5y = 2$

$\qquad y(x - 5) = 2$

$\qquad y = \dfrac{2}{x-5}$

$\qquad f^{-1}(x) = \dfrac{2}{x-5}$

**b.** $f(f^{-1}(x)) = \dfrac{2}{\frac{2}{x-5}} + 5$

$\qquad = \dfrac{2(x-5)}{2} + 5$

$\qquad = x - 5 + 5$

$\qquad = x$

$f^{-1}(f(x)) = \dfrac{2}{\frac{2}{x}+5-5}$

$\qquad = \dfrac{2}{\frac{2}{x}}$

$\qquad = \dfrac{2x}{2}$

$\qquad = x$

**90.** The inverse function exists.

**91.** The inverse function does not exist since it does not pass the horizontal line test.

**92.** The inverse function exists.

**93.** The inverse function does not exist since it does not pass the horizontal line test.

**94.**

![graph for problem 94]

**95.**
$$f(x) = 1 - x^2$$
$$y = 1 - x^2$$
$$x = 1 - y^2$$
$$y^2 = 1 - x$$
$$y = \sqrt{1-x}$$
$$f^{-1}(x) = \sqrt{1-x}$$

![graph for problem 95]

**96.**
$$f(x) = \sqrt{x} + 1$$
$$y = \sqrt{x} + 1$$
$$x = \sqrt{y} + 1$$
$$x - 1 = \sqrt{y}$$
$$(x-1)^2 = y$$
$$f^{-1}(x) = (x-1)^2, \quad x \ge 1$$

![graph for problem 96]

**97.**
$$d = \sqrt{[3-(-2)]^2 + [9-(-3)]^2}$$
$$= \sqrt{5^2 + 12^2}$$
$$= \sqrt{25 + 144}$$
$$= \sqrt{169}$$
$$= 13$$

**98.**
$$d = \sqrt{[-2-(-4)]^2 + (5-3)^2}$$
$$= \sqrt{2^2 + 2^2}$$
$$= \sqrt{4+4}$$
$$= \sqrt{8}$$
$$= 2\sqrt{2}$$
$$\approx 2.83$$

**99.** $\left( \dfrac{2+(-12)}{2}, \dfrac{6+4}{2} \right) = \left( \dfrac{-10}{2}, \dfrac{10}{2} \right) = (-5, 5)$

**100.** $\left( \dfrac{4+(-15)}{2}, \dfrac{-6+2}{2} \right) = \left( \dfrac{-11}{2}, \dfrac{-4}{2} \right) = \left( \dfrac{-11}{2}, -2 \right)$

**101.** $x^2 + y^2 = 3^2$
$$x^2 + y^2 = 9$$

**102.** $(x-(-2))^2 + (y-4)^2 = 6^2$
$$(x+2)^2 + (y-4)^2 = 36$$

**103.** center: (0, 0); radius: 1

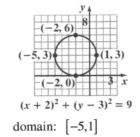

$$x^2 + y^2 = 1$$
domain: $[-1, 1]$
range: $[-1, 1]$

**104.** center: (−2, 3); radius: 3

![graph for problem 104]

$$(x+2)^2 + (y-3)^2 = 9$$
domain: $[-5, 1]$
range: $[0, 6]$

**105.**
$$x^2 + y^2 - 4x + 2y - 4 = 0$$
$$x^2 - 4x \quad + y^2 + 2y \quad = 4$$
$$x^2 - 4x + 4 + y^2 + 2y + 1 = 4 + 4 + 1$$
$$(x-2)^2 + (y+1)^2 = 9$$
center: (2, –1); radius: 3

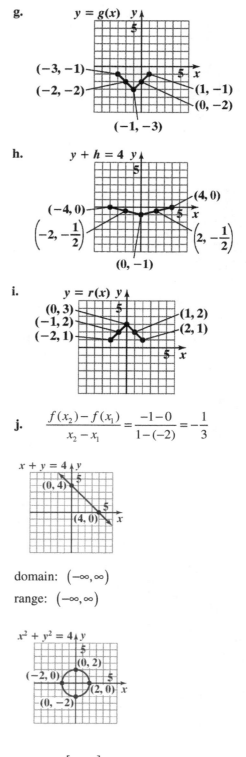

domain: $[-1, 5]$

range: $[-4, 2]$

## Chapter 2 Test

**1.** (b), (c), and (d) are not functions.

**2.**  **a.** $f(4) - f(-3) = 3 - (-2) = 5$

  **b.** domain: (–5, 6]

  **c.** range: [–4, 5]

  **d.** increasing: (–1, 2)

  **e.** decreasing: $(-5, -1)$ or $(2, 6)$

  **f.** $2, f(2) = 5$

  **g.** (–1, –4)

  **h.** *x*-intercepts: –4, 1, and 5.

  **i.** *y*-intercept: –3

**3.**  **a.** –2, 2

  **b.** –1, 1

  **c.** 0

  **d.** even;  $f(-x) = f(x)$

  **e.** no;  $f$  fails the horizontal line test

  **f.** $f(0)$  is a relative minimum.

**g.**

$y = g(x)$

(–3, –1), (–2, –2), (–1, –3), (1, –1), (0, –2)

**h.** $y + h = 4$

(–4, 0), $\left(-2, -\frac{1}{2}\right)$, (0, –1), (4, 0), $\left(2, -\frac{1}{2}\right)$

**i.** $y = r(x)$

(0, 3), (–1, 2), (–2, 1), (1, 2), (2, 1)

**j.** $\dfrac{f(x_2) - f(x_1)}{x_2 - x_1} = \dfrac{-1 - 0}{1 - (-2)} = -\dfrac{1}{3}$

**4.** $x + y = 4$

(0, 4), (4, 0)

domain: $(-\infty, \infty)$

range: $(-\infty, \infty)$

**5.** $x^2 + y^2 = 4$

(0, 2), (–2, 0), (2, 0), (0, –2)

domain: $[-2, 2]$

range: $[-2, 2]$

**6.**

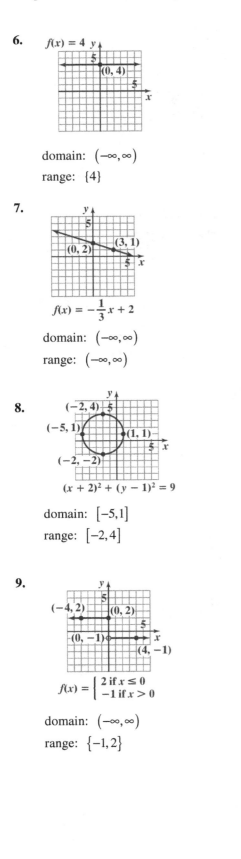

domain: $(-\infty, \infty)$

range: $\{4\}$

**7.**

domain: $(-\infty, \infty)$

range: $(-\infty, \infty)$

**8.**

domain: $[-5, 1]$

range: $[-2, 4]$

**9.**

domain: $(-\infty, \infty)$

range: $\{-1, 2\}$

**10.**

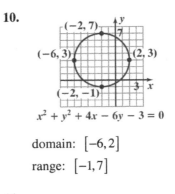

domain: $[-6, 2]$

range: $[-1, 7]$

**11.**

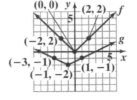

domain of $f$: $(-\infty, \infty)$

range of $f$: $[0, \infty)$

domain of $g$: $(-\infty, \infty)$

range of $g$: $[-2, \infty)$

**12.**

domain of $f$: $(-\infty, \infty)$

range of $f$: $[0, \infty)$

domain of $g$: $(-\infty, \infty)$

range of $g$: $(-\infty, 4]$

**13.**

domain of $f$: $(-\infty, \infty)$

range of $f$: $(-\infty, \infty)$

domain of $f^{-1}$: $(-\infty, \infty)$

range of $f^{-1}$: $(-\infty, \infty)$

**14.**

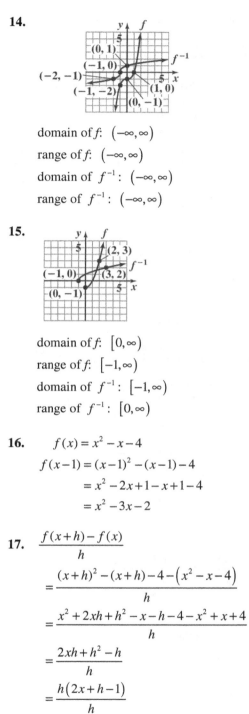

domain of $f$: $(-\infty, \infty)$

range of $f$: $(-\infty, \infty)$

domain of $f^{-1}$: $(-\infty, \infty)$

range of $f^{-1}$: $(-\infty, \infty)$

**15.**

domain of $f$: $[0, \infty)$

range of $f$: $[-1, \infty)$

domain of $f^{-1}$: $[-1, \infty)$

range of $f^{-1}$: $[0, \infty)$

**16.**   $f(x) = x^2 - x - 4$

$$f(x-1) = (x-1)^2 - (x-1) - 4$$
$$= x^2 - 2x + 1 - x + 1 - 4$$
$$= x^2 - 3x - 2$$

**17.**   $\dfrac{f(x+h) - f(x)}{h}$

$$= \dfrac{(x+h)^2 - (x+h) - 4 - (x^2 - x - 4)}{h}$$
$$= \dfrac{x^2 + 2xh + h^2 - x - h - 4 - x^2 + x + 4}{h}$$
$$= \dfrac{2xh + h^2 - h}{h}$$
$$= \dfrac{h(2x + h - 1)}{h}$$
$$= 2x + h - 1$$

**18.**   $(g - f)(x) = 2x - 6 - (x^2 - x - 4)$

$$= 2x - 6 - x^2 + x + 4$$
$$= -x^2 + 3x - 2$$

**19.**   $\left(\dfrac{f}{g}\right)(x) = \dfrac{x^2 - x - 4}{2x - 6}$

domain:  $(-\infty, 3) \cup (3, \infty)$

**20.**   $(f \circ g)(x) = f(g(x))$

$$= (2x - 6)^2 - (2x - 6) - 4$$
$$= 4x^2 - 24x + 36 - 2x + 6 - 4$$
$$= 4x^2 - 26x + 38$$

**21.**   $(g \circ f)(x) = g(f(x))$

$$= 2(x^2 - x - 4) - 6$$
$$= 2x^2 - 2x - 8 - 6$$
$$= 2x^2 - 2x - 14$$

**22.**   $g(f(-1)) = 2((-1)^2 - (-1) - 4) - 6$

$$= 2(1 + 1 - 4) - 6$$
$$= 2(-2) - 6$$
$$= -4 - 6$$
$$= -10$$

**23.**   $f(x) = x^2 - x - 4$

$$f(-x) = (-x)^2 - (-x) - 4$$
$$= x^2 + x - 4$$

$f$  is neither even nor odd.

**24.**   $m = \dfrac{-8 - 1}{-1 - 2} = \dfrac{-9}{-3} = 3$

point-slope form: $y - 1 = 3(x - 2)$
or $y + 8 = 3(x + 1)$
slope-intercept form: $y = 3x - 5$

**25.**   $y = -\dfrac{1}{4}x + 5$ so $m = 4$

point-slope form: $y - 6 = 4(x + 4)$
slope-intercept form: $y = 4x + 22$

**26.** Write $4x+2y-5=0$ in slope intercept form.

$4x+2y-5=0$

$2y=-4x+5$

$y=-2x+\dfrac{5}{2}$

The slope of the parallel line is –2, thus the slope of the desired line is $m=-2$.

$y-y_1=m(x-x_1)$

$y-(-10)=-2(x-(-7))$

$y+10=-2(x+7)$

$y+10=-2x-14$

$2x+y+24=0$

**27. a.** Find slope: $m=\dfrac{5870-4571}{4-1}=\dfrac{1299}{3}=433$

point-slope form:

$y-y_1=m(x-x_1)$

$y-4571=433(x-1)$

**b.** slope-intercept form:

$y-4571=433(x-1)$

$y-4571=433x-433$

$y=433x+4138$

$f(x)=433x+4138$

**c.** $f(x)=433x+4138$

$=433(10)+4138$

$=8468$

According to the model, 8468 fatalities will involve distracted driving in 2014.

**28.** $\dfrac{3(10)^2-5-[3(6)^2-5]}{10-6}$

$=\dfrac{205-103}{4}$

$=\dfrac{192}{4}$

$=48$

**29.** $g(-1)=3-(-1)=4$

$g(7)=\sqrt{7-3}=\sqrt{4}=2$

**30.** The denominator is zero when $x=1$ or $x=-5$.

domain: $(-\infty,-5)\cup(-5,1)\cup(1,\infty)$

**31.** The expressions under each radical must not be negative.

$x+5\geq0$ and $x-1\geq0$

$x\geq-5$ $\qquad x\geq1$

domain: $[1,\infty)$

**32.** $(f\circ g)(x)=\dfrac{7}{\dfrac{2}{x}-4}=\dfrac{7x}{2-4x}$

$x\neq0,\qquad 2-4x\neq0$

$x\neq\dfrac{1}{2}$

domain: $(-\infty,0)\cup\left(0,\dfrac{1}{2}\right)\cup\left(\dfrac{1}{2},\infty\right)$

**33.** $f(x)=x^7\qquad g(x)=2x+3$

**34.** $d=\sqrt{(x_2-x_1)^2+(y_2-y_1)^2}$

$d=\sqrt{(x_2-x_1)^2+(y_2-y_1)^2}$

$=\sqrt{(5-2)^2+(2-(-2))^2}$

$=\sqrt{3^2+4^2}$

$=\sqrt{9+16}$

$=\sqrt{25}$

$=5$

$\left(\dfrac{x_1+x_2}{2},\dfrac{y_1+y_2}{2}\right)=\left(\dfrac{2+5}{2},\dfrac{-2+2}{2}\right)$

$=\left(\dfrac{7}{2},0\right)$

The length is 5 and the midpoint is $\left(\dfrac{7}{2},0\right)$ or $(3.5,0)$.

## Cumulative Review Exercises (Chapters 1–2)

**1.** domain: $[0, 2)$

range: $[0, 2]$

**2.** $f(x) = 1$ at $\frac{1}{2}$ and $\frac{3}{2}$.

**3.** relative maximum: 2

**4.**

$g(x) = f(x-1) + 1$

**5.**

$h(x) = -2f\left(\frac{1}{2}x\right)$

**6.** $(x+3)(x-4) = 8$

$x^2 - x - 12 = 8$

$x^2 - x - 20 = 0$

$(x+4)(x-5) = 0$

$x + 4 = 0$ or $x - 5 = 0$

$x = -4$ or $x = 5$

**7.** $3(4x-1) = 4 - 6(x-3)$

$12x - 3 = 4 - 6x + 18$

$18x = 25$

$x = \dfrac{25}{18}$

**8.** $\sqrt{x} + 2 = x$

$\sqrt{x} = x - 2$

$(\sqrt{x})^2 = (x-2)^2$

$x = x^2 - 4x + 4$

$0 = x^2 - 5x + 4$

$0 = (x-1)(x-4)$

$x - 1 = 0$ or $x - 4 = 0$

$x = 1$ or $x = 4$

A check of the solutions shows that $x = 1$ is an extraneous solution.

The solution set is $\{4\}$.

**9.** $x^{2/3} - x^{1/3} - 6 = 0$

Let $u = x^{1/3}$. Then $u^2 = x^{2/3}$.

$u^2 - u - 6 = 0$

$(u+2)(u-3) = 0$

$u = -2$ or $u = 3$

$x^{1/3} = -2$ or $x^{1/3} = 3$

$x = (-2)^3$ or $x = 3^3$

$x = -8$ or $x = 27$

**10.** $\dfrac{x}{2} - 3 \le \dfrac{x}{4} + 2$

$4\left(\dfrac{x}{2} - 3\right) \le 4\left(\dfrac{x}{4} + 2\right)$

$2x - 12 \le x + 8$

$x \le 20$

The solution set is $(-\infty, 20]$.

**11.**

$3x - 6y - 12 = 0$

domain: $(-\infty, \infty)$

range: $(-\infty, \infty)$

**12.**

$(x-2)^2 + (y+1)^2 = 4$

domain: $[0, 4]$

range: $[-3, 1]$

**13.**

domain of $f$: $(-\infty, \infty)$

range of $f$: $(-\infty, \infty)$

domain of $g$: $(-\infty, \infty)$

range of $g$: $(-\infty, \infty)$

**14.**

domain of $f$: $[3, \infty)$

range of $f$: $[2, \infty)$

domain of $f^{-1}$: $[2, \infty)$

range of $f^{-1}$: $[3, \infty)$

**15.** $\dfrac{f(x+h) - f(x)}{h}$

$= \dfrac{\left(4 - (x+h)^2\right) - \left(4 - x^2\right)}{h}$

$= \dfrac{4 - (x^2 + 2xh + h^2) - \left(4 - x^2\right)}{h}$

$= \dfrac{4 - x^2 - 2xh - h^2 - 4 + x^2}{h}$

$= \dfrac{-2xh - h^2}{h}$

$= \dfrac{h(-2x - h)}{h}$

$= -2x - h$

**16.** $(f \circ g)(x) = f\left(g(x)\right)$

$(f \circ g)(x) = f\left(x + 5\right)$

$0 = 4 - \left(x + 5\right)^2$

$0 = 4 - (x^2 + 10x + 25)$

$0 = 4 - x^2 - 10x - 25$

$0 = -x^2 - 10x - 21$

$0 = x^2 + 10x + 21$

$0 = (x + 7)(x + 3)$

The value of $(f \circ g)(x)$ will be 0 when $x = -3$ or $x = -7$.

**17.** $y = -\dfrac{1}{4}x + \dfrac{1}{3}$, so $m = 4$.

point-slope form: $y - 5 = 4(x + 2)$

slope-intercept form: $y = 4x + 13$

general form: $4x - y + 13 = 0$

**18.** $0.07x + 0.09(6000 - x) = 510$

$0.07x + 540 - 0.09x = 510$

$-0.02x = -30$

$x = 1500$

$6000 - x = 4500$

$1500 was invested at 7% and $4500 was invested at 9%.

**19.** $200 + 0.05x = .15x$

$200 = 0.10x$

$2000 = x$

For $2000 in sales, the earnings will be the same.

**20.** width $= w$

length $= 2w + 2$

$2(2w + 2) + 2w = 22$

$4w + 4 + 2w = 22$

$6w = 18$

$w = 3$

$2w + 2 = 8$

The garden is 3 feet by 8 feet.

# Chapter 3
# Polynomial and Rational Functions

## Section 3.1

### Check Point Exercises

1.  $f(x) = -(x-1)^2 + 4$

    $f(x) = \overset{a=-1}{-}\left(x - \overset{h=1}{1}\right)^2 + \overset{k=4}{4}$

    Step 1: The parabola opens down because $a < 0$.
    Step 2: find the vertex: $(1, 4)$
    Step 3: find the $x$-intercepts:

    $0 = -(x-1)^2 + 4$

    $(x-1)^2 = 4$

    $x - 1 = \pm 2$

    $x = 1 \pm 2$

    $x = 3$ or $x = -1$

    Step 4: find the $y$-intercept:

    $f(0) = -(0-1)^2 + 4 = 3$

    Step 5: The axis of symmetry is $x = 1$.

    $f(x) = -(x - 1)^2 + 4$

2.  $f(x) = (x-2)^2 + 1$

    Step 1: The parabola opens up because $a > 0$.
    Step 2: find the vertex: $(2, 1)$
    Step 3: find the $x$-intercepts:

    $0 = (x-2)^2 + 1$

    $(x-2)^2 = -1$

    $x - 2 = \sqrt{-1}$

    $x = 2 \pm i$

    The equation has no real roots, thus the parabola has no $x$-intercepts.
    Step 4: find the $y$-intercept:

    $f(0) = (0-2)^2 + 1 = 5$

Step 5: The axis of symmetry is $x = 2$.

$f(x) = (x - 2)^2 + 1$

3.  $f(x) = -x^2 + 4x + 1$

    Step 1: The parabola opens down because $a < 0$.
    Step 2: find the vertex:

    $x = -\dfrac{b}{2a} = -\dfrac{4}{2(-1)} = 2$

    $f(2) = -2^2 + 4(2) + 1 = 5$

    The vertex is $(2, 5)$.
    Step 3: find the $x$-intercepts:

    $0 = -x^2 + 4x + 1$

    $x = \dfrac{-b \pm \sqrt{b^2 - 4ac}}{2a}$

    $x = \dfrac{-4 \pm \sqrt{4^2 - 4(-1)(1)}}{2(-1)}$

    $x = \dfrac{-4 \pm \sqrt{20}}{-2}$

    $x = 2 \pm \sqrt{5}$

    The $x$-intercepts are $x \approx -0.2$ and $x \approx -4.2$.

    Step 4: find the $y$-intercept: $f(0) = -0^2 + 4(0) + 1 = 1$

    Step 5: The axis of symmetry is $x = 2$.

    $f(x) = -x^2 + 4x + 1$

**4.** $f(x) = 4x^2 - 16x + 1000$

    **a.** $a = 4$. The parabola opens upward and has a minimum value.

    **b.** $x = \dfrac{-b}{2a} = \dfrac{16}{8} = 2$

        $f(2) = 4(2)^2 - 16(2) + 1000 = 984$

        The minimum point is 984 at $x = 2$.

    **c.** domain: $(-\infty, \infty)$ range: $[984, \infty)$

**5.** $f(x) = -0.005x^2 + 2x + 5$

    **a.** The information needed is found at the vertex.
        $x$-coordinate of vertex

        $x = \dfrac{-b}{2a} = \dfrac{-2}{2(-0.005)} = 200$

        $y$-coordinate of vertex

        $y = -0.005(200)^2 + 2(200) + 5 = 205$

        The vertex is (200, 205).
        The maximum height of the arrow is 205 feet.
        This occurs 200 feet from its release.

    **b.** The arrow will hit the ground when the height reaches 0.

        $f(x) = -0.005x^2 + 2x + 5$

        $0 = -0.005x^2 + 2x + 5$

        $x = \dfrac{-b \pm \sqrt{b^2 - 4ac}}{2a}$

        $x = \dfrac{-2 \pm \sqrt{2^2 - 4(-0.005)(5)}}{2(-0.005)}$

        $x \approx -2$ or $x \approx 402$
        The arrow travels 402 feet before hitting the ground.

    **c.** The starting point occurs when $x = 0$. Find the corresponding $y$-coordinate.

        $f(x) = -0.005(0)^2 + 2(0) + 5 = 5$

        Plot $(0, 5)$, $(402, 0)$, and $(200, 205)$, and connect them with a smooth curve.

**6.** Let $x$ = one of the numbers;
    $x - 8$ = the other number.
    The product is $f(x) = x(x - 8) = x^2 - 8x$
    The $x$-coordinate of the minimum is

    $x = -\dfrac{b}{2a} = -\dfrac{-8}{2(1)} = -\dfrac{-8}{2} = 4.$

    $f(4) = (4)^2 - 8(4)$
          $= 16 - 32 = -16$
    The vertex is $(4, -16)$.

    The minimum product is $-16$. This occurs when the two numbers are 4 and $4 - 8 = -4$.

**7.** Maximize the area of a rectangle constructed with 120 feet of fencing.
    Let $x$ = the length of the rectangle. Let $y$ = the width of the rectangle.
    Since we need an equation in one variable, use the perimeter to express $y$ in terms of $x$.
    $2x + 2y = 120$

        $2y = 120 - 2x$

        $y = \dfrac{120 - 2x}{2} = 60 - x$

    We need to maximize $A = xy = x(60 - x)$. Rewrite $A$ as a function of $x$.

    $A(x) = x(60 - x) = -x^2 + 60x$

    Since $a = -1$ is negative, we know the function opens downward and has a maximum at

    $x = -\dfrac{b}{2a} = -\dfrac{60}{2(-1)} = -\dfrac{60}{-2} = 30.$

    When the length $x$ is 30, the width $y$ is
    $y = 60 - x = 60 - 30 = 30.$
    The dimensions of the rectangular region with maximum area are 30 feet by 30 feet. This gives an area of $30 \cdot 30 = 900$ square feet.

**Concept and Vocabulary Check 3.1**

**1.** standard; parabola; $(h, k)$; $> 0$; $< 0$

**2.** $-\dfrac{b}{2a}$; $f\left(-\dfrac{b}{2a}\right)$; $-\dfrac{b}{2a}$; $f\left(-\dfrac{b}{2a}\right)$

**3.** true

**4.** false

**5.** true

**6.** $x-8$; $x^2-8x$

**7.** $40-x$; $-x^2+40x$

**Exercise Set 3.1**

**1.** vertex: (1, 1)
$h(x)=(x-1)^2+1$

**3.** vertex: (1, −1)
$j(x)=(x-1)^2-1$

**5.** The graph is $f(x)=x^2$ translated down one.
$h(x)=x^2-1$

**7.** The point (1, 0) is on the graph and
$g(1)=0$. $g(x)=x^2-2x+1$

**9.** $f(x)=2(x-3)^2+1$
$h=3, k=1$
The vertex is at (3, 1).

**11.** $f(x)=-2(x+1)^2+5$
$h=-1, k=5$
The vertex is at (−1, 5).

**13.** $f(x)=2x^2-8x+3$
$x=\dfrac{-b}{2a}=\dfrac{8}{4}=2$
$f(2)=2(2)^2-8(2)+3$
$=8-16+3=-5$
The vertex is at (2, −5).

**15.** $f(x)=-x^2-2x+8$
$x=\dfrac{-b}{2a}=\dfrac{2}{-2}=-1$
$f(-1)=-(-1)^2-2(-1)+8$
$=-1+2+8=9$
The vertex is at (−1, 9).

**17.** $f(x)=(x-4)^2-1$
vertex: (4, −1)
$x$-intercepts:
$0=(x-4)^2-1$
$1=(x-4)^2$
$\pm 1=x-4$
$x=3$ or $x=5$
$y$-intercept:
$f(0)=(0-4)^2-1=15$
The axis of symmetry is $x=4$.

$f(x)=(x-4)^2-1$

domain: $(-\infty,\infty)$

range: $[-1,\infty)$

**19.** $f(x)=(x-1)^2+2$
vertex: (1, 2)
$x$-intercepts:
$0=(x-1)^2+2$
$(x-1)^2=-2$
$x-1=\pm\sqrt{-2}$
$x=1\pm i\sqrt{2}$
No $x$-intercepts.
$y$-intercept:
$f(0)=(0-1)^2+2=3$
The axis of symmetry is $x=1$.

$f(x)=(x-1)^2+2$

domain: $(-\infty,\infty)$

range: $[2,\infty)$

**21.** $y - 1 = (x - 3)^2$

$y = (x - 3)^2 + 1$

vertex: (3, 1)

$x$-intercepts:

$0 = (x - 3)^2 + 1$

$(x - 3)^2 = -1$

$x - 3 = \pm i$

$x = 3 \pm i$

No $x$-intercepts.

$y$-intercept: 10

$y = (0 - 3)^2 + 1 = 10$

The axis of symmetry is $x = 3$.

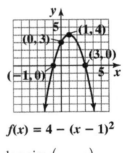

$y - 1 = (x - 3)^2$

domain: $(-\infty, \infty)$

range: $[1, \infty)$

**23.** $f(x) = 2(x + 2)^2 - 1$

vertex: (−2, −1)

$x$-intercepts:

$0 = 2(x + 2)^2 - 1$

$2(x + 2)^2 = 1$

$(x + 2)^2 = \dfrac{1}{2}$

$x + 2 = \pm \dfrac{1}{\sqrt{2}}$

$x = -2 \pm \dfrac{1}{\sqrt{2}} = -2 \pm \dfrac{\sqrt{2}}{2}$

$y$-intercept:

$f(0) = 2(0 + 2)^2 - 1 = 7$

The axis of symmetry is $x = -2$.

$f(x) = 2(x + 2)^2 - 1$

domain: $(-\infty, \infty)$

range: $[-1, \infty)$

**25.** $f(x) = 4 - (x - 1)^2$

$f(x) = -(x - 1)^2 + 4$

vertex: (1, 4)

$x$-intercepts:

$0 = -(x - 1)^2 + 4$

$(x - 1)^2 = 4$

$x - 1 = \pm 2$

$x = -1$ or $x = 3$

$y$-intercept:

$f(x) = -(0 - 1)^2 + 4 = 3$

The axis of symmetry is $x = 1$.

$f(x) = 4 - (x - 1)^2$

domain: $(-\infty, \infty)$

range: $(-\infty, 4]$

**27.** $f(x) = x^2 - 2x - 3$

$f(x) = (x^2 - 2x + 1) - 3 - 1$

$f(x) = (x-1)^2 - 4$

vertex: (1, –4)

*x*-intercepts:

$0 = (x-1)^2 - 4$

$(x-1)^2 = 4$

$x - 1 = \pm 2$

$x = -1$ or $x = 3$

*y*-intercept: –3

$f(0) = 0^2 - 2(0) - 3 = -3$

The axis of symmetry is $x = 1$.

$f(x) = x^2 + 3x - 10$

domain: $(-\infty, \infty)$

range: $[-4, \infty)$

**29.** $f(x) = x^2 + 3x - 10$

$f(x) = \left(x^2 + 3x + \dfrac{9}{4}\right) - 10 - \dfrac{9}{4}$

$f(x) = \left(x + \dfrac{3}{2}\right)^2 - \dfrac{49}{4}$

vertex: $\left(-\dfrac{3}{2}, -\dfrac{49}{4}\right)$

*x*-intercepts:

$0 = \left(x + \dfrac{3}{2}\right)^2 - \dfrac{49}{4}$

$\left(x + \dfrac{3}{2}\right)^2 = \dfrac{49}{4}$

$x + \dfrac{3}{2} = \pm \dfrac{7}{2}$

$x = -\dfrac{3}{2} \pm \dfrac{7}{2}$

$x = 2$ or $x = -5$

*y*-intercept:

$f(x) = 0^2 + 3(0) - 10 = -10$

The axis of symmetry is $x = -\dfrac{3}{2}$.

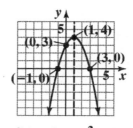

$f(x) = x^2 + 3x - 10$

domain: $(-\infty, \infty)$

range: $\left[-\dfrac{49}{4}, \infty\right)$

**31.** $f(x) = 2x - x^2 + 3$

$f(x) = -x^2 + 2x + 3$

$f(x) = -(x^2 - 2x + 1) + 3 + 1$

$f(x) = -(x-1)^2 + 4$

vertex: (1, 4)

*x*-intercepts:

$0 = -(x-1)^2 + 4$

$(x-1)^2 = 4$

$x - 1 = \pm 2$

$x = -1$ or $x = 3$

*y*-intercept:

$f(0) = 2(0) - (0)^2 + 3 = 3$

The axis of symmetry is $x = 1$.

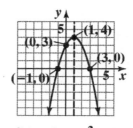

$f(x) = 2x - x^2 + 3$

domain: $(-\infty, \infty)$

range: $(-\infty, 4]$

**33.** $f(x) = x^2 + 6x + 3$

$f(x) = (x^2 + 6x + 9) + 3 - 9$

$f(x) = (x+3)^2 - 6$

vertex: $(-3, -6)$

$x$-intercepts:

$0 = (x+3)^2 - 6$

$(x+3)^2 = 6$

$x + 3 = \pm\sqrt{6}$

$x = -3 \pm \sqrt{6}$

$y$-intercept:

$f(0) = (0)^2 + 6(0) + 3$

$f(0) = 3$

The axis of symmetry is $x = -3$.

$f(x) = x^2 + 6x + 3$

domain: $(-\infty, \infty)$

range: $[-6, \infty)$

**35.** $f(x) = 2x^2 + 4x - 3$

$f(x) = 2(x^2 + 2x \quad) - 3$

$f(x) = 2(x^2 + 2x + 1) - 3 - 2$

$f(x) = 2(x+1)^2 - 5$

vertex: $(-1, -5)$

$x$-intercepts:

$0 = 2(x+1)^2 - 5$

$2(x+1)^2 = 5$

$(x+1)^2 = \dfrac{5}{2}$

$x + 1 = \pm\sqrt{\dfrac{5}{2}}$

$x = -1 \pm \dfrac{\sqrt{10}}{2}$

$y$-intercept:

$f(0) = 2(0)^2 + 4(0) - 3$

$f(0) = -3$

The axis of symmetry is $x = -1$.

$\left(-1 - \dfrac{\sqrt{10}}{2}, 0\right)$ $\left(-1 + \dfrac{\sqrt{10}}{2}, 0\right)$ $(0, -3)$ $(-1, -5)$

$f(x) = 2x^2 + 4x - 3$

domain: $(-\infty, \infty)$

range: $[-5, \infty)$

**37.** $f(x) = 2x - x^2 - 2$

$f(x) = -x^2 + 2x - 2$

$f(x) = -(x^2 - 2x + 1) - 2 + 1$

$f(x) = -(x-1)^2 - 1$

vertex: $(1, -1)$

$x$-intercepts:

$0 = -(x-1)^2 - 1$

$(x-1)^2 = -1$

$x - 1 = \pm i$

$x = 1 \pm i$

No $x$-intercepts.

$y$-intercept:

$f(0) = 2(0) - (0)^2 - 2 = -2$

The axis of symmetry is $x = 1$.

$(1, -1)$ $(0, -2)$

$f(x) = 2x - x^2 - 2$

domain: $(-\infty, \infty)$

range: $(-\infty, -1]$

**39.** $f(x) = 3x^2 - 12x - 1$

  **a.** $a = 3$. The parabola opens upward and has a minimum value.

  **b.** $x = \dfrac{-b}{2a} = \dfrac{12}{6} = 2$

   $f(2) = 3(2)^2 - 12(2) - 1$
   $= 12 - 24 - 1 = -13$
   The minimum is $-13$ at $x = 2$.

  **c.** domain: $(-\infty, \infty)$  range: $[-13, \infty)$

**41.** $f(x) = -4x^2 + 8x - 3$

  **a.** $a = -4$. The parabola opens downward and has a maximum value.

  **b.** $x = \dfrac{-b}{2a} = \dfrac{-8}{-8} = 1$

   $f(1) = -4(1)^2 + 8(1) - 3$
   $= -4 + 8 - 3 = 1$
   The maximum is $1$ at $x = 1$.

  **c.** domain: $(-\infty, \infty)$  range: $(-\infty, 1]$

**43.** $f(x) = 5x^2 - 5x$

  **a.** $a = 5$. The parabola opens upward and has a minimum value.

  **b.** $x = \dfrac{-b}{2a} = \dfrac{5}{10} = \dfrac{1}{2}$

   $f\left(\dfrac{1}{2}\right) = 5\left(\dfrac{1}{2}\right)^2 - 5\left(\dfrac{1}{2}\right)$

   $= \dfrac{5}{4} - \dfrac{5}{2} = \dfrac{5}{4} - \dfrac{10}{4} = \dfrac{-5}{4}$

   The minimum is $\dfrac{-5}{4}$ at $x = \dfrac{1}{2}$.

  **c.** domain: $(-\infty, \infty)$  range: $\left[\dfrac{-5}{4}, \infty\right)$

**45.** Since the parabola opens up, the vertex $(-1, -2)$ is a minimum point.
  domain: $(-\infty, \infty)$.  range: $[-2, \infty)$

**47.** Since the parabola has a maximum, it opens down from the vertex $(10, -6)$.

  domain: $(-\infty, \infty)$.  range: $(-\infty, -6]$

**49.** $(h, k) = (5, 3)$

  $f(x) = 2(x - h)^2 + k = 2(x - 5)^2 + 3$

**51.** $(h, k) = (-10, -5)$

  $f(x) = 2(x - h)^2 + k$

   $= 2[x - (-10)]^2 + (-5)$

   $= 2(x + 10)^2 - 5$

**53.** Since the vertex is a maximum, the parabola opens down and $a = -3$.
  $(h, k) = (-2, 4)$

  $f(x) = -3(x - h)^2 + k$

   $= -3[x - (-2)]^2 + 4$

   $= -3(x + 2)^2 + 4$

**55.** Since the vertex is a minimum, the parabola opens up and $a = 3$.
  $(h, k) = (11, 0)$

  $f(x) = 3(x - h)^2 + k$

   $= 3(x - 11)^2 + 0$

   $= 3(x - 11)^2$

**57.** **a.** $y = -0.01x^2 + 0.7x + 6.1$
   $a = -0.01, \ b = 0.7, \ c = 6.1$
   $x$-coordinate of vertex
   $= \dfrac{-b}{2a} = \dfrac{-0.7}{2(-0.01)} = 35$
   $y$-coordinate of vertex
   $y = -0.01x^2 + 0.7x + 6.1$
   $y = -0.01(35)^2 + 0.7(35) + 6.1 = 18.35$
   The maximum height of the shot is about 18.35 feet. This occurs 35 feet from its point of release.

**b.** The ball will reach the maximum horizontal distance when its height returns to 0.

$$y = -0.01x^2 + 0.7x + 6.1$$

$$0 = -0.01x^2 + 0.7x + 6.1$$

$$a = -0.01, \; b = 0.7, \; c = 6.1$$

$$x = \frac{-b \pm \sqrt{b^2 - 4ac}}{2a}$$

$$x = \frac{-0.7 \pm \sqrt{0.7^2 - 4(-0.01)(6.1)}}{2(-0.01)}$$

$$x \approx 77.8 \text{ or } x \approx -7.8$$

The maximum horizontal distance is 77.8 feet.

**c.** The initial height can be found at $x = 0$.

$$y = -0.01x^2 + 0.7x + 6.1$$

$$y = -0.01(0)^2 + 0.7(0) + 6.1 = 6.1$$

The shot was released at a height of 6.1 feet.

**59.** $y = -0.8x^2 + 2.4x + 6$

**a.** The information needed is found at the vertex.
$x$-coordinate of vertex

$$x = \frac{-b}{2a} = \frac{-2.4}{2(-0.8)} = 1.5$$

$y$-coordinate of vertex

$$y = -0.8(1.5)^2 + 2.4(1.5) + 6 = 7.8$$

The vertex is (1.5, 7.8).
The maximum height of the ball is 7.8 feet.
This occurs 1.5 feet from its release.

**b.** The ball will hit the ground when the height reaches 0.

$$y = -0.8x^2 + 2.4x + 6$$

$$0 = -0.8x^2 + 2.4x + 6$$

$$x = \frac{-b \pm \sqrt{b^2 - 4ac}}{2a}$$

$$x = \frac{-2.4 \pm \sqrt{2.4^2 - 4(-0.8)(6)}}{2(-0.8)}$$

$$x \approx -1.6 \text{ or } x \approx 4.6$$

The ball travels 4.6 feet before hitting the ground.

**c.** The starting point occurs when $x = 0$. Find the corresponding $y$-coordinate.

$$y = -0.8(0)^2 + 2.4(0) + 6 = 6$$

Plot $(0, 6)$, $(1.5, 7.8)$, and $(4.7, 0)$, and connect them with a smooth curve.

**61.** Let $x =$ one of the numbers;
$16 - x =$ the other number.

The product is $f(x) = x(16 - x)$
$$= 16x - x^2 = -x^2 + 16x$$

The $x$-coordinate of the maximum is

$$x = -\frac{b}{2a} = -\frac{16}{2(-1)} = -\frac{16}{-2} = 8.$$

$$f(8) = -8^2 + 16(8) = -64 + 128 = 64$$

The vertex is (8, 64). The maximum product is 64.
This occurs when the two numbers are 8 and $16 - 8 = 8$.

**63.** Let $x =$ one of the numbers;
$x - 16 =$ the other number.

The product is $f(x) = x(x - 16) = x^2 - 16x$

The $x$-coordinate of the minimum is

$$x = -\frac{b}{2a} = -\frac{-16}{2(1)} = -\frac{-16}{2} = 8.$$

$$f(8) = (8)^2 - 16(8) = 64 - 128 = -64$$

The vertex is $(8, -64)$. The minimum product is $-64$.
This occurs when the two numbers are 8 and $8 - 16 = -8$.

**65.** Maximize the area of a rectangle constructed along a river with 600 feet of fencing.

Let $x$ = the width of the rectangle;
$600 - 2x$ = the length of the rectangle
We need to maximize.

$A(x) = x(600 - 2x)$

$$= 600x - 2x^2 = -2x^2 + 600x$$

Since $a = -2$ is negative, we know the function opens downward and has a maximum at

$$x = -\frac{b}{2a} = -\frac{600}{2(-2)} = -\frac{600}{-4} = 150.$$

When the width is $x = 150$ feet, the length is
$600 - 2(150) = 600 - 300 = 300$ feet.

The dimensions of the rectangular plot with maximum area are 150 feet by 300 feet. This gives an area of $150 \cdot 300 = 45,000$ square feet.

**67.** Maximize the area of a rectangle constructed with 50 yards of fencing.

Let $x$ = the length of the rectangle. Let $y$ = the width of the rectangle.
Since we need an equation in one variable, use the perimeter to express $y$ in terms of $x$.

$2x + 2y = 50$

$2y = 50 - 2x$

$$y = \frac{50 - 2x}{2} = 25 - x$$

We need to maximize $A = xy = x(25 - x)$. Rewrite $A$ as a function of $x$.

$A(x) = x(25 - x) = -x^2 + 25x$

Since $a = -1$ is negative, we know the function opens downward and has a maximum at

$$x = -\frac{b}{2a} = -\frac{25}{2(-1)} = -\frac{25}{-2} = 12.5.$$

When the length $x$ is 12.5, the width $y$ is
$y = 25 - x = 25 - 12.5 = 12.5$.

The dimensions of the rectangular region with maximum area are 12.5 yards by 12.5 yards. This gives an area of $12.5 \cdot 12.5 = 156.25$ square yards.

**69.** Maximize the area of the playground with 600 feet of fencing.

Let $x$ = the length of the rectangle. Let $y$ = the width of the rectangle.
Since we need an equation in one variable, use the perimeter to express $y$ in terms of $x$.

$2x + 3y = 600$

$3y = 600 - 2x$

$$y = \frac{600 - 2x}{3}$$

$$y = 200 - \frac{2}{3}x$$

We need to maximize $A = xy = x\left(200 - \frac{2}{3}x\right)$.

Rewrite $A$ as a function of $x$.

$$A(x) = x\left(200 - \frac{2}{3}x\right) = -\frac{2}{3}x^2 + 200x$$

Since $a = -\frac{2}{3}$ is negative, we know the function opens downward and has a maximum at

$$x = -\frac{b}{2a} = -\frac{200}{2\left(-\frac{2}{3}\right)} = -\frac{200}{-\frac{4}{3}} = 150.$$

When the length $x$ is 150, the width $y$ is

$$y = 200 - \frac{2}{3}x = 200 - \frac{2}{3}(150) = 100.$$

The dimensions of the rectangular playground with maximum area are 150 feet by 100 feet. This gives an area of $150 \cdot 100 = 15,000$ square feet.

**71.** Maximize the cross-sectional area of the gutter:

$A(x) = x(20 - 2x)$

$$= 20x - 2x^2 = -2x^2 + 20x.$$

Since $a = -2$ is negative, we know the function opens downward and has a maximum at

$$x = -\frac{b}{2a} = -\frac{20}{2(-2)} = -\frac{20}{-4} = 5.$$

When the height $x$ is 5, the width is
$20 - 2x = 20 - 2(5) = 20 - 10 = 10$.

$A(5) = -2(5)^2 + 20(5)$

$$= -2(25) + 100 = -50 + 100 = 50$$

The maximum cross-sectional area is 50 square inches. This occurs when the gutter is 5 inches deep and 10 inches wide.

**73. a.** $C(x) = 0.55x + 525$

**b.** $P(x) = R - C$

$P(x) = -0.001x^2 + 3x - 0.55x - 525$

$P(x) = -0.001x^2 + 2.45x - 525$

**c.** $x = \dfrac{-2.45}{2(-0.001)} = 1225$

$P(x) = -0.001(1225)^2 + 2.45(1225) - 525$

$= 975.63$

The maximum profit will be $975.63 per week obtained by selling 1225 sandwiches.

**75. – 79.** Answers will vary.

**81.** $y = 2x^2 - 82x + 720$

**a.**

You can only see a little of the parabola.

**b.** $a = 2; b = -82$

$x = -\dfrac{b}{2a} = -\dfrac{-82}{4} = 20.5$

$y = 2(20.5)^2 - 82(20.5) + 720$

$= 840.5 - 1681 + 720$

$= -120.5$

vertex: $(20.5, -120.5)$

**c.** Ymax = 750

**d.** You can choose Xmin and Xmax so the $x$-value of the vertex is in the center of the graph. Choose Ymin to include the $y$-value of the vertex.

**83.** $y = -4x^2 + 20x + 160$

$x = \dfrac{-b}{2a} = \dfrac{-20}{-8} = 2.5$

$y = -4(2.5)^2 + 20(2.5) + 160$

$= -2.5 + 50 + 160 = 185$

The vertex is at $(2.5, 185)$.

**85.** $y = 0.01x^2 + 0.6x + 100$

$x = \dfrac{-b}{2a} = \dfrac{-0.6}{0.02} = -30$

$y = 0.01(-30)^2 + 0.6(-30) + 100$

$= 9 - 18 + 100 = 91$

The vertex is at $(-30, 91)$.

**87.** does not make sense; Explanations will vary. Sample explanation: Some parabolas have the $y$-axis as the axis of symmetry.

**89.** does not make sense; Explanations will vary. Sample explanation: If it is thrown vertically, its path will be a line segment.

**91.** true

**93.** false; Changes to make the statement true will vary. A sample change is: The graph has no $x$–intercepts. To find $x$–intercepts, set $y = 0$ and solve for $x$.

$0 = -2(x+4)^2 - 8$

$2(x+4)^2 = -8$

$(x+4)^2 = -4$

Because the solutions to the equation are imaginary, we know that there are no $x$–intercepts.

**95.** $f(x) = 3(x+2)^2 - 5; (-1, -2)$

axis: $x = -2$

$(-1, -2)$ is one unit right of $(-2, -2)$. One unit left of $(-2, -2)$ is $(-3, -2)$.

point: $(-3, -2)$

**97.** We start with the form $f(x) = a(x-h)^2 + k$.

Since we know the vertex is $(h, k) = (-3, -4)$, we have $f(x) = a(x+3)^2 - 4$. We also know that the graph passes through the point $(1, 4)$, which allows us to solve for $a$.

$4 = a(1+3)^2 - 4$

$8 = a(4)^2$

$8 = 16a$

$\dfrac{1}{2} = a$

Therefore, the function is $f(x) = \dfrac{1}{2}(x+3)^2 - 4$.

**99.** $3x + 4y = 1000$
$$4y = 1000 - 3x$$
$$y = 250 - 0.75x$$
$$A(x) = x(250 - 0.75x)$$
$$= 250x - 0.75x^2$$
$$x = \frac{-b}{2a} = \frac{-250}{-1.5} = 166\frac{2}{3}$$
$$y = 250 - 0.75\left(166\frac{2}{3}\right) = 125$$

The dimensions are $x = 166\frac{2}{3}$ ft, $y = 125$ ft.

The maximum area is about 20,833 ft$^2$.

**101.** Answers will vary.

**102.** $x^3 + 3x^2 - x - 3 = x^2(x+3) - 1(x+3)$
$$= (x+3)(x^2 - 1)$$
$$= (x+3)(x+1)(x-1)$$

**103.** $f(x) = x^3 - 2x - 5$
$$f(2) = (2)^3 - 2(2) - 5 = -1$$
$$f(3) = (3)^3 - 2(3) - 5 = 16$$
The graph passes through (2, –1), which is below the $x$-axis, and (3, 16), which is above the $x$-axis. Since the graph of $f$ is continuous, it must cross the $x$-axis somewhere between 2 and 3 to get from one of these points to the other.

**104.** $f(x) = x^4 - 2x^2 + 1$
$$f(-x) = (-x)^4 - 2(-x)^2 + 1$$
$$= x^4 - 2x^2 + 1$$
Since $f(-x) = f(x)$, the function is even.
Thus, the graph is symmetric with respect to the $y$-axis.

## Section 3.2

### Check Point Exercises

**1.** Since $n$ is even and $a_n > 0$, the graph rises to the left and to the right.

**2.** It is not necessary to multiply out the polynomial to determine its degree. We can find the degree of the polynomial by adding the degrees of each of its factors. $f(x) = 2 \overbrace{x^3}^{\text{degree 3}} \overbrace{(x-1)}^{\text{degree 1}}\overbrace{(x+5)}^{\text{degree 1}}$ has degree $3 + 1 + 1 = 5$.

$f(x) = 2x^3(x-1)(x+5)$ is of odd degree with a positive leading coefficient. Thus, the graph falls to the left and rises to the right.

**3.** Since $n$ is odd and the leading coefficient is negative, the function falls to the right. Since the ratio cannot be negative, the model won't be appropriate.

**4.** The graph does not show the function's end behavior. Since $a_n > 0$ and $n$ is odd, the graph should fall to the left but doesn't appear to do so.

**5.** $f(x) = x^3 + 2x^2 - 4x - 8$
$$0 = x^2(x+2) - 4(x+2)$$
$$0 = (x+2)(x^2 - 4)$$
$$0 = (x+2)^2(x-2)$$
$$x = -2 \text{ or } x = 2$$
The zeros are –2 and 2.

**6.** $f(x) = x^4 - 4x^2$
$$x^4 - 4x^2 = 0$$
$$x^2(x^2 - 4) = 0$$
$$x^2(x+2)(x-2) = 0$$
$$x = 0 \text{ or } x = -2 \text{ or } x = 2$$
The zeros are –2, 0, and 2.

**7.** $f(x) = -4\left(x+\dfrac{1}{2}\right)^2 (x-5)^3$

$-4\left(x+\dfrac{1}{2}\right)^2 (x-5)^3 = 0$

$x = -\dfrac{1}{2}$ or $x = 5$

The zeros are $-\dfrac{1}{2}$, with multiplicity 2, and 5, with multiplicity 3.

Because the multiplicity of $-\dfrac{1}{2}$ is even, the graph touches the $x$-axis and turns around at this zero. Because the multiplicity of 5 is odd, the graph crosses the $x$-axis at this zero.

**8.** $f(x) = 3x^3 - 10x + 9$

$f(-3) = 3(-3)^3 - 10(-3) + 9 = -42$

$f(-2) = 3(-2)^3 - 10(-2) + 9 = 5$

The sign change shows there is a zero between $-3$ and $-2$.

**9.** $f(x) = x^3 - 3x^2$

Since $a_n > 0$ and $n$ is odd, the graph falls to the left and rises to the right.

$x^3 - 3x^2 = 0$

$x^2(x-3) = 0$

$x = 0$ or $x = 3$

The $x$-intercepts are 0 and 3.

$f(0) = 0^3 - 3(0)^2 = 0$

The $y$-intercept is 0.

$f(-x) = (-x)^3 - 3(-x)^2 = -x^3 - 3x^2$

No symmetry.

$f(x) = x^3 - 3x^2$

**10.** $f(x) = 2(x+2)^2 (x-3)$

The leading term is $2 \cdot x^2 \cdot x$, or $2x^3$.

Since $a_n > 0$ and $n$ is odd, the graph falls to the left and rises to the right.

$2(x+2)^2 (x-3) = 0$

$x = -2$ or $x = 3$

The $x$-intercepts are $-2$ and 3.

$f(0) = 2(0+2)^2 (0-3) = -12$

The $y$-intercept is $-12$.

$f(-x) = 2\big((-x)+2\big)^2 \big((-x)-3\big)$

$= 2(-x+2)^2 (-x-3)$

No symmetry.

$f(x) = 2(x+2)^2 (x-3)$

**Concept and Vocabulary Check 3.2**

**1.** 5; $-2$

**2.** false

**3.** end; leading

**4.** falls; rises

**5.** rises; falls

**6.** rises; rises

**7.** falls; falls

**8.** true

**9.** true

**10.** $x$-intercept

**11.** turns around; crosses

**12.** 0; Intermediate Value

**13.** $n-1$

**Exercise Set 3.2**

**1.** polynomial function;
degree: 3

**3.** polynomial function;
degree: 5

**5.** not a polynomial function

**7.** not a polynomial function

**9.** not a polynomial function

**11.** polynomial function

**13.** Not a polynomial function because graph is not continuous.

**15.** (b)

**17.** (a)

**19.** $f(x) = 5x^3 + 7x^2 - x + 9$
Since $a_n > 0$ and $n$ is odd, the graph of $f(x)$ falls to the left and rises to the right.

**21.** $f(x) = 5x^4 + 7x^2 - x + 9$
Since $a_n > 0$ and $n$ is even, the graph of $f(x)$ rises to the left and to the right.

**23.** $f(x) = -5x^4 + 7x^2 - x + 9$
Since $a_n < 0$ and $n$ is even, the graph of $f(x)$ falls to the left and to the right.

**25.** $f(x) = 2(x-5)(x+4)^2$
$x = 5$ has multiplicity 1;
The graph crosses the $x$-axis.
$x = -4$ has multiplicity 2;
The graph touches the $x$-axis and turns around.

**27.** $f(x) = 4(x-3)(x+6)^3$
x = 3 has multiplicity 1;
The graph crosses the $x$-axis.
$x = -6$ has multiplicity 3;
The graph crosses the $x$-axis.

**29.** $f(x) = x^3 - 2x^2 + x$
$\quad = x\left(x^2 - 2x + 1\right)$
$\quad = x(x-1)^2$
$x = 0$ has multiplicity 1;
The graph crosses the $x$-axis.
$x = 1$ has multiplicity 2;
The graph touches the $x$-axis and turns around.

**31.** $f(x) = x^3 + 7x^2 - 4x - 28$
$\quad = x^2(x+7) - 4(x+7)$
$\quad = \left(x^2 - 4\right)(x+7)$
$\quad = (x-2)(x+2)(x+7)$
$x = 2$, $x = -2$ and $x = -7$ have multiplicity 1;
The graph crosses the $x$-axis.

**33.** $f(x) = x^3 - x - 1$
$f(1) = -1$
$f(2) = 5$
The sign change shows there is a zero between the given values.

**35.** $f(x) = 2x^4 - 4x^2 + 1$
$f(-1) = -1$
$f(0) = 1$
The sign change shows there is a zero between the given values.

**37.** $f(x) = x^3 + x^2 - 2x + 1$
$f(-3) = -11$
$f(-2) = 1$
The sign change shows there is a zero between the given values.

**39.** $f(x) = 3x^3 - 10x + 9$
$f(-3) = -42$
$f(-2) = 5$
The sign change shows there is a zero between the given values.

**41.** $f(x) = x^3 + 2x^2 - x - 2$

**a.** Since $a_n > 0$ and $n$ is odd, $f(x)$ rises to the right and falls to the left.

**b.** $\quad x^3 + 2x^2 - x - 2 = 0$
$\quad x^2(x+2) - (x+2) = 0$
$\quad (x+2)(x^2 - 1) = 0$
$\quad (x+2)(x-1)(x+1) = 0$
$\quad x = -2, x = 1, x = -1$
The zeros at $-2$, $-1$, and 1 have odd multiplicity so $f(x)$ crosses the $x$-axis at these points.

**c.** $f(0) = (0)^3 + 2(0)^2 - 0 - 2$
$\quad = -2$
The $y$-intercept is $-2$.

**d.** $f(-x)=(-x)+2(-x)^2-(-x)-2$
$\qquad = -x^3+2x^2+x-2$
$-f(x)=-x^3-2x^2+x+2$
The graph has neither origin symmetry nor *y*-axis symmetry.

**e.** The graph has 2 turning points and $2 \le 3-1$.

$f(x)=x^3+2x^2-x-2$

**43.** $f(x)=x^4-9x^2$

**a.** Since $a_n>0$ and *n* is even, *f*(*x*) rises to the left and the right.

**b.** $x^4-9x^2=0$
$x^2(x^2-9)=0$
$x^2(x-3)(x+3)=0$
$x=0, x=3, x=-3$

The zeros at –3 and 3 have odd multiplicity, so *f*(*x*) crosses the *x*-axis at these points. The root at 0 has even multiplicity, so *f*(*x*) touches the *x*-axis at 0.

**c.** $f(0)=(0)^4-9(0)^2=0$
The *y*-intercept is 0.

**d.** $f(-x)=x^4-9x^2$
$f(-x)=f(x)$
The graph has *y*-axis symmetry.

**e.** The graph has 3 turning points and $3 \le 4-1$.

$f(x)=x^4-9x^2$

**45.** $f(x)=-x^4+16x^2$

**a.** Since $a_n<0$ and *n* is even, *f*(*x*) falls to the left and the right.

**b.** $-x^4+16x^2=0$
$x^2(-x^2+16)=0$
$x^2(4-x)(4+x)=0$
$x=0, x=4, x=-4$
The zeros at –4 and 4 have odd multiplicity, so *f*(*x*) crosses the *x*-axis at these points. The root at 0 has even multiplicity, so *f*(*x*) touches the *x*-axis at 0.

**c.** $f(0)=(0)^4-9(0)^2=0$
The *y*-intercept is 0.

**d.** $f(-x)=-x^4+16x^2$
$f(-x)=f(x)$
The graph has *y*-axis symmetry.

**e.** The graph has 3 turning points and $3 \le 4-1$.

$f(x)=-x^4+16x^2$

**47.** $f(x)=x^4-2x^3+x^2$

**a.** Since $a_n>0$ and *n* is even, *f*(*x*) rises to the left and the right.

**b.** $x^4-2x^3+x^2=0$
$x^2(x^2-2x+1)=0$
$x^2(x-1)(x-1)=0$
$x=0, x=1$
The zeros at 1 and 0 have even multiplicity, so *f*(*x*) touches the *x*-axis at 0 and 1.

**c.** $f(0)=(0)^4-2(0)^3+(0)^2=0$
The *y*-intercept is 0.

**d.** $f(-x)=x^4+2x^3+x^2$
The graph has neither *y*-axis nor origin symmetry.

**e.**   The graph has 3 turning points and
$3 \leq 4 - 1$.

$f(x) = x^4 - 2x^3 + x^2$

**49.**   $f(x) = -2x^4 + 4x^3$

**a.**   Since $a_n < 0$ and $n$ is even, $f(x)$ falls to the left and the right.

**b.**   $-2x^4 + 4x^3 = 0$
$x^3(-2x + 4) = 0$
$x = 0, \ x = 2$
The zeros at 0 and 2 have odd multiplicity, so $f(x)$ crosses the $x$-axis at these points.

**c.**   $f(0) = -2(0)^4 + 4(0)^3 = 0$
The $y$-intercept is 0.

**d.**   $f(-x) = -2x^4 - 4x^3$
The graph has neither $y$-axis nor origin symmetry.

**e.**   The graph has 1 turning point and
$1 \leq 4 - 1$.

$f(x) = -2x^4 + 4x^3$

**51.**   $f(x) = 6x^3 - 9x - x^5$

**a.**   Since $a_n < 0$ and $n$ is odd, $f(x)$ rises to the left and falls to the right.

**b.**   $-x^5 + 6x^3 - 9x = 0$
$-x(x^4 - 6x^2 + 9) = 0$
$-x(x^2 - 3)(x^2 - 3) = 0$
$x = 0, \ x = \pm\sqrt{3}$
The root at 0 has odd multiplicity so $f(x)$ crosses the $x$-axis at (0, 0). The zeros at $-\sqrt{3}$ and $\sqrt{3}$ have even multiplicity so $f(x)$ touches the $x$-axis at $\sqrt{3}$ and $-\sqrt{3}$.

**c.**   $f(0) = -(0)^5 + 6(0)^3 - 9(0) = 0$
The $y$-intercept is 0.

**d.**   $f(-x) = x^5 - 6x^3 + 9x$
$f(-x) = -f(x)$
The graph has origin symmetry.

**e.**   The graph has 4 turning points and
$4 \leq 5 - 1$.

$f(x) = 6x^3 - 9x - x^5$

**53.**   $f(x) = 3x^2 - x^3$

**a.**   Since $a_n < 0$ and $n$ is odd, $f(x)$ rises to the left and falls to the right.

**b.**   $-x^3 + 3x^2 = 0$
$-x^2(x - 3) = 0$
$x = 0, \ x = 3$
The zero at 3 has odd multiplicity so $f(x)$ crosses the $x$-axis at that point. The root at 0 has even multiplicity so $f(x)$ touches the axis at (0, 0).

**c.**   $f(0) = -(0)^3 + 3(0)^2 = 0$
The $y$-intercept is 0.

**d.**   $f(-x) = x^3 + 3x^2$
The graph has neither $y$-axis nor origin symmetry.

**e.** The graph has 2 turning points and $2 \le 3 - 1$.

$$f(x) = 3x^2 - x^3$$

**e.** The graph has 2 turning points and $2 \le 6 - 1$.

$$f(x) = x^2(x-1)^3(x+2)$$

**55.** $f(x) = -3(x-1)^2(x^2-4)$

**a.** Since $a_n < 0$ and $n$ is even, $f(x)$ falls to the left and the right.

**b.** $-3(x-1)^2(x^2-4) = 0$

$x = 1, x = -2, x = 2$

The zeros at –2 and 2 have odd multiplicity, so $f(x)$ crosses the $x$-axis at these points. The root at 1 has even multiplicity, so $f(x)$ touches the $x$-axis at (1, 0).

**c.** $f(0) = -3(0-1)^2(0^2-4)^3$

$\quad = -3(1)(-4) = 12$

The $y$-intercept is 12.

**d.** $f(-x) = -3(-x-1)^2(x^2-4)$

The graph has neither $y$-axis nor origin symmetry.

**e.** The graph has 1 turning point and $1 \le 4 - 1$.

$$f(x) = -3(x-1)^2(x^2-4)$$

**57.** $f(x) = x^2(x-1)^3(x+2)$

**a.** Since $a_n > 0$ and $n$ is even, $f(x)$ rises to the left and the right.

**b.** $x = 0, x = 1, x = -2$

The zeros at 1 and –2 have odd multiplicity so $f(x)$ crosses the $x$-axis at those points. The root at 0 has even multiplicity so $f(x)$ touches the axis at (0, 0).

**c.** $f(0) = 0^2(0-1)^3(0+2) = 0$

The $y$-intercept is 0.

**d.** $f(-x) = x^2(-x-1)^3(-x+2)$

The graph has neither $y$-axis nor origin symmetry.

**59.** $f(x) = -x^2(x-1)(x+3)$

**a.** Since $a_n < 0$ and $n$ is even, $f(x)$ falls to the left and the right.

**b.** $x = 0, x = 1, x = -3$

The zeros at 1 and –3 have odd multiplicity so $f(x)$ crosses the $x$-axis at those points. The root at 0 has even multiplicity so $f(x)$ touches the axis at (0, 0).

**c.** $f(0) = -0^2(0-1)(0+3) = 0$

The $y$-intercept is 0.

**d.** $f(-x) = -x^2(-x-1)(-x+3)$

The graph has neither $y$-axis nor origin symmetry.

**e.** The graph has 3 turning points and $3 \le 4 - 1$.

$$f(x) = -x^2(x-1)(x+3)$$

**61.** $f(x) = -2x^3(x-1)^2(x+5)$

**a.** Since $a_n < 0$ and $n$ is even, $f(x)$ falls to the left and the right.

**b.** $x = 0, x = 1, x = -5$

The roots at 0 and –5 have odd multiplicity so $f(x)$ crosses the $x$-axis at those points. The root at 1 has even multiplicity so $f(x)$ touches the axis at (1, 0).

**c.** $f(0) = -2(0)^3(0-1)^2(0+5) = 0$

The $y$-intercept is 0.

**d.** $f(-x) = 2x^3(-x-1)^2(-x+5)$

The graph has neither $y$-axis nor origin symmetry.

**e.** The graph has 2 turning points and $2 \leq 6 - 1$.

$f(x) = -2x^3 (x - 1)^2 (x + 5)$

**63.** $f(x) = (x-2)^2 (x+4)(x-1)$

**a.** Since $a_n > 0$ and $n$ is even, $f(x)$ rises to the left and rises the right.

**b.** $x = 2, x = -4, x = 1$
The zeros at –4 and 1 have odd multiplicity so $f(x)$ crosses the $x$-axis at those points. The root at 2 has even multiplicity so $f(x)$ touches the axis at $(2, 0)$.

**c.** $f(0) = (0-2)^2 (0+4)(0-1) = -16$
The $y$-intercept is –16.

**d.** $f(-x) = (-x-2)^2 (-x+4)(-x-1)$
The graph has neither $y$-axis nor origin symmetry.

**e.** The graph has 3 turning points and $3 \leq 4 - 1$.

$f(x) = (x - 2)^2 (x + 4)(x - 1)$

**65. a.** The $x$-intercepts of the graph are $-2$, 1, and 4, so they are the zeros. Since the graph actually crosses the $x$-axis at all three places, all three have odd multiplicity.

**b.** Since the graph has two turning points, the function must be at least of degree 3. Since $-2$, 1, and 4 are the zeros, $x+2$, $x-1$, and $x-4$ are factors of the function. The lowest odd multiplicity is 1. From the end behavior, we can tell that the leading coefficient must be positive. Thus, the function is $f(x) = (x+2)(x-1)(x-4)$.

**c.** $f(0) = (0+2)(0-1)(0-4) = 8$

**67. a.** The $x$-intercepts of the graph are $-1$ and 3, so they are the zeros. Since the graph crosses the $x$-axis at $-1$, it has odd multiplicity. Since the graph touches the $x$-axis and turns around at 3, it has even multiplicity.

**b.** Since the graph has two turning points, the function must be at least of degree 3. Since $-1$ and 3 are the zeros, $x+1$ and $x-3$ are factors of the function. The lowest odd multiplicity is 1, and the lowest even multiplicity is 2. From the end behavior, we can tell that the leading coefficient must be positive. Thus, the function is $f(x) = (x+1)(x-3)^2$.

**c.** $f(0) = (0+1)(0-3)^2 = 9$

**69. a.** The $x$-intercepts of the graph are $-3$ and 2, so they are the zeros. Since the graph touches the $x$-axis and turns around at both $-3$ and 2, both have even multiplicity.

**b.** Since the graph has three turning points, the function must be at least of degree 4. Since $-3$ and 2 are the zeros, $x+3$ and $x-2$ are factors of the function. The lowest even multiplicity is 2. From the end behavior, we can tell that the leading coefficient must be negative. Thus, the function is $f(x) = -(x+3)^2 (x-2)^2$.

**c.** $f(0) = -(0+3)^2 (0-2)^2 = -36$

**71. a.** The $x$-intercepts of the graph are $-2$, $-1$, and 1, so they are the zeros. Since the graph crosses the $x$-axis at $-1$ and 1, they both have odd multiplicity. Since the graph touches the $x$-axis and turns around at $-2$, it has even multiplicity.

**b.** Since the graph has five turning points, the function must be at least of degree 6. Since $-2$, $-1$, and 1 are the zeros, $x+2$, $x+1$, and $x-1$ are factors of the function. The lowest even multiplicity is 2, and the lowest odd multiplicity is 1. However, to reach degree 6, one of the odd multiplicities must be 3. From the end behavior, we can tell that the leading coefficient must be positive. The function is $f(x) = (x+2)^2 (x+1)(x-1)^3$.

**c.** $f(0) = (0+2)^2 (0+1)(0-1)^3 = -4$

**73.** **a.** $f(x) = 0.76x^3 - 30x^2 - 882x + 37,807$

$f(40) = 0.76(40)^3 - 30(40)^2 - 882(40) + 37807$
$= 3167$

The world tiger population in 2010 (40 years after 1970) was about 3167.
This is represented by the point $(40, 3167)$.

**b.** This underestimates the actual data shown in the bar graph by 33.

**c.** The leading coefficient is positive, thus the graph rises to the right.
No, if conservation efforts fail, the model will not be useful. The model indicates an increasing world tiger population that will actually decrease without conservation efforts.

**75.** **a.** The woman's heart rate was increasing from 1 through 4 minutes and from 8 through 10 minutes.

**b.** The woman's heart rate was decreasing from 4 through 8 minutes and from 10 through 12 minutes.

**c.** There were 3 turning points during the 12 minutes.

**d.** Since there were 3 turning points, a polynomial of degree 4 would provide the best fit.

**e.** The leading coefficient should be negative. The graph falls to the left and to the right.

**f.** The woman's heart rate reached a maximum of about $116 \pm 1$ beats per minute. This occurred after 10 minutes.

**g.** The woman's heart rate reached a minimum of about $64 \pm 1$ beats per minute. This occurred after 8 minutes.

**77. – 93.** Answers will vary.

**95.**

**97.**

**99.**

**101.** does not make sense; Explanations will vary. Sample explanation: Since $(x + 2)$ is raised to an odd power, the graph crosses the *x*-axis at –2.

**103.** makes sense

**105.** false; Changes to make the statement true will vary. A sample change is: Such a function falls to the right and will eventually have negative values.

**107.** false; Changes to make the statement true will vary. A sample change is: A function with origin symmetry either falls to the left and rises to the right, or rises to the left and falls to the right.

**109.** $f(x) = x^3 - 2x^2$

**110.** $\dfrac{737}{21} = 35 + \dfrac{2}{21}$ or $35\dfrac{2}{21}$.

**111.** $6x^3 - x^2 - 5x + 4$

**112.** $2x^3 - x^2 - 11x + 6 = (x - 3)(2x^2 + 3x - 2)$
$\qquad\qquad\qquad\quad = (x - 3)(2x - 1)(x + 2)$

## Section 3.3

### Check Point Exercises

**1.**
$$
\begin{array}{r}
x+5 \\
x+9{\overline{\smash{\big)}\,x^2+14x+45}} \\
\underline{x^2+9x}\phantom{+45} \\
5x+45 \\
\underline{5x+45} \\
0
\end{array}
$$
The answer is $x + 5$.

**2.**
$$
\begin{array}{r}
2x^2+3x-2 \\
x-3{\overline{\smash{\big)}\,2x^3-3x^2-11x+7}} \\
\underline{2x^3-6x^2}\phantom{-11x+7} \\
3x^2-11x \\
\underline{3x^2-9x}\phantom{+7} \\
-2x+7 \\
\underline{-2x+6} \\
1
\end{array}
$$
The answer is $2x^2+3x-2+\dfrac{1}{x-3}$.

**3.**
$$
\begin{array}{r}
2x^2+7x+14 \\
x^2-2x{\overline{\smash{\big)}\,2x^4+3x^3+0x^2-7x-10}} \\
\underline{2x^4-4x^3}\phantom{+0x^2-7x-10} \\
7x^3+0x^2 \\
\underline{7x^3-14x^2}\phantom{-7x-10} \\
14x^2-7x \\
\underline{14x^2-28x}\phantom{-10} \\
21x-10
\end{array}
$$
The answer is $2x^2+7x+14+\dfrac{21x-10}{x^2-2x}$.

**4.**
$$
\begin{array}{r|rrrr}
-2 & 1 & 0 & -7 & -6 \\
   &   & -2 & 4 & 6 \\
\hline
   & 1 & -2 & -3 & 0
\end{array}
$$
The answer is $x^2-2x-3$.

**5.**
$$
\begin{array}{r|rrrr}
-4 & 3 & 4 & -5 & 3 \\
   &   & -12 & 32 & -108 \\
\hline
   & 3 & -8 & 27 & -105
\end{array}
$$
$f(-4) = -105$

**6.**
$$
\begin{array}{r|rrrr}
-1 & 15 & 14 & -3 & -2 \\
   &    & -15 & 1 & 2 \\
\hline
   & 15 & -1 & -2 & 0
\end{array}
$$
$15x^2 - x - 2 = 0$

$(3x+1)(5x-2) = 0$

$x = -\dfrac{1}{3}$ or $x = \dfrac{2}{5}$

The solution set is $\left\{-1, -\dfrac{1}{3}, \dfrac{2}{5}\right\}$.

### Concept and Vocabulary Check 3.3

**1.** $2x^3 + 0x^2 + 6x - 4$

**2.** $6x^3$; $3x$; $2x^2$; $7x^2$

**3.** $2x^2$; $5x-2$; $10x^3 - 4x^2$; $10x^3 + 6x^2$

**4.** $6x^2 - 10x$; $6x^2 + 8x$; $18x$; $-4$; $18x - 4$

**5.** 9; $3x-5$; 9; $3x-5+\dfrac{9}{2x+1}$

**6.** divisor; quotient; remainder; dividend

**7.** 4; 1; 5; $-7$; 1

**8.** $-5$; 4; 0; $-8$; $-2$

**9.** true

**10.** $f(c)$

**11.** $x-c$

### Exercise Set 3.3

**1.**
$$
\begin{array}{r}
x+3 \\
x+5{\overline{\smash{\big)}\,x^2+8x+15}} \\
\underline{x^2+5x}\phantom{+15} \\
3x+15 \\
\underline{3x+15} \\
0
\end{array}
$$
The answer is $x+3$.

**3.**
$$\begin{array}{r}
x^2+3x+1 \\
x+2\overline{\smash{\big)}\,x^3+5x^2+7x+2} \\
\underline{x^3+2x^2} \\
3x^2+7x \\
\underline{3x^2+6x} \\
x+2 \\
\underline{x+2} \\
0
\end{array}$$

The answer is $x^2+3x+1$.

**5.**
$$\begin{array}{r}
2x^2+3x+5 \\
3x-1\overline{\smash{\big)}\,6x^3+7x^2+12x-5} \\
\underline{6x^3-2x^2} \\
9x^2+12x \\
\underline{9x^2-3x} \\
15x-5 \\
\underline{15x-5} \\
0
\end{array}$$

The answer is $2x^2+3x+5$.

**7.**
$$\begin{array}{r}
4x+3+\dfrac{2}{3x-2} \\
3x-2\overline{\smash{\big)}\,12x^2+x-4} \\
\underline{12x^2-8x} \\
9x-4 \\
\underline{9x-6} \\
2
\end{array}$$

The answer is $4x+3+\dfrac{2}{3x-2}$.

**9.**
$$\begin{array}{r}
2x^2+x+6-\dfrac{38}{x+3} \\
x+3\overline{\smash{\big)}\,2x^3+7x^2+9x-20} \\
\underline{2x^3+6x^2} \\
x^2+9x \\
\underline{x^2+3x} \\
6x-20 \\
\underline{6x+18} \\
-38
\end{array}$$

The answer is $2x^2+x+6-\dfrac{38}{x+3}$.

**11.**
$$\begin{array}{r}
4x^3+16x^2+60x+246+\dfrac{984}{x-4} \\
x-4\overline{\smash{\big)}\,4x^4-4x^2+6x} \\
\underline{4x^4-16x^3} \\
16x^3-4x^2 \\
\underline{16x^3-64x^2} \\
60x^2+6x \\
\underline{60x^2-240x} \\
246x \\
\underline{246x-984} \\
984
\end{array}$$

The answer is
$$4x^3+16x^2+60x+246+\dfrac{984}{x-4}.$$

**13.**
$$\begin{array}{r}
2x+5 \\
3x^2-x-3\overline{\smash{\big)}\,6x^3+13x^2-11x-15} \\
\underline{6x^3-2x^2-6x} \\
15x^2-5x-15 \\
\underline{15x^2-5x-15} \\
0
\end{array}$$

The answer is $2x+5$.

**15.**
$$\begin{array}{r}
6x^2+3x-1 \\
3x^2+1\overline{\smash{\big)}\,18x^4+9x^3+3x^2} \\
\underline{18x^4+6x^2} \\
9x^3-3x^2 \\
\underline{9x^3+3x} \\
-3x^2-3x \\
\underline{-3x^2-1} \\
-3x+1
\end{array}$$

The answer is $6x^2+3x-1-\dfrac{3x-1}{3x^2+1}$.

**17.** $\left(2x^2+x-10\right)\div\left(x-2\right)$

$$\begin{array}{r|rrr}
2 & 2 & 1 & -10 \\
  &   & 4 & 10 \\
\hline
  & 2 & 5 & 0
\end{array}$$

The answer is $2x+5$.

**19.** $\left(3x^2 + 7x - 20\right) \div (x + 5)$

$$
\begin{array}{r|rrr}
-5 & 3 & 7 & -20 \\
   &   & -15 & 40 \\
\hline
   & 3 & -8 & 20
\end{array}
$$

The answer is $3x - 8 + \dfrac{20}{x + 5}$.

**21.** $\left(4x^3 - 3x^2 + 3x - 1\right) \div (x - 1)$

$$
\begin{array}{r|rrrr}
1 & 4 & -3 & 3 & -1 \\
  &   & 4 & 1 & 4 \\
\hline
  & 4 & 1 & 4 & 3
\end{array}
$$

The answer is $4x^2 + x + 4 + \dfrac{3}{x - 1}$.

**23.** $\left(6x^5 - 2x^3 + 4x^2 - 3x + 1\right) \div (x - 2)$

$$
\begin{array}{r|rrrrrr}
2 & 6 & 0 & -2 & 4 & -3 & 1 \\
  &   & 12 & 24 & 44 & 96 & 186 \\
\hline
  & 6 & 12 & 22 & 48 & 93 & 187
\end{array}
$$

The answer is

$6x^4 + 12x^3 + 22x^2 + 48x + 93 + \dfrac{187}{x - 2}$.

**25.** $\left(x^2 - 5x - 5x^3 + x^4\right) \div (5 + x) \Rightarrow$

$\left(x^4 - 5x^3 + x^2 - 5x\right) \div (x + 5)$

$$
\begin{array}{r|rrrrr}
-5 & 1 & -5 & 1 & -5 & 0 \\
   &   & -5 & 50 & -255 & 1300 \\
\hline
   & 1 & -10 & 51 & -260 & 1300
\end{array}
$$

The answer is

$x^3 - 10x^2 + 51x - 260 + \dfrac{1300}{x + 5}$.

**27.** $\dfrac{x^5 + x^3 - 2}{x - 1}$

$$
\begin{array}{r|rrrrrr}
1 & 1 & 0 & 1 & 0 & 0 & -2 \\
  &   & 1 & 1 & 2 & 2 & 2 \\
\hline
  & 1 & 1 & 2 & 2 & 2 & 0
\end{array}
$$

The answer is $x^4 + x^3 + 2x^2 + 2x + 2$.

**29.** $\dfrac{x^4 - 256}{x - 4}$

$$
\begin{array}{r|rrrrr}
4 & 1 & 0 & 0 & 0 & -256 \\
  &   & 4 & 16 & 64 & 256 \\
\hline
  & 1 & 4 & 16 & 64 & 0
\end{array}
$$

The answer is $x^3 + 4x^2 + 16x + 64$.

**31.** $\dfrac{2x^5 - 3x^4 + x^3 - x^2 + 2x - 1}{x + 2}$

$$
\begin{array}{r|rrrrrr}
-2 & 2 & -3 & 1 & -1 & 2 & -1 \\
   &   & -4 & 14 & -30 & 62 & -128 \\
\hline
   & 2 & -7 & 15 & -31 & 64 & -129
\end{array}
$$

The answer is

$2x^4 - 7x^3 + 15x^2 - 31x + 64 - \dfrac{129}{x + 2}$.

**33.** $f(x) = 2x^3 - 11x^2 + 7x - 5$

$$
\begin{array}{r|rrrr}
4 & 2 & -11 & 7 & -5 \\
  &   & 8 & -12 & -20 \\
\hline
  & 2 & -3 & -5 & -25
\end{array}
$$

$f(4) = -25$

**35.** $f(x) = 3x^3 - 7x^2 - 2x + 5$

$$
\begin{array}{r|rrrr}
-3 & 3 & -7 & -2 & 5 \\
   &   & -9 & 48 & -138 \\
\hline
   & 3 & -16 & 46 & -133
\end{array}
$$

$f(-3) = -133$

**37.** $f(x) = x^4 + 5x^3 + 5x^2 - 5x - 6$

$$
\begin{array}{r|rrrrr}
3 & 1 & 5 & 5 & -5 & -6 \\
  &   & 3 & 24 & 87 & 246 \\
\hline
  & 1 & 8 & 29 & 82 & 240
\end{array}
$$

$f(3) = 240$

**39.** $f(x) = 2x^4 - 5x^3 - x^2 + 3x + 2$

$$
\begin{array}{r|rrrrr}
-\tfrac{1}{2} & 2 & -5 & -1 & 3 & 2 \\
   &   & -1 & 3 & -1 & -1 \\
\hline
   & 2 & -6 & 2 & 2 & 1
\end{array}
$$

$f\left(-\dfrac{1}{2}\right) = 1$

**41.** Dividend: $x^3 - 4x^2 + x + 6$
Divisor: $x + 1$

$$
\begin{array}{r|rrrr}
-1 & 1 & -4 & 1 & 6 \\
   &   & -1 & 5 & -6 \\
\hline
   & 1 & -5 & 6 & 0
\end{array}
$$

The quotient is $x^2 - 5x + 6$.
$(x+1)(x^2 - 5x + 6) = 0$
$(x + 1)(x - 2)(x - 3) = 0$
$x = -1, x = 2, x = 3$
The solution set is $\{-1, 2, 3\}$.

**43.** $2x^3 - 5x^2 + x + 2 = 0$

$$
\begin{array}{r|rrrr}
2 & 2 & -5 & 1 & 2 \\
  &   & 4 & -2 & -2 \\
\hline
  & 2 & -1 & -1 & 0
\end{array}
$$

$(x - 2)(2x^2 - x - 1) = 0$
$(x - 2)(2x + 1)(x - 1) = 0$
$x = 2,\ x = -\dfrac{1}{2},\ x = 1$

The solution set is $\left\{-\dfrac{1}{2},\ 1,\ 2\right\}$.

**45.** $12x^3 + 16x^2 - 5x - 3 = 0$

$$
\begin{array}{r|rrrr}
-\frac{3}{2} & 12 & 16 & -5 & -3 \\
             &    & -18 & 3 & 3 \\
\hline
             & 12 & -2 & -2 & 0
\end{array}
$$

$\left(x + \dfrac{3}{2}\right)(12x^2 - 2x - 2) = 0$

$\left(x + \dfrac{3}{2}\right)2\left(6x^2 - x - 1\right) = 0$

$\left(x + \dfrac{3}{2}\right)2(3x + 1)(2x - 1) = 0$

$x = -\dfrac{3}{2},\ x = -\dfrac{1}{3},\ x = \dfrac{1}{2}$

The solution set is $\left\{-\dfrac{3}{2}, -\dfrac{1}{3}, \dfrac{1}{2}\right\}$.

**47.** The graph indicates that 2 is a solution to the equation.

$$
\begin{array}{r|rrrr}
2 & 1 & 2 & -5 & -6 \\
  &   & 2 & 8 & 6 \\
\hline
  & 1 & 4 & 3 & 0
\end{array}
$$

The remainder is 0, so 2 is a solution.
$x^3 + 2x^2 - 5x - 6 = 0$
$(x - 2)\left(x^2 + 4x + 3\right) = 0$
$(x - 2)(x + 3)(x + 1) = 0$
The solutions are 2, $-3$, and $-1$, or $\{-3, -1, 2\}$.

**49.** The table indicates that 1 is a solution to the equation.

$$
\begin{array}{r|rrrr}
1 & 6 & -11 & 6 & -1 \\
  &   & 6 & -5 & 1 \\
\hline
  & 6 & -5 & 1 & 0
\end{array}
$$

The remainder is 0, so 1 is a solution.
$6x^3 - 11x^2 + 6x - 1 = 0$
$(x - 1)\left(6x^2 - 5x + 1\right) = 0$
$(x - 1)(3x - 1)(2x - 1) = 0$

The solutions are 1, $\dfrac{1}{3}$, and $\dfrac{1}{2}$, or $\left\{\dfrac{1}{3}, \dfrac{1}{2}, 1\right\}$.

**51.** **a.** $14x^3 - 17x^2 - 16x - 177 = 0$

$$
\begin{array}{r|rrrr}
3 & 14 & -17 & -16 & -177 \\
  &    & 42 & 75 & 177 \\
\hline
  & 14 & 25 & 59 & 0
\end{array}
$$

The remainder is 0 so 3 is a solution.
$14x^3 - 17x^2 - 16x - 177$
$= (x - 3)\left(14x^2 + 25x + 59\right)$

**b.** $f(x) = 14x^3 - 17x^2 - 16x + 34$

We need to find $x$ when $f(x) = 211$.

$f(x) = 14x^3 - 17x^2 - 16x + 34$

$211 = 14x^3 - 17x^2 - 16x + 34$

$0 = 14x^3 - 17x^2 - 16x - 177$

This is the equation obtained in part **a.** One solution is 3. It can be used to find other solutions (if they exist).

$14x^3 - 17x^2 - 16x - 177 = 0$

$(x - 3)(14x^2 + 25x + 59) = 0$

The polynomial $14x^2 + 25x + 59$ cannot be factored, so the only solution is $x = 3$. The female moth's abdominal width is 3 millimeters.

**53.** $A = l \cdot w$ so

$$l = \frac{A}{w} = \frac{0.5x^3 - 0.3x^2 + 0.22x + 0.06}{x + 0.2}$$

$$\begin{array}{r|rrrr}
-0.2 & 0.5 & -0.3 & 0.22 & 0.06 \\
 & & -0.1 & 0.08 & -0.06 \\
\hline
 & 0.5 & -0.4 & 0.3 & 0
\end{array}$$

Therefore, the length of the rectangle is $0.5x^2 - 0.4x + 0.3$ units.

**55. a.**

$$f(30) = \frac{80(30) - 8000}{30 - 110} = 70$$

(30, 70) At a 30% tax rate, the government tax revenue will be $70 ten billion.

**b.**
$$\begin{array}{r|rr}
110 & 80 & -8000 \\
 & & 8800 \\
\hline
 & 80 & 800
\end{array}$$

$$f(x) = 80 + \frac{800}{x - 110}$$

$$f(30) = 80 + \frac{800}{80 - 110} = 70$$

(30, 70) same answer as in **a.**

**c.** $f(x)$ is not a polynomial function. It is a rational function because it is the quotient of two linear polynomials.

**57. – 65.** Answers will vary.

**67.** makes sense

**69.** does not make sense; Explanations will vary. Sample explanation: The zeros of $f$ are the same as the solutions of $f(x) = 0$.

**71.** true

**73.** false; Changes to make the statement true will vary. A sample change is: The divisor is a factor of the divided only if the remainder is the whole number 0.

**75.** $f(x) = d(x) \cdot q(x) + r(x)$

$2x^2 - 7x + 9 = d(x)(2x - 3) + 3$

$2x^2 - 7x + 6 = d(x)(2x - 3)$

$\dfrac{2x^2 - 7x + 6}{2x - 3} = d(x)$

$$\begin{array}{r}
x - 2 \\
2x-3\overline{)2x^2 - 7x + 6} \\
\underline{2x^2 - 3x} \\
-4x + 6 \\
\underline{-4x + 6} \\
\end{array}$$

The polynomial is $x - 2$.

**77.** $2x - 4 = 2(x - 2)$

Use synthetic division to divide by $x - 2$. Then divide the quotient by 2.

**79.** $x^2 + 4x - 1 = 0$

$$x = \frac{-b \pm \sqrt{b^2 - 4ac}}{2a}$$

$$x = \frac{-(4) \pm \sqrt{(4)^2 - 4(1)(-1)}}{2(1)}$$

$$x = \frac{-4 \pm \sqrt{20}}{2}$$

$$x = \frac{-4 \pm 2\sqrt{5}}{2}$$

$$x = -2 \pm \sqrt{5}$$

The solution set is $\left\{ -2 \pm \sqrt{5} \right\}$.

**80.** $x^2 + 4x + 6 = 0$

$$x = \frac{-b \pm \sqrt{b^2 - 4ac}}{2a}$$

$$x = \frac{-(4) \pm \sqrt{(4)^2 - 4(1)(6)}}{2(1)}$$

$$x = \frac{-4 \pm \sqrt{-8}}{2}$$

$$x = \frac{-4 \pm 2i\sqrt{2}}{2}$$

$$x = -2 \pm i\sqrt{2}$$

The solution set is $\left\{ -2 \pm i\sqrt{2} \right\}$.

**81.** $f(x) = a_n(x^4 - 3x^2 - 4)$

$$f(3) = -150$$

$$a_n \left( (3)^4 - 3(3)^2 - 4 \right) = -150$$

$$a_n (81 - 27 - 4) = -150$$

$$a_n (50) = -150$$

$$a_n = -3$$

**Section 3.4**

**Check Point Exercises**

1.  $p: \pm 1, \ \pm 2, \ \pm 3, \ \pm 6$

    $q: \pm 1$

    $\dfrac{p}{q}: \pm 1, \ \pm 2, \ \pm 3, \ \pm 6$

    are the possible rational zeros.

2.  $p: \pm 1, \ \pm 3$

    $q: \pm 1, \ \pm 2, \ \pm 4$

    $\dfrac{p}{q}: \pm 1, \ \pm 3, \ \pm \dfrac{1}{2}, \ \pm \dfrac{1}{4}, \ \pm \dfrac{3}{2}, \ \pm \dfrac{3}{4}$

    are the possible rational zeros.

3.  $\pm 1, \ \pm 2, \ \pm 4, \ \pm 5, \ \pm 10, \ \pm 20$ are possible rational zeros

| 1 | 1 | 8 | 11 | −20 |
|---|---|---|----|-----|
|   |   | 1 | 9  | 20  |
|   | 1 | 9 | 20 | 0   |

   1 is a zero.

   $x^2 + 9x + 20 = 0$

   $(x + 4)(x + 5) = 0$

   $x = -4$ or $x = -5$

   The zeros are $-5, -4,$ and $1$.

4.  $\pm 1, \ \pm 2$ are possible rational zeros

| 2 | 1 | 1 | -5 | −2 |
|---|---|---|----|-----|
|   |   | 2 | 6  | 2  |
|   | 1 | 3 | 1  | 0  |

   2 is a zero.

   $x^2 + 3x + 1 = 0$

   $$x = \frac{-b \pm \sqrt{b^2 - 4ac}}{2a}$$

   $$x = \frac{-3 \pm \sqrt{3^2 - 4(1)(1)}}{2(1)}$$

   $$= \frac{-3 \pm \sqrt{5}}{2}$$

   The zeros are $2, \dfrac{-3 - \sqrt{5}}{2}$, and $\dfrac{-3 + \sqrt{5}}{2}$..

**5.** $\pm 1, \pm 13$ are possible rational zeros.

$$
\begin{array}{r|rrrrr}
1 & 1 & -6 & 22 & -30 & 13 \\
  &   & 1 & -5 & 17 & -13 \\
\hline
  & 1 & -5 & 17 & -13 & 0
\end{array}
$$

1 is a zero.

$$
\begin{array}{r|rrrr}
1 & 1 & 5 & 17 & -13 \\
  &   & 1 & -4 & 13 \\
\hline
  & 1 & -4 & 13 & 0
\end{array}
$$

1 is a double root.

$x^2 - 4x + 13 = 0$

$x = \dfrac{4 \pm \sqrt{16-52}}{2} = \dfrac{4 \pm \sqrt{-36}}{2} = 2 + 3i$

The solution set is $\{1, 2 - 3i, 2 + 3i\}$.

**6.** $(x+3)(x-i)(x+i) = (x+3)(x^2+1)$

$f(x) = a_n(x+3)(x^2+1)$

$f(1) = a_n(1+3)(1^2+1) = 8a_n = 8$

$a_n = 1$

$f(x) = (x+3)(x^2+1)$

$f(x) = x^3 + 3x^2 + x + 3$

**7.** $f(x) = x^4 - 14x^3 + 71x^2 - 154x + 120$

$f(-x) = x^4 + 14x^3 + 71x^2 + 154x + 120$

Since $f(x)$ has 4 changes of sign, there are 4, 2, or 0 positive real zeros.
Since $f(-x)$ has no changes of sign, there are no negative real zeros.

## Concept and Vocabulary Check 3.4

**1.** $a_0$; $a_n$

**2.** true

**3.** false

**4.** $n$

**5.** $a - bi$

**6.** $-6$; $(x+6)(2x^2 - x - 1) = 0$

**7.** $n$; 1

**8.** false

**9.** true

**10.** true

## Exercise Set 3.4

**1.** $f(x) = x^3 + x^2 - 4x - 4$

$p : \pm 1, \pm 2, \pm 4$

$q : \pm 1$

$\dfrac{p}{q} : \pm 1, \pm 2, \pm 4$

**3.** $f(x) = 3x^4 - 11x^3 - x^2 + 19x + 6$

$p: \pm 1, \pm 2, \pm 3, \pm 6$

$q: \pm 1, \pm 3$

$\dfrac{p}{q} : \pm 1, \pm 2, \pm 3, \pm 6, \pm \dfrac{1}{3}, \pm \dfrac{2}{3}$

**5.** $f(x) = 4x^4 - x^3 + 5x^2 - 2x - 6$

$p: \pm 1, \pm 2, \pm 3, \pm 6$

$q: \pm 1, \pm 2, \pm 4$

$\dfrac{p}{q} : \pm 1, \pm 2, \pm 3, \pm 6, \pm \dfrac{1}{2}, \pm \dfrac{1}{4}, \pm \dfrac{3}{2}, \pm \dfrac{3}{4}$

**7.** $f(x) = x^5 - x^4 - 7x^3 + 7x^2 - 12x - 12$

$p : \pm 1, \pm 2, \pm 3 \pm 4 \pm 6 \pm 12$

$q : \pm 1$

$\dfrac{p}{q} : \pm 1, \pm 2, \pm 3 \pm 4 \pm 6 \pm 12$

**9.** $f(x) = x^3 + x^2 - 4x - 4$

**a.** $p : \pm 1, \pm 2, \pm 4$

$q : \pm 1$

$\dfrac{p}{q} : \pm 1, \pm 2, \pm 4$

**b.**
$$
\begin{array}{r|rrrr}
2 & 1 & 1 & -4 & -4 \\
  &   & 2 & 6 & 4 \\
\hline
  & 1 & 3 & 2 & 0
\end{array}
$$

2 is a zero.
2, –2, –1 are rational zeros.

**c.** $x^3 + x^2 - 4x - 4 = 0$

$(x-2)(x^2 + 3x + 2) = 0$

$(x-2)(x+2)(x+1) = 0$

$x - 2 = 0 \quad x + 2 = 0 \quad x + 1 = 0$

$x = 2, \ x = -2, \ x = -1$

The solution set is $\{2, -2, -1\}$.

**11.** $f(x) = 2x^3 - 3x^2 - 11x + 6$

    **a.** $p: \pm 1, \pm 2, \pm 3, \pm 6$

        $q: \pm 1, \pm 2$

        $\dfrac{p}{q}: \pm 1, \pm 2, \pm 3, \pm 6, \pm \dfrac{1}{2}, \pm \dfrac{3}{2}$

    **b.**

$$\begin{array}{r|rrrr} 3 & 2 & -3 & -11 & 6 \\ & & 6 & 9 & -6 \\ \hline & 2 & 3 & -2 & 0 \end{array}$$

        3 is a zero.

        $3, \dfrac{1}{2}, -2$ are rational zeros.

    **c.** $2x^3 - 3x^2 - 11x + 6 = 0$

        $(x - 3)(2x^2 + 3x - 2) = 0$

        $(x - 3)(2x - 1)(x + 2) = 0$

        $x = 3, \ x = \dfrac{1}{2}, \ x = -2$

        The solution set is $\left\{ 3, \dfrac{1}{2}, -2 \right\}$.

**13.** **a.** $f(x) = x^3 + 4x^2 - 3x - 6$

        $p: \pm 1, \pm 2, \pm 3, \pm 6$

        $q: \pm 1$

        $\dfrac{p}{q}: \pm 1, \pm 2, \pm 3, \pm 6$

    **b.**

$$\begin{array}{r|rrrr} -1 & 1 & 4 & -3 & -6 \\ & & -1 & -3 & 6 \\ \hline & 1 & 3 & -6 & 0 \end{array}$$

        $-1$ is a rational zero.

    **c.** $x^2 + 3x - 6 = 0$

        $x = \dfrac{-b \pm \sqrt{b^2 - 4ac}}{2a}$

        $x = \dfrac{-3 \pm \sqrt{3^2 - 4(1)(-6)}}{2(1)}$

        $= \dfrac{-3 \pm \sqrt{33}}{2}$

        The solution set is $\left\{ -1, \dfrac{-3 + \sqrt{33}}{2}, \dfrac{-3 - \sqrt{33}}{2} \right\}$.

**15.** **a.** $f(x) = 2x^3 + 6x^2 + 5x + 2$

        $p: \pm 1, \pm 2$

        $q: \pm 1, \pm 2$

        $\dfrac{p}{q}: \pm 1, \pm 2, \pm \dfrac{1}{2}$

    **b.**

$$\begin{array}{r|rrrr} -2 & 2 & 6 & 5 & 2 \\ & & -4 & -4 & -2 \\ \hline & 2 & 2 & 1 & 0 \end{array}$$

        $-2$ is a rational zero.

    **c.** $2x^2 + 2x + 1 = 0$

        $x = \dfrac{-b \pm \sqrt{b^2 - 4ac}}{2a}$

        $x = \dfrac{-2 \pm \sqrt{2^2 - 4(2)(1)}}{2(2)}$

        $= \dfrac{-2 \pm \sqrt{-4}}{4}$

        $= \dfrac{-2 \pm 2i}{4}$

        $= \dfrac{-1 \pm i}{2}$

        The solution set is $\left\{ -2, \dfrac{-1 + i}{2}, \dfrac{-1 - i}{2} \right\}$.

**17.** $x^3 - 2x^2 - 11x + 12 = 0$

    **a.** $p: \pm 1, \pm 2, \pm 3, \pm 4, \pm 6, \pm 12$

        $q: \pm 1$

        $\dfrac{p}{q}: \pm 1, \pm 2, \pm 3, \pm 4, \pm 6, \pm 12$

    **b.**

$$\begin{array}{r|rrrr} 4 & 1 & -2 & -11 & 12 \\ & & 4 & 8 & -12 \\ \hline & 1 & 2 & -3 & 0 \end{array}$$

        4 is a root.

        $-3, 1, 4$ are rational roots.

    **c.** $x^3 - 2x^2 - 11x + 12$

        $(x - 4)(x^2 + 2x - 3) = 0$

        $(x - 4)(x + 3)(x - 1) = 0$

        $x - 4 = 0 \quad x + 3 = 0 \quad x - 1 = 0$

        $x = 4 \qquad x = -3 \qquad x = 1$

        The solution set is $\{-3, 1, 4\}$.

**19.** $x^3 - 10x - 12 = 0$

  **a.** $p: \pm 1, \pm 2, \pm 3, \pm 4, \pm 6, \pm 12$
  $q: \pm 1$

  $\dfrac{p}{q}: \pm 1, \pm 2, \pm 3, \pm 4, \pm 6, \pm 12$

  **b.**

  $$\begin{array}{r|rrrr} -2 & 1 & 0 & -10 & -12 \\ & & -2 & 4 & 12 \\ \hline & 1 & -2 & -6 & 0 \end{array}$$

  $-2$ is a rational root.

  **c.** $x^3 - 10x - 12 = 0$
  $(x+2)(x^2 - 2x - 6) = 0$

  $x = \dfrac{2 \pm \sqrt{4+24}}{2} = \dfrac{2 \pm \sqrt{28}}{2}$

  $= \dfrac{2 \pm 2\sqrt{7}}{2} = 1 \pm \sqrt{7}$

  The solution set is $\left\{ -2, 1+\sqrt{7}, 1-\sqrt{7} \right\}$.

**21.** $6x^3 + 25x^2 - 24x + 5 = 0$

  **a.** $p: \pm 1, \pm 5$
  $q: \pm 1, \pm 2, \pm 3, \pm 6$
  $\dfrac{p}{q}: \pm 1, \pm 5, \pm \dfrac{1}{2}, \pm \dfrac{5}{2}, \pm \dfrac{1}{3}, \pm \dfrac{5}{3}, \pm \dfrac{1}{6}, \pm \dfrac{5}{6}$

  **b.**

  $$\begin{array}{r|rrrr} -5 & 6 & 25 & -24 & 5 \\ & & -30 & 25 & -5 \\ \hline & 6 & -5 & 1 & 0 \end{array}$$

  $-5$ is a root.

  $-5, \dfrac{1}{2}, \dfrac{1}{3}$ are rational roots.

  **c.** $6x^3 + 25x^2 - 24x + 5 = 0$
  $(x+5)(6x^2 - 5x + 1) = 0$
  $(x+5)(2x-1)(3x-1) = 0$
  $x + 5 = 0 \quad 2x - 1 = 0 \quad 3x - 1 = 0$

  $x = -5, \quad x = \dfrac{1}{2}, \quad x = \dfrac{1}{3}$

  The solution set is $\left\{ -5, \dfrac{1}{2}, \dfrac{1}{3} \right\}$.

**23.** $x^4 - 2x^3 - 5x^2 + 8x + 4 = 0$

  **a.** $p: \pm 1, \pm 2, \pm 4$
  $q: \pm 1$

  $\dfrac{p}{q}: \pm 1, \pm 2, \pm 4$

  **b.**

  $$\begin{array}{r|rrrrr} 2 & 1 & -2 & -5 & 8 & 4 \\ & & 2 & 0 & -10 & -4 \\ \hline & 1 & 0 & -5 & -2 & 0 \end{array}$$

  $2$ is a root.
  $-2, 2$ are rational roots.

  **c.** $x^4 - 2x^3 - 5x^2 + 8x + 4 = 0$
  $(x-2)(x^3 - 5x - 2) = 0$

  $$\begin{array}{r|rrrr} -2 & 1 & 0 & -5 & -2 \\ & & -2 & 4 & 2 \\ \hline & 1 & -2 & -1 & 0 \end{array}$$

  $-2$ is a zero of $x^3 - 5x - 2 = 0$.

  $(x-2)(x+2)\left(x^2 - 2x - 1\right) = 0$

  $x = \dfrac{2 \pm \sqrt{4+4}}{2} = \dfrac{2 \pm \sqrt{8}}{2} = \dfrac{2 \pm 2\sqrt{2}}{2}$

  $= 1 \pm \sqrt{2}$

  The solution set is
  $\left\{ -2, 2, 1+\sqrt{2}, 1-\sqrt{2} \right\}$.

**25.** $(x-1)(x+5i)(x-5i)$
  $= (x-1)\left(x^2 + 25\right)$
  $= x^3 + 25x - x^2 - 25$
  $= x^3 - x^2 + 25x - 25$
  $f(x) = a_n \left( x^3 - x^2 + 25x - 25 \right)$
  $f(-1) = a_n(-1 - 1 - 25 - 25)$
  $-104 = a_n(-52)$
  $a_n = 2$
  $f(x) = 2\left( x^3 - x^2 + 25x - 25 \right)$
  $f(x) = 2x^3 - 2x^2 + 50x - 50$

**27.** $(x+5)(x-4-3i)(x-4+3i)$

$\qquad = (x+5)\left(x^2 - 4x + 3ix - 4x + 16 - 12i\right.$

$\qquad\qquad \left. -3ix + 12i - 9i^2\right)$

$\qquad = (x+5)\left(x^2 - 8x + 25\right)$

$\qquad = \left(x^3 - 8x^2 + 25x + 5x^2 - 40x + 125\right)$

$\qquad = x^3 - 3x^2 - 15x + 125$

$\qquad f(x) = a_n(x^3 - 3x^2 - 15x + 125)$

$\qquad f(2) = a_n\left(2^3 - 3(2)^2 - 15(2) + 125\right)$

$\qquad 91 = a_n(91)$

$\qquad a_n = 1$

$\qquad f(x) = 1\left(x^3 - 3x^2 - 15x + 125\right)$

$\qquad f(x) = x^3 - 3x^2 - 15x + 125$

**29.** $(x-i)(x+i)(x-3i)(x+3i)$

$\qquad = \left(x^2 - i^2\right)\left(x^2 - 9i^2\right)$

$\qquad = \left(x^2 + 1\right)\left(x^2 + 9\right)$

$\qquad = x^4 + 10x^2 + 9$

$\qquad f(x) = a_n(x^4 + 10x^2 + 9)$

$\qquad f(-1) = a_n((-1)^4 + 10(-1)^2 + 9)$

$\qquad 20 = a_n(20)$

$\qquad a_n = 1$

$\qquad f(x) = x^4 + 10x^2 + 9$

**31.** $(x+2)\,(x-5)\,(x-3+2i)\,(x-3-2i)$

$\qquad = \left(x^2 - 3x - 10\right)\left(x^2 - 3x - 2ix - 3x + 9 + 6i + 2ix - 6i - 4i^2\right)$

$\qquad = \left(x^2 - 3x - 10\right)\left(x^2 - 6x + 13\right)$

$\qquad = x^4 - 6x + 13x^2 - 3x^3 + 18x^2 - 39x - 10x^2 + 60x - 130$

$\qquad = x^4 - 9x^3 + 21x^2 + 21x - 130$

$\qquad f(x) = a_n\left(x^4 - 9x^3 + 21x^2 + 21x - 130\right)$

$\qquad f(1) = a_n(1 - 9 + 21 + 21 - 130)$

$\qquad -96 = a_n(-96)$

$\qquad a_n = 1$

$\qquad f(x) = x^4 - 9x^3 + 21x^2 + 21x - 130$

**33.** $f(x) = x^3 + 2x^2 + 5x + 4$

Since $f(x)$ has no sign variations,
no positive real roots exist.

$f(-x) = -x^3 + 2x^2 - 5x + 4$

Since $f(-x)$ has 3 sign variations,
3 or 1 negative real roots exist.

**35.** $f(x) = 5x^3 - 3x^2 + 3x - 1$

Since $f(x)$ has 3 sign variations, 3 or 1 positive real roots exist.

$f(-x) = -5x^3 - 3x^2 - 3x - 1$

Since $f(-x)$ has no sign variations, no negative real roots exist.

**37.** $f(x) = 2x^4 - 5x^3 - x^2 - 6x + 4$

Since $f(x)$ has 2 sign variations, 2 or 0 positive real roots exist.

$f(-x) = 2x^4 + 5x^3 - x^2 + 6x + 4$

Since $f(-x)$ has 2 sign variations, 2 or 0 negative real roots exist.

**39.** $f(x) = x^3 - 4x^2 - 7x + 10$

$p: \pm 1, \pm 2, \pm 5, \pm 10$

$q: \pm 1$

$\dfrac{p}{q}: \pm 1, \pm 2, \pm 5, \pm 10$

Since $f(x)$ has 2 sign variations, 0 or 2 positive real zeros exist.

$f(-x) = -x^3 - 4x^2 + 7x + 10$

Since $f(-x)$ has 1 sign variation, exactly one negative real zeros exists.

$$\begin{array}{r|rrrr} -2 & 1 & -4 & -7 & 10 \\ & & -2 & 12 & -10 \\ \hline & 1 & -6 & 5 & 0 \end{array}$$

–2 is a zero.

$f(x) = (x+2)\left(x^2 - 6x + 5\right)$

$\quad = (x+2)(x-5)(x-1)$

$x = -2, x = 5, x = 1$

The solution set is $\{-2, 5, 1\}$.

**41.** $2x^3 - x^2 - 9x - 4 = 0$

$p: \pm 1, \pm 2, \pm 4$

$q: \pm 1, \pm 2$

$\dfrac{p}{q}: \pm 1, \pm 2, \pm 4 \pm \dfrac{1}{2}$

1 positive real root exists.

$f(-x) = -2x^3 - x^2 + 9x - 4$  2 or no negative real roots exist.

$$\begin{array}{r|rrrr} -\dfrac{1}{2} & 2 & -1 & -9 & -4 \\ & & & -1 & 1 & 4 \\ \hline & 2 & -2 & -8 & 0 \end{array}$$

$-\dfrac{1}{2}$ is a root.

$\left(x + \dfrac{1}{2}\right)\left(2x^2 - 2x - 8\right) = 0$

$2\left(x + \dfrac{1}{2}\right)\left(x^2 - x - 4\right) = 0$

$x = \dfrac{1 \pm \sqrt{1+16}}{2} = \dfrac{1 \pm \sqrt{17}}{2}$

The solution set is $\left\{ -\dfrac{1}{2}, \dfrac{1+\sqrt{17}}{2}, \dfrac{1-\sqrt{17}}{2} \right\}$.

**43.** $f(x) = x^4 - 2x^3 + x^2 + 12x + 8$

$p: \pm 1, \pm 2, \pm 4, \pm 8$

$q: \pm 1$

$\dfrac{p}{q}: \pm 1, \pm 2, \pm 4, \pm 8$

Since $f(x)$ has 2 sign changes, 0 or 2 positive roots exist.

$f(-x) = (-x)^4 - 2(-x)^3 + (-x)^2 - 12x + 8$

$\quad = x^4 + 2x^3 + x^2 - 12x + 8$

Since $f(-x)$ has 2 sign changes, 0 or 2 negative roots exist.

$$\begin{array}{r|rrrrr} -1 & 1 & -2 & 1 & 12 & 8 \\ & & -1 & 4 & -4 & -8 \\ \hline & 1 & -3 & 4 & 8 & 0 \end{array}$$

$$\begin{array}{r|rrrr} -1 & 1 & -3 & 4 & 8 \\ & & -1 & 4 & -8 \\ \hline & 1 & -4 & 8 & 0 \end{array}$$

$0 = x^2 - 4x + 8$

$x = \dfrac{-(-4) \pm \sqrt{(-4)^2 - 4(1)(8)}}{2(1)}$

$x = \dfrac{4 \pm \sqrt{16 - 32}}{2}$

$x = \dfrac{4 \pm \sqrt{-16}}{2}$

$x = \dfrac{4 \pm 4i}{2}$

$x = 2 \pm 2i$

The solution set is $\{-1, -1, 2 + 2i, 2 - 2i\}$.

**45.** $x^4 - 3x^3 - 20x^2 - 24x - 8 = 0$

$p : \pm 1, \pm 2, \pm 4, \pm 8$

$q : \pm 1$

$\dfrac{p}{q} : \pm 1, \pm 2, \pm 4 \pm 8$

1 positive real root exists.
3 or 1 negative real roots exist.

$$
\begin{array}{r|rrrrr}
-1 & 1 & -3 & -20 & -24 & -8 \\
   &   & -1 &  4  &  16 &  8 \\
\hline
   & 1 & -4 & -16 & -8  &  0
\end{array}
$$

$(x+1)\left(x^3 - 4x^2 - 16x - 8\right) = 0$

$$
\begin{array}{r|rrrr}
-2 & 1 & -4 & -16 & -8 \\
   &   & -2 &  12 &  8 \\
\hline
   & 1 & -6 & -4  &  0
\end{array}
$$

$(x+1)(x+2)\left(x^2 - 6x - 4\right) = 0$

$x = \dfrac{6 \pm \sqrt{36 + 16}}{2} = \dfrac{6 \pm \sqrt{52}}{2}$

$= \dfrac{6 \pm 2\sqrt{13}}{2} = \dfrac{3 \pm \sqrt{13}}{2}$

The solution set is

$\left\{-1, -2, 3 \pm \sqrt{13}, 3 - \sqrt{13}\right\}$.

**47.** $f(x) = 3x^4 - 11x^3 - x^2 + 19x + 6$

$p : \pm 1, \pm 2, \pm 3, \pm 6$

$q : \pm 1, \pm 3$

$\dfrac{p}{q} : \pm 1, \pm 2, \pm 3, \pm 6, \pm \dfrac{1}{3}, \pm \dfrac{2}{3}$

2 or no positive real zeros exists.
$f(-x) = 3x^4 + 11x^3 - x^2 - 19x + 6$
2 or no negative real zeros exist.

$$
\begin{array}{r|rrrrr}
-1 & 3 & -11 & -1 & 19 & 6 \\
   &   & -3  & 14 & -13 & -6 \\
\hline
   & 3 & -14 & 13 & 6  & 0
\end{array}
$$

$f(x) = (x+1)\left(3x^3 - 14x^2 + 13x + 6\right)$

$$
\begin{array}{r|rrrr}
2 & 3 & -14 & 13 & 6 \\
  &   &  6  & -16 & -6 \\
\hline
  & 3 & -8  & -3  & 0
\end{array}
$$

$f(x) = (x+1)(x-2)\left(3x^2 - 8x - 3\right)$

$\quad = (x+1)(x-2)(3x+1)(x-3)$

$x = -1, \; x = 2 \; x = -\dfrac{1}{3}, \; x = 3$

The solution set is $\left\{-1, 2, -\dfrac{1}{3}, 3\right\}$.

**49.** $4x^4 - x^3 + 5x^2 - 2x - 6 = 0$

$p : \pm 1, \pm 2, \pm 3, \pm 6$

$q : \pm 1, \pm 2, \pm 4$

$\dfrac{p}{q} : \pm 1, \pm 2, \pm 3, \pm 6, \pm \dfrac{1}{2}, \pm \dfrac{3}{2}, \pm \dfrac{1}{4}, \pm \dfrac{3}{4}$

3 or 1 positive real roots exists.
1 negative real root exists.

$$
\begin{array}{r|rrrrr}
1 & 4 & -1 & 5 & -2 & -6 \\
  &   &  4 & 3 &  8 &  6 \\
\hline
  & 4 &  3 & 8 &  6 &  0
\end{array}
$$

$(x-1)(4x^3 + 3x^2 + 8x + 6) = 0$

$4x^3 + 3x^2 + 8x + 6 = 0$ has no positive real roots.

$$\begin{array}{r|rrrr} -\frac{3}{4} & 4 & 3 & 8 & 6 \\ & & -3 & 0 & -6 \\ \hline & 4 & 0 & 8 & 0 \end{array}$$

$(x-1)\left(x+\frac{3}{4}\right)\left(4x^2+8\right)=0$

$4(x-1)\left(x+\frac{3}{4}\right)\left(x^2+2\right)=0$

$x^2+2=0$

$x^2=-2$

$x=\pm i\sqrt{2}$

The solution set is $\left\{1, -\frac{3}{4}, i\sqrt{2}, -i\sqrt{2}\right\}$.

**51.** $2x^5+7x^4-18x^2-8x+8=0$

$p: \pm 1, \pm 2, \pm 4, \pm 8$

$q: \pm 1, \pm 2$

$\frac{p}{q}: \pm 1, \pm 2, \pm 4, \pm 8, \pm\frac{1}{2}$

2 or no positive real roots exists.
3 or 1 negative real root exist.

$$\begin{array}{r|rrrrrr} -2 & 2 & 7 & 0 & -18 & -8 & 8 \\ & & -4 & -6 & 12 & 12 & -8 \\ \hline & 2 & 3 & -6 & -6 & 4 & 0 \end{array}$$

$(x+2)(2x^4+3x^3-6x^2-6x+4)=0$

$4x^3+3x^2+8x+6=0$ has no positive real roots.

$$\begin{array}{r|rrrrr} -2 & 2 & 3 & -6 & -6 & 4 \\ & & -4 & 2 & 8 & -4 \\ \hline & 2 & -1 & -4 & 2 & 0 \end{array}$$

$(x+2)^2(2x^3-x^2-4x+2)$

$$\begin{array}{r|rrrr} \frac{1}{2} & 2 & -1 & -4 & 2 \\ & & 1 & 0 & 2 \\ \hline & 2 & 0 & -4 & 0 \end{array}$$

$(x+2)^2\left(x-\frac{1}{2}\right)\left(2x^2-4\right)=0$

$2(x+2)^2\left(x-\frac{1}{2}\right)\left(x^2-2\right)=0$

$x^2-2=0$

$x^2=2$

$x=\pm\sqrt{2}$

The solution set is $\left\{-2, \frac{1}{2}, \sqrt{2}, -\sqrt{2}\right\}$.

**53.** $f(x)=-x^3+x^2+16x-16$

**a.** From the graph provided, we can see that $-4$ is an $x$-intercept and is thus a zero of the function. We verify this below:

$$\begin{array}{r|rrrr} -4 & -1 & 1 & 16 & -16 \\ & & 4 & -20 & 16 \\ \hline & -1 & 5 & -4 & 0 \end{array}$$

Thus, $\quad -x^3+x^2+16x-16=0$

$(x+4)\left(-x^2+5x-4\right)=0$

$-(x+4)\left(x^2-5x+4\right)=0$

$-(x+4)(x-1)(x-4)=0$

$x+4=0 \quad$ or $\quad x-1=0 \quad$ or $\quad x-4=0$

$x=-4 \qquad\qquad x=1 \qquad\qquad x=4$

The zeros are $-4$, 1, and 4.

**b.**

$$f(x)=-x^3+x^2+16x-16$$

**55.** $f(x) = 4x^3 - 8x^2 - 3x + 9$

**a.** From the graph provided, we can see that $-1$ is an $x$-intercept and is thus a zero of the function. We verify this below:

$$
\begin{array}{r|rrrr}
-1 & 4 & -8 & -3 & 9 \\
 & & -4 & 12 & -9 \\
\hline
 & 4 & -12 & 9 & 0 \\
\end{array}
$$

Thus, $4x^3 - 8x^2 - 3x + 9 = 0$

$$(x+1)(4x^2 - 12x + 9) = 0$$

$$(x+1)(2x-3)^2 = 0$$

$x + 1 = 0$ or $(2x-3)^2 = 0$

$x = -1$ $\quad\quad 2x - 3 = 0$

$\quad\quad\quad\quad 2x = 3$

$\quad\quad\quad\quad x = \dfrac{3}{2}$

The zeros are $-1$ and $\dfrac{3}{2}$.

**b.**

$f(x) = 4x^3 - 8x^2 - 3x + 9$

**57.** $f(x) = 2x^4 - 3x^3 - 7x^2 - 8x + 6$

**a.** From the graph provided, we can see that $\dfrac{1}{2}$ is an $x$-intercept and is thus a zero of the function. We verify this below:

$$
\begin{array}{r|rrrrr}
\frac{1}{2} & 2 & -3 & -7 & -8 & 6 \\
 & & 1 & -1 & -4 & -6 \\
\hline
 & 2 & -2 & -8 & -12 & 0 \\
\end{array}
$$

Thus, $2x^4 - 3x^3 - 7x^2 - 8x + 6 = 0$

$$\left(x - \dfrac{1}{2}\right)(2x^3 - 2x^2 - 8x - 12) = 0$$

$$2\left(x - \dfrac{1}{2}\right)(x^3 - x^2 - 4x - 6) = 0$$

To factor $x^3 - x^2 - 4x - 6$, we use the Rational Zero Theorem to determine possible rational zeros.

Factors of the constant term $-6$:
$\pm 1, \ \pm 2, \ \pm 3, \ \pm 6$

Factors of the leading coefficient 1: $\pm 1$

The possible rational zeros are:

$$\dfrac{\text{Factors of } -6}{\text{Factors of } 1} = \dfrac{\pm 1, \ \pm 2, \ \pm 3, \ \pm 6}{\pm 1}$$

$$= \pm 1, \ \pm 2, \ \pm 3, \ \pm 6$$

We test values from above until we find a zero. One possibility is shown next:

Test 3:

$$
\begin{array}{r|rrrr}
3 & 1 & -1 & -4 & -6 \\
 & & 3 & 6 & 6 \\
\hline
 & 1 & 2 & 2 & 0 \\
\end{array}
$$

The remainder is 0, so 3 is a zero of $f$.

$$2x^4 - 3x^3 - 7x^2 - 8x + 6 = 0$$

$$\left(x - \dfrac{1}{2}\right)(2x^3 - 2x^2 - 8x - 12) = 0$$

$$2\left(x - \dfrac{1}{2}\right)(x^3 - x^2 - 4x - 6) = 0$$

$$2\left(x - \dfrac{1}{2}\right)(x - 3)(x^2 + 2x + 2) = 0$$

Note that $x^2 + x + 1$ will not factor, so we use the quadratic formula:

$a = 1 \quad b = 2 \quad c = 2$

$$x = \dfrac{-2 \pm \sqrt{2^2 - 4(1)(2)}}{2(1)}$$

$$= \dfrac{-2 \pm \sqrt{-4}}{2} = \dfrac{-2 \pm 2i}{2} = -1 \pm i$$

The zeros are $\dfrac{1}{2}$, 3, and $-1 \pm i$.

**b.**

$f(x) = 2x^4 - 3x^3 - 7x^2 - 8x + 6$

**59.** $f(x) = 3x^5 + 2x^4 - 15x^3 - 10x^2 + 12x + 8$

  **a.** From the graph provided, we can see that 1 and 2 are *x*-intercepts and are thus zeros of the function. We verify this below:

```
1| 3   2   -15   -10    12    8
        3    5   -10   -20   -8
   ─────────────────────────────
   3   5   -10   -20    -8    0
```

Thus, $3x^5 + 2x^4 - 15x^3 - 10x^2 + 12x + 8$

$$= (x-1)\left(3x^4 + 5x^3 - 10x^2 - 20x - 8\right)$$

```
2| 3   5   -10   -20   -8
        6    22    24    8
   ───────────────────────
   3   11   12     4    0
```

Thus, $3x^5 + 2x^4 - 15x^3 - 10x^2 + 12x + 8$

$$= (x-1)\left(3x^4 + 5x^3 - 10x^2 - 20x - 8\right)$$

$$= (x-1)(x-2)\left(3x^3 + 11x^2 + 12x + 4\right)$$

To factor $3x^3 + 11x^2 + 12x + 4$, we use the Rational Zero Theorem to determine possible rational zeros.

Factors of the constant term 4: $\pm 1, \pm 2, \pm 4$
Factors of the leading coefficient 3: $\pm 1, \pm 3$

The possible rational zeros are:

$$\frac{\text{Factors of } 4}{\text{Factors of } 3} = \frac{\pm 1, \pm 2, \pm 4}{\pm 1, \pm 3}$$

$$= \pm 1, \pm 2, \pm 4, \pm\frac{1}{3}, \pm\frac{2}{3}, \pm\frac{4}{3}$$

We test values from above until we find a zero. One possibility is shown next:

Test $-1$:

```
-1| 3   11   12    4
        -3   -8   -4
    ─────────────────
    3   8    4    0
```

The remainder is 0, so $-1$ is a zero of *f*. We can now finish the factoring:

$$3x^5 + 2x^4 - 15x^3 - 10x^2 + 12x + 8 = 0$$

$$(x-1)\left(3x^4 + 5x^3 - 10x^2 - 20x - 8\right) = 0$$

$$(x-1)(x-2)\left(3x^3 + 11x^2 + 12x + 4\right) = 0$$

$$(x-1)(x-2)(x+1)\left(3x^2 + 8x + 4\right) = 0$$

$$(x-1)(x-2)(x+1)(3x+2)(x+2) = 0$$

$$x = 1, \; x = 2, \; x = -1, \; x = -\frac{2}{3}, \; x = -2$$

The zeros are $-2$, $-1$, $-\frac{2}{3}$, 1 and 2.

  **b.**

$f(x) = 3x^5 + 2x^4 - 15x^3 - 10x^2 + 12x + 8$

**61.** $V(x) = x(x+10)(30-2x)$

$$2000 = x(x+10)(30-2x)$$

$$2000 = -2x^3 + 10x^2 + 300x$$

$$2x^3 - 10x^2 - 300x + 2000 = 0$$

$$x^3 - 5x^2 - 150x + 1000 = 0$$

Find the roots.

```
10| 1   -5   -150   1000
         10    50   -1000
    ──────────────────────
    1    5   -100     0
```

Use the remaining quadratic to find the other 2 roots.

$$x = \frac{-b \pm \sqrt{b^2 - 4ac}}{2a}$$

$$x = \frac{-(5) \pm \sqrt{(5)^2 - 4(1)(-100)}}{2(1)}$$

$$x \approx -12.8, \; 7.8$$

Since the depth must be positive, reject the negative value.

The depth can be 10 inches or 7.8 inches to obtain a volume of 2000 cubic inches.

**63.**  **a.** The answers correspond to the points (7.8, 2000) and (10, 2000).

  **b.** The range is (0, 15).

**65. – 71.** Answers will vary.

**73.** $6x^3 - 19x^2 + 16x - 4 = 0$

$p$: $\pm 1, \pm 2, \pm 4$

$q$: $\pm 1, \pm 2, \pm 3, \pm 6$

$\dfrac{p}{q}$: $\pm 1, \pm 2, \pm 4, \pm\dfrac{1}{2}, \pm\dfrac{1}{3}, \pm\dfrac{2}{3}, \pm\dfrac{4}{3}, \pm\dfrac{1}{6}$

From the graph, we see that the solutions are $\dfrac{1}{2}, \dfrac{2}{3}$ and 2.

**75.** $4x^4 + 4x^3 + 7x^2 - x - 2 = 0$

$p$: ±1, ±2

$q$: ±1, ±2, ±4

$\dfrac{p}{q}$: ±1, ±2, ±$\dfrac{1}{2}$, ±$\dfrac{1}{4}$

From the graph, we see that the solutions are

$-\dfrac{1}{2}$ and $\dfrac{1}{2}$.

**77.** $f(x) = x^5 - x^4 + x^3 - x^2 + x - 8$

$f(x)$ has 5 sign variations, so either 5, 3, or 1 positive real roots exist.

$f(-x) = -x^5 - x^4 - x^3 - x^2 - x - 8$

$f(-x)$ has no sign variations, so no negative real roots exist.

**79.** $f(x) = x^3 - 6x - 9$

1 real zero

2 nonreal complex zeros

**81.** $f(x) = 3x^4 + 4x^3 - 7x^2 - 2x - 3$

**83.** makes sense

**85.** makes sense

**87.** false; Changes to make the statement true will vary. A sample change is: The equation has 0 sign variations, so no positive roots exist.

**89.** true

**91.** $(2x+1)(x+5)(x+2) - 3x(x+5) = 208$

$(2x^2 + 11x + 5)(x+2) - 3x^2 - 15x = 208$

$2x^3 + 4x^2 + 11x^2 + 22x + 5x + 10 - 3x^2 - 15x = 208$

$2x^3 + 15x^2 + 27x - 3x^2 - 15x - 198 = 0$

$2x^3 + 12x^2 + 12x - 198 = 0$

$2(x^3 + 6x^2 + 6x - 99) = 0$

$$\begin{array}{r|rrrr} 3 & 1 & 6 & 6 & -99 \\ & & 3 & 27 & 99 \\ \hline & 1 & 9 & 33 & 0 \end{array}$$

$x^2 + 9x + 33 = 0$

$b^2 - 4ac = -51$

$x = 3$ in.

**93.** Because the polynomial has two obvious changes of direction; the smallest degree is 3.

**95.** Because the polynomial has two obvious changes of direction and two roots have multiplicity 2, the smallest degree is 5.

**97.** Answers will vary.

**98.** The function is undefined at $x = 1$ and $x = 2$.

**99.** The equation of the vertical asymptote is $x = 1$.

**100.** The equation of the horizontal asymptote is $y = 0$.

**Mid-Chapter 3 Check Point**

**1.** $f(x) = (x-3)^2 - 4$

The parabola opens up because $a > 0$.
The vertex is $(3, -4)$.
$x$-intercepts:

$0 = (x-3)^2 - 4$

$(x-3)^2 = 4$

$x - 3 = \pm\sqrt{4}$

$x = 3 \pm 2$

The equation has $x$-intercepts at $x = 1$ and $x = 5$.
$y$-intercept:

$f(0) = (0-3)^2 - 4 = 5$

domain: $(-\infty, \infty)$  range: $[-4, \infty)$

$f(x) = (x - 3)^2 - 4$

**2.** $f(x) = 5 - (x+2)^2$

The parabola opens down because $a < 0$.
The vertex is $(-2, 5)$.
$x$-intercepts:

$0 = 5 - (x+2)^2$

$(x+2)^2 = 5$

$x + 2 = \pm\sqrt{5}$

$x = -2 \pm \sqrt{5}$

$y$-intercept:

$f(0) = 5 - (0+2)^2 = 1$

domain: $(-\infty, \infty)$  range: $(-\infty, 5]$

$f(x) = 5 - (x + 2)^2$

**3.** $f(x) = -x^2 - 4x + 5$

The parabola opens down because $a < 0$.

vertex: $x = -\dfrac{b}{2a} = -\dfrac{-4}{2(-1)} = -2$

$f(-2) = -(-2)^2 - 4(-2) + 5 = 9$

The vertex is $(-2, 9)$.
$x$-intercepts:

$0 = -x^2 - 4x + 5$

$x = \dfrac{-b \pm \sqrt{b^2 - 4ac}}{2a}$

$x = \dfrac{-(-4) \pm \sqrt{(-4)^2 - 4(-1)(5)}}{2(-1)}$

$x = \dfrac{4 \pm \sqrt{36}}{-2}$

$x = -2 \pm 3$

The $x$-intercepts are $x = 1$ and $x = -5$.
$y$-intercept:

$f(0) = -0^2 - 4(0) + 5 = 5$

domain: $(-\infty, \infty)$  range: $(-\infty, 9]$

$f(x) = -x^2 - 4x + 5$

**4.** $f(x) = 3x^2 - 6x + 1$

The parabola opens up because $a > 0$.

vertex: $x = -\dfrac{b}{2a} = -\dfrac{-6}{2(3)} = 1$

$f(1) = 3(1)^2 - 6(1) + 1 = -2$

The vertex is $(1, -2)$.
$x$-intercepts:

$0 = 3x^2 - 6x + 1$

$x = \dfrac{-b \pm \sqrt{b^2 - 4ac}}{2a}$

$x = \dfrac{-(-6) \pm \sqrt{(-6)^2 - 4(3)(1)}}{2(3)}$

$x = \dfrac{6 \pm \sqrt{24}}{6}$

$x = \dfrac{3 \pm \sqrt{6}}{3}$

*y*-intercept:

$$f(0) = 3(0)^2 - 6(0) + 1 = 1$$

domain: $(-\infty, \infty)$  range: $[-2, \infty)$

$$f(x) = 3x^2 - 6x + 1$$

**5.**  $f(x) = (x-2)^2 (x+1)^3$

$$(x-2)^2 (x+1)^3 = 0$$

Apply the zero-product principle:

$$(x-2)^2 = 0 \quad \text{or} \quad (x+1)^3 = 0$$

$$x - 2 = 0 \qquad\qquad x + 1 = 0$$

$$x = 2 \qquad\qquad\quad x = -1$$

The zeros are $-1$ and 2.

The graph of *f* crosses the *x*-axis at $-1$, since the zero has multiplicity 3. The graph touches the *x*-axis and turns around at 2 since the zero has multiplicity 2.

Since *f* is an odd-degree polynomial, degree 5, and since the leading coefficient, 1, is positive, the graph falls to the left and rises to the right.

Plot additional points as necessary and construct the graph.

$$f(x) = (x - 2)^2 (x + 1)^3$$

**6.**  $f(x) = -(x-2)^2 (x+1)^2$

$$-(x-2)^2 (x+1)^2 = 0$$

Apply the zero-product principle:

$$(x-2)^2 = 0 \quad \text{or} \quad (x+1)^2 = 0$$

$$x - 2 = 0 \qquad\qquad x + 1 = 0$$

$$x = 2 \qquad\qquad\quad x = -1$$

The zeros are $-1$ and 2.

The graph touches the *x*-axis and turns around both at $-1$ and 2 since both zeros have multiplicity 2.

Since *f* is an even-degree polynomial, degree 4, and since the leading coefficient, $-1$, is negative, the graph falls to the left and falls to the right.

Plot additional points as necessary and construct the graph.

$$f(x) = -(x - 2)^2 (x + 1)^2$$

**7.**  $f(x) = x^3 - x^2 - 4x + 4$

$$x^3 - x^2 - 4x + 4 = 0$$

$$x^2(x-1) - 4(x-1) = 0$$

$$(x^2 - 4)(x - 1) = 0$$

$$(x+2)(x-2)(x-1) = 0$$

Apply the zero-product principle:

$$x + 2 = 0 \quad \text{or} \quad x - 2 = 0 \quad \text{or} \quad x - 1 = 0$$

$$x = -2 \qquad\qquad x = 2 \qquad\qquad x = 1$$

The zeros are $-2$, 1, and 2.

The graph of *f* crosses the *x*-axis at all three zeros, $-2$, 1, and 2, since all have multiplicity 1.

Since *f* is an odd-degree polynomial, degree 3, and since the leading coefficient, 1, is positive, the graph falls to the left and rises to the right.

Plot additional points as necessary and construct the graph.

$$f(x) = x^3 - x^2 - 4x + 4$$

**8.** $f(x) = x^4 - 5x^2 + 4$

$$x^4 - 5x^2 + 4 = 0$$

$$(x^2 - 4)(x^2 - 1) = 0$$

$$(x+2)(x-2)(x+1)(x-1) = 0$$

Apply the zero-product principle,
$x = -2, \ x = 2, \ x = -1, \ x = 1$
The zeros are $-2$, $-1$, $1$, and $2$.

The graph crosses the $x$-axis at all four zeros, $-2$, $-1$, $1$, and $2$., since all have multiplicity 1.

Since $f$ is an even-degree polynomial, degree 4, and since the leading coefficient, 1, is positive, the graph rises to the left and rises to the right.

Plot additional points as necessary and construct the graph.

$f(x) = x^4 - 5x^2 + 4$

**9.** $f(x) = -(x+1)^6$

$$-(x+1)^6 = 0$$

$$(x+1)^6 = 0$$

$$x + 1 = 0$$

$$x = -1$$

The zero is are $-1$.

The graph touches the $x$-axis and turns around at $-1$ since the zero has multiplicity 6.

Since $f$ is an even-degree polynomial, degree 6, and since the leading coefficient, $-1$, is negative, the graph falls to the left and falls to the right.

Plot additional points as necessary and construct the graph.

$f(x) = -(x+1)^6$

**10.** $f(x) = -6x^3 + 7x^2 - 1$

To find the zeros, we use the Rational Zero Theorem:
List all factors of the constant term $-1$: $\pm 1$
List all factors of the leading coefficient $-6$:
$\pm 1, \ \pm 2, \ \pm 3, \ \pm 6$

The possible rational zeros are:

$$\frac{\text{Factors of } -1}{\text{Factors of } -6} = \frac{\pm 1}{\pm 1, \ \pm 2, \ \pm 3, \ \pm 6}$$

$$= \pm 1, \ \pm \frac{1}{2}, \ \pm \frac{1}{3}, \ \pm \frac{1}{6}$$

We test values from the above list until we find a zero. One is shown next:

Test 1:

$$\begin{array}{r|rrrr} 1 & -6 & 7 & 0 & -1 \\ & & -6 & 1 & 1 \\ \hline & -6 & 1 & 1 & 0 \end{array}$$

The remainder is 0, so 1 is a zero. Thus,

$$-6x^3 + 7x^2 - 1 = 0$$

$$(x-1)(-6x^2 + x + 1) = 0$$

$$-(x-1)(6x^2 - x - 1) = 0$$

$$-(x-1)(3x+1)(2x-1) = 0$$

Apply the zero-product property:

$$x = 1, \quad x = -\frac{1}{3}, \quad x = \frac{1}{2}$$

The zeros are $-\frac{1}{3}$, $\frac{1}{2}$, and 1.

The graph of $f$ crosses the $x$-axis at all three zeros, $-\frac{1}{3}$, $\frac{1}{2}$, and 1, since all have multiplicity 1.

Since $f$ is an odd-degree polynomial, degree 3, and since the leading coefficient, $-6$, is negative, the graph rises to the left and falls to the right.

Plot additional points as necessary and construct the graph.

$f(x) = -6x^3 + 7x^2 - 1$

**11.** $f(x) = 2x^3 - 2x$

$$2x^3 - 2x = 0$$

$$2x(x^2 - 1) = 0$$

$$2x(x+1)(x-1) = 0$$

Apply the zero-product principle:

$$x = 0, \qquad x = -1, \qquad x = 1$$

The zeros are $-1$, 0, and 1.

The graph of *f* crosses the *x*-axis at all three zeros, $-1$, 0, and 1, since all have multiplicity 1.

Since *f* is an odd-degree polynomial, degree 3, and since the leading coefficient, 2, is positive, the graph falls to the left and rises to the right.

Plot additional points as necessary and construct the graph.

$f(x) = 2x^3 - 2x$

**12.** $f(x) = x^3 - 2x^2 + 26x$

$$x^3 - 2x^2 + 26x = 0$$

$$x(x^2 - 2x + 26) = 0$$

Note that $x^2 - 2x + 26$ does not factor, so we use the quadratic formula:

$$x = 0 \quad \text{or} \quad x^2 - 2x + 26 = 0$$

$$a = 1, \ b = -2, \ c = 26$$

$$x = \frac{-(-2) \pm \sqrt{(-2)^2 - 4(1)(26)}}{2(1)}$$

$$= \frac{2 \pm \sqrt{-100}}{2} = \frac{2 \pm 10i}{2} = 1 \pm 5i$$

The zeros are 0 and $1 \pm 5i$.

The graph of *f* crosses the *x*-axis at 0 (the only real zero), since it has multiplicity 1.

Since *f* is an odd-degree polynomial, degree 3, and since the leading coefficient, 1, is positive, the graph falls to the left and rises to the right.

Plot additional points as necessary and construct the graph.

$f(x) = x^3 - 2x^2 + 26x$

**13.** $f(x) = -x^3 + 5x^2 - 5x - 3$

To find the zeros, we use the Rational Zero Theorem:

List all factors of the constant term $-3$: $\pm 1, \ \pm 3$

List all factors of the leading coefficient $-1$: $\pm 1$

The possible rational zeros are:

$$\frac{\text{Factors of } -3}{\text{Factors of } -1} = \frac{\pm 1, \ \pm 3}{\pm 1} = \pm 1, \ \pm 3$$

We test values from the previous list until we find a zero. One is shown next:

Test 3:

$$\begin{array}{r|rrrr} 3 & -1 & 5 & -5 & -3 \\ & & -3 & 6 & 3 \\ \hline & -1 & 2 & 1 & 0 \end{array}$$

The remainder is 0, so 3 is a zero. Thus,

$$-x^3 + 5x^2 - 5x - 3 = 0$$

$$(x - 3)(-x^2 + 2x + 1) = 0$$

$$-(x - 3)(x^2 - 2x - 1) = 0$$

Note that $x^2 - 2x - 1$ does not factor, so we use the quadratic formula:

$$x - 3 = 0 \quad \text{or} \quad x^2 - 2x - 1 = 0$$

$$x = 3 \qquad a = 1, \ b = -2, \ c = -1$$

$$x = \frac{-(-2) \pm \sqrt{(-2)^2 - 4(1)(-1)}}{2(1)}$$

$$= \frac{2 \pm \sqrt{8}}{2} = \frac{2 \pm 2\sqrt{2}}{2} = 1 \pm \sqrt{2}$$

The zeros are 3 and $1 \pm \sqrt{2}$.

The graph of *f* crosses the *x*-axis at all three zeros, 3 and $1 \pm \sqrt{2}$, since all have multiplicity 1.

Since *f* is an odd-degree polynomial, degree 3, and since the leading coefficient, $-1$, is negative, the graph rises to the left and falls to the right.

Plot additional points as necessary and construct the graph.

$f(x) = -x^3 + 5x^2 - 5x - 3$

**14.** $x^3 - 3x + 2 = 0$

We begin by using the Rational Zero Theorem to determine possible rational roots.

Factors of the constant term 2: $\pm 1, \pm 2$

Factors of the leading coefficient 1: $\pm 1$

The possible rational zeros are:

$$\frac{\text{Factors of } 2}{\text{Factors of } 1} = \frac{\pm 1, \pm 2}{\pm 1} = \pm 1, \pm 2$$

We test values from above until we find a root. One is shown next:

Test 1:

```
1| 1  0  -3   2
       1   1  -2
   _____
   1  1  -2   0
```

The remainder is 0, so 1 is a root of the equation. Thus,

$$x^3 - 3x + 2 = 0$$
$$(x-1)(x^2 + x - 2) = 0$$
$$(x-1)(x+2)(x-1) = 0$$
$$(x-1)^2(x+2) = 0$$

Apply the zero-product property:

$(x-1)^2 = 0$  or  $x + 2 = 0$

$x - 1 = 0$       $x = -2$

$x = 1$

The solutions are $-2$ and 1, and the solution set is $\{-2, 1\}$.

**15.** $6x^3 - 11x^2 + 6x - 1 = 0$

We begin by using the Rational Zero Theorem to determine possible rational roots.

Factors of the constant term $-1$: $\pm 1$
Factors of the leading coefficient 6:
$\pm 1, \pm 2, \pm 3, \pm 6$

The possible rational zeros are:

$$\frac{\text{Factors of } -1}{\text{Factors of } 6} = \frac{\pm 1}{\pm 1, \pm 2, \pm 3, \pm 6}$$

$$= \pm 1, \pm \frac{1}{2}, \pm \frac{1}{3}, \pm \frac{1}{6}$$

We test values from above until we find a root. One is shown next:

Test 1:

```
1| 6  -11   6  -1
        6  -5   1
   _____
   6  -5   1   0
```

The remainder is 0, so 1 is a root of the equation. Thus,

$$6x^3 - 11x^2 + 6x - 1 = 0$$
$$(x-1)(6x^2 - 5x + 1) = 0$$
$$(x-1)(3x-1)(2x-1) = 0$$

Apply the zero-product property:

$x - 1 = 0$   or   $3x - 1 = 0$   or   $2x - 1 = 0$

$x = 1$      $x = \dfrac{1}{3}$      $x = \dfrac{1}{2}$

The solutions are $\dfrac{1}{3}$, $\dfrac{1}{2}$ and 1, and the solution set is

$$\left\{\frac{1}{3}, \frac{1}{2}, 1\right\}.$$

**16.** $(2x+1)(3x-2)^3(2x-7) = 0$

Apply the zero-product property:

$2x + 1 = 0$   or   $(3x-2)^3 = 0$   or   $2x - 7 = 0$

$x = -\dfrac{1}{2}$      $3x - 2 = 0$      $x = \dfrac{7}{2}$   Th

                    $x = \dfrac{2}{3}$

e solutions are $-\dfrac{1}{2}$, $\dfrac{2}{3}$ and $\dfrac{7}{2}$, and the solution set

is $\left\{-\dfrac{1}{2}, \dfrac{2}{3}, \dfrac{7}{2}\right\}.$

**17.** $2x^3 + 5x^2 - 200x - 500 = 0$

We begin by using the Rational Zero Theorem to determine possible rational roots.

Factors of the constant term $-500$:
$\pm 1, \pm 2, \pm 4, \pm 5, \pm 10, \pm 20, \pm 25,$
$\pm 50, \pm 100, \pm 125, \pm 250, \pm 500$

Factors of the leading coefficient 2: $\pm 1, \pm 2$

The possible rational zeros are:

$$\frac{\text{Factors of } 500}{\text{Factors of } 2} = \pm 1, \pm 2, \pm 4, \pm 5,$$

$\pm 10, \pm 20, \pm 25, \pm 50, \pm 100, \pm 125,$

$\pm 250, \pm 500, \pm \dfrac{1}{2}, \pm \dfrac{5}{2}, \pm \dfrac{25}{2}, \pm \dfrac{125}{2}$

We test values from above until we find a root. One is shown next:

Test 10:

$$\begin{array}{r|rrrr} 10 & 2 & 5 & -200 & -500 \\ & & 20 & 250 & 500 \\ \hline & 2 & 25 & 50 & 0 \end{array}$$

The remainder is 0, so 10 is a root of the equation. Thus,

$$2x^3 + 5x^2 - 200x - 500 = 0$$
$$(x-10)(2x^2 + 25x + 50) = 0$$
$$(x-10)(2x+5)(x+10) = 0$$

Apply the zero-product property:

$x - 10 = 0$  or  $2x + 5 = 0$  or  $x + 10 = 0$

$x = 10$ $\qquad x = -\dfrac{5}{2} \qquad x = -10$

The solutions are $-10$, $-\dfrac{5}{2}$, and $10$, and the

solution set is $\left\{-10, \ -\dfrac{5}{2}, \ 10\right\}$.

**18.** $\qquad x^4 - x^3 - 11x^2 = x + 12$

$x^4 - x^3 - 11x^2 - x - 12 = 0$

We begin by using the Rational Zero Theorem to determine possible rational roots.

Factors of the constant term $-12$:
$\pm 1, \ \pm 2, \ \pm 3, \ \pm 4, \ \pm 6, \ \pm 12$

Factors of the leading coefficient 1: $\pm 1$

The possible rational zeros are:

$$\dfrac{\text{Factors of } -12}{\text{Factors of } 1}$$
$$= \dfrac{\pm 1, \ \pm 2, \ \pm 3, \ \pm 4, \ \pm 6, \ \pm 12}{\pm 1}$$
$$= \pm 1, \ \pm 2, \ \pm 3, \ \pm 4, \ \pm 6, \ \pm 12$$

We test values from this list we find a root. One possibility is shown next:

Test $-3$:

$$\begin{array}{r|rrrrr} -3 & 1 & -1 & -11 & -1 & -12 \\ & & -3 & 12 & -3 & 12 \\ \hline & 1 & -4 & 1 & -4 & 0 \end{array}$$

The remainder is 0, so $-3$ is a root of the equation. Using the Factor Theorem, we know that $x - 1$ is a factor. Thus,

$$x^4 - x^3 - 11x^2 - x - 12 = 0$$
$$(x+3)(x^3 - 4x^2 + x - 4) = 0$$
$$(x+3)[x^2(x-4) + 1(x-4)] = 0$$
$$(x+3)(x-4)(x^2+1) = 0$$

As this point we know that $-3$ and 4 are roots of the equation. Note that $x^2 + 1$ does not factor, so we use the square-root principle: $x^2 + 1 = 0$

$$x^2 = -1$$
$$x = \pm\sqrt{-1} = \pm i$$

The roots are $-3$, 4, and $\pm i$, and the solution set is $\{-3, \ 4, \ \pm i\}$.

**19.** $2x^4 + x^3 - 17x^2 - 4x + 6 = 0$

We begin by using the Rational Zero Theorem to determine possible rational roots.

Factors of the constant term 6: $\pm 1, \ \pm 2, \ \pm 3, \ \pm 6$
Factors of the leading coefficient 4: $\pm 1, \ \pm 2$

The possible rational roots are:

$$\dfrac{\text{Factors of } 6}{\text{Factors of } 2} = \dfrac{\pm 1, \ \pm 2, \ \pm 3, \ \pm 6}{\pm 1, \ \pm 2}$$
$$= \pm 1, \ \pm 2, \ \pm 3, \ \pm 6, \ \pm\dfrac{1}{2}, \ \pm\dfrac{3}{2}$$

We test values from above until we find a root. One possibility is shown next:

Test $-3$:

$$\begin{array}{r|rrrrr} -3 & 2 & 1 & -17 & -4 & 6 \\ & & -6 & 15 & 6 & -6 \\ \hline & 2 & -5 & -2 & 2 & 0 \end{array}$$

The remainder is 0, so $-3$ is a root. Using the Factor Theorem, we know that $x + 3$ is a factor of the polynomial. Thus,

$$2x^4 + x^3 - 17x^2 - 4x + 6 = 0$$
$$(x+3)(2x^3 - 5x^2 - 2x + 2) = 0$$

To solve the equation above, we need to factor $2x^3 - 5x^2 - 2x + 2$. We continue testing potential roots:

Test $\dfrac{1}{2}$ :

$$
\begin{array}{r|rrrr}
\frac{1}{2} & 2 & -5 & -2 & 2 \\
  &   & 1 & -2 & -2 \\
\hline
  & 2 & -4 & -4 & 0
\end{array}
$$

The remainder is 0, so $\dfrac{1}{2}$ is a zero and $x-\dfrac{1}{2}$ is a factor.

Summarizing our findings so far, we have

$$2x^4 + x^3 - 17x^2 - 4x + 6 = 0$$

$$(x+3)\left(2x^3 - 5x^2 - 2x + 2\right) = 0$$

$$(x+3)\left(x - \dfrac{1}{2}\right)\left(2x^2 - 4x - 4\right) = 0$$

$$2(x+3)\left(x - \dfrac{1}{2}\right)\left(x^2 - 2x - 2\right) = 0$$

At this point, we know that $-3$ and $\dfrac{1}{2}$ are roots of the equation. Note that $x^2 - 2x - 2$ does not factor, so we use the quadratic formula:

$x^2 - 2x - 2 = 0$

$a = 1, \; b = -2, \; c = -2$

$$x = \dfrac{-(-2) \pm \sqrt{(-2)^2 - 4(1)(-2)}}{2(1)}$$

$$= \dfrac{2 \pm \sqrt{4+8}}{2} = \dfrac{2 \pm \sqrt{12}}{2} = \dfrac{2 \pm 2\sqrt{3}}{2} = 1 \pm \sqrt{3}$$

The solutions are $-3$, $\dfrac{1}{2}$, and $1 \pm \sqrt{3}$, and the

solution set is $\left\{ -3, \; \dfrac{1}{2}, \; 1 \pm \sqrt{3} \right\}$.

**20.** $P(x) = -x^2 + 150x - 4425$

Since $a = -1$ is negative, we know the function opens down and has a maximum at

$x = -\dfrac{b}{2a} = -\dfrac{150}{2(-1)} = -\dfrac{150}{-2} = 75$.

$P(75) = -75^2 + 150(75) - 4425$

$\phantom{P(75)} = -5625 + 11{,}250 - 4425 = 1200$

The company will maximize its profit by manufacturing and selling 75 cabinets per day. The maximum daily profit is \$1200.

**21.** Let $x$ = one of the numbers;
$-18 - x$ = the other number

The product is $f(x) = x(-18 - x) = -x^2 - 18x$

The $x$-coordinate of the maximum is

$x = -\dfrac{b}{2a} = -\dfrac{-18}{2(-1)} = -\dfrac{-18}{-2} = -9$.

$f(-9) = -9\left[ -18 - (-9) \right]$

$\phantom{f(-9)} = -9(-18 + 9) = -9(-9) = 81$

The vertex is $(-9, 81)$. The maximum product is 81. This occurs when the two numbers are $-9$ and $-18 - (-9) = -9$.

**22.** Let $x$ = height of triangle;
$40 - 2x$ = base of triangle

$A = \dfrac{1}{2} bh = \dfrac{1}{2} x(40 - 2x)$

$A(x) = 20x - x^2$

The height at which the triangle will have

maximum area is $x = -\dfrac{b}{2a} = -\dfrac{20}{2(-1)} = 10$.

$A(10) = 20(10) - (10)^2 = 100$

The maximum area is 100 squares inches.

**23.**

$$
\begin{array}{r}
2x^2 - x - 3 \\
3x^2 - 1 \overline{\smash{\big)}\, 6x^4 - 3x^3 - 11x^2 + 2x + 4} \\
\underline{6x^4 \phantom{{}- 3x^3} - 2x^2 \phantom{{}+ 2x + 4}} \\
-3x^3 - 9x^2 + 2x \phantom{{}+ 4} \\
\underline{-3x^3 \phantom{{}- 9x^2} + x \phantom{{}+ 4}} \\
-9x^2 + x + 4 \\
\underline{-9x^2 \phantom{{}+ x} + 3} \\
x + 1
\end{array}
$$

$2x^2 - x - 3 + \dfrac{x+1}{3x^2 - 1}$

**24.** $\left( 2x^4 - 13x^3 + 17x^2 + 18x - 24 \right) \div (x - 4)$

$$
\begin{array}{r|rrrrr}
4 & 2 & -13 & 17 & 18 & -24 \\
  &   & 8 & -20 & -12 & 24 \\
\hline
  & 2 & -5 & -3 & 6 & 0
\end{array}
$$

The quotient is $2x^3 - 5x^2 - 3x + 6$.

**25.** $(x-1)(x-i)(x+i) = (x-1)(x^2+1)$

$f(x) = a_n(x-1)(x^2+1)$

$f(-1) = a_n(-1-1)((-1)^2+1) = -4a_n = 8$

$a_n = -2$

$f(x) = -2(x-1)(x^2+1)$ or $-2x^3+2x^2-2x+2$

**26.** $(x-2)(x-2)(x-3i)(x+3i)$

$= (x-2)(x-2)(x^2+9)$

$f(x) = a_n(x-2)(x-2)(x^2+9)$

$f(0) = a_n(0-2)(0-2)(0^2+9)$

$36 = 36a_n$

$a_n = 1$

$f(x) = 1(x-2)(x-2)(x^2+9)$

$f(x) = x^4-4x^3+13x^2-36x+36$

**27.** $f(x) = x^3-x-5$

$f(1) = 1^3-1-5 = -5$

$f(2) = 2^3-2-5 = 1$

Yes, the function must have a real zero between 1 and 2 because $f(1)$ and $f(2)$ have opposite signs.

## Section 3.5

### Check Point Exercises

**1.** Because division by 0 is undefined, we must exclude from the domain of each function values of $x$ that cause the polynomial function in the denominator to be 0.

**a.** $x-5 = 0$

$x = 5$

$\{x | x \neq 5\}$ or $(-\infty, 5) \cup (5, \infty)$.

**b.** $x^2-25 = 0$

$x^2 = 25$

$x = \pm 5$

$\{x | x \neq 5, x \neq -5\}$ or $(-\infty, -5) \cup (-5, 5) \cup (5, \infty)$.

**c.** The denominator cannot equal zero.
All real numbers or $(-\infty, \infty)$. .

**2.** **a.** $x^2-1 = 0$

$x^2 = 1$

$x = 1, x = -1$

**b.** $g(x) = \dfrac{x-1}{x^2-1} = \dfrac{x-1}{(x-1)(x+1)} = \dfrac{1}{x+1}$

$x = -1$

**c.** The denominator cannot equal zero.
No vertical asymptotes.

**3.** **a.** The degree of the numerator, 2, is equal to the degree of the denominator, 2. Thus, the leading coefficients of the numerator and denominator, 9 and 3, are used to obtain the equation of the horizontal asymptote.

$y = \dfrac{9}{3} = 3$

$y = 3$ is a horizontal asymptote.

**b.** The degree of the numerator, 1, is less than the degree of the denominator, 2. Thus, the graph has the $x$-axis as a horizontal asymptote
$y = 0$ is a horizontal asymptote.

**c.** The degree of the numerator, 3, is greater than the degree of the denominator, 2. Thus, the graph has no horizontal asymptote.

**4.** Begin with the graph of $f(x) = \dfrac{1}{x}$.

$g(x) = \dfrac{1}{x+2} - 1$

Shift the graph 2 units to the left by subtracting 2 from each $x$-coordinate. Shift the graph 1 unit down by subtracting 1 from each $y$-coordinate.

**5.** $f(x) = \dfrac{3x-3}{x-2}$

$f(-x) = \dfrac{3(-x)-3}{-x-2} = \dfrac{-3x-3}{-x-2} = \dfrac{3x+3}{x+2}$

no symmetry

$f(0) = \dfrac{3(0)-3}{0-2} = \dfrac{3}{2}$

The $y$-intercept is $\dfrac{3}{2}$.

$3x-3 = 0$

$3x = 3$

$x = 1$

The $x$-intercept is 1.

Vertical asymptote:

$x - 2 = 0$

$x = 2$

Horizontal asymptote:

$y = \dfrac{3}{1} = 3$

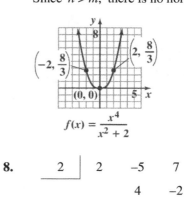

**6.** $f(x) = \dfrac{2x^2}{x^2-9}$

$f(-x) = \dfrac{2(-x)^2}{(-x)^2-9} = \dfrac{2x^2}{x^2-9} = f(x)$

The $y$-axis symmetry.

$f(0) = \dfrac{2(0)^2}{0^2-9} = 0$

The $y$-intercept is 0.

$2x^2 = 0$

$x = 0$

The $x$-intercept is 0.

vertical asymptotes:

$x^2 - 9 = 0$

$x = 3, x = -3$

horizontal asymptote:

$y = \dfrac{2}{1} = 2$

**7.** $f(x) = \dfrac{x^4}{x^2+2}$

$f(-x) = \dfrac{(-x)^4}{(-x)^2+2} = \dfrac{x^4}{x^2+2} = f(x)$

$y$-axis symmetry

$f(0) = \dfrac{0^4}{0^2+2} = 0$

The $y$-intercept is 0.

$x^4 = 0$

$x = 0$

The $x$-intercept is 0.

vertical asymptotes:

$x^2 + 2 = 0$

$x^2 = -2$

no vertical asymptotes

horizontal asymptote:

Since $n > m$, there is no horizontal asymptote.

**8.**

|   | 2 | 2 | –5 | 7 |
|---|---|---|----|---|
|   |   |   | 4  | –2 |
|   |   | 2 | –1 | 5 |

the equation of the slant asymptote is
$y = 2x - 1$.

**9. a.** $C(x) = 500,000 + 400x$

**b.** $\overline{C}(x) = \dfrac{500,000 + 400x}{x}$

**c.** $\overline{C}(1000) = \dfrac{500,000 + 400(1000)}{1000}$

$= 900$

$\overline{C}(10,000) = \dfrac{500,000 + 400(10,000)}{10,000}$

$= 450$

$\overline{C}(100,000) = \dfrac{500,000 + 400(100,000)}{100,000}$

$= 405$

The average cost per wheelchair of producing 1000, 10,000, and 100,000 wheelchairs is \$900, \$450, and \$405, respectively.

**d.** $y = \dfrac{400}{1} = 400$

The cost per wheelchair approaches \$400 as more wheelchairs are produced.

## Concept and Vocabulary Check 3.5

**1.** polynomial

**2.** false

**3.** true

**4.** vertical asymptote; $x = -5$

**5.** horizontal asymptote; $y = 0$; $y = \dfrac{1}{3}$

**6.** true

**7.** left; down

**8.** one more than

**9.** $y = 3x + 5$

## Exercise Set 3.5

**1.** $f(x) = \dfrac{5x}{x-4}$

$\{x | x \neq 4\}$

**3.** $g(x) = \dfrac{3x^2}{(x-5)(x+4)}$

$\{x | x \neq 5, x \neq -4\}$

**5.** $h(x) = \dfrac{x+7}{x^2 - 49}$

$x^2 - 49 = (x-7)(x+7)$

$\{x | x \neq 7, x \neq -7\}$

**7.** $f(x) = \dfrac{x+7}{x^2 + 49}$

all real numbers

**9.** $-\infty$

**11.** $-\infty$

**13.** $0$

**15.** $+\infty$

**17.** $-\infty$

**19.** $1$

**21.** $f(x) = \dfrac{x}{x+4}$

$x + 4 = 0$

$x = -4$

vertical asymptote: $x = -4$

There are no holes.

**23.** $g(x) = \dfrac{x+3}{x(x+4)}$

$x(x+4) = 0$

$x = 0, x = -4$

vertical asymptotes: $x = 0, x = -4$

There are no holes.

**25.** $h(x) = \dfrac{x}{x(x+4)} = \dfrac{1}{x+4}$

$x + 4 = 0$

$x = -4$

vertical asymptote: $x = -4$

There is a hole at $x = 0$.

**27.** $r(x) = \dfrac{x}{x^2 + 4}$

$x^2 + 4$ has no real zeros
There are no vertical asymptotes.
There are no holes.

**29.** $f(x) = \dfrac{x^2 - 9}{x - 3}$

$\quad = \dfrac{(x+3)(x-3)}{x-3}$

$\quad = x + 3$

There are no vertical asymptotes.
There is a hole at $x = 3$.

**31.** $g(x) = \dfrac{x - 3}{x^2 - 9}$

$\quad = \dfrac{x - 3}{(x+3)(x-3)}$

$\quad = \dfrac{1}{x + 3}$

vertical asymptote: $x = -3$
There is a hole at $x = 3$.

**33.** $h(x) = \dfrac{x + 7}{x^2 + 4x - 21}$

$\quad = \dfrac{x + 7}{(x+7)(x-3)}$

$\quad = \dfrac{1}{x - 3}$

vertical asymptote: $x = 3$
There is a hole at $x = -7$.

**35.** $r(x) = \dfrac{x^2 + 4x - 21}{x + 7}$

$\quad = \dfrac{(x+7)(x-3)}{x+7}$

$\quad = x - 3$

There are no vertical asymptotes.
There is a hole at $x = -7$.

**37.** $f(x) = \dfrac{12x}{3x^2 + 1}$

$n < m$
horizontal asymptote: $y = 0$

**39.** $g(x) = \dfrac{12x^2}{3x^2 + 1}$

$n = m$,

horizontal asymptote: $y = \dfrac{12}{3} = 4$

**41.** $h(x) = \dfrac{12x^3}{3x^2 + 1}$

$n > m$
no horizontal asymptote

**43.** $f(x) = \dfrac{-2x + 1}{3x + 5}$

$n = m$

horizontal asymptote: $y = -\dfrac{2}{3}$

**45.** $g(x) = \dfrac{1}{x - 1}$

Shift the graph of $f(x) = \dfrac{1}{x}$ 1 unit to the right.

**47.** $h(x) = \dfrac{1}{x} + 2$

Shift the graph of $f(x) = \dfrac{1}{x}$ 2 units up.

**49.** $g(x) = \dfrac{1}{x + 1} - 2$

Shift the graph of $f(x) = \dfrac{1}{x}$ 1 unit left and 2 units down.

**51.** $g(x) = \dfrac{1}{(x+2)^2}$

Shift the graph of $f(x) = \dfrac{1}{x^2}$ 2 units left.

$$g(x) = \dfrac{1}{(x+2)^2}$$

**53.** $h(x) = \dfrac{1}{x^2} - 4$

Shift the graph of $f(x) = \dfrac{1}{x^2}$ 4 units down.

$$h(x) = \dfrac{1}{x^2} - 4$$

**55.** $h(x) = \dfrac{1}{(x-3)^2} + 1$

Shift the graph of $f(x) = \dfrac{1}{x^2}$ 3 units right and 1 unit up.

$$h(x) = \dfrac{1}{(x-3)^2} + 1$$

**57.** $f(x) = \dfrac{4x}{x-2}$

$f(-x) = \dfrac{4(-x)}{(-x)-2} = \dfrac{4x}{x+2}$

$f(-x) \neq f(x), f(-x) \neq -f(x)$

no symmetry

$y$-intercept: $y = \dfrac{4(0)}{0-2} = 0$

$x$-intercept: $4x = 0$
$x = 0$

vertical asymptote:
$x - 2 = 0$
$\qquad x = 2$

horizontal asymptote:

$n = m$, so $y = \dfrac{4}{1} = 4$

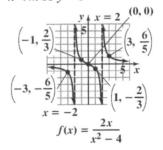

$$f(x) = \dfrac{4x}{x-2}$$

**59.** $f(x) = \dfrac{2x}{x^2 - 4}$

$f(-x) = \dfrac{2(-x)}{(-x)^2 - 4} = -\dfrac{2x}{x^2 - 4} = -f(x)$

Origin symmetry

$y$-intercept: $\dfrac{2(0)}{0^2 - 4} = \dfrac{0}{-4} = 0$

$x$-intercept:
$2x = 0$
$x = 0$

vertical asymptotes:
$x^2 - 4 = 0$
$x = \pm 2$

horizontal asymptote:
$n < m$ so $y = 0$

$$f(x) = \dfrac{2x}{x^2 - 4}$$

**61.** $f(x) = \dfrac{2x^2}{x^2 - 1}$

$f(-x) = \dfrac{2(-x)^2}{(-x)^2 - 1} = \dfrac{2x^2}{x^2 - 1} = f(x)$

$y$-axis symmetry

$y$-intercept: $y = \dfrac{2(0)^2}{0^2 - 1} = \dfrac{0}{1} = 0$

$x$-intercept:

$2x^2 = 0$

$x = 0$

vertical asymptote:

$x^2 - 1 = 0$

$x^2 = 1$

$x = \pm 1$

horizontal asymptote:

$n = m$, so $y = \dfrac{2}{1} = 2$

$$f(x) = \frac{2x^2}{x^2 - 1}$$

**63.** $f(x) = \dfrac{-x}{x+1}$

$f(-x) = \dfrac{-(-x)}{(-x)+1} = \dfrac{x}{-x+1}$

$f(-x) \neq f(x), f(-x) \neq -f(x)$

no symmetry

$y$-intercept: $y = \dfrac{-(0)}{0+1} = \dfrac{0}{1} = 0$

$x$-intercept:

$-x = 0$

$x = 0$

vertical asymptote:

$x + 1 = 0$

$x = -1$

horizontal asymptote:

$n = m$, so $y = \dfrac{-1}{1} = -1$

$$f(x) = \frac{-x}{x+1}$$

**65.** $f(x) = -\dfrac{1}{x^2 - 4}$

$f(-x) = -\dfrac{1}{(-x)^2 - 4} = -\dfrac{1}{x^2 - 4} = f(x)$

$y$-axis symmetry

$y$-intercept: $y = -\dfrac{1}{0^2 - 4} = \dfrac{1}{4}$

$x$-intercept: $-1 \neq 0$

no $x$-intercept

vertical asymptotes:

$x^2 - 4 = 0$

$x^2 = 4$

$x = \pm 2$

horizontal asymptote:

$n < m$ or $y = 0$

$$f(x) = -\frac{1}{x^2 - 4}$$

**67.** $f(x) = \dfrac{2}{x^2 + x - 2}$

$f(-x) = -\dfrac{2}{(-x)^2 - x - 2} = \dfrac{2}{x^2 - x - 2}$

$f(-x) \neq f(x), f(-x) \neq -f(x)$

no symmetry

y-intercept: $y = \dfrac{2}{0^2 + 0 - 2} = \dfrac{2}{-2} = -1$

x-intercept: none

vertical asymptotes:

$x^2 + x - 2 = 0$

$(x + 2)(x - 1) = 0$

$x = -2, x = 1$

horizontal asymptote:

$n < m$ so $y = 0$

$f(x) = \dfrac{2}{x^2 + x - 2}$

**69.** $f(x) = \dfrac{2x^2}{x^2 + 4}$

$f(-x) = \dfrac{2(-x)^2}{(-x)^2 + 4} = \dfrac{2x^2}{x^2 + 4} = f(x)$

y axis symmetry

y-intercept: $y = \dfrac{2(0)^2}{0^2 + 4} = 0$

x-intercept: $2x^2 = 0$

$x = 0$

vertical asymptote: none

horizontal asymptote:

$n = m$, so $y = \dfrac{2}{1} = 2$

$f(x) = \dfrac{2x^2}{x^2 + 4}$

**71.** $f(x) = \dfrac{x + 2}{x^2 + x - 6}$

$f(-x) = \dfrac{-x + 2}{(-x)^2 - (-x) - 6} = \dfrac{-x + 2}{x^2 + x - 6}$

$f(-x) \neq f(x), f(-x) \neq -f(x)$

no symmetry

y-intercept: $y = \dfrac{0 + 2}{0^2 + 0 - 6} = -\dfrac{2}{6} = -\dfrac{1}{3}$

x-intercept:

$x + 2 = 0$

$x = -2$

vertical asymptotes:

$x^2 + x - 6 = 0$

$(x + 3)(x - 2)$

$x = -3, x = 2$

horizontal asymptote:

$n < m$, so $y = 0$

$f(x) = \dfrac{x + 2}{x^2 + x - 6}$

**73.** $f(x) = \dfrac{x - 2}{x^2 - 4}$

$f(-x) = \dfrac{-x - 2}{(-x)^2 - 4} = \dfrac{-x - 2}{x^2 - 4}$

$f(-x) \neq f(x), f(-x) \neq -f(x)$

no symmetry

y-intercept: $y = \dfrac{0 - 2}{0^2 - 4} = \dfrac{-2}{-4} = \dfrac{1}{2}$

x-intercept:

$x - 2 = 0, x = 2$

vertical asymptotes:

$f(x) = \dfrac{x - 2}{x^2 - 4}$

$= \dfrac{x - 2}{(x - 2)(x + 2)}$

$= \dfrac{1}{x + 2}$

$x = -2$ is a vertical asymptote.

Furthermore, the value 2 causes the original denominator to be zero, but the reduced form of the function's equation does not cause the denominator to be zero. Thus, there is a hole at $x = 2$.

horizontal asymptote:

$n < m$, so $y = 0$

$$f(x) = \frac{x - 2}{x^2 - 4}$$

**75.** $f(x) = \dfrac{x^4}{x^2 + 2}$

$$f(-x) = \frac{(-x)^4}{(-x)^2 + 2} = \frac{x^4}{x^2 + 2} = f(x)$$

$y$-axis symmetry

$y$-intercept: $y = \dfrac{0^4}{0^2 + 2} = 0$

$x$-intercept: $x^4 = 0$
$x = 0$

vertical asymptote: none
horizontal asymptote:
$n > m$, so none

$$f(x) = \frac{x^4}{x^2 + 2}$$

**77.** $f(x) = \dfrac{x^2 + x - 12}{x^2 - 4}$

$$f(-x) = \frac{(-x)^2 - x - 12}{(-x)^2 - 4} = \frac{x^2 - x - 12}{x^2 - 4}$$

$f(-x) \neq f(x), f(-x) \neq -f(x)$
no symmetry

$y$-intercept: $y = \dfrac{0^2 + 0 - 12}{0^2 - 4} = 3$

$x$-intercept: $x^2 + x - 12 = 0$
$$(x - 3)(x + 4) = 0$$
$$x = 3, x = -4$$

vertical asymptotes:
$$x^2 - 4 = 0$$
$$(x - 2)(x + 2) = 0$$
$$x = 2, x = -2$$

horizontal asymptote:

$n = m$, so $y = \dfrac{1}{1} = 1$

$$f(x) = \frac{x^2 + x - 12}{x^2 - 4}$$

**79.** $f(x) = \dfrac{3x^2 + x - 4}{2x^2 - 5x}$

$$f(-x) = \frac{3(-x)^2 - x - 4}{2(-x)^2 + 5x} = \frac{3x^2 - x - 4}{2x^2 + 5x}$$

$f(-x) \neq f(x), f(-x) \neq -f(x)$
no symmetry

$y$-intercept: $y = \dfrac{3(0)^2 + 0 - 4}{2(0)^2 - 5(0)} = \dfrac{-4}{0}$

no $y$-intercept
$x$-intercepts:
$$3x^2 + x - 4 = 0$$
$$(3x + 4)(x - 1) = 0$$
$$3x + 4 = 0 \quad x - 1 = 0$$
$$3x = -4$$
$$x = -\frac{4}{3}, x = 1$$

vertical asymptotes:
$$2x^2 - 5x = 0$$
$$x(2x - 5) = 0$$
$$x = 0, 2x = 5$$
$$x = \frac{5}{2}$$

horizontal asymptote:

$n = m$, so $y = \dfrac{3}{2}$

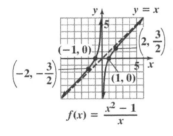

$f(x) = \dfrac{3x^2 + x - 4}{2x^2 - 5x}$

**81. a.** Slant asymptote:

$f(x) = x - \dfrac{1}{x}$

$y = x$

**b.** $f(x) = \dfrac{x^2 - 1}{x}$

$f(-x) = \dfrac{(-x)^2 - 1}{(-x)} = \dfrac{x^2 - 1}{-x} = -f(x)$

Origin symmetry

$y$-intercept: $y = \dfrac{0^2 - 1}{0} = \dfrac{-1}{0}$

no $y$-intercept

$x$-intercepts: $x^2 - 1 = 0$

$x = \pm 1$

vertical asymptote: $x = 0$


horizontal asymptote:

$n < m$, so none exist.

$f(x) = \dfrac{x^2 - 1}{x}$

**83. a.** Slant asymptote:

$f(x) = x + \dfrac{1}{x}$

$y = x$

**b.** $f(x) = \dfrac{x^2 + 1}{x}$

$f(-x) = \dfrac{(-x)^2 + 1}{-x} = \dfrac{x^2 + 1}{-x} = -f(x)$

Origin symmetry

$y$-intercept: $y = \dfrac{0^2 + 1}{0} = \dfrac{1}{0}$

no $y$-intercept

$x$-intercept:

$x^2 + 1 = 0$

$x^2 = -1$

no $x$-intercept

vertical asymptote: $x = 0$

horizontal asymptote:

$n > m$, so none exist.

$f(x) = \dfrac{x^2 + 1}{x}$

**85. a.** Slant asymptote:

$f(x) = x + 4 + \dfrac{6}{x - 3}$

$y = x + 4$

**b.** $f(x) = \dfrac{x^2 + x - 6}{x - 3}$

$f(-x) = \dfrac{(-x)^2 + (-x) - 6}{-x - 3} = \dfrac{x^2 - x - 6}{-x - 3}$

$f(-x) \neq g(x), \; g(-x) \neq -g(x)$

No symmetry

$y$-intercept: $y = \dfrac{0^2 + 0 - 6}{0 - 3} = \dfrac{-6}{-3} = 2$

$x$-intercept:

$x^2 + x - 6 = 0$

$(x + 3)(x - 2) = 0$

$x = -3$ and $x = 2$

vertical asymptote:
$x - 3 = 0$
$x = 3$
horizontal asymptote:
$n > m$, so none exist.

$$f(x) = \frac{x^2 + x - 6}{x - 3}$$

**87.** $f(x) = \dfrac{x^3 + 1}{x^2 + 2x}$

**a.** slant asymptote:

$$x^2 + 2x \overline{\smash{\big)}\, x^3 \qquad + 1}$$
$$\underline{x^3 + 2x^2}$$
$$-2x^2$$
$$\underline{-2x^2 + 4x}$$
$$-4x + 1$$

$y = x - 2$

**b.**
$$f(-x) = \frac{(-x)^3 + 1}{(-x)^2 + 2(-x)} = \frac{-x^3 + 1}{x^2 - 2x}$$
$$f(-x) \neq f(x), \ f(-x) \neq -f(x)$$

no symmetry

$y$-intercept: $y = \dfrac{0^3 + 1}{0^2 + 2(0)} = \dfrac{1}{0}$

no $y$-intercept
$x$-intercept: $x^3 + 1 = 0$
$x^3 = -1$
$x = -1$
vertical asymptotes:
$x^2 + 2x = 0$
$x(x + 2) = 0$
$x = 0, x = -2$

horizontal asymptote:
$n > m$, so none

$$f(x) = \frac{x^3 + 1}{x^2 + 2x}$$

**89.** $\dfrac{5x^2}{x^2 - 4} \cdot \dfrac{x^2 + 4x + 4}{10x^3}$

$$= \frac{\cancel{5}\, \cancel{x^2}}{(x+2)(x-2)} \cdot \frac{(x+2)^{\cancel{2}}}{\underset{2}{\cancel{10}\, x^{\cancel{3}\,1}}}$$

$$= \frac{x + 2}{2x(x - 2)}$$

So, $f(x) = \dfrac{x + 2}{2x(x - 2)}$

$$f(x) = \frac{x + 2}{2x(x - 2)}$$

**91.** $\dfrac{x}{2x + 6} - \dfrac{9}{x^2 - 9}$

$$\frac{x}{2x + 6} - \frac{9}{x^2 - 9} = \frac{x}{2(x + 3)} - \frac{9}{(x + 3)(x - 3)}$$

$$= \frac{x(x - 3) - 9(2)}{2(x + 3)(x - 3)}$$

$$= \frac{x^2 - 3x - 18}{2(x + 3)(x - 3)}$$

$$= \frac{(x - 6)(x + 3)}{2(x + 3)(x - 3)} = \frac{x - 6}{2(x - 3)}$$

So, $f(x) = \dfrac{x-6}{2(x-3)}$

$f(x) = \dfrac{x-6}{2(x-3)}$

**93.**
$$\dfrac{1-\dfrac{3}{x+2}}{1+\dfrac{1}{x-2}} = \dfrac{1-\dfrac{3}{x+2}}{1+\dfrac{1}{x-2}} \cdot \dfrac{(x+2)(x-2)}{(x+2)(x-2)}$$

$$= \dfrac{(x+2)(x-2)-3(x-2)}{(x+2)(x-2)+(x+2)}$$

$$= \dfrac{x^2-4-3x+6}{x^2-4+x+2}$$

$$= \dfrac{x^2-3x+2}{x^2+x-2}$$

$$= \dfrac{(x-2)\cancel{(x-1)}}{(x+2)\cancel{(x-1)}} = \dfrac{x-2}{x+2}$$

So, $f(x) = \dfrac{x-2}{x+2}$

$f(x) = \dfrac{x-2}{x+2}$

**95.** $g(x) = \dfrac{2x+7}{x+3} = \dfrac{1}{x+3} + 2$

$g(x) = \dfrac{1}{x+3} + 2$

**97.** $g(x) = \dfrac{3x-7}{x-2} = \dfrac{-1}{x-2} + 3$

$g(x) = \dfrac{-1}{x-2} + 3$

**99. a.** $C(x) = 100x + 100,000$

**b.** $\overline{C}(x) = \dfrac{100x+100,000}{x}$

**c.** $\overline{C}(500) = \dfrac{100(500)+100,000}{500} = \$300$
When 500 bicycles are manufactured, it costs $300 to manufacture each.

$\overline{C}(1000) = \dfrac{100(1000)+100,000}{1000} = \$200$
When 1000 bicycles are manufactured, it costs $200 to manufacture each.

$\overline{C}(2000) = \dfrac{100(2000)+100,000}{2000} = \$150$
When 2000 bicycles are manufactured, it costs $150 to manufacture each.

$\overline{C}(4000) = \dfrac{100(4000)+100,000}{4000} = \$125$
When 4000 bicycles are manufactured, it costs $125 to manufacture each.
The average cost decreases as the number of bicycles manufactured increases.

**d.** $n = m$, so $y = \dfrac{100}{1} = 100$.

As greater numbers of bicycles are manufactured, the average cost approaches $100.

**101. a.** From the graph the pH level of the human mouth 42 minutes after a person eats food containing sugar will be about 6.0.

**b.** From the graph, the pH level is lowest after about 6 minutes.

$$f(6) = \dfrac{6.5(6)^2 - 20.4(6) + 234}{6^2 + 36}$$
$$= 4.8$$
The pH level after 6 minutes (i.e. the lowest pH level) is 4.8.

**c.** From the graph, the pH level appears to approach 6.5 as time goes by. Therefore, the normal pH level must be 6.5.

**d.** $y = 6.5$

Over time, the pH level rises back to the normal level.

**e.** During the first hour, the pH level drops quickly below normal, and then slowly begins to approach the normal level.

**103.** $P(10) = \dfrac{100(10-1)}{10} = 90 \quad (10, 90)$

For a disease that smokers are 10 times more likely to contact than non-smokers, 90% of the deaths are smoking related.

**105.** $y = 100$ As incidence of the diseases increases, the percent of death approaches, but never gets to be, 100%.

**107. a.** $f(x) = \dfrac{p(x)}{q(x)} = \dfrac{1.75x^2 - 15.9x + 160}{2.1x^2 - 3.5x + 296}$

**b.** According to the graph, $\dfrac{2504.0}{3720.7} \approx 0.67$ or 67% of federal expenditures were spent on human resources in 2010.

**c.** According to the function,

$f(x) = \dfrac{1.75(40)^2 - 15.9(40) + 160}{2.1(40)^2 - 3.5(40) + 296} \approx 0.66$ or

66% of federal expenditures were spent on human resources in 2010.

**d.** The degree of the numerator, 2, is equal to the degree of the denominator, 2. The leading coefficients of the numerator and denominator are 1.75 and 2.1, respectively. The equation of the horizontal asymptote is $y = \dfrac{1.75}{2.1}$ which is about 83%

Thus, about 83% of federal expenditures will be spent on human resources over time.

**109. – 117.** Answers will vary.

**119.**

The graph approaches the horizontal asymptote faster and the vertical asymptote slower as $n$ increases.

**121. a.** $f(x) = \dfrac{27725(x-14)}{x^2 + 9} - 5x$

**b.** The graph increases from late teens until about the age of 25, and then the number of arrests decreases.

**c.** At age 25 the highest number arrests occurs. There are about 356 arrests for every 100,000 drivers.

**123.** does not make sense; Explanations will vary. Sample explanation: The function has one vertical asymptote, $x = 2$.

**125.** does not make sense; Explanations will vary. Sample explanation: As production level increases, the average cost for a company to produce each unit of its product decreases.

**127.** true

**129.** true

**131. – 133.** Answers will vary.

**134.**
$$2x^2 + x = 15$$
$$2x^2 + x - 15 = 0$$
$$(2x - 5)(x + 3) = 0$$
$$2x - 5 = 0 \quad \text{or} \quad x + 3 = 0$$
$$x = \frac{5}{2} \qquad\qquad x = -3$$

The solution set is $\left\{ -3, \dfrac{5}{2} \right\}$.

**135.**
$$x^3 + x^2 = 4x + 4$$
$$x^3 + x^2 - 4x - 4 = 0$$
$$x^2(x+1) - 4(x+1) = 0$$
$$(x+1)(x^2 - 4) = 0$$
$$(x+1)(x+2)(x-2) = 0$$

The solution set is $\{-2, -1, 2\}$.

**136.** $\dfrac{x+1}{x+3} - 2 = \dfrac{x+1}{x+3} - \dfrac{2(x+3)}{x+3}$

$\quad\quad\quad\quad\quad = \dfrac{x+1}{x+3} - \dfrac{2x+6}{x+3}$

$\quad\quad\quad\quad\quad = \dfrac{x+1-2x-6}{x+3}$

$\quad\quad\quad\quad\quad = \dfrac{-x-5}{x+3}$ or $-\dfrac{x+5}{x+3}$

## Section 3.6

### Check Point Exercises

**1.**
$$x^2 - x > 20$$
$$x^2 - x - 20 > 0$$
$$(x+4)(x-5) > 0$$

Solve the related quadratic equation.
$$(x+4)(x-5) = 0$$

Apply the zero product principle.
$$x+4=0 \quad \text{or} \quad x-5=0$$
$$x=-4 \quad\quad\quad x=5$$

The boundary points are −2 and 4.

| Test Interval | Test Number | Test | Conclusion |
|---|---|---|---|
| $(-\infty,-4)$ | −5 | $(-5)^2 - (-5) > 20$ <br> $30 > 20$, true | $(-\infty,-4)$ belongs to the solution set. |
| $(-4,5)$ | 0 | $(0)^2 - (0) > 20$ <br> $0 > 20$, false | $(-4,5)$ does not belong to the solution set. |
| $(5,\infty)$ | 10 | $(10)^2 - (10) > 20$ <br> $90 > 20$, true | $(5,\infty)$ belongs to the solution set. |

The solution set is $(-\infty,-4)\cup(5,\infty)$ or $\{x|x<-4 \text{ or } x>5\}$.

**2.** $$2x^2 \le -6x - 1$$

$2x^2 + 6x + 1 \le 0$

Solve the related quadratic equation to find the boundary points.

$2x^2 + 6x + 1 = 0$

$a = 2 \quad b = 6 \quad c = 1$

$$x = \frac{-(6) \pm \sqrt{(6)^2 - 4(2)(1)}}{2(2)}$$

$$= \frac{-6 \pm \sqrt{36 - 8}}{4}$$

$$= \frac{-6 \pm \sqrt{28}}{4}$$

$$= \frac{-6 \pm 2\sqrt{7}}{4}$$

$$= \frac{-3 \pm \sqrt{7}}{2}$$

$x = \dfrac{-3 + \sqrt{7}}{2} \quad \text{or} \quad x = \dfrac{-3 - \sqrt{7}}{2}$

$x \approx -0.2 \qquad\qquad x \approx -2.8$

| Interval | Test Value | Test | Conclusion |
|---|---|---|---|
| $\left(-\infty, \dfrac{-3 - \sqrt{7}}{2}\right)$ | $-10$ | $2(-10)^2 \le -6(-10) - 1$ <br> $200 \le 59$, false | $\left(-\infty, \dfrac{-3 - \sqrt{7}}{2}\right)$ is not part of the solution set |
| $\left(\dfrac{-3 - \sqrt{7}}{2}, \dfrac{-3 + \sqrt{7}}{2}\right)$ | $-1$ | $2(-1)^2 \le -6(-1) - 1$ <br> $2 \le 5$, true | $\left(\dfrac{-3 - \sqrt{7}}{2}, \dfrac{-3 + \sqrt{7}}{2}\right)$ is part of the solution set |
| $\left(\dfrac{-3 + \sqrt{7}}{2}, \infty\right)$ | $0$ | $2(0)^2 \le -6(0) - 1$ <br> $0 \le -1$, false | $\left(\dfrac{-3 + \sqrt{7}}{2}, \infty\right)$ is not part of the solution set |

The solution set is $\left(\dfrac{-3 - \sqrt{7}}{2}, \dfrac{-3 + \sqrt{7}}{2}\right)$.

**3.**
$$x^3 + 3x^2 \leq x + 3$$
$$x^3 + 3x^2 - x - 3 \leq 0$$
$$(x+1)(x-1)(x+3) \leq 0$$
$$(x+1)(x-1)(x+3) = 0$$
$$x+1=0 \quad \text{or} \quad x-1=0 \quad \text{or} \quad x+3=0$$
$$x=-1 \qquad x=1 \qquad x=-3$$

| Test Interval | Test Number | Test | Conclusion |
|---|---|---|---|
| $(-\infty,-3)$ | -4 | $(-4)^3 + 3(-4)^2 \leq (-4)+3$ <br> $-16 \leq -1$ true | $(-\infty,-3)$ belongs to the solution set. |
| $(-3,-1]$ | $-2$ | $(-2)^3 + 3(-2)^2 \leq (-2)+3$ <br> $4 \leq 1$ false | $(-3,-1]$ does not belong to the solution set. |
| $[-1,1]$ | 0 | $(0)^3 + 3(0)^2 \leq (0)+3$ <br> $0 \leq 3$ true | $[-1,1]$ belongs to the solution set. |
| $[1,\infty)$ | 2 | $(6+3)(6-5) > 0$ <br> true | $[1,\infty)$ does not belong to the solution set. |

The solution set is $(-\infty,-3]\cup[-1,1]$ or $\{x \mid x \leq -3 \text{ or } -1 \leq x \leq 1\}$.

**4.**
$$\frac{2x}{x+1} \geq 1$$
$$\frac{2x}{x+1} - 1 \geq 0$$
$$\frac{x-1}{x+1} \geq 0$$
$$x-1=0 \quad \text{or} \quad x+1=0$$
$$x=1 \qquad x=-1$$

| Test Interval | Test Number | Test | Conclusion |
|---|---|---|---|
| $(-\infty,-1)$ | $-2$ | $\dfrac{2(-2)}{-2+1} \geq 1$ <br> $4 \geq 1$, true | $(-\infty,-1)$ belongs to the solution set. |
| $(-1,1]$ | 0 | $\dfrac{2(0)}{0+1} \geq 1$ <br> $0 \geq 1$, false | $(-1,1]$ does not belong to the solution set. |
| $[1,\infty)$ | 2 | $\dfrac{2(2)}{2+1} \geq 1$ <br> $\dfrac{4}{3} \geq 1$, true | $[1,\infty)$ belongs to the solution set. |

The solution set is $(-\infty,-1)\cup[1,\infty)$ or $\{x \mid x < -1 \text{ or } x \geq 1\}$.

**5.**

$$-16t^2 + 80t > 64$$
$$-16t^2 + 80t - 64 > 0$$
$$-16(t-1)(t-4) > 0$$
$$t - 1 = 0 \quad \text{or} \quad t - 4 = 0$$
$$t = 1 \qquad\qquad t = 4$$

| Test Interval | Test Number | Test | Conclusion |
|---|---|---|---|
| $(-\infty, 1)$ | 0 | $-16(0)^2 + 80(0) > 64$ <br> $0 > 64$, false | $(-\infty, 1)$ does not belong to the solution set. |
| $(1, 4)$ | 2 | $-16(2)^2 + 80(2) > 64$ <br> $96 > 64$, true | $(1, 4)$ belongs to the solution set. |
| $(4, \infty)$ | 5 | $-16(5)^2 + 80(5) > 64$ <br> $0 > 64$, false | $(4, \infty)$ does not belong to the solution set. |

The object will be more than 64 feet above the ground between 1 and 4 seconds.

## Concept and Vocabulary Check 3.6

**1.** $x^2 + 8x + 15 = 0$; boundary

**2.** $(-\infty, -5)$; $(-5, -3)$; $(-3, \infty)$

**3.** true

**4.** true

**5.** $[-\infty, -2) \cup [1, \infty)$

## Exercise Set 3.6

**1.** $(x-4)(x+2) > 0$
$x = 4$ or $x = -2$

```
    T   |   F   |   T
 _____
       -2       4
```

Test $-3$: $(-3 - 4)(-3 + 2) > 0$
$\qquad\qquad\qquad\qquad 7 > 0$ True
Test $0$: $(0 - 4)(0 + 2) > 0$
$\qquad\qquad\qquad\qquad -8 > 0$ False
Test $5$: $(5 - 4)(5 + 2) > 0$
$\qquad\qquad\qquad\qquad 7 > 0$ True
$(-\infty, -2)$ or $(4, \infty)$

**3.** $(x-7)(x+3) \le 0$
$x = 7$ or $x = -3$

| F | T | F |
|---|---|---|
| | −3 | 7 |

Test −4: $(-4-7)(-4+3) \le 0$
$\qquad 11 \le 0$ False
Test 0: $(0-7)(0+3) \le 0$
$\qquad -21 \le 0$ True
Test 8: $(8-7)(8+3) \le 0$
$\qquad 11 \le 0$ False
The solution set is $[-3, 7]$.

**5.** $x^2 - 5x + 4 > 0$
$(x-4)(x-1) > 0$
$x = 4 \quad$ or $\quad x = 1$

| T | F | T |
|---|---|---|
| | 1 | 4 |

Test 0: $0^2 - 5(0) + 4 > 0$
$\qquad 4 > 0$ True
Test 2: $2^2 - 5(2) + 4 > 0$
$\qquad -2 > 0$ False
Test 5: $5^2 - 5(5) + 4 > 0$
$\qquad 4 > 0$ True
The solution set is $(-\infty, 1)$ or $(4, \infty)$.

**7.** $x^2 + 5x + 4 > 0$
$(x+1)(x+4) > 0$
$x = -1$ or $x = -4$

| T | F | T |
|---|---|---|
| | −4 | −1 |

Test −5: $(-5)^2 + 5(-5) + 4 > 0$
$\qquad 4 > 0$ True
Test −3: $(-3)^2 + 5(-3) + 4 > 0$
$\qquad -2 > 0$ False
Test 0: $0^2 + 5(0) + 4 > 0$
$\qquad 4 > 0$ True
The solution set is $(-\infty, -4)$ or $(-1, \infty)$.

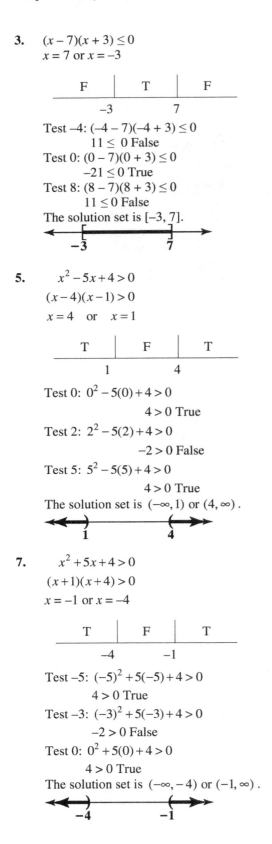

**9.** $x^2 - 6x + 9 < 0$
$(x-3)(x-3) < 0$
$x = 3$

| F | F |
|---|---|
| | 3 |

Test 0: $0^2 - 6(0) + 9 < 0$
$\qquad 9 < 0$ False
Test 4: $4^2 - 6(4) + 9 < 0$
$\qquad 1 < 0$ False
The solution set is the empty set, $\varnothing$.

**11.** $3x^2 + 10x - 8 \le 0$
$(3x-2)(x+4) \le 0$
$x = \dfrac{2}{3}$ or $x = -4$

| F | T | F |
|---|---|---|
| | −4 | $\frac{2}{3}$ |

Test −5: $3(-5)^2 + 10(-5) - 8 \le 0$
$\qquad 17 \le 0$ False
Test 0: $3(0)^2 + 10(0) - 8 \le 0$
$\qquad -8 \le 0$ True
Test 1: $3(1)^2 + 10(1) - 8 \le 0$
$\qquad 5 \le 0$ False
The solution set is $\left[-4, \dfrac{2}{3}\right]$.

**13.**
$$2x^2 + x < 15$$
$$2x^2 + x - 15 < 0$$
$$(2x-5)(x+3) < 0$$
$$2x - 5 = 0 \quad \text{or} \quad x+3=0$$
$$2x = 5$$
$$x = \frac{5}{2} \quad \text{or} \quad x = -3$$

| F | | T | | F |
|---|---|---|---|---|
| | $-3$ | | $\frac{5}{2}$ | |

Test $-4$: $2(-4)^2 + (-4) < 15$
$28 < 15$ False
Test 0: $2(0)^2 + 0 < 15$
$0 < 15$ True
Test 3: $2(3)^2 + 3 < 15$
$21 < 15$ False

The solution set is $\left(-3, \frac{5}{2}\right)$.

**15.**
$$4x^2 + 7x < -3$$
$$4x^2 + 7x + 3 < 0$$
$$(4x+3)(x+1) < 0$$
$$4x + 3 = 0 \quad \text{or} \quad x+1=0$$
$$4x-3=0$$
$$x = -\frac{3}{4} \quad \text{or} \quad x = -1$$

| F | | T | | F |
|---|---|---|---|---|
| | $-1$ | | $-\frac{3}{4}$ | |

Test $-2$: $4(-2)^2 + 7(-2) < -3$
$2 < -3$ False

Test $-\frac{7}{8}$: $4\left(-\frac{7}{8}\right)^2 + 7\left(-\frac{7}{8}\right) < -3$
$$\frac{49}{16} - \frac{49}{8} < -3$$
$$-\frac{49}{16} < -3 \text{ True}$$

Test 0: $4(0)^2 + 7(0) < -3$
$0 < -3$ False

The solution set is $\left(-1, -\frac{3}{4}\right)$.

**17.**
$$5x \leq 2 - 3x^2$$
$$3x^2 + 5x - 2 \leq 0$$
$$(3x-1)(x+2) \leq 0$$
$$3x - 1 = 0 \text{ or } x + 2 = 0$$
$$3x = 1$$
$$3x - 1 = 0 \quad \text{or} \quad x + 2 = 0$$
$$3x = 1$$
$$x = \frac{1}{3} \quad \text{or} \quad x = -2$$

| F | | T | | F |
|---|---|---|---|---|
| | $-2$ | | $\frac{1}{3}$ | |

Test $-3$: $5(-3) \leq 2 - 3(-3)^2$
$-15 \leq -25$ False
Test 0: $5(0) \leq 2 - 3(0)^2$
$0 \leq 2$ True
Test 1: $5(1) \leq 2 - 3(1)^2$
$5 \leq -1$ False

The solution set is $\left[-2, \frac{1}{3}\right]$.

**19.** $x^2 - 4x \geq 0$
$x(x-4) \geq 0$
$x = 0$ or $x - 4 = 0$
$x = 4$

| T | | F | | T |
|---|---|---|---|---|
| | 0 | | 4 | |

Test $-1$: $(-1)^2 - 4(-1) \geq 0$
$5 \geq 0$ True
Test 1: $(1)^2 - 4(1) \geq 0$
$-3 \geq 0$ False
$0 \leq 2$ True
Test 5: $5^2 - 4(5) \geq 0$
$5 \geq 0$ True
The solution set is $(-\infty, 0]$ or $[4, \infty)$.

**21.** $2x^2 + 3x > 0$

$x(2x+3) > 0$

$x = 0 \text{ or } x = -\dfrac{3}{2}$

| T | | F | | T |
|---|---|---|---|---|

$-\dfrac{3}{2}$ 0

Test $-2$: $2(-2)^2 + 3(-2) > 0$

$2 > 0$ True

Test $-1$: $2(-1)^2 + 3(-1) > 0$

$-1 > 0$ False

Test 1: $2(1)^2 + 3(1) > 0$

$5 > 0$ True

The solution set is $\left(-\infty, -\dfrac{3}{2}\right)$ or $(0, \infty)$.

**23.** $-x^2 + x \geq 0$

$x^2 - x \leq 0$

$x(x-1) \leq 0$

$x = 0 \quad \text{or} \quad x = 1$

| F | | T | | F |
|---|---|---|---|---|

0   1

Test $-1$: $-(-1)^2 + (-1) \geq 0$

$-2 \geq 0$ False

Test $\dfrac{1}{2}$: $-\left(\dfrac{1}{2}\right)^2 + \left(\dfrac{1}{2}\right) \geq 0$

$\dfrac{1}{4} \geq 0$ True

Test 2: $-(2)^2 + 2 \geq 0$

$-2 \geq 0$ False

The solution set is $[0, 1]$.

**25.** $x^2 \leq 4x - 2$

$x^2 - 4x + 2 \leq 0$

Solve $x^2 - 4x + 2 = 0$

$x = \dfrac{-b \pm \sqrt{b^2 - 4ac}}{2a}$

$x = \dfrac{-(-4) \pm \sqrt{(-4)^2 - 4(1)(2)}}{2(1)}$

$= \dfrac{4 \pm \sqrt{8}}{2}$

$= 2 \pm \sqrt{2}$

$x \approx 0.59 \text{ or } x \approx 3.41$

| F | | T | | F |
|---|---|---|---|---|

0.59   3.41

The solution set is $\left[2 - \sqrt{2}, 2 + \sqrt{2}\right]$ or $[0.59, 3.41]$.

$2 - \sqrt{2}$    $2 + \sqrt{2}$

**27.** $x^2 - 6x + 9 < 0$

Solve $x^2 - 6x + 9 = 0$

$(x-3)(x-3) = 0$

$(x-3)^2 = 0$

$x = 3$

| F | | F |
|---|---|---|

3

The solution set is the empty set, $\varnothing$.

**29.** $(x-1)(x-2)(x-3) \geq 0$

Boundary points: 1, 2, and 3

Test one value in each interval.

| F | | T | | F | | T |
|---|---|---|---|---|---|---|

1   2   3

The solution set is $[1, 2] \cup [3, \infty)$.

1   2   3

**31.** $x(3-x)(x-5) \le 0$

Boundary points: 0, 3, and 5
Test one value in each interval.

| F | | T | | F | | T |
|---|---|---|---|---|---|---|
| | 0 | | 3 | | 5 | |

The solution set is $[0, 3] \cup [5, \infty)$.

**33.** $(2-x)^2 \left(x - \frac{7}{2}\right) < 0$

Boundary points: 2, and $\frac{7}{2}$
Test one value in each interval.

| T | | T | | F |
|---|---|---|---|---|
| | 2 | | $\frac{7}{2}$ | |

The solution set is $(-\infty, 2) \cup \left(2, \frac{7}{2}\right)$.

**35.** $x^3 + 2x^2 - x - 2 \ge 0$

$x^2(x+2) - 1(x+2) \ge 0$

$(x+2)(x^2-1) \ge 0$

$(x+2)(x-1)(x+1) \ge 0$

Boundary points: $-2$, $-1$, and 2
Test one value in each interval.

| F | | T | | F | | T |
|---|---|---|---|---|---|---|
| | $-2$ | | $-1$ | | 2 | |

The solution set is $[-2, -1] \cup [1, \infty)$.

**37.** $x^3 + 2x^2 - x - 2 \ge 0$

$x^2(x-3) - 9(x-3) \ge 0$

$(x-3)(x^2-9) \ge 0$

$(x-3)(x+3)(x-3) \ge 0$

Boundary points: $-3$ and 3
Test one value in each interval.

| T | | F | | F |
|---|---|---|---|---|
| | $-3$ | | 3 | |

The solution set is $(-\infty, -3]$.

**39.** $x^3 + x^2 + 4x + 4 > 0$

$x^2(x+1) + 4(x+1) \ge 0$

$(x+1)(x^2+4) \ge 0$

Boundary point: $-1$
Test one value in each interval.

| F | | T |
|---|---|---|
| | $-1$ | |

The solution set is $(-1, \infty)$.

**41.** $x^3 - 9x^2 \ge 0$

$x^2(x-9) \ge 0$

Boundary points: 0 and 9
Test one value in each interval.

| F | | F | | T |
|---|---|---|---|---|
| | 0 | | 9 | |

The solution set is $[0,0] \cup [9, \infty)$.

**43.** $\dfrac{x-4}{x+3} > 0$

$x-4 = 0 \quad x+3 = 0$

$x = 4 \qquad x = -3$

| T | | F | | T |
|---|---|---|---|---|
| | $-3$ | | 4 | |

The solution set is $(-\infty, -3) \cup (4, \infty)$.

**45.** $\dfrac{x+3}{x+4} < 0$

$x = -3 \quad \text{or} \quad x = -4$

| F | | T | | F |
|---|---|---|---|---|
| | $-4$ | | $-3$ | |

The solution set is $(-4, -3)$.

**47.** $\dfrac{-x+2}{x-4} \ge 0$

$x = 2 \text{ or } x = 4$

| F | | T | | F |
|---|---|---|---|---|
| | 2 | | 4 | |

The solution set is $[2, 4)$.

**49.** $\dfrac{4-2x}{3x+4} \le 0$

$x = 2$ or $x = -\dfrac{4}{3}$

| T | F | T |
|---|---|---|

$-\dfrac{4}{3}$     $2$

The solution set is $\left(-\infty, \dfrac{-4}{3}\right) \cup [2, \infty)$.

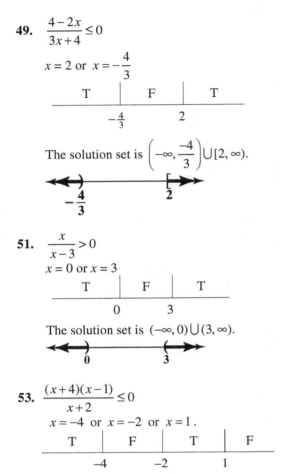

$-\dfrac{4}{3}$    $2$

**51.** $\dfrac{x}{x-3} > 0$

$x = 0$ or $x = 3$

| T | F | T |
|---|---|---|

$0$     $3$

The solution set is $(-\infty, 0) \cup (3, \infty)$.

$0$    $3$

**53.** $\dfrac{(x+4)(x-1)}{x+2} \le 0$

$x = -4$ or $x = -2$ or $x = 1$.

| T | F | T | F |
|---|---|---|---|

$-4$    $-2$    $1$

Values of $x = -4$ or $x = 1$ result in $f(x) = 0$ and, therefore must be included in the solution set.
The solution set is $(-\infty, -4] \cup (-2, 1]$

| T | F | T |
|---|---|---|

$\dfrac{1}{2}$     $\dfrac{7}{5}$

$-4$   $-2$   $1$

**55.** $\dfrac{x+1}{x+3} < 2$

$\dfrac{x+1}{x+3} - 2 < 0$

$\dfrac{x+1-2(x+3)}{x+3} < 0$

$\dfrac{x+1-2x-6}{x+3} < 0$

$\dfrac{-x-5}{x+3} < 0$

$x = $ or $x = -3$

| T | F | T |
|---|---|---|

$-5$     $-3$

The solution set is $(-\infty, -5) \cup (-3, \infty)$.

$-5$    $-3$

**57.** $\dfrac{x+4}{2x-1} \le 3$

$\dfrac{x+4}{2x-1} - 3 \le 0$

$\dfrac{x+4-3(2x-1)}{2x-1} \le 0$

$\dfrac{x+4-6x+3}{2x-1} \le 0$

$\dfrac{-5x+7}{2x-1} \le 0$

$x = \dfrac{7}{5}$   or   $x = \dfrac{1}{2}$

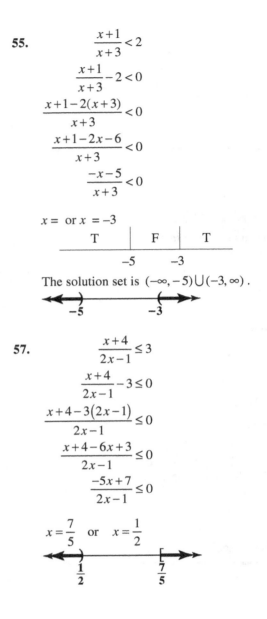

$\dfrac{1}{2}$     $\dfrac{7}{5}$

**59.**
$$\frac{x-2}{x+2} \leq 2$$

$$\frac{x-2}{x+2} - 2 \leq 0$$

$$\frac{x-2-2(x+2)}{x+2} \leq 0$$

$$\frac{x-2-2x-4}{x+2} \leq 0$$

$$\frac{-x-6}{x+2} \leq 0$$

$$x = -6 \quad \text{or} \quad x = -2$$

| T | F | T |
|---|---|---|
| | -6 | -2 |

The solution set is $(-\infty, -6] \cup (-2, \infty)$.

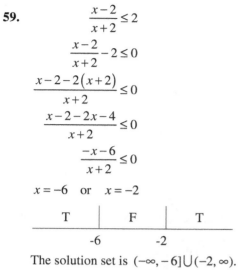

**61.** $f(x) = \sqrt{2x^2 - 5x + 2}$

The domain of this function requires that $2x^2 - 5x + 2 \geq 0$

Solve $2x^2 - 5x + 2 = 0$

$(x-2)(2x-1) = 0$

$x = \dfrac{1}{2}$ or $x = 2$

| T | F | T |
|---|---|---|
| $\dfrac{1}{2}$ | 2 | |

The domain is $\left(-\infty, \dfrac{1}{2}\right] \cup [2, \infty)$.

**63.** $f(x) = \sqrt{\dfrac{2x}{x+1} - 1}$

The domain of this function requires that $\dfrac{2x}{x+1} - 1 \geq 0$ or $\dfrac{x-1}{x+1} \geq 0$

$x = -1$ or $x = 1$

The value $x = 1$ results in 0 and, thus, it must be included in the domain.
The domain is $(-\infty, -1) \cup [1, \infty)$.

**65.** $\left|x^2 + 2x - 36\right| > 12$

Express the inequality without the absolute value symbol:

$x^2 + 2x - 36 < -12$ or $x^2 + 2x - 36 > 12$

$x^2 + 2x - 24 < 0$ $\qquad$ $x^2 + 2x - 48 > 0$

Solve the related quadratic equations.

$x^2 + 2x - 24 = 0$ or $x^2 + 2x - 48 = 0$

$(x+6)(x-4) = 0$ $\qquad$ $(x+8)(x-6) = 0$

Apply the zero product principle.

$x + 6 = 0$ or $x - 4 = 0$ or $x + 8 = 0$ or $x - 6 = 0$

$\quad x = -6 \qquad\quad x = 4 \qquad\quad x = -8 \qquad\quad x = 6$

The boundary points are $-8, \ -6, \ 4$ and $6$.

| Test Interval | Test Number | Test | Conclusion |
|---|---|---|---|
| $(-\infty, -8)$ | $-9$ | $\left\|(-9)^2 + 2(-9) - 36\right\| > 12$ <br><br> $27 > 12$, True | $(-\infty, -8)$ belongs to the solution set. |
| $(-8, -6)$ | $-7$ | $\left\|(-7)^2 + 2(-7) - 36\right\| > 12$ <br><br> $1 > 12$, False | $(-8, -6)$ does not belong to the solution set. |
| $(-6, 4)$ | $0$ | $\left\|0^2 + 2(0) - 36\right\| > 12$ <br><br> $36 > 12$, True | $(-6, 4)$ belongs to the solution set. |
| $(4, 6)$ | $5$ | $\left\|5^2 + 2(5) - 36\right\| > 12$ <br><br> $1 > 12$, False | $(4, 6)$ does not belong to the solution set. |
| $(6, \infty)$ | $7$ | $\left\|7^2 + 2(7) - 36\right\| > 12$ <br><br> $27 > 12$, True | $(6, \infty)$ belongs to the solution set. |

The solution set is $(-\infty, -8) \cup (-6, 4) \cup (6, \infty)$ or $\left\{x \mid x < -8 \text{ or } -6 < x < 4 \text{ or } x > 6\right\}$.

**67.** $\dfrac{3}{x+3} > \dfrac{3}{x-2}$

Express the inequality so that one side is zero.

$$\frac{3}{x+3} - \frac{3}{x-2} > 0$$

$$\frac{3(x-2)}{(x+3)(x-2)} - \frac{3(x+3)}{(x+3)(x-2)} > 0$$

$$\frac{3x-6-3x-9}{(x+3)(x-2)} < 0$$

$$\frac{-15}{(x+3)(x-2)} < 0$$

Find the values of $x$ that make the denominator zero.

$$x+3=0 \qquad x-2=0$$
$$x=-3 \qquad x=2$$

The boundary points are $-3$ and 2.

| Test Interval | Test Number | Test | Conclusion |
|---|---|---|---|
| $(-\infty,-3)$ | $-4$ | $\dfrac{3}{-4+3} > \dfrac{3}{-4-2}$ <br> $-3 > \dfrac{1}{2}$, False | $(-\infty,-3)$ does not belong to the solution set. |
| $(-3,2)$ | $0$ | $\dfrac{3}{0+3} > \dfrac{3}{0-2}$ <br> $1 > -\dfrac{3}{2}$, True | $(-3,2)$ belongs to the solution set. |
| $(2,\infty)$ | $3$ | $\dfrac{3}{3+3} > \dfrac{3}{3-2}$ <br> $\dfrac{1}{2} > 3$, False | $(2,\infty)$ does not belong to the solution set. |

The solution set is $(-3,2)$ or $\{x | -3 < x < 2\}$.

**69.** $\dfrac{x^2 - x - 2}{x^2 - 4x + 3} > 0$

Find the values of $x$ that make the numerator and denominator zero.

$x^2 - x - 2 = 0$      $x^2 - 4x + 3 = 0$

$(x-2)(x+1) = 0$     $(x-3)(x-1) = 0$

Apply the zero product principle.

$x - 2 = 0$   or   $x + 1 = 0$     $x - 3 = 0$   or   $x - 1 = 0$

     $x = 2$        $x = -1$       $x = 3$        $x = 1$

The boundary points are $-1$, 1, 2 and 3.

| Test Interval | Test Number | Test | Conclusion |
|---|---|---|---|
| $(-\infty, -1)$ | $-2$ | $\dfrac{(-2)^2 - (-2) - 2}{(-2)^2 - 4(-2) + 3} > 0$ <br><br> $\dfrac{4}{15} > 0$, True | $(-\infty, -1)$ belongs to the solution set. |
| $(-1, 1)$ | $0$ | $\dfrac{0^2 - 0 - 2}{0^2 - 4(0) + 3} > 0$ <br><br> $-\dfrac{2}{3} > 0$, False | $(-1, 1)$ does not belong to the solution set. |
| $(1, 2)$ | $1.5$ | $\dfrac{1.5^2 - 1.5 - 2}{1.5^2 - 4(1.5) + 3} > 0$ <br><br> $\dfrac{5}{3} > 0$, True | $(1, 2)$ belongs to the solution set. |
| $(2, 3)$ | $2.5$ | $\dfrac{2.5^2 - 2.5 - 2}{2.5^2 - 4(2.5) + 3} > 0$ <br><br> $-\dfrac{7}{3} > 0$, False | $(2, 3)$ does not belong to the solution set. |
| $(3, \infty)$ | $4$ | $\dfrac{4^2 - 4 - 2}{4^2 - 4(4) + 3} > 0$ <br><br> $\dfrac{10}{3} > 0$, True | $(3, \infty)$ belongs to the solution set. |

The solution set is $(-\infty, -1) \cup (1, 2) \cup (3, \infty)$ or $\{x \mid x < -1 \text{ or } 1 < x < 2 \text{ or } x > 3\}$.

**71.**
$$2x^3 + 11x^2 \geq 7x + 6$$
$$2x^3 + 11x^2 - 7x - 6 \geq 0$$

The graph of $f(x) = 2x^3 + 11x^2 - 7x - 6$ appears to cross the $x$-axis at $-6$, $-\dfrac{1}{2}$, and 1. We verify this numerically by substituting these values into the function:

$$f(-6) = 2(-6)^3 + 11(-6)^2 - 7(-6) - 6 = 2(-216) + 11(36) - (-42) - 6 = -432 + 396 + 42 - 6 = 0$$

$$f\left(-\frac{1}{2}\right) = 2\left(-\frac{1}{2}\right)^3 + 11\left(-\frac{1}{2}\right)^2 - 7\left(-\frac{1}{2}\right) - 6 = 2\left(-\frac{1}{8}\right) + 11\left(\frac{1}{4}\right) - \left(-\frac{7}{2}\right) - 6 = -\frac{1}{4} + \frac{11}{4} + \frac{7}{2} - 6 = 0$$

$$f(1) = 2(1)^3 + 11(1)^2 - 7(1) - 6 = 2(1) + 11(1) - 7 - 6 = 2 + 11 - 7 - 6 = 0$$

Thus, the boundaries are $-6$, $-\dfrac{1}{2}$, and 1. We need to find the intervals on which $f(x) \geq 0$. These intervals are indicated on the graph where the curve is above the $x$-axis. Now, the curve is above the $x$-axis when $-6 < x < -\dfrac{1}{2}$ and when $x > 1$. Thus, the solution set is $\left\{ x \,\middle|\, -6 \leq x \leq -\dfrac{1}{2} \text{ or } x \geq 1 \right\}$ or $\left[-6, -\dfrac{1}{2}\right] \cup [1, \infty)$.

**73.**
$$\frac{1}{4(x+2)} \leq -\frac{3}{4(x-2)}$$
$$\frac{1}{4(x+2)} + \frac{3}{4(x-2)} \leq 0$$

Simplify the left side of the inequality:

$$\frac{x-2}{4(x+2)} + \frac{3(x+2)}{4(x-2)} = \frac{x-2+3x+6}{4(x+2)(x-2)} = \frac{4x+4}{4(x+2)(x-2)} = \frac{4(x+1)}{4(x+2)(x-2)} = \frac{x+1}{x^2-4}.$$

The graph of $f(x) = \dfrac{x+1}{x^2-4}$ crosses the $x$-axis at $-1$, and has vertical asymptotes at $x = -2$ and $x = 2$. Thus, the boundaries are $-2$, $-1$, and 1. We need to find the intervals on which $f(x) \leq 0$. These intervals are indicated on the graph where the curve is below the $x$-axis. Now, the curve is below the $x$-axis when $x < -2$ and when $-1 < x < 2$. Thus, the solution set is $\left\{ x \,\middle|\, x < -2 \text{ or } -1 \leq x < 2 \right\}$ or $(-\infty, -2) \cup [-1, 2)$.

**75.** $s(t) = -16t^2 + 8t + 87$

The diver's height will exceed that of the cliff when $s(t) > 87$

$$-16t^2 + 8t + 87 > 87$$
$$-16t^2 + 8t > 0$$
$$-8t(2t - 1) > 0$$

The boundaries are 0 and $\dfrac{1}{2}$. Testing each interval shows that the diver will be higher than the cliff for the first half second after beginning the jump. The interval is $\left(0, \dfrac{1}{2}\right)$.

**77.** $f(x) = 0.0875x^2 - 0.4x + 66.6$

$g(x) = 0.0875x^2 + 1.9x + 11.6$

**a.** $f(35) = 0.0875(35)^2 - 0.4(35) + 66.6 \approx 160$ feet

$g(35) = 0.0875(35)^2 + 1.9(35) + 11.6 \approx 185$ feet

**b.** Dry pavement: graph (b)

Wet pavement: graph (a)

**c.** The answers to part (a) model the actual stopping distances shown in the figure extremely well. The function values and the data are identical.

**d.** $0.0875x^2 - 0.4x + 66.6 > 540$

$0.0875x^2 - 0.4x + 473.4 > 0$

Solve the related quadratic equation.

$0.0875x^2 - 0.4x + 473.4 = 0$

$x = \dfrac{-b \pm \sqrt{b^2 - 4ac}}{2a}$

$x = \dfrac{-(-0.4) \pm \sqrt{(-0.4)^2 - 4(0.0875)(473.4)}}{2(0.0875)}$

$x \approx -71$ or $76$

Since the function's domain is $x \geq 30$, we must test the following intervals.

| Interval | Test Value | Test | Conclusion |
|---|---|---|---|
| $(30, 76)$ | 50 | $0.0875(50)^2 - 0.4(50) + 66.6 > 540$ $265.35 > 540$, False | $(30, 76)$ does not belong to the solution set. |
| $(76, \infty)$ | 100 | $0.0875(100)^2 - 0.4(100) + 66.6 > 540$ $901.6 > 540$, True | $(76, \infty)$ belongs to the solution set. |

On dry pavement, stopping distances will exceed 540 feet for speeds exceeding 76 miles per hour. This is represented on graph (b) to the right of point (76, 540).

**79.** Let $x$ = the length of the rectangle.

Since Perimeter $= 2(\text{length}) + 2(\text{width})$, we know

$50 = 2x + 2(\text{width})$

$50 - 2x = 2(\text{width})$

$\text{width} = \dfrac{50 - 2x}{2} = 25 - x$

Now, $A = (\text{length})(\text{width})$, so we have that

$A(x) \leq 114$

$x(25 - x) \leq 114$

$25x - x^2 \leq 114$

Solve the related equation

$25x - x^2 = 114$

$0 = x^2 - 25x + 114$

$0 = (x - 19)(x - 6)$

Apply the zero product principle:

$x - 19 = 0$    or    $x - 6 = 0$

$\qquad x = 19 \qquad\qquad x = 6$

The boundary points are 6 and 19.

| Test Interval | Test Number | Test | Conclusion |
|---|---|---|---|
| $(-\infty, 6)$ | 0 | $25(0) - 0^2 \le 114$ <br> $0 \le 114$, True | $(-\infty, 6)$ belongs to the solution set. |
| $(6, 19)$ | 10 | $25(10) - 10^2 \le 114$ <br> $150 \le 114$, False | $(6, 19)$ does not belong to the solution set. |
| $(19, \infty)$ | 20 | $25(20) - 20^2 \le 114$ <br> $100 \le 114$, True | $(19, \infty)$ belongs to the solution set. |

If the length is 6 feet, then the width is 19 feet. If the length is less than 6 feet, then the width is greater than 19 feet. Thus, if the area of the rectangle is not to exceed 114 square feet, the length of the shorter side must be 6 feet or less.

**81. – 85.** Answers will vary.

**87.** Graph $y_1 = 2x^2 + 5x - 3$ in a standard window. The graph is below or equal to the $x$-axis for $-3 \le x \le \dfrac{1}{2}$.

The solution set is $\left\{ x \left| -3 \le x \le \dfrac{1}{2} \right. \right\}$ or $\left[ -3, \dfrac{1}{2} \right]$.

**89.** Graph $y_1 = \dfrac{x-4}{x-1}$ in a standard viewing window. The graph is below the $x$-axis for

$1 < x \le 4$.

The solution set is $(1, 4]$.

**91.** Graph $y_1 = \dfrac{1}{x+1}$ and $y_2 = \dfrac{2}{x+4}$

$y_1$ less than or equal to $y_2$ for $-4 < x < -1$ or $x \ge 2$.

The solution set is $(-4, -1) \cup [2, \infty)$

**93.** **a.** $f(x) = 0.1375x^2 + 0.7x + 37.8$

    **b.** $0.1375x^2 + 0.7x + 37.8 > 446$

$0.1375x^2 + 0.7x + 408.2 > 0$

Solve the related quadratic equation.

$0.1375x^2 + 0.7x + 408.2 = 0$

$$x = \frac{-b \pm \sqrt{b^2 - 4ac}}{2a}$$

$$x = \frac{-(0.7) \pm \sqrt{(0.7)^2 - 4(0.1375)(408.2)}}{2(0.1375)}$$

$x \approx -57 \text{ or } 52$

Since the function's domain must be $x \geq 0$, we must test the following intervals.

| Interval | Test Value | Test | Conclusion |
|---|---|---|---|
| $(0, 52)$ | 10 | $0.1375(10)^2 + 0.7(10) + 37.8 > 446$ <br> $58.55 > 446$, False | $(0, 52)$ does not belong to the solution set. |
| $(52, \infty)$ | 100 | $0.1375(100)^2 + 0.7(100) + 37.8 > 446$ <br> $1482.8 > 446$, True | $(52, \infty)$ belongs to the solution set. |

On wet pavement, stopping distances will exceed 446 feet for speeds exceeding 52 miles per hour.

**95.** does not make sense; Explanations will vary. Sample explanation: Polynomials are defined for all values.

**97.** does not make sense; Explanations will vary. Sample explanation: To solve this inequality you must first subtract 2 from both sides.

**99.** false; Changes to make the statement true will vary. A sample change is: The inequality cannot be solved by multiplying both sides by $x + 3$. We do not know if $x + 3$ is positive or negative. Thus, we would not know whether or not to reverse the order of the inequality.

**101.** true

**103.** One possible solution: $\dfrac{x-3}{x+4} \geq 0$

**105.** Because any number squared other than zero is positive, the solution includes only 2.

**107.** Because any number squared other than zero is positive, and the reciprocal of zero is undefined, the solution is all real numbers except 2.

**109.** $\sqrt{27-3x^2} \geq 0$

$27-3x^2 \geq 0$

$9-x^2 \geq 0$

$(3-x)(3+x) \geq 0$

$3-x=0 \qquad 3+x=0$

$x=3$ or $\qquad x=-3$

Test –4: $\sqrt{27-3(-4)^2} \geq 0$

$\sqrt{27-48} \geq 0$

$\sqrt{-21} \geq 0$

no graph- imaginary

Test 0: $\sqrt{27-3(0)^2} \geq 0$

$\sqrt{27} \geq 0$ True

Test 4: $\sqrt{27-3(4)^2} \geq 0$

$\sqrt{27-48} \geq 0$

$\sqrt{-21} \geq 0$

no graph -imaginary

The solution set is [–3, 3].

**110. a.** $y=kx^2$

$64 = k \cdot 2^2$

$64 = 4k$

$16 = k$

**b.** $y=kx^2$

$y=16x^2$

**c.** $y=kx^2$

$y=16x^2$

$y=16 \cdot 5^2$

$y=400$

**111. a.** $y=\dfrac{k}{x}$

$12=\dfrac{k}{8}$

$96=k$

**b.** $y=\dfrac{k}{x}$

$y=\dfrac{96}{x}$

**c.** $y=\dfrac{96}{x}$

$y=\dfrac{96}{3}$

$y=32$

**112.** $S=\dfrac{kA}{P}$

$12,000=\dfrac{k \cdot 60,000}{40}$

$\dfrac{12,000 \cdot 40}{60,000}=k$

$8=k$

**Section 3.7**

**Check Point Exercises**

**1.** $y$ varies directly as $x$ is expressed as $y=kx$.

The volume of water, $W$, varies directly as the time, $t$ can be expressed as $W=kt$.

Use the given values to find $k$.

$W=kt$

$30=k(5)$

$6=k$

Substitute the value of $k$ into the equation.

$W=kt$

$W=6t$

Use the equation to find $W$ when $t=11$.

$W=6t$

$=6(11)$

$=66$

A shower lasting 11 minutes will use 66 gallons of water.

**2.** $y$ varies directly as the cube of $x$ is expressed as $y = kx^3$.

The weight, $w$, varies directly as the cube of the length, $l$ can be expressed as $w = kl^2$.

Use the given values to find $k$.

$w = kl^3$

$2025 = k(15)^3$

$0.6 = k$

Substitute the value of $k$ into the equation.

$w = kl^3$

$w = 0.6l^3$

Use the equation to find $w$ when $l = 25$.

$w = 0.6l^3$

$= 0.6(25)^3$

$= 9375$

The 25-foot long shark was 9375 pounds.

**3.** $y$ varies inversely as $x$ is expressed as $y = \dfrac{k}{x}$.

The length, $L$, varies inversely as the frequency, $f$ can be expressed as $L = \dfrac{k}{f}$.

Use the given values to find $k$.

$L = \dfrac{k}{f}$

$8 = \dfrac{k}{640}$

$5120 = k$

Substitute the value of $k$ into the equation.

$L = \dfrac{k}{f}$

$L = \dfrac{5120}{f}$

Use the equation to find $f$ when $L = 10$.

$L = \dfrac{5120}{f}$

$10 = \dfrac{5120}{f}$

$10f = 5120$

$f = 512$

A 10 inch violin string will have a frequency of 512 cycles per second.

**4.** let $M$ represent the number of minutes
let $Q$ represent the number of problems
let $P$ represent the number of people
$M$ varies directly as $Q$ and inversely as $P$ is expressed as $M = \dfrac{kQ}{P}$.

Use the given values to find $k$.

$M = \dfrac{kQ}{P}$

$32 = \dfrac{k(16)}{4}$

$8 = k$

Substitute the value of $k$ into the equation.

$M = \dfrac{kQ}{P}$

$M = \dfrac{8Q}{P}$

Use the equation to find $M$ when $P = 8$ and $Q = 24$.

$M = \dfrac{8Q}{P}$

$M = \dfrac{8(24)}{8}$

$M = 24$

It will take 24 minutes for 8 people to solve 24 problems.

**5.** $V$ varies jointly with $h$ and $r^2$ and can be modeled as $V = khr^2$.

Use the given values to find $k$.

$V = khr^2$

$120\pi = k(10)(6)^2$

$\dfrac{\pi}{3} = k$

Therefore, the volume equation is $V = \dfrac{1}{3}hr^2$.

$V = \dfrac{\pi}{3}(2)(12)^2 = 96\pi$ cubic feet

**Concept and Vocabulary Check 3.7**

**1.** $y - kx$; constant of variation

**2.** $y = kx^n$

**3.** $y = \dfrac{k}{x}$

**4.** $y = \dfrac{kx}{z}$

**5.** $y = kxz$

**6.** directly; inversely

**7.** jointly; inversely

## Exercise Set 3.7

**1.** Use the given values to find $k$.

$$y = kx$$
$$65 = k \cdot 5$$
$$\frac{65}{5} = \frac{k \cdot 5}{5}$$
$$13 = k$$

The equation becomes $y = 13x$.

When $x = 12$, $y = 13x = 13 \cdot 12 = 156$.

**3.** Since $y$ varies inversely with $x$, we have $y = \dfrac{k}{x}$.

Use the given values to find $k$.

$$y = \frac{k}{x}$$
$$12 = \frac{k}{5}$$
$$5 \cdot 12 = 5 \cdot \frac{k}{5}$$
$$60 = k$$

The equation becomes $y = \dfrac{60}{x}$.

When $x = 2$, $y = \dfrac{60}{2} = 30$.

**5.** Since $y$ varies inversely as $x$ and inversely as the square of $z$, we have $y = \dfrac{kx}{z^2}$.

Use the given values to find $k$.

$$y = \frac{kx}{z^2}$$
$$20 = \frac{k(50)}{5^2}$$
$$20 = \frac{k(50)}{25}$$
$$20 = 2k$$
$$10 = k$$

The equation becomes $y = \dfrac{10x}{z^2}$.

When $x = 3$ and $z = 6$,

$$y = \frac{10x}{z^2} = \frac{10(3)}{6^2} = \frac{10(3)}{36} = \frac{30}{36} = \frac{5}{6}.$$

**7.** Since $y$ varies jointly as $x$ and $z$, we have $y = kxz$.

Use the given values to find $k$.

$$y = kxz$$
$$25 = k(2)(5)$$
$$25 = k(10)$$
$$\frac{25}{10} = \frac{k(10)}{10}$$
$$\frac{5}{2} = k$$

The equation becomes $y = \dfrac{5}{2}xz$.

When $x = 8$ and $z = 12$, $y = \dfrac{5}{2}(8)(12) = 240$.

**9.** Since $y$ varies jointly as $a$ and $b$ and inversely as the square root of $c$, we have $y = \dfrac{kab}{\sqrt{c}}$.

Use the given values to find $k$.

$$y = \dfrac{kab}{\sqrt{c}}$$

$$12 = \dfrac{k(3)(2)}{\sqrt{25}}$$

$$12 = \dfrac{k(6)}{5}$$

$$12(5) = \dfrac{k(6)}{5}(5)$$

$$60 = 6k$$

$$\dfrac{60}{6} = \dfrac{6k}{6}$$

$$10 = k$$

The equation becomes $y = \dfrac{10ab}{\sqrt{c}}$.

When $a = 5$, $b = 3$, $c = 9$,

$$y = \dfrac{10ab}{\sqrt{c}} = \dfrac{10(5)(3)}{\sqrt{9}} = \dfrac{150}{3} = 50.$$

**11.** $x = kyz$ ;

Solving for $y$:

$$x = kyz$$

$$\dfrac{x}{kz} = \dfrac{kyz}{yz}.$$

$$y = \dfrac{x}{kz}$$

**13.** $x = \dfrac{kz^3}{y}$;

Solving for $y$

$$x = \dfrac{kz^3}{y}$$

$$xy = y \cdot \dfrac{kz^3}{y}$$

$$xy = kz^3$$

$$\dfrac{xy}{x} = \dfrac{kz^3}{x}$$

$$y = \dfrac{kz^3}{x}$$

**15.** $x = \dfrac{kyz}{\sqrt{w}}$ ;

Solving for $y$:

$$x = \dfrac{kyz}{\sqrt{w}}$$

$$x\left(\sqrt{w}\right) = \left(\sqrt{w}\right)\dfrac{kyz}{\sqrt{w}}$$

$$x\sqrt{w} = kyz$$

$$\dfrac{x\sqrt{w}}{kz} = \dfrac{kyz}{kz}$$

$$y = \dfrac{x\sqrt{w}}{kz}$$

**17.** $x = kz(y+w)$ ;

Solving for $y$:

$$x = kz(y+w)$$

$$x = kzy + kzw$$

$$x - kzw = kzy$$

$$\dfrac{x - kzw}{kz} = \dfrac{kzy}{kz}$$

$$y = \dfrac{x - kzw}{kz}$$

**19.** $x = \dfrac{kz}{y-w}$ ;

Solving for $y$:

$$x = \dfrac{kz}{y-w}$$

$$(y-w)x = (y-w)\dfrac{kz}{y-w}$$

$$xy - wx = kz$$

$$xy = kz + wx$$

$$\dfrac{xy}{x} = \dfrac{kz + wx}{x}$$

$$y = \dfrac{xw + kz}{x}$$

21. Since $T$ varies directly as $B$, we have $T = kB$.
Use the given values to find $k$.

$T = kB$

$3.6 = k(4)$

$\dfrac{3.6}{4} = \dfrac{k(4)}{4}$

$0.9 = k$

The equation becomes $T = 0.9B$.
When $B = 6$, $T = 0.9(6) = 5.4$.
The tail length is 5.4 feet.

23. Since $B$ varies directly as $D$, we have $B = kD$.
Use the given values to find $k$.

$B = kD$

$8.4 = k(12)$

$\dfrac{8.4}{12} = \dfrac{k(12)}{12}$

$k = \dfrac{8.4}{12} = 0.7$

The equation becomes $B = 0.7D$.
When $B = 56$,

$56 = 0.7D$

$\dfrac{56}{0.7} = \dfrac{0.7D}{0.7}$

$D = \dfrac{56}{0.7} = 80$

It was dropped from 80 inches.

25. Since a man's weight varies directly as the
cube of his height, we have $w = kh^3$.
Use the given values to find $k$.

$w = kh^3$

$170 = k(70)^3$

$170 = k(343,000)$

$\dfrac{170}{343,000} = \dfrac{k(343,000)}{343,000}$

$0.000496 = k$

The equation becomes $w = 0.000496h^3$.
When $h = 107$,

$w = 0.000496(107)^3$

$\quad = 0.000496(1,225,043) \approx 607$.

Robert Wadlow's weight was approximately
607 pounds.

27. Since the banking angle varies inversely as
the turning radius, we have $B = \dfrac{k}{r}$.
Use the given values to find $k$.

$B = \dfrac{k}{r}$

$28 = \dfrac{k}{4}$

$28(4) = 28\left(\dfrac{k}{4}\right)$

$112 = k$

The equation becomes $B = \dfrac{112}{r}$.

When $r = 3.5$, $B = \dfrac{112}{r} = \dfrac{112}{3.5} = 32$.

The banking angle is $32°$ when the turning
radius is 3.5 feet.

29. Since intensity varies inversely as the square
of the distance, we have pressure, we have
$I = \dfrac{k}{d}$.
Use the given values to find $k$.

$I = \dfrac{k}{d^2}$.

$62.5 = \dfrac{k}{3^2}$

$62.5 = \dfrac{k}{9}$

$9(62.5) = 9\left(\dfrac{k}{9}\right)$

$562.5 = k$

The equation becomes $I = \dfrac{562.5}{d^2}$.

When $d = 2.5$, $I = \dfrac{562.5}{2.5^2} = \dfrac{562.5}{6.25} = 90$

The intensity is 90 milliroentgens per hour.

**31.** Since index varies directly as weight and inversely as the square of one's height, we have $I = \dfrac{kw}{h^2}$.

Use the given values to find $k$.

$$I = \frac{kw}{h^2}$$

$$35.15 = \frac{k(180)}{60^2}$$

$$35.15 = \frac{k(180)}{3600}$$

$$(3600)35.15 = \frac{k(180)}{3600}$$

$$126540 = k(180)$$

$$k = \frac{126540}{180} = 703$$

The equation becomes $I = \dfrac{703w}{h^2}$.

When $w = 170$ and $h = 70$,

$$I = \frac{703(170)}{(70)^2} \approx 24.4.$$

This person has a BMI of 24.4 and is not overweight.

**33.** Since heat loss varies jointly as the area and temperature difference, we have $L = kAD$.
Use the given values to find $k$.

$$L = kAD$$

$$1200 = k(3 \cdot 6)(20)$$

$$1200 = 360k$$

$$\frac{1200}{360} = \frac{360k}{360}$$

$$k = \frac{10}{3}$$

The equation becomes $L = \dfrac{10}{3}AD$

When $A = 6 \cdot 9 = 54$, $D = 10$,

$$L = \frac{10}{3}(9 \cdot 6)(10) = 1800.$$

The heat loss is 1800 Btu.

**35.** Since intensity varies inversely as the square of the distance from the sound source, we have $I = \dfrac{k}{d^2}$. If you move to a seat twice as far, then $d = 2d$. So we have

$$I = \frac{k}{(2d)^2} = \frac{k}{4d^2} = \frac{1}{4} \cdot \frac{k}{d^2}.$$ The intensity

will be multiplied by a factor of $\dfrac{1}{4}$. So the

sound intensity is $\dfrac{1}{4}$ of what it was

originally.

**37. a.** Since the average number of phone calls varies jointly as the product of the populations and inversely as the square of the distance, we have

$$C = \frac{kP_1P_2}{d^2}.$$

**b.** Use the given values to find $k$.

$$C = \frac{kP_1P_2}{d^2}$$

$$326,000 = \frac{k(777,000)(3,695,000)}{(420)^2}$$

$$326,000 = \frac{k(2.87 \times 10^{12})}{176,400}$$

$$326,000 = 16269841.27k$$

$$0.02 \approx k$$

The equation becomes $C = \dfrac{0.02P_1P_2}{d^2}$.

**c.** $C = \dfrac{0.02(650,000)(490,000)}{(400)^2}$

$$\approx 39,813$$

There are approximately 39,813 daily phone calls.

**39. a.**

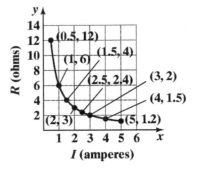

**b.** Current varies inversely as resistance. Answers will vary.

**c.** Since the current varies inversely as resistance we have $R = \dfrac{k}{I}$. Using one of the given ordered pairs to find $k$.

$$12 = \frac{k}{0.5}$$

$$12(0.5) = \frac{k}{0.5}(0.5)$$

$$k = 6$$

The equation becomes $R = \dfrac{6}{I}$.

**41. – 47.** Answers will vary.

**49.** does not make sense; Explanations will vary. Sample explanation: For an inverse variation, the independent variable can not be zero.

**51.** makes sense

**53.** Pressure, $P$, varies directly as the square of wind velocity, $v$, can be modeled as $P = kv^2$.
If $v = x$ then $P = k(x)^2 = kx^2$
If $v = 2x$ then $P = k(2x)^2 = 4kx^2$
If the wind speed doubles the pressure is 4 times more destructive.

**55.** The Heat, $H$, varies directly as the square of the voltage, $v$, and inversely as the resistance, $r$.

$$H = \frac{kv^2}{r}$$

If the voltage remains constant, to triple the heat the resistant must be reduced by a multiple of 3.

**57.** Answers will vary.

**58.**

$f(x) = 2^x$

**59.**

$g(x) = f(-x) = 2^{-x}$

**60.**

$h(x) = f(x) + 1 = 2^x + 1$

## Chapter 3 Review Exercises

**1.** $f(x) = -(x+1)^2 + 4$
vertex: $(-1, 4)$
$x$-intercepts:

$$0 = -(x+1)^2 + 4$$

$$(x+1)^2 = 4$$

$$x + 1 = \pm 2$$

$$x = -1 \pm 2$$

$$x = -3 \text{ or } x = 1$$

$y$-intercept:

$$f(0) = -(0+1)^2 + 4 = 3$$

The axis of symmetry is $x = -1$.

$f(x) = -(x + 1)^2 + 4$

domain: $(-\infty, \infty)$   range: $(-\infty, 4]$

**2.** $f(x) = (x+4)^2 - 2$
vertex: $(-4, -2)$
$x$-intercepts:

$$0 = (x+4)^2 - 2$$

$$(x+4)^2 = 2$$

$$x + 4 = \pm\sqrt{2}$$

$$x = -4 \pm \sqrt{2}$$

*y*-intercept:

$$f(0) = (0+4)^2 - 2 = 14 = -1$$

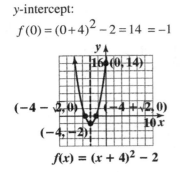

$$f(x) = (x + 4)^2 - 2$$

The axis of symmetry is $x = -4$.

domain: $(-\infty, \infty)$  range: $[-2, \infty)$

3.  $f(x) = -x^2 + 2x + 3$

$$= -(x^2 - 2x + 1) + 3 + 1$$

$$f(x) = -(x-1)^2 + 4$$

$$f(x) = -x^2 + 2x + 3$$

domain: $(-\infty, \infty)$  range: $(-\infty, 4]$

4.  $f(x) = 2x^2 - 4x - 6$

$$f(x) = 2(x^2 - 2x + 1) - 6 - 2$$

$$2(x-1)^2 - 8$$

$$f(x) = 2x^2 - 4x - 6$$

axis of symmetry: $x = 1$

domain: $(-\infty, \infty)$  range: $[-8, \infty)$

5.  $f(x) = -x^2 + 14x - 106$

a.  Since $a < 0$ the parabola opens down with the maximum value occurring at

$$x = -\frac{b}{2a} = -\frac{14}{2(-1)} = 7.$$

The maximum value is $f(7)$.

$$f(7) = -(7)^2 + 14(7) - 106 = -57$$

b.  domain: $(-\infty, \infty)$  range: $(-\infty, -57]$

6.  $f(x) = 2x^2 + 12x + 703$

a.  Since $a > 0$ the parabola opens up with the minimum value occurring at

$$x = -\frac{b}{2a} = -\frac{12}{2(2)} = -3.$$

The minimum value is $f(-3)$.

$$f(-3) = 2(-3)^2 + 12(-3) + 703 = 685$$

b.  domain: $(-\infty, \infty)$  range: $[685, \infty)$

7.  a.  The maximum height will occur at the vertex.

$$f(x) = -0.025x^2 + x + 6$$

$$x = -\frac{b}{2a} = -\frac{1}{2(-0.025)} = 20$$

$$f(20) = -0.025(20)^2 + (20) + 6 = 16$$

The maximum height of 16 feet occurs when the ball is 20 yards downfield.

b.  $f(x) = -0.025x^2 + x + 6$

$$f(0) = -0.025(0)^2 + (0) + 6 = 6$$

The ball was tossed at a height of 6 feet.

c.  The ball is at a height of 0 when it hits the ground.

$$f(x) = -0.025x^2 + x + 6$$

$$0 = -0.025x^2 + x + 6$$

$$x = \frac{-b \pm \sqrt{b^2 - 4ac}}{2a}$$

$$x = \frac{-(1) \pm \sqrt{(1)^2 - 4(-0.025)(6)}}{2(-0.025)}$$

$$x \approx 45.3, \ -5.3 (\text{reject})$$

The ball will hit the ground 45.3 yards downfield.

**d.** The football's path:

$$f(x) = -0.025x^2 + x + 6$$

Ball's Height (feet)

(20, 16)

(0, 6)

(45.3, 0)

Ball's Horizontal Distance (yards)

**8.** Maximize the area using $A = lw$.

$$A(x) = x(1000 - 2x)$$

$$A(x) = -2x^2 + 1000x$$

Since $a = -2$ is negative, we know the function opens downward and has a maximum at

$$x = -\frac{b}{2a} = -\frac{1000}{2(-2)} = -\frac{1000}{-4} = 250.$$

The maximum area is achieved when the width is 250 yards. The maximum area is

$$A(250) = 250(1000 - 2(250))$$

$$= 250(1000 - 500)$$

$$= 250(500) = 125,000.$$

The area is maximized at 125,000 square yards when the width is 250 yards and the length is $1000 - 2 \cdot 250 = 500$ yards.

**9.** Let $x =$ one of the numbers
Let $14 + x =$ the other number

We need to minimize the function

$$P(x) = x(14 + x)$$

$$= 14x + x^2$$

$$= x^2 + 14x.$$

The minimum is at

$$x = -\frac{b}{2a} = -\frac{14}{2(1)} = -\frac{14}{2} = -7.$$

The other number is $14 + x = 14 + (-7) = 7$.

The numbers which minimize the product are 7 and $-7$. The minimum product is $-7 \cdot 7 = -49$.

**10.** $f(x) = -x^3 + 12x^2 - x$

The graph rises to the left and falls to the right and goes through the origin, so graph (c) is the best match.

**11.** $g(x) = x^6 - 6x^4 + 9x^2$

The graph rises to the left and rises to the right, so graph (b) is the best match.

**12.** $h(x) = x^5 - 5x^3 + 4x$

The graph falls to the left and rises to the right and crosses the $y$-axis at zero, so graph (a) is the best match.

**13.** $f(x) = -x^4 + 1$

$f(x)$ falls to the left and to the right so graph (d) is the best match.

**14. a.** Since $n$ is odd and $a_n > 0$, the graph rises to the right.

**b.** No, the model will not be useful. The model indicates increasing deforestation despite a declining rate in which the forest is being cut down.

**c.** The graph of function $g$ falls to the right.

**d.** No, the model will not be useful. The model indicates the amount of forest cleared, in square kilometers, will eventually be negative, which

**15.** In the polynomial, $f(x) = -x^4 + 21x^2 + 100$, the leading coefficient is $-1$ and the degree is 4. Applying the Leading Coefficient Test, we know that even-degree polynomials with negative leading coefficient will fall to the left and to the right. Since the graph falls to the right, we know that the elk population will die out over time.

**16.** $f(x) = -2(x-1)(x+2)^2(x+5)^3$

$x = 1$, multiplicity 1, the graph crosses the $x$-axis
$x = -2$, multiplicity 2, the graph touches the $x$-axis
$x = -5$, multiplicity 5, the graph crosses the $x$-axis

**17.** $f(x) = x^3 - 5x^2 - 25x + 125$

$$= x^2(x-5) - 25(x-5)$$

$$= (x^2 - 25)(x-5)$$

$$= (x+5)(x-5)^2$$

$x = -5$, multiplicity 1, the graph crosses the $x$-axis
$x = 5$, multiplicity 2, the graph touches the $x$-axis

**18.** $f(x) = x^3 - 2x - 1$

$f(1) = (1)^3 - 2(1) - 1 = -2$

$f(2) = (2)^3 - 2(2) - 1 = 3$

The sign change shows there is a zero between the given values.

**19.** $f(x) = x^3 - x^2 - 9x + 9$

**a.** Since $n$ is odd and $a_n > 0$, the graph falls to the left and rises to the right.

**b.** $f(-x) = (-x)^3 - (-x)^2 - 9(-x) + 9$

$\quad = -x^3 - x^2 + 9x + 9$

$f(-x) \neq f(x), f(-x) \neq -f(x)$

no symmetry

**c.** $f(x) = (x-3)(x+3)(x-1)$

zeros: 3, −3, 1

$f(x) = x^3 - x^2 - 9x + 9$

**20.** $f(x) = 4x - x^3$

**a.** Since $n$ is odd and $a_n < 0$, the graph rises to the left and falls to the right.

**b.** $f(-x) = -4x + x^3$

$f(-x) = -f(x)$

origin symmetry

**c.** $f(x) = x(x^2 - 4) = x(x-2)(x+2)$

zeros: $x = 0, 2, -2$

$f(x) = 4x - x^3$

**21.** $f(x) = 2x^3 + 3x^2 - 8x - 12$

**a.** Since $h$ is odd and $a_n > 0$, the graph falls to the left and rises to the right.

**b.** $f(-x) = -2x^3 + 3x^2 + 8x - 12$

$f(-x) \neq f(x), f(-x) = -f(x)$

no symmetry

**c.** $f(x) = (x-2)(x+2)(2x+3)$

zeros: $x = 2, -2, -\dfrac{3}{2}$

$f(x) = 2x^3 + 3x^2 - 8x - 12$

**22.** $g(x) = -x^4 + 25x^2$

**a.** The graph falls to the left and to the right.

**b.** $f(-x) = -(-x)^4 + 25(-x)^2$

$\quad = -x^4 + 25x^2 = f(x)$

$y$-axis symmetry

**c.** $-x^4 + 25x^2 = 0$

$-x^2(x^2 - 25) = 0$

$-x^2(x-5)(x+5) = 0$

zeros: $x = -5, 0, 5$

$f(x) = -x^4 + 25x^2$

**23.** $f(x) = -x^4 + 6x^3 - 9x^2$

**a.** The graph falls to the left and to the right.

**b.** $f(-x) = -(-x)^4 + 6(-x)^3 - 9(-x)$

$\quad = -x^4 - 6x^3 - 9x^2 f(-x) \neq f(x)$

$f(-x) \neq -f(x)$

no symmetry

**c.**   $=-x^2\left(x^2-6x+9\right)=0$

$-x^2(x-3)(x-3)=0$

zeros: $x=0,\,3$

$f(x)=-x^4+6x^3-9x^2$

**24.**   $f(x)=3x^4-15x^3$

**a.**   The graph rises to the left and to the right.

**b.**   $f(-x)=3(-x)^4-15(-x)^2=3x^4+15x^3$

$f(-x)\neq f(x),\,f(-x)\neq -f(x)$

no symmetry

**c**   $3x^4-15x^3=0$

$3x^3(x-5)=0$

zeros: $x=0,\,5$

$f(x)=3x^4-15x^3$

**25.**   $f(x)=2x^2(x-1)^3(x+2)$

Since $a_n>0$ and $n$ is even, $f(x)$ rises to the left and the right.

$x=0,\,x=1,\,x=-2$

The zeros at 1 and –2 have odd multiplicity so $f(x)$ crosses the $x$-axis at those points. The root at 0 has even multiplicity so $f(x)$ touches the axis at $(0,\,0)$

$f(0)=2(0)^2(0-1)^3(0+2)=0$

The $y$-intercept is 0.

$f(x)=2x^2(x-1)^3(x+2)$

**26.**   $f(x)=-x^3(x+4)^2(x-1)$

Since $a_n<0$ and $n$ is even, $f(x)$ falls to the left and the right.

$x=0,\,x=-4,\,x=1$

The roots at 0 and 1 have odd multiplicity so $f(x)$ crosses the $x$-axis at those points. The root at –4 has even multiplicity so $f(x)$ touches the axis at $(-4,\,0)$

$f(0)=-(0)^3(0+4)^2(0-1)=0$

The $y$-intercept is 0.

$f(x)=-x^3(x+4)^2(x-1)$

**27.**

$$
\begin{array}{r}
4x^2-7x+5 \\
x+1\overline{\smash{\big)}\,4x^3-3x^2-2x+1} \\
\underline{4x^3+4x^2}\phantom{} \\
-7x^2-2x \\
\underline{-7x^2-7x} \\
5x+1 \\
\underline{5x+5} \\
-4
\end{array}
$$

Quotient: $4x^2-7x+5-\dfrac{4}{x+1}$

**28.**

$$
\begin{array}{r}
2x^2-4x+1 \\
5x-3\overline{\smash{\big)}\,10x^3-26x^2+17x-13} \\
\underline{10x^3+6x^2}\phantom{} \\
-20x^2+17x \\
\underline{-20x^2+12x} \\
5x-13 \\
\underline{5x-\phantom{0}3} \\
-10
\end{array}
$$

Quotient: $2x^2-4x+1-\dfrac{10}{5x-3}$

**29.**

$$
\begin{array}{r}
2x^2+3x-1 \\
2x^2+1\overline{)4x^4+6x^3+3x-1} \\
\underline{4x^2+2x^2} \\
6x^3-2x^2+3x \\
\underline{6x^2+3x} \\
-2x^2-1 \\
\underline{-2x^2-1} \\
0
\end{array}
$$

**30.** $(3x^4+11x^3-20x^3+7x+35)\div(x+5)$

$$
\begin{array}{r|rrrrr}
-5 & 3 & 11 & -20 & 7 & 35 \\
   &   & -15 & 20 & 0 & -35 \\
\hline
   & 3 & -4 & 0 & 7 & 0
\end{array}
$$

Quotient: $3x^3-4x^2+7$

**31.** $(3x^4-2x^2-10x)\div(x-2)$

$$
\begin{array}{r|rrrrr}
2 & 3 & 0 & -2 & -10 & 0 \\
  &   & 6 & 12 & 20 & 20 \\
\hline
  & 3 & 6 & 10 & 10 & 20
\end{array}
$$

Quotient: $3x^3+6x^2+10x+10+\dfrac{20}{x-2}$

**32.** $f(x)=2x^3-7x^2+9x-3$

$$
\begin{array}{r|rrrr}
-13 & 2 & -7 & 9 & -3 \\
    &   & -26 & 429 & -5694 \\
\hline
    & 2 & -33 & 438 & -5697
\end{array}
$$

Quotient: $f(-13)=-5697$

**33.** $f(x)=2x^3+x^2-13x+6$

$$
\begin{array}{r|rrrr}
2 & 2 & 1 & -13 & 6 \\
  &   & 4 & 10 & -6 \\
\hline
  & 2 & 5 & -3 & 0
\end{array}
$$

$f(x)=(x-2)(2x^2+5x-3)$

$\phantom{f(x)}=(x-2)(2x-1)(x+3)$

Zeros: $x=2,\dfrac{1}{2},-3$

**34.** $x^3-17x+4=0$

$$
\begin{array}{r|rrrr}
4 & 1 & 0 & -17 & 4 \\
  &   & 4 & 16 & -4 \\
\hline
  & 1 & 4 & -1 & 0
\end{array}
$$

$(x-4)(x^2+4x-1)=0$

$x=\dfrac{-4\pm\sqrt{16+4}}{2}=\dfrac{-4\pm2\sqrt{5}}{2}=-2\pm\sqrt{5}$

The solution set is $\left\{4,-2+\sqrt{5},-2-\sqrt{5}\right\}$.

**35.** $f(x)=x^4-6x^3+14x^2-14x+5$

$p:\pm1,\pm5$

$q:\pm1$

$\dfrac{p}{q}:\pm1,\pm5$

**36.** $f(x)=3x^5-2x^4-15x^3+10x^2+12x-8$

$p:\pm1,\pm2,\pm4,\pm8$

$q:\pm1,\pm3$

$\dfrac{p}{q}:\pm1,\pm2,\pm4,\pm8,\pm\dfrac{8}{3},\pm\dfrac{4}{3},\pm\dfrac{2}{3},\pm\dfrac{1}{3}$

**37.** $f(x)=3x^4-2x^3-8x+5$

$f(x)$ has 2 sign variations, so $f(x)=0$ has 2 or 0 positive solutions.

$f(-x)=3x^4+2x^3+x+5$

$f(-x)$ has no sign variations, so $f(x)=0$ has no negative solutions.

**38.** $f(x)=2x^5-3x^3-5x^2+3x-1$

$f(x)$ has 3 sign variations, so $f(x)=0$ has 3 or 1 positive real roots.

$f(-x)=-2x^5+3x^3-5x^2-3x-1$

$f(-x)$ has 2 sign variations, so $f(x)=0$ has 2 or 0 negative solutions.

**39.** $f(x)=f(-x)=2x^4+6x^2+8$

No sign variations exist for either $f(x)$ or $f(-x)$, so no real roots exist.

**40.** $f(x) = x^3 + 3x^2 - 4$

  **a.**  $p: \pm 1, \pm 2, \pm 4$

      $q: \pm 1$

      $\dfrac{p}{q}: \pm 1, \pm 2, \pm 4$

  **b.**  1 sign variation $\Rightarrow$ 1 positive real zero

      $f(-x) = -x^3 + 3x^2 - 4$

      2 sign variations $\Rightarrow$ 2 or no negative real zeros

  **c.**

$$\begin{array}{c|cccc}
1 & 1 & 3 & 0 & -4 \\
  &   & 1 & 4 & -4 \\
\hline
  & 1 & 4 & 4 & 0
\end{array}$$

      1 is a zero.

      1, –2 are rational zeros.

  **d.**  $(x-1)(x^2 + 4x + 4) = 0$

      $(x-1)(x+2)^2 = 0$

      $x = 1$ or $x = -2$

      The solution set is $\{1, -2\}$.

**41.** $f(x) = 6x^3 + x^2 - 4x + 1$

  **a.**  $p: \pm 1$

      $q: \pm 1, \pm 2, \pm 3, \pm 6$

      $\dfrac{p}{q}: \pm 1, \pm \dfrac{1}{2}, \pm \dfrac{1}{3}, \pm \dfrac{1}{6}$

  **b.**  $f(x) = 6x^3 + x^2 - 4x + 1$

      2 sign variations; 2 or 0 positive real zeros.

      $f(-x) = -6x^3 + x^2 + 4x + 1$

      1 sign variation; 1 negative real zero.

  **c.**

$$\begin{array}{c|cccc}
-1 & 6 & 1 & -4 & 1 \\
   &   & -6 & 5 & -1 \\
\hline
   & 6 & -5 & 1 & 0
\end{array}$$

      –1 is a zero.

      $-1, \dfrac{1}{3}, \dfrac{1}{2}$ are rational zeros.

  **d.**  $6x^3 + x^2 - 4x + 1 = 0$

      $(x+1)(6x^2 - 5x + 1) = 0$

      $(x+1)(3x-1)(2x-1) = 0$

      $x = -1$ or $x = \dfrac{1}{3}$ or $x = \dfrac{1}{2}$

      The solution set is $\left\{-1, \dfrac{1}{3}, \dfrac{1}{2}\right\}$.

**42.** $f(x) = 8x^3 - 36x^2 + 46x - 15$

  **a.**  $p: \pm 1, \pm 3, \pm 5, \pm 15$

      $q: \pm 1, \pm 2, \pm 4, \pm 8$

      $\dfrac{p}{q}: \pm 1, \pm 3, \pm 5, \pm 15, \pm \dfrac{1}{2}, \pm \dfrac{1}{4}, \pm \dfrac{1}{8},$

          $\pm \dfrac{3}{2}, \pm \dfrac{3}{4}, \pm \dfrac{3}{8}, \pm \dfrac{5}{2}, \pm \dfrac{5}{4},$

          $\pm \dfrac{5}{8}, \pm \dfrac{15}{2}, \pm \dfrac{15}{4}, \pm \dfrac{15}{8}$

  **b.**  $f(x) = 8x^3 - 36x^2 + 46x - 15$

      3 sign variations; 3 or 1 positive real solutions.

      $f(-x) = -8x^3 - 36x^2 - 46x - 15$

      0 sign variations; no negative real solutions.

  **c.**

$$\begin{array}{c|cccc}
\frac{1}{2} & 8 & -36 & 46 & -15 \\
            &   & 4 & -16 & 15 \\
\hline
            & 8 & -32 & 30 & 0
\end{array}$$

      $\dfrac{1}{2}$ is a zero.

      $\dfrac{1}{2}, \dfrac{3}{2}, \dfrac{5}{2}$ are rational zeros.

  **d.**

$$8x^3 - 36x^2 + 46x - 15 = 0$$

$$\left(x - \dfrac{1}{2}\right)(8x^2 - 32x + 30) = 0$$

$$2\left(x - \dfrac{1}{2}\right)(4x^2 - 16x + 15) = 0$$

$$2\left(x - \dfrac{1}{2}\right)(2x - 5)(2x - 3) = 0$$

$$x = \dfrac{1}{2} \text{ or } x = \dfrac{5}{2} \text{ or } x = \dfrac{3}{2}$$

      The solution set is $\left\{\dfrac{1}{2}, \dfrac{3}{2}, \dfrac{5}{2}\right\}$.

**43.**  $2x^3 + 9x^2 - 7x + 1 = 0$

   **a.**  $p$: $\pm 1$

       $q$: $\pm 1, \pm 2$

       $\dfrac{p}{q}$: $\pm 1, \pm \dfrac{1}{2}$

   **b.**  $f(x) = 2x^3 + 9x^2 - 7x + 1$

       2 sign variations;  2 or 0 positive real zeros.

       $f(-x) = -2x^3 + 9x^2 + 7x + 1$

       1 sign variation;  1 negative real zero.

   **c.**

$$\begin{array}{r|rrrr}
\frac{1}{2} & 2 & 9 & -7 & 1 \\
  &   & 1 & 5 & -1 \\
\hline
  & 2 & 10 & -2 & 0
\end{array}$$

       $\dfrac{1}{2}$ is a rational zero.

   **d.**  $2x^3 + 9x^2 - 7x + 1 = 0$

       $\left(x - \dfrac{1}{2}\right)(2x^2 + 10x - 2) = 0$

       $2\left(x - \dfrac{1}{2}\right)(x^2 + 5x - 1) = 0$

       Solving $x^2 + 5x - 1 = 0$ using the quadratic

       formula gives $x = \dfrac{-5 \pm \sqrt{29}}{2}$

       The solution set is $\left\{ \dfrac{1}{2}, \dfrac{-5 + \sqrt{29}}{2}, \dfrac{-5 - \sqrt{29}}{2} \right\}$.

**44.**  $x^4 - x^3 - 7x^2 + x + 6 = 0$

   **a.**  $p = \dfrac{p}{q}$: $\pm 1, \pm 2, \pm 3, \pm 6$

   **b.**  $f(x) = x^4 - x^3 - 7x^2 + x + 6$

       2 sign variations;  2 or 0 positive real zeros.

       $f(-x) = x^4 + x^3 - 7x^2 - x + 6$

       2 sign variations;  2 or 0 negative real zeros.

   **c.**

$$\begin{array}{r|rrrrr}
1 & 1 & -1 & -7 & 1 & 6 \\
  &   & 1 & 0 & -7 & -6 \\
\hline
  & 1 & 0 & -7 & -6 & 0
\end{array}$$

$$\begin{array}{r|rrrrr}
-1 & 1 & 0 & -7 & -6 \\
   &   & -1 & 1 & 6 \\
\hline
   & 1 & -1 & -6 & 0
\end{array}$$

$-2, -1, 1, 3$ are rational zeros.

   **d.**  $x^4 - x^3 - 7x^2 + x + 6 = 0$

       $(x-1)(x+1)(x^2 - x + 6) = 0$

       $(x-1)(x+1)(x-3)(x+2) = 0$

       The solution set is $\{-2, -1, 1, 3\}$.

**45.**  $4x^4 + 7x^2 - 2 = 0$

   **a.**  $p$: $\pm 1, \pm 2$

       $q$: $\pm 1, \pm 2, \pm 4$

       $\dfrac{p}{q}$: $\pm 1, \pm 2, \pm \dfrac{1}{2}, \pm \dfrac{1}{4}$

   **b.**  $f(x) = 4x^4 + 7x^2 - 2$

       1 sign variation;  1 positive real zero.

       $f(-x) = 4x^4 + 7x^2 - 2$

       1 sign variation;  1 negative real zero.

   **c.**

$$\begin{array}{r|rrrrr}
\frac{1}{2} & 4 & 0 & 7 & 0 & -2 \\
  &   & 2 & 1 & 4 & 2 \\
\hline
  & 4 & 2 & 8 & 4 & 0
\end{array}$$

$$\begin{array}{r|rrrrr}
-\frac{1}{2} & 4 & 2 & 8 & 4 \\
  &   & -2 & 0 & -4 \\
\hline
  & 4 & 0 & 8 & 0
\end{array}$$

$-\dfrac{1}{2}, \dfrac{1}{2}$ are rational zeros.

   **d.**  $4x^4 + 7x^2 - 2 = 0$

       $\left(x - \dfrac{1}{2}\right)\left(x + \dfrac{1}{2}\right)(4x^2 + 8) = 0$

       $4\left(x - \dfrac{1}{2}\right)\left(x + \dfrac{1}{2}\right)(x^2 + 2) = 0$

       Solving $x^2 + 2 = 0$ using the quadratic formula

       gives $x = \pm 2i$

       The solution set is $\left\{ -\dfrac{1}{2}, \dfrac{1}{2}, 2i, -2i \right\}$.

**46.** $f(x) = 2x^4 + x^3 - 9x^2 - 4x + 4$

a. $p: \pm 1, \pm 2, \pm 4$
$q: \pm 1, \pm 2$

$\dfrac{p}{q}: \pm 1, \pm 2, \pm 4, \pm \dfrac{1}{2}$

b. $f(x) = 2x^4 + x^3 - 9x^2 - 4x + 4$
2 sign variations; 2 or 0 positive real zeros.
$f(-x) = 2x^4 - x^3 - 9x^2 + 4x + 4$
2 sign variations; 2 or 0 negative real zeros.

c.

$$\begin{array}{r|rrrrr} 2 & 2 & 1 & -9 & -4 & 4 \\ & & 4 & 10 & 2 & -4 \\ \hline & 2 & 5 & 1 & -2 & 0 \end{array}$$

$$\begin{array}{r|rrrr} -1 & 2 & 5 & 1 & -2 \\ & & -2 & -3 & 2 \\ \hline & 2 & 3 & -2 & 0 \end{array}$$

$-2, -1, \dfrac{1}{2}, 2$ are rational zeros.

d. $2x^2 + 3x - 2 = 0$
$(2x - 1)(x + 2) = 0$

$x = -2$ or $x = \dfrac{1}{2}$

The solution set is $\left\{ -2, -1, \dfrac{1}{2}, 2 \right\}$.

**47.** $f(x) = a_n(x-2)(x-2+3i)(x-2-3i)$
$f(x) = a_n(x-2)\left(x^2 - 4x + 13\right)$
$f(1) = a_n(1-2)\left[1^2 - 4(1) + 13\right]$
$-10 = -10a_n$
$a_n = 1$
$f(x) = 1(x-2)\left(x^2 - 4x + 13\right)$
$f(x) = x^3 - 4x^2 + 13x - 2x^2 + 8x - 26$
$f(x) = x^3 - 6x^2 + 21x - 26$

**48.** $f(x) = a_n(x-i)(x+i)(x+3)^2$
$f(x) = a_n\left(x^2 + 1\right)\left(x^2 + 6x + 9\right)$
$f(-1) = a_n\left[(-1)^2 + 1\right]\left[(-1)^2 + 6(-1) + 9\right]$
$16 = 8a_n$
$a_n = 2$
$f(x) = 2\left(x^2 + 1\right)\left(x^2 + 6x + 9\right)$
$f(x) = 2\left(x^4 + 6x^3 + 9x^2 + x^2 + 6x + 9\right)$
$f(x) = 2x^4 + 12x^3 + 20x^2 + 12x + 18$

**49.** $f(x) = 2x^4 + 3x^3 + 3x - 2$
$p: \pm 1, \pm 2$
$q: \pm 1, \pm 2$

$\dfrac{p}{q}: \pm 1, \pm 2, \pm \dfrac{1}{2}$

$$\begin{array}{r|rrrrr} -2 & 2 & 3 & 0 & 3 & -2 \\ & & -4 & 2 & -4 & 2 \\ \hline & 2 & -1 & 2 & -1 & 0 \end{array}$$

$2x^4 + 3x^3 + 3x - 2 = 0$
$(x+2)(2x^3 - x^2 + 2x - 1) = 0$
$(x+2)[x^2(2x-1) + (2x-1)] = 0$
$(x+2)(2x-1)(x^2 + 1) = 0$

$x = -2,\ x = \dfrac{1}{2}$ or $x = \pm i$

The zeros are $-2,\ \dfrac{1}{2},\ \pm i$.
$f(x) = (x-i)(x+i)(x+2)(2x-1)$

**50.** $g(x) = x^4 - 6x^3 + x^2 + 24x + 16$
$p: \pm 1, \pm 2, \pm 4, \pm 8, \pm 16$
$q: \pm 1$

$\dfrac{p}{q}: \pm 1, \pm 2, \pm 4, \pm 8, \pm 16$

$$\begin{array}{r|rrrrr} -1 & 1 & -6 & 1 & 24 & 16 \\ & & -1 & 7 & -8 & -16 \\ \hline & 1 & -7 & 8 & 16 & 0 \end{array}$$

$x^4 - 6x^3 + x^2 + 24x + 16 = 0$
$(x+1)(x^3 - 7x^2 + 8x + 16) = 0$

$$-1 \,\big|\; 1 \quad -7 \quad 8 \quad 16$$
$$\phantom{-1 \,\big|\; 1 \quad} -1 \quad 8 \quad -16$$
$$\overline{\phantom{-1 \,\big|\;} 1 \quad -8 \quad 16 \quad 0}$$

$$(x+1)^2(x^2 - 8x + 16) = 0$$
$$(x+1)^2(x-4)^2 = 0$$
$$x = -1 \text{ or } x = 4$$
$$g(x) = (x+1)^2(x-4)^2$$

**51.** 4 real zeros, one with multiplicity two

**52.** 3 real zeros; 2 nonreal complex zeros

**53.** 2 real zeros, one with multiplicity two; 2 nonreal complex zeros

**54.** 1 real zero; 4 nonreal complex zeros

**55.** $g(x) = \dfrac{1}{(x+2)^2} - 1$

**56.** $h(x) = \dfrac{1}{x-1} + 3$

**57.** $f(x) = \dfrac{2x}{x^2 - 9}$

Symmetry: $f(-x) = -\dfrac{2x}{x^2 - 9} = -f(x)$

origin symmetry
$x$-intercept:
$$0 = \frac{2x}{x^2 - 9}$$
$$2x = 0$$
$$x = 0$$

$y$-intercept: $y = \dfrac{2(0)}{0^2 - 9} = 0$

Vertical asymptote:
$$x^2 - 9 = 0$$
$$(x-3)(x+3) = 0$$
$$x = 3 \text{ and } x = -3$$

Horizontal asymptote:
$n < m$, so $y = 0$

**58.** $g(x) = \dfrac{2x - 4}{x + 3}$

Symmetry: $g(-x) = \dfrac{-2x - 4}{x + 3}$

$g(-x) \neq g(x)$, $g(-x) \neq -g(x)$
No symmetry
$x$-intercept:
$$2x - 4 = 0$$
$$x = 2$$

$y$-intercept: $y = \dfrac{2(0) - 4}{(0) + 3} = -\dfrac{4}{3}$

Vertical asymptote:
$x + 3 = 0$
$x = -3$

Horizontal asymptote:
$n = m$, so $y = \dfrac{2}{1} = 2$

$$f(x) = \frac{2x - 4}{x + 3}$$

**59.** $h(x) = \dfrac{x^2 - 3x - 4}{x^2 - x - 6}$

Symmetry: $h(-x) = \dfrac{x^2 + 3x - 4}{x^2 + x - 6}$

$h(-x) \neq h(x),\ h(-x) \neq -h(x)$
No symmetry
$x$-intercepts:
$x^2 - 3x - 4 = 0$
$(x - 4)(x + 1)$
$x = 4 \quad x = -1$

$y$-intercept: $y = \dfrac{0^2 - 3(0) - 4}{0^2 - 0 - 6} = \dfrac{2}{3}$

Vertical asymptotes:
$x^2 - x - 6 = 0$
$(x - 3)(x + 2) = 0$
$x = 3, -2$

Horizontal asymptote:

$n = m$, so $y = \dfrac{1}{1} = 1$

$$h(x) = \frac{x^2 - 3x - 4}{x^2 - x - 6}$$

**60.** $r(x) = \dfrac{x^2 + 4x + 3}{(x + 2)^2}$

Symmetry: $r(-x) = \dfrac{x^2 - 4x + 3}{(-x + 2)^2}$

$r(-x) \neq r(x),\ r(-x) \neq -r(x)$
No symmetry
$x$-intercepts:
$x^2 + 4x + 3 = 0$
$(x + 3)(x + 1) = 0$
$x = -3, -1$

$y$-intercept: $y = \dfrac{0^2 + 4(0) + 3}{(0 + 2)^2} = \dfrac{3}{4}$

Vertical asymptote:
$x + 2 = 0$
$x = -2$

Horizontal asymptote:

$n = m$, so $y = \dfrac{1}{1} = 1$

$$r(x) = \frac{x^2 + 4x + 3}{(x + 2)^2}$$

**61.** $y = \dfrac{x^2}{x + 1}$

Symmetry: $f(-x) = \dfrac{x^2}{-x + 1}$

$f(-x) \neq f(x),\ f(-x) \neq -f(x)$
No symmetry
$x$-intercept:
$x^2 = 0$
$x = 0$

$y$-intercept: $y = \dfrac{0^2}{0 + 1} = 0$

Vertical asymptote:
$x + 1 = 0$
$x = -1$
$n > m$, no horizontal asymptote.
Slant asymptote:
$$y = x - 1 + \frac{1}{x + 1}$$
$y = x - 1$

$$y = \frac{x^2}{x + 1}$$

**62.** $y = \dfrac{x^2 + 2x - 3}{x - 3}$

Symmetry: $f(-x) = \dfrac{x^2 - 2x - 3}{-x - 3}$

$f(-x) \neq f(x),\ f(-x) \neq -f(x)$
No symmetry
$x$-intercepts:
$x^2 + 2x - 3 = 0$
$(x + 3)(x - 1) = 0$
$x = -3, 1$

y-intercept: $y = \dfrac{0^2 + 2(0) - 3}{0 - 3} = \dfrac{-3}{-3} = 1$

Vertical asymptote:
$x - 3 = 0$
$x = 3$
Horizontal asymptote:
$n > m$, so no horizontal asymptote.

Slant asymptote:

$y = x + 5 + \dfrac{12}{x - 3}$

$y = x + 5$

$$y = \dfrac{x^2 + 2x - 3}{x - 3}$$

**63.** $f(x) = \dfrac{-2x^3}{x^2 + 1}$

Symmetry: $f(-x) = \dfrac{2}{x^2 + 1} = -f(x)$

Origin symmetry
x-intercept:
$-2x^3 = 0$
$x = 0$

y-intercept: $y = \dfrac{-2(0)^3}{0^2 + 1} = \dfrac{0}{1} = 0$

Vertical asymptote:
$x^2 + 1 = 0$
$x^2 = -1$
No vertical asymptote.
Horizontal asymptote:
$n > m$, so no horizontal asymptote.
Slant asymptote:

$f(x) = -2x + \dfrac{2x}{x^2 + 1}$

$y = -2x$

$$f(x) = \dfrac{-2x^3}{x^2 + 1}$$

**64.** $g(x) = \dfrac{4x^2 - 16x + 16}{2x - 3}$

Symmetry: $g(-x) = \dfrac{4x^2 + 16x + 16}{-2x - 3}$

$g(-x) \neq g(x)$, $g(-x) \neq -g(x)$
No symmetry
x-intercept:
$4x^2 - 16x + 16 = 0$
$4(x - 2)^2 = 0$
$x = 2$
y-intercept:

$y = \dfrac{4(0)^2 - 16(0) + 16}{2(0) - 3} = -\dfrac{16}{3}$

Vertical asymptote:
$2x - 3 = 0$

$x = \dfrac{3}{2}$

Horizontal asymptote:
$n > m$, so no horizontal asymptote.
Slant asymptote:

$g(x) = 2x - 5 + \dfrac{1}{2x - 3}$

$y = 2x - 5$

$$g(x) = \dfrac{4x^2 - 16x + 16}{2x - 3}$$

**65. a.** $C(x) = 50{,}000 + 25x$

**b.** $\overline{C}(x) = \dfrac{25x + 50{,}000}{x}$

**c.** $\overline{C}(50) = \dfrac{25(50) + 50{,}000}{50} = 1025$

When 50 calculators are manufactured, it costs $1025 to manufacture each.

$\overline{C}(100) = \dfrac{25(100) + 50{,}000}{100} = 525$

When 100 calculators are manufactured, it costs $525 to manufacture each.

$\overline{C}(1000) = \dfrac{25(1000) + 50{,}000}{1000} = 75$

When 1,000 calculators are manufactured, it costs $75 to manufacture each.

$$\overline{C}(100,000) = \frac{25(100,000) + 50,000}{100,000} = 25.5 \text{ Wh}$$

en 100,000 calculators are manufactured, it costs $25.50 to manufacture each.

**d.** $n = m$, so $y = \dfrac{25}{1} = 25$ is the horizontal asymptote. Minimum costs will approach $25.

**66.** $f(x) = \dfrac{150x + 120}{0.05x + 1}$

$n = m$, so $y = \dfrac{150}{0.05} = 3000$

The number of fish available in the pond approaches 3000.

**67.** $P(x) = \dfrac{72,900}{100x^2 + 729}$

$n < m$ so $y = 0$

As the number of years of education increases the percentage rate of unemployment approaches zero.

**68. a.** 
$$\begin{aligned} P(x) &= M(x) + F(x) \\ &= 1.53x + 114.8 + 1.46x + 120.7 \\ &= 2.99x + 235.5 \end{aligned}$$

**b.** $R(x) = \dfrac{M(x)}{P(x)} = \dfrac{1.53x + 114.8}{2.99x + 235.5}$

**c.** $y = \dfrac{1.53}{2.99} \approx 0.51$

Over time, the percentage of men in the U.S. population will approach 51%.

**69.** $2x^2 + 5x - 3 < 0$

Solve the related quadratic equation.

$$2x^2 + 5x - 3 = 0$$
$$(2x - 1)(x + 3) = 0$$

The boundary points are $-3$ and $\dfrac{1}{2}$.

Testing each interval gives a solution set of $\left( -3, \dfrac{1}{2} \right)$

**70.** $2x^2 + 9x + 4 \geq 0$

Solve the related quadratic equation.

$$2x^2 + 9x + 4 = 0$$
$$(2x + 1)(x + 4) = 0$$

The boundary points are $-4$ and $-\dfrac{1}{2}$.

Testing each interval gives a solution set of

$$\left( -\infty, -4 \right] \cup \left[ -\dfrac{1}{2}, \infty \right)$$

**71.** $x^3 + 2x^2 > 3x$

Solve the related equation.

$$x^3 + 2x^2 = 3x$$
$$x^3 + 2x^2 - 3x = 0$$
$$x\left( x^2 + 2x - 3 \right) = 0$$
$$x(x + 3)(x - 1) = 0$$

The boundary points are $-3$, 0, and 1.

Testing each interval gives a solution set of

$$(-3, 0) \cup (1, \infty)$$

**72.** $\dfrac{x - 6}{x + 2} > 0$

Find the values of $x$ that make the numerator and denominator zero.

The boundary points are $-2$ and 6.

Testing each interval gives a solution set of

$$(-\infty, -2) \cup (6, \infty).$$

**73.** $\dfrac{(x + 1)(x - 2)}{x - 1} \geq 0$

Find the values of $x$ that make the numerator and denominator zero.

The boundary points are $-1$, 1 and 2. We exclude 1 from the solution set, since this would make the denominator zero.

Testing each interval gives a solution set of

$$[-1, 1) \cup [2, \infty).$$

**74.** $\dfrac{x+3}{x-4} \le 5$

Express the inequality so that one side is zero.

$$\dfrac{x+3}{x-4} - 5 \le 0$$

$$\dfrac{x+3}{x-4} - \dfrac{5(x-4)}{x-4} \le 0$$

$$\dfrac{-4x+23}{x-4} \le 0$$

Find the values of $x$ that make the numerator and denominator zero.

The boundary points are $4$ and $\dfrac{23}{4}$. We exclude $4$ from the solution set, since this would make the denominator zero.

Testing each interval gives a solution set of

$$(-\infty, 4) \cup \left[\dfrac{23}{4}, \infty\right).$$

**75.** **a.** $g(x) = 0.125x^2 + 2.3x + 27$

$g(35) = 0.125(35)^2 + 2.3(35) + 27 \approx 261$

The stopping distance on wet pavement for a motorcycle traveling 35 miles per hour is about 261 feet. This overestimates the distance shown in the graph by 1 foot.

**b.** $f(x) = 0.125x^2 - 0.8x + 99$

$0.125x^2 - 0.8x + 99 > 267$

$0.125x^2 - 0.8x - 168 > 0$

Solve the related quadratic equation.

$0.125x^2 - 0.8x - 168 = 0$

$$x = \dfrac{-b \pm \sqrt{b^2 - 4ac}}{2a}$$

$$x = \dfrac{-(-0.8) \pm \sqrt{(-0.8)^2 - 4(0.125)(-168)}}{2(0.125)}$$

$x = -33.6, \ 40$

Testing each interval gives a solution set of $(-\infty, -33.6) \cup (40, \infty)$.

Thus, speeds exceeding 40 miles per hour on dry pavement will require over 267 feet of stopping distance.

**76.** $s = -16t^2 + v_0 t + s_0$

$32 < -16t^2 + 48t + 0$

$0 < -16t^2 + 48t - 32$

$0 < -16(t^2 - 3t + 2)$

$0 < -16(t-2)(t-1)$

| | F | | T | | F | |
|---|---|---|---|---|---|---|
| | | 1 | | 2 | | |

The projectile's height exceeds 32 feet during the time period from 1 to 2 seconds.

**77.** $w = ks$

$28 = k \cdot 250$

$0.112 = k$

Thus, $w = 0.112s$.

$w = 0.112(1200) = 134.4$

1200 cubic centimeters of melting snow will produce 134.4 cubic centimeters of water.

**78.** $d = kt^2$

$144 = k(3)^2$

$k = 16$

$d = 16t^2$

$d = 16(10)^2 = 1,600 \text{ ft}$

**79.** $p = \dfrac{k}{w}$

$660 = \dfrac{k}{1.6}$

$1056 = k$

Thus, $p = \dfrac{1056}{w}$.

$p = \dfrac{1056}{2.4} = 440$

The pitch is 440 vibrations per second.

**80.** $l = \dfrac{k}{d^2}$

$28 = \dfrac{k}{8^2}$

$k = 1792$

$l = \dfrac{1792}{d^2}$

$l = \dfrac{1792}{4^2} = 112 \text{ decibels}$

**81.** $t = \dfrac{kc}{w}$

$10 = \dfrac{k \cdot 30}{6}$

$10 = 5h$

$h = 2$

$t = \dfrac{2c}{w}$

$t = \dfrac{2(40)}{5} = 16$ hours

**82.** $V = khB$

$175 = k \cdot 15 \cdot 35$

$k = \dfrac{1}{3}$

$V = \dfrac{1}{3}hB$

$V = \dfrac{1}{3} \cdot 20 \cdot 120 = 800 \text{ ft}^3$

**83.** **a.** Use $L = \dfrac{k}{R}$ to find $k$.

$L = \dfrac{k}{R}$

$30 = \dfrac{k}{63}$

$63 \cdot 30 = 63 \cdot \dfrac{k}{63}$

$1890 = k$

Thus, $L = \dfrac{1890}{R}$.

**b.** This is an approximate model.

**c.** $L = \dfrac{1890}{R}$

$L = \dfrac{1890}{27} = 70$

The average life span of an elephant is 70 years.

## Chapter 3 Test

**1.** $f(x) = (x+1)^2 + 4$

vertex: $(-1, 4)$
axis of symmetry: $x = -1$
$x$-intercepts:

$(x+1)^2 + 4 = 0$

$x^2 + 2x + 5 = 0$

$x = \dfrac{-2 \pm \sqrt{4 - 20}}{2} = -1 \pm 2i$

no $x$-intercepts
$y$-intercept:

$f(0) = (0+1)^2 + 4 = 5$

$f(x) = (x + 1)^2 + 4$
domain: $(-\infty, \infty)$ ; range: $[4, \infty)$

**2.** $f(x) = x^2 - 2x - 3$

$x = \dfrac{-b}{2a} = \dfrac{2}{2} = 1$

$f(1) = 1^2 - 2(1) - 3 = -4$

vertex: $(1, -4)$
axis of symmetry $x = 1$
$x$-intercepts:

$x^2 - 2x - 3 = 0$

$(x-3)(x+1) = 0$

$x = 3$ or $x = -1$

$y$-intercept:

$f(0) = 0^2 - 2(0) - 3 = -3$

$f(x) = x^2 - 2x - 3$
domain: $(-\infty, \infty)$ ; range: $[-4, \infty)$

**3.**  $f(x) = -2x^2 + 12x - 16$

Since the coefficient of $x^2$ is negative, the graph of $f(x)$ opens down and $f(x)$ has a maximum point.

$$x = \frac{-12}{2(-2)} = 3$$

$$f(3) = -2(3)^2 + 12(3) - 16$$
$$= -18 + 36 - 16$$
$$= 2$$

Maximum point: (3, 2)

domain: $(-\infty, \infty)$ ; range: $(-\infty, 2]$

**4.**  $f(x) = -x^2 + 46x - 360$

$$x = -\frac{b}{2a} = \frac{-46}{-2} = 23$$

23 computers will maximize profit.

$$f(23) = -(23)^2 + 46(23) - 360 = 169$$

Maximum daily profit = \$16,900.

**5.**  Let $x =$ one of the numbers;
$14 - x =$ the other number.

The product is $f(x) = x(14 - x)$

$$f(x) = x(14 - x) = -x^2 + 14x$$

The $x$-coordinate of the maximum is

$$x = -\frac{b}{2a} = -\frac{14}{2(-1)} = -\frac{14}{-2} = 7.$$

$$f(7) = -7^2 + 14(7) = 49$$

The vertex is (7, 49).  The maximum product is 49. This occurs when the two numbers are 7 and $14 - 7 = 7$.

**6.**  **a.**  $f(x) = x^3 - 5x^2 - 4x + 20$

$$x^3 - 5x^2 - 4x + 20 = 0$$
$$x^2(x - 5) - 4(x - 5) = 0$$
$$(x - 5)(x - 2)(x + 2) = 0$$
$$x = 5, 2, -2$$

The solution set is $\{5, 2, -2\}$.

**b.**  The degree of the polynomial is odd and the leading coefficient is positive. Thus the graph falls to the left and rises to the right.

$$f(x) = x^3 - 5x^2 - 4x + 20$$

**7.**  $f(x) = x^5 - x$

Since the degree of the polynomial is odd and the leading coefficient is positive, the graph of $f$ should fall to the left and rise to the right. The $x$-intercepts should be $-1$ and 1.

**8.**  **a.**  The integral root is 2.

**b.**

$$
\begin{array}{r|rrrr}
2 & 6 & -19 & 16 & -4 \\
  &   & 12  & -14 & 4 \\
\hline
  & 6 & -7  & 2  & 0
\end{array}
$$

$$6x^2 - 7x + 2 = 0$$
$$(3x - 2)(2x - 1) = 0$$
$$x = \frac{2}{3} \text{ or } x = \frac{1}{2}$$

The other two roots are $\frac{1}{2}$ and $\frac{2}{3}$.

**9.**  $2x^3 + 11x^2 - 7x - 6 = 0$

$p$: ±1, ±2, ±3, ±6
$q$: ±1, ±2

$$\frac{p}{q}: \pm 1, \pm 2, \pm 3, \pm 6, \pm \frac{1}{2}, \pm \frac{3}{2}$$

**10.**  $f(x) = 3x^5 - 2x^4 - 2x^2 + x - 1$

$f(x)$ has 3 sign variations.

$$f(-x) = -3x^5 - 2x^4 - 2x^2 - x - 1$$

$f(-x)$ has no sign variations.

There are 3 or 1 positive real solutions and no negative real solutions.

**11.**  $x^3 + 9x^2 + 16x - 6 = 0$

Since the leading coefficient is 1, the possible rational zeros are the factors of 6

$$p = \frac{p}{q}: \pm 1, \pm 2, \pm 3, \pm 6$$

$$
\begin{array}{r|rrrr}
-3 & 1 & 9 & 16 & -6 \\
   &   & -3 & -18 & 6 \\
\hline
   & 1 & 6 & -2 & 0
\end{array}
$$

Thus  $x = 3$  is a root.

Solve the quotient  $x^2 + 6x - 2 = 0$  using the quadratic formula to find the remaining roots.

$$x = \frac{-b \pm \sqrt{b^2 - 4ac}}{2a}$$

$$x = \frac{-(6) \pm \sqrt{(6)^2 - 4(1)(-2)}}{2(1)}$$

$$= \frac{-6 \pm \sqrt{44}}{2}$$

$$= -3 \pm \sqrt{11}$$

The zeros are $-3$, $-3+\sqrt{11}$, and $-3-\sqrt{11}$.

**12.** $f(x) = 2x^4 - x^3 - 13x^2 + 5x + 15$

**a.** Possible rational zeros are:

$p$: $\pm 1$, $\pm 3$, $\pm 5$, $\pm 15$

$q$: $\pm 1$, $\pm 2$

$\dfrac{p}{q}$: $\pm 1, \pm 3, \pm 5, \pm 15, \pm \dfrac{1}{2}, \pm \dfrac{3}{2}, \pm \dfrac{5}{2}, \pm \dfrac{15}{2}$

**b.** Verify that $-1$ and $\dfrac{3}{2}$ are zeros as it appears in the graph:

```
 -1|  2   -1   -13    5    15
         -2    3   10   -15
     ─────────────────────────
      2   -3   -10   15    0
```

```
  3|
  2|  2   -3   -10   15
             3    0   -15
     ─────────────────────
      2    0   -10    0
```

Thus, $-1$ and $\dfrac{3}{2}$ are zeros, and the polynomial factors as follows:

$$2x^4 - x^3 - 13x^2 + 5x + 15 = 0$$

$$(x+1)(2x^3 - 3x^2 - 10x + 15) = 0$$

$$(x+1)\left(x - \frac{3}{2}\right)(2x^2 - 10) = 0$$

Find the remaining zeros by solving:

$$2x^2 - 10 = 0$$

$$2x^2 = 10$$

$$x^2 = 5$$

$$x = \pm\sqrt{5}$$

The zeros are $-1$, $\dfrac{3}{2}$, and $\pm\sqrt{5}$.

**13.** $f(x)$ has zeros at $-2$ and $1$. The zero at $-2$ has multiplicity of 2.

$$x^3 + 3x^2 - 4 = (x-1)(x+2)^2$$

**14.** $f(x) = a_0(x+1)(x-1)(x+i)(x-i)$

$$= a_0(x^2 - 1)(x^2 + 1)$$

$$= a_0(x^4 - 1)$$

Since $f(3) = 160$, then

$$a_0(3^4 - 1) = 160$$

$$a_0(80) = 160$$

$$a_0 = \frac{160}{80}$$

$$a_0 = 2$$

$$f(x) = 2(x^4 - 1) = 2x^4 - 2$$

**15.** $f(x) = -3x^3 - 4x^2 + x + 2$

The graph shows a root at $x = -1$.
Use synthetic division to verify this root.

```
 -1|  -3   -4    1    2
            3    1    4
     ─────────────────────
      -3   -1    2    0
```

Factor the quotient to find the remaining zeros.

$$-3x^2 - x + 2 = 0$$

$$-(3x - 2)(x + 1) = 0$$

The zeros (x-intercepts) are $-1$ and $\dfrac{2}{3}$.

The y-intercept is $f(0) = 2$

$$f(x) = -3x^3 - 4x^2 + x + 2$$

**16.** $f(x) = \dfrac{1}{(x+3)^2}$

domain: $\{x \mid x \neq -3\}$ or $(-\infty, -3) \cup (-3, \infty)$

$f(x) = \dfrac{1}{(x+3)^2}$

**17.** $f(x) = \dfrac{1}{x-1} + 2$

domain: $\{x \mid x \neq 1\}$ or $(-\infty, 1) \cup (1, \infty)$

$f(x) = \dfrac{1}{x-1} + 2$

**18.** $f(x) = \dfrac{x}{x^2 - 16}$

domain: $\{x \mid x \neq 4, x \neq -4\}$

Symmetry: $f(-x) = \dfrac{-x}{x^2 - 16} = -f(x)$

$y$-axis symmetry
$x$-intercept: $x = 0$

$y$-intercept: $y = \dfrac{0}{0^2 - 16} = 0$

Vertical asymptotes:
$x^2 - 16 = 0$
$(x-4)(x+4) = 0$
$x = 4, -4$
Horizontal asymptote:
$n < m$, so $y = 0$ is the horizontal asymptote.

$f(x) = \dfrac{x}{x^2 - 16}$

**19.** $f(x) = \dfrac{x^2 - 9}{x - 2}$

domain: $\{x \mid x \neq 2\}$

Symmetry: $f(-x) = \dfrac{x^2 - 9}{-x - 2}$

$f(-x) \neq f(x), f(-x) \neq -f(x)$
No symmetry

$x$-intercepts:
$x^2 - 9 = 0$
$(x-3)(x+3) = 0$
$x = 3, -3$

$y$-intercept: $y = \dfrac{0^2 - 9}{0 - 2} = \dfrac{9}{2}$

Vertical asymptote:
$x - 2 = 0$
$x = 2$
Horizontal asymptote:
$n > m$, so no horizontal asymptote exists.

Slant asymptote: $f(x) = x + 2 - \dfrac{5}{x-2}$

$y = x + 2$

$f(x) = \dfrac{x^2 - 9}{x - 2}$

**20.** $f(x) = \dfrac{x+1}{x^2 + 2x - 3}$

$x^2 + 2x - 3 = (x+3)(x-1)$

domain: $\{x \mid x \neq -3, x \neq 1\}$

Symmetry: $f(-x) = \dfrac{-x+1}{x^2 - 2x - 3}$

$f(-x) \neq f(x), f(-x) \neq -f(x)$
No symmetry
$x$-intercept:
$x + 1 = 0$
$x = -1$

$y$-intercept: $y = \dfrac{0+1}{0^2 + 2(0) - 3} = -\dfrac{1}{3}$

Vertical asymptotes:
$x^2 + 2x - 3 = 0$
$(x+3)(x-1) = 0$
$x - 3, 1$

Horizontal asymptote:
$n < m$, so $y = 0$ is the horizontal asymptote.

$$f(x) = \frac{x+1}{x^2 + 2x - 3}$$

**21.** $f(x) = \dfrac{4x^2}{x^2 + 3}$

domain: all real numbers

Symmetry: $f(-x) = \dfrac{4x^2}{x^2 + 3} = f(x)$

$y$-axis symmetry

$x$-intercept:

$4x^2 = 0$

$x = 0$

$y$-intercept: $y = \dfrac{4(0)^2}{0^2 + 3} = 0$

Vertical asymptote:

$x^2 + 3 = 0$

$x^2 = -3$

No vertical asymptote.

Horizontal asymptote:

$n = m$, so $y = \dfrac{4}{1} = 4$ is the horizontal asymptote.

$$f(x) = \frac{4x^2}{x^2 + 3}$$

**22. a.** $\overline{C}(x) = \dfrac{300,000 + 10x}{x}$

**b.** Since the degree of the numerator equals the degree of the denominator, the horizontal asymptote is $x = \dfrac{10}{1} = 10$.

This represents the fact that as the number of satellite radio players produced increases, the production cost approaches $10 per radio.

**23.**
$$x^2 < x + 12$$
$$x^2 - x - 12 < 0$$
$$(x + 3)(x - 4) < 0$$
Boundary values: $-3$ and $4$
Solution set: $(-3, 4)$

**24.**
$$\frac{2x+1}{x-3} \le 3$$
$$\frac{2x+1}{x-3} - 3 \le 0$$
$$\frac{10-x}{x-3} \le 0$$
Boundary values: 3 and 10
Solution set: $(-\infty, 3) \cup [10, \infty)$

**25.** $i = \dfrac{k}{d^2}$

$20 = \dfrac{k}{15^2}$

$4500 = k$

$i = \dfrac{4500}{d^2} = \dfrac{4500}{10^2} = 45$ foot-candles

**Cumulative Review Exercises (Chapters 1–3)**

**1.** domain: $(-2, 2)$  range: $[0, \infty)$

**2.** The zero at $-1$ touches the $x$-axis at turns around so it must have a minimum multiplicity of 2.
The zero at 1 touches the $x$-axis at turns around so it must have a minimum multiplicity of 2.

**3.** There is a relative maximum at the point $(0, 3)$.

**4.** $(f \circ f)(-1) = f(f(-1)) = f(0) = 3$

**5.** $f(x) \to \infty$ as $\underline{x \to -2^+}$ or as $\underline{x \to 2^-}$

**6.**

$x = -4$

(−2, 4)

(−3, 1)   (−1, 1)

$g(x) = f(x + 2) + 1$

**7.**  $|2x - 1| = 3$

$2x - 1 = 3$

$2x = 4$

$x = 2$

$2x - 1 = -3$

$2x = -2$

$x = -1$

The solution set is {−1, 2}.

**8.**  $3x^2 - 5x + 1 = 0$

$x = \dfrac{5 \pm \sqrt{25 - 12}}{6} = \dfrac{5 \pm \sqrt{13}}{6}$

The solution set is $\left\{ \dfrac{5 + \sqrt{13}}{6}, \dfrac{5 - \sqrt{13}}{6} \right\}$.

**9.**  $9 + \dfrac{3}{x} = \dfrac{2}{x^2}$

$9x^2 + 3x = 2$

$9x^2 + 3x - 2 = 0$

$(3x - 1)(3x + 2) = 0$

$3x - 1 = 0 \qquad 3x + 2 = 0$

$x = \dfrac{1}{3} \qquad \text{or} \qquad x = -\dfrac{2}{3}$

The solution set is $\left\{ -\dfrac{2}{3}, \dfrac{1}{3} \right\}$.

**10.**  $x^3 + 2x^2 - 5x - 6 = 0$

$p: \pm 1, \pm 2, \pm 3, \pm 6$

$q: \pm 1$

$\dfrac{p}{q}: \pm 1, \pm 2, \pm 3, \pm 6$

$$
\begin{array}{r|rrrr}
-3 & 1 & 2 & -5 & -6 \\
   &   & -3 & 3 & 6 \\
\hline
   & 1 & -1 & -2 & 0 \\
\end{array}
$$

$x^3 + 2x^2 - 5x - 6 = 0$

$(x + 3)(x^2 - x - 2) = 0$

$(x + 3)(x + 1)(x - 2) = 0$

$x = -3$ or $x = -1$ or $x = 2$

The solution set is {−3, −1, 2}.

**11.**  $|2x - 5| > 3$

$2x - 5 > 3$

$2x > 8$

$x > 4$

$2x - 5 < -3$

$2x < 2$

$x < 1$

$(-\infty, 1)$ or $(4, \infty)$

**12.**  $3x^2 > 2x + 5$

$3x^2 - 2x - 5 > 0$

$3x^2 - 2x - 5 = 0$

$(3x - 5)(x + 1) = 0$

$x = \dfrac{5}{3}$ or $x = -1$

Test intervals are $(-\infty, -1)$, $\left( -1, \dfrac{5}{3} \right)$, $\left( \dfrac{5}{3}, \infty \right)$.

Testing points, the solution is $(-\infty, -1)$ or $\left( \dfrac{5}{3}, \infty \right)$.

**13.**  $f(x) = x^3 - 4x^2 - x + 4$

$x$-intercepts:

$x^3 - 4x^2 - x + 4 = 0$

$x^2(x - 4) - 1(x - 4) = 0$

$(x - 4)(x^2 - 1) = 0$

$(x - 4)(x + 1)(x - 1) = 0$

$x = -1, 1, 4$

$x$-intercepts:

$f(0) = 0^3 - 4(0)^2 - 0 + 4 = 4$

The degree of the polynomial is odd and the leading coefficient is positive. Thus the graph falls to the left and rises to the right.

(0, 4)   (1, 0)

(−1, 0)   (4, 0)

(3, −8)

$f(x) = x^3 - 4x^2 - x + 4$

**14.**  $f(x) = x^2 + 2x - 8$

$$x = \frac{-b}{2a} = \frac{-2}{2} = -1$$

$$f(-1) = (-1)^2 + 2(-1) - 8$$
$$= 1 - 2 - 8 = -9$$

vertex: $(-1, -9)$

$x$-intercepts:

$$x^2 + 2x - 8 = 0$$

$$(x + 4)(x - 2) = 0$$

$x = -4$ or $x = 2$

$y$-intercept: $f(0) = -8$

**15.**  $f(x) = x^2(x - 3)$

zeros: $x = 0$ (multiplicity 2) and $x = 3$

$y$-intercept: $y = 0$

$$f(x) = x^3 - 3x^2$$

$n = 3, a_n = 0$ so the graph falls to the left and rises to the right.

**16.**  $f(x) = \dfrac{x - 1}{x - 2}$

vertical asymptote: $x = 2$

horizontal asymptote: $y = 1$

$x$-intercept: $x = 1$

$y$-intercept: $y = \dfrac{1}{2}$

**17.**

**18.**

$$x^2 + y^2 - 2x + 4y - 4 = 0$$

**19.**  $(f \circ g)(x) = f\big(g(x)\big)$

$$= 2(4x - 1)^2 - (4x - 1) - 1$$
$$= 32x^2 - 20x + 2$$

**20.**  $\dfrac{f(x + h) - f(x)}{h}$

$$= \frac{\left[2(x + h)^2 - (x + h) - 1\right] - \left[2x^2 - x - 1\right]}{h}$$

$$= \frac{2x^2 + 4hx - x + 2h^2 - h - 1 - 2x^2 + x + 1}{h}$$

$$= \frac{4hx + 2h^2 - h}{h}$$

$$= 4x + 2h - 1$$

## Section 4.1

## Check Point Exercises

1.  $f(x) = 42.2(1.56)^x$

    $f(3) = 42.2(1.56)^3 \approx 160.20876 \approx 160$

    According to the function, the average amount spent after three hours of shopping at the mall is $160. This overestimates the actual amount shown by $11.

2.  Begin by setting up a table of coordinates.

    | $x$ | $f(x) = 3^x$ |
    |---|---|
    | $-3$ | $f(-3) = 3^{-3} = \frac{1}{27}$ |
    | $-2$ | $f(-2) = 3^{-2} = \frac{1}{9}$ |
    | $-1$ | $f(-1) = 3^{-1} = \frac{1}{3}$ |
    | $0$ | $f(0) = 3^0 = 1$ |
    | $1$ | $f(1) = 3^1 = 3$ |
    | $2$ | $f(2) = 3^2 = 9$ |
    | $3$ | $f(3) = 3^3 = 27$ |

    Plot a few of these points, connecting them with a continuous curve.

    $f(x) = 3^x$

3.  Begin by setting up a table of coordinates.

    | $x$ | $f(x) = \left(\frac{1}{3}\right)^x$ |
    |---|---|
    | $-2$ | $\left(\frac{1}{3}\right)^{-2} = 9$ |
    | $-1$ | $\left(\frac{1}{3}\right)^{-1} = 3$ |
    | $0$ | $\left(\frac{1}{3}\right)^0 = 1$ |
    | $1$ | $\left(\frac{1}{3}\right)^1 = \frac{1}{3}$ |
    | $2$ | $\left(\frac{1}{3}\right)^2 = \frac{1}{9}$ |

Plot a few of these points, connecting them with a continuous curve.

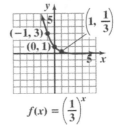

$f(x) = \left(\frac{1}{3}\right)^x$

4.  Note that the function $g(x) = 3^{x-1}$ has the general form $g(x) = b^{x+c}$ where $c = -1$. Because $c < 0$, we graph $g(x) = 3^{x-1}$ by shifting the graph of $f(x) = 3^x$ one unit to the right. Construct a table showing some of the coordinates for $f$ and $g$.

    | $x$ | $f(x) = 3^x$ | $g(x) = 3^{x-1}$ |
    |---|---|---|
    | $-2$ | $3^{-2} = \frac{1}{9}$ | $3^{-2-1} = 3^{-3} = \frac{1}{27}$ |
    | $-1$ | $3^{-1} = \frac{1}{3}$ | $3^{-1-1} = 3^{-2} = \frac{1}{9}$ |
    | $0$ | $3^0 = 1$ | $3^{0-1} = 3^{-1} = \frac{1}{3}$ |
    | $1$ | $3^1 = 3$ | $3^{1-1} = 3^0 = 1$ |
    | $2$ | $3^2 = 9$ | $3^{2-1} = 3^1 = 3$ |

    $f(x) = 3^x$
    $g(x) = 3^{x-1}$

**5.** Note that the function $g(x) = 2^x + 1$ has the general form $g(x) = b^x + c$ where $c = 1$. Because $c > 0$, we graph $g(x) = 2^x + 1$ by shifting the graph of $f(x) = 2^x$ up one unit. Construct a table showing some of the coordinates for $f$ and $g$.

| $x$ | $f(x) = 2^x$ | $g(x) = 2^x + 1$ |
|---|---|---|
| $-2$ | $2^{-2} = \frac{1}{4}$ | $2^{-2} + 1 = \frac{1}{4} + 1 = \frac{5}{4}$ |
| $-1$ | $2^{-1} = \frac{1}{2}$ | $2^{-1} + 1 = \frac{1}{2} + 1 = \frac{3}{2}$ |
| $0$ | $2^0 = 1$ | $2^0 + 1 = 1 + 1 = 2$ |
| $1$ | $2^1 = 2$ | $2^1 + 1 = 2 + 1 = 3$ |
| $2$ | $2^2 = 4$ | $2^2 + 1 = 4 + 1 = 5$ |

$$f(x) = 2^x$$
$$g(x) = 2^x + 1$$

**6.** 2012 is 34 years after 1978.

$$f(x) = 1066e^{0.042x}$$
$$f(34) = 1066e^{0.042(34)} \approx 4446$$

In 2012 the gray wolf population of the Western Great Lakes is projected to be about 4446.

**7. a.** $A = P\left(1 + \dfrac{r}{n}\right)^{nt}$

$$A = 10,000\left(1 + \dfrac{0.08}{4}\right)^{4(5)}$$
$$= \$14,859.47$$

**b.** $A = Pe^{rt}$
$$A = 10,000e^{0.08(5)}$$
$$= \$14,918.25$$

**Concept and Vocabulary Check 4.1**

**1.** $b^x$; $(-\infty, \infty)$; $(0, \infty)$

**2.** $x$; $y = 0$; horizontal

**3.** $e$; natural; 2.72

**4.** $A$; $P$; $r$; $n$

**5.** semiannually; quarterly; continuous

**Exercise Set 4.1**

**1.** $2^{3.4} \approx 10.556$

**3.** $3^{\sqrt{5}} \approx 11.665$

**5.** $4^{-1.5} = 0.125$

**7.** $e^{2.3} \approx 9.974$

**9.** $e^{-0.95} \approx 0.387$

**11.**

| $x$ | $f(x) = 4^x$ |
|---|---|
| $-2$ | $4^{-2} = \frac{1}{16}$ |
| $-1$ | $4^{-1} = \frac{1}{4}$ |
| $0$ | $4^0 = 1$ |
| $1$ | $4^1 = 4$ |
| $2$ | $4^2 = 16$ |

$$f(x) = 4^x$$

**13.**

| $x$ | $g(x) = \left(\frac{3}{2}\right)^x$ |
|---|---|
| $-2$ | $\left(\frac{3}{2}\right)^{-2} = \frac{4}{9}$ |
| $-1$ | $\left(\frac{3}{2}\right)^{-1} = \frac{2}{3}$ |
| $0$ | $\left(\frac{3}{2}\right)^0 = 1$ |
| $1$ | $\left(\frac{3}{2}\right)^1 = \frac{3}{2}$ |
| $2$ | $\left(\frac{3}{2}\right)^2 = \frac{9}{4}$ |

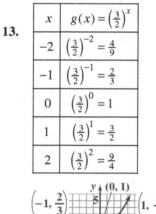

$$g(x) = \left(\frac{3}{2}\right)^x$$

**15.**

| $x$ | $h(x) = \left(\frac{1}{2}\right)^x$ |
|---|---|
| $-2$ | $\left(\frac{1}{2}\right)^{-2} = 4$ |
| $-1$ | $\left(\frac{1}{2}\right)^{-1} = 2$ |
| $0$ | $\left(\frac{1}{2}\right)^{0} = 1$ |
| $1$ | $\left(\frac{1}{2}\right)^{1} = \frac{1}{2}$ |
| $2$ | $\left(\frac{1}{2}\right)^{2} = \frac{1}{4}$ |

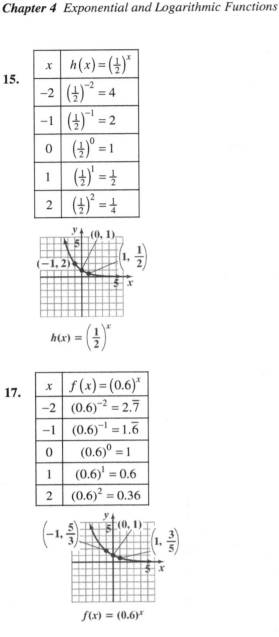

$$h(x) = \left(\frac{1}{2}\right)^x$$

**17.**

| $x$ | $f(x) = (0.6)^x$ |
|---|---|
| $-2$ | $(0.6)^{-2} = 2.\overline{7}$ |
| $-1$ | $(0.6)^{-1} = 1.\overline{6}$ |
| $0$ | $(0.6)^{0} = 1$ |
| $1$ | $(0.6)^{1} = 0.6$ |
| $2$ | $(0.6)^{2} = 0.36$ |

$$f(x) = (0.6)^x$$

**19.** This is the graph of $f(x) = 3^x$ reflected about the *x*-axis and about the *y*-axis, so the function is $H(x) = -3^{-x}$.

**21.** This is the graph of $f(x) = 3^x$ reflected about the *x*-axis, so the function is $F(x) = -3^x$.

**23.** This is the graph of $f(x) = 3^x$ shifted one unit downward, so the function is $h(x) = 3^x - 1$.

**25.** The graph of $g(x) = 2^{x+!}$ can be obtained by shifting the graph of $f(x) = 2^x$ one unit to the left.

$$f(x) = 2^x$$
$$g(x) = 2^{x+1}$$

asymptote: $y = 0$

domain: $(-\infty, \infty)$ ; range: $(0, \infty)$

**27.** The graph of $g(x) = 2^x - 1$ can be obtained by shifting the graph of $f(x) = 2^x$ downward one unit.

$$f(x) = 2^x$$
$$g(x) = 2^x - 1$$

asymptote: $y = -1$

domain: $(-\infty, \infty)$ ; range: $(-1, \infty)$

**29.** The graph of $h(x) = 2^{x+1} - 1$ can be obtained by shifting the graph of $f(x) = 2^x$ one unit to the left and one unit downward.

$$f(x) = 2^x$$
$$h(x) = 2^{x+1} - 1$$

asymptote: $y = -1$

domain: $(-\infty, \infty)$ ; range: $(-1, \infty)$

**31.** The graph of $g(x) = -2^x$ can be obtained by reflecting the graph of $f(x) = 2^x$ about the $x$-axis.

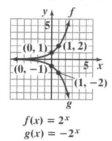

$f(x) = 2^x$
$g(x) = -2^x$

asymptote: $y = 0$

domain: $(-\infty, \infty)$ ; range: $(-\infty, 0)$

**33.** The graph of $g(x) = 2 \cdot 2^x$ can be obtained by vertically stretching the graph of $f(x) = 2^x$ by a factor of two.

$f(x) = 2^x$
$g(x) = 2 \cdot 2^x$

asymptote: $y = 0$

domain: $(-\infty, \infty)$ ; range: $(0, \infty)$

**35.** The graph of $g(x) = e^{x-1}$ can be obtained by moving $f(x) = e^x$ 1 unit right.

$g(x) = e^{x-1}$

asymptote: $y = 0$

domain: $(-\infty, \infty)$ ; range: $(0, \infty)$

**37.** The graph of $g(x) = e^x + 2$ can be obtained by moving $f(x) = e^x$ 2 units up.

$g(x) = e^x + 2$

asymptote: $y = 2$

domain: $(-\infty, \infty)$ ; range: $(2, \infty)$

**39.** The graph of $h(x) = e^{x-1} + 2$ can be obtained by moving $f(x) = e^x$ 1 unit right and 2 units up.

$h(x) = e^{x-1} + 2$

asymptote: $y = 2$

domain: $(-\infty, \infty)$ ; range: $(2, \infty)$

**41.** The graph of $h(x) = e^{-x}$ can be obtained by reflecting $f(x) = e^x$ about the $y$-axis.

$h(x) = e^{-x}$

asymptote: $y = 0$

domain: $(-\infty, \infty)$ ; range: $(0, \infty)$

**43.** The graph of $g(x) = 2e^x$ can be obtained by stretching $f(x) = e^x$ vertically by a factor of 2.

$$g(x) = 2e^x$$

asymptote: $y = 0$

domain: $(-\infty, \infty)$; range: $(0, \infty)$

**45.** The graph of $h(x) = e^{2x} + 1$ can be obtained by stretching $f(x) = e^x$ horizontally by a factor of 2 and then moving the graph up 1 unit.

$$h(x) = e^{2x} + 1$$

asymptote: $y = 1$

domain: $(-\infty, \infty)$; range: $(1, \infty)$

**47.** The graph of $g(x)$ can be obtained by reflecting $f(x)$ about the *y*-axis.

$$f(x) = 3^x$$
$$g(x) = 3^{-x}$$

asymptote of $f(x)$: $y = 0$
asymptote of $g(x)$: $y = 0$

**49.** The graph of $g(x)$ can be obtained by vertically shrinking $f(x)$ by a factor of $\frac{1}{3}$.

$$f(x) = 3^x$$
$$g(x) = \frac{1}{3} \cdot 3^x$$

asymptote of $f(x)$: $y = 0$
asymptote of $g(x)$: $y = 0$

**51.** The graph of $g(x)$ can be obtained by moving the graph of $f(x)$ one space to the right and one space up.

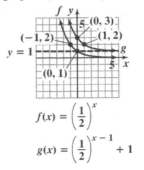

$$f(x) = \left(\frac{1}{2}\right)^x$$
$$g(x) = \left(\frac{1}{2}\right)^{x-1} + 1$$

asymptote of $f(x)$: $y = 0$
asymptote of $g(x)$: $y = 1$

**53.** **a.** $A = 10,000\left(1 + \dfrac{0.055}{2}\right)^{2(5)}$

$\approx \$13,116.51$

**b.** $A = 10,000\left(1 + \dfrac{0.055}{4}\right)^{4(5)}$

$\approx \$13,140.67$

**c.** $A = 10,000\left(1 + \dfrac{0.055}{12}\right)^{12(5)}$

$\approx \$13,157.04$

**d.** $A = 10,000e^{0.055(5)}$

$\approx \$13,165.31$

**55.** $A = 12,000\left(1 + \dfrac{0.07}{12}\right)^{12(3)}$

$\approx 14,795.11$ (7% yield)

$A = 12,000e^{0.0685(3)}$

$\approx 14,737.67$ (6.85% yield)

Investing \$12,000 for 3 years at 7% compounded monthly yields the greater return.

**57.**

| $x$ | $f(x) = 2^x$ | $g(x) = 2^{-x}$ |
|-----|------|------|
| $-2$ | $\dfrac{1}{4}$ | $4$ |
| $-1$ | $\dfrac{1}{2}$ | $2$ |
| $0$ | $1$ | $1$ |
| $1$ | $2$ | $\dfrac{1}{2}$ |
| $2$ | $4$ | $\dfrac{1}{4}$ |

The point of intersection is $(0,1)$.

**59.**

| $x$ | $y = 2^x$ |
|-----|------|
| $-2$ | $\dfrac{1}{4}$ |
| $-1$ | $\dfrac{1}{2}$ |
| $0$ | $1$ |
| $1$ | $2$ |
| $2$ | $4$ |

| $y$ | $x = 2^y$ |
|-----|------|
| $-2$ | $\dfrac{1}{4}$ |
| $-1$ | $\dfrac{1}{2}$ |
| $0$ | $1$ |
| $1$ | $2$ |
| $2$ | $4$ |

**61.** The graph is of the form $y = b^x$.

Substitute values from the point (1, 4) to find b.

$y = b^x$

$4 = b^1$

$4 = b$

The equation of the graph is $y = 4^x$

**63.** The graph is of the form $y = -b^x$.

Substitute values from the point (1, $-e$) to find b.

$y = -b^x$

$-e = -b^1$

$e = b$

The equation of the graph is $y = -e^x$

**65.** **a.** $f(0) = 574(1.026)^0$

$= 574(1) = 574$

India's population in 1974 was 574 million.

**b.** $f(27) = 574(1.026)^{27} \approx 1148$

India's population in 2001 will be 1148 million.

**c.** Since $2028 - 1974 = 54$, find

$f(54) = 574(1.026)^{54} \approx 2295$.

India's population in 2028 will be 2295 million.

**d.** $2055 - 1974 = 81$, find

$f(54) = 574(1.026)^{81} \approx 4590$.

India's population in 2055 will be 4590 million.

**e.** India's population appears to be doubling every 27 years.

**67.** $S = 465,000(1 + 0.06)^{10}$

$= 465,000(1.06)^{10} \approx \$832,744$

**69.** $2^{1.7} \approx 3.249009585$

$2^{1.73} \approx 3.317278183$

$2^{1.732} \approx 3.321880096$

$2^{1.73205} \approx 3.321995226$

$2^{1.7320508} \approx 3.321997068$

$2^{\sqrt{3}} \approx 3.321997085$

The closer the exponent is to $\sqrt{3}$, the closer the value is to $2^{\sqrt{3}}$.

**71. a.** 2011 is 11 years after 2000.

$$f(x) = 782x + 6564$$

$$f(11) = 782(11) + 6564 = 15,166$$

According to the linear model, the average cost of a family health insurance plan in 2011 was $15,166.

**b.** 2011 is 11 years after 2000.

$$g(x) = 6875e^{0.077x}$$

$$g(11) = 6875e^{0.077(11)} \approx 16,037$$

According to the exponential model, the average cost of a family health insurance plan in 2011 was $16,037.

**c.** The linear model is the better model for the data in 2011.

**73. a.** $f(0) = 80e^{-0.5(0)} + 20$

$$= 80e^0 + 20$$

$$= 80(1) + 20$$

$$= 100$$

100% of the material is remembered at the moment it is first learned.

**b.** $f(1) = 80e^{-0.5(1)} + 20 \approx 68.5$

68.5% of the material is remembered 1 week after it is first learned.

**c.** $f(4) = 80e^{-0.5(4)} + 20 \approx 30.8$

30.8% of the material is remembered 4 weeks after it is first learned.

**d.** $f(52) = 80e^{-0.5(52)} + 20 \approx 20$

20% of the material is remembered 1 year after it is first learned.

**75.** $f(x) = 6.25(1.029)^x$

$$f(59) = 6.25(1.029)^{59} \approx 33.8$$

$$g(x) = \frac{38.8}{1 + 6.3e^{-0.051x}}$$

$$g(59) = \frac{38.8}{1 + 6.3e^{-0.051(59)}} \approx 29.6$$

Function $g(x)$ is a better model for the graph's value of 29.5 in 2009.

**77. – 79.** Answers will vary.

**81. a.** $A = 10,000\left(1 + \dfrac{0.05}{4}\right)^{4t}$

$$A = 10,000\left(1 + \dfrac{0.045}{12}\right)^{12t}$$

**b.** 5% compounded quarterly offers the better return.

**83.** does not make sense; Explanations will vary. Sample explanation: The horizontal asymptote is $y = 0$.

**85.** does not make sense; Explanations will vary. Sample explanation: An exponential model appears better than a linear model.

**87.** false; Changes to make the statement true will vary. A sample change is: The amount of money will not increase without bound.

**89.** false; Changes to make the statement true will vary. A sample change is: If $f(x) = 2^x$ then $f(a+b) = f(a) \cdot f(b)$.

**91.** $y = 3^x$ is (d). $y$ increases as $x$ increases, but not as quickly as $y = 5^x$. $y = 5^x$ is (c). $y = \left(\dfrac{1}{3}\right)^x$ is (a).

$y = \left(\dfrac{1}{3}\right)^x$ is the same as $y = 3^{-x}$, so it is (d) reflected about the $y$-axis. $y = \left(\dfrac{1}{5}\right)^x$ is (b). $y = \left(\dfrac{1}{5}\right)^x$ is the same as $y = 5^{-x}$, so it is (c) reflected about the $y$-axis.

**93.** **a.**
$$\cosh(-x) = \frac{e^{-x} + e^{-(-x)}}{2}$$
$$= \frac{e^{-x} + e^{x}}{2}$$
$$= \frac{e^{x} + e^{-x}}{2}$$
$$= \cosh x$$

**b.**
$$\sinh(-x) = \frac{e^{-x} - e^{-(-x)}}{2}$$
$$= \frac{e^{-x} - e^{x}}{2}$$
$$= \frac{-\left(-e^{-x} + e^{x}\right)}{2}$$
$$= -\frac{e^{x} - e^{-x}}{2}$$
$$= -\sinh x$$

**c.**
$$(\cosh x)^2 - (\sinh x)^2 \overset{?}{=} 1$$
$$\left(\frac{e^{x} + e^{-x}}{2}\right)^2 - \left(\frac{e^{x} - e^{-x}}{2}\right)^2 \overset{?}{=} 1$$
$$\frac{e^{2x} + 2 + e^{-2x}}{4} - \frac{e^{2x} - 2 + e^{-2x}}{4} \overset{?}{=} 1$$
$$\frac{e^{2x} + 2 + e^{-2x} - e^{2x} + 2 - e^{-2x}}{4} \overset{?}{=} 1$$
$$\frac{4}{4} \overset{?}{=} 1$$
$$1 = 1$$

**94.** We do not know how to solve $x = 2^y$ for $y$.

**95.** $\dfrac{1}{2}$; i.e. $25^{1/2} = 5$

**96.** $(x - 3)^2 > 0$

Solving the related equation, $(x - 3)^2 = 0$, gives $x = 3$.
Note that the boundary value $x = 3$ does not satisfy
the inequality.
Testing each interval gives a solution set of
$(-\infty, 3) \cup (3, \infty)$.

**Section 4.2**

**Check Point Exercises**

**1.** **a.** $3 = \log_7 x$ written in exponential form is
$7^3 = x$.

**b.** $2 = \log_b 25$ written in exponential form is
$b^2 = 25$.

**c.** $\log_4 26 = y$ written in exponential form is
$4^y = 26$.

**2.** **a.** $2^5 = x$ written in logarithmic form is
$5 = \log_2 x$.

**b.** $b^3 = 27$ written in logarithmic form
is $3 = \log_b 27$.

**c.** $e^y = 33$ written in logarithmic form
is $y = \log_e 33$.

**3.** **a.** Question: 10 to what power gives 100?
$\log_{10} 100 = 2$ because $10^2 = 100$.

**b.** Question: 5 to what power gives $\dfrac{1}{125}$?
$\log_5 \dfrac{1}{125} = -3$ because $5^{-3} = \dfrac{1}{5^3} = \dfrac{1}{125}$.

**c.** Question: 36 to what power gives 6?
$\log_{36} 6 = \dfrac{1}{2}$ because $36^{1/2} = \sqrt{36} = 6$

**d.** Question: 3 to what power gives $\sqrt[7]{3}$?
$\log_3 \sqrt[7]{3} = \dfrac{1}{7}$ because $3^{1/7} = \sqrt[7]{3}$.

**4.** **a.** Because $\log_b b = 1$, we conclude $\log_9 9 = 1$.

**b.** Because $\log_b 1 = 0$, we conclude $\log_8 1 = 0$.

**5.** **a.** Because $\log_b b^x = x$, we conclude $\log_7 7^8 = 8$.

**b.** Because $b^{\log_b x} = x$, we conclude $3^{\log_3 17} = 17$.

**6.** First, set up a table of coordinates for $f(x) = 3^x$.

| $x$ | $-2$ | $-1$ | $0$ | $1$ | $2$ | $3$ |
|---|---|---|---|---|---|---|
| $f(x) = 3^x$ | $\frac{1}{9}$ | $\frac{1}{3}$ | $1$ | $3$ | $9$ | $27$ |

Reversing these coordinates gives the coordinates for the inverse function $g(x) = \log_3 x$.

| $x$ | $\frac{1}{9}$ | $\frac{1}{3}$ | $1$ | $3$ | $9$ | $27$ |
|---|---|---|---|---|---|---|
| $g(x) = \log_3 x$ | $-2$ | $-1$ | $0$ | $1$ | $2$ | $3$ |

The graph of the inverse can also be drawn by reflecting the graph of $f(x) = 3^x$ about the line $y = x$.

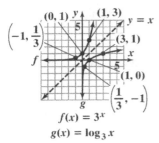

$f(x) = 3^x$
$g(x) = \log_3 x$

**7.** The domain of $h$ consists of all $x$ for which $x - 5 > 0$. Solving this inequality for $x$, we obtain $x > 5$. Thus, the domain of $h$ is $(5, \infty)$.

**8.** Substitute the boy's age, 10, for $x$ and evaluate the function at 10.
$$f(10) = 29 + 48.8\log(10 + 1)$$
$$= 29 + 48.8\log(11)$$
$$\approx 80$$
A 10-year-old boy is approximately 80% of his adult height.

**9.** Because $I = 10,000\, I_0$,
$$R = \log\frac{10,000 I_0}{I_0}$$
$$= \log 10,000$$
$$= 4$$
The earthquake registered 4.0 on the Richter scale.

**10. a.** The domain of $f$ consists of all $x$ for which $4 - x > 0$. Solving this inequality for $x$, we obtain $x < 4$.
The domain of $f$ is $(-\infty, 4)$

   **b.** The domain of $g$ consists of all $x$ for which $x^2 > 0$. Solving this inequality for $x$, we obtain $x < 0$ or $x > 0$.
The domain of $g$ is $(-\infty, 0) \cup (0, \infty)$.

**11.** Find the temperature increase after 30 minutes by substituting 30 for $x$ and evaluating the function at 30.
$$f(x) = 13.4\ln x - 11.6$$
$$f(30) = 13.4\ln 30 - 11.6$$
$$\approx 34$$
The temperature increase after 30 minutes is $34°$. The function models the actual increase shown in the graph extremely well.

**Concept and Vocabulary Check 4.2**

1. $b^y = x$

2. logarithmic; $b$

3. 1

4. 0

5. $x$

6. $x$

7. $(0, \infty)$; $(-\infty, \infty)$

8. $y$; $x = 0$; vertical

9. up 5 units

10. to the left 5 units

11. $x$-axis

12. $y$-axis

13. $5 - x > 0$

14. common; $\log x$

15. natural; $\ln x$

**Exercise Set 4.2**

1. $2^4 = 16$

3. $3^2 = x$

5. $b^5 = 32$

7. $6^y = 216$

9. $\log_2 8 = 3$

**11.** $\log_2 \dfrac{1}{16} = -4$

**13.** $\log_8 2 = \dfrac{1}{3}$

**15.** $\log_{13} x = 2$

**17.** $\log_b 1000 = 3$

**19.** $\log_7 200 = y$

**21.** $\log_4 16 = 2$ because $4^2 = 16.$

**23.** $\log_2 64 = 6$ because $2^6 = 64.$

**25.** $\log_5 \dfrac{1}{5} = -1$ because $5^{-1} = \dfrac{1}{5}.$

**27.** $\log_2 \dfrac{1}{8} = -3$ because $2^{-3} = \dfrac{1}{8}.$

**29.** $\log_7 \sqrt{7} = \dfrac{1}{2}$ because $7^{\frac{1}{2}} = \sqrt{7}.$

**31.** $\log_2 \dfrac{1}{\sqrt{2}} = -\dfrac{1}{2}$ because $2^{-\frac{1}{2}} = \dfrac{1}{\sqrt{2}}.$

**33.** $\log_{64} 8 = \dfrac{1}{2}$ because $64^{\frac{1}{2}} = \sqrt{64} = 8.$

**35.** Because $\log_b b = 1$, we conclude $\log_5 5 = 1.$

**37.** Because $\log_b 1 = 0$, we conclude $\log_4 1 = 0.$

**39.** Because $\log_b b^x = x$, we conclude $\log_5 5^7 = 7.$

**41.** Because $b^{\log_b x} = x$, we conclude $8^{\log_8 19} = 19.$

**43.** First, set up a table of coordinates for $f(x) = 4^x.$

| $x$ | $-2$ | $-1$ | $0$ | $1$ | $2$ | $3$ |
|---|---|---|---|---|---|---|
| $f(x) = 4x$ | $\frac{1}{16}$ | $\frac{1}{4}$ | $1$ | $4$ | $16$ | $64$ |

Reversing these coordinates gives the coordinates for the inverse function $g(x) = \log_4 x.$

| $x$ | $\frac{1}{16}$ | $\frac{1}{4}$ | $1$ | $4$ | $16$ | $64$ |
|---|---|---|---|---|---|---|
| $g(x) = \log_{4x}$ | $-2$ | $-1$ | $0$ | $1$ | $2$ | $3$ |

$f(x) = 4^x$

$g(x) = \log_4 x$

**45.** First, set up a table of coordinates for $f(x) = \left(\dfrac{1}{2}\right)^x.$

| $x$ | $-2$ | $-1$ | $0$ | $1$ | $2$ | $3$ |
|---|---|---|---|---|---|---|
| $f(x) = \left(\frac{1}{2}\right)^x$ | $4$ | $2$ | $1$ | $\frac{1}{2}$ | $\frac{1}{4}$ | $\frac{1}{8}$ |

Reversing these coordinates gives the coordinates for the inverse function $g(x) = \log_{1/2} x.$

| $x$ | $4$ | $2$ | $1$ | $\frac{1}{2}$ | $\frac{1}{4}$ | $\frac{1}{8}$ |
|---|---|---|---|---|---|---|
| $g(x) = \log_{1/2} x$ | $-2$ | $-1$ | $0$ | $1$ | $2$ | $3$ |

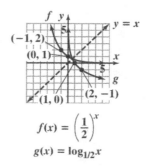

$f(x) = \left(\dfrac{1}{2}\right)^x$

$g(x) = \log_{1/2} x$

**47.** This is the graph of $f(x) = \log_3 x$ reflected about the *x*-axis and shifted up one unit, so the function is $H(x) = 1 - \log_3 x$.

**49.** This is the graph of $f(x) = \log_3 x$ shifted down one unit, so the function is $h(x) = \log_3 x - 1$.

**51.** This is the graph of $f(x) = \log_3 x$ shifted right one unit, so the function is $g(x) = \log_3(x-1)$.

**53.**

$$f(x) = \log_2 x$$
$$g(x) = \log_2(x+1)$$

vertical asymptote: $x = -1$
domain: $(-1, \infty)$ ; range: $(-\infty, \infty)$

**55.**

$$f(x) = \log_2 x$$
$$h(x) = 1 + \log_2 x$$

vertical asymptote: $x = 0$
domain: $(0, \infty)$ ; range: $(-\infty, \infty)$

**57.**

$$f(x) = \log_2 x$$
$$g(x) = \frac{1}{2}\log_2 x$$

vertical asymptote: $x = 0$
domain: $(0, \infty)$ ; range: $(-\infty, \infty)$

**59.**

$$g(x) = \log(x-1)$$

vertical asymptote: $x = 1$
domain: $(1, \infty)$ ; range: $(-\infty, \infty)$

**61.**

$$h(x) = \log x - 1$$

vertical asymptote: $x = 0$
domain: $(0, \infty)$ ; range: $(-\infty, \infty)$

**63.**

$$g(x) = 1 - \log x$$

vertical asymptote: $x = 0$
domain: $(0, \infty)$ ; range: $(-\infty, \infty)$

**65.**

$$g(x) = \ln(x+2)$$

vertical asymptote: $x = -2$
domain: $(-2, \infty)$ ; range: $(-\infty, \infty)$

**67.**

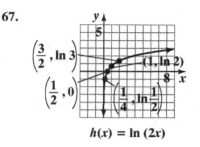

$h(x) = \ln(2x)$

vertical asymptote: $x = 0$
domain: $(0, \infty)$ ; range: $(-\infty, \infty)$

**69.**

$g(x) = 2 \ln x$

vertical asymptote: $x = 0$
domain: $(0, \infty)$ ; range: $(-\infty, \infty)$

**71.**

$h(x) = -\ln x$

vertical asymptote: $x = 0$
domain: $(0, \infty)$ ; range: $(-\infty, \infty)$

**73.**

$g(x) = 2 - \ln x$

vertical asymptote: $x = 0$
domain: $(0, \infty)$ ; range: $(-\infty, \infty)$

**75.** The domain of $f$ consists of all $x$ for which $x + 4 > 0$. Solving this inequality for $x$, we obtain $x > -4$. Thus, the domain of $f$ is $(-4, \infty)$.

**77.** The domain of $f$ consists of all $x$ for which $2 - x > 0$. Solving this inequality for $x$, we obtain $x < 2$. Thus, the domain of $f$ is $(-\infty, 2)$.

**79.** The domain of $f$ consists of all $x$ for which $(x - 2)^2 > 0$. Solving this inequality for $x$, we obtain $x < 2$ or $x > 2$. Thus, the domain of $f$ is $(-\infty, 2)$ or $(2, \infty)$.

**81.** $\log 100 = \log_{10} 100 = 2$
because $10^2 = 100$.

**83.** Because $\log 10^x = x$, we
conclude $\log 10^7 = 7$.

**85.** Because $10^{\log x} = x$, we
conclude $10^{\log 33} = 33$.

**87.** $\ln 1 = 0$ because $e^0 = 1$.

**89.** Because $\ln e^x = x$, we
conclude $\ln e^6 = 6$.

**91.** $\ln \dfrac{1}{e^6} = \ln e^{-6}$

Because $\ln e^x = x$ we conclude
$\ln e^{-6} = -6$, so $\ln \dfrac{1}{e^6} = -6$.

**93.** Because $e^{\ln x} = x$, we conclude $e^{\ln 125} = 125$.

**95.** Because $\ln e^x = x$, we conclude $\ln e^{9x} = 9x$.

**97.** Because $e^{\ln x} = x$, we conclude $e^{\ln 5x^2} = 5x^2$.

**99.** Because $10^{\log x} = x$, we conclude $10^{\log \sqrt{x}} = \sqrt{x}$.

**101.** $\log_3(x - 1) = 2$
$$3^2 = x - 1$$
$$9 = x - 1$$
$$10 = x$$
The solution is 10, and the solution set is $\{10\}$.

**103.** $\log_4 x = -3$
$$4^{-3} = x$$
$$x = \frac{1}{4^3} = \frac{1}{64}$$

The solution is $\dfrac{1}{64}$ , and the solution set is $\left\{\dfrac{1}{64}\right\}$ .

**105.** $\log_3\left(\log_7 7\right) = \log_3 1 = 0$

**107.** $\log_2\left(\log_3 81\right) = \log_2\left(\log_3 3^4\right)$
$$= \log_2 4 = \log_2 2^2 = 2$$

**109.** For $f(x) = \ln(x^2 - x - 2)$ to be real, $x^2 - x - 2 > 0$.
Solve the related equation to find the boundary points:
$$x^2 - x - 2 = 0$$
$$(x+1)(x-2) = 0$$
The boundary points are –1 and 2. Testing each interval gives a domain of $(-\infty, -1) \cup (2, \infty)$.

**111.** For $f(x) = \ln\left(\dfrac{x+1}{x-5}\right)$ to be real, $\dfrac{x+1}{x-5} > 0$.
The boundary points are –1 and 5. Testing each interval gives a domain of $(-\infty, -1) \cup (5, \infty)$.

**113.** $f(13) = 62 + 35\log(13-4) \approx 95.4$
She is approximately 95.4% of her adult height.

**115. a.** 2008 is 39 years after 1969.
$$f(x) = -7.52\ln x + 53$$
$$f(39) = -7.52\ln 39 + 53 \approx 25.5$$

According to the function, 25.5% of first-year college men expressed antifeminist views in 2008. This underestimates the value in the graph by 0.7%.

**b.** 2015 is 46 years after 1969.
$$f(x) = -7.52\ln x + 53$$
$$f(46) = -7.52\ln 46 + 53 \approx 24.2$$

According to the function, 24.2% of first-year college men will express antifeminist views in 2015.

**117.** $D = 10\log\left[10^{12}(6.3\times10^6)\right] \approx 188$

Yes, the sound can rupture the human eardrum.

**119. a.** $f(0) = 88 - 15\ln(0 + 1) = 88$
The average score on the original exam was 88.

**b.** $f(2) = 88 - 15\ln(2 + 1) = 71.5$
$f(4) = 88 - 15\ln(4 + 1) = 63.9$
$f(6) = 88 - 15\ln(6 + 1) = 58.8$
$f(8) = 88 - 15\ln(8 + 1) = 55$

$f(10) = 88 - 15\ln(10 + 1) = 52$
$f(12) = 88 - 15\ln(12 + 1) = 49.5$

The average score after 2 months was about 71.5, after 4 months was about 63.9, after 6 months was about 58.8, after 8 months was about 55, after 10 months was about 52, and after one year was about 49.5.

**121. – 127.** Answers will vary.

**129.**

$g(x)$ is $f(x)$ shifted 3 units upward.

**131.**

$g(x)$ is $f(x)$ shifted right 2 units and upward 1 unit.

**133. a.**

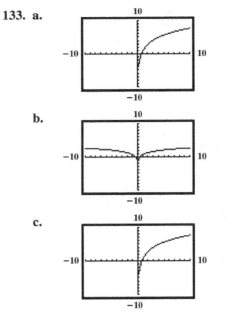

**b.**

**c.**

**d** They are the same.
$\log_b MN = \log_b M + \log_b N$

**e.** The sum of the logarithms of its factors.

**135.** makes sense

**137.** makes sense

**139.** false; Changes to make the statement true will vary.

A sample change is: $\dfrac{\log_2 8}{\log_2 4} = \dfrac{3}{2}$

**141.** false; Changes to make the statement true will vary. A sample change is: The domain of $f(x) = \log_2 x$ is $(0, \infty)$.

**143.** $\dfrac{\log_3 81 - \log_\pi 1}{\log_{2\sqrt{2}} 8 - \log 0.001} = \dfrac{4-0}{2-(-3)} = \dfrac{4}{5}$

**145.** $\log_4 60 < \log_4 64 = 3$ so $\log_4 60 < 3$.

$\log_3 40 > \log_3 27 = 3$ so $\log_3 40 > 3$.

$\log_4 60 < 3 < \log_3 40$

$\log_3 40 > \log_4 60$

**147. a.** $\log_2 32 = \log_2 2^5 = 5$

   **b.** $\log_2 8 + \log_2 4 = \log_2 2^3 + \log_2 2^2 = 3 + 2 = 5$

   **c.** $\log_2(8 \cdot 4) = \log_2 8 + \log_2 4$

**148. a.** $\log_2 16 = \log_2 2^4 = 4$

   **b.** $\log_2 32 - \log_2 2 = \log_2 2^5 - \log_2 2 = 5 - 1 = 4$

   **c.** $\log_2\left(\dfrac{32}{2}\right) = \log_2 32 - \log_2 2$

**149. a.** $\log_3 81 = \log_3 3^4 = 4$

   **b.** $2\log_3 9 = 2\log_3 3^2 = 2 \cdot 2 = 4$

   **c.** $\log_3 9^2 = 2\log_3 9$

## Section 4.3

### Check Point Exercises

**1. a.** $\log_6(7 \cdot 11) = \log_6 7 + \log_6 11$

   **b.** $\log(100x) = \log 100 + \log x$
$$= 2 + \log x$$

**2. a.** $\log_8\left(\dfrac{23}{x}\right) = \log_8 23 - \log_8 x$

   **b.** $\ln\left(\dfrac{e^5}{11}\right) = \ln e^5 - \ln 11$
$$= 5 - \ln 11$$

**3. a.** $\log_6 3^9 = 9\log_6 3$

   **b.** $\ln \sqrt[3]{x} = \ln x^{1/3} = \dfrac{1}{3}\ln x$

   **c.** $\log(x+4)^2 = 2\log(x+4)$

**4. a.** $\log_b x^4 \sqrt[3]{y}$
$$= \log_b x^4 y^{1/3}$$
$$= \log_b x^4 + \log_b y^{1/3}$$
$$= 4\log_b x + \dfrac{1}{3}\log_b y$$

   **b.** $\log_5 \dfrac{\sqrt{x}}{25y^3}$
$$= \log_5 \dfrac{x^{1/2}}{25y^3}$$
$$= \log_5 x^{1/2} - \log_5 25y^3$$
$$= \log_5 x^{1/2} - \left(\log_5 5^2 + \log_5 y^3\right)$$
$$= \tfrac{1}{2}\log_5 x - \log_5 5^2 - \log_5 y^3$$
$$= \tfrac{1}{2}\log_5 x - 2\log_5 5 - 3\log_5 y$$
$$= \tfrac{1}{2}\log_5 x - 2 - 3\log_5 y$$

**5. a.** $\log 25 + \log 4 = \log(25 \cdot 4) = \log 100 = 2$

   **b.** $\log(7x+6) - \log x = \log \dfrac{7x+6}{x}$

**6. a.** $\ln x^2 + \dfrac{1}{3}\ln(x+5)$
$$= \ln x^2 + \ln(x+5)^{1/3}$$
$$= \ln x^2 (x+5)^{1/3}$$
$$= \ln x^2 \sqrt[3]{x+5}$$

**b.** $2\log(x-3)-\log x$

$=\log(x-3)^2-\log x$

$=\log\dfrac{(x-3)^2}{x}$

**c.** $\dfrac{1}{4}\log_b x-2\log_b 5-10\log_b y$

$=\log_b x^{1/4}-\log_b 5^2-\log_b y^{10}$

$=\log_b x^{1/4}-\left(\log_b 25-\log_b y^{10}\right)$

$=\log_b x^{1/4}-\log_b 25y^{10}$

$=\log_b\dfrac{x^{1/4}}{25y^{10}}$ or $\log_b\dfrac{\sqrt[4]{x}}{25y^{10}}$

**7.** $\log_7 2506=\dfrac{\log 2506}{\log 7}\approx 4.02$

**8.** $\log_7 2506=\dfrac{\ln 2506}{\ln 7}\approx 4.02$

**Concept and Vocabulary Check 4.3**

**1.** $\log_b M+\log_b N$ ; sum

**2.** $\log_b M-\log_b N$ ; difference

**3.** $p\log_b M$ ; product

**4.** $\log_a M$ ; $\log_a b$

**Exercise Set 4.3**

**1.** $\log_5(7\cdot 3)=\log_5 7+\log_5 3$

**3.** $\log_7(7x)=\log_7 7+\log_7 x=1+\log_7 x$

**5.** $\log(1000x)=\log 1000+\log x=3+\log x$

**7.** $\log_7\left(\dfrac{7}{x}\right)=\log_7 7-\log_7 x=1-\log_7 x$

**9.** $\log\left(\dfrac{x}{100}\right)=\log x-\log 100=\log x-2$

**11.** $\log_4\left(\dfrac{64}{y}\right)=\log_4 64-\log_4 y$

$=3-\log_4 y$

**13.** $\ln\left(\dfrac{e^2}{5}\right)=\ln e^2-\ln 5=2\ln e-\ln 5=2-\ln 5$

**15.** $\log_b x^3=3\log_b x$

**17.** $\log N^{-6}=-6\log N$

**19.** $\ln\sqrt[5]{x}=\ln x^{(1/5)}=\dfrac{1}{5}\ln x$

**21.** $\log_b x^2 y=\log_b x^2+\log_b y=2\log_b x+\log_b y$

**23.** $\log_4\left(\dfrac{\sqrt{x}}{64}\right)=\log_4 x^{1/2}-\log_4 64=\dfrac{1}{2}\log_4 x-3$

**25.** $\log_6\left(\dfrac{36}{\sqrt{x+1}}\right)=\log_6 36-\log_6(x+1)^{1/2}$

$=2-\dfrac{1}{2}\log_6(x+1)$

**27.** $\log_b\left(\dfrac{x^2 y}{z^2}\right)=\log_b\left(x^2 y\right)-\log_b z^2$

$=\log_b x^2+\log_b y-\log_b z^2$

$=2\log_b x+\log_b y-2\log_b z$

**29.** $\log\sqrt{100x}=\log(100x)^{1/2}$

$=\dfrac{1}{2}\log(100x)$

$=\dfrac{1}{2}(\log 100+\log x)$

$=\dfrac{1}{2}(2+\log x)$

$=1+\dfrac{1}{2}\log x$

**31.** $\log\sqrt[3]{\dfrac{x}{y}}=\log\left(\dfrac{x}{y}\right)^{1/3}$

$=\dfrac{1}{3}\left[\log\left(\dfrac{x}{y}\right)\right]$

$=\dfrac{1}{3}(\log x-\log y)$

$=\dfrac{1}{3}\log x-\dfrac{1}{3}\log y$

**33.** $\log_b \dfrac{\sqrt{x}\,y^3}{z^3}$

$= \log_b x^{1/2} + \log_b y^3 - \log_b z^3$

$= \dfrac{1}{2}\log_b x + 3\log_b y - 3\log_b z$

**35.** $\log_5 \sqrt[3]{\dfrac{x^2 y}{25}}$

$= \log_5 x^{2/3} + \log_5 y^{1/3} - \log_5 25^{1/3}$

$= \dfrac{2}{3}\log_5 x + \dfrac{1}{3}\log_5 y - \log_5 5^{2/3}$

$= \dfrac{2}{3}\log_5 x + \dfrac{1}{3}\log_5 y - \dfrac{2}{3}$

**37.** $\ln\left[\dfrac{x^3\sqrt{x^2+1}}{(x+1)^4}\right]$

$= \ln x^3 + \ln\sqrt{x^2+1} - \ln(x+1)^4$

$= 3\ln x + \dfrac{1}{2}\ln(x^2+1) - 4\ln(x+1)$

**39.** $\log\left[\dfrac{10x^2\sqrt[3]{1-x}}{7(x+1)^2}\right]$

$= \log 10 + \log x^2 + \log\sqrt[3]{1-x} - \log 7 - \log(x+1)^2$

$= 1 + 2\log x + \dfrac{1}{3}\log(1-x) - \log 7 - 2\log(x+1)$

**41.** $\log 5 + \log 2 = \log(5\cdot 2) = \log 10 = 1$

**43.** $\ln x + \ln 7 = \ln(7x)$

**45.** $\log_2 96 - \log_2 3 = \log_2\left(\dfrac{96}{3}\right) = \log_2 32 = 5$

**47.** $\log(2x+5) - \log x = \log\left(\dfrac{2x+5}{x}\right)$

**49.** $\log x + 3\log y = \log x + \log y^3 = \log(xy^3)$

**51.** $\dfrac{1}{2}\ln x + \ln y = \ln x^{1/2} + \ln y$

$= \ln\left(x^{\frac{1}{2}}y\right)$ or $\ln\left(y\sqrt{x}\right)$

**53.** $2\log_b x + 3\log_b y = \log_b x^2 + \log_b y^3$

$= \log_b(x^2 y^3)$

**55.** $5\ln x - 2\ln y = \ln x^5 - \ln y^2 = \ln\left(\dfrac{x^5}{y^2}\right)$

**57.** $3\ln x - \dfrac{1}{3}\ln y = \ln x^3 - \ln y^{1/3}$

$= \ln\left(\dfrac{x^3}{y^{1/3}}\right)$ or $\ln\left(\dfrac{x^3}{\sqrt[3]{y}}\right)$

**59.** $4\ln(x+6) - 3\ln x = \ln(x+6)^4 - \ln x^3$

$= \ln\dfrac{(x+6)^4}{x^3}$

**61.** $3\ln x + 5\ln y - 6\ln z$

$= \ln x^3 + \ln y^5 - \ln z^6$

$= \ln\dfrac{x^3 y^5}{z^6}$

**63.** $\dfrac{1}{2}\left(\log x + \log y\right)$

$= \dfrac{1}{2}(\log xy)$

$= \log(xy)^{1/2}$

$= \log\sqrt{xy}$

**65.** $\dfrac{1}{2}(\log_5 x + \log_5 y) - 2\log_5(x+1)$

$= \dfrac{1}{2}\log_5 xy - \log_5(x+1)^2$

$= \log_5(xy)^{1/2} - \log_5(x+1)^2$

$= \log_5\dfrac{(xy)^{1/2}}{(x+1)^2}$

$= \log_5\dfrac{\sqrt{xy}}{(x+1)^2}$

**67.** $\dfrac{1}{3}[2\ln(x+5)-\ln x - \ln(x^2-4)]$

$=\dfrac{1}{3}[\ln(x+5)^2 - \ln x - \ln(x^2-4)]$

$=\dfrac{1}{3}\left[\ln\dfrac{(x+5)^2}{x(x^2-4)}\right]$

$=\ln\left[\dfrac{(x+5)^2}{x(x^2-4)}\right]^{1/3}$

$=\ln\sqrt[3]{\dfrac{(x+5)^2}{x(x^2-4)}}$

**69.** $\log x + \log\left(x^2-1\right) - \log 7 - \log(x+1)$

$=\log x + \log\left(x^2-1\right) - \left(\log 7 + \log(x+1)\right)$

$=\log\left(x\left(x^2-1\right)\right) - \log\left(7(x+1)\right)$

$=\log\dfrac{x\left(x^2-1\right)}{7(x+1)}$

$=\log\dfrac{x(x+1)(x-1)}{7(x+1)}$

$=\log\dfrac{x(x-1)}{7}$

**71.** $\log_5 13 = \dfrac{\log 13}{\log 5} \approx 1.5937$

**73.** $\log_{14} 87.5 = \dfrac{\ln 87.5}{\ln 14} \approx 1.6944$

**75.** $\log_{0.1} 17 = \dfrac{\log 17}{\log 0.1} \approx -1.2304$

**77.** $\log_{\pi} 63 = \dfrac{\ln 63}{\ln \pi} \approx 3.6193$

**79.** $y = \log_3 x = \dfrac{\log x}{\log 3}$

**81.** $y = \log_2(x+2) = \dfrac{\log(x+2)}{\log 2}$

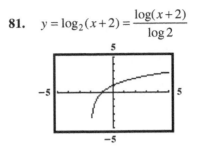

**83.** $\log_b \dfrac{3}{2} = \log_b 3 - \log_b 2 = C - A$

**85.** $\log_b 8 = \log_b 2^3 = 3\log_b 2 = 3A$

**87.** $\log_b \sqrt{\dfrac{2}{27}} = \log_b\left(\dfrac{2}{27}\right)^{\frac{1}{2}}$

$=\dfrac{1}{2}\log_b\left(\dfrac{2}{3^3}\right)$

$=\dfrac{1}{2}\left(\log_b 2 - \log_b 3^3\right)$

$=\dfrac{1}{2}\left(\log_b 2 - 3\log_b 3\right)$

$=\dfrac{1}{2}\log_b 2 - \dfrac{3}{2}\log_b 3$

$=\dfrac{1}{2}A - \dfrac{3}{2}C$

**89.** false; $\ln e = 1$

**91.** false; $\log_4 (2x)^3 = 3\log_4 (2x)$

**93.** true; $x\log 10^x = x\cdot x = x^2$

**95.** true; $\ln(5x) + \ln 1 = \ln 5x + 0 = \ln 5x$

**97.** false; $\log(x+3) - \log(2x) = \log\dfrac{x+3}{2x}$

**99.** true; quotient rule

**101.** true; $\log_3 7 = \dfrac{\log 7}{\log 3} = \dfrac{1}{\frac{\log 3}{\log 7}} = \dfrac{1}{\log_7 3}$

**103. a.** $D = 10\log\dfrac{I}{I_0}$

**b.** $D_1 = 10\log\left(\dfrac{100I}{I_0}\right)$

$= 10\log\left(100I - I_0\right)$

$= 10\log 100 + 10\log I - 10\log I_0$

$= 10(2) + 10\log I - 10\log I_0$

$= 20 + 10\log\dfrac{I}{I_0}$

This is 20 more than the loudness level of the softer sound. This means that the 100 times louder sound will be 20 decibels louder.

**105. – 111.** Answers will vary.

**113.** **a.** $y = \log_3 x = \dfrac{\ln x}{\ln 3}$

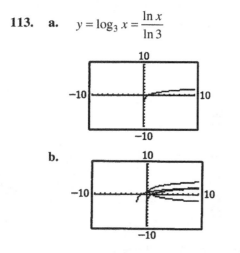

**b.**

To obtain the graph of $y = 2 + \log_3 x$, shift the graph of $y = \log_3 x$ two units upward. To obtain the graph of $y = \log_3(x + 2)$, shift the graph of $y = \log_3 x$ two units left. To obtain the graph of $y = -\log_3 x$, reflect the graph of $y = \log_3 x$ about the $x$-axis.

**115.** $\log_3 x = \dfrac{\log x}{\log 3}$;

$\log_{25} x = \dfrac{\log x}{\log 25}$;

$\log_{100} x = \dfrac{\log x}{\log 100}$

**a.** top graph: $y = \log_{100} x$
bottom graph: $y = \log_3 x$

**b.** top graph: $y = \log_3 x$
bottom graph: $y = \log_{100} x$

**c.** Comparing graphs of $\log_b x$ for $b > 1$, the graph of the equation with the largest $b$ will be on the top in the interval $(0, 1)$ and on the bottom in the interval $(1, \infty)$.

**117. – 119.** Answers will vary.

**121.** makes sense

**123.** makes sense

**125.** true

**127.** false; Changes to make the statement true will vary. A sample change is: $\log_b\left(x^3 + y^3\right)$ cannot be simplified. If we were taking the logarithm of a product and not a sum, we would have been able to simplify as follows.

$\log_b\left(x^3 y^3\right) = \log_b x^3 + \log_b y^3$

$= 3\log_b x + 3\log_b y$

**129.** $\log e = \log_{10} e = \dfrac{\ln e}{\ln 10} = \dfrac{1}{\ln 10}$

**131.** $e^{\ln 8x^5 - \ln 2x^2} = e^{\ln\left(\frac{8x^5}{2x^2}\right)} = e^{\ln\left(4x^3\right)} = 4x^3$

**133.** Let $\log_b M = R$ and $\log_b N = S$.

Then $\log_b M = R$ means $b^R = M$ and

$\log_b N = S$ means $b^S = N$.

Thus, $\dfrac{M}{N} = \dfrac{b^R}{b^S}$

$\log_b \dfrac{M}{N} = \log_b b^{R-S}$

$\log_b \dfrac{M}{N} = R - S$

$\log_b \dfrac{M}{N} = \log_b M - \log_b N$

**134.** $a(x-2) = b(2x+3)$

$ax - 2a = 2bx + 3b$

$ax - 2bx = 2a + 3b$

$x(a - 2b) = 2a + 3b$

$x = \dfrac{2a + 3b}{a - 2b}$

**135.** $x(x-7) = 3$

$x^2 - 7x = 3$

$x^2 - 7x - 3 = 0$

$x = \dfrac{-b \pm \sqrt{b^2 - 4ac}}{2a}$

$x = \dfrac{-(-7) \pm \sqrt{(-7)^2 - 4(1)(-3)}}{2(1)}$

$x = \dfrac{7 \pm \sqrt{61}}{2}$

The solution set is $\left\{ \dfrac{7 \pm \sqrt{61}}{2} \right\}$.

**136.** $\dfrac{x+2}{4x+3} = \dfrac{1}{x}$

$x(4x+3)\left( \dfrac{x+2}{4x+3} \right) = x(4x+3)\left( \dfrac{1}{x} \right)$

$x(x+2) = 4x+3$

$x^2 + 2x = 4x + 3$

$x^2 - 2x - 3 = 0$

$(x+1)(x-3) = 0$

$x+1 = 0 \quad$ or $\quad x-3 = 0$

$x = -1 \qquad\qquad x = 3$

The solution set is $\{-1, 3\}$.

**Mid-Chapter 4 Check Point**

**1.**

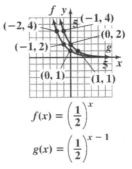

$f(x) = 2^x$
$g(x) = 2^x - 3$

asymptote of $f$: $\ y = 0$
asymptote of $g$: $\ y = -3$
domain of $f$ = domain of $g = (-\infty, \infty)$
range of $f = (0, \infty)$
range of $g = (-3, \infty)$

**2.**

$f(x) = \left( \dfrac{1}{2} \right)^x$

$g(x) = \left( \dfrac{1}{2} \right)^{x-1}$

asymptote of $f$: $\ y = 0$
asymptote of $g$: $\ y = 0$
domain of $f$ = domain of $g = (-\infty, \infty)$
range of $f$ = range of $g = (0, \infty)$

**3.**

$f(x) = e^x$
$g(x) = \ln x$

asymptote of $f$: $\ y = 0$
asymptote of $g$: $\ x = 0$
domain of $f$ = range of $g = (-\infty, \infty)$
range of $f$ = domain of $g = (0, \infty)$

**4.**

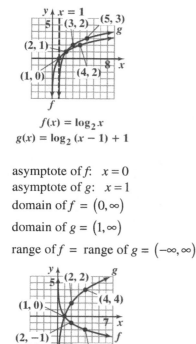

$$f(x) = \log_2 x$$
$$g(x) = \log_2(x-1) + 1$$

asymptote of $f$: $x = 0$
asymptote of $g$: $x = 1$
domain of $f = (0, \infty)$
domain of $g = (1, \infty)$
range of $f$ = range of $g = (-\infty, \infty)$

**5.**

$$f(x) = \log_{1/2} x$$
$$g(x) = -2\log_{1/2} x$$

asymptote of $f$: $x = 0$
asymptote of $g$: $x = 0$
domain of $f$ = domain of $g = (0, \infty)$
range of $f$ = range of $g = (-\infty, \infty)$

**6.** $f(x) = \log_3(x+6)$

The argument of the logarithm must be positive:
$x + 6 > 0$
$\quad x > -6$
domain: $\{x \mid x > -6\}$ or $(-6, \infty)$.

**7.** $f(x) = \log_3 x + 6$

The argument of the logarithm must be positive:
$x > 0$
domain: $\{x \mid x > 0\}$ or $(0, \infty)$.

**8.** $\log_3(x+6)^2$

The argument of the logarithm must be positive.
Now $(x+6)^2$ is always positive, except when
$x = -6$
domain: $\{x \mid x \neq 0\}$ or $(-\infty, -6) \cup (-6, \infty)$.

**9.** $f(x) = 3^{x+6}$

domain: $\{x \mid x \text{ is a real number}\}$ or $(-\infty, \infty)$.

**10.** $\log_2 8 + \log_5 25 = \log_2 2^3 + \log_5 5^2$
$$= 3 + 2 = 5$$

**11.** $\log_3 \dfrac{1}{9} = \log_3 \dfrac{1}{3^2} = \log_3 3^{-2} = -2$

**12.** Let $\log_{100} 10 = y$
$$100^y = 10$$
$$\left(10^2\right)^y = 10^1$$
$$10^{2y} = 10^1$$
$$2y = 1$$
$$y = \frac{1}{2}$$

**13.** $\log \sqrt[3]{10} = \log 10^{\frac{1}{3}} = \dfrac{1}{3}$

**14.** $\log_2(\log_3 81) = \log_2\left(\log_3 3^4\right)$
$$= \log_2 4 = \log_2 2^2 = 2$$

**15.** $\log_3\left(\log_2 \dfrac{1}{8}\right) = \log_3\left(\log_2 \dfrac{1}{2^3}\right)$
$$= \log_3\left(\log_2 2^{-3}\right)$$
$$= \log_3(-3)$$
$$= \text{not possible}$$

This expression is impossible to evaluate because $\log_3(-3)$ is undefined.

**16.** $6^{\log_6 5} = 5$

**17.** $\ln e^{\sqrt{7}} = \sqrt{7}$

**18.** $10^{\log 13} = 13$

**19.** $\log_{100} 0.1 = y$
$$100^y = 0.1$$
$$\left(10^2\right)^y = \frac{1}{10}$$
$$10^{2y} = 10^{-1}$$
$$2y = -1$$
$$y = -\frac{1}{2}$$

**20.** $\log_\pi \pi^{\sqrt{\pi}} = \sqrt{\pi}$

**21.** $\log\left(\dfrac{\sqrt{xy}}{1000}\right) = \log\left(\sqrt{xy}\right) - \log 1000$

$$= \log(xy)^{\frac{1}{2}} - \log 10^3$$

$$= \frac{1}{2}\log(xy) - 3$$

$$= \frac{1}{2}(\log x + \log y) - 3$$

$$= \frac{1}{2}\log x + \frac{1}{2}\log y - 3$$

**22.** $\ln\left(e^{19}x^{20}\right) = \ln e^{19} + \ln x^{20}$

$$= 19 + 20\ln x$$

**23.** $8\log_7 x - \dfrac{1}{3}\log_7 y = \log_7 x^8 - \log_7 y^{\frac{1}{3}}$

$$= \log_7\left(\frac{x^8}{y^{\frac{1}{3}}}\right)$$

$$= \log_7\left(\frac{x^8}{\sqrt[3]{y}}\right)$$

**24.** $7\log_5 x + 2\log_5 x = \log_5 x^7 + \log_5 x^2$

$$= \log_5\left(x^7 \cdot x^2\right)$$

$$= \log_5 x^9$$

**25.** $\dfrac{1}{2}\ln x - 3\ln y - \ln(z-2)$

$$= \ln x^{\frac{1}{2}} - \ln y^3 - \ln(z-2)$$

$$= \ln\sqrt{x} - \left[\ln y^3 + \ln(z-2)\right]$$

$$= \ln\sqrt{x} - \ln\left[y^3(z-2)\right]$$

$$= \ln\left[\frac{\sqrt{x}}{y^3(z-2)}\right]$$

**26.** Continuously: $A = 8000e^{0.08(3)}$

$$\approx 10,170$$

Monthly: $A = 8000\left(1 + \dfrac{0.08}{12}\right)^{12\cdot3}$

$$\approx 10,162$$

$10,170 - 10,162 = 8$

Interest returned will be \$8 more if compounded continuously.

**Section 4.4**

**Check Point Exercises**

1.  **a.**  $5^{3x-6} = 125$

$5^{3x-6} = 5^3$

$3x - 6 = 3$

$3x = 9$

$x = 3$

The solution set is $\{3\}$.

    **b.**  $8^{x+2} = 4^{x-3}$

$\left(2^3\right)^{x+2} = \left(2^2\right)^{x-3}$

$2^{3x+6} = 2^{2x-6}$

$3x + 6 = 2x - 6$

$x = -12$

The solution set is $\{-12\}$.

2.  **a.**  $5^x = 134$

$\ln 5^x = \ln 134$

$x \ln 5 = \ln 134$

$x = \dfrac{\ln 134}{\ln 5} \approx 3.04$

The solution set is $\left\{\dfrac{\ln 134}{\ln 5}\right\}$,

approximately 3.04.

    **b.**  $10^x = 8000$

$\log 10^x = \log 8000$

$x \log 10 = \log 8000$

$x = \log 8000 \approx 3.90$

The solution set is $\{\log 8000\}$, approximately 3.90.

3.  $7e^{2x} = 63$

$e^{2x} = 9$

$\ln e^{2x} = \ln 9$

$2x = \ln 9$

$x = \dfrac{\ln 9}{2} \approx 1.10$

The solution set is $\left\{\dfrac{\ln 9}{2}\right\}$,

approximately 1.10.

4.  $3^{2x-1} = 7^{x+1}$

$\ln 3^{2x-1} = \ln 7^{x+1}$

$(2x-1)\ln 3 = (x+1)\ln 7$

$2x \ln 3 - \ln 3 = x \ln 7 + \ln 7$

$2x \ln 3 - x \ln 7 = \ln 3 + \ln 7$

$x(2\ln 3 - \ln 7) = \ln 3 + \ln 7$

$x = \dfrac{\ln 3 + \ln 7}{2\ln 3 - \ln 7}$

$x \approx 12.11$

The solution set is $\left\{\dfrac{\ln 3 + \ln 7}{2\ln 3 - \ln 7}\right\}$,

approximately 12.11.

5.  $e^{2x} - 8e^x + 7 = 0$

$\left(e^x - 7\right)\left(e^x - 1\right) = 0$

$e^x - 7 = 0 \qquad \text{or} \qquad e^x - 1 = 0$

$e^x = 7 \qquad\qquad\qquad e^x = 1$

$\ln e^x = \ln 7 \qquad\qquad \ln e^x = \ln 1$

$x = \ln 7 \qquad\qquad\quad x = 0$

The solution set is {0, ln7}. The solutions are 0 and (approximately) 1.95.

6.  **a.**  $\log_2(x-4) = 3$

$2^3 = x - 4$

$8 = x - 4$

$12 = x$

Check:

$\log_2(x-4) = 3$

$\log_2(12-4) = 3$

$\log_2 8 = 3$

$3 = 3$

The solution set is $\{12\}$.

**b.**
$$4\ln 3x = 8$$
$$\ln 3x = 2$$
$$e^{\ln 3x} = e^2$$
$$3x = e^2$$
$$x = \frac{e^2}{3} \approx 2.46$$

Check
$$4\ln 3x = 8$$
$$4\ln 3\left(\frac{e^2}{3}\right) = 8$$
$$4\ln e^2 = 8$$
$$4(2) = 8$$
$$8 = 8$$

The solution set is $\left\{\dfrac{e^2}{3}\right\}$,
approximately 2.46.

**7.**
$$\log x + \log(x-3) = 1$$
$$\log x(x-3) = 1$$
$$10^1 = x(x-3)$$
$$10 = x^2 - 3x$$
$$0 = x^2 - 3x - 10$$
$$0 = (x-5)(x+2)$$
$$x - 5 = 0 \quad \text{or} \quad x + 2 = 0$$
$$x = 5 \quad \text{or} \qquad x = -2$$

Check
Checking 5:
$$\log 5 + \log(5-3) = 1$$
$$\log 5 + \log 2 = 1$$
$$\log(5 \cdot 2) = 1$$
$$\log 10 = 1$$
$$1 = 1$$

Checking –2:
$$\log x + \log(x-3) = 1$$
$$\log(-2) + \log(-2-3) \; \mathbf{0} \; 1$$

Negative numbers do not have logarithms so
–2 does not check.
The solution set is $\{5\}$.

**8.**
$$\ln(x-3) = \ln(7x-23) - \ln(x+1)$$
$$\ln(x-3) = \ln\frac{7x-23}{x+1}$$
$$x - 3 = \frac{7x-23}{x+1}$$
$$(x-3)(x+1) = 7x - 23$$
$$x^2 - 2x - 3 = 7x - 23$$
$$x^2 - 9x + 20 = 0$$
$$(x-4)(x-5) = 0$$
$$x = 4 \quad \text{or} \quad x = 5$$

Both values produce true statements.
The solution set is $\{4, 5\}$.

**9.** For a risk of 7%, let $R = 7$ in
$$R = 6e^{12.77x}$$
$$6e^{12.77x} = 7$$
$$e^{12.77x} = \frac{7}{6}$$
$$\ln e^{12.77x} = \ln\left(\frac{7}{6}\right)$$
$$12.77x = \ln\left(\frac{7}{6}\right)$$
$$x = \frac{\ln\left(\frac{7}{6}\right)}{12.77} \approx 0.01$$

For a blood alcohol concentration of 0.01, the risk
of a car accident is 7%.

**10.**
$$A = P\left(1 + \frac{r}{n}\right)^{nt}$$
$$3600 = 1000\left(1 + \frac{0.08}{4}\right)^{4t}$$
$$1000\left(1 - \frac{0.08}{4}\right)^{4t} = 3600$$
$$1000(1 + 0.02)^{4t} = 3600$$
$$1000(1.02)^{4t} = 3600$$
$$(1.02)^{4t} = \ln 3.6$$
$$4t\ln(1.02) = \ln 3.6$$
$$t = \frac{\ln 3.6}{4\ln 1.02}$$
$$\approx 16.2$$

After approximately 16.2 years, the $1000 will
grow to an accumulated value of $3600.

**11.** $f(x) = 62 + 35\log(x-4)$

Solve the equation when $f(x) = 97$.

$62 + 35\log(x-4) = 97$

$35\log(x-4) = 35$

$\log(x-4) = \dfrac{35}{35}$

$\log(x-4) = 1$

$x - 4 = 10^1$

$x = 10 + 4$

$x = 14$

At age 14, a girl will attain 97% of her adult height.

**Concept and Vocabulary Check 4.4**

**1.** $M = N$

**2.** $4x - 1$

**3.** $\dfrac{\ln 20}{\ln 9}$

**4.** $\ln 6$

**5.** $5^3$

**6.** $(x^2 + x)$

**7.** $\dfrac{7x - 23}{x + 1}$

**8.** false

**9.** true

**10.** false

**11.** true

**Exercise Set 4.4**

**1.** $2^x = 64$

$2^x = 2^6$

$x = 6$

The solution is 6, and the solution set is $\{6\}$.

**3.** $5^x = 125$

$5^x = 5^3$

$x = 3$

The solution is 3, and the solution set is $\{3\}$.

**5.** $2^{2x-1} = 32$

$2^{2x-1} = 2^5$

$2x - 1 = 5$

$2x = 6$

$x = 3$

The solution is 3, and the solution set is $\{3\}$.

**7.** $4^{2x-1} = 64$

$4^{2x-1} = 4^3$

$2x - 1 = 3$

$2x = 4$

$x = 2$

The solution is 2, and the solution set is $\{2\}$.

**9.** $32^x = 8$

$\left(2^5\right)^x = 2^3$

$2^{5x} = 2^3$

$5x = 3$

$x = \dfrac{3}{5}$

The solution is $\dfrac{3}{5}$, and the solution set is $\left\{\dfrac{3}{5}\right\}$.

**11.** $9^x = 27$

$\left(3^2\right)^x = 3^3$

$3^{2x} = 3^3$

$2x = 3$

$x = \dfrac{3}{2}$

The solution is $\dfrac{3}{2}$, and the solution set is $\left\{\dfrac{3}{2}\right\}$.

13. $3^{1-x} = \dfrac{1}{27}$

$3^{1-x} = \dfrac{1}{3^3}$

$3^{1-x} = 3^{-3}$

$1 - x = -3$

$-x = -4$

$x = 4$

The solution set is $\{4\}$.

15. $6^{\frac{x-3}{4}} = \sqrt{6}$

$6^{\frac{x-3}{4}} = 6^{\frac{1}{2}}$

$\dfrac{x-3}{4} = \dfrac{1}{2}$

$2(x-3) = 4(1)$

$2x - 6 = 4$

$2x = 10$

$x = 5$

The solution is 5, and the solution set is $\{5\}$.

17. $4^x = \dfrac{1}{\sqrt{2}}$

$\left(2^2\right)^x = \dfrac{1}{2^{\frac{1}{2}}}$

$2^{2x} = 2^{-\frac{1}{2}}$

$2x = -\dfrac{1}{2}$

$x = \dfrac{1}{2}\left(-\dfrac{1}{2}\right) = -\dfrac{1}{4}$

The solution is $-\dfrac{1}{4}$, and the solution set is $\left\{-\dfrac{1}{4}\right\}$.

19. $8^{x+3} = 16^{x-1}$

$\left(2^3\right)^{x+3} = \left(2^4\right)^{x-1}$

$2^{3x+9} = 2^{4x-4}$

$3x + 9 = 4x - 4$

$13 = x$

The solution set is $\{13\}$.

21. $e^{x+1} = \dfrac{1}{e}$

$e^{x+1} = e^{-1}$

$x + 1 = -1$

$x = -2$

The solution set is $\{-2\}$.

23. $10^x = 3.91$

$\ln 10^x = \ln 3.91$

$x \ln 10 = \ln 3.91$

$x = \dfrac{\ln 3.91}{\ln 10} \approx 0.59$

25. $e^x = 5.7$

$\ln e^x = 5.7$

$x = \ln 5.7 \approx 1.74$

27. $5^x = 17$

$\ln 5^x = \ln 17$

$x \ln 5 = \ln 17$

$x = \dfrac{\ln 17}{\ln 5} \approx 1.76$

29. $5e^x = 23$

$e^x = \dfrac{23}{5}$

$\ln e^x = \ln \dfrac{23}{5}$

$x = \ln \dfrac{23}{5} \approx 1.53$

31. $3e^{5x} = 1977$

$e^{5x} = 659$

$\ln e^{5x} = \ln 659$

$x = \dfrac{\ln 659}{5} \approx 1.30$

33. $e^{1-5x} = 793$

$\ln e^{1-5x} = \ln 793$

$(1-5x)(\ln e) = \ln 793$

$1 - 5x = \ln 793$

$5x = 1 - \ln 793$

$x = \dfrac{1 - \ln 793}{5} \approx -1.14$

**35.**
$$e^{5x-3} - 2 = 10,476$$
$$e^{5x-3} = 10,478$$
$$\ln e^{5x-3} = \ln 10,478$$
$$(5x-3)\ln e = \ln 10,478$$
$$5x - 3 = \ln 10,478$$
$$5x = \ln 10,478 + 3$$
$$x = \frac{\ln 10,478 + 3}{5} \approx 2.45$$

**37.**
$$7^{x+2} = 410$$
$$\ln 7^{x+2} = \ln 410$$
$$(x+2)\ln 7 = \ln 410$$
$$x + 2 = \frac{\ln 410}{\ln 7}$$
$$x = \frac{\ln 410}{\ln 7} - 2 \approx 1.09$$

**39.**
$$7^{0.3x} = 813$$
$$\ln 7^{0.3x} = \ln 813$$
$$0.3x \ln 7 = \ln 813$$
$$x = \frac{\ln 813}{0.3 \ln 7} \approx 11.48$$

**41.**
$$5^{2x+3} = 3^{x-1}$$
$$\ln 5^{2x+3} = \ln 3^{x-1}$$
$$(2x+3)\ln 5 = (x-1)\ln 3$$
$$2x \ln 5 + 3 \ln 5 = x \ln 3 - \ln 3$$
$$3\ln 5 + \ln 3 = x \ln 3 - 2x \ln 5$$
$$3\ln 5 + \ln 3 = x(\ln 3 - 2\ln 5)$$
$$\frac{3\ln 5 + \ln 3}{\ln 3 - 2\ln 5} = x$$
$$-2.80 \approx x$$

**43.**
$$e^{2x} - 3e^x + 2 = 0$$
$$(e^x - 2)(e^x - 1) = 0$$
$$e^x - 2 = 0 \quad \text{or} \quad e^x - 1 = 0$$
$$e^x = 2 \qquad\qquad e^x = 1$$
$$\ln e^x = \ln 2 \qquad \ln e^x = \ln 1$$
$$x = \ln 2 \qquad\qquad x = 0$$
The solution set is {0, ln 2}. The solutions are 0 and approximately 0.69.

**45.**
$$e^{4x} + 5e^{2x} - 24 = 0$$
$$\left(e^{2x} + 8\right)\left(e^{2x} - 3\right) = 0$$
$$e^{2x} + 8 = 0 \qquad \text{or} \quad e^{2x} - 3 = 0$$
$$e^{2x} = -8 \qquad\qquad e^{2x} = 3$$
$$\ln e^{2x} = \ln(-8) \qquad \ln e^{2x} = \ln 3$$
$$2x = \ln(-8) \qquad\qquad 2x = \ln 3$$
$$\ln(-8) \text{ does not exist} \qquad x = \frac{\ln 3}{2}$$
$$x = \frac{\ln 3}{2} \approx 0.55$$

**47.**
$$3^{2x} + 3^x - 2 = 0$$
$$(3^x + 2)(3^x - 1) = 0$$
$$3^x + 2 = 0 \qquad\qquad 3^x - 1 = 0$$
$$3^x = -2 \qquad\qquad 3^x = 1$$
$$\log 3^x = \log(-2) \qquad \log 3^x = \log 1$$
$$\text{does not exist} \qquad \log 3 = 0$$
$$x = \frac{0}{\log 3}$$
$$x = 0$$
The solution set is {0}.

**49.**
$$\log_3 x = 4$$
$$3^4 = x$$
$$81 = x$$

**51.**
$$\ln x = 2$$
$$e^2 = x$$
$$7.39 \approx x$$

**53.**
$$\log_4(x+5) = 3$$
$$4^3 = x + 5$$
$$59 = x$$

**55.**
$$\log_2(x+25) = 4$$
$$2^4 = x + 25$$
$$16 = x + 25$$
$$-9 = x$$

**57.** $\log_3(x+4) = -3$

$3^{-3} = x+4$

$\dfrac{1}{27} = x+4$

$\dfrac{1}{27} - 4 = x$

$-\dfrac{107}{27} = x$

$-3.96 \approx x$

**59.** $\log_4(3x+2) = 3$

$4^3 = 3x+2$

$64 = 3x+2$

$62 = 3x$

$\dfrac{62}{3} = x$

$20.67 \approx x$

**61.** $5\ln(2x) = 20$

$\ln(2x) = 4$

$e^{\ln(2x)} = e^4$

$2x = e^4$

$x = \dfrac{e^4}{2} \approx 27.30$

**63.** $6 + 2\ln x = 5$

$2\ln x = -1$

$\ln x = -\dfrac{1}{2}$

$e^{\ln x} = e^{-1/2}$

$x = e^{-1/2} \approx 0.61$

**65.** $\ln\sqrt{x+3} = 1$

$e^{\ln\sqrt{x+3}} = e^1$

$\sqrt{x+3} = e$

$x+3 = e^2$

$x = e^2 - 3 \approx 4.39$

**67.** $\log_5 x + \log_5(4x-1) = 1$

$\log_5(4x^2 - x) = 1$

$4x^2 - x = 5$

$4x^2 - x - 5 = 0$

$(4x-5)(x+1) = 0$

$x = \dfrac{5}{4}$ or $x = -1$

$x = -1$ does not check because $\log_5(-1)$ does not exist.

The solution set is $\left\{ \dfrac{5}{4} \right\}$.

**69.** $\log_3(x+6) + \log_3(x+4) = 1$

$\log_3\left[(x+6)(x+4)\right] = 1$

$(x+6)(x+4) = 3^1$

$x^2 + 10x + 24 = 3$

$x^2 + 10x + 21 = 0$

$(x+7)(x+3) = 0$

$x = -7$ or $x = -3$

$x = -7$ does not check because $\log_3(-7+6)$ does not exist. The solution set is $\{-3\}$.

**71.** $\log_2(x+2) - \log_2(x-5) = 3$

$\log_2\left(\dfrac{x+2}{x-5}\right) = 3$

$\dfrac{x+2}{x-5} = 2^3$

$\dfrac{x+2}{x-5} = 8$

$x+2 = 8(x-5)$

$x+2 = 8x - 40$

$7x = 42$

$x = 6$

**73.** $2\log_3(x+4) = \log_3 9 + 2$

$2\log_3(x+4) = 2 + 2$

$2\log_3(x+4) = 4$

$\log_3(x+4) = 2$

$3^2 = x+4$

$9 = x+4$

$5 = x$

**75.** $\log_2(x-6)+\log_2(x-4)-\log_2 x = 2$

$$\log_2 \frac{(x-6)(x-4)}{x} = 2$$

$$\frac{(x-6)(x-4)}{x} = 2^2$$

$$x^2 - 10x + 24 = 4x$$

$$x^2 - 14x + 24 = 0$$

$$(x-12)(x-2) = 0$$

$x - 12 = 0$    or    $x - 2 = 0$

$\quad\quad x = 12 \quad\quad\quad\quad x = 2$

The solution set is $\{12\}$ since $\log_2 (2-6) = \log_2 (-4)$ is not possible.

**77.** $\log(x+4) = \log x + \log 4$

$\log(x+4) = \log 4x$

$\quad\quad x + 4 = 4x$

$\quad\quad\quad 4 = 3x$

$\quad\quad\quad x = \frac{4}{3}$

This value is rejected. The solution set is $\left\{\frac{4}{3}\right\}$.

**79.** $\log(3x-3) = \log(x+1) + \log 4$

$\log(3x-3) = \log(4x+4)$

$\quad\quad 3x - 3 = 4x + 4$

$\quad\quad\quad -7 = x$

This value is rejected. The solution set is $\{\ \}$.

**81.** $2\log x = \log 25$

$\log x^2 = \log 25$

$\quad\quad x^2 = 25$

$\quad\quad x = \pm 5$

$-5$ is rejected. The solution set is $\{5\}$.

**83.** $\log(x+4) - \log 2 = \log(5x+1)$

$$\log \frac{x+4}{2} = \log(5x+1)$$

$$\frac{x+4}{2} = 5x+1$$

$$x + 4 = 10x + 2$$

$$-9x = -2$$

$$x = \frac{2}{9}$$

$$x \approx 0.22$$

**85.** $2\log x - \log 7 = \log 112$

$\log x^2 - \log 7 = \log 112$

$$\log \frac{x^2}{7} = \log 112$$

$$\frac{x^2}{7} = 112$$

$$x^2 = 784$$

$$x = \pm 28$$

$-28$ is rejected. The solution set is $\{28\}$.

**87.** $\log x + \log(x+3) = \log 10$

$\log(x^2 + 3x) = \log 10$

$\quad\quad x^2 + 3x = 10$

$\quad x^2 + 3x - 10 = 0$

$\quad (x+5)(x-2) = 0$

$x = -5$ or $x = 2$

$-5$ is rejected. The solution set is $\{2\}$.

**89.** $\ln(x-4) + \ln(x+1) = \ln(x-8)$

$\ln(x^2 - 3x - 4) = \ln(x-8)$

$\quad\quad x^2 - 3x - 4 = x - 8$

$\quad\quad x^2 - 4x + 4 = 0$

$\quad\quad (x-2)(x-2) = 0$

$\quad\quad x = 2$

2 is rejected. The solution set is $\{\ \}$.

**91.** $\ln(x-2) - \ln(x+3) = \ln(x-1) - \ln(x+7)$

$$\ln \frac{x-2}{x+3} = \ln \frac{x-1}{x+7}$$

$$\frac{x-2}{x+3} = \frac{x-1}{x+7}$$

$$(x-2)(x+7) = (x+3)(x-1)$$

$$x^2 + 5x - 14 = x^2 + 2x - 3$$

$$3x = 11$$

$$x = \frac{11}{3}$$

$$x \approx 3.67$$

**93.** $5^{2x} \cdot 5^{4x} = 125$

$$5^{2x+4x} = 5^3$$

$$5^{6x} = 5^3$$

$$6x = 3$$

$$x = \frac{1}{2}$$

**95.**
$$2|\ln x| - 6 = 0$$
$$2|\ln x| = 6$$
$$|\ln x| = 3$$
$$\ln x = 3 \quad \text{or} \quad \ln x = -3$$
$$x = e^3 \qquad\qquad x = e^{-3}$$
$$x \approx 20.09 \qquad\quad x \approx 0.05$$

**97.**
$$3^{x^2} = 45$$
$$\ln 3^{x^2} = \ln 45$$
$$x^2 \ln 3 = \ln 45$$
$$x^2 = \frac{\ln 45}{\ln 3}$$
$$x = \pm\sqrt{\frac{\ln 45}{\ln 3}} \approx \pm 1.86$$

**99.**
$$\ln(2x+1) + \ln(x-3) - 2\ln x = 0$$
$$\ln(2x+1) + \ln(x-3) - \ln x^2 = 0$$
$$\ln\frac{(2x+1)(x-3)}{x^2} = 0$$
$$\frac{(2x+1)(x-3)}{x^2} = e^0$$
$$\frac{2x^2 - 5x - 3}{x^2} = 1$$
$$2x^2 - 5x - 3 = x^2$$
$$x^2 - 5x - 3 = 0$$
$$x = \frac{-b \pm \sqrt{b^2 - 4ac}}{2a}$$
$$x = \frac{-(-5) \pm \sqrt{(-5)^2 - 4(1)(-3)}}{2(1)}$$
$$x = \frac{5 \pm \sqrt{37}}{2}$$
$$x = \frac{5 + \sqrt{37}}{2} \approx 5.54$$
$$x = \frac{5 - \sqrt{37}}{2} \approx -0.54 \ \text{(rejected)}$$

The solution set is $\left\{\dfrac{5+\sqrt{37}}{2}\right\}$.

**101.**
$$5^{x^2 - 12} = 25^{2x}$$
$$5^{x^2 - 12} = \left(5^2\right)^{2x}$$
$$5^{x^2 - 12} = 5^{4x}$$
$$x^2 - 12 = 4x$$
$$x^2 - 4x - 12 = 0$$
$$(x-6)(x+2) = 0$$
Apply the zero product property:
$$x - 6 = 0 \quad \text{or} \quad x + 2 = 0$$
$$x = 6 \qquad\qquad x = -2$$
The solutions are $-2$ and $6$, and the solution set is $\{-2,\ 6\}$.

**103. a.**  2010 is 0 years after 2010.
$$A = 37.3e^{0.0095t}$$
$$A = 37.3e^{0.0095(0)} = 37.3$$
The population of California was 37.3 million in 2010.

**b.**
$$A = 37.3e^{0.0095t}$$
$$40 = 37.3e^{0.0095t}$$
$$\frac{40}{37.3} = e^{0.0095t}$$
$$\ln\frac{40}{37.3} = \ln e^{0.0095t}$$
$$0.0095t = \ln\frac{40}{37.3}$$
$$t = \frac{\ln\dfrac{40}{37.3}}{0.0095} \approx 7$$

The population of California will reach 40 million about 7 years after 2010, or 2017

**105.**
$$f(x) = 20(0.975)^x$$
$$1 = 20(0.975)^x$$
$$\frac{1}{20} = 0.975^x$$
$$\ln\frac{1}{20} = \ln 0.975^x$$
$$\ln\frac{1}{20} = x \ln 0.975$$
$$x = \frac{\ln\dfrac{1}{20}}{\ln 0.975}$$
$$x \approx 118$$

There is 1% of surface sunlight at 118 feet. This is represented by the point $(118, 1)$.

**107.** $20,000 = 12,500\left(1 + \dfrac{0.0575}{4}\right)^{4t}$

$12,500(1.014375)^{4t} = 20,000$

$(1.014375)^{4t} = 1.6$

$\ln(1.014375)^{4t} = \ln 1.6$

$4t\ln(1.014375) = \ln 1.6$

$t = \dfrac{\ln 1.6}{4\ln 1.014375} \approx 8.2$

8.2 years

**109.**

$1400 = 1000\left(1 + \dfrac{0.168}{360}\right)^{360 \cdot t}$

$\dfrac{1400}{1000} = \left(1 + \dfrac{0.168}{360}\right)^{360 \cdot t}$

$\ln\left(1 + \dfrac{0.168}{360}\right)^{360 \cdot t} = \ln\left(\dfrac{1400}{1000}\right)$

$360 \cdot t \ln\left(1 + \dfrac{0.168}{360}\right) = \ln 1.4$

$t = \dfrac{\ln 1.4}{360\ln\left(1 + \dfrac{0.168}{360}\right)}$

$t \approx 2.0$

**111.** accumulated amount = 2(8000) = 16,000

$16,000 = 8000e^{0.08t}$

$e^{0.08t} = 2$

$\ln e^{0.08t} = \ln 2$

$0.08t = \ln 2$

$t = \dfrac{\ln 2}{0.08}$

$t \approx 8.7$

The amount would double in 8.7 years.

**113.** accumulated amount = 3(2350) = 7050

$7050 = 2350e^{0.157t}$

$e^{0.157t} = \dfrac{7050}{2350}$

$\ln e^{0.157t} = \ln 3$

$0.157t = \ln 3$

$t = \dfrac{\ln 3}{0.157}$

$t \approx 7.0$

About 7 years.

**115. a.** 2009 is 3 years after 2006.

$f(x) = 1.2\ln x + 15.7$

$f(3) = 1.2\ln 3 + 15.7 \approx 17.0$

According to the function, 17.0% of the of the U.S. gross domestic product went toward healthcare in 2009. This underestimates the value shown in the graph by 0.3%.

**b.** $f(x) = 1.2\ln x + 15.7$

$18.5 = 1.2\ln x + 15.7$

$2.8 = 1.2\ln x$

$\dfrac{2.8}{1.2} = \dfrac{1.2\ln x}{1.2}$

$\dfrac{2.8}{1.2} = \ln x$

$x = e^{\frac{2.8}{1.2}}$

$x \approx 10$

If the trend continues, 18.5% of the U.S. gross domestic product will go toward healthcare 10 years after 2006, or 2016.

**117.** $P(x) = 95 - 30\log_2 x$

$40 = 95 - 30\log_2 x$

$30\log_2 x = 45$

$\log_2 x = 1.5$

$x = 2^{1.5} \approx 2.8$

Only half the students recall the important features of the lecture after 2.8 days.
This is represented by the point (2.8, 50).

**119. a.** $\text{pH} = -\log x$

$5.6 = -\log x$

$-5.6 = \log x$

$x = 10^{-5.6}$

The hydrogen ion concentration is $10^{-5.6}$ mole per liter.

**b.** $\text{pH} = -\log x$

$2.4 = -\log x$

$-2.4 = \log x$

$x = 10^{-2.4}$

The hydrogen ion concentration is $10^{-2.4}$ mole per liter.

**c.** $\dfrac{10^{-2.4}}{10^{-5.6}} = 10^{-2.4-(-5.6)} = 10^{3.2}$

The concentration of the acidic rainfall in part (b) is $10^{3.2}$ times greater than the normal rainfall in part (a).

**121. – 123.** Answers will vary.

**125.**

The intersection point is (2, 8).
Verify: $x = 2$

$$2^{x+1} = 8$$
$$2^{2+1} = 2$$
$$2^3 = 8$$
$$8 = 8$$

The solution set is $\{2\}$.

**127.**

The intersection point is (4, 2).
Verify: $x = 4$

$$\log_3(4 \cdot 4 - 7) = 2$$
$$\log_3 9 = 2$$
$$2 = 2$$

The solution set is $\{4\}$.

**129.**

The intersection point is (2, 1).
Verify: $x = 2$

$$\log(2+3) + \log 2 = 1$$
$$\log 5 + \log 2 = 1$$
$$\log(5 \cdot 2) = 1$$
$$\log 10 = 1$$
$$1 = 1$$

The solution set is $\{2\}$.

**131.**

There are 2 points of intersection, approximately
(−1.391606, 0.21678798) and
(1.6855579, 6.3711158).
Verify $x \approx -1.391606$
$$3^x = 2x + 3$$
$$3^{-1.391606} \approx 2(-1.391606) + 3$$
$$0.2167879803 \approx 0.216788$$
Verify $x \approx 1.6855579$
$$3^x = 2x + 3$$
$$3^{1.6855579} \approx 2(1.6855579) + 3$$
$$6.37111582 \approx 6.371158$$
The solution set is $\{-1.391606, 1.6855579\}$.

**133.**

As the distance from the eye increases, barometric air pressure increases, leveling off at about 30 inches of mercury.

**135.**

When $P = 70$, $t \approx 7.9$, so it will take about 7.9 minutes.
Verify:
$$70 = 45e^{-0.092(7.9)}$$
$$70 \approx 70.10076749$$
The runner's pulse will be 70 beats per minute after about 7.9 minutes.

**137.** does not make sense; Explanations will vary. Sample explanation: $2^x = 15$ requires logarithms. $2^x = 16$ can be solved by rewriting 16 as $2^4$.

$$2^x = 15$$
$$\ln 2^x = \ln 15$$
$$x \ln 2 = \ln 15$$
$$x = \frac{\ln 15}{\ln 2}$$

$$2^x = 16$$
$$2^x = 2^4$$
$$x = 4$$

**139.** makes sense

**141.** false; Changes to make the statement true will vary. A sample change is: If $\log(x+3) = 2$, then $10^2 = x + 3$.

**143.** true

**145.** Account paying 3% interest:

$$A = 4000\left(1 + \frac{0.03}{1}\right)^{1 \cdot t}$$

Account paying 5% interest:

$$A = 2000\left(1 + \frac{0.05}{1}\right)^{1 \cdot t}$$

The two accounts will have the same balance when

$$4000(1.03)^t = 2000(1.05)^t$$
$$(1.03)^t = 0.5(1.05)^t$$
$$\left(\frac{1.03}{1.05}\right)^t = 0.5$$
$$\ln\left(\frac{1.03}{1.05}\right)^t = \ln 0.5$$
$$t \ln\left(\frac{1.03}{1.05}\right) = \ln 0.5$$
$$t = \frac{\ln 0.5}{\ln\left(\dfrac{1.03}{1.05}\right)} \approx 36$$

The accounts will have the same balance in about 36 years.

**147.** $(\log x)(2\log x + 1) = 6$

$$2(\log x)^2 + \log x - 6 = 0$$
$$(2\log x - 3)(\log x + 2) = 0$$
$$2\log x - 3 = 0 \text{ or } \log x + 2 = 0$$
$$2\log x = 3 \qquad \log x = -2$$
$$\log x = \frac{3}{2} \qquad x = 10^{-2}$$
$$x = 10^{3/2} \qquad x = \frac{1}{100}$$
$$x = 10\sqrt{10}$$

The solution set is $\left\{\dfrac{1}{100},\ 10\sqrt{10}\right\}$.

Check by direct substitution:

$$\text{Check:} x = 10\sqrt{10} = 10^{3/2}$$
$$(\log x)(2\log x + 1) = 6$$
$$\left(\log 10^{3/2}\right)(2\log 10^{3/2} + 1) = 6$$
$$\left(\frac{3}{2}\right)\left(2 \cdot \frac{3}{2} + 1\right) = 6$$
$$\left(\frac{3}{2}\right)(3 + 1) = 6$$
$$\left(\frac{3}{2}\right)(4) = 6$$
$$6 = 6$$

**149.** Answers will vary.

**150.** $A = 10e^{-0.003t}$

**a.** 2006: $A = 10e^{-0.003(0)} = 10$ million
2007: $A = 10e^{-0.003(1)} \approx 9.97$ million
2008: $A = 10e^{-0.003(2)} \approx 9.94$ million
2009: $A = 10e^{-0.003(3)} \approx 9.91$ million

**b.** The population is decreasing.

**151. a.**    $e^{\ln 3} = 3$

**b.**    $e^{\ln 3} = 3$

$\left(e^{\ln 3}\right)^x = 3^x$

$e^{(\ln 3)x} = 3^x$

**152.** An exponential function is the best choice.

**Annual Retail Sales of** *Call of Duty* **Games**

**Section 4.5**

**Check Point Exercises**

**1.    a.**    $A_0 = 807$. Since 2011 is 11 years after 2000, when $t = 11$, $A = 1052$.

$A = A_0 e^{kt}$

$1052 = 807 e^{k(11)}$

$\dfrac{1052}{807} = e^{11k}$

$\ln\left(\dfrac{1052}{807}\right) = \ln e^{11k}$

$\ln\left(\dfrac{1052}{807}\right) = 11k$

$k = \dfrac{\ln\left(\dfrac{1052}{807}\right)}{11} \approx 0.024$

Thus, the growth function is $A = 807e^{0.024t}$.

**b.**    $A = 807e^{0.024t}$

$2000 = 807e^{0.024t}$

$\dfrac{2000}{807} = e^{0.024t}$

$\ln\left(\dfrac{2000}{807}\right) = \ln e^{0.024t}$

$\ln\left(\dfrac{2000}{807}\right) = 0.024t$

$t = \dfrac{\ln\left(\dfrac{2000}{807}\right)}{0.024} \approx 38$

Africa's population will reach 2000 million approximately 38 years after 2000, or 2038.

**2.    a.**    In the exponential decay model $A = A_0 e^{kt}$, substitute $\dfrac{A_0}{2}$ for $A$ since the amount present after 28 years is half the original amount.

$\dfrac{A_0}{2} = A_0 e^{k \cdot 28}$

$e^{28k} = \dfrac{1}{2}$

$\ln e^{28k} = \ln \dfrac{1}{2}$

$28k = \ln \dfrac{1}{2}$

$k = \dfrac{\ln^{1/2}}{28} \approx -0.0248$

So the exponential decay model is $A = A_0 e^{-0.0248t}$

**b.**    Substitute 60 for $A_0$ and 10 for $A$ in the model from part (a) and solve for $t$.

$10 = 60e^{-0.0248t}$

$e^{-0.0248t} = \dfrac{1}{6}$

$\ln e^{-0.0248t} = \ln \dfrac{1}{6}$

$-0.0248t = \ln \dfrac{1}{6}$

$t = \dfrac{\ln \frac{1}{6}}{-0.0248} \approx 72$

The strontium-90 will decay to a level of 10 grams about 72 years after the accident.

**3. a.** The time prior to learning trials corresponds to $t = 0$.

$$f(0) = \frac{0.8}{1 + e^{-0.2(0)}} = 0.4$$

The proportion of correct responses prior to learning trials was 0.4.

**b.** Substitute 10 for $t$ in the model:

$$f(10) = \frac{0.8}{1 + e^{-0.2(10)}} \approx 0.7$$

The proportion of correct responses after 10 learning trials was 0.7.

**c.** In the logistic growth model, $f(t) = \dfrac{c}{1 + ae^{-bt}}$,

the constant $c$ represents the limiting size that $f(t)$ can attain. The limiting size of the proportion of correct responses as continued learning trials take place is 0.8.

**4.** A logarithmic function would be a good choice for modeling the data.

**5.** An exponential function would be a good choice for modeling the data although model choices may vary.

**6. a.** 1970 is 21 years after 1949.
$$f(x) = 0.074x + 2.294$$
$$f(21) = 0.074(21) + 2.294$$
$$\approx 3.8$$
$$g(x) = 2.577(1.017)^x$$
$$g(21) = 2.577(1.017)^{21}$$
$$\approx 3.7$$
The exponential function $g$ serves as a better model for 1970.

**b.** 2050 is 101 years after 1949.
$$f(x) = 0.074x + 2.294$$
$$f(101) = 0.074(101) + 2.294$$
$$\approx 9.8$$
$$g(x) = 2.577(1.017)^x$$
$$g(101) = 2.577(1.017)^{101}$$
$$\approx 14.1$$
The linear function $f$ serves as a better model for 2050.

**7.** $y = 4(7.8)^x$
$$= 4e^{(\ln 7.8)x}$$
Rounded to three decimal places:
$$y = 4e^{(\ln 7.8)x}$$
$$= 4e^{2.054x}$$

**Concept and Vocabulary Check 4.5**

**1.** $> 0$; $< 0$

**2.** $A_0$; $A$

**3.** $A$; $c$

**4.** logarithmic

**5.** exponential

**6.** linear

**7.** $\ln 5$

**Exercise Set 4.5**

**1.** Since 2010 is 0 years after 2010, find $A$ when $t = 0$:

$A = 127.3e^{-0.006t}$

$A = 127.3e^{-0.006(0)}$

$A = 127.3e^0$

$A = 127.3(1)$

$A = 127.3$

In 2010, the population of Japan was 127.3 million.

**3.** Since $k = 0.019$, Iraq has the greatest growth rate at 1.9% per year.

**5.** Substitute $A = 1377$ into the model for India and solve for $t$:

$1377 = 1173.1e^{0.008t}$

$\dfrac{1377}{1173.1} = e^{0.008t}$

$\ln\dfrac{1377}{1173.1} = \ln e^{0.008t}$

$\ln\dfrac{1377}{1173.1} = 0.008t$

$t = \dfrac{\ln\dfrac{1377}{1173.1}}{0.008} \approx 20$

The population of India will be 1377 million approximately 20 years after 2010, or 2030.

**7. a.** $A_0 = 6.04$. Since 2050 is 50 years after 2000, when $t = 50$, $A = 10$.

$A = A_0 e^{kt}$

$10 = 6.04e^{k(50)}$

$\dfrac{10}{6.04} = e^{50k}$

$\ln\left(\dfrac{10}{6.04}\right) = \ln e^{50k}$

$\ln\left(\dfrac{10}{6.04}\right) = 50k$

$k = \dfrac{\ln\left(\dfrac{10}{6.04}\right)}{50} \approx 0.01$

Thus, the growth function is $A = 6.04e^{0.01t}$.

**b.**

$9 = 6.04e^{0.01t}$

$\dfrac{9}{6.04} = e^{0.01t}$

$\ln\left(\dfrac{9}{6.04}\right) = \ln e^{0.01t}$

$\ln\left(\dfrac{9}{6.04}\right) = 0.01t$

$t = \dfrac{\ln\left(\dfrac{9}{6.04}\right)}{0.01} \approx 40$

Now, 2000 + 40 = 2040, so the population will be 9 million is approximately the year 2040.

**9.** $P(x) = 99.9e^{0.0095t}$

$P(40) = 99.9e^{0.0095(40)}$

$P(40) = 99.9e^{0.0095(40)} \approx 146.1$

The population is projected to be 146.1 million in 2050.

**11.** $P(x) = 44.2e^{kt}$

$62.9 = 44.2e^{40k}$

$\dfrac{62.9}{44.2} = e^{40k}$

$\ln\left(\dfrac{62.9}{44.2}\right) = \ln e^{40k}$

$\ln\left(\dfrac{62.9}{44.2}\right) = 40k$

$\dfrac{\ln\left(\dfrac{62.9}{44.2}\right)}{40} = k$

$k \approx 0.0088$

The growth rate is 0.0088.

**13.** $P(x) = 82.3e^{kt}$

$70.5 = 82.3e^{40k}$

$\dfrac{70.5}{82.3} = e^{40k}$

$\ln\left(\dfrac{70.5}{82.3}\right) = \ln e^{40k}$

$\ln\left(\dfrac{70.5}{82.3}\right) = 40k$

$\dfrac{\ln\left(\dfrac{70.5}{82.3}\right)}{40} = k$

$k \approx -0.0039$

The growth rate is −0.0039.

**15.**  $A = 16e^{-0.000121t}$

$A = 16e^{-0.000121(5715)}$

$A = 16e^{-0.691515}$

$A \approx 8.01$

Approximately 8 grams of carbon-14 will be present in 5715 years.

**17.**  After 10 seconds, there will be $16 \cdot \dfrac{1}{2} = 8$ grams present. After 20 seconds, there will be $8 \cdot \dfrac{1}{2} = 4$ grams present. After 30 seconds, there will be $4 \cdot \dfrac{1}{2} = 2$ grams present. After 40 seconds, there will be $2 \cdot \dfrac{1}{2} = 1$ grams present. After 50 seconds, there will be $1 \cdot \dfrac{1}{2} = \dfrac{1}{2}$ gram present.

**19.**  $A = A_0 e^{-0.000121t}$

$15 = 100e^{-0.000121t}$

$\dfrac{15}{100} = e^{-0.000121t}$

$\ln 0.15 = \ln e^{-0.000121t}$

$\ln 0.15 = -0.000121t$

$t = \dfrac{\ln 0.15}{-0.000121} \approx 15,679$

The paintings are approximately 15,679 years old.

**21.**  $0.5 = e^{kt}$

$0.5 = e^{-0.055t}$

$\ln 0.5 = \ln e^{-0.055t}$

$\ln 0.5 = -0.055t$

$\dfrac{\ln 0.5}{-0.055} = t$

$t \approx 12.6$

The half-life is 12.6 years.

**23.**  $0.5 = e^{kt}$

$0.5 = e^{1620k}$

$\ln 0.5 = \ln e^{1620k}$

$\ln 0.5 = 1620k$

$\dfrac{\ln 0.5}{1620} = k$

$k \approx -0.000428$

The decay rate is 0.0428% per year.

**25.**  $0.5 = e^{kt}$

$0.5 = e^{17.5k}$

$\ln 0.5 = \ln e^{17.5k}$

$\ln 0.5 = 17.5k$

$\dfrac{\ln 0.5}{17.5} = k$

$k \approx -0.039608$

The decay rate is 3.9608% per day.

**27. a.**  $\dfrac{1}{2} = 1e^{k1.31}$

$\ln \dfrac{1}{2} = \ln e^{1.31k}$

$\ln \dfrac{1}{2} = 1.31k$

$k = \dfrac{\ln \dfrac{1}{2}}{1.31} \approx -0.52912$

The exponential model is given by $A = A_0 e^{-0.52912t}$.

**b.**  $A = A_0 e^{-0.52912t}$

$0.945A_0 = A_0 e^{-0.52912t}$

$0.945 = e^{-0.52912t}$

$\ln 0.945 = \ln e^{-0.52912t}$

$\ln 0.945 = -0.52912t$

$t = \dfrac{\ln 0.945}{-0.52912} \approx 0.1069$

The age of the dinosaur ones is approximately 0.1069 billion or 106,900,000 years old.

**29.**  First find the decay equation.

$0.5 = e^{kt}$

$0.5 = e^{22k}$

$\ln 0.5 = \ln e^{22k}$

$\ln 0.5 = 22k$

$\dfrac{\ln 0.5}{22} = k$

$k \approx -0.031507$

$A = e^{-0.031507t}$

Next use the decay equation answer question.

$$A = e^{-0.031507t}$$

$$0.8 = e^{-0.031507t}$$

$$\ln 0.8 = \ln e^{-0.031507t}$$

$$\ln 0.8 = -0.031507t$$

$$\frac{\ln 0.8}{-0.031507} = t$$

$$t \approx 7.1$$

It will take 7.1 years.

**31.** First find the decay equation.

$$0.5 = e^{kt}$$

$$0.5 = e^{36k}$$

$$\ln 0.5 = \ln e^{36k}$$

$$\ln 0.5 = 36k$$

$$\frac{\ln 0.5}{36} = k$$

$$k \approx -0.019254$$

$$A = e^{-0.019254t}$$

Next use the decay equation answer question.

$$A = e^{-0.019254t}$$

$$0.9 = e^{-0.019254t}$$

$$\ln 0.9 = \ln e^{-0.019254t}$$

$$\ln 0.9 = -0.019254t$$

$$\frac{\ln 0.9}{-0.019254} = t$$

$$t \approx 5.5$$

It will take 5.5 hours.

**33.**
$$2A_0 = A_0 e^{kt}$$

$$2 = e^{kt}$$

$$\ln 2 = \ln e^{kt}$$

$$\ln 2 = kt$$

$$t = \frac{\ln 2}{k}$$

The population will double in $t = \dfrac{\ln 2}{k}$ years.

**35.** $A = 4.3e^{0.01t}$

**a.** $k = 0.01$, so New Zealand's growth rate is 1%.

**b.**
$$A = 4.3e^{0.01t}$$

$$2 \cdot 4.3 = 4.3e^{0.01t}$$

$$2 = e^{0.01t}$$

$$\ln 2 = \ln e^{0.01t}$$

$$\ln 2 = 0.01t$$

$$t = \frac{\ln 2}{0.01} \approx 69$$

New Zealand's population will double in approximately 69 years.

**37. a.** When the epidemic began, $t = 0$.
$$f(0) = \frac{100,000}{1 + 5000e^0} \approx 20$$

Twenty people became ill when the epidemic began.

**b.** $f(4) = \dfrac{100,000}{1 + 5,000e^{-4}} \approx 1080$

About 1080 people were ill at the end of the fourth week.

**c.** In the logistic growth model,
$$f(t) = \frac{c}{1 + ae^{-bt}},$$
the constant $c$ represents the limiting size that $f(t)$ can attain. The limiting size of the population that becomes ill is 100,000 people.

**39.** $f(x) = \dfrac{12.57}{1 + 4.11e^{-0.026(x)}}$

$$f(61) = \frac{12.57}{1 + 4.11e^{-0.026(61)}} \approx 6.8$$

The function models the data quite well.

**41.**

$$f(x) = \frac{12.57}{1 + 4.11e^{-0.026(x)}}$$

$$8 = \frac{12.57}{1 + 4.11e^{-0.026(x)}}$$

$$8\left(1 + 4.11e^{-0.026(x)}\right) = 12.57$$

$$8 + 32.88e^{-0.026(x)} = 12.57$$

$$32.88e^{-0.026(x)} = 4.57$$

$$e^{-0.026(x)} = \frac{4.57}{32.88}$$

$$\ln e^{-0.026(x)} = \ln \frac{4.57}{32.88}$$

$$-0.026x = \ln \frac{4.57}{32.88}$$

$$x = \frac{\ln \frac{4.57}{32.88}}{-0.026}$$

$$x \approx 76$$

The world population will reach 8 billion 76 years after 1949, or 2025.

**43.** $P(20) = \dfrac{90}{1 + 271e^{-0.122(20)}} \approx 3.7$

The probability that a 20-year-old has some coronary heart disease is about 3.7%.

**45.**

$$0.5 = \frac{0.9}{1 + 271e^{-0.122t}}$$

$$0.5\left(1 + 271e^{-0.122t}\right) = 0.9$$

$$1 + 271e^{-0.122t} = 1.8$$

$$271e^{-0.122t} = 0.8$$

$$e^{-0.122t} = \frac{0.8}{271}$$

$$\ln e^{-0.122t} = \ln \frac{0.8}{271}$$

$$-0.122t = \ln \frac{0.8}{271}$$

$$t = \frac{\ln \frac{0.8}{271}}{-0.122} \approx 48$$

The probability of some coronary heart disease is 50% at about age 48.

**47. a.** Scatter plot:

**b.** An exponential function appears to be the best choice for modeling the data.

**49. a.** Scatter plot:

**b.** A logarithmic function appears to be the best choice for modeling the data.

**51. a.** Scatter plot:

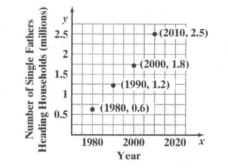

**b.** A linear function appears to be the best choice for modeling the data.

**53.** $y = 100(4.6)^x$ is equivalent to
$y = 100e^{(\ln 4.6)x}$;
Using $\ln 4.6 \approx 1.526$,
$y = 100e^{1.526x}$.

**55.** $y = 2.5(0.7)^x$ is equivalent to

$y = 2.5e^{(\ln 0.7)x}$;

Using $\ln 0.7 \approx -0.357$, $y = 2.5e^{-0.357x}$.

**57. – 65.** Answers will vary.

**67.** The logarithmic model is $y = 191.9 + 24.569\ln x$.
Since $r = 0.870$ is fairly close to 1, the model fits the data okay, but not great.

**69.** The power regression model is $y = 195.05x^{0.100}$.
Since $r = 0.896$, the model fits the data fairly well.

**71. a.** Exponential Regression:
$y = 3.46(1.02)^x$; $r \approx 0.994$
Logarithmic Regression:
$y = 14.752\ln x - 26.512$; $r \approx 0.673$
Linear Regression:
$y = 0.557x - 10.972$; $r \approx 0.947$
The exponential model has an $r$ value closer to 1. Thus, the better model is $y = 3.46(1.02)^x$.

**b.** $y = 3.46(1.02)^x$

$y = 3.46e^{(\ln 1.02)x}$

$y = 3.46e^{0.02x}$

The 65-and-over population is increasing by approximately 2% each year.

**73.** does not make sense; Explanations will vary. Sample explanation: Since the car's value is decreasing (depreciating), the growth rate is negative.

**75.** makes sense

**77.** true

**79.** true

**81.** Use $T_0 = 210$, $C = 70$, $t = 30$, and $T = 140$ to determine the constant $k$:

$$T = C + \left(T_0 - C\right)e^{-kt}$$

$$140 = 70 + \left(210 - 70\right)e^{-k(30)}$$

$$140 = 70 + 140e^{-30k}$$

$$70 = 140e^{-30k}$$

$$\frac{70}{140} = \frac{140e^{-30k}}{140}$$

$$0.5 = e^{-30k}$$

$$\ln 0.5 = \ln e^{-30k}$$

$$\ln 0.5 = -30k$$

$$k = \frac{\ln 0.5}{-30} \approx 0.0231$$

Thus, the model for these conditions is
$T = 70 + 140e^{-0.0231t}$.
Evaluate the model for $t = 40$:
$T = 70 + 140e^{-0.0231(40)} \approx 126$
Thus, the temperature of the cake after 40 minutes will be approximately $126°F$.

**83. a.** $x + 2y = 2$
$4 + 2(-1) = 2$
$4 - 2 = 2$
$2 = 2$, true
Yes, $(4, -1)$ satisfies the equation $x + 2y = 2$.

**b.** $x - 2y = 6$
$4 - 2(-1) = 6$
$4 + 2 = 6$
$6 = 6$, true
Yes, $(4, -1)$ satisfies the equation $x - 2y = 6$.

**84.** The graphs intersect at the point $(4, -1)$.

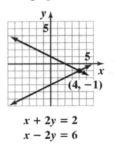

$x + 2y = 2$
$x - 2y = 6$

**85.**  $5(2x-3)-4x=9$

$10x-15-4x=9$

$6x-15=9$

$6x=24$

$x=4$

The solution set is $\{4\}$.

## Chapter 4 Review Exercises

**1.** This is the graph of $f(x)=4^x$ reflected about the $y$-axis, so the function is $g(x)=4^{-x}$.

**2.** This is the graph of $f(x)=4^x$ reflected about the $x$-axis and about the $y$-axis, so the function is $h(x)=-4^{-x}$.

**3.** This is the graph of $f(x)=4^x$ reflected about the $x$-axis and about the $y$-axis then shifted upward 3 units, so the function is $r(x)=-4^{-x}+3$.

**4.** This is the graph of $f(x)=4^x$.

**5.** The graph of $g(x)$ shifts the graph of $f(x)$ one unit to the right.

$f(x)=2^x$

$g(x)=2^{x-1}$

asymptote of $f$: $y=0$

asymptote of $g$: $y=0$

domain of $f$ = domain of $g$ = $(-\infty,\infty)$

range of $f$ = range of $g$ = $(0,\infty)$

**6.** The graph of $g(x)$ shifts the graph of $f(x)$ one unit down.

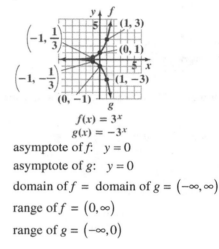

$f(x)=3^x$

$g(x)=3^x-1$

asymptote of $f$: $y=0$

asymptote of $g$: $y=-1$

domain of $f$ = domain of $g$ = $(-\infty,\infty)$

range of $f$ = $(0,\infty)$

range of $g$ = $(-1,\infty)$

**7.** The graph of $g(x)$ reflects the graph of $f(x)$ about the $y-$ axis.

$f(x)=3^x$

$g(x)=-3^x$

asymptote of $f$: $y=0$

asymptote of $g$: $y=0$

domain of $f$ = domain of $g$ = $(-\infty,\infty)$

range of $f$ = $(0,\infty)$

range of $g$ = $(-\infty,0)$

**8.** The graph of $g(x)$ reflects the graph of $f(x)$ about the $x-$ axis.

$f(x)=\left(\dfrac{1}{2}\right)^x$

$g(x)=\left(\dfrac{1}{2}\right)^{-x}$

asymptote of $f$: $y=0$

asymptote of $g$: $y=0$

domain of $f$ = domain of $g$ = $(-\infty,\infty)$

range of $f$ = range of $g$ = $(0,\infty)$

**9.** The graph of $g(x)$ vertically stretches the graph of $f(x)$ by a factor of 2.

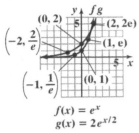

$$f(x) = e^x$$
$$g(x) = 2e^{x/2}$$

asymptote of $f$:   $y = 0$

asymptote of $g$:   $y = 0$

domain of $f$ = domain of $g = (-\infty, \infty)$

range of $f$ = range of $g = (0, \infty)$

**10.** 5.5% compounded semiannually:

$$A = 5000\left(1 + \frac{0.055}{2}\right)^{2 \cdot 5} \approx 6558.26$$

5.25% compounded monthly:

$$A = 5000\left(1 + \frac{0.0525}{12}\right)^{12 \cdot 5} \approx 6497.16$$

5.5% compounded semiannually yields the greater return.

**11.** 7% compounded monthly:

$$A = 14{,}000\left(1 + \frac{0.07}{12}\right)^{12 \cdot 10} \approx 28{,}135.26$$

6.85% compounded continuously:

$$A = 14{,}000e^{0.0685(10)} \approx 27{,}772.81$$

7% compounded monthly yields the greater return.

**12. a.** When first taken out of the microwave, the temperature of the coffee was 200°.

**b.** After 20 minutes, the temperature of the coffee was about 120°.
$$T = 70 + 130e^{-0.04855(20)} \approx 119.23$$
Using a calculator, the temperature is about 119°.

**c.** The coffee will cool to about 70°;
The temperature of the room is 70°.

**13.** $49^{1/2} = 7$

**14.** $4^3 = x$

**15.** $3^y = 81$

**16.** $\log_6 216 = 3$

**17.** $\log_b 625 = 4$

**18.** $\log_{13} 874 = y$

**19.** $\log_4 64 = 3$ because $4^3 = 64$.

**20.** $\log_5 \dfrac{1}{25} = -2$ because $5^{-2} = \dfrac{1}{25}$.

**21.** $\log_3(-9)$ is undefined and cannot be evaluated since $\log_b x$ is defined only for $x > 0$.

**22.** $\log_{16} 4 = \dfrac{1}{2}$ because $16^{1/2} = \sqrt{16} = 4$.

**23.** Because $\log_b b = 1$,
we conclude $\log_{17} 17 = 1$.

**24.** Because $\log_b b^x = x$,
we conclude $\log_3 3^8 = 8$.

**25.** Because $\ln e^x = x$,
we conclude $\ln e^5 = 5$.

**26.** $\log_3 \dfrac{1}{\sqrt{3}} = \log_3 \dfrac{1}{3^{\frac{1}{2}}} = \log_3 3^{-\frac{1}{2}} = -\dfrac{1}{2}$

**27.** $\ln \dfrac{1}{e^2} = \ln e^{-2} = -2$

**28.** $\log \dfrac{1}{1000} = \log \dfrac{1}{10^3} = \log 10^{-3} = -3$

**29.** Because $\log_b b = 1$,
we conclude $\log_8 8 = 1$.
So, $\log_3(\log_8 8) = \log_3 1$.
Because $\log_b 1 = 0$
we conclude $\log_3 1 = 0$.
Therefore, $\log_3(\log_8 8) = 0$.

**30.**

$f(x) = 2^x$
$g(x) = \log_2 x$

domain of $f$ = range of $g$ = $(-\infty, \infty)$

range of $f$ = domain of $g$ = $(0, \infty)$

**31.**

$f(x) = \left(\dfrac{1}{3}\right)^x$

$g(x) = \log_{1/3} x$

domain of $f$ = range of $g$ = $(-\infty, \infty)$

range of $f$ = domain of $g$ = $(0, \infty)$

**32.** This is the graph of $f(x) = \log x$ reflected about the
y-axis, so the function is
$g(x) = \log(-x)$.

**33.** This is the graph of $f(x) = \log x$
shifted left 2 units, reflected about the
y-axis, then shifted upward one unit, so the function
is $r(x) = 1 + \log(2 - x)$.

**34.** This is the graph of $f(x) = \log x$
shifted left 2 units then reflected about the
y-axis, so the function is $h(x) = \log(2 - x)$ .

**35.** This is the graph of $f(x) = \log x$.

**36.**

$x = 2$
$f(x) = \log_2 x$
$g(x) = \log_2(x - 2)$

x-intercept: (3, 0)
vertical asymptote: $x = 2$
domain: $(2, \infty)$
range: $(-\infty, \infty)$

**37.**

$f(x) = \log_2 x$
$h(x) = -1 + \log_2 x$

x-intercept: (2, 0)
vertical asymptote: $x = 0$
domain: $(0, \infty)$
range: $(-\infty, \infty)$

**38.**

$f(x) = \log_2 x$
$r(x) = \log_2(-x)$

x-intercept: (−1, 0)
vertical asymptote: $x = 0$
domain: $(-\infty, 0)$
range: $(-\infty, \infty)$

**39.**

asymptote of $f$: $x = 0$
asymptote of $g$: $x = -3$
domain of $f = (0, \infty)$
domain of $g = (-3, \infty)$
range of $f$ = range of $g = (-\infty, \infty)$

**40.**

asymptote of $f$: $x = 0$
asymptote of $g$: $x = 0$
domain of $f$ = domain of $g = (0, \infty)$
range of $f$ = range of $g = (-\infty, \infty)$

**41.** The domain of $f$ consists of all $x$ for which $x + 5 > 0$. Solving this inequality for $x$, we obtain $x > -5$. Thus the domain of $f$ is $(-5, \infty)$

**42.** The domain of $f$ consists of all $x$ for which $3 - x > 0$. Solving this inequality for $x$, we obtain $x < 3$. Thus, the domain of $f$ is $(-\infty, 3)$.

**43.** The domain of $f$ consists of all $x$ for which $(x - 1)^2 > 0$.
Solving this inequality for $x$, we obtain $x < 1$ or $x > 1$.
Thus, the domain of $f$ is $(-\infty, 1) \cup (1, \infty)$.

**44.** Because $\ln e^x = x$, we conclude $\ln e^{6x} = 6x$.

**45.** Because $e^{\ln x} = x$, we conclude $e^{\ln \sqrt{x}} = \sqrt{x}$.

**46.** Because $10^{\log x} = x$, we conclude $10^{\log 4x^2} = 4x^2$.

**47.** $R = \log \dfrac{1000 I_0}{I_0} = \log 1000 = 3$
The Richter scale magnitude is 3.0.

**48. a.** $f(0) = 76 - 18\log(0 + 1) = 76$
When first given, the average score was 76.

**b.** $f(2) = 76 - 18\log(2 + 1) \approx 67$
$f(4) = 76 - 18\log(4 + 1) \approx 63$
$f(6) = 76 - 18\log(6 + 1) \approx 61$
$f(8) = 76 - 18\log(8 + 1) \approx 59$
$f(12) = 76 - 18\log(12 + 1) \approx 56$
After 2, 4, 6, 8, and 12 months, the average scores are about 67, 63, 61, 59, and 56, respectively.

**c.**

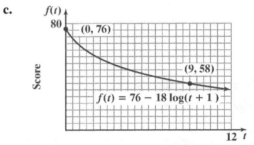

Retention decreases as time passes.

**49.** $t = \dfrac{1}{0.06}\ln\left(\dfrac{12}{12 - 5}\right) \approx 8.98$
It will take about 9 weeks.

**50.** $\log_6\left(36 x^3\right)$
$= \log_6 36 + \log_6 x^3$
$= \log_6 36 + 3\log_6 x$
$= 2 + 3\log_6 x$

**51.** $\log_4 \dfrac{\sqrt{x}}{64} = \log_4 x^{1/2} - \log_4 64$
$= \dfrac{1}{2}\log_4 x - 3$

**52.** $\log_2 \dfrac{xy^2}{64} = \log_2 xy^2 - \log_2 64$
$= \log_2 x + \log_2 y^2 - \log_2 64$
$= \log_2 x + 2\log_2 y - 6$

**53.** $\ln \sqrt[3]{\dfrac{x}{e}}$

$= \ln \left(\dfrac{x}{e}\right)^{\!\frac{1}{3}}$

$= \dfrac{1}{3}\left[\ln x - \ln e\right]$

$= \dfrac{1}{3}\ln x - \dfrac{1}{3}\ln e$

$= \dfrac{1}{3}\ln x - \dfrac{1}{3}$

**54.** $\log_b 7 + \log_b 3$

$= \log_b (7 \cdot 3)$

$= \log_b 21$

**55.** $\log 3 - 3\log x$

$= \log 3 - \log x^3$

$= \log \dfrac{3}{x^3}$

**56.** $3\ln x + 4\ln y$

$= \ln x^3 + \ln y^4$

$= \ln\left(x^3 y^4\right)$

**57.** $\dfrac{1}{2}\ln x - \ln y$

$= \ln x^{\frac{1}{2}} - \ln y$

$= \ln \dfrac{\sqrt{x}}{y}$

**58.** $\log_6 72{,}348 = \dfrac{\log 72{,}348}{\log 6} \approx 6.2448$

**59.** $\log_4 0.863 = \dfrac{\ln 0.863}{\ln 4} \approx -0.1063$

**60.** true; $(\ln x)(\ln 1) = (\ln x)(0) = 0$

**61.** false; $\log(x+9) - \log(x+1) = \log \dfrac{(x+9)}{(x+1)}$

**62.** false; $\log_2 x^4 = 4\log_2 x$

**63.** true; $\ln e^x = x\ln e$

**64.** $2^{4x-2} = 64$

$2^{4x-2} = 2^6$

$4x - 2 = 6$

$4x = 8$

$x = 2$

**65.** $125^x = 25$

$\left(5^3\right)^x = 5^2$

$5^{3x} = 5^2$

$3x = 2$

$x = \dfrac{2}{3}$

**66.** $10^x = 7000$

$\log 10^x = \log 7000$

$x\log 10 = \log 7000$

$x = \log 7000$

$x \approx 3.85$

**67.** $9^{x+2} = 27^{-x}$

$\left(3^2\right)^{x+2} = \left(3^3\right)^{-x}$

$3^{2x+4} = 3^{-3x}$

$2x + 4 = -3x$

$5x = -4$

$x = -\dfrac{4}{5}$

**68.** $8^x = 12{,}143$

$\ln 8^x = \ln 12{,}143$

$x\ln 8 = \ln 12{,}143$

$x = \dfrac{\ln 12{,}143}{\ln 8} \approx 4.52$

**69.** $9e^{5x} = 1269$

$e^{5x} = 141$

$\ln e^{5x} = \ln 141$

$5x = \ln 141$

$x = \dfrac{\ln 141}{5}$

$x = \dfrac{1}{5}\ln 141 \approx 0.99$

**70.** $e^{12-5x} - 7 = 123$

$\qquad e^{12-5x} = 130$

$\quad \ln e^{12-5x} = \ln 130$

$\qquad 12 - 5x = \ln 130$

$\qquad\quad 5x = 12 - \ln 130$

$\qquad\quad x = \dfrac{12 - \ln 130}{5} \approx 1.43$

**71.** $\qquad\qquad 5^{4x+2} = 37,500$

$\qquad\qquad \ln 5^{4x+2} = \ln 37,500$

$\qquad (4x+2)\ln 5 = \ln 37,500$

$\quad 4x \ln 5 + 2 \ln 5 = \ln 37,500$

$\qquad\qquad 4x \ln 5 = \ln 37,500 - 2 \ln 5$

$\qquad\qquad\qquad x = \dfrac{\ln 37,500 - 2 \ln 5}{4 \ln 5} \approx 1.14$

**72.** $\qquad\qquad 3^{x+4} = 7^{2x-1}$

$\qquad\qquad \ln 3^{x+4} = \ln 7^{2x-1}$

$\quad (x+4)\ln 3 = (2x-1)\ln 7$

$\quad x \ln 3 + 4 \ln 3 = 2x \ln 7 - \ln 7$

$\quad x \ln 3 - 2x \ln 7 = -4 \ln 3 - \ln 7$

$\quad x(\ln 3 - 2 \ln 7) = -4 \ln 3 - \ln 7$

$\qquad\qquad\quad x = \dfrac{-4 \ln 3 - \ln 7}{\ln 3 - 2 \ln 7}$

$\qquad\qquad\quad x = \dfrac{4 \ln 3 + \ln 7}{2 \ln 7 - \ln 3}$

$\qquad\qquad\quad x \approx 2.27$

**73.** $e^{2x} - e^x - 6 = 0$

$\quad \left(e^x - 3\right)\left(e^x + 2\right) = 0$

$\quad e^x - 3 = 0 \ \text{ or } \ e^x + 2 = 0$

$\quad e^x = 3 \qquad\quad e^x = -2$

$\quad \ln e^x = \ln 3 \quad \ln e^x - \ln(-2)$

$\quad x = \ln 3 \qquad\quad x = \ln(-2)$

$\quad x = \ln 3 \approx 1.099 \ \ \ln(-2)$ does not exist.

The solution set is $\left\{\ln 3\right\}$,

approximately 1.10.

**74.** $\log_4 (3x - 5) = 3$

$\quad 3x - 5 = 4^3$

$\quad 3x - 5 = 64$

$\quad 3x = 69$

$\quad x = 23$

The solutions set is $\{23\}$.

**75.** $3 + 4 \ln(2x) = 15$

$\qquad 4 \ln(2x) = 12$

$\qquad\quad \ln(2x) = 3$

$\qquad\qquad 2x = e^3$

$\qquad\qquad\quad x = \dfrac{e^3}{2}$

$\qquad\qquad\quad x \approx 10.04$

The solutions set is $\left\{\dfrac{e^3}{2}\right\}$.

**76.** $\log_2 (x+3) + \log_2 (x-3) = 4$

$\quad \log_2 (x+3)(x-3) = 4$

$\quad \log_2 (x^2 - 9) = 4$

$\quad x^2 - 9 = 2^4$

$\quad x^2 - 9 = 16$

$\quad x^2 = 25$

$\quad x = \pm 5$

$x = -5$ does not check because $\log_2 (-5 + 3)$ does not exist.

The solution set is $\{5\}$.

**77.** $\log_3 (x-1) - \log_3 (x+2) = 2$

$\quad \log_3 \dfrac{x-1}{x+2} = 2$

$\quad \dfrac{x-1}{x-2} = 3^2$

$\quad \dfrac{x-1}{x+2} = 9$

$\quad x - 1 = 9(x+2)$

$\quad x - 1 = 9x + 18$

$\quad 8x = -19$

$\quad x = -\dfrac{19}{8}$

$x = -\dfrac{19}{8}$ does not check because $\log_3 \left(-\dfrac{19}{8} - 1\right)$

does not exist.

The solution set is $\varnothing$.

**78.** $\ln(x+4)-\ln(x+1)=\ln x$

$$\ln\frac{x+4}{x+1}=\ln x$$

$$\frac{x+4}{x+1}=x$$

$$x(x+1)=x+4$$

$$x^2+x=x+4$$

$$x^2=4$$

$$x=\pm 2$$

$x=-2$ does not check and must be rejected.
The solution set is $\{2\}$.

**79.** $\log_4(2x+1)=\log_4(x-3)+\log_4(x+5)$

$$\log_4(2x+1)=\log_4(x-3)+\log_4(x+5)$$

$$\log_4(2x+1)=\log_4\left(x^2+2x-15\right)$$

$$2x+1=x^2+2x-15$$

$$16=x^2$$

$$x^2=16$$

$$x=\pm 4$$

$x=-4$ does not check and must be rejected.
The solution set is $\{4\}$.

**80.** $P(x)=14.7e^{-0.21x}$

$$4.6=14.7e^{-0.21x}$$

$$\frac{4.6}{14.7}=e^{-0.21x}$$

$$\ln\frac{4.6}{14.7}=\ln e^{-0.21x}$$

$$\ln\frac{4.6}{14.7}=-0.21x$$

$$t=\frac{\ln\frac{4.6}{14.7}}{-0.21}\approx 5.5$$

The peak of Mt. Everest is about 5.5 miles above sea level.

**81.** $S(t)=21.4(1.56)^t$

$$1170.9=21.4(1.56)^t$$

$$\frac{1170.9}{21.4}=(1.56)^t$$

$$\ln\frac{1170.9}{21.4}=\ln(1.56)^t$$

$$\ln\frac{1170.9}{21.4}=t\ln 1.56$$

$$t=\frac{\ln\frac{1170.9}{21.4}}{\ln 1.56}\approx 9$$

This model projects that 1170.9 million smartphones will be sold 9 years after 2004, or 2013.

**82. a.** $f(x)=-6.2\ln x+40.5$

$$16=-6.2\ln x+40.5$$

$$-24.5=-6.2\ln x$$

$$\frac{-24.5}{-6.2}=\ln x$$

$$\frac{24.5}{6.2}=\ln x$$

$$e^{\frac{24.5}{6.2}}=e^{\ln x}$$

$$x=e^{\frac{24.5}{6.2}}\approx 52$$

This model projects that opposition to homosexual relationships among first-year college women will diminish to 16% 52 years after 1979, or 2031.

**b.** $g(x)=-7\ln x+59$

$$40=-7\ln x+59$$

$$-19=-7\ln x$$

$$\frac{-19}{-7}=\ln x$$

$$e^{\frac{19}{7}}=e^{\ln x}$$

$$x=e^{\frac{19}{7}}\approx 15$$

This model projects that opposition to homosexual relationships among first-year college men diminish to 40% 15 years after 1979, or 1994.

**83.** $20,000 = 12,500\left(1+\dfrac{0.065}{4}\right)^{4t}$

$12,500(1.01625)^{4t} = 20,000$

$(1.01625)^{4t} = 1.6$

$\ln(1.01625)^{4t} = \ln 1.6$

$4t\ln 1.01625 = \ln 1.6$

$t = \dfrac{\ln 1.6}{4\ln 1.01625} \approx 7.3$

It will take about 7.3 years.

**84.** $3 \cdot 50,000 = 50,000e^{0.075t}$

$50,000e^{0.075t} = 150,000$

$e^{0.075} = 3$

$\ln e^{0.075t} = \ln 3$

$0.075t = \ln 3$

$t = \dfrac{\ln 3}{0.075} \approx 14.6$

It will take about 14.6 years.

**85.** When an investment value triples, $A = 3P$.

$3P = Pe^{5r}$

$e^{5r} = 3$

$\ln e^{5r} = \ln 3$

$5r = \ln 3$

$r = \dfrac{\ln 3}{5} \approx 0.2197$

The interest rate would need to be about 22%

**86. a.** $50.5 = 35.3e^{k10}$

$\dfrac{50.5}{35.3} = e^{10k}$

$\ln \dfrac{50.5}{35.3} = \ln e^{10k}$

$\ln \dfrac{50.5}{35.3} = 10k$

$\dfrac{\ln \dfrac{50.5}{35.3}}{10} = k$

$0.036 \approx k$

$A = 35.3e^{0.036t}$

**b.** $A = 35.3e^{0.036(15)} \approx 60.6$

In 2015, the population will be about 60.6 million.

**c.** $70 = 35.3e^{0.036t}$

$\dfrac{70}{35.3} = e^{0.036t}$

$\ln \dfrac{70}{35.3} = \ln e^{0.036t}$

$\ln \dfrac{70}{35.3} = 0.036t$

$\dfrac{\ln \dfrac{70}{35.3}}{0.036} = t$

$19 \approx t$

The population will reach 70 million about 19 years after 2000, in 2019.

**87.** Use the half-life of 140 days to find $k$.

$A = A_0e^{kt}$

$\tfrac{1}{2} = e^{k \cdot 140}$

$\tfrac{1}{2} = e^{140k}$

$\ln\tfrac{1}{2} = \ln e^{140k}$

$\ln\tfrac{1}{2} = 140k$

$\dfrac{\ln\tfrac{1}{2}}{140} = k$

$k \approx -0.004951$

Use $A = A_0e^{kt}$ to find $t$.

$A = A_0e^{-0.004951t}$

$0.2 = e^{-0.004951t}$

$\ln 0.2 = \ln e^{-0.004951t}$

$\ln 0.2 = -0.004951t$

$t = \dfrac{\ln 0.2}{-0.004951}$

$t \approx 325$

It will take about 325 days for the substance to decay to 20% of its original amount.

**88. a.** $f(0) = \dfrac{500,000}{1+2499e^{-0.92(0)}} = 200$

200 people became ill when the epidemic began.

**b.** $f(6) = \dfrac{500,000}{1+2499e^{-0.92(6)}} = 45,411$

45,410 were ill after 6 weeks.

**c.** 500,000 people

**89. a.** Scatter plot:

**b.** A linear function appears to be the better choice for modeling the data.

**90. a.** Scatter plot:

**b.** An logarithmic function appears to be the better choice for modeling the data.

**91. a.** Scatter plot:

**b.** An exponential function appears to be the better choice for modeling the data.

**92.** $y = 73(2.6)^x$

$y = 73e^{(\ln 2.6)x}$

$y = 73e^{0.956x}$

**93.** $y = 6.5(0.43)^x$

$y = 6.5e^{(\ln 0.43)x}$

$y = 6.5e^{-0.844x}$

**94.** Answers will vary.

**Chapter 4 Test**

**1.**

$f(x) = 2^x$
$g(x) = 2^{x+1}$

**2.**

$f(x) = \log_2 x$
$g(x) = \log_2(x-1)$

**3.** $5^3 = 125$

**4.** $\log_{36} 6 = \dfrac{1}{2}$

**5.** The domain of $f$ consists of all $x$ for which $3 - x > 0.$ Solving this inequality for $x$, we obtain $x < 3.$
Thus, the domain of $f$ is $(-\infty, 3).$

**6.** $\log_4\left(64x^5\right) = \log_4 64 + \log_4 x^5$
$= 3 + 5\log_4 x$

**7.** $\log_3 \dfrac{\sqrt[3]{x}}{81} = \log_3 x^{\frac{1}{3}} - \log_3 81$
$= \dfrac{1}{3}\log_3 x - 4$

**8.** $6\log x + 2\log y = \log x^6 + \log y^2$
$= \log\left(x^6 y^2\right)$

**9.** $\ln 7 - 3\ln x = \ln 7 - \ln x^3$
$= \ln \dfrac{7}{x^3}$

**10.** $\log_{15} 71 = \dfrac{\log 71}{\log 15} \approx 1.5741$

**11.** $3^{x-2} = 9^{x+4}$

$3^{x-2} = \left(3^2\right)^{x+4}$

$3^{x-2} = 3^{2x+8}$

$x - 2 = 2x + 8$

$-x = 10$

$x = -10$

**12.** $5^x = 1.4$

$\ln 5^x = \ln 1.4$

$x \ln 5 = \ln 1.4$

$x = \dfrac{\ln 1.4}{\ln 5} \approx 0.2091$

**13.** $400e^{0.005x} = 1600$

$e^{0.005x} = 4$

$\ln e^{0.005x} = \ln 4$

$0.005x = \ln 4$

$x = \dfrac{\ln 4}{0.005} \approx 277.2589$

**14.** $e^{2x} - 6e^x + 5 = 0$

$\left(e^x - 5\right)\left(e^x - 1\right) = 0$

$e^x - 5 = 0$     or    $e^x - 1 = 0$

$e^x = 5$           $e^x = 1$

$\ln e^x = \ln 5$     $\ln e^x = \ln 1$

$x = \ln 5$        $x = \ln 1$

$x \approx 1.6094$      $x = 0$

The solution set is $\{0, \ln 5\}$; $\ln \approx 1.6094$.

**15.** $\log_6 (4x-1) = 3$

$4x - 1 = 6^3$

$4x - 1 = 216$

$4x = 217$

$x = \dfrac{217}{4} = 54.25$

**16.** $2\ln 3x = 8$

$\ln 3x = 4$

$3x = e^4$

$x = \dfrac{e^4}{3} \approx 18.1994$

**17.** $\log x + \log(x+15) = 2$

$\log\left(x^2 + 15x\right) = 2$

$x^2 + 15x = 10^2$

$x^2 + 15x - 100 = 0$

$(x+20)(x-5) = 0$

$x + 20 = 0$ or $x - 5 = 0$

$x = -20$      $x = 5$

$x = -20$ does not check because $\log(-20)$ does not exist.

The solution set is $\{5\}$.

**18.** $\ln(x-4) - \ln(x+1) = \ln 6$

$\ln \dfrac{x-4}{x+1} = \ln 6$

$\dfrac{x-4}{x+1} = 6$

$6(x+1) = x - 4$

$6x + 6 = x - 4$

$5x = -10$

$x = -2$

$x = -2$ does not check and must be rejected.

The solution set is $\varnothing$.

**19.** $D = 10\log \dfrac{10^{12} I_0}{I_0}$

$= 10\log 10^{12}$

$= 10 \cdot 12$

$= 120$

The loudness of the sound is 120 decibels.

**20.** Since $\ln e^x = x$, $\ln e^{5x} = 5x$.

**21.** $\log_b b = 1$ because $b^1 = b$.

**22.** $\log_6 1 = 0$ because $6^0 = 1$.

**23.** 6.5% compounded semiannually:

$A = 3{,}000\left(1 + \dfrac{0.065}{2}\right)^{2(10)} \approx \$5{,}687.51$

6% compounded continuously:

$A = 3{,}000e^{0.06(10)} \approx \$5{,}466.36$

6.5% compounded semiannually yields about $221 more than 6% compounded continuously.

**24.**

$$8000 = 4000\left(1 + \frac{0.05}{4}\right)^{4t}$$

$$\frac{8000}{4000} = (1 + 0.0125)^{4t}$$

$$2 = (1.0125)^{4t}$$

$$\ln 2 = \ln(1.0125)^{4t}$$

$$\ln 2 = 4t\ln(1.0125)$$

$$\frac{\ln 2}{4\ln(1.0125)} = \frac{4t\ln(1.0125)}{4\ln(1.0125)}$$

$$t = \frac{\ln 2}{4\ln(1.0125)} \approx 13.9$$

It will take approximately 13.9 years for the money to grow to $8000.

**25.**

$$2 = 1e^{r10}$$

$$2 = e^{10r}$$

$$\ln 2 = \ln e^{10r}$$

$$\ln 2 = 10r$$

$$r = \frac{\ln 2}{10} \approx 0.069$$

The money will double in 10 years with an interest rate of approximately 6.9%.

**26. a.**

$$A = 82.3e^{-0.004(x)}$$

$$A = 82.3e^{-0.004(0)} \approx 82.3$$

In 2010, the population of Germany was 82.3 million.

**b.** The population of Germany is decreasing. We can tell because the model has a negative $k = -0.004$.

**c.**

$$79.1 = 82.3e^{-0.002t}$$

$$\frac{79.1}{82.3} = e^{-0.004t}$$

$$\ln\frac{79.1}{82.3} = \ln e^{-0.004t}$$

$$\ln\frac{79.1}{82.3} = -0.004t$$

$$t = \frac{\ln\frac{79.1}{82.3}}{-0.004} \approx 10$$

The population of Germany will be 79.1 million approximately 10 years after 2010 in the year 2020.

**27.** In 2010, $t = 0$ and $A_0 = 4121$
In 2050, $t = 2050 - 2010 = 40$ and $A = 5231$.

$$5231 = 4121e^{k40}$$

$$\frac{5231}{4121} = e^{40k}$$

$$\ln\frac{5231}{4121} = \ln e^{40k}$$

$$\ln\frac{5231}{4121} = 40k$$

$$\frac{\ln\frac{5231}{4121}}{40} = k$$

$$0.006 \approx k$$

The exponential growth function is
$A = 4121e^{0.006t}$.

**28.** First find the decay equation.

$$0.5 = e^{kt}$$

$$0.5 = e^{7.2k}$$

$$\ln 0.5 = \ln e^{7.2k}$$

$$\ln 0.5 = 7.2k$$

$$\frac{\ln 0.5}{7.2} = k$$

$$k \approx -0.096270$$

$$A = e^{-0.096270t}$$

Next use the decay equation answer question.

$$A = e^{-0.096270t}$$

$$0.3 = e^{-0.096270t}$$

$$\ln 0.3 = \ln e^{-0.096270t}$$

$$\ln 0.3 = -0.096270t$$

$$\frac{\ln 0.3}{-0.096270} = t$$

$$t \approx 12.5$$

It will take 12.5 days.

**29. a.** $f(0) = \dfrac{140}{1 + 9e^{-0.165(0)}} = 14$

Fourteen elk were initially introduced to the habitat.

**b.** $f(10) = \dfrac{140}{1 + 9e^{-0.165(10)}} \approx 51$

After 10 years, about 51 elk are expected.

**c.** In the logistic growth model,

$$f(t) = \frac{c}{1 + ae^{-bt}},$$

the constant $c$ represents the limiting size that $f(t)$ can attain. The limiting size of the elk population is 140 elk.

**30.** Plot the ordered pairs.

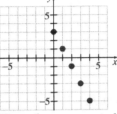

The values appear to belong to a linear function.

**31.** Plot the ordered pairs.

The values appear to belong to a logarithmic function.

**32.** Plot the ordered pairs.

The values appear to belong to an exponential function.

**33.** Plot the ordered pairs.

The values appear to belong to a quadratic function.

**34.** $y = 96(0.38)^x$

$y = 96e^{(\ln 0.38)x}$

$y = 96e^{-0.968x}$

**Cumulative Review Exercises (Chapters 1–4)**

**1.** $|3x - 4| = 2$

$3x - 4 = 2$ or $\quad 3x - 4 = -2$

$3x = 6 \qquad\qquad 3x = 2$

$x = 2 \qquad\quad x = \dfrac{2}{3}$

The solution set is $\left\{\dfrac{2}{3}, 2\right\}$.

**2.** $\sqrt{2x - 5} - \sqrt{x - 3} = 1$

$\sqrt{2x - 5} = 1 + \sqrt{x - 3}$

$\left(\sqrt{2x - 5}\right)^2 = \left(1 + \sqrt{x - 3}\right)^2$

$2x - 5 = 1 + 2\sqrt{x - 3} + x - 3$

$2x - 5 = x - 2 + 2\sqrt{x - 3}$

$x - 3 = 2\sqrt{x - 3}$

$(x - 3)^2 = \left(2\sqrt{x - 3}\right)^2$

$(x - 3)^2 = 4(x - 3)$

$x^2 - 6x + 9 = 4x - 12$

$x^2 - 10x + 21 = 0$

$(x - 3)(x - 7) = 0$

$x = 3$ or $x = 7$

Both solutions satisfy the original equation when checked.

The solution set is $\{3, 7\}$.

**3.** $x^4 + x^3 - 3x^2 - x + 2 = 0$

$p: \pm 1, \pm 2$

$q: \pm 1$

$\dfrac{p}{q}: \pm 1, \pm 2$

$$
\begin{array}{r|rrrrr}
-2 & 1 & 1 & -3 & -1 & 2 \\
& & -2 & 2 & 2 & -2 \\
\hline
& 1 & -1 & -1 & 1 & 0
\end{array}
$$

$(x+2)(x^3 - x^2 - x + 1) = 0$

$(x+2)[x^2(x-1) - (x-1)] = 0$

$(x+2)(x^2 - 1)(x - 1) = 0$

$(x+2)(x+1)(x-1)(x-1) = 0$

$(x+2)(x+1)(x-1)^2 = 0$

$x + 2 = 0 \quad$ or $\quad x + 1 = 0 \quad$ or $\quad x - 1 = 0$

$x = -2 \qquad\qquad x = -1 \qquad\qquad x = 1$

The solution set is $\{-2, -1, 1\}$.

**4.** $e^{5x} - 32 = 96$

$e^{5x} = 128$

$\ln e^{5x} = \ln 128$

$5x = \ln 128$

$x = \dfrac{\ln 128}{5} \approx 0.9704$

The solution set is $\left\{ \dfrac{\ln 128}{5} \right\}$, approximately 0.9704.

**5.** $\log_2(x+5) + \log_2(x-1) = 4$

$\log_2[(x+5)(x-1)] = 4$

$(x+5)(x-1) = 2^4$

$x^2 + 4x - 5 = 16$

$x^2 + 4x - 21 = 0$

$(x+7)(x-3) = 0$

$x + 7 = 0 \quad$ or $\quad x - 3 = 0$

$x = -7 \qquad\qquad x = 3$

$x = -7$ does not check because $\log_2(-7 + 5)$ does not exist.

The solution set is $\{3\}$.

**6.** $\ln(x+4) + \ln(x+1) = 2\ln(x+3)$

$\ln\big((x+4)(x+1)\big) = \ln(x+3)^2$

$(x+4)(x+1) = (x+3)^2$

$x^2 + 5x + 4 = x^2 + 6x + 9$

$5x + 4 = 6x + 9$

$-x = 5$

$x = -5$

$x = -5$ does not check and must be rejected.

The solution set is $\varnothing$.

**7.** $14 - 5x \geq -6$

$-5x \geq -20$

$x \leq 4$

The solution set is $(-\infty, 4]$.

**8.** $|2x - 4| \leq 2$

$2x - 4 \leq 2$ and $2x - 4 \geq -2$

$2x \leq 6 \qquad\qquad 2x \geq 2$

$x \leq 3 \qquad$ and $\quad x \geq 1$

The solution set is $[1, 3]$.

**9.** Circle with center: (3, –2) and radius of 2

$(x - 3)^2 + (y + 2)^2 = 4$

**10.** Parabola with vertex: (2, –1)

$f(x) = (x - 2)^2 - 1$

**11.** *x*-intercepts:

$$x^2 - 1 = 0$$
$$x^2 = 1$$
$$x = \pm 1$$

The *x*-intercepts are $(1, 0)$ and $(-1, 0)$.

vertical asymptotes:

$$x^2 - 4 = 0$$
$$x^2 = 4$$
$$x = \pm 2$$

The vertical asymptotes are $x = 2$ and $x = -2$.

Horizontal asymptote: $y = 1$

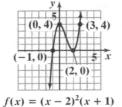

$$f(x) = \frac{x^2 - 1}{x^2 - 4}$$

**12.** *x*-intercepts:

$$x - 2 = 0 \quad \text{or} \quad x + 1 = 0$$
$$x = 2 \quad \text{or} \quad x = -1$$

The *x*-intercepts are $(2, 0)$ and $(-1, 0)$.

$$f(x) = (x - 2)^2(x + 1)$$

**13.**

$$f(x) = 2x - 4$$
$$f^{-1}(x) = \frac{x + 4}{2}$$

**14.**

$$f(x) = \ln x$$
$$g(x) = \ln(x - 2) + 1$$

**15.** $m = \dfrac{3 - (-3)}{1 - 3} = \dfrac{6}{-2} = -3$

Using (1, 3) point-slope form:

$$y - 3 = -3(x - 1)$$

slope-intercept form:

$$y - 3 = -3(x - 1)$$
$$y - 3 = -3x + 3$$
$$y = -3x + 6$$

**16.** $(f \circ g)(x) = f(x + 2)$

$$= (x + 2)^2$$
$$= x^2 + 4x + 4$$

$(g \circ f)(x) = g(x^2)$

$$= x^2 + 2$$

**17.** *y* varies inversely as the square of *x* is expressed as $y = \dfrac{k}{x^2}$.

The hours, *H* , vary inversely as the square of the number of cups of coffee, *C* can be expressed as $H = \dfrac{k}{C^2}$ .

Use the given values to find *k* .

$$H = \frac{k}{C^2}$$
$$8 = \frac{k}{2^2}$$
$$32 = k$$

Substitute the value of *k* into the equation.

$$H = \frac{k}{C^2}$$
$$H = \frac{32}{C^2}$$

Use the equation to find $H$ when $C = 4$.

$$H = \frac{32}{C^2}$$

$$H = \frac{32}{4^2}$$

$$H = 2$$

If 4 cups of coffee are consumed you should expect to sleep 2 hours.

**18.**  $s(t) = -16t^2 + 64t + 5$

The ball reaches its maximum height at

$$t = \frac{-b}{2a} = \frac{-(64)}{2(-16)} = 2 \text{ seconds.}$$

The maximum height is $s(2)$.

$$s(2) = -16(2)^2 + 64(2) + 5 = 69 \text{ feet.}$$

**19.**  $s(t) = -16t^2 + 64t + 5$

Let $s(t) = 0$:

$$0 = -16t^2 + 64t + 5$$

Use the quadratic formula to solve.

$$t = \frac{-b \pm \sqrt{b^2 - 4ac}}{2a}$$

$$t = \frac{-(64) \pm \sqrt{(64)^2 - 4(-16)(5)}}{2(-16)}$$

$t \approx 4.1, \quad t \approx -0.1$

The negative value is rejected.

The ball hits the ground after about 4.1 seconds.

**20.**  $40x + 10(1.5x) = 660$

$$40x + 15x = 660$$

$$55x = 660$$

$$x = 12$$

Your normal hourly salary is $12 per hour.

# Chapter 5
# Systems of Equations and Inequalities

## Section 5.1

### Check Point Exercises

**1. a.**
$$2x = 3y = -4$$
$$2(1) - 3(2) = -4$$
$$2 - 6 = -4$$
$$-4 = -4 \text{ true}$$
$$2x + y = 4$$
$$2(1) + 2 = 4$$
$$2 + 2 = 4$$
$$4 = 4 \text{ true}$$
(1, 2) is a solution of the system.

**b.**
$$2x = 3y = -4$$
$$2(7) - 3(6) = -4$$
$$14 - 18 = -4$$
$$-4 = -4 \text{ true}$$
$$2x + y = 4$$
$$2(7) + 6 = 4$$
$$14 + 6 = 4$$
$$20 = 4 \text{ false}$$
(7, 6) is not a solution of the system.

**2.**
$$3x + 2y = 4$$
$$2x + y = 1$$
Solve $2x + y = 1$ for $y$.
$$2x + y = 1$$
$$y = 1 - 2x$$
Substitute $1 - 2x$ for $y$ in the other equation and solve.
$$3x + 2\overbrace{(1 - 2x)}^{y} = 4$$
$$3x + 2 - 4x = 4$$
$$-x = 2$$
$$x = -2$$
Back-substitute the obtained value:
$$3x + 2y = 4$$
$$3(-2) + 2y = 4$$
$$-6 + 2y = 4$$
$$2y = 10$$
$$y = 5$$
Checking confirms the solution set is $\{(-2, 5)\}$.

**3.** Rewrite one or both equations:
$$4x + 5y = 3 \xrightarrow{\text{No change}} 4x + 5y = 3$$
$$2x - 3y = 7 \xrightarrow{\text{Mult. by } -2} \underline{-4x + 6y = -14}$$
$$11y = -11$$
$$y = -1$$
Back-substitute into either equation:
$$4x + 5y = 3$$
$$4x + 5(-1) = 3$$
$$4x - 5 = 3$$
$$4x = 8$$
$$x = 2$$
Checking confirms the solution set is $\{(2, -1)\}$.

**4.** Rewrite both equations in the form $Ax + By = C$:
$$2x = 9 + 3y \rightarrow 2x - 3y = 9$$
$$4y = 8 - 3x \rightarrow 3x + 4y = 8$$
Rewrite with opposite coefficients, then add and solve:
$$2x - 3y = 9 \xrightarrow{\text{Mult. by } 4} 8x - 12y = 36$$
$$3x + 4y = 8 \xrightarrow{\text{Mult. by } 3} \underline{9x + 12y = 24}$$
$$17x = 60$$
$$x = \frac{60}{17}$$
Back-substitute into either equation:
$$4y = 8 - 3x$$
$$4y = 8 - 3\left(\frac{60}{17}\right)$$
$$4y = -\frac{44}{17}$$
$$y = -\frac{11}{17}$$
Checking confirms the solution set is
$$\left\{\left(\frac{60}{17}, -\frac{11}{17}\right)\right\}.$$

**5.** Rewrite with a pair of opposite coefficients, then add:
$$5x - 2y = 4 \xrightarrow{\text{Mult. by } 2} 10x - 4y = 8$$
$$-10x + 4y = 7 \xrightarrow{\text{No change}} \underline{-10x + 4y = 7}$$
$$0 = 15$$
The statement $0 = 15$ is false which indicates that the system has no solution. The solution set is the empty set, $\varnothing$.

**6.** Substitute $4y - 8$ for $x$ in the other equation:

$$5\overbrace{(4y - 8)}^{x} - 20y = -40$$
$$20y - 40 - 20y = -40$$
$$-40 = -40$$

The statement $-40 = -40$ is true which indicates that the system has infinitely many solutions. The solution set is $\{(x, y) | x = 4y - 8\}$ or $\{(x, y) | 5x - 20y = -40\}$.

**7.**  **a.**  $C(x) = 300,000 + 30x$

**b.**  $R(x) = 80x$

**c.**  $R(x) = C(x)$
$$80x = 300,000 + 30x$$
$$50x = 300,000$$
$$x = 6000$$
$$C(6000) = 300,000 + 30(6000) = 480,000$$

The break-even point is (6000, 480,000).
The company will need to make 6000 pairs of shoes and earn $480,000 to break even.

## Concept and Vocabulary Check 5.1

**1.** satisfies both equations in the system

**2.** the intersection point

**3.** $\left\{ \left( \dfrac{1}{3}, -2 \right) \right\}$

**4.** $-2$

**5.** $-3$

**6.** $\varnothing$; inconsistent; parallel

**7.** $\{(x, y) | x = 3y + 2\}$ or $\{(x, y) | 5x - 15y = 10\}$;
dependent; are identical or coincide

**8.** revenue; profit

**9.** break-even point

## Exercise Set 5.1

**1.** $x + 3y = 11$
$2 + 3(3) = 11$
$2 + 9 = 11$
$11 = 11$ true
$x - 5y = -13$
$2 - 5(3) = -13$
$2 - 15 = -13$
$-13 = -13$ true
(2, 3) is a solution.

**3.** $2x + 3y = 17$
$2(2) + 3(5) = 17$
$4 + 15 = 17$
$19 = 17$ false
(2, 5) is not a solution.

**5.** $x + y = 4$
$y = 3x$
Substitute the expression $3x$ for $y$ in the first equation and solve for $x$.
$x + 3x = 4$
$4x = 4$
$x = 1$
Substitute 1 for $x$ in the second equation.
$y = 3(1) = 3$
The solution set is $\{(1, 3)\}$.

**7.** $x + 3y = 8$
$y = 2x - 9$
Substitute the expression $2x - 9$ for $y$ in the first equation and solve for $x$.
$x + 3(2x - 9) = 8$
$x + 6x - 27 = 8$
$7x = 35$
$x = 5$
Substitute 5 for $x$ in the second equation.
$y = 2(5) - 9 = 10 - 9 = 1$
The solution set is $\{(5, 1)\}$.

**9.** $x = 4y - 2$
$x = 6y + 8$
Substitute the expression $4y - 2$ for $x$ in the second equation and solve for $y$.
$4y - 2 = 6y + 8$
$-10 = 2y$
$-5 = y$
Substitute $-5$ for $y$ in the equation $x = 4y - 2$.
$x = 4(-5) - 2 = -22$
The solution set is $\{(-22, -5)\}$.

**11.** $5x + 2y = 0$
$x - 3y = 0$
Solve the second equation for $x$.
$x = 3y$
Substitute the expression $3y$ for $x$ in the first equation and solve for $y$.
$5(3y) + 2y = 0$
$15y + 2y = 0$
$17y = 0$
$y = 0$
Substitute 0 for $y$ in the equation $x = 3y$
$y = 3(0) = 0$
The solution set is $\{(0, 0)\}$.

**13.** $2x + 5y = -4$
$3x - y = 11$
Solve the second equation for $y$.
$-y = -3x + 11$
$y = 3x - 11$
Substitute the expression $3x - 11$ for $y$ in the first equation and solve for $x$.
$2x + 5(3x - 11) = -4$
$2x + 15x - 55 = -4$
$17x = 51$
$x = 3$
Substitute 3 for $x$ in the equation $y = 3x - 11$.
$y = 3(3) - 11 = 9 - 11 = -2$
The solution set is $\{(3, -2)\}$.

**15.** $2x - 3y = 8 - 2x$
$2x + 4y = x + 3y + 14$
Solve the second equation for $y$.
$y = -2x + 14$
Substitute the expression $-2x + 14$ for $y$ in the first equation and solve for $x$.
$2x - 3(-2x + 14) = 8 - 2x$
$2x + 6x - 42 = 8 - 2x$
$8x - 42 = 8 - 2x$
$10x = 50$
$x = 5$
Substitute 5 for $x$ in the equation $y = -2x + 14$.
$y = -2(5) + 14 = -10 + 14 = 4$
The solution set is $\{(5, 4)\}$.

**17.** $y = \frac{1}{3}x + \frac{2}{3}$
$y = \frac{5}{7}x - 2$
Substitute the expression $y = \frac{1}{3}x + \frac{2}{3}$ for $y$ in the second equation and solve for $x$.
$\frac{1}{3}x + \frac{2}{3} = \frac{5}{7}x - 2$
$7x + 14 = 15x - 42$
$56 = 8x$
$7 = x$
Substitute 7 for $x$ in the equation $y = \frac{1}{3}x + \frac{2}{3}$ and solve for $y$.
$y = \frac{1}{3}(7) + \frac{2}{3} = \frac{7}{3} + \frac{2}{3} = \frac{9}{3} = 3$
The solution set is $\{(7, 3)\}$.

**19.** Eliminate $y$ by adding the equations.
$x + y = 1$
$\underline{x - y = 3}$
$2x = 4$
$x = 2$
Substitute 2 for $x$ in the first equation.
$2 + y = 1$
$y = -1$
The solution set is $\{(2, -1)\}$.

**21.** Eliminate $y$ by adding the equations.
$2x + 3y = 6$
$\underline{2x - 3y = 6}$
$4x = 12$
$x = 3$
Substitute 3 for $x$ in the first equation.
$2(3) + 3y = 6$
$6 + 3y = 6$
$3y = 0$
$y = 0$
The solution set is $\{(3, 0)\}$.

**23.** $x + 2y = 2$
$-4x + 3y = 25$
Eliminate $x$ by multiplying the first equation by 4 and adding the resulting equations.
$4x + 8y = 8$
$\underline{-4x + 3y = 25}$
$11y = 33$
$y = 3$

Substitute 3 for y in the first equation.
$$x + 2(3) = 2$$
$$x + 6 = 2$$
$$x = -4$$
The solution set is $\{(-4, 3)\}$.

**25.** $4x + 3y = 15$
$2x - 5y = 1$
Eliminate $x$ by multiplying the second equation by $-2$ and adding the resulting equations.
$$4x + 3y = 15$$
$$\underline{-4x + 10y = -2}$$
$$13y = 13$$
$$y = 1$$
Substitute 1 for y in the second equation.
$$2x - 5(1) = 1$$
$$2x = 6$$
$$x = 3$$
The solution set is $\{(3, 1)\}$.

**27.** $3x - 4y = 11$
$2x + 3y = -4$
Eliminate $x$ by multiplying the first equation by 2 and the second equation by $-3$. Add the resulting equations.
$$6x - 8y = 22$$
$$\underline{-6x - 9y = 12}$$
$$-17y = 34$$
$$y = -2$$
Substitute $-2$ for y in the second equation.
$$2x + 3(-2) = -4$$
$$2x - 6 = -4$$
$$2x = 2$$
$$x = 1$$
The solution set is $\{(1, -2)\}$.

**29.** $3x = 4y + 1$
$3y = 1 - 4x$
Arrange the system so that variable terms appear on the left and constants appear on the right.
$3x - 4y = 1$
$4x + 3y = 1$
Eliminate $y$ by multiplying the first equation by 3 and the second equation by 4. Add the resulting equations.
$$9x - 12y = 3$$
$$\underline{16x + 12y = 4}$$
$$25x = 7$$
$$x = \frac{7}{25}$$

Substitute $\frac{7}{25}$ for $x$ in the second equation.
$$3y = 1 - 4\left(\frac{7}{25}\right)$$
$$3y = \frac{-3}{25}$$
$$y = \frac{-1}{25}$$
The solution set is $\left\{\left(\frac{7}{25}, -\frac{1}{25}\right)\right\}$.

**31.** The substitution method is used here to solve the system.
$$x = 9 - 2y$$
$$x + 2y = 13$$
Substitute the expression $9 - 2y$ for $x$ in the second equation and solve for $y$.
$$9 - 2y + 2y = 13$$
$$9 = 13$$
The false statement $9 = 13$ indicates that the system has no solution.
The solution set is the empty set, $\varnothing$.

**33.** The substitution method is used here to solve the system.
$$y = 3x - 5$$
$$21x - 35 = 7y$$
Substitute the expression $3x - 5$ for $y$ in the second equation and solve for $x$.
$$21x - 35 = 7(3x - 5)$$
$$21x - 35 = 21x - 35$$
$$-35 = -35$$
This true statement indicates that the system has infinitely many solutions.
The solution set is $\left\{(x, y) \mid y = 3x - 5\right\}$

**35.** The elimination method is used here to solve the system.
$3x - 2y = -5$
$4x + y = 8$
Eliminate $y$ by multiplying the second equation by 2 and adding the resulting equations.
$$3x - 2y = -5$$
$$\underline{8x + 2y = 16}$$
$$11x = 11$$
$$x = 1$$
Substitute 1 for $x$ in the second equation.
$$4(1) + y = 8$$
$$y = 4$$
The solution set is $\{(1, 4)\}$.

**37.** The elimination method is used here to solve the system.

$x + 3y = 2$

$3x + 9y = 6$

Eliminate $x$ by multiplying the first equation by $-3$ and adding the resulting equations.

$-3x - 9y = -6$

$\underline{3x + 9y = 6}$

$0 = 0$

This true statement indicates that the system has infinitely many solutions.

The solution set is $\left\{(x, y) \mid x + 3y = 2\right\}$.

**39.** First multiply each term in the first equation by 4 to eliminate the fractions.

$\dfrac{x}{4} - \dfrac{y}{4} = -1$

$x - y = -4$

Multiply the first equation by $-1$ and add to the second equation and solve for $y$.

$-x + y = 4$

$x + 4y = -9$

$5y = -5$

$y = -1$

Substitute $-1$ for $y$ in the equation $x - y = -4$ and solve for $x$.

$x - (-1) = -4$

$x + 1 = -4$

$x = -5$

The solution set is $\{(-5, -1)\}$.

**41.** Rearrange the equations to get in the standard form.

$2x - 3y = 4$

$4x + 5y = 3$

Multiply the first equation by $-2$ and add to the second equation. Solve for $y$.

$-4x + 6y = -8$

$4x + 5y = 3$

$11y = -5$

$y = -\dfrac{5}{11}$

Multiply the first equation by 5 and the second equation by 3 and add the equations. Solve for $x$.

$10x - 15y = 20$

$12x + 15y = 9$

$22x = 29$

$x = \dfrac{29}{22}$

The solution set is $\left\{\left(\dfrac{29}{22}, -\dfrac{5}{11}\right)\right\}$.

**43.** Add the equations to eliminate $y$.

$x + y = 7$

$\underline{x - y = -1}$

$2x = 6$

$x = 3$

Substitute 3 for $x$ in the first equation.

$3 + y = 7$

$y = 4$

The numbers are 3 and 4.

**45.** $3x - y = 1$

$x + 2y = 12$

Eliminate $y$ by multiplying the first equation by 2 and adding the resulting equations.

$6x - 2y = 2$

$\underline{x + 2y = 12}$

$7x = 14$

$x = 2$

Substitute 2 for $x$ in the first equation.

$3(2) - y = 1$

$6 - y = 1$

$-y = -5$

$y = 5$

The numbers are 2 and 5.

**47.**
$$\frac{x+2}{2} - \frac{y+4}{3} = 3$$
$$\frac{x+y}{5} = \frac{x-y}{2} - \frac{5}{2}$$

Start by multiplying each equation by its LCD and simplifying to clear the fractions.

$$\frac{x+2}{2} - \frac{y+4}{3} = 3$$

$$\frac{x+y}{5} = \frac{x-y}{2} - \frac{5}{2}$$

Start by multiplying each equation by its LCD and simplifying to clear the fractions.

$$6\left(\frac{x+2}{2} - \frac{y+4}{3}\right) = 6(3)$$
$$3(x+2) - 2(y+4) = 18$$
$$3x+6-2y-8 = 18$$
$$3x-2y = 20$$

$$10\left(\frac{x+y}{5}\right) = 10\left(\frac{x-y}{2} - \frac{5}{2}\right)$$
$$2(x+y) = 5(x-y) - 5(5)$$
$$2x+2y = 5x-5y-25$$
$$3x-7y = 25$$

We now need to solve the equivalent system of equations:
$$3x-2y = 20$$
$$3x-7y = 25$$

Subtract the two equations:
$$3x-2y = 20$$
$$\underline{-(3x-7y = 25)}$$
$$5y = -5$$
$$y = -1$$

Back-substitute this value for *y* and solve for *x*.
$$3x-2y = 20$$
$$3x-2(-1) = 20$$
$$3x+2 = 20$$
$$3x = 18$$
$$x = 6$$

The solution is $(6, -1)$.

**49.**
$$5ax+4y = 17$$
$$ax+7y = 22$$

Multiply the second equation by $-5$ and add the equations.

$$5ax+4y = 17$$
$$\underline{-5ax-35y = -110}$$
$$-31y = -93$$
$$y = 3$$

Back-substitute into one of the original equations to solve for *x*.
$$ax+7y = 22$$
$$ax+7(3) = 22$$
$$ax+21 = 22$$
$$ax = 1$$
$$x = \frac{1}{a}$$

The solution is $\left(\frac{1}{a}, 3\right)$.

**51.**
$$f(-2) = 11 \quad \rightarrow \quad -2m+b = 11$$
$$f(3) = -9 \quad \rightarrow \quad 3m+b = -9$$

We need to solve the resulting system of equations:
$$-2m+b = 11$$
$$3m+b = -9$$

Subtract the two equations:
$$-2m+b = 11$$
$$\underline{3m+b = -9}$$
$$-5m = 20$$
$$m = -4$$

Back-substitute into one of the original equations to solve for *b*.
$$-2m+b = 11$$
$$-2(-4)+b = 11$$
$$8+b = 11$$
$$b = 3$$

Therefore, $m = -4$ and $b = 3$.

**53.** The solution to a system of linear equations is the point of intersection of the graphs of the equations in the system. If $(6,2)$ is a solution, then we need to find the lines that intersect at that point.
Looking at the graph, we see that the graphs of $x+3y=12$ and $x-y=4$ intersect at the point $(6,2)$. Therefore, the desired system of equations is

$$x+3y=12 \quad \text{or} \quad y=-\frac{1}{3}x+4$$
$$x-y=4 \qquad\qquad y=x-4$$

**55.** At the break-even point, $R(x)=C(x)$.

$$10000+30x=50x$$
$$10000=20x$$
$$10000=20x$$
$$500=x$$

Five hundred radios must be produced and sold to break-even.

**57.** $R(x)=50x$

$$R(200)=50(200)=10000$$

$$C(x)=10000+30x$$
$$C(200)=10000+30(200)$$
$$=10000+6000=16000$$
$$R(200)-C(200)=10000-16000$$
$$=-6000$$

This means that if 200 radios are produced and sold the company will lose $6,000.

**59. a.**
$$P(x)=R(x)-C(x)$$
$$=50x-(10,000+30x)$$
$$=50x-10,000-30x$$
$$=20x-10,000$$
$$P(x)=20x-10,000$$

**b.**
$$P(10,000)=20(10,000)-10,000$$
$$=200,000-10,000=190,000$$

If 10,0000 radios are produced and sold the profit will be $190,000.

**61. a.** The cost function is:
$$C(x)=18,000+20x$$

**b.** The revenue function is:
$$R(x)=80x$$

**c.** At the break-even point, $R(x)=C(x)$.

$$80x=18,000+20x$$
$$60x=18,000$$
$$x=300$$
$$R(x)=80x$$
$$R(300)=80(300)$$
$$=24,000$$

When approximately 300 canoes are produced the company will break-even with cost and revenue at $24,000.

**63. a.** The cost function is:
$$C(x)=30000+2500x$$

**b.** The revenue function is:
$$R(x)=3125x$$

**c.** At the break-even point, $R(x)=C(x)$.

$$3125x=30000+2500x$$
$$625x=30000$$
$$x=48$$

After 48 sold out performances, the investor will break-even. ($150,000)

**65. a.** Substitute $0.375x+3$ for $p$ in the first equation.
$$p=-0.325x+5.8$$

$$\overbrace{0.375x+3}^{p}=-0.325x+5.8$$
$$0.375x+3=-0.325x+5.8$$
$$0.375x+0.325x+3=-0.325x+0.325x+5.8$$
$$0.7x+3=5.8$$
$$0.7x+3-3=5.8-3$$
$$0.7x=2.8$$
$$\frac{0.7x}{0.7}=\frac{2.8}{0.7}$$
$$x=4$$

Back-substitute to find $p$.
$$p=-0.325x+5.8$$
$$p=-0.325(4)+5.8=4.5$$

The ordered pair is (4,4.5).
Equilibrium number of workers: 4 million
Equilibrium hourly wage: $4.50

**b.** If workers are paid $4.50 per hour, there will be 4 million available workers and 4 million workers will be hired. In this state of market equilibrium, there is no unemployment.

**c.**
$$p = -0.325x + 5.8$$
$$5.15 = -0.325x + 5.8$$
$$0.65 = -0.325x$$
$$\frac{-0.65}{-0.325} = \frac{-0.325x}{-0.325}$$
$$2 = x$$

At $5.15 per hour, 2 million workers will be hired.

**d.**
$$p = 0.375x + 3$$
$$5.15 = 0.375x + 3$$
$$2.15 = 0.375x$$
$$\frac{2.15}{0.375} = \frac{0.375x}{0.375}$$
$$x \approx 5.7$$

At $5.15 per hour, there will be about 5.7 million available workers.

**e.** $5.7 - 2 = 3.7$

At $5.15 per hour, there will be about 3.7 million more people looking for work than employers are willing to hire.

**67.** **a.**
$$-3x + 10y = 160$$
$$x + 2y = 142$$

Multiply the second equation by -3 and then add the equations.
$$-3x + 10y = 160$$
$$\underline{3x + 6y = 426}$$
$$16y = 586$$
$$y = \frac{586}{16}$$
$$y = \frac{293}{8}$$
$$y \approx 37$$

Back-substitute $\frac{293}{8}$ for $y$ and find $x$.
$$x + 2y = 142$$
$$x + 2\left(\frac{293}{8}\right) = 142$$
$$x + \frac{293}{4} = 142$$
$$x = \frac{275}{4}$$
$$x \approx 69$$

About 69 years after 1970, or 2039, the percentage of never-married adults will be the same as the percentage of married adults, about 37%.

**b.** The answer from part (a) is represented by the intersection point $(69, 37)$.

**69.** **a.** $y = 0.45x + 0.8$

**b.** $y = 0.15x + 2.6$

**c.** To find the week in the semester when both groups report the same number of symptoms, we set the two equations equal to each other and solve for $x$.
$$0.45x + 0.8 = 0.15x + 2.6$$
$$0.3x = 1.8$$
$$x = 6$$

The number of symptoms will be the same in week 6.
$$y = 0.15x + 2.6$$
$$y = 0.15(6) + 2.6$$
$$y = 3.5$$

The number of symptoms in week 6 will be 3.5 for both groups. This is shown in the graph by the intersection point $(6, 3.5)$.

**71.** **a.** $y = 0.2x + 5$

**b.** $y = -0.4x + 47$

**c.** To project the year in which the percentage of adults living alone will be the same as the percentage of married adults living with kids, we set the two equations equal to each other and solve for $x$.
$$0.2x + 5 = -0.4x + 47$$
$$0.6x = 42$$
$$x = 70$$

The percentage will be the same 70 years after 1960, or 2030.

$$y = 0.2x + 5$$
$$y = 0.2(70) + 5$$
$$= 19$$

At that time about 19% will belong to each group.

**73.** Let $x$ = the number of calories in a Mr. Goodbar.
Let $y$ = the number of calories in a Mounds bar.
$$x + 2y = 780$$
$$2x + y = 786$$
Multiply the bottom equation by –2 and then add the equations to eliminate $y$.

$$x + 2y = \phantom{-}780$$
$$\underline{-4x - 2y = -1572}$$
$$-3x \phantom{-2y} = -792$$
$$x = \phantom{-}264$$

Back-substitute to find $y$.
$$x + 2y = 780$$
$$264 + 2y = 780$$
$$2y = 516$$
$$y = 258$$

There are 264 calories in a Mr. Goodbar and 258 calories in a Mounds bar.

**75.** Let $x$ = the number of Mr. Goodbars.
Let $y$ = the number of Mounds bars.
$$x + y = 5$$
$$16.3x + 14.1y - 70 = 7.1$$
Solve the first equation for $y$ in terms of $x$.
$$x + y = 5$$
$$y = -x + 5$$
Substitute $-x + 5$ for $y$ in the second equation.
$$16.3x + 14.1y - 70 = 7.1$$

$$16.3x + 14.1\overbrace{(-x + 5)}^{y} - 70 = 7.1$$
$$16.3x + 14.1(-x + 5) - 70 = 7.1$$
$$16.3x - 14.1x + 70.5 - 70 = 7.1$$
$$2.2x + 0.5 = 7.1$$
$$2.2x = 6.6$$
$$x = 3$$

Back-substitute to find $y$.
$$x + y = 5$$
$$3 + y = 5$$
$$y = 2$$
There are 3 Mr. Goodbars and 2 Mounds bars.

**77.**
$$x + y = 200$$
$$100x + 80y = 17000$$
Multiply the first equation by –100 and add to the second equation. Solve for $y$.
$$-100x - 100y = -20000$$
$$100x + 80y = 17000$$
$$-20y = -3000$$
$$y = 150$$
Substitute 150 for $y$ in the first equation and solve for $x$.
$$x + 150 = 200$$
$$x = 50$$
There are 50 rooms with kitchenettes and 150 rooms without.

**79.**
$$2x + 2y = 360$$
$$20x + 8(2y) = 3280$$
Multiply the first equation by –10 and add to the second equation. Solve for $y$.
$$-20x - 20y = -3600$$
$$20x + 16y = 3280$$
$$-4y = -320$$
$$y = 80$$
Substitute 80 for $y$ in the first equation and solve for $x$.
$$2x + 2(80) = 360$$
$$2x + 160 = 360$$
$$2x = 200$$
$$x = 100$$
The lot is 100 feet long and 80 feet wide.

**81.**
$$(x + y)2 = 16$$
$$(x - y)2 = 8$$
Multiply to remove the parentheses and then add the two equations together. Solve for $x$.
$$2x + 2y = 16$$
$$2x - 2y = 8$$
$$4x = 24$$
$$x = 6$$
Substitute 6 for $x$ in the first equation and solve for $y$.
$$2(6) + 2y = 16$$
$$12 + 2y = 16$$
$$2y = 4$$
$$y = 2$$
The crew rows 6 mph and the current is 2 mph.

**83.**
$$x + 2y = 180$$
$$(2x - 30) + y = 180$$
Rewrite the second equation in standard form.
$$x + 2y = 180$$
$$2x + y = 210$$
Multiply the first equation by $-2$ and add the equations.

$$\begin{array}{r} -2x - 4y = -360 \\ 2x + y = 210 \\ \hline -3y = -150 \\ y = 50 \end{array}$$

Back-substitute to solve for $x$.
$$x + 2y = 180$$
$$x + 2(50) = 180$$
$$x + 100 = 180$$
$$x = 80$$
The three interior angles measure $80°$, $50°$, and $50°$.

**85. – 93.** Answers will vary.

**95.** makes sense

**97.** makes sense

**99.** $a_1 x + b_1 y = c_1$
$a_2 x + b_2 y = c_2$
Solve the first equation for $x$.
$$x = \frac{c_1 - b_1 y}{a_1}$$

Substitute the expression $\dfrac{c_1 - b_1 y}{a_1}$ for $x$ in the second

equation and solve for $y$.

$$a_2 \left( \frac{c_1 - b_1 y}{a_1} \right) + b_2 y = c_2$$

$$a_2 \left( \frac{c_1 - b_1 y}{a_1} \right) + \frac{a_1 b_2 y}{a_1} = c_2$$

$$\frac{a_2 c_1 - a_2 b_1 y + a_1 b_2 y}{a_1} = c_2$$

$$a_2 c_1 - a_2 b_1 y + a_1 b_2 y = a_1 c_2$$
$$y(a_1 b_2 - a_2 b_1) = a_1 c_2 - a_2 c_1$$
$$y = \frac{a_1 c_2 - a_2 c_1}{a_1 b_2 - a_2 b_1}$$

Substitute the expression $\dfrac{a_1 c_2 - a_2 c_1}{a_1 b_2 - a_2 b_1}$ for $y$ in the first

equation and solve for $x$.

$$a_1 x + b_1 \left( \frac{a_1 c_2 - a_2 c_1}{a_1 b_2 - a_2 b_1} \right) = c_1$$

$$a_1 x + \frac{a_1 b_1 c_2 - a_2 b_1 c_1}{a_1 b_2 - a_2 b_1} = c_1$$

$$a_1 x = c_1 - \frac{a_1 b_1 c_2 - a_2 b_1 c_1}{a_1 b_2 - a_2 b_1}$$

$$= \frac{c_1 (a_1 b_2 - a_2 b_1)}{a_1 b_2 - a_2 b_1} - \frac{a_1 b_1 c_2 - a_2 b_1 c_1}{a_1 b_2 - a_2 b_1}$$

$$= \frac{a_1 b_2 c_1 - a_1 b_1 c_2}{a_1 b_2 - a_2 b_1}$$

$$x = \frac{a_1 b_2 c_1 - a_1 b_1 c_2}{a_1 b_2 - a_2 b_1} \div a_1$$

$$= \frac{a_1 b_2 c_1 - a_1 b_1 c_2}{a_1 (a_1 b_2 - a_2 b_1)} = \frac{a_1 (b_2 c_1 - b_1 c_2)}{a_1 (a_1 b_2 - a_2 b_1)}$$

$$x = \frac{b_2 c_1 - b_1 c_2}{a_1 b_2 - a_2 b_1}$$

**101.** $x =$ number of hexagons formed
$y =$ number of squares formed
$$6x + y = 52$$
$$x + 4y = 24$$
Eliminate $x$ by multiplying the second equation by $-6$ and adding the resulting equations.

$$\begin{array}{r} 6x + y = 52 \\ -6x - 24y = -144 \\ \hline -23y = -92 \\ y = 4 \end{array}$$

Substitute 4 for $y$ in the second equation.
$$x + 4(4) = 24$$
$$x + 16 = 24$$
$$x = 8$$
Yes, they should make 8 hexagons and 4 squares.

**103.**
$$2x - y + 4z = -8$$
$$2(3) - (2) + 4(-3) = -8$$
$$-8 = -8, \text{ true}$$
Yes, the ordered triple satisfies the equation.

**104.** $5x - 2y - 4z = 3$
$3x + 3y + 2z = -3$
Multiply Equation 2 by 2.
$5x - 2y - 4z = 3$
$6x + 6y + 4z = -6$
Then add to eliminate $z$.

$$\begin{array}{r} 5x - 2y - 4z = 3 \\ 6x + 6y + 4z = -6 \\ \hline 11x + 4y \quad\quad = -3 \end{array}$$

**105.** $y = ax^2 + bx + c$
$$1682 = a(4)^2 + b(4) + c$$
$$1682 = 16a + 4b + c$$

## Section 5.2

### Check Point Exercises

**1.**
$$x - 2y + 3z = 22$$
$$-1 - 2(-4) + 3(5) = 22$$
$$-1 + 8 + 15 = 22$$
$$22 = 22 \text{ true}$$
$$2x - 3y - z = 5$$
$$2(-1) - 3(-4) - 5 = 5$$
$$-2 + 12 - 5 = 5$$
$$5 = 5 \text{ true}$$
$$3x + y - 5z = -32$$
$$3(-1) - 4 - 5(5) = -32$$
$$-3 - 4 - 25 = -32$$
$$-32 = -32 \text{ true}$$
$(-1, -4, 5)$ is a solution of the system.

**2.**
$$x + 4y - z = 20$$
$$3x + 2y + z = 8$$
$$2x - 3y + 2z = -16$$
Eliminate $z$ from Equations 1 and 2 by adding Equation 1 and Equation 2.

$$\begin{array}{r} x + 4y - z = 20 \\ 3x + 2y + z = 8 \\ \hline 4x + 6y = 28 \quad \text{Equation 4} \end{array}$$

Eliminate $z$ from Equations 2 and 3 by multiplying Equation 2 by $-2$ and adding the resulting equation to Equation 3.

$$\begin{array}{r} -6x - 4y - 2z = -16 \\ 2x - 3y + 2z = -16 \\ \hline -4x - 7y = -32 \quad \text{Equation 5} \end{array}$$

Solve Equations 4 and 5 for $x$ and $y$ by adding Equation 4 and Equation 5.

$$\begin{array}{r} 4x + 6y = 28 \\ -4x - 7y = -32 \\ \hline -y = -4 \\ y = 4 \end{array}$$

Substitute 4 for $y$ in Equation 4 and solve for $x$.
$$4x + 6(4) = 28$$
$$4x + 24 = 28$$
$$4x = 4$$
$$x = 1$$
Substitute 1 for $x$ and 4 for $y$ in Equation 2 and solve for $z$.
$$3(1) + 2(4) + z = 8$$
$$3 + 8 + z = 8$$
$$11 + z = 8$$
$$z = -3$$
The solution set is $\{(1, 4, -3)\}$.

**3.**
$$2y - z = 7$$
$$x + 2y + z = 17$$
$$2x - 3y + 2z = -1$$
Eliminate $x$ and $z$ from Equations 2 and 3 by multiplying Equation 2 by $-2$ and adding the resulting equation to Equation 3.

$$\begin{array}{r} -2x - 4y - 2z = -34 \\ 2x - 3y + 2z = -1 \\ \hline -7y = -35 \\ y = 5 \end{array}$$

Substitute 5 for $y$ in Equation 1 and solve for $z$.

$$2(5) - z = 7$$
$$10 - z = 7$$
$$-z = -3$$
$$z = 3$$

Substitute 5 for $y$ and 3 for $z$ in Equation 2 and solve for $x$.

$$x + 2(5) + 3 = 17$$
$$x + 10 + 3 = 17$$
$$x + 13 = 17$$
$$x = 4$$

The solution set is $\{(4, 5, 3)\}$.

4. $(1, 4), (2, 1), (3, 4)$

$$y = ax^2 + bx + c$$

Substitute 1 for $x$ and 4 for $y$ in
$$y = ax^2 + bx + c.$$
$$4 = a(1)^2 + b(1) + c$$
$$4 = a + b + c \quad \text{Equation 1}$$

Substitute 2 for $x$ and 1 for $y$ in
$$y = ax^2 + bx + c.$$
$$1 = a(2)^2 + b(2) + c$$
$$1 = 4a + 2b + c \quad \text{Equation 2}$$

Substitute 3 for $x$ and 4 for $y$ in
$$y = ax^2 + bx + c.$$
$$4 = a(3)^2 + b(3) + c$$
$$4 = 9a + 3b + c \quad \text{Equation 3}$$

Eliminate $c$ from Equations 1 and 2 by multiplying Equation 2 by –1 and adding the resulting equation to Equation 1.

$$4 = a + b + c$$
$$\underline{-1 = -4a - 2b - c}$$
$$3 = -3a - b \qquad \text{Equation 4}$$

Eliminate $c$ from Equation 2 and 3 by multiplying Equation 3 by –1 and adding the resulting equation to Equation 2.

$$1 = 4a + 2b + c$$
$$\underline{-4 = -9a - 3b - c}$$
$$-3 = -5a - b \qquad \text{Equation 5}$$

Solve Equations 4 and 5 for $a$ and $b$ by multiplying Equation 5 by –1 and adding the resulting equation to Equation 4.

$$3 = -3a - b$$
$$\underline{3 = 5a + b}$$
$$6 = 2a$$
$$a = 3$$

Substitute 3 for $a$ in Equation 4 and solve for $b$.

$$3 = -3(3) - b$$
$$3 = -9 - b$$
$$12 = -b$$
$$b = -12$$

Substitute 3 for $a$ and –12 for $b$ in Equation 1 and solve for $c$.

$$4 = 3 - 12 + c$$
$$4 = -9 + c$$
$$c = 13$$

Substituting 3 for $a$, –12 for $b$, and 13 for $c$ in the quadratic equation $y = ax^2 + bx + c$ gives

$$y = 3x^2 - 12x + 13 .$$

### Concept and Vocabulary Check 5.2

1. triple; all

2. $-2$; $-4$

3. $z$; add Equations 1 and 3

### Exercise Set 5.2

1. $x + y + z = 4$
   $2 - 1 + 3 = 4$
   $\qquad 4 = 4$ true
   $\qquad x - 2y - z = 1$
   $2(2) - 2(-1) - 3 = 1$
   $\qquad 4 + 2 - 3 = 1$
   $\qquad\qquad 1 = 1$ true
   $\qquad\qquad 2x - y - 2z = -1$
   $2(2) - (-1) - 2(3) = -1$
   $\qquad\quad 4 + 1 - 6 = -1$
   $\qquad\qquad\qquad -1 = -1$ false

   $(2, -1, 3)$ is a solution.

3. $x - 2y = 2$
   $4 - 2(1) = 2$
   $\qquad 4 - 2 = 2$
   $\qquad\quad 2 = 2$ true
   $\qquad 2x + 3y = 11$
   $2(4) + 3(1) = 11$
   $\qquad 8 + 3 = 11$
   $\qquad\quad 11 = 11$ true
   $\qquad y - 4z = -7$
   $1 - 4(2) = -7$
   $\qquad 1 - 8 = -7$
   $\qquad -7 = -7$ true

   $(4, 1, 2)$ is a solution.

**5.** $x + y + 2z = 11$
$x + y + 3z = 14$
$x + 2y - z = 5$

Eliminate $x$ and $y$ from Equations 1 and 2 by multiplying Equation 2 by –1 and adding the resulting equation to Equation 1.

$-x - y - 3z = -14$
$\underline{x + y + 2z = \phantom{0}11}$
$\phantom{x + y}-z = -3$
$\phantom{x + y - }z = 3$

Substitute 3 for $z$ in Equations 1 and 3.
$x + y + 2(3) = 11$
$x + 2y - (3) = 5$
Simplify:
$x + y = 5$ \qquad Equation 4
$x + 2y = 8$ \qquad Equation 5

Solve Equations 4 and 5 for $x$ and $y$ by multiplying Equation 5 by –1 and adding the resulting equation to Equation 4.

$x + y = 5$
$\underline{-x - 2y = -8}$
$\phantom{x + }-y = -3$
$\phantom{x + -}y = 3$

Substitute 3 for $z$ and 3 for $y$ in Equation 2 and solve for $x$.
$x + 3 + 3(3) = 14$
$\phantom{x + 3}x + 12 = 14$
$\phantom{x + 3 + 3(}x = 2$
The solution set is {(2, 3, 3)}.

**7.** $4x - y + 2z = 11$
$x + 2y - z = -1$
$2x + 2y - 3z = -1$

Eliminate $y$ from Equation 1 and 2 by multiplying Equation 1 by 2 and adding the resulting equation to Equation 2 and 3.

$8x - 2y + 4z = 22$
$\underline{x + 2y - z = -1}$
$9x + 3z = 21$ \qquad Equation 4

Eliminate $y$ from Equations 1 and 3 by multiplying Equation 1 by 2 and adding the resulting equation to Equation 3.

$8x - 2y + 4z = 22$
$\underline{2x + 2y - 3z = -1}$
$10x + z = 21$ \qquad Equation 5

Solve Equations 4 and 5 for $x$ and $z$ by multiplying Equation 5 by –3 and adding the resulting equation to Equation 4.

$9x + 3z = 21$
$\underline{-30x - 3z = -63}$
$\phantom{9x}-21x = -42$
$\phantom{9x - 2}x = 2$

Substitute 2 for $x$ in Equation 5 and solve for $z$. $10(2) + z = 21$
$\phantom{10(2}20 + z = 21$
$\phantom{10(2)+}z = 1$

Substitute 2 for $x$ and 1 for $z$ in Equation 2 and solve for $y$.
$2 + 2y - 1 = -1$
$\phantom{2 + }2y + 1 = -1$
$\phantom{2 + }2y = -2$
$\phantom{2 + 2}y = -1$
The solution set is $\{(2, -1, 1)\}$.

**9.** $3x + 2y - 3z = -2$
$2x - 5y + 2z = -2$
$4x - 3y + 4z = 10$

Eliminate $z$ from Equations 1 and 2 by multiplying Equation 1 by 2 and Equation 2 by 3. Add the resulting equations.

$6x + 4y - 6z = -4$
$\underline{6x - 15y + 6z = -6}$
$12x - 11y = -10$ \quad Equation 4

Eliminate $z$ from Equations 2 and 3 by multiplying Equation 2 by –2.

$-4x + 10y - 4z = 4$
$\underline{4x - 3y + 4z = 10}$
$\phantom{-4x + 10y}7y = 14$ \quad Equation 5

Solve Equation 5 for $y$
$7y = 14$
$\phantom{7}y = 2$

Solve for $x$ by substituting 7 for $y$ in Equation 4.
$12x - 11y = -10$
$12x - 11(2) = -10$
$\phantom{12x }12x - 22 = -10$
$\phantom{12x - 2}12x = 12$
$\phantom{12x - 22}x = 1$

Substitute 2 for $y$ and 1 for $x$ in Equation 2 and solve for $z$.
$2x - 5y + 2z = -2$
$2(1) - 5(2) + 2z = -2$
$\phantom{2(1)}2 - 10 + 2z = -2$
$\phantom{2(1) - 1}2z = 6$
$\phantom{2(1) - 10}z = 3$
The solution set is {(1, 2, 3)}.

**11.** $2x - 4y + 3z = 17$
$x + 2y - z = 0$
$4x - y - z = 6$

Eliminate $z$ from Equations 1 and 2 by multiplying Equation 2 by 3 and adding the resulting equation to Equation 1.

$2x - 4y + 3z = 17$
$\underline{3x + 6y - 3z = 0}$
$\quad\; 5x + 2y = 17$     Equation 4

Eliminate $z$ from Equations 2 and 3 by multiplying Equation 2 by −1 and adding the resulting equation to Equation 3.

$-x - 2y + z = 0$
$\underline{4x - y - z = 6}$
$\quad 3x - 3y = 6$        Equation 5

Solve Equations 4 and 5 for $x$ and $y$ by multiplying

Equation 5 by $\dfrac{2}{3}$ and adding the resulting equation to

Equation 4.

$5x + 2y = 17$
$\underline{2x - 2y = 4}$
$\quad 7x = 21$
$\quad\; x = 3$

Substitute 3 for $x$ in Equation 4 and solve for $y$.

$5(3) + 2y = 17$
$\quad 15 + 2y = 17$
$\qquad 2y = 2$
$\qquad\; y = 1$

Substitute 3 for $x$ and 1 for $y$ in Equation 2 and solve for $z$.

$3 + 2(1) - z = 0$
$\quad 3 + 2 - z = 0$
$\qquad 5 - z = 0$
$\qquad\quad 5 = z$

The solution set is $\{(3, 1, 5)\}$.

**13.** $2x + y = 2$
$x + y - z = 4$
$3x + 2y + z = 0$

Eliminate $z$ from Equations 2 and 3 by adding Equation 2 and Equation 3.

$x + y - z = 4$
$\underline{3x + 2y + z = 0}$
$\quad 4x + 3y = 4$     Equation 4

Solve Equations 1 and 4 for $x$ and $y$ by multiplying Equation 1 by −3 and adding the resulting equation to Equation 4.

$-6x - 3y = -6$
$\underline{4x + 3y = 4}$
$\quad -2x = -2$
$\qquad x = 1$

Substitute 1 for $x$ in Equation 1 and solve for $y$.

$2(1) + y = 2$
$\quad 2 + y = 2$
$\qquad y = 0$

Substitute 1 for $x$ and 0 for $y$ in Equation 2 and solve for $z$.

$1 + 0 - z = 4$
$\quad 1 - z = 4$
$\quad\; -z = 3$
$\qquad z = -3$

The solution set is $\{(1, 0, -3)\}$.

**15.** $\quad\; x + y = -4$
$\qquad y - z = 1$
$2x + y + 3z = -21$

Eliminate $y$ from Equations 1 and 2 by multiplying Equation 1 by −1 and adding the resulting equation to Equation 2.

$-x - y = 4$
$\underline{y - z = 1}$
$-x - z = 5$     Equation 4

Eliminate $y$ from Equations 2 and 3 by multiplying Equation 2 by −1 and adding the resulting equation to Equation 3.

$\quad -y + z = -1$
$\underline{2x + y + 3z = -21}$
$\quad 2x + 4z = -22$     Equation 5

Solve Equations 4 and 5 for $x$ and $z$ by multiplying Equation 4 by 2 and adding the resulting equation to Equation 5.

$-2x - 2z = 10$
$\underline{2x + 4z = -22}$
$\quad 2z = -12$
$\qquad z = -6$

Substitute −6 for $z$ in Equation 2 and solve for $y$.

$y - (-6) = 1$
$\quad y + 6 = 1$
$\qquad y = -5$

Substitute −5 for $y$ in Equation 1 and solve for $x$

$x + (-5) = -4$
$\qquad x = 1$

The solution set is $\{(1, -5, -6)\}$.

**17.** $3(2x + y) + 5z = -1$

$2(x - 3y + 4z) = -9$

$4(1 + x) = -3(z - 3y)$

Simplify each equation.

$6x + 3y + 5z = -1$   Equation 4

$2x - 6y + 8z = -9$   Equation 5

$4 + 4x = -3z + 9y$

$4x - 9y + 3z = -4$   Equation 6

Eliminate *x* from Equations 4 and 5 by multiplying Equation 5 by –3 and adding
the resulting equation to Equation 4.

$-6x + 3y + 5z = -1$

$\underline{-6x + 18y - 24z = 27}$

$\phantom{-6x+}21y - 19z = 26$   Equation 7

Eliminate *x* from Equations 5 and 6 by multiplying Equation 5 by –2 and adding
the resulting equation to Equation 6.

$-4x + 12y - 16z = 18$

$\underline{4x - 9y + 3z = -4}$

$\phantom{-4x+}3y - 13z = 14$   Equation 8

Solve Equations 7 and 8 for *y* and *z* by multiplying Equation 8 by –7 and adding
the resulting equation to Equation 7.

$21y - 19z = 26$

$\underline{-21y + 91z = -98}$

$\phantom{21y+}72z = -72$

$\phantom{21y+}z = -1$

Substitute –1 for *z* in Equation 8 and solve for *y*.

$3y - 13(-1) = 14$

$3y + 13 = 14$

$3y = 1$

$y = \dfrac{1}{3}$

Substitute $\dfrac{1}{3}$ for *y* and –1 for *z* in Equation 5 and

solve for *x*.

$2x - 6\left(\dfrac{1}{3}\right) + 8(-1) = -9$

$2x - 2 - 8 = -9$

$2x - 10 = -9$

$2x = 1$

$x = \dfrac{1}{2}$

The solution set is $\left\{ \left( \dfrac{1}{2}, \dfrac{1}{3}, -1 \right) \right\}$.

**19.** $(-1, 6), (1, 4), (2, 9)$

$y = ax^2 + bx + c$

Substitute –1 for *x* and 6 for *y* in $y = ax^2 + bx + c$.

$6 = a(-1)^2 + b(-1) + c$

$6 = a - b + c$   Equation 1

Substitute 1 for *x* and 4 for *y* in $y = ax^2 + bx + c$.

$4 = a(1)^2 + b(1) + c$

$4 = a + b + c$   Equation 2

Substitute 2 for *x* and 9 for *y* in

$y = ax^2 + bx + c$.

$9 = a(2)^2 + b(2) + c$

$9 = 4a + 2b + c$   Equation 3

Eliminate *b* from Equations 1 and 2 by adding Equation 1 and Equation 2.

$6 = a - b + c$

$\underline{4 = a + b + c}$

$10 = 2a + 2c$   Equation 4

Eliminate *b* from Equations 1 and 3 by multiplying Equation 1 by 2 and adding the resulting equation to Equation 3.

$12 = 2a - 2b + 2c$

$\underline{9 = 4a + 2b + c}$

$21 = 6a + 3c$   Equation 5

Solve Equations 4 and 5 for *a* and *c* by multiplying Equation 4 by –3 and adding the resulting equation to Equation 5.

$-30 = -6a - 6c$

$\underline{21 = 6a + 3c}$

$-9 = -3c$

$c = 3$

Substitute 3 for *c* in Equation 4 and solve for *a*.

$10 = 2a + 2(3)$

$10 = 2a + 6$

$4 = 2a$

$a = 2$

Substitute 2 for *a* and 3 for *c* in Equation 2 and solve for *b*.

$4 = 2 + b + 3$

$4 = b + 5$

$b = -1$

Substituting 2 for *a*, –1 for *b*, and 3 for *c* in the quadratic equation $y = ax^2 + bx + c$ gives

$y = 2x^2 - x + 3$.

**21.** $(-1, -4), (1, -2), (2, 5)$

Substitute $-1$ for $x$ and $-4$ for $y$ in $y = ax^2 + bx + c$.

$-4 = a(-1)^2 + b(-1) + c$

$-4 = a - b + c$        Equation 1

Substitute $1$ for $x$ and $-2$ for $y$ in $y = ax^2 + bx + c$.

$-2 = a(1)^2 + b(1) + c$

$-2 = a + b + c$        Equation 2

Substitute $2$ for $x$ and $5$ for $y$ in $y = ax^2 + bx + c$.

$5 = a(2)^2 + b(2) + c$

$5 = 4a + 2b + c$        Equation 3

Eliminate $a$ and $b$ from Equations 1 and 2 by multiplying Equation 1 by $-1$ and adding the resulting equation to Equation 2.

$4 = -a + b - c$

$\underline{-2 = a + b + c}$

$2 = 2b$

$b = 1$

Eliminate $c$ from Equations 1 and 3 by multiplying Equation 1 by $-1$ and adding the resulting equation to Equation 3.

$4 = -a + b - c$

$\underline{5 = 4a + 2b + c}$

$9 = 3a + 3b$        Equation 4

Substitute $1$ for $b$ in Equation 4 and solve for $a$.

$9 = 3a + 3(1)$

$9 = 3a + 3$

$6 = 3a$

$a = 2$

Substitute $2$ for $a$ and $1$ for $b$ in Equation 2 and solve for $c$.

$-2 = 2 + 1 + c$

$-2 = c + 3$

$c = -5$

Substituting $2$ for $a$, $1$ for $b$, and $-5$ for $c$ in quadratic equation $y = ax^2 + bx + c$ gives $y = 2x^2 + x - 5$.

**23.** $x + y + z = 16$

$2x + 3y + 4z = 46$

$5x - y = 31$

Eliminate $z$ from Equations 1 and 2 by multiplying Equation 1 by $-4$ and adding the resulting equation to Equation 2.

$-4x - 4y - 4z = -64$

$\underline{2x + 3y + 4z = 46}$

$-2x - y = -18$        Equation 4

Solve Equations 3 and 4 for $x$ and $y$ by multiplying

Equation 4 by $-1$ and adding the resulting equation to Equation 3.

$5x - y = 31$

$\underline{2x + y = 18}$

$7x = 49$

$x = 7$

Substitute $7$ for $x$ in Equation 3 and solve for $y$.

$5(7) - y = 31$

$35 - y = 31$

$-y = -4$

$y = 4$

Substitute $7$ for $x$ and $4$ for $y$ in Equation 1 and solve for $z$.

$7 + 4 + z = 16$

$z + 11 = 16$

$z = 5$

The numbers are 7, 4 and 5.

**25.**

$$\frac{x+2}{6} - \frac{y+4}{3} + \frac{z}{2} = 0$$

$$6\left(\frac{x+2}{6} - \frac{y+4}{3} + \frac{z}{2}\right) = 6(0)$$

$$(x+2) - 2(y+4) + 3z = 0$$

$$x + 2 - 2y - 8 + 3z = 0$$

$$x - 2y + 3z = 6$$

$$\frac{x+1}{2} + \frac{y-1}{2} - \frac{z}{4} = \frac{9}{2}$$

$$4\left(\frac{x+1}{2} + \frac{y-1}{2} - \frac{z}{4}\right) = 4\left(\frac{9}{2}\right)$$

$$2(x+1) + 2(y-1) - z = 18$$

$$2x + 2 + 2y - 2 - z = 18$$

$$2x + 2y - z = 18$$

$$\frac{x-5}{4} + \frac{y+1}{3} + \frac{z-2}{2} = \frac{19}{4}$$

$$12\left(\frac{x-5}{4} + \frac{y+1}{3} + \frac{z-2}{2}\right) = 12\left(\frac{19}{4}\right)$$

$$3(x-5) + 4(y+1) + 6(z-2) = 57$$

$$3x - 15 + 4y + 4 + 6z - 12 = 57$$

$$3x + 4y + 6z = 80$$

We need to solve the equivalent system:

$x - 2y + 3z = 6$

$2x + 2y - z = 18$

$3x + 4y + 6z = 80$

Add the first two equations together.

$x - 2y + 3z = 6$
$\underline{2x + 2y - z = 18}$
$\quad 3x + 2z = 24$

Multiply the second equation by $-2$ and add it to the third equation.

$-4x - 4y + 2z = -36$
$\underline{\quad 3x + 4y + 6z = 80}$
$\quad -x + 8z = 44$

Using the two reduced equations, we solve the system

$3x + 2z = 24$

$-x + 8z = 44$

Multiply the second equation by 3 and add the equations.

$\quad 3x + 2z = 24$
$\underline{-3x + 24z = 132}$
$\quad 26z = 156$
$\quad z = 6$

Back-substitute to find $x$.

$-x + 8(6) = 44$
$-x + 48 = 44$
$-x = -4$
$x = 4$

Back substitute to find $y$.

$x - 2y + 3z = 6$
$4 - 2y + 3(6) = 6$
$-2y = -16$
$y = 8$

The solution is $(4, 8, 6)$.

**27.** Selected points may vary, but the equation will be the same.

$y = ax^2 + bx + c$

Use the points $(2, -2)$, $(4, 1)$, and $(6, -2)$ to get the system

$4a + 2b + c = -2$

$16a + 4b + c = 1$

$36a + 6b + c = -2$

Multiply the first equation by $-1$ and add to the second equation.

$-4a - 2b - c = 2$
$\underline{16a + 4b + c = 1}$
$\quad 12a + 2b = 3$

Multiply the first equation by $-1$ and add to the third equation.

$-4a - 2b - c = 2$
$\underline{36a + 6b + c = -2}$
$\quad 32a + 4b = 0$

Using the two reduced equations, we get the system

$12a + 2b = 3$

$32a + 4b = 0$

Multiply the first equation by $-2$ and add to the second equation.

$-24a - 4b = -6$
$\underline{\quad 32a + 4b = 0}$
$\quad 8a = -6$
$\quad a = -\dfrac{3}{4}$

Back-substitute to solve for $b$.

$12a + 2b = 3$

$12\left(-\dfrac{3}{4}\right) + 2b = 3$

$-9 + 2b = 3$

$2b = 12$

$b = 6$

Back-substitute to solve for $c$.

$4a + 2b + c = -2$

$4\left(-\dfrac{3}{4}\right) + 2(6) + c = -2$

$-3 + 12 + c = -2$

$c = -11$

The equation is:

$y = -\dfrac{3}{4}x^2 + 6x - 11$

**29.** $ax - by - 2cz = 21$

$\quad ax + by + cz = 0$

$\quad 2ax - by + cz = 14$

Add the first two equations.

$ax - by - 2cz = 21$
$\underline{ax + by + cz = 0}$
$\quad 2ax - cz = 21$

Multiply the first equation by $-1$ and add to the third equation.

$-ax + by + 2cz = -21$

$\underline{2ax - by + cz = 14}$

$ax + 3cz = -7$

Use the two reduced equations to get the following system:

$2ax - cz = 21$

$ax + 3cz = -7$

Multiply the second equation by $-2$ and add the equations.

$2ax - cz = 21$

$\underline{-2ax - 6cz = 14}$

$-7cz = 35$

$z = -\dfrac{5}{c}$

Back-substitute to solve for $x$.

$ax + 3cz = -7$

$ax + 3c\left(-\dfrac{5}{c}\right) = -7$

$ax - 15 = -7$

$ax = 8$

$x = \dfrac{8}{a}$

Back-substitute to solve for $y$.

$ax + by + cz = 0$

$a\left(\dfrac{8}{a}\right) + by + c\left(-\dfrac{5}{c}\right) = 0$

$8 + by - 5 = 0$

$by = -3$

$y = -\dfrac{3}{b}$

The solution is $\left(\dfrac{8}{a}, -\dfrac{3}{b}, -\dfrac{5}{c}\right)$.

**31. a.** Substitute the values for $x$ and $y$ into the quadratic form.

$224 = a(1)^2 + b(1) + c$

$a + b + c = 224$

$176 = a(3)^2 + b(3) + c$

$9a + 3b + c = 176$

$104 = a(4)^2 + b(4) + c$

$16a + 4b + c = 104$

Multiply the first equation by $-1$ and add to both the second and the third equations to obtain 2 new equations with 2 variables.

$-a - b - c = -224$

$9a + 3b + c = 176$

$8a + 2b = -48$

$-a - b - c = -224$

$16a + 4b + c = 104$

$15a + 3b = -120$

Use the two new equations to solve for $a$ and $b$. Multiply the first equation by $-3$ and the second equation by 2 and add the results together. Solve for $a$. Substitute that value in $8a + 2b = -48$ and solve for $b$.

$-24a - 6b = 144$

$30a + 6b = -240$

$6a = -96$

$a = -16$

$8(-16) + 2b = -48$

$-128 + 2b = -48$

$2b = 80$

$b = 40$

Substitute $-16$ for $a$ and 40 for $b$ into the equation $a + b + c = 224$ and solve for $c$.

$-16 + 40 + c = 224$

$c = 200$

The equation is $y = -16x^2 + 40x + 200$.

**b.** $y = -16(5)^2 + 40(5) + 200 = 0$

The ball hit the ground after 5 seconds.

**33.** Let $x$ = annual spending in 2010 per person on housing.
Let $y$ = annual spending in 2010 per person on vehicles/gas.
Let $z$ = annual spending in 2010 per person on health care.

$x + y + z = 13,840$

$x - y = 3864$

$x - z = 695$

Solve the second equation for $y$.

$x - y = 3864$

$-y = -x + 3864$

$y = x - 3864$

Solve the third equation for $z$.

$$x - z = 695$$
$$-z = -x + 695$$
$$z = x - 695$$

Substitute the expressions for $x$ and $z$ into the first equation and solve for $y$.

$$x + y + z = 13,840$$

$$x + \overbrace{(x - 3864)}^{y} + \overbrace{(x - 695)}^{z} = 13,840$$
$$x + x - 3864 + x - 695 = 13,840$$
$$3x - 4559 = 13,840$$
$$3x = 18,399$$
$$x = 6133$$

Back-substitute to solve for $y$ and $z$.

$$y = x - 3864$$
$$= 6133 - 3864$$
$$= 2269$$
$$z = x - 695$$
$$= 6133 - 695$$
$$= 5438$$

The annual spending in 2010 per person on housing is \$6133, on vehicles/gas is \$2269, and on health care is \$5438.

**35.** Let $m =$ the cost per gallon of milk.
Let $w =$ the cost per bottle of water.
Let $c =$ the cost per bag of chips.
The information is represented by the following system of equations.

$$2m + 5w + 6c = 19$$
$$w = 2c$$
$$m = w + 2$$

Use substitution to find $m$ in terms of $c$:

$$m = \overset{2c}{\overbrace{w}} + 2$$
$$m = 2c + 2$$

Use substitution to find $c$.

$$2m + 5w + 6c = 19$$
$$2(2c + 2) + 5(2c) + 6c = 19$$
$$4c + 4 + 10c + 6c = 19$$
$$20c + 4 = 19$$
$$20c = 15$$
$$c = 0.75$$

Find $w$.

$$w = 2c$$
$$w = 2(0.75)$$
$$= 1.50$$

Find $m$.

$$m = w + 2$$
$$m = 1.50 + 2$$
$$= 3.50$$

The cost per gallon of milk is \$3.50. The cost per bottle of water is \$1.50. The cost per bag of chips is \$0.75.

**37.** $x =$ number of \$8 tickets sold
$y =$ number of \$10 tickets sold
$z =$ number of \$12 tickets sold
From the given conditions we have the following system of equations.

$$x + y + z = 400$$
$$8x + 10y + 12z = 3700$$
$$x + y = 7z \text{ or } x + y - 7z = 0$$

Eliminate $z$ from Equations 1 and 2 multiplying Equation 1 by $-12$ and adding the resulting equation to Equation 2.

$$-12x - 12y - 12z = -4800$$
$$\underline{8x + 10y + 12z = 3700}$$
$$-4x - 2y = -1100 \quad \text{Equation 4}$$

Eliminate $z$ from Equations 1 and 3 by multiplying Equation 1 by 7 and adding the resulting equation to Equation 3.

$$7x + 7y + 7z = 2800$$
$$\underline{x + y - 7z = 0}$$
$$8x + 8y = 2800 \quad \text{Equation 5}$$

Solve Equations 4 and 5 for $x$ and $y$ by multiplying Equation 4 by 2 and adding the resulting equation to Equation 5.

$$-8x - 4y = -2200$$
$$\underline{8x + 8y = 2800}$$
$$4y = 600$$
$$y = 150$$

Substitute 150 for $y$ in Equation 5 and solve for $x$.

$$8x + 8(150) = 2800$$
$$8x = 2800 - 1200$$
$$8x = 1600$$
$$x = 200$$

Substitute 200 for $x$ and 150 for $y$ in Equation 1 and solve for $z$.

$$200 + 150 + z = 400$$
$$350 + z = 400$$
$$z = 50$$

The number of \$8 tickets sold was 200.
The number of \$10 tickets sold was 150.
The number of \$12 tickets sold was 50.

**39.** $x$ = amount of money invested at 10%
$y$ = amount of money invested at 12%
$z$ = amount of money invested at 15%

$$x + y + z = 6700$$

$$0.08x + 0.10y + 0.12z = 716$$

$$z = x + y + 300$$

Arrange Equation 3 so that variable terms appear on the left and constants appear on the right.

$$-x - y + z = 300 \quad \text{Equation 4}$$

Eliminate $x$ and $y$ from Equations 1 and 4 by adding Equations 1 and 4.

$$x + y + z = 6700$$
$$\underline{-x - y + z = 300}$$
$$2z = 7000$$
$$z = 3500$$

Substitute 3500 for $z$ in Equation 1 and Equation 2 and simplify.

$$x + y + 3500 = 6700$$

$$x + y = 3200 \quad \text{Equation 5}$$

$$0.08x + 0.10y + 0.12(3500) = 716$$

$$0.08x + 0.10y + 420 = 716 \qquad \text{Solve}$$

$$0.08x + 10y = 296 \quad \text{Equation 6}$$

Equations 5 and 6 for $x$ and $y$ by multiplying Equation 5 by $-0.10$ and adding the resulting equation to Equation 6.

$$-0.10x - 0.10y = -320$$
$$\underline{0.08x + 0.10y = 296}$$
$$-0.02x = 24$$
$$x = 1200$$

Substitute 1200 for $x$ and 3,500 for $z$ in Equation 1 and solve for $y$.

$$1200 + y + 3500 = 6700$$

$$y + 4700 = 6700$$

$$y = 2000$$

The person invested $1200 at 8%, $2000 at 10%, and $3500 at 12%.

**41.** $x + y + z = 180$

$$2x - 5 + z = 180$$
$$2x + z = 185$$

$$2x + 5 + y = 180$$
$$2x + y = 175$$

Multiply the second equation by $-1$ and add to the first equation. Use the new equation and the third equation to solve for $x$ and $z$.

$$-2x - z = -185$$
$$\underline{x + y + z = 180}$$
$$-x + y = -5$$

Multiply the new equation by -1.

$$x - y = 5$$
$$\underline{2x + y = 175}$$
$$3x = 180$$
$$x = 60$$

$$60 - y = 5$$
$$-y = -55$$
$$y = 55$$

Substitute 60 for $x$ and 55 for $y$ in the first equation and solve for $z$.

$$60 + 55 + z = 180$$
$$z = 65$$

**43. – 47.** Answers will vary.

**49.** does not make sense; Explanations will vary. Sample explanation: A system of linear equations in three variables can contain an equation of the form $y = mx + b$. For this equation, the coefficient of $z$ is 0.

**51.** makes sense

**53.** $x$ = number of triangles
$y$ = number of rectangles
$z$ = number of pentagons

$x + y + z = 40$

$3x + 4y + 5z = 153$

$2y + 5z = 72$

Eliminate $x$ from Equations 1 and 2 by multiplying Equation 1 by –3 and adding the resulting equation to Equation 2.

$-3x - 3y - 3z = -120$

$\underline{3x + 4y + 5z = 153}$

$\qquad y + 2z = 33 \qquad$ Equation 4

Solve for $z$ by multiplying Equation 4 by –2 and adding the resulting equation to Equation 3.

$2y + 5z = 72$

$\underline{-2y - 4z = -66}$

$\qquad z = 6$

Substitute 6 for $z$ in Equation 4 and solve for $y$.

$y + 2(6) = 33$

$y + 12 = 33$

$y = 21$

Substitute 21 for $y$ and 6 for $z$ in Equation 1 and solve for $x$.

$x + 21 + 6 = 40$

$x + 27 = 40$

$x = 13$

The painting has 13 triangles, 21 rectangles, and 6 pentagons.

**55.**
$$\frac{3}{x-4} - \frac{2}{x+2} = \frac{3(x+2)}{(x-4)(x+2)} - \frac{2(x-4)}{(x-4)(x+2)}$$
$$= \frac{3x+6}{(x-4)(x+2)} - \frac{2x-8}{(x-4)(x+2)}$$
$$= \frac{3x+6-2x+8}{(x-4)(x+2)}$$
$$= \frac{x+14}{(x-4)(x+2)}$$

**56.**
$$\frac{5x-3}{x^2+1} + \frac{2x}{(x^2+1)^2} = \frac{(5x-3)(x^2+1)}{(x^2+1)(x^2+1)} + \frac{2x}{(x^2+1)^2}$$
$$= \frac{5x^3 - 3x^2 + 5x - 3}{(x^2+1)^2} + \frac{2x}{(x^2+1)^2}$$
$$= \frac{5x^3 - 3x^2 + 5x - 3 + 2x}{(x^2+1)^2}$$
$$= \frac{5x^3 - 3x^2 + 7x - 3}{(x^2+1)^2}$$

**57.** $\qquad A + B = 3$

$\qquad 2A - 2B + C = 17$

$\qquad 4A - 2C = 14$

Solving $A + B = 3$ for $B$ gives $B = 3 - A$.

Solving $4A - 2C = 14$ for $A$ gives $C = 2A - 7$.

Use substitution to find $f$.

$$2A - 2B + C = 17$$

$$2A - 2\overbrace{(3-A)}^{B} + \overbrace{2A - 7}^{C} = 17$$

$$2A - 2(3 - A) + 2A - 7 = 17$$

$$2A - 6 + 2A + 2A - 7 = 17$$

$$6A - 13 = 17$$

$$6A = 30$$

$$A = 5$$

Find $B$.

$B = 3 - A$

$B = 3 - 5 = -2$

Find $C$.

$C = 2A - 7$

$C = 2(5) - 7 = 3$

The solution set is $\{(5, -2, 3)\}$.

**Section 5.3**

**Check Point Exercises**

**1.** $\dfrac{5x-1}{(x-3)(x+4)} = \dfrac{A}{x-3} + \dfrac{B}{x+4}$

Multiply both sides of the equation by the least common denominator $(x-3)(x+4)$ and divide out common factors.

$5x-1 = A(x+4) + B(x-3)$

$5x-1 = Ax + 4A + Bx - 3B$

$5x-1 = (A+B)x + 4A - 3B$

Equate coefficients of like powers of $x$ and equate constant terms.

$A+B = 5$

$4A - 3B = -1$

Solving the above system for $A$ and $B$ we find $A = 2$ and $B = 3$.

$\dfrac{5x-1}{(x-3)(x+4)} = \dfrac{2}{x-3} + \dfrac{3}{x+4}$

**2.** $\dfrac{x+2}{x(x-1)^2} = \dfrac{A}{x} + \dfrac{B}{x-1} + \dfrac{C}{(x-1)^2}$

Multiply both sides of the equation by the least common denominator $x(x-1)^2$ and divide out common factors.

$x+2 = A(x-1)^2 + Bx(x-1) + Cx$

$x+2 = A\left(x^2 - 2x + 1\right) + Bx^2 - Bx + Cx$

$x+2 = Ax^2 - 2Ax + A + Bx^2 - Bx + Cx$

$x+2 = Ax^2 + Bx^2 - 2Ax - Bx + Cx + A$

$x+2 = (A+B)x^2 + (-2A - B + C)x + A$

Equate coefficients of like powers of $x$ and equate constant terms.

$A+B = 0$

$-2A - B + C = 1$

$A = 2$

Since $A = 2$, we find that $B = -2$ and $C = 3$ by substitution.

$\dfrac{x+2}{x(x-1)^2} = \dfrac{2}{x} - \dfrac{2}{x-1} + \dfrac{3}{(x-1)^2}$

**3.** $\dfrac{8x^2 + 12x - 20}{(x+3)\left(x^2 + x + 2\right)} = \dfrac{A}{x+3} + \dfrac{Bx+C}{x^2 + x + 2}$

Multiply both sides of the equation by the least common denominator $(x+3)\left(x^2 + x + 2\right)$ and divide out common factors.

$8x^2 + 12x - 20 = A\left(x^2 + x + 2\right) + (Bx + C)(x+3)$

$8x^2 + 12x - 20 = Ax^2 + Ax + 2A + Bx^2 + 3Bx + Cx + 3C$

$8x^2 + 12x - 20 = Ax^2 + Bx^2 + Ax + 3Bx + Cx + 2A + 3C$

$8x^2 + 12x - 20 = (A+B)x^2 + (A + 3B + C)x + 2A + 3C$

Equate coefficients of like powers of $x$ and equate constant terms.

$A+B = 8$

$A + 3B + C = 12$

$2A + 3C = -20$

Solving the above system for $A$, $B$, and $C$ we find $A = 2$, $B = 6$, and $C = -8$.

$\dfrac{8x^2 + 12x - 20}{(x+3)\left(x^2 + x + 2\right)} = \dfrac{2}{x+3} + \dfrac{6x-8}{x^2 + x + 2}$

**4.** $\dfrac{2x^3 + x + 3}{\left(x^2 + 1\right)^2} = \dfrac{Ax+B}{x^2 + 1} + \dfrac{Cx+D}{\left(x^2 + 1\right)^2}$

Multiply both sides of the equation by the common denominator $\left(x^2 + 1\right)^2$ and divide out common factors.

$2x^3 + x + 3 = (Ax + B)(x^2 + 1) + Cx + D$

$2x^3 + x + 3 = Ax^3 + Bx^2 + Ax + B + Cx + D$

$2x^3 + x + 3 = Ax^3 + Bx^2 + Ax + Cx + B + D$

$2x^3 + x + 3 = Ax^3 + Bx^2 + (A+C)x + B + D$

Equate coefficients of like powers of $x$ and equate constant terms.

$A = 2$

$B = 0$

$A + C = 1$

$B + D = 3$

Since $A = 2$ and $B = 0$ we find that $C = -1$ and $D = 3$ by substitution.

$\dfrac{2x^3 + x + 3}{\left(x^2 + 1\right)^2} = \dfrac{2x}{x^2 + 1} + \dfrac{-x+3}{\left(x^2 + 1\right)^2}$ or $\dfrac{2x}{x^2 + 1} - \dfrac{x-3}{\left(x^2 + 1\right)^2}$

**Concept and Vocabulary Check 5.3**

1. correct

2. incorrect

3. incorrect

4. correct

**Exercise Set 5.3**

1. $\dfrac{11x-10}{(x-2)(x+1)} = \dfrac{A}{x-2} + \dfrac{B}{x+1}$

3. $\dfrac{6x^2 - 14x - 27}{(x+2)(x-3)^2} = \dfrac{A}{x+2} + \dfrac{B}{x-3} + \dfrac{C}{(x-3)^2}$

5. $\dfrac{5x^2 - 6x + 7}{(x-1)(x^2+1)} = \dfrac{A}{x-1} + \dfrac{Bx+C}{x^2+1}$

7. $\dfrac{x^3 + x^2}{(x^2+4)^2} = \dfrac{Ax+B}{x^2+4} + \dfrac{Cx+D}{(x^2+4)^2}$

9. $\dfrac{x}{(x-3)(x-2)} = \dfrac{A}{x-3} + \dfrac{B}{x-2}$

Multiply both sides of the equation by the least common denominator $(x-3)(x-2)$ and divide out common factors.

$x = A(x-2) + B(x-3)$

$x = A\,x - 2A + Bx - 3B$

$x = Ax + Bx - 2A - 3B$

$x = (A+B)x - (2A+3B)$

Equate coefficients of like powers of $x$, and equate constant terms.

$A+B = 1$

$2A + 3B = 0$

Solving the above system for $A$ and $B$, we find $A = 3$ and $B = -2$.

$\dfrac{x}{(x-3)(x-2)} = \dfrac{3}{x-3} - \dfrac{2}{x-2}$

11. $\dfrac{3x+50}{(x-9)(x+2)} = \dfrac{A}{x-9} + \dfrac{B}{x+2}$

Multiply both sides of the equation by the least common denominator $(x-9)(x+2)$ and divide out common factors.

$3x + 50 = A(x+2) + B(x-9)$

$3x + 50 = Ax + 2A + Bx - 9B$

$3x + 50 = Ax + Bx + 2A - 9B$

$3x + 50 = (A+B)x + (2A - 9B)$

Equate coefficients of like powers of $x$, and equate constant terms.

$A+B = 3$

$2A - 9B = 50$

Solving the above system for $A$ and $B$, we find $A = 7$ and $B = -4$.

$\dfrac{3x+50}{(x-9)(x+2)} = \dfrac{7}{x-9} - \dfrac{4}{x+2}$

13. $\dfrac{7x-4}{x^2 - x - 12} = \dfrac{7x-4}{(x-4)(x+3)} = \dfrac{A}{x-4} + \dfrac{B}{x+3}$

Multiply both sides of the last equation by the least common denominator $(x-4)(x-3)$ and divide out common factors.

$7x - 4 = A(x+3) + B(x-4)$

$7x - 4 = A\,x + 3A + Bx - 4B$

$7x - 4 = Ax + Bx + 3A - 4B$

$7x - 4 = (A+B)x + (3A - 4B)$

Equate coefficients of like powers of $x$, and equate constant terms.

$A+B = 7$

$3A - 4B = -4$

Solving the above system for $A$ and $B$, we find $A = \dfrac{24}{7}$ and $B = \dfrac{25}{7}$.

$\dfrac{7x-4}{x^2 - x - 12} = \dfrac{24}{7(x-4)} + \dfrac{25}{7(x+3)}$

15. $\dfrac{4}{(2x+1)(x-3)} = \dfrac{A}{2x+1} + \dfrac{B}{x-3}$

Multiply both sides of the equation by the least common denominator $(2x + 1)(x - 3)$ and divide out common factors.

$4 = A(x-3) + B(2x+1)$

$4 = Ax - A3 + B2x + B$

$4 = (A+2B)x + (-3A + B)$

Equate coefficients of like powers of $x$ and equate the constant terms. Solve for $A$ and $B$.

$A + 2B = 0$

$-3A + B = 4$

$3A + 6B = 0$

$-3A + B = 4$

$7B = 4$

$B = \dfrac{4}{7}$

$A + 2B = 0$

$6A - 2B = -8$

$7A = -8$

$A = -\dfrac{8}{7}$

$\dfrac{4}{(2x+1)(x-3)} = \dfrac{-8}{7(2x+1)} + \dfrac{4}{7(x-3)}$

**17.** $\dfrac{4x^2 + 13x - 9}{x(x-1)(x+3)} = \dfrac{A}{x} + \dfrac{B}{x-3} + \dfrac{C}{x+3}$

Multiply both sides of the equation by the least common denominator $x(x-1)(x+3)$ and divide out common factors.

$4x^2 + 13x - 9 = A(x\text{-}1)(x+3) + Bx(x+3) + Cx(x-1)$

$4x^2 + 13x - 9 = A\left(x^2 + 2x - 3\right) + Bx^2 + 3Bx + Cx^2 - Cx$

$4x^2 + 13x - 9 = Ax^2 + 2Ax - 3A + Bx^2 + 3Bx + Cx^2 - Cx$

$4x^2 + 13x - 9 = Ax^2 + Bx^2 + Cx^2 + 2Ax + 3Bx - Cx - 3A$

$4x^2 + 13x - 9 = \left(A + B + C\right)x^2 + \left(2A + 3B - C\right)x - 3A$

Equate coefficients of like powers of $x$, and equate constant terms.

$A + B + C = 4$

$2A + 3B - C = 13$

$-3A = -9$

Solving the above system for $A$, $B$, and $C$, we find $A = 3$ and $B = 2$, and $C = -1$.

$\dfrac{4x^2 + 13x - 9}{x(x-1)(x+3)} = \dfrac{3}{x} + \dfrac{2}{x-1} - \dfrac{1}{x+3}$

**19.** $\dfrac{4x^2 - 7x - 3}{x^3 - x} = \dfrac{4x^2 - 7x - 3}{x(x+1)(x-1)} = \dfrac{A}{x} + \dfrac{B}{x+1} + \dfrac{C}{x-1}$

Multiply both sides of the last equation by the least common denominator $x(x+1)(x-1)$ and divide out common factors.

$4x^2 - 7x - 3 = A(x+1)(x-1) + Bx(x-1) + Cx(x+1)$

$4x^2 - 7x - 3 = A(x^2 - 1) + Bx^2 - Bx + Cx^2 + Cx$

$4x^2 - 7x - 3 = Ax^2 - A + Bx^2 - Bx + Cx^2 + Cx$

$4x^2 - 7x - 3 = Ax^2 + Bx^2 + Cx^2 - Bx + Cx - A$

$4x^2 - 7x - 3 = (A + B + C)x^2 + (-B + C)x - A$

Equate coefficients of like powers of $x$, and equate constant terms.

$A + B + C = 4$

$-B + C = -7$

$-A = -3$

Solving the above system for $A$, $B$, and $C$, we find $A = 3$ and $B = 4$, and $C = -3$.

$\dfrac{4x^2 - 7x - 3}{x^3 - x} = \dfrac{3}{x} + \dfrac{4}{x+1} - \dfrac{3}{x-1}$

**21.** $\dfrac{6x - 11}{(x-1)^2} = \dfrac{A}{x-1} + \dfrac{B}{(x-1)^2}$

Multiply both sides of the equation by the least common denominator $(x-1)^2$ and divide out common factors.

$6x - 11 = A(x-1) + B$

$6x - 11 = Ax - A + B$

Equate coefficients of like powers of $x$, and equate constant terms.

$A = 6$

$-A + B = -11$

Since $A = 6$, we find that $B = -5$ by substitution.

$\dfrac{6x - 11}{(x-1)^2} = \dfrac{6}{x-1} - \dfrac{5}{(x-1)^2}$

**23.** $\dfrac{x^2 - 6x + 3}{(x-2)^3} = \dfrac{A}{x-2} + \dfrac{B}{(x-2)^2} + \dfrac{C}{(x-2)^3}$

Multiply both sides of the equation by the least common denominator $(x-2)^3$ and divide out common factors.

$x^2 - 6x + 3 = A(x-2)^2 + B(x-2) + C$

$x^2 - 6x + 3 = A(x^2 - 4x + 4) + Bx - 2B + C$

$x^2 - 6x + 3 = Ax^2 - 4Ax + 4A + Bx - 2B + C$

$x^2 - 6x + 3 = Ax^2 - 4Ax + Bx + 4A - 2B + C$

$x^2 - 6x + 3 = Ax^2 + (-4A + B)x + 4A - 2B + C$

Equate coefficients of like powers of $x$, and equate constant terms.

$$A = 1$$
$$-4A + B = -6$$
$$4A - 2B + C = 3$$

Since $A = 1$, we find that $B = -2$ and $C = -5$ by substitution. $\dfrac{x^2 - 6x + 3}{(x-2)^3} = \dfrac{1}{x-2} - \dfrac{2}{(x-2)^2} - \dfrac{5}{(x-2)^3}$

**25.** $\dfrac{x^2 + 2x + 7}{x(x-1)^2} = \dfrac{A}{x} + \dfrac{B}{x-1} + \dfrac{C}{(x-1)^2}$

Multiply both sides of the equation by the least common denominator $x(x-1)^2$ and divide out common factors.

$x^2 + 2x + 7 = A(x-1)^2 + Bx(x-1) + Cx$

$x^2 + 2x + 7 = A(x^2 - 2x + 1) + Bx^2 - Bx + Cx$

$x^2 + 2x + 7 = Ax^2 - 2Ax + A + Bx^2 - Bx + Cx$

$x^2 + 2x + 7 = Ax^2 + Bx^2 - 2Ax - Bx + Cx + A$

$x^2 + 2x + 7 = (A+B)x^2 + (-2A - B + C)x + A$

$$A + B = 1$$
$$-2A - B + C = 2$$
$$A = 7$$

Since $A = 7$, we find that $B = -6$ and $C = 10$ by substitution. $\dfrac{x^2 + 2x + 7}{x(x-1)^2} = \dfrac{7}{x} - \dfrac{6}{x-1} + \dfrac{10}{(x-1)^2}$

**27.** $\dfrac{x^2}{(x+1)(x-1)^2} = \dfrac{A}{x+1} + \dfrac{B}{x-1} + \dfrac{C}{(x-1)^2}$

Multiply both sides of the equation by the least common denominator $(x+1)(x-1)^2$ and divide out common factors.

$x^2 = A(x-1)^2 + B(x+1)(x-1) + C(x+1)$

$x^2 = x^2 A - 2xA + A + Bx^2 - B + Cx + C$

$x^2 = (A+B)x^2 + (-2A+C)x + (A-B+C)$

Equate coefficients of like powers of $x$, and equate constant terms.

$$A + B = 1$$
$$-2A + C = 0$$
$$A - B + C = 0$$

Solving the above system for $A$, $B$, and $C$, we find $A = \dfrac{1}{4}$, $B = \dfrac{3}{4}$, and $C = \dfrac{1}{2}$.

$$\dfrac{x^2}{(x+1)(x-1)^2} = \dfrac{1}{4(x+1)} + \dfrac{3}{4(x-1)} + \dfrac{1}{2(x-1)^2}$$

**29.** $\dfrac{5x^2-6x+7}{(x-1)(x^2+1)} = \dfrac{A}{x-1} + \dfrac{Bx+C}{x^2+1}$

Multiply both sides of the equation by the least common denominator $(x-1)(x^2+1)$ and divide out common factors.

$5x^2-6x+7 = A(x^2+1)+(Bx+C)(x-1)$

$5x^2-6x+7 = Ax^2+A+Bx^2-Bx+Cx-C$

$5x^2-6x+7 = Ax^2+Bx^2-Bx+Cx+A-C$

$5x^2-6x+7 = (A+B)x^2+(-B+C)x+A-C$

Equate coefficients of like powers of $x$, and equate constant terms.

$A+B=5$
$-B+C=-6$
$A-C=7$

Solving the above system for $A$, $B$, and $C$, we find $A=3$, $B=2$, and $C=-4$.

$\dfrac{5x^2-6x+7}{(x-1)(x^2+1)} = \dfrac{3}{x-1} + \dfrac{2x-4}{x^2+1}$

**31.** $\dfrac{5x^2+6x+3}{(x+1)(x^2+2x+2)} = \dfrac{A}{x+1} + \dfrac{Bx+C}{x^2+2x+2}$

Multiply both sides of the equation by the least common denominator $(x+1)(x^2+2x+2)$ and divide out common factors.

$5x^2+6x+3 = A(x^2+2x+2)+(Bx+C)(x+1)$

$5x^2+6x+3 = Ax^2+2Ax+2A+Bx^2+Bx+Cx+C$

$5x^2+6x+3 = Ax^2+Bx^2+2Ax+Bx+Cx+2A+C$

$5x^2+6x+3 = (A+B)x^2+(2A+B+C)x+2A+C$

Equate coefficients of like powers of $x$, and equate constant terms.

$A+B=5$
$2A+B+C=6$
$2A+C=3$

Solving the above system for $A$, $B$, and $C$, we find $A=2$, $B=3$, and $C=-1$.

$\dfrac{5x^2+6x+3}{(x+1)(x^2+2x+2)} = \dfrac{2}{x+1} + \dfrac{3x-1}{x^2+2x+2}$

**33.** $\dfrac{x+4}{x^2(x^2+4)} = \dfrac{A}{x} + \dfrac{B}{x^2} + \dfrac{Cx+D}{x^2+4}$

Multiply both sides of the equation by the least common denominator $x^2(x^2+4)$ and divide out common factors.

$x+4 = Ax(x^2+4)+B(x^2+4)+(Cx+D)x^2$

$x+4 = Ax^3+4Ax+Bx^2+4B+Cx^3+Dx^2$

$x+4 = (A+C)x^3+(B+D)x^2+4Ax+4B$

Equate coefficients of like powers of $x$, and equate constant terms

$A + C = 0$

$B + D = 0$

$4A = 1$

$4B = 4$

Solving the above system for $A$, $B$, and $C$, we find $A = \dfrac{1}{4}$, $B = 1$, $C = -\dfrac{1}{4}$, and $D = -1$.

$$\frac{x+4}{x^2(x^2+4)} = \frac{1}{4x} + \frac{1}{x^2} + \frac{-1x-4}{4\left(x^2+4\right)}$$

**35.** $\dfrac{6x^2 - x + 1}{x^3 + x^2 + x + 1} = \dfrac{6x^2 - x + 1}{(x+1)(x^2+1)} = \dfrac{A}{x+1} + \dfrac{Bx+C}{x^2+1}$

Multiply both sides of the last equation by the least common denominator $(x+1)(x^2+1)$ and divide out common factors.

$6x^2 - x + 1 = A(x^2+1) + (Bx+C)(x+1)$

$6x^2 - x + 1 = Ax^2 + A + Bx^2 + Bx + Cx + C$

$6x^2 - x + 1 = Ax^2 + Bx^2 + Bx + Cx + A + C$

$6x^2 - x + 1 = (A+B)x^2 + (B+C)x + A + C$

Equate coefficients of like powers of $x$, and equate constant terms.

$A + B = 6$

$B + C = -1$

$A + C = 1$

Solving the above system for $A$, $B$, and $C$, we find $A = 4$, $B = 2$, and $C = -3$.

$$\frac{6x^2 - x + 1}{x^3 + x^2 + x + 1} = \frac{4}{x+1} + \frac{2x-3}{x^2+1}$$

**37.** $\dfrac{x^3 + x^2 + 2}{\left(x^2+2\right)^2} = \dfrac{Ax+B}{x^2+2} + \dfrac{Cx+D}{\left(x^2+2\right)^2}$

Multiply both sides of the last equation by the least common denominator $(x^2+2)^2$ and divide out common factors.

$x^3 + x^2 + 2 = \left(Ax+B\right)\left(x^2+2\right) + Cx + D$

$x^3 + x^2 + 2 = Ax^3 + Bx^2 + 2Ax + 2B + Cx + D$

$x^3 + x^2 + 2 = Ax^3 + Bx^2 + 2Ax + Cx + 2B + D$

$x^3 + x^2 + 2 = Ax^3 + Bx^2 + \left(2A+C\right)x + \left(2B+D\right)$

Equate coefficients of like powers of $x$, and equate constant terms.

$A = 1$

$B = 1$

$2A + C = 0$

$2B + D = 2$

Since $A = 1$ and $B = 1$, we find that $C = -2$ and $D = 0$ by substitution.

$$\frac{x^3 + x^2 + 2}{\left(x^2+2\right)^2} = \frac{x+1}{x^2+2} - \frac{2x}{\left(x^2+2\right)^2}$$

**39.** $\dfrac{x^3 - 4x^2 + 9x - 5}{(x^2 - 2x + 3)^2} = \dfrac{Ax + B}{x^2 - 2x + 3} + \dfrac{Cx + D}{(x^2 - 2x + 3)^2}$

Multiply both sides of the equation by the least common denominator $(x^2 - 2x + 3)^2$ and divide out common factors.

$x^3 - 4x^2 + 9x - 5 = (Ax + B)(x^2 - 2x + 3) + Cx + D$

$x^3 - 4x^2 + 9x - 5 = Ax^3 - 2Ax^2 + 3Ax + Bx^2 - 2Bx + 3B + Cx + D$

$x^3 - 4x^2 + 9x - 5 = Ax^3 - 2Ax^2 + Bx^2 + 3Ax - 2Bx + Cx + 3B + D$

$x^3 - 4x^2 + 9x - 5 = Ax^3 + (-2A + B)x^2 + (3A - 2B + C)x + 3B + D$

Equate coefficients of like powers of $x$, and equate constant terms.

$A = 1$
$-2A + B = -4$
$3A - 2B + C = 9$
$3B + D = -5$

Since $A = 1$, we find that $B = -2$, $C = 2$, and $D = 1$ by substitution.

$\dfrac{x^3 - 4x^2 + 9x - 5}{(x^2 - 2x + 3)^2} = \dfrac{x - 2}{x^2 - 2x + 3} + \dfrac{2x + 1}{(x^2 - 2x + 3)^2}$

**41.** $\dfrac{4x^2 + 3x + 14}{x^3 - 8} = \dfrac{4x^2 + 3x + 14}{(x - 2)(x^2 + 2x + 4)} = \dfrac{A}{x - 2} + \dfrac{Bx + C}{x^2 + 2x + 4}$

Multiply both sides of the last equation by the least common denominator $(x - 2)(x^2 + 2x + 4)$ and divide out common factors.

$4x^2 + 3x + 14 = A\left(x^2 + 2x + 4\right) + (Bx + C)(x - 2)$

$4x^2 + 3x + 14 = A^2 + 2Ax + 4A + Bx^2 - 2Bx + Cx - 2C$

$4x^2 + 3x + 14 = Ax^2 + Bx^2 + 2Ax - 2Bx + Cx + 4A - 2C$

$4x^2 + 3x + 14 = (A + B)x^2 + (2A - 2B + C)x + (4A - 2C)$

Equate coefficients of like powers of $x$, and equate constant terms.

$A + B = 4$

$2A - 2B + C = 3$

$4A - 2C = 14$

Solving the above system for $A$, $B$, and $C$, we find $A = 3$, $B = 1$, and $C = -1$.

$\dfrac{4x^2 + 3x + 4}{x^3 - 8} = \dfrac{3}{x - 2} + \dfrac{x - 1}{x^2 + 2x + 4}$

**43.** Divide $x^5 + 2$ by $x^2 - 1$.

$$
\begin{array}{r}
x^3 + x \phantom{+2} \\
x^2 - 1\overline{) x^5 \phantom{aaaaaa} + 2} \\
\underline{x^5 - x^3} \phantom{aaaa} \\
x^3 \phantom{aaaa} \\
\underline{x^3 - x} \phantom{aa} \\
x + 2
\end{array}
$$

$\dfrac{x^5 + 2}{x^2 - 1} = x^3 + x + \dfrac{x + 2}{x^2 - 1}$

Decompose $\dfrac{x + 2}{x^2 - 1}$.

$$\frac{x+2}{x^2-1} = \frac{x+2}{(x+1)(x-1)} = \frac{A}{x+1} + \frac{B}{x-1}$$

$$(x+1)(x-1)\frac{x+2}{(x+1)(x-1)} = (x+1)(x-1)\left(\frac{A}{x+1} + \frac{B}{x-1}\right)$$

$$x+2 = (x-1)A + (x+1)B$$

$$x+2 = Ax - A + Bx + B$$

$$x+2 = (A+B)x + (-A+B)$$

Equate coefficients:

$$A+B=1$$

$$-A+B=2$$

Solving this system results in $A = -\dfrac{1}{2}$ and $B = \dfrac{3}{2}$.

$$\frac{x^5+2}{x^2-1} = x^3 + x + \frac{-\dfrac{1}{2}}{x+1} + \frac{\dfrac{3}{2}}{x-1} \quad \text{or} \quad x^3 + x + \frac{-1}{2(x+1)} + \frac{3}{2(x-1)}$$

**45.** Divide $x^4 - x^2 + 2$ by $x^3 - x^2$.

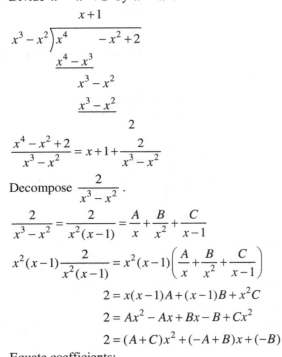

$$\frac{x^4 - x^2 + 2}{x^3 - x^2} = x + 1 + \frac{2}{x^3 - x^2}$$

Decompose $\dfrac{2}{x^3 - x^2}$.

$$\frac{2}{x^3 - x^2} = \frac{2}{x^2(x-1)} = \frac{A}{x} + \frac{B}{x^2} + \frac{C}{x-1}$$

$$x^2(x-1)\frac{2}{x^2(x-1)} = x^2(x-1)\left(\frac{A}{x} + \frac{B}{x^2} + \frac{C}{x-1}\right)$$

$$2 = x(x-1)A + (x-1)B + x^2C$$

$$2 = Ax^2 - Ax + Bx - B + Cx^2$$

$$2 = (A+C)x^2 + (-A+B)x + (-B)$$

Equate coefficients:

$$A+C=0$$

$$-A+B=0$$

$$-B=2$$

Solving this system results in $A = -2$, $B = -2$, and $C = 2$.

$$\frac{x^4 - x^2 + 2}{x^3 - x^2} = x + 1 + \frac{-2}{x} + \frac{-2}{x^2} + \frac{2}{x-1}$$

**47.** $\dfrac{1}{x^2 - c^2} = \dfrac{1}{(x+c)(x-c)} = \dfrac{A}{x+c} + \dfrac{B}{x-c}$

$(x+c)(x-c)\dfrac{1}{(x+c)(x-c)} = (x+c)(x-c)\left(\dfrac{A}{x+c} + \dfrac{B}{x-c}\right)$

$$1 = (x-c)A + (x+c)B$$
$$1 = Ax - Ac + Bx + Bc$$
$$1 = (A+B)x + (-Ac + Bc)$$

Equate coefficients:

$A + B = 0$

$-Ac + Bc = 1$

Solving this system results in $A = \dfrac{-1}{2c}$ and $B = \dfrac{1}{2c}$.

$\dfrac{1}{x^2 - c^2} = \dfrac{\frac{-1}{2c}}{x+c} + \dfrac{\frac{1}{2c}}{x-c}$

**49.** $\dfrac{ax+b}{(x-c)^2} = \dfrac{D}{x-c} + \dfrac{E}{(x-c)^2}$

$\dfrac{ax+b}{(x-c)^2} = \dfrac{D}{x-c} + \dfrac{E}{(x-c)^2}$

$(x-c)^2 \dfrac{ax+b}{(x-c)^2} = (x-c)^2\left(\dfrac{D}{x-c} + \dfrac{E}{(x-c)^2}\right)$

$$ax + b = (x-c)D + E$$
$$ax + b = Dx - Dc + E$$
$$ax + b = (D)x + (-Dc + E)$$

Equate coefficients:

$D = a$

$-Dc + E = b$

Solving this system results in $D = a$ and $E = ac + b$.

$\dfrac{ax+b}{(x-c)^2} = \dfrac{a}{x-c} + \dfrac{ac+b}{(x-c)^2}$

**51.** $\dfrac{1}{x(x+1)} = \dfrac{A}{x} + \dfrac{B}{x+1}$

Multiply both sides of the equation by the least common denominator $x(x+1)$ and divide out common factors.

$1 = A(x+1) + Bx$

$1 = Ax + A + Bx$

$1 = Ax + Bx + A$

$1 = (A+B)x + A$

Equate coefficients of like powers of $x$, and equate constant terms.

$A + B = 0$

$\quad A = 1$

Since $A = 1$ we find that $B = -1$ by substitution.

$$\frac{1}{x(x+1)} = \frac{1}{x} - \frac{1}{x+1}$$

$$\frac{1}{1 \cdot 2} + \frac{1}{2 \cdot 3} + \frac{1}{3 \cdot 4} + \cdots \frac{1}{99 \cdot 100} = \left(\frac{1}{1} - \frac{1}{2}\right) + \left(\frac{1}{2} - \frac{1}{3}\right) + \left(\frac{1}{3} - \frac{1}{4}\right) + \cdots \left(\frac{1}{99} - \frac{1}{100}\right)$$

$$= \frac{1}{1} - \frac{1}{100}$$

$$= \frac{99}{100}$$

**53. – 59.** Answers will vary.

**61.** does not make sense; Explanations will vary. Sample explanation: To perform partial fraction decomposition, the degree in the numerator must be less than the degree in the denominator.

**63.** does not make sense; Explanations will vary. Sample explanation: The second denominator should be $(x+3)^2$.

**65.** $\dfrac{4x^2 + 5x - 9}{x^3 - 6x - 9} = \dfrac{4x^2 + 5x - 9}{(x-3)(x^2 + 3x + 3)} = \dfrac{A}{x-3} + \dfrac{Bx + C}{x^2 + 3x + 3}$

Multiply both sides of the last equation by the common denominator $(x-3)(x^2 + 3x + 3)$ and divide out common factors.

$4x^2 + 5x - 9 = A(x^2 + 3x + 3) + (Bx + C)(x - 3)$

$4x^2 + 5x - 9 = Ax^2 + 3Ax + 3A + Bx^2 - 3Bx + Cx - 3C$

$4x^2 + 5x - 9 = Ax^2 + Bx^2 + 3Ax - 3Bx + Cx + 3A - 3C$

$4x^2 + 5x - 9 = (A+B)x^2 + (3A - 3B + C)x + 3A - 3C$

Equate coefficients of like powers of $x$ and equate constant terms.

$A + B = 4$

$3A - 3B + C = 5$

$3A - 3C = -9$

Solving the above system for $A$, $B$, and $C$, we find $A = 2$, and $B = 2$, and $C = 5$.

$$\frac{4x^2 + 5x - 9}{x^3 - 6x - 9} = \frac{2}{x-3} + \frac{2x+5}{x^2 + 3x + 3}$$

**66.** $4x + 3y = 4$

$y = 2x - 7$

Substitute.

$4x + 3y = 4$

$4x + 3\overbrace{(2x-7)}^{y} = 4$

$4x + 3(2x - 7) = 4$

$4x + 6x - 21 = 4$

$10x = 25$

$x = 2.5$

Back-substitute.

$y = 2x - 7$

$y = 2(2.5) - 7 = -2$

The solution set is $\{(2.5, -2)\}$.

**67.** $2x + 4y = -4$

$3x + 5y = -3$

Multiply the first equation by –5 and the second equation by 4.

$-10x - 20y = 20$

$12x + 20y = -12$

Add the equations and solve for $x$.

$2x = 8$

$x = 4$

Back-substitute to find $y$.

$2x + 4y = -4$

$2(4) + 4y = -4$

$8 + 4y = -4$

$4y = -12$

$y = -3$

The solution set is $\{(4, -3)\}$.

**68.** The points of intersection are $(0, -3)$ and $(2, -1)$.

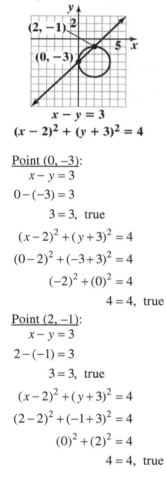

$x - y = 3$

$(x - 2)^2 + (y + 3)^2 = 4$

Point $(0, -3)$:

$x - y = 3$

$0 - (-3) = 3$

$3 = 3$, true

$(x - 2)^2 + (y + 3)^2 = 4$

$(0 - 2)^2 + (-3 + 3)^2 = 4$

$(-2)^2 + (0)^2 = 4$

$4 = 4$, true

Point $(2, -1)$:

$x - y = 3$

$2 - (-1) = 3$

$3 = 3$, true

$(x - 2)^2 + (y + 3)^2 = 4$

$(2 - 2)^2 + (-1 + 3)^2 = 4$

$(0)^2 + (2)^2 = 4$

$4 = 4$, true

**Section 5.4**

**Check Point Exercises**

**1.** $x^2 = y - 1$

$4x - y = -1$

Solve the first equation for $y$.

$y = x^2 + 1$

Substitute the expression $x^2 + 1$ for $y$ in the second equation and solve for $x$.

$4x - (x^2 + 1) = -1$

$4x - x^2 - 1 = -1$

$x^2 - 4x = 0$

$x(x - 4) = 0$

$x = 0$ or $x - 4 = 0$

$x = 4$

If $x = 0$, $y = (0)^2 + 1 = 1$.

If $x = 4$, $y = (4)^2 + 1 = 17$.

The solution set is $\{(0, 1), (4, 17)\}$.

**2.** $x + 2y = 0$

$(x - 1)^2 + (y - 1)^2 = 5$

Solve the first equation for $x$.

$x = -2y$

Substitute the expression $-2y$ for $x$ in the second equation and solve for $y$.

$(-2y - 1)^2 + (y - 1)^2 = 5$

$4y^2 + 4y + 1 + y^2 - 2y + 1 = 5$

$5y^2 + 2y - 3 = 0$

$(5y - 3)(y + 1) = 0$

$5y - 3 = 0$ or $y + 1 = 0$

$y = \dfrac{3}{5}$ or $y = -1$

If $y = \dfrac{3}{5}$, $x = -2\left(\dfrac{3}{5}\right) = -\dfrac{6}{5}$.

If $y = -1$, $x = -2(-1) = 2$.

The solution set is $\left\{\left(-\dfrac{6}{5}, \dfrac{3}{5}\right), (2, -1)\right\}$.

**3.** $3x^2 + 2y^2 = 35$

$4x^2 + 3y^2 = 48$

Eliminate the $y^2$-term by multiplying the first equation by –3 and the second equation by 2. Add the resulting equations.

$-9x^2 - 6y = -105$

$\underline{8x^2 + 6y^2 = 96}$

$-x^2 = -9$

$x^2 = 9$

$x = \pm 3$

If $x = 3$,

$3(3)^2 + 2y^2 = 35$

$y^2 = 4$

$y = \pm 2$

If $x = -3$,

$3(-3)^2 + 2y^2 = 35$

$y^2 = 4$

$y = \pm 2$

The solution set is $\{(3,2),(3,-2),(-3,2),(-3,-2)\}$.

**4.** $y = x^2 + 5$

$x^2 + y^2 = 25$

Arrange the first equation so that variable terms appear on the left, and constants appear on the right.

Add the resulting equations to eliminate the $x^2$-terms and solve for $y$.

$-x^2 + y = 5$

$\underline{x^2 + y^2 = 25}$

$y^2 + y = 30$

$y^2 + y - 30 = 0$

$(y+6)(y-5) = 0$

$y + 6 = 0 \quad$ or $\quad y - 5 = 0$

$y = -6 \quad$ or $\quad y = 5$

If $y = -6$,

$x^2 + (-6)^2 = 25$

$x^2 = -11$

no real solution

If $y = 5$,

$x^2 + (5)^2 = 25$

$x^2 = 0$

$x = 0$

The solution set is $\{(0, 5)\}$.

**5.** $2x + 2y = 20$

$xy = 21$

Solve the second equation for $x$.

$x = \dfrac{21}{7}$

Substitute the expression $\dfrac{21}{y}$ for $x$ in the first equation and solve for $y$.

$2\left(\dfrac{21}{y}\right) + 2y = 20$

$\dfrac{42}{y} + 2y = 20$

$y^2 - 10y + 21 = 0$

$(y-7)(y-3) = 0$

$y - 7 = 0 \quad$ or $\quad y - 3 = 0$

$y = 7 \quad$ or $\quad\quad y = 3$

If $y = 7$, $x = \dfrac{21}{7} = 3$.

If $y = 3$, $x = \dfrac{21}{3} = 7$.

The dimensions are 7 feet by 3 feet.

**Concept and Vocabulary Check 5.4**

**1.** nonlinear

**2.** $\{(-4,\ 3),\ (0,\ -1)\}$

**3.** 3

**4.** $\left\{(2,\sqrt{3}),\ (2,-\sqrt{3}),\ (-2,\sqrt{3}),\ (-2,-\sqrt{3})\right\}$

**5.** $-1$; $y^2 + y = 6$

**6.** $\dfrac{4}{x}$; $\dfrac{4}{x}$; $y$

**Exercise Set 5.4**

**1.** $x + y = 2$

$\qquad y = x^2 - 4$

Solve the first equation for $y$. $y = 2 - x$.

Substitute the expression $2 - x$ for $y$ in the second equation and solve for $x$.

$\qquad 2 - x = x^2 - 4$

$\qquad x^2 + x - 6 = 0$

$\qquad (x + 3)(x - 2) = 0$

$\qquad x + 3 = 0 \quad\text{or}\quad x - 2 = 0$

$\qquad\quad x = -3 \quad\text{or}\quad\quad x = 2$

If $x = -3$, $y = 2 - (-3) = 5$.

If $x = 2$, $y = 2 - 2 = 0$.

The solution set is $\{(-3, 5), (2, 0)\}$.

**3.** $x + y = 2$

$\qquad y = x^2 - 4x + 4$

Substitute the expression $x^2 - 4x + 4$ for $y$ in the first equation and solve for $x$.

$x + x^2 - 4x + 4 = 2$

$\qquad x^2 - 3x + 2 = 0$

$\qquad (x - 1)(x - 2) = 0$

$\qquad\qquad x - 1 = 0 \quad x - 2 = 0$

$\qquad\qquad\quad x = 1 \qquad x = 2$

Substitute $x = 1$ and then $x = 2$ into the equation $x + y = 2$ and solve for each value of $y$.

$\quad 1 + y = 2 \quad 2 + y = 2$

$\qquad y = 1 \qquad\qquad y = 0$

The solution set is $\{(1, 1), (2, 0)\}$.

**5.** $y = x^2 - 4x - 10$

$\quad y = -x^2 - 2x + 14$

Substitute the expression $x^2 - 4x - 10$ for $y$ in the second equation and solve for $x$.

$\quad x^2 - 4x - 10 = -x^2 - 2x + 14$

$2x^2 - 2x - 24 = 0$

$\qquad x^2 - x - 12 = 0$

$\big(x - 4\big)\big(x + 3\big) = 0$

$x - 4 = 0 \quad\text{or}\quad x + 3 = 0$

$\quad x = 4 \quad\text{or}\quad\quad x = -3$

If $x = 4$, $y = \big(4\big)^2 - 4\big(4\big) - 10 = -10$.

If $x = -3$, $y = \big(-3\big)^2 - 4\big(-3\big) - 10 = 11$.

The solution set is $\big\{(4, -10), (-3, 11)\big\}$.

**7.** $x^2 + y^2 = 25$

$\quad x - y = 1$

Solve the second equation for $y$. $y = x - 1$

Substitute the expression $x - 1$ for $y$ in the first equation and solve for $x$.

$\qquad x^2 + (x - 1)^2 = 25$

$\quad x^2 + x^2 - 2x + 1 = 25$

$\qquad 2x^2 - 2x - 24 = 0$

$\qquad\quad x^2 - x - 12 = 0$

$\qquad\quad (x - 4)(x + 3) = 0$

$\quad x - 4 = 0 \quad\text{or}\quad x + 3 = 0$

$\qquad x = 4 \quad\text{or}\quad\quad x = -3$

If $x = 4$, $y = 4 - 1 = 3$.

If $x = -3$, $y = -3 - 1 = -4$.

The solution set is $\{(4, 3), (-3, -4)\}$.

**9.** $\qquad xy = 6$

$2x - y = 1$

Solve the first equation for $y$.

$y = \dfrac{6}{x}$

Substitute the expression $\dfrac{6}{x}$ for $y$ in the second equation and solve for $x$.

$\qquad 2x - \dfrac{6}{x} = 1$

$\qquad 2x^2 - 6 = x$

$\quad 2x^2 - x - 6 = 0$

$(2x + 3)(x - 2) = 0$

$2x + 3 = 0 \quad\text{or}\quad x - 2 = 0$

$\quad x = -\dfrac{3}{2} \quad\text{or}\quad\quad x = 2$

If $x = -\dfrac{3}{2}$, $y = \dfrac{6}{-\frac{3}{2}} = -4$.

If $x = 2$, $y = \dfrac{6}{2} = 3$.

The solution set is $\left\{\left(-\dfrac{3}{2}, -4\right), (2, 3)\right\}$.

**11.** $y^2 = x^2 - 9$
$2y = x - 3$

Solve the second equation for $y$.

$y = \dfrac{x-3}{2}$

Substitute the expression $\dfrac{x-3}{2}$ for $y$ in the first equation and solve for $x$.

$\left(\dfrac{x-3}{2}\right)^2 = x^2 - 9$

$\dfrac{x^2 - 6x + 9}{4} = x^2 - 9$

$x^2 - 6x + 9 = 4x^2 - 36$

$3x^2 + 6x - 45 = 0$

$x^2 + 2x - 15 = 0$

$(x+5)(x-3) = 0$

$x + 5 = 0$  or  $x - 3 = 0$

$x = -5$  or  $x = 3$

If $x = -5$, $y = \dfrac{-5-3}{2} = -4$.

If $x = 3$, $y = \dfrac{3-3}{2} = 0$.

The solution set is $\{(-5,-4),(3,0)\}$.

**13.** $xy = 3$
$x^2 + y^2 = 10$

Solve the second equation for $y$.

$y = \dfrac{3}{x}$

Substitute the expression $\dfrac{3}{x}$ for $y$ in the second equation and solve for $x$.

$x^2 + \left(\dfrac{3}{x}\right)^2 = 10$

$x^2 + \dfrac{9}{x^2} - 10 = 0$

$x^4 - 10x^2 + 9 = 0$

$(x^2 - 9)(x^2 - 1) = 0$

$(x-3)(x+3)(x-1)(x+1) = 0$

$x - 3 = 0$  or  $x + 3 = 0$  or  $x - 1 = 0$  or  $x + 1 = 0$

$x = 3$  or  $x = -3$  or  $x = 1$  or  $x = -1$

If $x = 3$, $y = \dfrac{3}{3} = 1$.

If $x = -3$, $y = \dfrac{3}{-3} = -1$.

If $x = 1$, $y = \dfrac{3}{1} = 3$.

If $x = -1$, $y = \dfrac{3}{-1} = -3$.

The solution set is $\{(3,1),(-3,-1),(1,3),(-1,-3)\}$.

**15.** $x + y = 1$
$x^2 + xy - y^2 = -5$

Solve the first equation for $y$.  $y = 1 - x$

Substitute the expression $1 - x$ for $y$ in the second equation and solve for $x$.

$x^2 + x(1-x) - (1-x)^2 = -5$

$x^2 + x - x^2 - (1 - 2x + x^2) = -5$

$x - 1 + 2x - x^2 = -5$

$x^2 - 3x - 4 = 0$

$(x-4)(x+1) = 0$

$x - 4 = 0$  or  $x + 1 = 0$

$x = 4$  or  $x = -1$

If $x = 4$, $y = 1 - 4 = -3$.

If $x = -1$, $y = 1 - (-1) = 2$.

The solution set is $\{(4,-3),(-1,2)\}$.

**17.** $x + y = 1$

$(x-1)^2 + (y+2)^2 = 10$

Solve the first equation for $y$.

$y = 1 - x$

Substitute the expression $1 - x$ for $y$ in the second equation and solve for $x$.

$(x-1)^2 + (1-x+2)^2 = 10$

$(x-1)^2 + (3-x)^2 = 10$

$x^2 - 2x + 1 + 9 - 6x + x^2 - 10 = 0$

$2x^2 - 8x = 0$

$x^2 - 4x = 0$

$x(x-4) = 0$

$x = 0 \quad \text{or} \quad x - 4 = 0$

$x = 4$

If $x = 0$, $y = 1 - 0 = 1$.

If $x = 4$, $y = 1 - 4 = -3$.

The solution set is $\{(0,1),(4,-3)\}$.

**19.** Eliminate the $y^2$ –terms by adding the equations.

$x^2 + y^2 = 13$

$\underline{x^2 - y^2 = 5}$

$2x^2 = 18$

$x^2 = 9$

$x = \pm 3$

If $x = 3$,

$(3)^2 + y^2 = 13$

$y^2 = 4$

$y = \pm 2$

If $x = -3$,

$(-3)^2 + y^2 = 13$

$y^2 = 4$

$y = \pm 2$

The solution set is
$\{(3, 2), (3, -2), (-3, 2), (-3, -2)\}$.

**21.** $x^2 - 4y^2 = -7$

$3x^2 + y^2 = 31$

Eliminate the $x^2$ –terms by multiplying the first equation by –3 and adding the resulting equations.

$-3x^2 + 12y^2 = 21$

$\underline{3x^2 + y^2 = 31}$

$13y^2 = 52$

$y^2 = 4$

$y = \pm 2$

If $y = 2$,

$x^2 - 4(2)^2 = -7$

$x^2 = 9$

$x = \pm 3$

If $y = -2$,

$x^2 - 4(-2)^2 = -7$

$x^2 = 9$

$x = \pm 3$

The solution set is $\{(3,2),(3,-2),(-3,2),(-3,-2)\}$.

**23.** Arrange the equations so that variable terms appear on the left and constants appear on the right.

$3x^2 + 4y^2 = 16$

$2x^2 - 3y^2 = 5$

Eliminate the $y^2$ –terms by multiplying the first equation by 3 and the second equation by 4. Add the resulting equations.

$9x^2 + 12y^2 = 48$

$\underline{8x^2 - 12y^2 = 20}$

$17x^2 = 68$

$x^2 = 4$

$x = \pm 2$

If $x = 2$,

$3(2)^2 + 4y^2 = 16$

$y^2 = 1$

$y = \pm 1$

If $x = -2$,

$3(-2)^2 + 4y^2 = 16$

$y = \pm 1$

The solution set is
$\{(2, 1), (2, -1), (-2, 1), (-2, -1)\}$.

**25.**
$$x^2 + y^2 = 25$$
$$(x-8)^2 + y^2 = 41$$

Expand the second equation and eliminate $x^2$ and $y^2$–terms by multiplying the first equation by –1 and adding the resulting equations.

$$x^2 - 16x + 64 + y^2 = 41$$
$$\underline{-x^2 - y^2 = -25}$$
$$-16x + 64 = 16$$
$$-16x = -48$$
$$x = 3$$

If $x = 3$,
$$(3)^2 + y^2 = 25$$
$$y^2 = 16$$
$$y = \pm 4$$

The solution set is $\{(3,4), (3,-4)\}$.

**27.**
$$y^2 - x = 4$$
$$x^2 + y^2 = 4$$

Eliminate the $y^2$–terms by multiplying the first equation by –1 and adding the resulting equations.

$$x - y^2 = -4$$
$$\underline{x^2 + y^2 = 4}$$
$$x^2 + x = 0$$
$$x(x+1) = 0$$
$$x = 0 \quad \text{or} \quad x + 1 = 0$$
$$x = -1$$

If $x = 0$,
$$y^2 = 4$$
$$y = \pm 2$$

If $x = -1$,
$$y^2 - (-1) = 4$$
$$y^2 = 3$$
$$y = \pm\sqrt{3}$$

The solution set is
$$\left\{(0,2),(0,-2),\left(-1,\sqrt{3}\right),\left(-1,-\sqrt{3}\right)\right\}.$$

$$\left\{(0,-4),\left(2\sqrt{3},2\right),\left(-2\sqrt{3},2\right)\right\}.$$

**29.** The addition method is used here to solve the system.
$$3x^2 + 4y^2 = 16$$
$$2x^2 - 3y^2 = 5$$

Eliminate the $y^2$–terms by multiplying the first equation by 3 and the second equation by 4. Add the resulting equations.

$$9x^2 + 12y^2 = 48$$
$$\underline{8x^2 - 12y^2 = 20}$$
$$17x^2 = 68$$
$$x^2 = 4$$
$$x = \pm 2$$

If $x = 2$,
$$3(2)^2 + 4y^2 = 16$$
$$y^2 = 1$$
$$y = \pm 1$$

If $x = -2$,
$$3(-2)^2 + 4y^2 = 16$$
$$y = \pm 1$$

The solution set is
$$\{(2, 1), (2, -1), (-2, 1), (-2, -1)\}.$$

**31.** The substitution method is used here to solve the system.
$$2x^2 + y^2 = 18$$
$$xy = 4$$

Solve the second equation for $y$.

$$y = \frac{4}{x}$$

Substitute the expression $\frac{4}{x}$ for $y$ in the first equation and solve for $x$.

$$2x^2 + \left(\frac{4}{x}\right)^2 = 18$$
$$2x^2 + \frac{16}{x^2} = 18$$
$$2x^4 + 16 = 18x^2$$
$$x^4 - 9x^2 + 8 = 0$$
$$\left(x^2 - 8\right)\left(x^2 - 1\right) = 0$$
$$x^2 - 8 = 0 \quad \text{or} \quad x^2 - 1 = 0$$
$$x^2 = 8 \quad \text{or} \quad x^2 = 1$$
$$x = \pm 2\sqrt{2} \quad \text{or} \quad x = \pm 1$$

If $x = 2\sqrt{2}$, $y = \dfrac{4}{2\sqrt{2}} = \sqrt{2}$.

If $x = -2\sqrt{2}$, $y = \dfrac{4}{-2\sqrt{2}} = -\sqrt{2}$.

If $x = 1$, $y = \dfrac{4}{1} = 4$.

If $x = -1$, $y = \dfrac{4}{-1} = -4$.

The solution set is

$\left\{\left(2\sqrt{2}, \sqrt{2}\right), \left(-2\sqrt{2}, -\sqrt{2}\right), (1,4), (-1,-4)\right\}$.

**33.** The substitution method is used here to solve the system.
$$x^2 + 4y^2 = 20$$
$$x + 2y = 6$$
Solve the second equation for $x$.
$$x = 6 - 2y$$
Substitute the expression $6 - 2y$ for $x$ in the first equation and solve for $y$.
$$(6 - 2y)^2 + 4y^2 = 20$$
$$36 - 24y + 4y^2 + 4y^2 - 20 = 0$$
$$8y^2 - 24y + 16 = 0$$
$$y^2 - 3y + 2 = 0$$
$$(y - 2)(y - 1) = 0$$
$$y - 2 = 0 \quad \text{or} \quad y - 1 = 0$$
$$y = 2 \quad \text{or} \quad y = 1$$
If $y = 2$, $x = 6 - 2(2) = 2$.

If $y = 1$, $x = 6 - 2(1) = 4$.

The solution set is $\left\{(2,2), (4,1)\right\}$.

**35.** Eliminate $y$ by adding the equations.
$$x^3 + y = 0$$
$$\underline{x^2 - y = 0}$$
$$x^3 + x^2 = 0$$
$$x^2(x + 1) = 0$$
$$x^2 = 0 \quad \text{or} \quad x + 1 = 0$$
$$x = 0 \quad \text{or} \quad x = -1$$
If $x = 0$,
$$(0)^3 + y = 0$$
$$y = 0$$
If $x = -1$,
$$(-1)^3 + y = 0$$
$$y = 1$$
The solution set is $\left\{(0,0), (-1,1)\right\}$.

**37.** The substitution method is used here to solve the system.
$$x^2 + (y - 2)^2 = 4$$
$$x^2 - 2y = 0$$
Solve the second equation for $x^2$.
$$x^2 = 2y$$
Substitute the expression $2y$ for $x^2$ in the first equation and solve for $y$.
$$2y + (y - 2)^2 = 4$$
$$2y + y^2 - 4y + 4 = 4$$
$$y^2 - 2y = 0$$
$$y(y - 2) = 0$$
$$y = 0 \quad \text{or} \quad y - 2 = 0$$
$$y = 2$$
If $y = 0$,
$$x^2 = 2(0)$$
$$x^2 = 0$$
$$x = 0$$
If $y = 2$,
$$x^2 = 2(2)$$
$$x^2 = 4$$
$$x = \pm 2$$
The solution set is $\{(0, 0), (-2, 2), (2, 2)\}$.

**39.** The substitution method is used here to solve the system.
$$y = (x + 3)^2$$
$$x + 2y = -2$$
Solve the first equation for $x$.
$$x = -2y - 2$$
Substitute the expression $-2y-2$ for $x$ in the first equation and solve for $y$.
$$y = (-2y - 2 + 3)^2 = (-2y + 1)^2$$
$$y = 4y^2 - 4y + 1$$
$$4y^2 - 5y + 1 = 0$$
$$(4y - 1)(y - 1) = 0$$
$$4y - 1 = 0 \quad \text{or} \quad y - 1 = 0$$
$$y = \dfrac{1}{4} \quad \text{or} \quad y = 1$$
If $y = \dfrac{1}{4}$, $x = -2\left(\dfrac{1}{4}\right) - 2 = -\dfrac{5}{2}$.

If $y = 1$, $x = -2(1) - 2 = -4$.

The solution set is $\left\{(-4, 1), \left(-\dfrac{5}{2}, \dfrac{1}{4}\right)\right\}$.

**41.** The substitution method is used here to solve the system.

$x^2 + y^2 + 3y = 22$

$2x + y = -1$

Solve the second equation for $y$.

$y = -2x - 1$

Substitute the expression $-2x-1$ for $y$ in the first equation and solve for $x$.

$x^2 + (-2x-1)^2 + 3(-2x-1) - 22 = 0$

$x^2 + 4x^2 + 4x + 1 - 6x - 3 - 22 = 0$

$$5x^2 - 2x - 24 = 0$$

$$(5x - 12)(x + 2) = 0$$

$5x - 12 = 0$  or  $x + 2 = 0$

$x = \dfrac{12}{5}$  or  $x = -2$

If $x = \dfrac{12}{5}$, $y = -2\left(\dfrac{12}{5}\right) - 1 = -\dfrac{29}{5}$.

If $x = -2$, $y = -2(-2) - 1 = 3$.

The solution set is $\left\{ \left( \dfrac{12}{5}, -\dfrac{29}{5} \right), (-2, 3) \right\}$.

**43.** The substitution method is used here to solve the system.

$x + y = 10$

$xy = 24$

Solve the first equation for $y$.

$y = 10 - x$

Substitute the expression $10 - x$ for $y$ in the second equation and solve for $x$.

$x(10 - x) = 24$

$10x - x^2 = 24$

$x^2 - 10x + 24 = 0$

$(x - 4)(x - 6) = 0$

$x - 4 = 0$  or  $x - 6 = 0$

$x = 4$  or  $x = 6$

If $x = 4$, $y = 10 - 4 = 6$.

If $x = 6$, $y = 10 - 6 = 4$.

The numbers are 4 and 6.

**45.** Eliminate the $y^2$–terms by adding the equations.

$x^2 - y^2 = 3$

$\underline{2x^2 + y^2 = 9}$

$3x^2 = 12$

$x^2 = 4$

$x = \pm 2$

If $x = 2$,

$2(2)^2 + y^2 = 9$

$y^2 = 1$

$y = \pm 1$

If $x = -2$,

$2(-2)^2 + y^2 = 9$

$y^2 = 1$

$y = \pm 1$

The numbers are 2 and 1, 2 and $-1$, $-2$ and 1, or $-2$ and $-1$.

**47.**  $2x^2 + xy = 6$

$x^2 + 2xy = 0$

Multiply the first equation by $-2$ and add the two equations.

$-4x^2 - 2xy = -12$

$\underline{x^2 + 2xy = 0}$

$-3x^2 = -12$

$x^2 = 4$

$x = \pm 2$

Back-substitute these values for $x$ in the second equation and solve for $y$.

For $x = -2$:  $(-2)^2 + 2(-2)y = 0$

$4 - 4y = 0$

$-4y = -4$

$y = 1$

For $x = 2$:  $(2)^2 + 2(2)y = 0$

$4 + 4y = 0$

$4y = -4$

$y = -1$

The solution set is $\{(-2, 1), (2, -1)\}$.

**49.** $-4x + y = 12$

$y = x^3 + 3x^2$

Substitute $x^3 + 3x^2$ for $y$ in the first equation and solve for $x$.

$-4x + \left(x^3 + 3x^2\right) = 12$

$x^3 + 3x^2 - 4x - 12 = 0$

$x^2(x+3) - 4(x+3) = 0$

$(x+3)\left(x^2 - 4\right) = 0$

$(x+3)(x-2)(x+2) = 0$

$x = -3$, $x = 2$, or $x = -2$

Substitute these values for $x$ in the second equation and solve for $y$.

For $x = -3$:  $y = (-3)^3 + 3(-3)^2 = -27 + 27$

$= 0$

For $x = 2$:  $y = (2)^3 + 3(2)^2 = 8 + 12$

$= 20$

For $x = -2$:  $y = (-2)^3 + 3(-2)^2 = -8 + 12$

$= 4$

The solution set is $\{(-3,0),(2,20),(-2,4)\}$.

**51.** $\dfrac{3}{x^2} + \dfrac{1}{y^2} = 7$

$\dfrac{5}{x^2} - \dfrac{2}{y^2} = -3$

Multiply the first equation by 2 and add the equations.

$\dfrac{6}{x^2} + \dfrac{2}{y^2} = 14$

$\dfrac{5}{x^2} - \dfrac{2}{y^2} = -3$

_____

$\dfrac{11}{x^2} = 11$

$x^2 = 1$

$x = \pm 1$

Back-substitute these values for $x$ in the first equation and solve for $y$.

For $x = -1$:

$\dfrac{3}{(-1)^2} + \dfrac{1}{y^2} = 7$

$3 + \dfrac{1}{y^2} = 7$

$\dfrac{1}{y^2} = 4$

$y^2 = \dfrac{1}{4}$

$y = \pm\dfrac{1}{2}$

For $x = 1$:

$\dfrac{3}{(1)^2} + \dfrac{1}{y^2} = 7$

$3 + \dfrac{1}{y^2} = 7$

$\dfrac{1}{y^2} = 4$

$y^2 = \dfrac{1}{4}$

$y = \pm\dfrac{1}{2}$

The solution set is

$\left\{\left(-1,-\dfrac{1}{2}\right),\left(-1,\dfrac{1}{2}\right),\left(1,-\dfrac{1}{2}\right),\left(1,\dfrac{1}{2}\right)\right\}$.

**53.**

$y = x^2 - 4$
$y = 2x - 4$

**55.** $16x^2 + 4y^2 = 64$

$y = x^2 - 4$

Substitute the expression $x^2 - 4$ for $y$ in the first equation and solve for $x$.

$16x^2 + 4\left(x^2 - 4\right)^2 = 64$

$16x^2 + 4\left(x^4 - 8x^2 + 16\right) = 64$

$16x^2 + 4x^4 - 32x^2 + 64 = 64$

$4x^4 - 16x^2 = 0$

$x^4 - 4x^2 = 0$

$x^2\left(x^2 - 4\right) = 0$

$x^2 = 0$  or  $x^2 - 4 = 0$

$x = 0$  or  $x^2 = 4$

$x = \pm 2$

If $x = 0$, $y = (0)^2 - 4 = -4$.

If $x = 2$, $y = (2)^2 - 4 = 0$.

If $x = -2$, $y = (-2)^2 - 4 = 0$.

It is possible for the comet to intersect the orbiting body at $(0, -4)$, $(-2, 0)$, $(2, 0)$.

**57.** $2L + 2W = 36$
$\qquad LW = 77$

Divide each term in the first equation by 2 and solve *L*.

$L + W = 18$

$\qquad L = 18 - W$

Substitute the expression $18 - W$ for *L* in the second equation and solve for *W*.

$$(18 - W)W = 77$$

$$18W - W^2 = 77$$

$$W^2 - 18W + 77 = 0$$

$$(W - 11)(W - 7) = 0$$

$$W - 11 = 0 \quad \text{or} \quad W - 7 = 0$$

$$W = 11 \quad \text{or} \qquad W = 7$$

If $W = 11, L = 18 - 11 = 7$.

If $W = 7, L = 18 - 7 = 11$.

The dimensions are 11 feet by 7 feet.

**59.** $L^2 + W^2 = 10^2 = 100$

$\qquad LW = 48$

Solve the second equation for *L*. $L = \dfrac{48}{W}$

Substitute the expression $\dfrac{48}{W}$ for *L* in the first equation and solve for *W*.

$$\left(\frac{48}{W}\right)^2 + W^2 = 100$$

$$\frac{2304}{W^2} + W^2 - 100 = 0$$

$$2304 + W^4 - 100W^2 = 0$$

$$W^4 - 100W^2 + 2304 = 0$$

$$\left(W^2 - 36\right)\left(W^2 - 64\right) = 0$$

$$W^2 - 36 = 0 \quad \text{or} \quad W^2 - 64 = 0$$

$$W^2 = 36 \quad \text{or} \qquad W^2 = 64$$

$$W = \pm 6 \quad \text{or} \qquad W = \pm 8$$

The width cannot be –6 or –8 inches.

If $W = 6$,

$$L = \frac{48}{6} = 8$$

If $W = 8$,

$$L = \frac{48}{8} = 6$$

The dimensions are 8 inches by 6 inches.

**61.** $x^2 - y^2 = 21$
$\qquad 4x + 2y = 24$

Divide each term in the second equation by 2 and solve for *y*.

$2x + y = 12$

$y = 12 - 2x$

Substitute the expression $12 - 2x$ for *y* in the first equation and solve for *x*.

$$x^2 - (12 - 2x)^2 = 21$$

$$x^2 - \left(144 - 48x + 4x^2\right) = 21$$

$$3x^2 - 48x + 165 = 0$$

$$x^2 - 16x + 55 = 0$$

$$(x - 5)(x - 11) = 0$$

$$x - 5 = 0 \quad \text{or} \quad x - 11 = 0$$

$$x = 5 \quad \text{or} \qquad x = 11$$

If $x = 11, \ y = 12 - 2(11) = -10$.

If $x = 5, \ y = 12 - 2(5) = 2$.

The dimensions of the floor are 5 meters by 5 meters and the dimensions of the square that will accommodate the pool are 2 meters by 2 meters.

**63.** **a.** It appears from the graphs that the percentage of white-collar workers was the same as blue-collar workers between the 1940s and 1960s.

**b.** $0.5x - y = -18$

$\qquad\quad y = -0.004x^2 + 0.23x + 41$

Substitute $-0.004x^2 + 0.23x + 41$ in the first equation and solve for *x*.

$$0.5x - y = -18$$

$$0.5x - \overbrace{(-0.004x^2 + 0.23x + 41)}^{y} = -18$$

$$0.5x + 0.004x^2 - 0.23x - 41 = -18$$

$$0.004x^2 + 0.27x - 23 = 0$$

Use the quadratic formula.

$$x = \frac{-b \pm \sqrt{b^2 - 4ac}}{2a}$$

$$x = \frac{-(0.27) \pm \sqrt{0.27^2 - 4(0.004)(-23)}}{2(0.004)}$$

$$x \approx 49 \quad \text{or} \ -117$$

The percentage of white-collar workers was the same as blue-collar workers 49 years after 1900, or 1949.

Let $x = 49$ and solve for $y$ in the white-collar model.

$$0.5x - y = -18$$
$$0.5(49) - y = -18$$
$$24.5 - y = -18$$
$$-y = -42.5$$
$$y \approx 43$$

The percentage of white-collar workers in 1949 was about 43%

Let $x = 49$ and solve for $y$ in the blue-collar model.

$$y = -0.004(49)^2 + 0.23(49) + 41 \approx 43\%$$

The percentage of blue-collar workers in 1949 was about 43%.

**c.** According to the graph, the percentage of white-collar workers was the same as farmers in 1920. The percentages of white-collar workers and farmers in 1920 were both 28%.

**d.**
$$0.5x - y = -18$$
$$\underline{0.4x + y = \phantom{-}35}$$
$$0.9x \phantom{aaaa} = 17$$
$$x = \frac{17}{0.9}$$
$$x \approx 19$$

According to the models, the percentage of white-collar workers was the same as farmers 19 years after 1900, or 1919.

Let $x = 19$ and solve for $y$ in the white-collar model.

$$0.5x - y = -18$$
$$0.5(19) - y = -18$$
$$9.5 - y = -18$$
$$-y = -27.5$$
$$y = 27.5$$

The percentage of white-collar workers in 1919 was about 27.5%

Let $x = 19$ and solve for $y$ in the farming model.

$$0.4x + y = 35$$
$$0.4(19) + y = 35$$
$$0.4(19) + y = 35$$
$$7.6 + y = 35$$
$$y = 27.4$$

The percentage of farm workers in 1919 was about 27.4%

These answers model the actual data from part c (the graph) fairly well.

**65. – 67.** Answers will vary.

**69.** makes sense

**71.** makes sense

**73.** false; Changes to make the statement true will vary. A sample change is: A circle and a line can intersect in at most two points, and therefore such a system has at most two real solutions.

**75.** false; Changes to make the statement true will vary. A sample change is: It is possible that a system of two equations in two variables whose graphs represent circles do not intersect, or intersect in a single point. This means that the system would have no solution, or a single solution, respectively.

**77.** First determine the solution to the following system of equations.

$$xy = 20$$
$$x^2 + y^2 = 41$$

Solve the first equation for $x$.

$$x = \frac{20}{y}$$

Substitute the expression $\dfrac{20}{y}$ for $x$ in the second equation and solve for  in the second equation and solve for $y$.

$$\left(\frac{20}{y}\right)^2 + y^2 = 41$$
$$\frac{400}{y^2} + y^2 - 41 = 0$$
$$y^4 - 41y^2 + 400 = 0$$
$$\left(y^2 - 25\right)\left(y^2 - 16\right) = 0$$
$$y^2 - 25 = 0 \quad \text{or} \quad y^2 - 16 = 0$$
$$y^2 = 25 \quad \text{or} \quad y^2 = 16$$
$$y = \pm 5 \quad \text{or} \quad y = \pm 4$$

If $y = 5$, $x = \dfrac{20}{5} = 4$.

If $y = -5$, $x = \dfrac{20}{-5} = -4$.

If $y = 4$, $x = \dfrac{20}{4} = 5$.

If $y = -4$, $x = \dfrac{20}{-4} = -5$.

The rectangle formed by joining the points of intersection has sides $a$ and $b$. The length of $a$ is $\sqrt{(5-4)^2+(4-5)^2}=\sqrt{2}$. The length of $b$ is $\sqrt{(4-(-5))^2+(5-(-4))^2}=9\sqrt{2}$.

The area of the rectangle is $a \cdot b = \left(\sqrt{2}\right)\left(9\sqrt{2}\right)=18$ square units.

**79.** $\log_y x = 3$

$\log_y(4x)=5$

$x = y^3$

$4x = y^5$

Substitute $y^3$ for $x$ in the equation $4x = y^5$ and solve for $y$.

$4(y^3)=y^5$

$4 = y^2$

$y = \pm 2$

Keep the positive base.

$x = 2^3 = 8$

The solution set is $\{(8, 2)\}$.

**81.**

**82.**

**83.**

**Mid-Chapter 5 Check Point**

**1.**
$$x = 3y - 7$$
$$4x + 3y = 2$$

Since the first equation is solved for $x$ already, we will use substitution.

Let $x = 3y - 7$ in the second equation and solve for $y$.

$4(3y-7)+3y=2$

$12y - 28 + 3y = 2$

$15y = 30$

$y = 2$

Substitute this value for $y$ in the first equation.

$x = 3(2) - 7 = 6 - 7 = -1$

The solution is $(-1, 2)$.

**2.** $3x + 4y = -5$

$2x - 3y = 8$

Multiply the first equation by 3 and the second equation by 4, then add the equations.

$9x + 12y = -15$

$\underline{8x - 12y = 32}$

$17x = 17$

$x = 1$

Back-substitute to solve for $y$.

$3x + 4y = -5$

$3(1) + 4y = -5$

$3 + 4y = -5$

$4y = -8$

$y = -2$

The solution is $(1, -2)$.

**3.**
$$\frac{2x}{3} + \frac{y}{5} = 6$$

$$\frac{x}{6} - \frac{y}{2} = -4$$

Multiply the first equation by 15 and the second equation by 6 to eliminate the fractions.

$$15\left(\frac{2x}{3} + \frac{y}{5}\right) = 15(6)$$

$$10x + 3y = 90$$

$$6\left(\frac{x}{6} - \frac{y}{2}\right) = 6(-4)$$

$$x - 3y = -24$$

We now need to solve the equivalent system
$$10x + 3y = 90$$
$$x - 3y = -24$$

Add the two equations to eliminate $y$.
$$10x + 3y = 90$$
$$\underline{x - 3y = -24}$$
$$11x = 66$$
$$x = 6$$

Back-substitute to solve for $y$.
$$x - 3y = -24$$
$$6 - 3y = -24$$
$$-3y = -30$$
$$y = 10$$

The solution is $(6, 10)$.

**4.**
$$y = 4x - 5$$
$$8x - 2y = 10$$

Since the first equation is already solved for $y$, we will use substitution.
Let $y = 4x - 5$ in the second equation and solve for $x$.

$$8x - 2(4x - 5) = 10$$
$$8x - 8x + 10 = 10$$
$$10 = 10$$

This statement is an identity. The system is dependent so there are an infinite number of solutions. The solution set is $\{(x, y) \mid y = 4x - 5\}$.

**5.**
$$2x + 5y = 3$$
$$3x - 2y = 1$$

Multiply the first equation by 3 and the second equation by $-2$, then add the equations.

$$6x + 15y = 9$$
$$\underline{-6x + 4y = -2}$$
$$19y = 7$$
$$y = \frac{7}{19}$$

Back-substitute to solve for $x$.

$$2x + 5y = 3$$
$$2x + 5\left(\frac{7}{19}\right) = 3$$
$$2x + \frac{35}{19} = 3$$
$$2x = \frac{22}{19}$$
$$x = \frac{11}{19}$$

The solution is $\left(\frac{11}{19}, \frac{7}{19}\right)$.

**6.**
$$\frac{x}{12} - y = \frac{1}{4}$$
$$4x - 48y = 16$$

Solve the first equation for $y$.

$$\frac{x}{12} - y = \frac{1}{4}$$
$$-y = -\frac{x}{12} + \frac{1}{4}$$
$$y = \frac{x}{12} - \frac{1}{4}$$

Let $y = \frac{x}{12} - \frac{1}{4}$ in the second equation and solve for $x$.

$$4x - 48\left(\frac{x}{12} - \frac{1}{4}\right) = 16$$
$$4x - 4x + 12 = 16$$
$$12 = 16$$

This statement is a contradiction. The system is inconsistent so there is no solution. The solution is $\{\ \}$ or $\varnothing$.

**7.**
$$2x - y + 2z = -8$$
$$x + 2y - 3z = 9$$
$$3x - y - 4z = 3$$

Multiply the first equation by 2 and add to the second equation.
$$4x - 2y + 4z = -16$$
$$\underline{x + 2y - 3z = 9}$$
$$5x + z = -7$$

Multiply the first equation by $-1$ and add to the third equation.
$$-2x + y - 2z = 8$$
$$\underline{3x - y - 4z = 3}$$
$$x - 6z = 11$$

Use the two reduced equations to get the following system:
$$5x + z = -7$$
$$x - 6z = 11$$

Multiply the first equation by 6 and add to the second equation.
$$30x + 6z = -42$$
$$\underline{x - 6z = 11}$$
$$31x = -31$$
$$x = -1$$

Back-substitute to solve for z.
$$5x + z = -7$$
$$5(-1) + z = -7$$
$$-5 + z = -7$$
$$z = -2$$

Back-substitute to solve for y.
$$2x - y + 2z = -8$$
$$2(-1) - y + 2(-2) = -8$$
$$-2 - y - 4 = -8$$
$$-y = -2$$
$$y = 2$$

The solution is $(-1, 2, -2)$.

**8.**
$$x \quad - 3z = -5$$
$$2x - y + 2z = 16$$
$$7x - 3y - 5z = 19$$

Multiply the second equation by $-3$ and add to the third equation.
$$-6x + 3y - 6z = -48$$
$$\underline{7x - 3y - 5z = 19}$$
$$x - 11z = -29$$

Use this reduced equation and the original first equation to obtain the following system:
$$x - 3z = -5$$
$$x - 11z = -29$$

Multiply the second equation by $-1$ and add to the first equation.
$$x - 3z = -5$$
$$\underline{-x + 11z = 29}$$
$$8z = 24$$
$$z = 3$$

Back-substitute to solve for x.
$$x - 3z = -5$$
$$x - 3(3) = -5$$
$$x - 9 = -5$$
$$x = 4$$

Back-substitute to solve for y.
$$2x - y + 2z = 16$$
$$2(4) - y + 2(3) = 16$$
$$8 - y + 6 = 16$$
$$-y = 2$$
$$y = -2$$

The solution is $(4, -2, 3)$.

**9.** Solve $x + 2y - 3 = 0$ for x and substitute into the other equation.
$$(\overset{x}{\overbrace{-2y+3}})^2 + y^2 = 9$$
$$4y^2 - 12y + 9 + y^2 = 9$$
$$5y^2 - 12y = 0$$
$$y(5y - 12) = 0$$
$$y = 0 \text{ or } y = \frac{12}{5}$$

Back-substitute these values to find x.
When $y = 0$, $x = -2(0) + 3 = 3$.
When $y = \frac{12}{5}$, $x = -2\left(\frac{12}{5}\right) + 3 = -\frac{9}{5}$.

The solution set is $\left\{(3,0), \left(-\frac{9}{5}, \frac{12}{5}\right)\right\}$.

**10.** $3x^2 + 2y^2 = 14$

$2x^2 - y^2 = 7$

Multiply the second equation by 2 and add.

$3x^2 + 2y^2 = 14$

$\underline{4x^2 - 2y^2 = 14}$

$7x^2 = 28$

$x^2 = 4$

$x = \pm 2$

Back-substitute these values to find $y$.

$3(2)^2 + 2y^2 = 14$

$12 + 2y^2 = 14$

$2y^2 = 2$

$y^2 = 1$

$y = \pm 1$

$3(-2)^2 + 2y^2 = 14$

$12 + 2y^2 = 14$

$2y^2 = 2$

$y^2 = 1$

$y = \pm 1$

The solution set is $\{(2,1),(2,-1),(-2,1),(-2,-1)\}$.

**11.** Use the first equation to substitute for $y$ in the second equation.

$$x^2 + \overbrace{(x^2-6)}^{y}{}^2 = 8$$

$$x^2 + (x^2-6)^2 = 8$$

$$x^2 + x^4 - 12x^2 + 36 = 8$$

$$x^4 - 11x^2 + 28 = 0$$

$$(x^2 - 4)(x^2 - 7) = 0$$

$x = \pm 2$ or $x = \pm\sqrt{7}$

Back-substitute these values to find $x$.

$y = (2)^2 - 6 \qquad y = (-2)^2 - 6$

$y = -2 \qquad\qquad y = -2$

$y = (\sqrt{7})^2 - 6 \quad y = (-\sqrt{7})^2 - 6$

$y = 1 \qquad\qquad y = 1$

The solution set is $\left\{(2,-2),(-2,-2),\left(\sqrt{7},1\right),\left(-\sqrt{7},1\right)\right\}$.

**12.** Use the first equation to substitute for $x$ in the second equation.

$$2y^2 + (\overbrace{2y+4}^{x})y = 8$$
$$2y^2 + (2y+4)y = 8$$
$$2y^2 + 2y^2 + 4y = 8$$
$$4y^2 + 4y - 8 = 0$$
$$y^2 + y - 2 = 0$$
$$(y+2)(y-1) = 0$$
$$y = -2 \text{ or } y = 1$$

Back-substitute these values to find $x$.

$$x = 2(-2) + 4 \quad x = 2(1) + 4$$
$$x = 0 \qquad\qquad y = 6$$

The solution set is $\{(0,-2),(6,1)\}$.

**13.** $\dfrac{x^2 - 6x + 3}{(x-2)^3} = \dfrac{A}{x-2} + \dfrac{B}{(x-2)^2} + \dfrac{C}{(x-2)^3}$

Multiply both sides of the equation by the common denominator $(x-2)^3$.

$$(x-2)^3 \frac{x^2 - 6x + 3}{(x-2)^3} = (x-2)^3 \left( \frac{A}{x-2} + \frac{B}{(x-2)^2} + \frac{C}{(x-2)^3} \right)$$
$$x^2 - 6x + 3 = A(x-2)^2 + B(x-2) + C$$
$$x^2 - 6x + 3 = A(x^2 - 4x + 4) + Bx - 2B + C$$
$$x^2 - 6x + 3 = Ax^2 - 4Ax + 4A + Bx - 2B + C$$
$$x^2 - 6x + 3 = (A)x^2 + (-4A + B)x + (4A - 2B + C)$$

Equate coefficients of like powers of $x$ and equate constant terms.

$$A = 1$$
$$-4A + B = -6$$
$$4A - 2B + C = 3$$

Since $A = 1$, we find that $B = -2$ and $C = -5$ by substitution.

$$\frac{2x^2 - 6x + 3}{(x-2)^3} = \frac{1}{x-2} + \frac{-2}{(x-2)^2} + \frac{-5}{(x-2)^3}$$

**14.** $\dfrac{10x^2 + 9x - 7}{(x+2)(x^2-1)} = \dfrac{10x^2 + 9x - 7}{(x+2)(x+1)(x-1)} = \dfrac{A}{x+2} + \dfrac{B}{x+1} + \dfrac{C}{x-1}$

Multiply both sides of the equation by the least common denominator.

$$(x+2)(x+1)(x-1)\frac{10x^2 + 9x - 7}{(x+2)(x+1)(x-1)} = (x+2)(x+1)(x-1)\left( \frac{A}{x+2} + \frac{B}{x+1} + \frac{C}{x-1} \right)$$
$$10x^2 + 9x - 7 = (x+1)(x-1)A + (x+2)(x-1)B + (x+2)(x+1)C$$
$$10x^2 + 9x - 7 = (x^2 - 1)A + (x^2 + x - 2)B + (x^2 + 3x + 2)C$$
$$10x^2 + 9x - 7 = x^2A - A + x^2B + xB - 2B + x^2C + 3xC + 2C$$
$$10x^2 + 9x - 7 = (A + B + C)x^2 + (B + 3C)x + (-A - 2B + 2C)$$

Equate coefficients of like powers of $x$, and equate constant terms.

$$A + B + C = 10$$
$$B + 3C = 9$$
$$-A - 2B + 2C = -7$$

Solving the above system for $A$, $B$, and $C$, we find $A = 5$, $B = 3$, and $C = 2$.

$$\frac{10x^2 + 9x - 7}{(x+2)(x^2-1)} = \frac{5}{x+2} + \frac{3}{x+1} + \frac{2}{x-1}$$

**15.** $\dfrac{x^2 + 4x - 23}{(x+3)(x^2+4)} = \dfrac{A}{x+3} + \dfrac{Bx+C}{x^2+4}$

Multiply both sides of the equation by the least common denominator.

$$(x+3)(x^2+4)\frac{x^2+4x-23}{(x+3)(x^2+4)} = (x+3)(x^2+4)\left(\frac{A}{x+3} + \frac{Bx+C}{x^2+4}\right)$$

$$x^2 + 4x - 23 = (x^2+4)A + (x+3)(Bx+C)$$

$$x^2 + 4x - 23 = x^2A + 4A + x^2B + 3xB + xC + 3C$$

$$x^2 + 4x - 23 = (A+B)x^2 + (3B+C)x + (4A+3C)$$

Equate coefficients of like powers of $x$, and equate constant terms.

$$A + B = 1$$
$$3B + C = 4$$
$$4A + 3C = -23$$

Solving the above system for $A$, $B$, and $C$, we find $A = -2$, $B = 3$, and $C = -5$.

$$\frac{x^2 + 4x - 23}{(x+3)(x^2+4)} = \frac{-2}{x+3} + \frac{3x-5}{x^2+4}$$

**16.** $\dfrac{x^3}{\left(x^2+4\right)^2} = \dfrac{Ax+B}{x^2+4} + \dfrac{Cx+D}{\left(x^2+4\right)^2}$

Multiply both sides of the equation by the least common denominator.

$$\left(x^2+4\right)^2\frac{x^3}{\left(x^2+4\right)^2} = \left(x^2+4\right)^2\left(\frac{Ax+B}{x^2+4} + \frac{Cx+D}{\left(x^2+4\right)^2}\right)$$

$$x^3 = \left(x^2+4\right)(Ax+B) + (Cx+D)$$

$$x^3 = x^3A + x^2B + 4xA + 4B + Cx + D$$

$$x^3 = (A)x^3 + (B)x^2 + (4A+C)x + (4B+D)$$

Equate coefficients of like powers of $x$, and equate constant terms.

$$A = 1$$
$$B = 0$$
$$4A + C = 0$$
$$4B + D = 0$$

Since $A = 1$ and $B = 0$, we find that $C = -4$ and $D = 0$ by substitution.

$$\frac{x^3}{\left(x^2+4\right)^2} = \frac{1x+0}{x^2+4} + \frac{-4x+0}{\left(x^2+4\right)^2} = \frac{x}{x^2+4} + \frac{-4x}{\left(x^2+4\right)^2}$$

**17.**  **a.**  $C(x) = 400,000 + 20x$

**b.**  $R(x) = 100x$

**c.**  $\begin{aligned} P(x) &= R(x) - C(x) \\ &= 100x - (400,000 + 20x) \\ &= 80x - 400,000 \end{aligned}$

**d.**  The break even point is the point where cost and revenue are the same. We need to solve the following system.

$y = 400,000 + 20x$

$y = 100x$

Let $y = 400,000 + 20x$ in the second equation and solve for $x$.

$400,000 + 20x = 100x$

$\qquad 400,000 = 80x$

$\qquad\quad 5000 = x$

Back-substitute to solve for $y$.

$\begin{aligned} y &= 100x \\ &= 100(5000) \\ &= 500,000 \end{aligned}$

Thus, the break-even point is $(5000,\ 500,000)$. The company will break even when it produces and sells 5000 PDAs. At this level, the cost and revenue will both be \$500,000.

**18.**  Let $x =$ the number of roses.

Let $y =$ the number of carnations.

$\qquad x + y = 20$

$3x + 1.5y = 39$

Solve the first equation for $x$.

$\quad x + y = 20$

$\qquad x = 20 - y$

Substitute this expression for $x$ in the second equation and solve for $y$.

$3(20 - y) + 1.5y = 39$

$\quad 60 - 3y + 1.5y = 39$

$\qquad\qquad -1.5y = -21$

$\qquad\qquad\quad\ y = 14$

Back-substitute to solve for $x$.

$x = 20 - y = 20 - 14 = 6$

There are 6 roses and 14 carnations in the bouquet.

**19.** Because the sum of the measures of the angles of any triangle is 180°, or $x + y + 90 = 180$

$$x + y = 90.$$

Because the angle with measures $x$ and $(3y + 20)$ are supplementary, $x + (3y + 20) = 180$.

Simplify this equation.

$$x + 3y + 20 = 180$$
$$x + 3y = 160$$

We now have the system

$$x + y = 90$$
$$x + 3y = 160.$$

To solve this system by the addition method, multiply the first equation by $-1$ and add to the second equation.

$$\begin{array}{r} -x - y = -90 \\ x + 3y = 160 \\ \hline 2y = 70 \\ y = 35 \end{array}$$

Back-substitute.

$$x + y = 90$$
$$x + 35 = 90$$
$$x = 55$$

Thus, $x = 55°$, $y = 35°$, and $(3y + 20) = 125°$.

**20.** Using the points $(-1, 0)$, $(1, 4)$, and $(2, 3)$ in the equation $y = ax^2 + bx + c$, we get the following system of equations:

$$a - b + c = 0$$
$$a + b + c = 4$$
$$4a + 2b + c = 3$$

Add the first two equations.

$$\begin{array}{r} a - b + c = 0 \\ a + b + c = 4 \\ \hline 2a + 2c = 4 \end{array}$$

Multiply the first equation by 2 and add to the third equation.

$$\begin{array}{r} 2a - 2b + 2c = 0 \\ 4a + 2b + c = 3 \\ \hline 6a + 3c = 3 \end{array}$$

Using the two reduced equations, we get the following system of equations:

$$2a + 2c = 4$$
$$6a + 3c = 3$$

Multiply the first equation by $-3$ and add to the second equation.

$$\begin{array}{r} -6a - 6c = -12 \\ 6a + 3c = 3 \\ \hline -3c = -9 \\ c = 3 \end{array}$$

Back-substitute to solve for $a$.

$$2a + 2c = 4$$
$$2a + 2(3) = 4$$
$$2a + 6 = 4$$
$$2a = -2$$
$$a = -1$$

Back-substitute to solve for $b$.

$$a + b + c = 4$$
$$-1 + b + 3 = 4$$
$$b = 2$$

The equation is $y = -x^2 + 2x + 3$.

**21.**
$$2l + 2w = 21$$
$$lw = 20$$

Solving $lw = 20$ for $l$ gives $l = \dfrac{20}{w}$.

Substitute into the other equation:

$$2\underbrace{\left(\dfrac{20}{w}\right)}_{l} + 2w = 21$$

$$\dfrac{40}{w} + 2w = 21$$

$$w\left(\dfrac{40}{w} + 2w\right) = w(21)$$

$$40 + 2w^2 = 21w$$

$$2w^2 - 21w + 40 = 0$$

$$(2w - 5)(w - 8) = 0$$

$$w = \dfrac{5}{2} = 2.5 \text{ or } w = 8.$$

Back-substitute to find $l$.

When $w = 2.5$, $l = \dfrac{20}{2.5} = 8$.

When $w = 8$, $l = \dfrac{20}{8} = 2.5$.

The dimensions of the rectangle are 2.5 m by 8.

**Section 5.5**

**Check Point Exercises**

**1.** $4x - 2y \geq 8$

Graph the equation $4x - 2y = 8$ as a solid line.
Choose a test point that is not on the line.

$\qquad$ Test $(0,0)$

$$4x - 2y \geq 8$$

$$4(0) - 2(0) \geq 8$$

$$0 \geq 8, \text{ false}$$

Since the statement is false, shade the other half-plane.

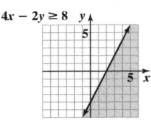

**2.** $y > -\dfrac{3}{4}x$

Graph the equation $y = -\dfrac{3}{4}x$ as a dashed line.

Choose a test point that is not on the line.

$\qquad$ Test $(1,1)$

$$y > -\dfrac{3}{4}x$$

$$1 > -\dfrac{3}{4}(1)$$

$$1 > -\dfrac{3}{4}, \text{ true}$$

Since the statement is true, shade the half-plane containing the point.

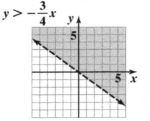

**3.** **a.** $y > 1$

Graph the equation $y = 1$ as a dashed line.
Choose a test point that is not on the line.

$\qquad$ Test $(0,0)$

$$y > 1$$

$$0 > 1, \text{ false}$$

Since the statement is false, shade the other half-plane.

**b.** Graph the equation $x = -2$ as a solid line.
Choose a test point that is not on the line.

$\qquad$ Test $(0,0)$

$$x \leq -2$$

$$0 \leq -2, \text{ false}$$

Since the statement is false, shade the other half-plane.

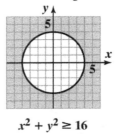

**4.** Graph $x^2 + y^2 = 16$ as a solid circle with radius 4 and center (0, 0).

$\qquad$ Test $(0,0)$

$$x^2 + y^2 \geq 16$$

$$(0)^2 + (0)^2 \geq 16$$

$$0 \geq 16, \text{ false}$$

Shade the region not containing (0, 0).

$$x^2 + y^2 \geq 16$$

**5.** Point $B = (66, 130)$

$$4.9x - y \geq 165$$

$$4.9(66) - 130 \geq 165$$

$$193.4 \geq 165, \text{ true}$$

$$3.7x - y \leq 125$$

$$3.7(66) - 130 \leq 125$$

$$114.2 \leq 125, \text{ true}$$

Point $B$ is a solution of the system.

**6.** Graph the equation $x - 3y = 6$ as a dashed line.

Test $(0, 0)$
———————

$$x - 3y < 6$$

$$(0) - 3(0) < 6$$

$$0 < 6, \text{ true}$$

Since the statement is true, shade the half-plane containing the point.
Graph the equation $2x + 3y = -6$ as a solid line.

Test $(0, 0)$
———————

$$2x + 3y \geq -6$$

$$2(0) + 3(0) \geq -6$$

$$0 \geq -6, \text{ true}$$

Since the statement is true, shade the half-plane containing the point.
For the solution graph, place an open circle at the point of intersection and shade the region that satisfies both inequalities.

$$\begin{cases} x - 3y < 6 \\ 2x + 3y \geq -6 \end{cases}$$

**7.** Begin by graphing $y = x^2 - 4$ as a solid parabola with vertex $(0, -4)$ and $x$-intercepts $(-2, 0)$ and $(2, 0)$. Since $(0, 0)$ makes the inequality $y \geq x^2 - 4$ true, shade the region containing $(0, 0)$. Graph $x + y = 2$ as a solid line by using its $x$-intercept, $(2, 0)$, and its $y$-intercept $(0, 2)$. Since $(0, 0)$ makes the inequality $x + y \leq 2$ true, shade the region containing $(0, 0)$.

$$\begin{cases} y \geq x^2 - 4 \\ x + y \leq 2 \end{cases}$$

**8.** Graph the lines $x + y = 2$, $x = 1$, and $y = -3$ with dashed lines.
Graph the line $x = -2$ with a solid line.
Test points indicate that the solution contains the region to the right of $-2$, to the left of 1, above $-3$, and below the line $x + y = 2$. The corner points are represented as open circles because none satisfy all three inequalities.

$$\begin{cases} x + y < 2 \\ -2 \leq x < 1 \\ y > -3 \end{cases}$$

**Concept and Vocabulary Check 5.5**

**1.** solution; $x$; $y$; $5 > 1$

**2.** graph

**3.** half-plane

**4.** false

**5.** true

**6.** false

**7.** $(0, 0)$ answers will vary

**8.** $x - y < 1$; $2x + 3y \geq 12$

**9.** false

**Exercise Set 5.5**

**1.** Graph $x + 2y = 8$ as a solid line using its $x$-intercept, $(8, 0)$, and its $y$-intercept, $(0, 4)$.
Test $(0, 0)$:
$0 + 2(0) \leq 8$?
$0 \leq 8$ true
Shade the half-plane containing $(0, 0)$.

$$x + 2y \leq 8$$

**3.** Graph $x - 2y = 10$ as a dashed line using its $x$-intercept, $(10, 0)$, and its $y$-intercept, $(0, -5)$.
Test $(0, 0)$:
$0 - 2(0) > 10$?
$\qquad 0 > 10$ false
Shade the half-plane not containing $(0, 0)$.

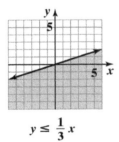

$x - 2y > 10$

**5.** Graph $y = \frac{1}{3}x$ as a solid line using its slope, $\frac{1}{3}$, and its $y$-intercept $(0, 0)$.
Test $(1, 1)$:
$1 \le \frac{1}{3}(1)$?
$1 \le \frac{1}{3}$ false
Shade the half-plane not containing $(1, 1)$.

$y \le \frac{1}{3}x$

**7.** Graph $y = 2x - 1$ as a dashed line using its $x$-intercept, $\left(\frac{1}{2}, 0\right)$ and its $y$-intercept, $(0, -1)$.
Test $(0, 0)$:
$0 > 2(0) - 1$?
$0 > -1$ true
Shade the half-plane containing $(0, 0)$.

$y > 2x - 1$

**9.** Graph $x = 1$ as a solid vertical line.
Test $(0, 0)$:
$0 \le 1$ true
Shade the half-plane containing $(0, 0)$.

$x \le 1$

**11.** Graph $y = 1$ as a dashed horizontal line.
Test $(0, 0)$:
$0 > 1$ false
Shade the half-plane not containing $(0, 0)$.

$y > 1$

**13.** Graph $x^2 + y^2 = 1$ as a solid circle with radius 1 and center $(0, 0)$.
Test $(0, 0)$:
$(0)^2 + (0)^2 \le 1$?
$\qquad 0 \le 1$ true
Shade the region containing $(0, 0)$.

$x^2 + y^2 \le 1$

**15.** Graph $x^2 + y^2 = 25$ as a dashed circle with radius 5 and center (0, 0).
Test (0, 0):
$(0)^2 + (0)^2 > 25?$
$\quad 0 > 25$ false
Shade the region not containing (0, 0).

$x^2 + y^2 > 25$

**17.** Graph $(x-2)^2 + (y+1)^2 = 9$ as a dashed circle.
Test (0,0)
$(x-2)^2 + (y+1)^2 < 9$
$(0-2)^2 + (0+1)^2 < 9$
$\quad\quad 5 < 9,\quad$ true
Shade the region containing (0, 0).

$(x-2)^2 + (y+1)^2 < 9$

**19.** Graph $y = x^2 - 1$ as a dashed parabola with vertex (0, −1) and x-intercepts (1, 0) and (−1, 0).
Test (0, 0):
$0 < (0)^2 - 1?$
$0 < -1$ false
Shade the region not containing (0, 0).

$y < x^2 - 1$

**21.** Graph $y = x^2 - 9$ as a solid parabola with vertex (0, −9) and x-intercepts (3, 0) and (−3, 0).
Test (0, 0):
$0 \geq (0)^2 - 9?$
$0 \geq -9$ true
Shade the region containing (0, 0).

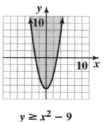

$y \geq x^2 - 9$

**23.** Graph $y = 2^x$ as a dashed exponential function with base 2 that passes through the point (0, 1).
Test (0, 0):
$0 > 2^0?$
$0 > 1$ false
Shade the region not containing (0, 0).

$y > 2^x$

**25.** Graph $y = \log_2(x+1)$ as a solid logarithmic function.
Test (0, 0):
$y \geq \log_2(x+1)$
$0 \geq \log_2(0+1)$
$0 \geq 0,\quad$ true
Shade the region containing (0, 0).

$y \geq \log_2(x+1)$

**27.** Begin by graphing $3x + 6y = 6$ as a solid line using its $x$-intercept, (2, 0), and its $y$-intercept, (0, 1). Since (0, 0) makes the inequality $3x + 6y \le 6$ true, shade the half-plane containing (0, 0). Graph $2x + y = 8$ as a solid line using its $x$-intercept, (4, 0), and its $y$-intercept, (0, 8). Since (0, 0) makes the inequality $2x + y \le 8$ true, shade the half-plane containing (0, 0). The solution set of the system is the intersection of the above shaded half-planes, and is shown as the shaded region in the following graph.

$\begin{cases} 3x + 6y \le 6 \\ 2x + y \le 8 \end{cases}$

**29.** Begin by graphing $2x - 5y = 10$ as a solid line using its $x$-intercept, (5, 0), and its $y$-intercept, (0, –2). Since (0, 0) makes the inequality $2x - 5y \le 10$ true, shade the half-plane containing (0, 0). Graph $3x - 2y = 6$ as a dashed line using its $x$-intercept, (2, 0), and its $y$-intercept, (0, –3). Since (0, 0) makes the inequality $3x - 2y > 6$ false, shade the half-plane containing (0, 0). The solution set of the system is the intersection of the above shaded half-planes, and is shown as the shaded region in the following graph.

$\begin{cases} 2x - 5y \le 10 \\ 3x - 2y > 6 \end{cases}$

**31.** Begin by graphing $y = 2x - 3$ as a dashed line using its slope, 2, and its $y$-intercept, (0, –3). Since (0, 0) makes the inequality $y > 2x - 3$ true, shade the half-plane containing (0, 0). Graph $y = -x + 6$ as a dashed line using its slope, –1, and its $y$-intercept, (0, 6). Since (0, 0) makes the inequality $y < -x + 6$ true, shade the half-plane containing (0, 0). The solution set of the system is the intersection of the above shaded half-planes, and is shown as the shaded region in the following graph.

$\begin{cases} y > 2x - 3 \\ y < -x + 6 \end{cases}$

**33.** Begin by graphing $x + 2y = 4$ as a solid line using its $x$-intercept, (4, 0), and its $y$-intercept, (0, 2). Since (0, 0) makes the inequality $x + 2y \le 4$ true, shade the half-plane containing (0, 0). Graph $y = x - 3$ as a solid line using its slope, 1, and its $y$-intercept, (0, –3). Since (0, 0) makes the inequality $y \ge x - 3$ true, shade the half-plane containing (0, 0). The solution set of the system is the intersection of the above shaded half-planes, and is shown as the shaded region in the following graph.

$\begin{cases} x + 2y \le 4 \\ y \ge x - 3 \end{cases}$

**35.** Begin by graphing $x = 2$ as a solid vertical line . Since (0, 0) makes the inequality $x \leq 2$ true, shade the half-plane containing (0, 0). Graph $y = 1$ as a solid horizontal line. Since (0, 0) makes the inequality $y \geq -1$ true, shade the half-plane containing (0, 0). The solution set of the system is the intersection of the above shaded half-planes, and is shown as the shaded region in the following graph.

$\begin{cases} x \leq 2 \\ y \geq -1 \end{cases}$

**37.** Graph $x = -2$ as a solid vertical line and $x = 5$ as a dashed vertical line. Since (0, 0) makes the inequality $-2 \leq x < 5$ true, shade the region between the two lines.

$-2 \leq x < 5$

**39.** Begin by graphing $x - y = 1$ as a solid line using its x-intercept, (1, 0), and its y-intercept (0, -1). Since (0, 0) makes the inequality $x - y \leq 1$ true, shade the half-plane containing (0, 0). Graph $x = 2$ as a solid horizontal line. Since (0, 0) makes the inequality $x \geq 2$ false, shade the half-plane not containing (0, 0). The solution set of the system is the intersection of the above shaded half-planes, and is shown as the shaded region in the following graph.

$\begin{cases} x - y \leq 1 \\ x \geq 2 \end{cases}$

**41.** $x + y > 4$
$x + y < -1$
Begin by graphing $x + y = 4$ as a dashed line using its x-intercept, (4, 0), and its y-intercept (0, 4). Since (0, 0) makes the inequality $x + y > 4$ false, shade the half-plane not containing (0, 0). Graph $x + y = -1$ as a dashed line using its x-intercept, (-1, 0), and its y-intercept, (0, -1). Since (0, 0) makes the inequality $x + y < -1$ false, shade the half-plane not containing (0, 0). Since these half-planes do not intersect the system has no solution.

**43.** Begin by graphing $x + y = 4$ as a dashed line using its x-intercept, (4, 0), and its y-intercept, (0, 4). Since (0, 0) makes the inequality $x + y > 4$ false, shade the half-plane not containing (0, 0). Graph $x + y = -1$ as a dashed line using its x-intercept, (-1, 0), and its y-intercept, (0, -1). Since (0, 0) makes the inequality $x + y > -1$ true, shade the half-plane containing (0, 0). The solution set of the system is the intersection of the above half-planes, and is shown as the shaded region in the following graph.

$\begin{cases} x + y > 4 \\ x + y > -1 \end{cases}$

**45.** Begin by graphing $y = x^2 - 1$ as a solid parabola with vertex (0, -1) and x-intercepts, (-1, 0), and (1, 0). Since (0, 0) makes the inequality $y \geq x^2 - 1$ true, shade the half-plane containing (0, 0). Graph $x - y = -1$ as a solid line using its x-intercept, (-1, 0), and its y-intercept, (0, 1). Since (0, 0) makes the inequality $x - y \geq -1$ true, shade the half-plane containing (0, 0). The solution set of the system is the intersection of the above half-planes, and is shown as the shaded region in the following graph.

$\begin{cases} y \geq x^2 - 1 \\ x - y \geq -1 \end{cases}$

**47.** Begin by graphing $x^2 + y^2 = 16$ as a solid circle with radius 4 and center, $(0, 0)$. Since $(0, 0)$ makes the inequality $x^2 + y^2 \le 16$ true, shade the half-plane containing $(0, 0)$. Graph $x + y = 2$ as a dashed line using its $x$-intercept, $(2, 0)$, and its $y$-intercept, $(0, 2)$. Since $(0, 0)$ makes the inequality $x + y > 2$ false, shade the half-plane not containing $(0, 0)$. The solution set of the system is the intersection of the above half-planes, and is shown as the shaded region in the following graph.

$$\begin{cases} x^2 + y^2 \le 16 \\ x + y > 2 \end{cases}$$

**49.** Begin by graphing $x^2 + y^2 = 1$ as a dashed circle with radius 1 and center, $(0, 0)$. Since $(0, 0)$ makes the inequality $x^2 + y^2 > 1$ false, shade the half-plane not containing $(0, 0)$. Graph $x^2 + y^2 = 16$ as a dashed circle with radius 4 and center $(0, 0)$. Since $(0, 0)$ makes the inequality $x^2 + y^2 < 16$ true, shade the half-plane containing $(0, 0)$. The solution set of the system is the intersection of the above half-planes, and is shown as the shaded region in the following graph.

$$\begin{cases} x^2 + y^2 > 1 \\ x^2 + y^2 < 16 \end{cases}$$

**51.**

$$\begin{cases} (x - 1)^2 + (y + 1)^2 < 25 \\ (x - 1)^2 + (y + 1)^2 \ge 16 \end{cases}$$

**53.**

$$\begin{cases} x^2 + y^2 \le 1 \\ y - x^2 > 0 \end{cases}$$

**55.**

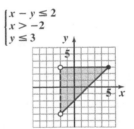

$$\begin{cases} x^2 + y^2 < 16 \\ y \ge 2^x \end{cases}$$

**57.** Begin by graphing $x - y = 2$ as a solid line using its $x$-intercept, $(2, 0)$, and its $y$-intercept, $(0, -2)$. Since $(0, 0)$ makes the inequality $x - y \le 2$ true, shade the half-plane containing $(0, 0)$. Graph $x = -2$ as a solid vertical line. Since $(0, 0)$ makes the inequality $x \ge -2$ true, shade the half-plane containing $(0, 0)$. Graph $y = 3$ as a solid horizontal line. Since $(0, 0)$ makes the inequality $y \le 3$ true, shade the half-plane containing $(0, 0)$. The solution set of the system is the intersection of the above half-planes, and is shown as the shaded region in the following graph.

$$\begin{cases} x - y \le 2 \\ x > -2 \\ y \le 3 \end{cases}$$

**59.** Since $x \geq 0$ and $y \geq 0$ the solution to the system lies in the first quadrant. Graph $2x + 5y = 10$ as a solid line using its $x$-intercept, (5, 0), and its $y$-intercept, (0, 2). Since (0, 0) makes the inequality $2x + 5y \leq 10$ true, shade the half-plane containing (0, 0). Graph $3x + 4y = 12$ as a solid line by using its $x$-intercept, (4, 0), and its $y$-intercept, (0, 3). Since (0, 0) makes the inequality $3x + 4y \leq 12$ true, shade the half-plane containing (0, 0). The solution set of the system is the intersection of the above half-planes which lies in the first quadrant, and is shown as the shaded region in the following graph.

$$\begin{cases} x \geq 0 \\ y \geq 0 \\ 2x + 5y < 10 \\ 3x + 4y \leq 12 \end{cases}$$

**61.** Begin by graphing $3x + y = 6$ as a solid line using its $x$-intercept, (2, 0), and its $y$-intercept, (0, 6). Since (0, 0) makes the inequality $3x + y \leq 6$ true, shade the half-plane containing (0, 0). Graph $2x - y = -1$ as a solid line using its $x$-intercept, $\left(-\frac{1}{2}, 0\right)$, and its $y$-intercept, (0, 1). Since (0, 0) makes the inequality $2x - y \leq -1$ false, shade the half-plane not containing (0, 0). Graph $x = -2$ as a solid vertical line. Since (0, 0) makes the inequality $x \geq -2$ true, shade the half-plane containing (0, 0). Graph $y = 4$ as a solid horizontal line. Since (0, 0) makes the inequality $y \leq 4$ true, shade the half-plane containing (0, 0). The solution set of the system is the intersection of the above half-planes, and is shown as the shaded region in the following graph.

$$\begin{cases} 3x + y \leq 6 \\ 2x - y \leq -1 \\ x > -2 \\ y < 4 \end{cases}$$

**63.** $y \geq -2x + 4$

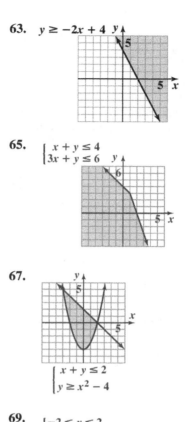

**65.** $\begin{cases} x + y \leq 4 \\ 3x + y \leq 6 \end{cases}$

**67.** $\begin{cases} x + y \leq 2 \\ y \geq x^2 - 4 \end{cases}$

**69.** $\begin{cases} -2 \leq x \leq 2 \\ -3 \leq y \leq 3 \end{cases}$

**71.** Find the union of solutions of $y > \frac{3}{2}x - 2$ and $y < 4$.

**73.** The system $\begin{array}{c} 3x+3y<9 \\ 3x+3y>9 \end{array}$ has no solution.

The number $3x+3y$ cannot both be less than 9 and greater than 9 at the same time.

**75.** The system has an infinite number of solutions. The solution is the set of points that make up the circle $(x+4)^2+(y-3)^2=9$

**77.** Point $A=(66,160)$

$5.3x-y\ge180$

$5.3(66)-160\ge180$

$189.8\ge180,\ \text{true}$

$4.1x-y\le14$

$4.1(66)-160\le140$

$110.6\le140,\ \text{true}$

Point $A$ is a solution of the system.

**79.** Point $=(72,205)$

$5.3x-y\ge180$

$5.3(72)-205\ge180$

$176.6\ge180,\ \text{false}$

$4.1x-y\le14$

$4.1(72)-205\le140$

$90.2\le140,\ \text{true}$

The data does not satisfy both inequalities. The person is not within the healthy weight region.

**81. a.** $50x+150y>2000$

**b.** Graph $50x+150y$ as a dashed line using its $x$-intercept, (40, 0), and its $y$-intercept, $\left(0,\dfrac{40}{3}\right)$.

Test (0, 0):

$50(0)+150(0)>2000$?

$0>2000$ false

Shade the half-plane not containing (0, 0).

$50x+150y>2000$

**c.** Ordered pairs may vary.

**83. a.** $\begin{cases} y\ge0 \\ x+y\ge5 \\ x\ge1 \\ 200x+100y\le700 \end{cases}$

**b.** $\begin{cases} y\ge0 \\ x+y\ge5 \\ x\ge1 \\ 200x+100y\le700 \end{cases}$

**c.** 2 nights

**85. a.** $\text{BMI}=\dfrac{703W}{H^2}=\dfrac{703(200)}{72^2}\approx27.1$

**b.** A 20 year old man with a BMI of 27.1 is classified as overweight.

**87. – 95.** Answers will vary.

**97.**

**99.** $y\ge x^2-4$

**101.** $2x + y \le 6$

**103. – 105.** Answers will vary.

**107.** does not make sense; Explanations will vary. Sample explanation: It is necessary to graph the linear equation with a dashed line to represent its role as a borderline.

**109.** makes sense

**111.** $y > x - 3$
$y \le x$

**113.** $x + 2y \le 6$ or $2x + y \le 6$

**115.** $y \ge nx + b$ $(n < 0,\ b > 0)$
$y \le mx + b$ $(m > 0,\ b > 0)$

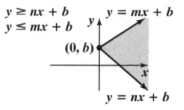

**116. a.** $x + y \ge 6$
$x \le 8$
$y \le 5$

**b.** The corner points are $(1,5)$, $(8,5)$, and $(8,-2)$.

**c.** At $(1,5)$, $3x + 2y = 3(1) + 2(5) = 13$.
At $(8,5)$, $3x + 2y = 3(8) + 2(5) = 34$.
At $(8,-2)$, $3x + 2y = 3(8) + 2(-2) = 20$.

**117. a.** $x \ge 0$
$y \ge 0$
$3x - 2y \le 6$
$y \le -x + 7$

**b.** The corner points are $(0,0)$, $(2,0)$, $(4,3)$, and $(0,7)$.

**c.** At $(0,0)$, $2x + 5y = 2(0) + 5(0) = 0$.
At $(2,0)$, $2x + 5y = 2(2) + 5(0) = 4$.
At $(4,3)$, $2x + 5y = 2(4) + 5(3) = 23$.
At $(0,7)$, $2x + 5y = 2(0) + 5(7) = 35$.

**118.** $20x + 10y \le 80{,}000$

**Section 5.6**

**Check Point Exercises**

**1.** The total profit is 25 times the number of bookshelves, $x$, plus 55 times the number of desks, $y$. The objective function is $z = 25x + 55y$

**2.** Not more than a total of 80 bookshelves and desks can be manufactured per day. This is represented by the inequality $x + y \le 80$.

**3.** Objective function: $z = 25x + 55y$
Constraints: $\begin{cases} x + y \le 80 \\ 30 \le x \le 80 \\ 10 \le y \le 30 \end{cases}$

**4.** Graph the constraints and find the corners, or vertices, of the region of intersection.

Find the value of the objective function at each corner of the graphed region.

| Corner $(x, y)$ | Objective Function $z = 25x + 55y$ z |
|---|---|
| (30, 10) | $z = 25(30) + 55(10)$ $= 750 + 550 = 1300$ |
| (30, 30) | $z = 25(30) + 55(30)$ $= 750 + 1650 = 2400$ |
| (50, 30) | $z = 25(50) + 55(30)$ $= 1250 + 1650 = 2900 \leftarrow$ Maximum |
| (70, 10) | $z = 25(70) + 55(10)$ $= 1750 + 550 = 2300$ |

The maximum value of $z$ is 2900 and it occurs at the point (50, 30).
In order to maximize profit, 50 bookshelves and 30 desks must be produced each day for a profit of $2900.

**5.** objective function: $z = 3x + 5y$
constraints: $x \geq 0$, $y \geq 0$
$$x + y \geq 1$$
$$x + y \leq 6$$

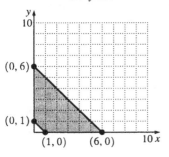

Evaluate the objective function at the four vertices of
the region shown:

(1, 0): $3(1) + 5(0) = 3$
(0, 1): $3(0) + 5(1) = 5$
(0, 6): $3(0) + 5(6) = 30$
(6, 0): $3(6) + 5(0) = 18$
The maximum value of $z$ is 30 and this occurs when
$x = 0$ and $y = 6$.

**Concept and Vocabulary Check 5.6**

**1.** linear programming

**2.** objective

**3.** constraints; corner

Copyright © 2014 Pearson Education, Inc.

**Exercise Set 5.6**

**1.**  $z = 5x + 6y$

(1, 2): 5(1) + 6(2) = 5 + 12 = 17
(2, 10): 5(2) + 6(10) = 10 + 60 = 70
(7, 5): 5(7) + 6(5) = 35 + 30 = 65
(8, 3): 5(8) + 6(3) = 40 + 18 = 58
The maximum value is $z = 70$; the minimum value is $z = 17$.

**3.**  $z = 40x + 50y$

(0, 0): 40(0) + 50(0) = 0 + 0 = 0
(0, 8): 40(0) + 50(8) = 0 + 400 = 400
(4, 9): 40(4) + 50(9) = 160 + 450 = 610
(8, 0): 40(8) + 50(0) = 320 + 0 = 320
The maximum value is $z = 610$; the minimum value is $z = 0$.

**5.**  $z = 3x + 2y$

$x \geq 0, y \geq 0$

$2x + y \leq 8$

$x + y \geq 4$

**a.**

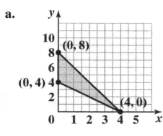

**b.**  $(0,8): z = 3(0) + 2(8) = 16$

$(0,4): z = 3(0) + 2(4) = 8$

$(4,0): z = 3(4) + 2(0) = 12$

**c.**  The maximum value is 16 at $x = 0$ and $y = 8$.

**7.**  $z = 4x + y$
$x \geq 0, y \geq 0$
$2x + 3y \leq 12$
$x + y \geq 3$

**a.**

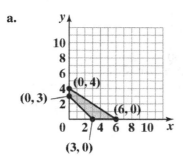

**b.**  (0, 4): $z = 4(0) + 4 = 4$
(0, 3): $z = 4(0) + 3 = 3$
(3, 0): $z = 4(3) + 0 = 12$
(6, 0): $z = 4(6) + 0 = 24$

**c.**  The maximum value is 24 at $x = 6$ and $y = 0$.

**9.**  $z = 3x - 2y$

$1 \leq x \leq 5$

$y \geq 2$

$x - y \geq -3$

**a.**

**b.**  (1, 2): $z = 3(1) - 2(2) = -1$
(1, 4): $z = 3(1) - 2(4) = -5$
(5, 8): $z = 3(5) - 2(8) = -1$
(5, 2): $z = 3(5) - 2(2) = 11$

**c.**  Maximum value is 11 at $x = 5$ and $y = 2$.

**11.**  $z = 4x + 2y$
$x \geq 0, y \geq 0$
$2x + 3y \leq 12$
$3x + 2y \leq 12$
$x + y \geq 2$

**a.**

**b.**  (0, 4): $z = 4(0) + 2(4) = 8$
(0, 2): $z = 4(0) + 2(2) = 4$
(2, 0): $z = 4(2) + 2(0) = 8$
(4, 0): $z = 4(4) + 2(0) = 16$
$\left(\dfrac{12}{5}, \dfrac{12}{5}\right): z = 4\left(\dfrac{12}{5}\right) + 2\left(\dfrac{12}{5}\right)$

$\qquad = \dfrac{48}{5} + \dfrac{24}{5} = \dfrac{72}{5}$

**c.**  The maximum value is 16 at $x = 4$ and $y = 0$.

**13.** $z = 10x + 12y$
$x \geq 0, y \geq 0$
$x + y \leq 7$
$2x + y \leq 10$
$2x + 3y \leq 18$

**a.**

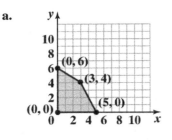

**b.** $(0, 6)$: $z = 10(0) + 12(6) = 72$
$(0, 0)$: $z = 10(0) + 12(0) = 0$
$(5, 0)$: $z = 10(5) + 12(0) = 50$
$(3, 4)$: $z = 10(3) + 12(4)$
$= 30 + 48 = 78$

**c.** The maximum value is 78 at $x = 3$ and $y = 4$.

**15. a.** $z = 125x + 200y$

**b.** $x \leq 450$
$y \leq 200$
$600x + 900y \leq 360,000$

**c.** Simplify the third inequality by dividing by 300 to get $2x + 3y \leq 1200$.

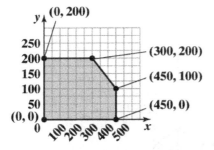

**d.** $(0, 0)$: $125(0) + 200(0) = 0 + 0 = 0$
$(0, 200)$: $125(0) + 200(200)$
$= 0 + 40,000 = 40,000$
$(300, 200)$: $125(300) + 200(200)$
$= 37,500 + 40,000 = 77,500$
$(450, 100)$: $125(450) + 200(100)$
$= 56,250 + 20,000 = 76,250$
$(450, 0)$: $125(450) + 200(0)$
$= 56,250 + 0 = 56,250$

**e.** The television manufacturer will make the greatest profit by manufacturing <u>300</u> rear-projection televisions each month and <u>200</u> plasma televisions each month. The maximum monthly profit is <u>$77,500</u>.

**17.** Let $x$ = number of model *A* bicycles and $y$ = number of model *B* bicycles.
The constraints are
$5x + 4y \leq 200$
$2x + 3y \leq 108$
Graph these inequalities in the first quadrant, since $x$ and $y$ cannot be negative.

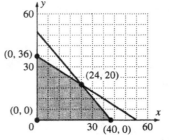

The quantity to be maximized is the profit, which is $25x + 15y$.
$(0, 0)$: $25(0) + 15(0) = 0 + 0 = 0$
$(0, 36)$: $25(0) + 15(36) = 0 + 540 = 540$
$(24, 20)$: $25(24) + 15(20) = 600 + 300 = 900$
$(40, 0)$: $25(40) + 15(0) = 1000 + 0 = 1000$
40 model *A* bicycles and 0 model *B* bicycles should be produced.

**19.** Let $x$ = the number of cartons of food and $y$ = the number of cartons of clothing.
The constraints are:
$20x + 10y \leq 8,000$ or $2x + y \leq 8000$
$50x + 20y \leq 19,000$ or $5x + 2y \leq 1900$
Graph these inequalities in the first quadrant, since $x$ and $y$ cannot be negative.

The quantity to be maximized is the number of people helped, which is $12x + 5y$.
$(0, 0)$: $12(0) + 5(0) = 0 + 0 = 0$
$(0, 800)$: $12(0) + 5(800) = 0 + 4000 = 4000$
$(300, 200)$: $12(300) + 5(200) = 4600$
$(380, 0)$: $12(380) + 5(0) = 4500$
300 cartons of food and 200 cartons of clothing should be shipped. This will help 4600 people.

**21.** Let $x$ = number of students attending and
$y$ = number of parents attending.
The constraints are
$x + y \le 150$
$\quad 2x \ge y$
or
$\quad x + y \le 150$
$2x - y \ge 0$
Graph these inequalities in the first quadrant, since $x$
and $y$ cannot be negative.

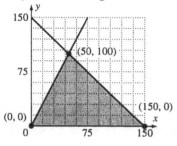

The quantity to be maximized is the amount of
money raised, which is $x + 2y$.
$(0, 0)$: $0 + 2(0) = 0 + 0 = 0$
$(50, 100)$: $50 + 2(100) = 50 + 200 = 250$
$(150, 0)$: $150 + 2(0) = 150 + 0 = 150$
50 students and 100 parents should attend.

**23.** Let $x$ = number of Boeing 727s, $y$ = number of
Falcon 20s.
Maximize $z = x + y$ with the following constraints:
$1400x + 500y \le 35,000$ or $14x + 5y \le 350$

$42,000x + 6000y \ge 672,000$ or $7x + y \ge 112$

$x \le 20$

$x \ge 0, y \ge 0$

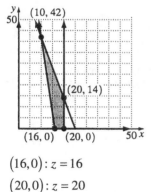

$(16, 0)$: $z = 16$

$(20, 0)$: $z = 20$

$(20, 14)$: $z = 34$

$(10, 42)$: $z = 52$

Federal Express should have purchased 10 Boeing
727s and 42 Falcon 20s.

**25. – 27.** Answers will vary.

**29.** does not make sense; Explanations will vary.
Sample explanation: Solving a linear programming
problem does not require graphing the objective
function.

**31.** makes sense

**33.** Let $x$ = amount invested in stocks and
$y$ = amount invested in bonds.
The constraints are:
$x + y \le 10,000$
$\quad y \ge 3000$
$\quad x \ge 2000$
$\quad y \ge x$
Graph these inequalities in the first quadrant, since $x$
and $y$ cannot be negative.

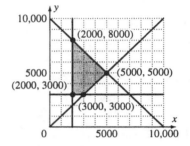

The quantity to be maximized is the return on the
investment, which is $0.12x + 0.08y$.
$(2000, 3000)$:
$0.12(2000) + 0.08(3000) = 240 + 240 = 480$
$(2000, 8000)$:
$0.12(2000) + 0.08(8000) = 240 + 640 = 880$
$(5000, 5000)$:
$0.12(5000) + 0.08(5000) = 600 + 400 = 1000$
$(3000, 3000)$:
$0.12(3000) + 0.08(3000) = 360 + 240 = 600$
The greatest return occurs when \$5000 is invested in
stocks and \$5000 is invested in bonds.

**35.** Answers will vary.

**37.** Back-substitute $z = 5$ to find $y$ in the second
equation.
$$y + 2z = 13$$
$$y + 2(5) = 13$$
$$y + 10 = 13$$
$$y = 3$$
Back-substitute to find $x$ in the first equation.
$$x + y + 2z = 19$$
$$x + 3 + 2(5) = 19$$
$$x + 13 = 19$$
$$x = 6$$
The solution set is $\{(6, 3, 5)\}$.
Explanations may vary.

**38.** Back-substitute $z = 3$ to find $y$ in the third equation.

$$y - z = 1$$
$$y - 3 = 1$$
$$y = 4$$

Back-substitute to find $x$ in the second equation.

$$x - \frac{1}{3}y + z = \frac{8}{3}$$
$$x - \frac{1}{3}(4) + 3 = \frac{8}{3}$$
$$x - \frac{4}{3} + \frac{9}{3} = \frac{8}{3}$$
$$x + \frac{5}{3} = \frac{8}{3}$$
$$x = \frac{3}{3}$$
$$x = 1$$

Back-substitute to find $w$ in the first equation.

$$w - x + 2y - 2z = -1$$
$$w - 1 + 2(4) - 2(3) = -1$$
$$w - 1 + 8 - 6 = -1$$
$$w + 1 = -1$$
$$w = -2$$

The solution set is $\{(-2, 1, 4, 3)\}$.
Explanations may vary.

**39.** $\begin{bmatrix} 1 & 2 & -1 \\ 1(-4)+4 & 2(-4)+(-3) & -1(-4)+(-15) \end{bmatrix}$

$= \begin{bmatrix} 1 & 2 & -1 \\ 0 & -11 & -11 \end{bmatrix}$

## Chapter 5 Review Exercises

**1.** $$y = 4x + 1$$
$$3x + 2y = 13$$

Substitute $4x + 1$ for $y$ in the second equation:

$$3x + 2(4x + 1) = 13$$
$$3x + 8x + 2 = 13$$
$$11x = 11$$
$$x = 1$$
$$y = 4(1) + 1 = 5$$

The solution set is $\{(1,5)\}$.

**2.** $$x + 4y = 14$$
$$2x - y = 1$$

Multiply the second equation by 4 and add to the first equation.

$$x + 4y = 14$$
$$\underline{8x - 4y = 4}$$
$$9x = 18$$
$$x = 2$$
$$2(2) - y = 1$$
$$-y = -3$$
$$y = 3$$

The solution set is $\{(2,3)\}$.

**3.** $$5x + 3y = 1$$
$$3x + 4y = -6$$

Multiply the first equation by 4 and the second equation by $-3$.
Then add.

$$20x + 12y = 4$$
$$\underline{-9x - 12y = 18}$$
$$11x = 22$$
$$x = 2$$
$$5(2) + 3y = 1$$
$$3y = -9$$
$$y = -3$$

The solution set is $\{(2, -3)\}$.

**4.** $$2y - 6x = 7$$
$$3x - y = 9$$

The second equation can be written as $y = 3x - 9$.
Substitute:

$$2(3x - 9) - 6x = 7$$
$$6x - 18 - 6x = 7$$
$$-18 = 7$$

Since this is false, the system has no solution.
The solution set is the empty set, $\varnothing$.

**5.** $$4x - 8y = 16$$
$$3x - 6y = 12$$

Divide the first equation by 4 and the second equation by 3.

$$x - 2y = 4$$
$$x - 2y = 4$$

Since these equations are identical, the system has an infinite number of solutions.
The solution set is $\{(x, y) \mid 3x - 6y = 12\}$.

**6. a.** $C(x) = 60,000 + 200x$

**b.** $R(x) = 450x$

**c.** $450x = 60000 + 200x$
$250x = 60000$
$x = 240$
$450(240) = 108,000$
The company must make 240 desks at a cost of $108,000 to break even.

**7.** Let $x =$ the selling price for Klint's work.
Let $y =$ the selling price for Picasso's work.
$x + y = 239$
$x - y = 31$
Add the equations to eliminate $y$ and solve for $x$.
$x + y = 239$
$\underline{x - y = \phantom{0}31}$
$2x \phantom{00} = 270$
$x = 135$
Back-substitute to find $y$.
$x + y = 239$
$135 + y = 239$
$y = 104$
Klint's work sold for $135 million and Picasso's work sold for $104 million.

**8. a.** $y = 219x + 142$

**b.** Using substitution,
$95x + y = 770$
$95x + \overbrace{219x + 142}^{y} = 770$
$314x + 142 = 770$
$314x = 628$
$x = 2$
Digital track sales caught up with album sales 2 years after 2004, or 2006.
$y = 219x + 142$
$y = 219(2) + 142$
$= 580$
In 2006 there were 580 million digital tracks and 580 million albums sold.

**9.** Let $l =$ the length of the table.
Let $w =$ the width of the table.
$2l + 2w = 34$
$4l - 3w = 33$
Multiply the first equation by $-2$ and solve by addition.
$-4l - 4w = -68$
$\underline{4l - 3w = \phantom{0}33}$
$-7w = -35$
$w = 5$
Back-substitute 5 for $w$ to find $l$.
$2l + 2w = 34$
$2l + 2(5) = 34$
$2l + 10 = 34$
$2l = 24$
$l = 12$
The dimensions of the table are 12 feet by 5 feet.

**10.** Let $x =$ the cost of the hotel
$y =$ the cost of the car
$3x + 2y = 360$
$4x + 3y = 500$
Solve the system.
$12x + 8y = 1440$
$-12x - 9y = -1500$
$-y = -60$
$y = 60$

$3x + 2(60) = 360$
$3x = 240$
$x = 80$
The room costs $80 a day and the car rents for $60 a day.

**11.** $x =$ number of apples
$y =$ number of avocados
$100x + 350y = 1000$
$24x + 14y = 100$
$100x + 350y = 1000$
$\underline{-600x - 350y = -2500}$
$-500x = -1500$
$x = 3$
$100(3) + 350y = 1000$
$350y = 700$
$y = 2$
3 apples and 2 avocados supply 1000 calories and 100 grams of carbohydrates.

**12.**
$$2x - y + z = 1 \,(1)$$
$$3x - 3y + 4z = 5 \,(2)$$
$$4x - 2y + 3z = 4 \,(3)$$
Eliminate $y$ from (1) and (2) by multiplying (1) by –3 and adding the result to (2).
$$-6x + 3y - 3z = -3$$
$$\underline{3x - 3y + 4z = 5}$$
$$-3x + z = 2 \ (4)$$
Eliminate $y$ from (1) and (3) by multiplying (1) by –2 and adding the result to (3).
$$-4x + 2y - 2z = -2$$
$$\underline{4x - 2y + 3z = 4}$$
$$z = 2$$
Substituting $z = 2$ into (4), we get:
$$-3x + 2 = 2$$
$$-3x = 0$$
$$x = 0$$
Substituting $x = 0$ and $z = 2$ into (1), we have:
$$2(0) - y + 2 = 1$$
$$-y = -1$$
$$y = 1$$
The solution set is $\{(0, 1, 2)\}$.

**13.**
$$x + 2y - z = 5 \ (1)$$
$$2x - y + 3z = 0 \ (2)$$
$$2y + z = 1 \ (3)$$
Eliminate $x$ from (1) and (2) by multiplying (1) by –2 and adding the result to (2).
$$-2x - 4y + 2z = -10$$
$$\underline{2x - y + 3z = 0}$$
$$-5y + 5z = -10$$
$$y - z = 2 \ (4)$$
Adding (3) and (4), we get:
$$2y + z = 1$$
$$\underline{y - z = 2}$$
$$3y = 3$$
$$y = 1$$
Substituting $y = 1$ into (3), we have:
$$2(1) + z = 1$$
$$z = -1$$
Substituting $y = 1$ and $z = -1$ into (1), we obtain:
$$x + 2(1) - (-1) = 5$$
$$x + 3 = 5$$
$$x = 2$$
The solution set is $\{(2, 1, -1)\}$.

**14.**
$$y = ax^2 + bx + c$$
$$(1,4): 4 = a + b + c \qquad (1)$$
$$(3,20): 20 = 9a + 3b + c \qquad (2)$$
$$(-2,25): 25 = 4a - 2b + c \quad (3)$$
Multiply (1) by –1 and add to (2).
$$20 = 9a + 3b + c$$
$$\underline{-4 = -a - b - c}$$
$$16 = 8a + 2b$$
$$8 = 4a + b$$
$$8 = 4a + b \qquad\qquad (4)$$
Multiply (1) by –1 and add to (3).
$$25 = 4a - 2b + c$$
$$\underline{-4 = -a - b - c}$$
$$21 = 3a - 3b$$
$$7 = a - b \qquad\qquad (5)$$
Add (4) and (5).
$$8 = 4a + b$$
$$\underline{7 = a - b}$$
$$15 = 5a$$
$$a = 3$$
$$8 = 4(3) + b$$
$$b = -4$$
$$3 - 4 + c = 4$$
$$c = 5$$
Hence, the quadratic function is $y = 3x^2 - 4x + 5$.

**15.** Let $x$ = number of deaths, in millions, in the 20$^{\text{th}}$ century from war.
Let $y$ = number of deaths, in millions, in the 20$^{\text{th}}$ century from famine.
Let $z$ = number of deaths, in millions, in the 20$^{\text{th}}$ century from tobacco.
$$x + y + z = 306$$
$$x - y = 13$$
$$x - z = 53$$
Solve the second equation for $y$.
$$x - y = 13$$
$$-y = -x + 13$$
$$y = x - 13$$
Solve the third equation for $z$.
$$x - z = 53$$
$$-z = -x + 53$$
$$z = x - 53$$
Substitute the expressions for $x$ and $z$ into the first equation and solve for $y$.

$$x \ + \ y \ + \ z \ = 306$$

$$x+\overbrace{(x-13)}^{y}+\overbrace{(x-53)}^{z}=306$$

$$x+x-13+x-53=306$$

$$3x-66=306$$

$$3x=372$$

$$x=124$$

Back-substitute to solve for $y$ and $z$.

$$y=x-13$$
$$=124-13$$
$$=111$$
$$z=x-53$$
$$=124-53$$
$$=71$$

The number of deaths in the 20th century from war is 124 million, from famine is 111 million, and from tobacco is 71 million.

**16.** $$\frac{x}{(x-3)(x+2)}=\frac{A}{x-3}+\frac{B}{x+2}$$

$$x=A(x+2)+B(x-3)$$
$$=(A+B)x+(2A-3B)$$
$$A+B=1$$
$$2A-3B=0$$

Multiply first equation by 3, then add to second equation.

$$3A+3B=3$$
$$\underline{2A-3B=0}$$
$$5A=3$$

$$A=\frac{3}{5},\ B=\frac{2}{5}$$

$$\frac{x}{(x-3)(x+2)}=\frac{3}{5(x-3)}+\frac{2}{5(x+2)}$$

**17.** $$\frac{11x-2}{x^2-x-12}=\frac{11x-2}{(x-4)(x+3)}=\frac{A}{x-4}+\frac{B}{x+3}$$

$$11x-2=A(x+3)+B(x-4)$$
$$=Ax+3A+Bx-4B$$
$$=(A+B)x+(3A-4B)$$
$$A+B=11$$
$$3A-4B=-2$$

Multiply first equation by 4, then add to second equation.

$$3A-4B=-2$$
$$\underline{4A+4B=44}$$
$$7A=42$$

$$A=6,\ B=5$$

$$\frac{11x-2}{x^2-x-12}=\frac{6}{x-4}+\frac{5}{x+3}$$

**18.** $$\frac{4x^2-3x-4}{x^3+x^2-2x}=\frac{4x^2-3x-4}{x(x+2)(x-1)}$$

$$=\frac{A}{x}+\frac{B}{x+2}+\frac{C}{x-1}$$

$$4x^2-3x-4=A(x+2)(x-1)+Bx(x-1)+Cx(x+2)$$
$$=A(x^2+x-2)+Bx^2-Bx+Cx^2+2Cx$$
$$=Ax^2+Ax-2A+Bx^2-Bx+Cx^2+2Cx$$
$$=(A+B+C)x^2+(A-B+2C)x-2A$$

$$A+B+C=4$$
$$A-B+2C=-3$$
$$-2A=-4$$
$$A=2$$
$$B+C=2$$
$$\underline{-B+2C=-5}$$
$$3C=-3$$
$$C=-1$$
$$B-1=2$$
$$B=3$$

$$\frac{4x^2-3x-4}{x^3+x^2-2x}=\frac{2}{x}+\frac{3}{x+2}-\frac{1}{x-1}$$

**19.** $\dfrac{2x+1}{(x-2)^2} = \dfrac{A}{x-2} + \dfrac{B}{(x-2)^2}$

$2x+1 = A(x-2)+B = Ax-2A+B$

$A = 2$

$-2A+B = 1$

$-2(2)+B = 1$

$B = 5$

$\dfrac{2x+1}{(x-2)^2} = \dfrac{2}{x-2} + \dfrac{5}{(x-2)^2}$

**20.** $\dfrac{2x-6}{(x-1)(x-2)^2} = \dfrac{A}{x-1} + \dfrac{B}{x-2} + \dfrac{C}{(x-2)^2}$

$2x-6 = A(x-2)^2 + B(x-1)(x-2) + C(x-1)$

$\quad = A(x^2-4x+4)+B(x^2-3x+2)+C(x-1)$

$\quad = Ax^2-4Ax+4A+Bx^2-3Bx+2B+Cx-C$

$\quad = (A+B)x^2 + (-4A-3B+C)x + (4A+2B-C)$

$A+B = 0$

$-4A-3B+C = 2$

$\underline{4A+2B-C = -6}$

$\quad -B = -4$

$\quad B = 4$

$\quad A = -4$

$4(-4)+2(4)-C = -6$

$-16+8-C = -6$

$-C-8 = -6$

$-C = 2$

$C = -2$

$\dfrac{2x-6}{(x-1)(x-2)^2} = -\dfrac{4}{x-1} + \dfrac{4}{x-2} - \dfrac{2}{(x-2)^2}$

**21.** $\dfrac{3x}{(x-2)(x^2+1)} = \dfrac{A}{x-2} + \dfrac{Bx+C}{x^2+1}$

$3x = A(x^2+1)+(Bx+C)(x-2)$

$\quad = Ax^2+A+Bx^2-2Bx+Cx-2C$

$\quad = (A+B)x^2+(-2B+C)x-(2C-A)$

$A+B = 0$

$-2B+C = 3$

$2C-A = 0$

$A = 2C$

$B+2C = 0$

$\underline{4B-2C = -6}$

$5B = -6$

$B = -\dfrac{6}{5}$

$A = \dfrac{6}{5}$

$C = \dfrac{6}{10} = \dfrac{3}{5}$

$\dfrac{3x}{(x-2)(x^2+1)} = \dfrac{6}{5(x-2)} + \dfrac{-6x+3}{5(x^2+1)}$

**22.** $\dfrac{7x^2-7x+23}{(x-3)(x^2+4)} = \dfrac{A}{x-3} + \dfrac{Bx+C}{x^2+4}$

$7x^2-7x+23 = A(x^2+4)+(Bx+C)(x-3)$

$\quad = Ax^2+4A+Bx^2-3Bx+Cx-3C$

$\quad = (A+B)x^2+(-3B+C)x+(4A-3C)$

$A+B = 7$

$-3B+C = -7$

$4A-3C = 23$

$3A+3B = 21$

$\underline{-3B+C = -7}$

$3A+C = 14$

$9A+3C = 42$

$\underline{4A-3C = 23}$

$13A = 65$

$A = 5$

$5+B = 7$

$B = 7-5 = 2$

$-3(2)+C = -7$

$C = -7+6 = -1$

$\dfrac{7x^2-7x+23}{(x-3)(x^2+4)} = \dfrac{5}{x-3} + \dfrac{2x-1}{x^2+4}$

**23.** $\dfrac{x^3}{(x^2+4)^2} = \dfrac{Ax+B}{x^2+4} + \dfrac{Cx+D}{(x^2+4)^2}$

$x^3 = (Ax+B)(x^2+4)+Cx+D$

$\quad = Ax^3+4Ax+Bx^2+4B+Cx+D$

$\quad = Ax^3+Bx^2+(4A+C)x+(4B+D)$

$\quad A=1$

$\quad B=0$

$4A+C=0$

$4B+D=0$

$\quad C=-4$

$0+D=0, D=0$

$\dfrac{x^2}{(x^2+4)^2} = \dfrac{x}{x^2+4} - \dfrac{4x}{(x^2+4)^2}$

**24.** $\dfrac{4x^3+5x^2+7x-1}{(x^2+x+1)^2} = \dfrac{Ax+B}{x^2+x+1} + \dfrac{Cx+D}{(x^2+x+1)^2}$

$4x^3+5x^2+7x-1$

$= (Ax+B)(x^2+x+1)+Cx+D$

$= Ax^3+Ax^2+Ax+Bx^2+Bx+B+Cx+D$

$= Ax^3+(A+B)x^2(A+B+C)x+(B+D)$

$\quad A=4$

$\quad A+B=5$

$\quad A+B+C=7$

$\quad B+D=-1$

$\quad 4+B=5, B=1$

$\quad 4+1+C=7, C=2$

$\quad 1+D=-1, D=-2$

$\dfrac{4x^3+5x^2+7x-1}{(x^2+x+1)^2} = \dfrac{4x+1}{x^2+x+1} + \dfrac{2x-2}{(x^2+x+1)^2}$

**25.** $\quad\quad 5y=x^2-1$

$\quad\quad x-y=1$

$\quad\quad y=x-1$

$\quad 5(x-1)=x^2-1$

$\quad 5x-5=x^2-1$

$x^2-5x+4=0$

$(x-4)(x-1)=0$

$\quad\quad x=4,1$

If $x=4$, $y=4-1=3$.

If $x=1$, $y=1-1=0$.

The solution set is $\{(4,3),(1,0)\}$.

**26.** $\quad y=x^2+2x+1$

$\quad x+y=1$

$\quad\quad y=1-x$

$\quad 1-x=x^2+2x+1$

$\quad x^2+3x=0$

$\quad x(x+3)=0$

$\quad\quad x=0,-3$

If $x=0$, $y=1-0=1$.

If $x=-3$, $y=1-(-3)=4$.

The solution set is $\{(0,1),(-3,4)\}$.

**27.** $\quad x^2+y^2=2$

$\quad x+y=0$

$\quad\quad x=-y$

$(-y)^2+y^2=2$

$\quad 2y^2=2$

$\quad y^2=1$

$\quad y=1,-1$

If $y=1, x=-1$.

If $y=-1, x=1$.

The solution set is $\{(1,-1),(-1,1)\}$.

**28.** $\quad 2x^2+y^2=24$

$\quad x^2+y^2=15$

$\quad 2x^2+y^2=24$

$\quad \underline{-x^2-y^2=-15}$

$\quad\quad x^2=9$

$\quad\quad x=3,-3$

If $x=3, 3^2+y^2=15$, $y^2=6$ and $y=\pm\sqrt{6}$.

If $x=-3, y=\pm\sqrt{6}$.

The solution set is

$\left\{\left(3,\sqrt{6}\right),\left(3,-\sqrt{6}\right),\left(-3,\sqrt{6}\right),\left(-3,-\sqrt{6}\right)\right\}$.

**29.** $xy-4=0$

$\quad y-x=0$

$\quad\quad y=x$

$\quad\quad xy=4$

$\quad\quad x^2=4$

$\quad\quad x=2,-2$

If $x=2, y=2$.

If $x=-2, y=-2$.

The solution set is $\{(2,2),(-2,-2)\}$.

**30.**
$$y^2 = 4x$$
$$x - 2y + 3 = 0$$
$$x = \frac{y^2}{4}$$
$$\frac{y^2}{4} - 2y + 3 = 0$$
$$y^2 - 8y + 12 = 0$$
$$(y - 6)(y - 2) = 0$$
$$y = 6, 2$$

If $y = 6, x = \frac{36}{4} = 9.$

If $y = 2, x = \frac{4}{4} = 1.$

The solution set is $\{(9,6),(1,2)\}$.

**31.** $x^2 + y^2 = 10$
$$y = x + 2$$
$$x^2 + (x + 2)^2 = 10$$
$$x^2 + x^2 + 4x + 4 - 10 = 0$$
$$2x^2 + 4x - 6 = 0$$
$$x^2 + 2x - 3 = 0$$
$$(x + 3)(x - 1) = 0$$
$$x = -3, 1$$

If $x = -3, y = -3 + 2 = -1.$
If $x = 1, y = 1 + 2 = 3.$
The solution set is $\{(-3,-1),(1,3)\}$.

**32.** $xy = 1$
$$y = 2x + 1$$
$$x(2x + 1) = 1$$
$$2x^2 + x - 1 = 0$$
$$(2x - 1)(x + 1) = 0$$
$$x = \frac{1}{2}, -1$$

If $x = \frac{1}{2}, y = 2\left(\frac{1}{2}\right) + 1 = 2.$

If $x = -1, y = 2(-1) + 1 = -1.$

The solution set is $\left\{\left(\frac{1}{2}, 2\right), (-1, -1)\right\}$.

**33.**
$$x + y + 1 = 0$$
$$x^2 + y^2 + 6y - x = -5$$
$$x = -y - 1$$
$$(-y - 1)^2 + y^2 + 6y - (-y - 1) + 5 = 0$$
$$y^2 + 2y + 1 + y^2 + 6y + y + 1 + 5 = 0$$
$$2y^2 + 9y + 7 = 0$$
$$(2y + 7)(y + 1) = 0$$
$$y = -\frac{7}{2}, -1$$

If $y = -\frac{7}{2}, x = \frac{7}{2} - 1 = \frac{5}{2}.$

If $y = -1, x = 1 - 1 = 0.$

The solution set is $\left\{\left(\frac{5}{2}, -\frac{7}{2}\right), (0, -1)\right\}$.

**34.** $x^2 + y^2 = 13$
$$x^2 - y = 7$$
$$x^2 + y^2 = 13$$
$$\underline{-x^2 + y = -7}$$
$$y^2 + y = 6$$
$$y^2 + y - 6 = 0$$
$$(y + 3)(y - 2) = 0$$
$$y = -3, 2$$

If $y = -3, x^2 + 3 = 7$
$$x^2 = 4, x = 2, -2$$

If $y = 2, x^2 - 2 = 7, x^2 = 9, x = 3, -3.$
The solution set is $\{(2,-3),(-2,-3),(3,2),(-3,2)\}$.

**35.** $2x^2 + 3y^2 = 21$
$$3x^2 - 4y^2 = 23$$
$$8x^2 + 12y^2 = 84$$
$$\underline{9x^2 - 12y^2 = 69}$$
$$17x^2 = 153$$
$$x^2 = \frac{153}{17} = 9$$
$$x = 3, -3$$

If $x = 3, 2(3)^2 + 3y^2 = 21.$
$$3y^2 = 21 - 18 = 3$$
$$y^2 = 1, y = 1, -1$$

If $x = -3, y = 1, -1.$
The solution set is $\{(3,1),(3,-1),(-3,1),(-3,-1)\}$.

**36.**
$$2L + 2W = 26$$
$$LW = 40$$
$$L = \frac{40}{W}$$
$$2\left(\frac{40}{W}\right) + 2W = 26$$
$$\frac{80}{W} + 2W = 26$$
$$80 + 2W^2 = 26W$$
$$2W^2 - 26W + 80 = 0$$
$$W^2 - 13W + 40 = 0$$
$$(W - 8)(W - 5) = 0$$
$$W = 8, 5$$
If $W = 5, L = \frac{40}{5} = 8$

The dimensions are 8 m by 5 m.

**37.**
$$xy = 6$$
$$y = \frac{6}{x}$$
$$2x + y = 8$$
$$2x + \frac{6}{x} = 8$$
$$2x^2 + 6 = 8x$$
$$2x^2 - 8x + 6 = 0$$
$$x^2 - 4x + 3 = 0$$
$$(x - 1)(x - 3) = 0$$
$$x = 1, 3$$
If $x = 1, y = 6.$
If $x = 3, y = 2.$
The solution set is $\{(1, 6), (3, 2)\}$.

**38.**
$$x^2 + y^2 = 2900$$
$$4x + 2y = 240$$
$$2x + y = 120$$
$$y = 120 - 2x$$
$$x^2 + (120 - 2x)^2 = 2900$$
$$x^2 + 14,400 - 480x + 4x^2 - 2900 = 0$$
$$5x^2 - 480x + 11,500 = 0$$
$$x^2 - 96x + 2300 = 0$$
$$(x - 46)(x - 50) = 0$$
$$x = 46, 50$$

If $x = 46, y = 120 - 2(46) = 28.$
If $x = 50, y = 120 - 2(50) = 20.$
$x = 46$ ft and $y = 28$ ft  or
$x = 50$ ft  and $y = 20$ ft

**39.**

$3x - 4y > 12$

**40.**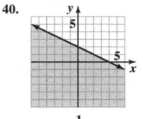

$y \le -\frac{1}{2}x + 2$

**41.**

$x < -2$

**42.**

$y \ge 3$

**43.**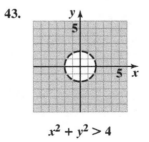

$x^2 + y^2 > 4$

429

**44.**

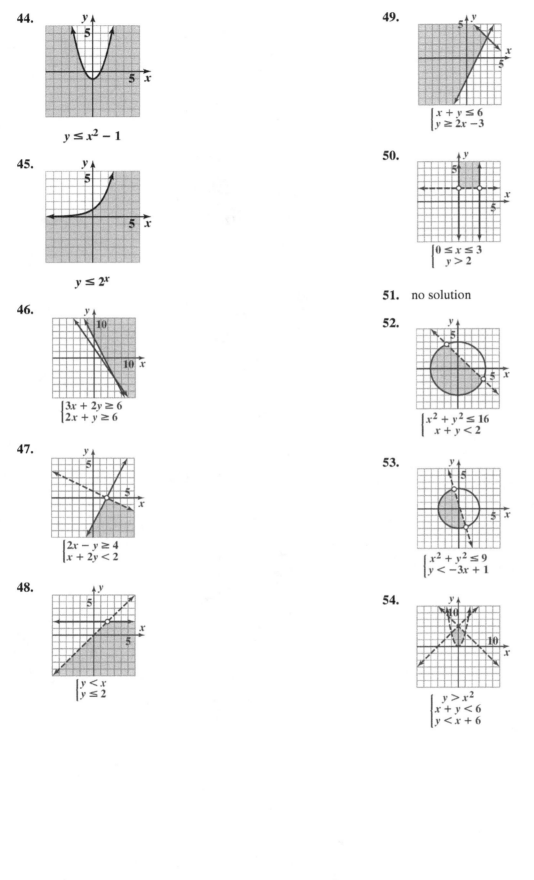

$y \le x^2 - 1$

**45.**

$y \le 2^x$

**46.**

$\begin{cases} 3x + 2y \ge 6 \\ 2x + y \ge 6 \end{cases}$

**47.**

$\begin{cases} 2x - y \ge 4 \\ x + 2y < 2 \end{cases}$

**48.**

$\begin{cases} y < x \\ y \le 2 \end{cases}$

**49.**

$\begin{cases} x + y \le 6 \\ y \ge 2x - 3 \end{cases}$

**50.**

$\begin{cases} 0 \le x \le 3 \\ y > 2 \end{cases}$

**51.** no solution

**52.**

$\begin{cases} x^2 + y^2 \le 16 \\ x + y < 2 \end{cases}$

**53.**

$\begin{cases} x^2 + y^2 \le 9 \\ y < -3x + 1 \end{cases}$

**54.**

$\begin{cases} y > x^2 \\ x + y < 6 \\ y < x + 6 \end{cases}$

**55.**

$$\begin{cases} y \ge 0 \\ 3x + 2y \ge 4 \\ x - y \le 3 \end{cases}$$

**56.** $z = 2x + 3y$

$(2,2): z = 2(2) + 3(2) = 10$

$(4,0): z = 2(4) + 3(0) = 8$

$\left(\frac{1}{2}, \frac{1}{2}\right): z = 2\left(\frac{1}{2}\right) + 3\left(\frac{1}{2}\right) = \frac{5}{2}$

$(1,0): z = 2(1) + 3(0) = 2$

The maximum value is 10 and the minimum value is 2.

**57.** $z = 2x + 3y$

$x \ge 0, \ y \ge 0$
$x + y \le 8$
$3x + 2y \ge 6$

$(0,8): z = 2(0) + 3(8) = 24$

$(8,0): z = 2(8) + 3(0) = 16$

$(0,3): z = 2(0) + 3(3) = 9$

$(2,0): z = 2(2) + 3(0) = 6$

Maximum value is 24 at $(0, 8)$.

**58.** $z = x + 4y$

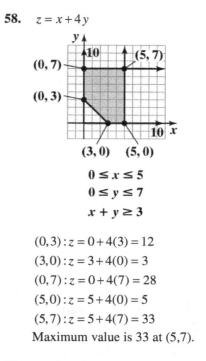

$0 \le x \le 5$
$0 \le y \le 7$
$x + y \ge 3$

$(0,3): z = 0 + 4(3) = 12$

$(3,0): z = 3 + 4(0) = 3$

$(0,7): z = 0 + 4(7) = 28$

$(5,0): z = 5 + 4(0) = 5$

$(5,7): z = 5 + 4(7) = 33$

Maximum value is 33 at $(5,7)$.

**59.** $z = 5x + 6y$

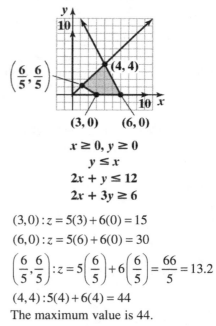

$x \ge 0, \ y \ge 0$
$y \le x$
$2x + y \le 12$
$2x + 3y \ge 6$

$(3,0): z = 5(3) + 6(0) = 15$

$(6,0): z = 5(6) + 6(0) = 30$

$\left(\frac{6}{5}, \frac{6}{5}\right): z = 5\left(\frac{6}{5}\right) + 6\left(\frac{6}{5}\right) = \frac{66}{5} = 13.2$

$(4,4): 5(4) + 6(4) = 44$

The maximum value is 44.

**60.** **a.** $z = 500x + 350y$

**b.** $x + y \leq 200$
$\phantom{x+y}x \geq 10$
$\phantom{x+y}y \geq 80$

**c.**

$x + y \leq 200$
$x \geq 10, y \geq 80$

**d.**

| Vertex | Objective Function |
|---|---|
| | $z = 500x + 350y$ |
| (10, 80) | $z = 500(10) + 350(80)$ |
| | $= 33,000$ |
| (10, 190) | $z = 500(10) + 350(190)$ |
| | $= 71,500$ |
| (120, 80) | $z = 500(120) + 350(80)$ |
| | $= 88,000$ |

**e.** The company will make the greatest profit by producing 120 units of writing paper and 80 units of newsprint each day. The maximum daily profit is \$88,000.

**61.** Let $x$ = number of model $A$ tents produced and $y$ = number of model $B$ tents produced.
The constraints are:
$0.9x + 1.8y \leq 864$
$0.8x + 1.2y \leq 672$
$\phantom{0.8x+1.2}x \geq 0$
$\phantom{0.8x+1.2}y \geq 0$

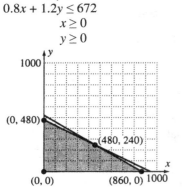

The vertices of the region are (0, 0), (0, 480), (480, 240), and (840, 0).
The objective is to maximize $25x + 40y$.
(0, 0): $25(0) + 40(0) = 0 + 0 = 0$
(0, 480): $25(0) + 40(480) = 0 + 19,200 = 19,200$
(480, 240): $25(480) + 40(240) = 12,000 + 9600$
$\phantom{(480,240): 25(480)} = 21,600$
(840, 0): $25(840) + 40(0) = 21,000 + 0 = 21,000$
The manufacturer should make 480 of model $A$ and 240 of model $B$.

**Chapter 5 Test**

**1.** $x = y + 4$
$3x + 7y = -18$
Substitute $y + 4$ for $x$ into second equation.
$3(y + 4) + 7y = -18$
$3y + 12 + 7y = -18$
$\phantom{3y + 12 + 7}10y = -30$
$\phantom{3y + 12 + 7 10}y = -3$
$\phantom{3y + 12}x = -3 + 4 = 1$
The solution set to the system is $\{(1, -3)\}$.

**2.** $2x + 5y = -2$
$3x - 4y = 20$
Multiply the first equation by 3 and the second equation by –2 and add the result.
$\phantom{-}6 + 15y = -6$
$-6x + 8y = -40$
$\overline{\phantom{-6x+}23y = -46}$
$\phantom{-6x+23}y = -2$
Substitute $y = -2$ into the first equation:
$2x + 5(-2) = -2$
$\phantom{2x}2x - 10 = -2$
$\phantom{2x-1}2x = 8$
$\phantom{2x-1}x = 4$
The solution to the system is $\{(4, -2)\}$.

**3.** $\phantom{3}x + y + z = 6 \,(1)$
$3x + 4y - 7z = 1 \,(2)$
$2x - y + 3z = 5 \,(3)$
Eliminate $x$ by multiplying (1) by –3 and adding the result to (2) and by multiplying (1) by –2 and adding the result to (3).
$-3x - 3y - 3z = -18$
$\underline{\phantom{-}3x + 4y - 7z = 1}$
$\phantom{-3x-3}y - 10z = -17 \,(4)$

$-2x - 2y - 2z = -12$
$\underline{\phantom{-}2x - y + 3z = 5}$
$\phantom{-2x-}-3y + z = -7 \,(5)$

Multiply (4) by 3 and add the result to (5) to eliminate $y$.

$$3y - 30z = -51$$
$$\underline{-3y + z = -7}$$
$$-29z = -58$$
$$z = 2$$

Substitute $z = 2$ into (5).
$$-3y + 2 = -7$$
$$-3y = -9$$
$$y = 3$$

Substitute $z = 2$ and $y = 3$ into (1).
$$x + 3 + 2 = 6$$
$$x = 1$$

The solution to the system is $\{(1, 3, 2)\}$.

**4.**   $$x^2 + y^2 = 25$$
$$x + y = 1$$
$$y = 1 - x$$

Substitute $1 - x$ for $y$ in the first equation.
$$x^2 + (1-x)^2 = 25$$
$$x^2 + 1 - 2x + x^2 = 25$$
$$2x^2 - 2x - 24 = 0$$
$$x^2 - x - 12 = 0$$
$$(x-4)(x+3) = 0$$
$$x = 4, -3$$

If $x = 4$, $y = 1 - 4 = -3$.
If $x = -3$, $y = 1 - (-3) = 4$.
The solution set is $\{(4,-3),(-3,4)\}$.

**5.**   $$2x^2 - 5y^2 = -2$$
$$3x^2 + 2y^2 = 35$$

Multiply first equation by 2 and the second equation by 5. Then add.
$$4x^2 - 10y^2 = -4$$
$$\underline{15x^2 + 10y^2 = 175}$$
$$19x^2 = 171$$
$$x^2 = 9$$
$$x = 3, -3$$

If $x = 3, 2(3)^2 - 5y^2 = -2$.
$$18 - 5y^2 = -2$$
$$-5y^2 = -20$$
$$y^2 = 4$$
$$y = 2, -2$$

If $x = -3$, $y = -2$.
The solution to the system is
$\{(3, 2),(3, -2),(-3, 2),(-3, -2)\}$.

**6.**   $$\frac{x}{(x+1)(x^2+9)} = \frac{A}{x+1} + \frac{Bx+C}{x^2+9}$$
$$x = A(x^2+9) + (Bx+C)(x+1)$$
$$= Ax^2 + 9A + Bx^2 + Bx + Cx + C$$
$$= (A+B)x^2 + (B+C)x + (9A+C)$$
$$A + B = 0 \rightarrow A = -B$$
$$B + C = 1$$
$$9A + C = 0$$
$$-9B + C = 0$$
$$9B - C = 0$$
$$\underline{B + C = 1}$$
$$10B = 1$$
$$B = \frac{1}{10}$$
$$A = -\frac{1}{10}$$
$$\frac{1}{10} + C = 1, \ C = \frac{9}{10}$$
$$\frac{x}{(x+1)(x^2+9)} = \frac{-1}{10(x+1)} + \frac{x+9}{10(x^2+9)}$$

**7.**

$x - 2y < 8$

**8.**

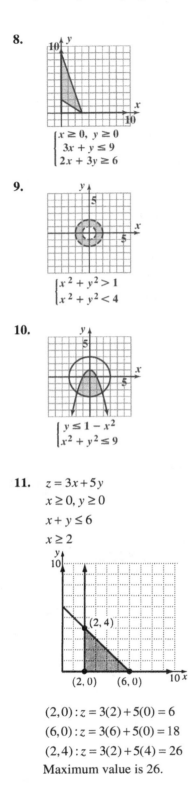

$$\begin{cases} x \geq 0, \ y \geq 0 \\ 3x + y \leq 9 \\ 2x + 3y \geq 6 \end{cases}$$

**9.**

$$\begin{cases} x^2 + y^2 > 1 \\ x^2 + y^2 < 4 \end{cases}$$

**10.**

$$\begin{cases} y \leq 1 - x^2 \\ x^2 + y^2 \leq 9 \end{cases}$$

**11.** $z = 3x + 5y$

$x \geq 0, \ y \geq 0$

$x + y \leq 6$

$x \geq 2$

$(2,0): z = 3(2) + 5(0) = 6$

$(6,0): z = 3(6) + 5(0) = 18$

$(2,4): z = 3(2) + 5(4) = 26$

Maximum value is 26.

**12.** $x =$ mg of cholesterol in one ounce of shrimp
$y =$ mg of cholesterol in one ounce of scallops
$3x + 2y = 156$
$5x + 3y = 255$
Multiply the first equation by –3 and multiply the
second equation by 2.
Add the resulting equations together.

$-9x - 6y = -468$

$\underline{10x + 6y = \ 510}$

$\phantom{-9x-6y} x = \ 42$

$3(42) + 2y = 156$
$126 + 2y = 156$
$2y = 30$
$y = 15$
$3(42) + 2y = 156$
$126 + 2y = 156$
$2y = 30$
$y = 15$

Shrimp: 42 mg of cholesterol per ounce
Scallops: 15 mg of cholesterol per ounce

**13. a.** $C(x) = 360,000 + 850x$

**b.** $R(x) = 1150x$

**c.** $1150x = 360000 + 850x$
$300x = 360000$
$x = 1200$
$1150(1200) = 1,380,000$
The break-even point is $(1200, 1, 380, 000)$.

This means that 1200 computers need to be sold to
make \$1,380,00 and for the company to break even.

**14.** $y = ax^2 + bx + c$

$(-1, -2) : -2 = a - b + c$

$(2, 1) : 1 = 4a + 2b + c$

$(-2, 1) : 1 = 4a - 2b + c$

$4a + 2b + c = 1$

$\underline{-4a + 2b - c = -1}$

$\phantom{4a+2b+c=1} 4b = 0$

$\phantom{4a+2b+c=1} b = 0$

$a + c = -2$

$4a + c = 1$

$\underline{-a - c = 2}$

$\phantom{4a} 3a = 3$

$\phantom{4a} a = 1$

$a + c = -2$

$c = -3$

The quadratic function is $y = x^2 - 3$.

**15.** $2x + y = 39$

$xy = 180$

$y = 39 - 2x$

$\quad x(39 - 2x) = 180$

$\quad 39x - 2x^2 = 180$

$2x^2 - 39x + 180 = 0$

$(2x - 15)(x - 12) = 0$

$\quad\quad x = \dfrac{15}{2}, 12$

If $x = \dfrac{15}{2}, \dfrac{15}{2} y = 180$ and $y = 24$.

If $x = 12, 12 y = 180$ and $y = 15$.

The dimensions are 7.5 ft by 24 ft or 12 ft by 15 ft

**16.** Let $x$ = regular, $y$ = deluxe.

objective function: $z = 200x + 250y$

constraints: $x \geq 50, y \geq 75$

$\quad\quad x + y \leq 150$

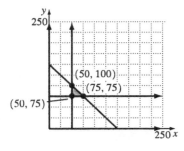

$(50, 75): z = 200(50) + 250(75) = 28,750$

$(50, 100): z = 200(50) + 250(100) = 35,000$

$(75, 75): z = 200(75) + 250(75) = 33,750$

For a maximum profit of \$35,000 a week, the company should manufacture 50 regular and 100 deluxe jet skis.

**Cumulative Review Exercises (Chapters 1–5)**

**1.** Domain: $(-\infty, \infty)$  Range: $[-\infty, 3)$

**2.** $-1$ and 1 are the zeros.

**3.** The relative maximum is $y = 3$ and it occurs at $x = 0$.

**4.** $f(x)$ is decreasing on the interval $(0, 2)$.

**5.** At $x = -0.7$, the curve is above the $x$-axis and thus $f(x)$ is positive.

**6.** $(f \circ f)(-1) = f\big(f(-1)\big) = f(0) = 3$

**7.** $f(x) \to -\infty$ as $x \to -2^+$ or as $x \to 2^-$.

**8.** $f(-x) = f(x)$ thus the function is even.

**9.** The graph of $g(x) = f(x+2) - 1$ can be obtained by shifting $f(x)$ 2 units left and 1 unit down.

$g(x) = f(x+2) - 1$

**10.** The graph of $h(x) = \frac{1}{2} f\left(\frac{1}{2} x\right)$ can be obtained by shrinking the graph of $f(x)$ horizontally by a factor of $\frac{1}{2}$ and vertically by a factor of $\frac{1}{2}$.

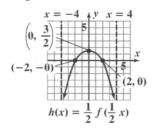

$h(x) = \dfrac{1}{2} f\left(\dfrac{1}{2} x\right)$

**11.** $\quad\quad \sqrt{x^2 - 3x} = 2x - 6$

$\quad\quad x^2 - 3x = 4x^2 - 24x + 36$

$3x^2 - 21x + 36 = 0$

$\quad x^2 - 7x + 12 = 0$

$\quad (x - 3)(x - 4) = 0$

$\quad\quad\quad x = 3, 4$

The solution set is $\{3, 4\}$.

**12.**
$$4x^2 = 8x - 7$$
$$4x^2 - 8x + 7 = 0$$
$$x = \frac{8 \pm \sqrt{64 - 112}}{8} = \frac{8 \pm \sqrt{-48}}{8}$$
$$= \frac{8 \pm 4\sqrt{3}i}{8} = \frac{2 \pm \sqrt{3}i}{2}$$

The solution set is $\left\{ \frac{2 + i\sqrt{3}}{2}, \frac{2 - i\sqrt{3}}{2} \right\}$.

**13.** $\left| \frac{x}{3} + 2 \right| < 4$

$$-4 < \frac{x}{3} + 2 < 4$$
$$-6 < \frac{x}{3} < 2$$
$$-18 < x < 6$$

The solution is $\left\{ x \mid -18 < x < 6 \right\}$ or $(-18, 6)$.

**14.**
$$\frac{x+5}{x-1} > 2$$
$$\frac{x+5}{x-1} - 2 > 0$$
$$\frac{x + 5 - 2(x-1)}{x-1} > 0$$
$$\frac{x + 5 - 2x + 2}{x-1} > 0$$
$$\frac{-x + 7}{x-1} > 0$$

$\frac{-x+7}{x-1} = 0$ when $x = 7$ and is undefined when $x = 1$.

Test $x = 0$:
$$\frac{0+5}{0-1} > 2?$$
$$\frac{5}{-1} > 2?$$
$$-5 \not> 2$$

Test $x = 2$:
$$\frac{2+5}{2-1} > 2?$$
$$\frac{7}{1} > 2?$$
$$7 > 2$$

Test $x = 8$:
$$\frac{8+5}{8-1} > 2?$$
$$\frac{13}{7} > 2?$$
$$\frac{13}{7} \geq \frac{14}{7}$$

The solution is $\left\{ x \mid 1 < x < 7 \right\}$ or $(1, 7)$.

**15.** $2x^3 + x^2 - 13x + 6 = 0$

$f(x) = 2x^3 + x^2 - 13x + 6$ has 2 sign changes: 2 or 0 positive real roots.

$f(-x) = -2x^3 + x^2 + 13x + 6$ has 1 sign change: 1 negative real root.

$p$: $\pm 1, \pm 2, \pm 3, \pm 6$

$q$: $\pm 1, \pm 2$

$\frac{p}{q}$: $\pm 1, \pm \frac{1}{2}, \pm 2, \pm 3, \pm \frac{3}{2}, \pm 6$

$$
\begin{array}{r|rrrr}
-3 & 2 & 1 & -13 & 6 \\
   &   & -6 & 15 & -6 \\
\hline
   & 2 & -5 & 2 & 0 \\
\end{array}
$$

$$2x^3 + x^2 - 13x + 6 = (x+3)(2x^2 - 5x + 2)$$
$$= (x+3)(2x-1)(x-2)$$

$x = -3$, $x = \frac{1}{2}$, $x = 2$

The solution set is $\left\{ -3, \frac{1}{2}, 2 \right\}$.

**16.** $6x - 3(5x + 2) = 4(1 - x)$
$$6x - 15x - 6 = 4 - 4x$$
$$-9x - 6 = 4 - 4x$$
$$-5x = 10$$
$$x = -2$$

The solution set is $\{-2\}$.

**17.** $\log(x+3) + \log x = 1$
$$\log x(x+3) = 1$$
$$x(x+3) = 10$$
$$x^2 + 3x - 10 = 0$$
$$(x+5)(x-2) = 0$$
$$x = -5 \text{ or } x = 2$$
$$x = -5 \text{ is extraneous.}$$
$$x = 2$$

The solution set is $\{2\}$.

**18.** $3^{x+2} = 11$

$\log_3 3^{x+2} = \log_3 11$

$x + 2 = \log_3 11$

$x = -2 + \log_3 11$

$x = -2 + \dfrac{\log 11}{\log 3} \approx 0.18$

The solution set is $\{-2 + \log_3 11\}$.

**19.** $x^{\frac{1}{2}} - 2x^{\frac{1}{4}} - 15 = 0$

$\sqrt[4]{x}^2 - 2\sqrt[4]{x} - 15 = 0$

$\left(\sqrt[4]{x} + 3\right)\left(\sqrt[4]{x} - 5\right) = 0$

$\sqrt[4]{x} + 3 = 0 \qquad\qquad \sqrt[4]{x} - 5 = 0$

$\sqrt[4]{x} = -3 \qquad\quad\text{or}\quad\qquad \sqrt[4]{x} = 5$

$x = (-3)^4 \qquad\qquad\qquad x = 5^4$

$x = 81 \qquad\qquad\qquad\quad x = 625$

81 does not check. The solution set is $\{625\}$.

**20.** $3x - y = -2$

$2x^2 - y = 0$

Solve the first equation for $y$.

$3x - y = -2$

$y = 3x + 2$

Use this equation to substitute into the other equation.

$2x^2 - \overbrace{(3x + 2)}^{y} = 0$

$2x^2 - 3x - 2 = 0$

$(2x + 1)(x - 2) = 0$

$x = -\dfrac{1}{2}$ or $x = 2$.

Back-substitute to find $y$.

$y = 3x + 2 \qquad$ or $\qquad y = 3x + 2$

$y = 3(2) + 2$

$y = 8 \qquad\qquad\qquad y = 3\left(-\dfrac{1}{2}\right) + 2$

$\qquad\qquad\qquad\qquad y = \dfrac{1}{2}$

$y = 3x + 2$

The solution set is $\left\{(2, 8), \left(-\dfrac{1}{2}, \dfrac{1}{2}\right)\right\}$.

**21.** $x + 2y + 3z = -2$

$3x + 3y + 10z = -2$

$2y - 5z = 6$

Multiply equation 1 by –3 and add to equation 2.

$-3x - 6y - 9z = 6$

$\underline{3x + 3y + 10z = -2}$

$\qquad -3y + z = 4 \quad$ Equation 4

Multiply equation 4 by 5 and add to equation 3 and solve for $y$.

$-15y + 5z = 20$

$\underline{2y - 5z = 6}$

$-13y = 26$

$y = -2$

Back-substitute to find $z$.

$-3y + z = 4$

$-3(-2) + z = 4$

$z = -2$

Back-substitute to find $x$.

$x + 2y + 3z = -2$

$x + 2(-2) + 3(-2) = -2$

$x = 8$

The solution set is $\{8, -2, -2\}$.

**22.** vertex: (–2, –4)

$y$-intercept:

$f(0) = (0 + 2)^2 - 4 = 0$

$x$-intercepts:

$(x + 2)^2 - 4 = 0$

$x^2 + 4x + 4 - 4 = 0$

$x^2 + 4x = 0$

$x(x + 4) = 0$

$x = 0, x - 4$

$f(x) = (x + 2)^2 - 4$

**23.**

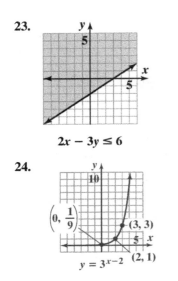

$2x - 3y \le 6$

**24.**

$y = 3^{x-2}$

**25.** vertical asymptote: $x = -1$
horizontal asymptote: $m > n$, none
$x$-intercepts:

$$x^2 - x - 6 = 0$$
$$(x-3)(x+2) = 0$$
$$x = 3, x = -2$$

$y$-intercept:

$$f(0) = \frac{0^2 - 0 - 6}{0+1} = -6$$

$$f(x) = \frac{x^2 - x - 6}{x + 1}$$

**26.**

$$f(x) = 2x - 4$$
$$f^{-1}(x) = \frac{x+4}{2}$$

**27.**

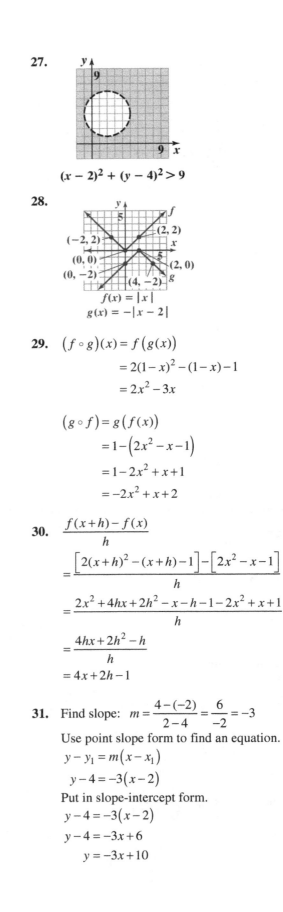

$(x - 2)^2 + (y - 4)^2 > 9$

**28.**

$$f(x) = |x|$$
$$g(x) = -|x - 2|$$

**29.** $(f \circ g)(x) = f(g(x))$
$$= 2(1-x)^2 - (1-x) - 1$$
$$= 2x^2 - 3x$$

$$(g \circ f) = g(f(x))$$
$$= 1 - (2x^2 - x - 1)$$
$$= 1 - 2x^2 + x + 1$$
$$= -2x^2 + x + 2$$

**30.** $\dfrac{f(x+h) - f(x)}{h}$

$$= \frac{\left[2(x+h)^2 - (x+h) - 1\right] - \left[2x^2 - x - 1\right]}{h}$$

$$= \frac{2x^2 + 4hx + 2h^2 - x - h - 1 - 2x^2 + x + 1}{h}$$

$$= \frac{4hx + 2h^2 - h}{h}$$

$$= 4x + 2h - 1$$

**31.** Find slope: $m = \dfrac{4 - (-2)}{2 - 4} = \dfrac{6}{-2} = -3$
Use point slope form to find an equation.
$$y - y_1 = m(x - x_1)$$
$$y - 4 = -3(x - 2)$$
Put in slope-intercept form.
$$y - 4 = -3(x - 2)$$
$$y - 4 = -3x + 6$$
$$y = -3x + 10$$

**32.** Find the slope of the perpendicular line by putting in slope-intercept form.

$x + 3y - 6 = 0$

$$3y = -x + 6$$

$$y = -\frac{1}{3}x + 2$$

The slope of the perpendicular line is $-\frac{1}{3}$ so the slope of the desired line is the negative reciprocal, or 3.

Use point slope form to find an equation.

$y - y_1 = m(x - x_1)$

$$y - 0 = 3(x + 1)$$

Put in slope-intercept form.

$y - 0 = 3(x + 1)$

$$y = 3x + 3$$

**33.** Let $x$ = the amount invested at 12%
Let $4000 - x$ = the amount invested at 14%

$0.12x + 0.14(4000 - x) = 508$

$$0.12x + 560 - 0.14x = 508$$

$$-0.02x = -52$$

$$x = \frac{-52}{-0.02}$$

$$x = 2600$$

$4000 - x = 4000 - 2600 = 1400$

Thus, $2600 was invested at 12% and $1400 was invested at 14%.

**34.**
$$L = 2W + 1$$
$$LW = 36$$
$$W(2W + 1) = 36$$
$$2W^2 + W - 36 = 0$$
$$(2W + 9)(W - 4) = 0$$
$$W = -\frac{9}{2} \text{ or } 4$$

Length cannot be negative. If $W = 4$, $4L = 36$, $L = 9$. The dimensions are 4 m by 9 m.

**35.** $A = Pe^{rt}$

$18,000 = 6000e^{10r}$

$$3 = e^{10r}$$

$$\ln 3 = \ln e^{10r}$$

$$\ln 3 = 10r$$

$$r = \frac{\ln 3}{10} \approx 0.1099$$

10.99%

# Chapter 6
# Matrices and Determinants

## Section 6.1

### Check Point Exercises

**1.** **a.** The notation $R_1 \leftrightarrow R_2$ means to interchange the elements in row 1 and row 2. This results in the row-equivalent matrix

$$\begin{bmatrix} 1 & 6 & -3 & | & 7 \\ 4 & 12 & -20 & | & 8 \\ -3 & -2 & 1 & | & -9 \end{bmatrix}.$$

**b.** The notation $\frac{1}{4}R_1$ means to multiply each element in row 1 by $\frac{1}{4}$. This results in the row-equivalent matrix

$$\begin{bmatrix} \frac{1}{4}(4) & \frac{1}{4}(12) & \frac{1}{4}(-20) & | & \frac{1}{4}(8) \\ 1 & 6 & -3 & | & 7 \\ -3 & -2 & 1 & | & -9 \end{bmatrix} = \begin{bmatrix} 1 & 3 & -5 & | & 2 \\ 1 & 6 & -3 & | & 7 \\ -3 & -2 & 1 & | & -9 \end{bmatrix}.$$

**c.** The notation $3R_2 + R_3$ means to add 3 times the elements in row 2 to the corresponding elements in row 3. Replace the elements in row 3 by these sums. First, we find 3 times the elements in row 2:
$3(1) = 3$, $3(6) = 18$, $3(-3) = -9$, $3(7) = 21$. Now we add these products to the corresponding elements in row 3. This results in the row equivalent matrix

$$\begin{bmatrix} 4 & 12 & -20 & | & 8 \\ 1 & 6 & -3 & | & 7 \\ -3+3=0 & -2+18=16 & 1-9=-8 & | & -9+21=12 \end{bmatrix} = \begin{bmatrix} 4 & 12 & -20 & | & 8 \\ 1 & 6 & -3 & | & 7 \\ 0 & 16 & -8 & | & 12 \end{bmatrix}.$$

**2.** $\begin{cases} 2x + y + 2z = 18 \\ x - y + 2z = 9 \\ x + 2y - z = 6 \end{cases} \rightarrow \begin{bmatrix} 2 & 1 & 2 & | & 18 \\ 1 & -1 & 2 & | & 9 \\ 1 & 2 & -1 & | & 6 \end{bmatrix}$

Interchange row 1 with row 2 to get 1 in the top position of the first column.

$$\begin{bmatrix} 1 & -1 & 2 & | & 9 \\ 2 & 1 & 2 & | & 18 \\ 1 & 2 & -1 & | & 6 \end{bmatrix}$$

Multiply the first row by –2 and add these products to row 2.

$$\begin{bmatrix} 1 & -1 & 2 & | & 9 \\ 2+-2=0 & 1+2=3 & 2+-4=-2 & | & -18+18=0 \\ 1 & 2 & -1 & | & 6 \end{bmatrix} = \begin{bmatrix} 1 & -1 & 2 & | & 9 \\ 0 & 3 & -2 & | & 0 \\ 1 & 2 & -1 & | & 6 \end{bmatrix}$$

Next, multiply the top row by –1 and add these products to row 3.

$$\begin{bmatrix} 1 & -1 & 2 & | & 9 \\ 0 & 3 & -2 & | & 0 \\ 1+-1=0 & 2+1=3 & -1-2=-3 & | & 6-9=-3 \end{bmatrix} = \begin{bmatrix} 1 & -1 & 2 & | & 9 \\ 0 & 3 & -2 & | & 0 \\ 0 & 3 & -3 & | & -3 \end{bmatrix}$$

Next, to obtain a 1 in the second row, second column, multiply 3 by its reciprocal, $\frac{1}{3}$. Therefore, we multiply all the numbers in the second row by $\frac{1}{3}$ to get

$$\begin{bmatrix} 1 & -1 & 2 & | & 9 \\ 0 & 1 & -\frac{2}{3} & | & 0 \\ 0 & 3 & -3 & | & -3 \end{bmatrix}.$$

Next, to obtain a 0 in the third row, second column, multiply the second row by $-3$ and add the products to row three. The resulting matrix is

$$\begin{bmatrix} 1 & -1 & 2 & | & 9 \\ 0 & 1 & -\frac{2}{3} & | & 0 \\ 0 & 0 & -1 & | & -3 \end{bmatrix}.$$

To get 1 in the third row, third column, multiply $-1$ by its reciprocal, $-1$. Multiply all numbers in the third row by $-1$ to obtain the resulting matrix

$$\begin{bmatrix} 1 & -1 & 2 & | & 9 \\ 0 & 1 & -\frac{2}{3} & | & 0 \\ 0 & 0 & 1 & | & 3 \end{bmatrix}.$$

The system represented by this matrix is:

$$x - y + 2z = 9$$
$$y - \frac{2}{3}z = 0$$
$$z = 3$$

Use back substitution to find $y$ and $x$.

$$y - \frac{2}{3}(3) = 0 \qquad\qquad x - 2 + 6 = 9$$
$$y - 2 = 0 \qquad\qquad\qquad x + 4 = 9$$
$$y = 2 \qquad\qquad\qquad\quad x = 5$$

The solution set for the original system is $\{(5, 2, 3)\}$.

**3.**
$$w - 3x - 2y + z = -3$$
$$2w - 7x - y + 2z = 1$$
$$3w - 7x - 3y + 3z = -5$$
$$5w + x + 4y - 2z = 18$$

The augmented matrix is

$$\begin{bmatrix} 1 & -3 & -2 & 1 & | & -3 \\ 2 & -7 & -1 & 2 & | & 1 \\ 3 & -7 & -3 & 3 & | & -5 \\ 5 & 1 & 4 & -2 & | & 18 \end{bmatrix}.$$

Multiply the top row by $-2$ and add the products to the second row. Multiply the top row by $-3$ and add the products to the third row. Multiply the top row by $-5$ and add the products to the fourth row. The resulting matrix is

$$\begin{bmatrix} 1 & -3 & -2 & 1 & | & -3 \\ 0 & -1 & 3 & 0 & | & 7 \\ 0 & 2 & 3 & 0 & | & 4 \\ 0 & 16 & 14 & -7 & | & 33 \end{bmatrix}.$$

Next, multiply the second row by –1 to obtain a 1 in the second row, second column.

$$\begin{bmatrix} 1 & -3 & -2 & 1 & | & -3 \\ 0 & 1 & -3 & 0 & | & -7 \\ 0 & 2 & 3 & 0 & | & 4 \\ 0 & 16 & 14 & -7 & | & 33 \end{bmatrix}$$

Next, multiply the second row by –2 and add the products to the third row. Multiply the second row by –16 and add the products to the fourth row. The resulting matrix is

$$\begin{bmatrix} 1 & -3 & -2 & 1 & | & -3 \\ 0 & 1 & -3 & 0 & | & -7 \\ 0 & 0 & 9 & 0 & | & 18 \\ 0 & 0 & 62 & -7 & | & 145 \end{bmatrix}.$$

Next, multiply the third row by $\frac{1}{9}$ to obtain a 1 in the third row, third column. The resulting matrix is

$$\begin{bmatrix} 1 & -3 & -2 & 1 & | & -3 \\ 0 & 1 & -3 & 0 & | & -7 \\ 0 & 0 & 1 & 0 & | & 2 \\ 0 & 0 & 62 & -7 & | & 145 \end{bmatrix}.$$

Multiply the third row by –62 and add the products to the fourth row to obtain the resulting matrix

$$\begin{bmatrix} 1 & -3 & -2 & 1 & | & -3 \\ 0 & 1 & -3 & 0 & | & -7 \\ 0 & 0 & 1 & 0 & | & 2 \\ 0 & 0 & 0 & -7 & | & 21 \end{bmatrix}.$$

Multiply the fourth row by $-\frac{1}{7}$, the reciprocal of –7. The resulting matrix is

$$\begin{bmatrix} 1 & -3 & -2 & 1 & | & -3 \\ 0 & 1 & -3 & 0 & | & -7 \\ 0 & 0 & 1 & 0 & | & 2 \\ 0 & 0 & 0 & 1 & | & -3 \end{bmatrix}.$$

The system of linear equations corresponding to the resulting matrix is

$$w - 3x - 2y + z = -3$$
$$x - 3y = -7$$
$$y = 2$$
$$z = -3$$

Using back-substitution solve for $x$ and $w$.

$$x - 3(2) = -7$$
$$x = -1$$

$$w - 3(-1) - 2(2) - 3 = -3$$
$$w - 4 = -3$$
$$w = 1$$

The solution set is $\{(1, -1, 2, -3)\}$.

**4.** The matrix obtained in 3 will be the starting point.

$$\begin{bmatrix} 1 & -1 & 2 & 9 \\ 0 & 1 & -\frac{2}{3} & 0 \\ 0 & 0 & 1 & 3 \end{bmatrix}$$

Next, multiply the third row by $\frac{2}{3}$ and add the products to the second row. Multiply the third row by 2 and add the products to the first row. The resulting matrix is

$$\begin{bmatrix} 1 & -1 & 0 & 3 \\ 0 & 1 & 0 & 2 \\ 0 & 0 & 1 & 3 \end{bmatrix}.$$

Add the second row to the first row and replace the first row.

$$\begin{bmatrix} 1 & 0 & 0 & 5 \\ 0 & 1 & 0 & 2 \\ 0 & 0 & 1 & 3 \end{bmatrix}$$

This matrix corresponds to $x = 5$, $y = 2$ and $z = 3$. The solution set is $\{(5, 2, 3)\}$.

## Concept and Vocabulary Check 6.1

**1.** matrix; elements

**2.** −3; 2; 3; second; −2; third

**3.** $\begin{bmatrix} 2 & 1 & 4 & -4 \\ 3 & 0 & 1 & 1 \\ 4 & 3 & 1 & 8 \end{bmatrix}$

**4.** $\{(1, 3, -2)\}$

**5.** true

**6.** false

## Exercise Set 6.1

**1.** $\begin{bmatrix} 2 & 1 & 2 & 2 \\ 3 & -5 & -1 & 4 \\ 1 & -2 & -3 & -6 \end{bmatrix}$

**3.** $\begin{bmatrix} 1 & -1 & 1 & 8 \\ 0 & 1 & -12 & -15 \\ 0 & 0 & 1 & 1 \end{bmatrix}$

**5.** $\begin{bmatrix} 5 & -2 & -3 & 0 \\ 1 & 1 & 0 & 5 \\ 2 & 0 & -3 & 4 \end{bmatrix}$

**7.** $\begin{bmatrix} 2 & 5 & -3 & 1 & | & 2 \\ 0 & 3 & 1 & 0 & | & 4 \\ 1 & -1 & 5 & 0 & | & 9 \\ 5 & -5 & -2 & 0 & | & 1 \end{bmatrix}$

**9.** $\begin{aligned} 5x + 3z &= -11 \\ y - 4z &= 12 \\ 7x + 2y &= 3 \end{aligned}$

**11.** $\begin{aligned} w + x + 4y + z &= 3 \\ -w + x - y &= 7 \\ 12w + 5z &= 11 \\ 12y + 4z &= 5 \end{aligned}$

**13.** $\begin{bmatrix} 2\left(\frac{1}{2}\right) & -6\left(\frac{1}{2}\right) & 4\left(\frac{1}{2}\right) & | & 10\left(\frac{1}{2}\right) \\ 1 & 5 & -5 & | & 0 \\ 3 & 0 & 4 & | & 7 \end{bmatrix} \frac{1}{2}R_1$

$\begin{bmatrix} 1 & -3 & 2 & | & 5 \\ 1 & 5 & -5 & | & 0 \\ 3 & 0 & 4 & | & 7 \end{bmatrix}$

**15.** $\begin{bmatrix} 1 & -3 & 2 & | & 0 \\ -3(1)+3 & -3(-3)+1 & -3(2)+-1 & | & -3(0)+7 \\ 2 & -2 & 1 & | & 3 \end{bmatrix} -3R_1 + R_2$

$\begin{bmatrix} 1 & -3 & 2 & | & 0 \\ 0 & 10 & -7 & | & 7 \\ 2 & -2 & 1 & | & 3 \end{bmatrix}$

**17.** $\begin{bmatrix} 1 & -1 & 1 & 1 & | & 3 \\ 0 & 1 & -2 & -1 & | & 0 \\ 2 & 0 & 3 & 4 & | & 11 \\ 5 & 1 & 2 & 4 & | & 6 \end{bmatrix} \begin{matrix} \\ \\ -2R_1 + R_3 \\ -5R_1 + R_4 \end{matrix}$

$\begin{bmatrix} 1 & -1 & 1 & 1 & | & 3 \\ 0 & 1 & -2 & -1 & | & 0 \\ -2(1)+2 & -2(-1)+0 & -2(1)+3 & -2(1)+4 & | & -2(3)+11 \\ -5(1)+5 & -5(-1)+1 & -5(1)+2 & -5(1)+4 & | & -5(3)+6 \end{bmatrix} = \begin{bmatrix} 1 & -1 & 1 & 1 & | & 3 \\ 0 & 1 & -2 & -1 & | & 0 \\ 0 & 2 & 1 & 2 & | & 5 \\ 0 & 6 & -3 & -1 & | & -9 \end{bmatrix}$

**19.** $\begin{bmatrix} 1 & -1 & 1 & | & 8 \\ 2 & 3 & -1 & | & -2 \\ 3 & -2 & -9 & | & 9 \end{bmatrix}$

$\begin{bmatrix} 1 & -1 & 1 & | & 8 \\ -2(1)+2 & -2(-1)+3 & -2(1)-1 & | & -2(8)-2 \\ -3(1)+3 & -3(-1)-2 & -3(1)-9 & | & -3(8)+9 \end{bmatrix}$

$\begin{bmatrix} 1 & -1 & 1 & | & 8 \\ 0 & 5 & \boxed{-3} & | & \boxed{-18} \\ 0 & 1 & \boxed{-12} & | & \boxed{-15} \end{bmatrix}$

$\begin{bmatrix} 1 & -1 & 1 & | & 8 \\ 0(\frac{1}{5}) & 1(\frac{1}{5}) & -3(\frac{1}{5}) & | & -18(\frac{1}{5}) \\ 0 & 1 & -12 & | & -15 \end{bmatrix}$

$\begin{bmatrix} 1 & -1 & 1 & | & 8 \\ 0 & 1 & \boxed{-\frac{3}{5}} & | & \boxed{-\frac{18}{5}} \\ 0 & 1 & \boxed{-12} & | & \boxed{-15} \end{bmatrix}$

**21.**
$x+y-z=-2$
$2x-y+z=5$
$-x+2y+2z=1$

$\begin{bmatrix} 1 & 1 & -1 & | & -2 \\ 2 & -1 & 1 & | & 5 \\ -1 & 2 & 2 & | & 1 \end{bmatrix} -2R_1+R_2$

$\begin{bmatrix} 1 & 1 & -1 & | & -2 \\ 0 & -3 & 3 & | & 9 \\ -1 & 2 & 2 & | & 1 \end{bmatrix} 1R_1+R_3$

$\begin{bmatrix} 1 & 1 & -1 & | & -2 \\ 0 & -3 & 3 & | & 9 \\ 0 & 3 & 1 & | & -1 \end{bmatrix} -\frac{1}{3}R_2$

$\begin{bmatrix} 1 & 1 & -1 & | & -2 \\ 0 & 1 & -1 & | & -3 \\ 0 & 3 & 1 & | & -1 \end{bmatrix} -3R_2+R_3$

$=\begin{bmatrix} 1 & 1 & -1 & | & -2 \\ 0 & 1 & -1 & | & -3 \\ 0 & 0 & 4 & | & 8 \end{bmatrix}$

$4z=8$
$z=2$
$y-z=-3$
$y-2=-3$
$y=-1$

$x+y-z=-2$
$x-1-2=-2$
$x-3=-2$
$x=1$
The solution set is $\{(1,-1,2)\}$.

**23.**
$x+3y=0$
$x+y+z=1$
$3x-y-z=11$

$\begin{bmatrix} 1 & 3 & 0 & | & 0 \\ 1 & 1 & 1 & | & 1 \\ 3 & -1 & -1 & | & 11 \end{bmatrix} -1R_1+R_2$

$\begin{bmatrix} 1 & 3 & 0 & | & 0 \\ 0 & -2 & 1 & | & 1 \\ 3 & -1 & -1 & | & 11 \end{bmatrix} -3R_1+R_3$

$\begin{bmatrix} 1 & 3 & 0 & | & 0 \\ 0 & -2 & 1 & | & 1 \\ 0 & -10 & -1 & | & 11 \end{bmatrix} -\frac{1}{2}R_2$

$\begin{bmatrix} 1 & 3 & 0 & | & 0 \\ 0 & 1 & -\frac{1}{2} & | & -\frac{1}{2} \\ 0 & -10 & -1 & | & 11 \end{bmatrix} 10R_2+R_3$

$\begin{bmatrix} 1 & 3 & 0 & | & 0 \\ 0 & 1 & -\frac{1}{2} & | & -\frac{1}{2} \\ 0 & 0 & -6 & | & 6 \end{bmatrix} -\frac{1}{6}R_3$

$\begin{bmatrix} 1 & 3 & 0 & | & 0 \\ 0 & 1 & -\frac{1}{2} & | & -\frac{1}{2} \\ 0 & 0 & 1 & | & -1 \end{bmatrix}$

$z=-1$

$y-\frac{1}{2}z=-\frac{1}{2}$
$y-\frac{1}{2}(-1)=-\frac{1}{2}$
$y+\frac{1}{2}=-\frac{1}{2}$
$y=-1$

Interchange row one and row two.
$x+3y=0$
$x+3(-1)=0$
$x=3$
The solution set is $\{(3,-1,-1)\}$.

**25.** $2x - y - z = 4$
$x + y - 5z = -4$
$x - 2y = 4$

$$\begin{bmatrix} 2 & -1 & -1 & 4 \\ 1 & 1 & -5 & -4 \\ 1 & -2 & 0 & 4 \end{bmatrix}$$

Interchange rows one and two.

$$\begin{bmatrix} 1 & 1 & -5 & -4 \\ 2 & -1 & -1 & 4 \\ 1 & -2 & 0 & 4 \end{bmatrix}$$

Replace row two with $-2R_1 + R_2$.
Replace row three with $-R_1 + R_3$.

$$\begin{bmatrix} 1 & 1 & -5 & -4 \\ 0 & -3 & 9 & 12 \\ 0 & -3 & 5 & 8 \end{bmatrix}$$

Replace row two with $-\dfrac{1}{3}R_2$.

$$\begin{bmatrix} 1 & 1 & -5 & -4 \\ 0 & 1 & -3 & -4 \\ 0 & -3 & 5 & 8 \end{bmatrix}$$

Replace row three with $3R_2 + R_3$.

$$\begin{bmatrix} 1 & 1 & -5 & -4 \\ 0 & 1 & -3 & -4 \\ 0 & 0 & -4 & -4 \end{bmatrix}$$

Replace row three with $-\dfrac{1}{4}R_3$.

$$\begin{bmatrix} 1 & 1 & -5 & -4 \\ 0 & 1 & -3 & -4 \\ 0 & 0 & 1 & 1 \end{bmatrix}$$

$z = 1$
$y - 3z = -4$
$y - 3(1) = -4$
$y = -1$
$x + y - 5z = -4$
$x - 1 - 5(1) = -4$
$x - 6 = -4$
$x = 2$

The solution set is $\{(2, -1, 1)\}$.

**27.** $x + y + z = 4$
$x - y - z = 0$
$x - y + z = 2$

$$\begin{bmatrix} 1 & 1 & 1 & 4 \\ 1 & -1 & -1 & 0 \\ 1 & -1 & 1 & 2 \end{bmatrix}$$

Replace row two with $-R_1 + R_2$.
Replace row three with $-R_1 + R_3$.

$$\begin{bmatrix} 1 & 1 & 1 & 4 \\ 0 & -2 & -2 & -4 \\ 0 & -2 & 0 & -2 \end{bmatrix}$$

Replace row two with $-\dfrac{1}{2}R_2$.

$$\begin{bmatrix} 1 & 1 & 1 & 4 \\ 0 & 1 & 1 & 2 \\ 0 & -2 & 0 & -2 \end{bmatrix}$$

Replace row 3 with $2R_2 + R_3$.

$$\begin{bmatrix} 1 & 1 & 1 & 4 \\ 0 & 1 & 1 & 2 \\ 0 & 0 & 2 & 2 \end{bmatrix}$$

Replace row 3 with $\dfrac{1}{2}R_3$.

$$\begin{bmatrix} 1 & 1 & 1 & 4 \\ 0 & 1 & 1 & 2 \\ 0 & 0 & 1 & 1 \end{bmatrix}$$

$z = 1$
$y + 1 = 2$
$y = 1$
$x + 1 + 1 = 4$
$x = 2$

The solution set is $\{(2, 1, 1)\}$.

**29.** Write the equations in standard form.
$x + 2y - z = -1$
$x - y + z = 4$
$x + y - 3z = -2$

$$\begin{bmatrix} 1 & 2 & -1 & -1 \\ 1 & -1 & 1 & 4 \\ 1 & 1 & -3 & -2 \end{bmatrix}$$

Replace row two with $-R_1 + R_2$.
Replace row three with $-R_1 + R_3$.

$$\begin{bmatrix} 1 & 2 & -1 & -1 \\ 0 & -3 & 2 & 5 \\ 0 & -1 & -2 & -1 \end{bmatrix}$$

Replace row two with $-R_3$.
Replace row three with $R_2$.

$$\begin{bmatrix} 1 & 2 & -1 & -1 \\ 0 & 1 & 2 & 1 \\ 0 & -3 & 2 & 5 \end{bmatrix}$$

Replace row 3 with $3R_2 + R_3$.

$$\begin{bmatrix} 1 & 2 & -1 & -1 \\ 0 & 1 & 2 & 1 \\ 0 & 0 & 8 & 8 \end{bmatrix}$$

Replace row 3 with $\dfrac{1}{8}R_3$.

$$\begin{bmatrix} 1 & 2 & -1 & -1 \\ 0 & 1 & 2 & 1 \\ 0 & 0 & 1 & 1 \end{bmatrix}$$

$z = 1$

$y + 2(1) = 1$

$\quad y = -1$

$x + 2(-1) - 1 = -1$

$\quad\quad x = 2$

The solution set is $\{(2, -1, 1)\}$.

**31.** $3a - b - 4c = 3$

$\quad 2a - b + 2c = -8$

$\quad a + 2b - 3c = 9$

Interchange equations 1 and 3.

$$\begin{bmatrix} 1 & 2 & -3 & 9 \\ 2 & -1 & 2 & -8 \\ 3 & -1 & -4 & 3 \end{bmatrix}$$

Replace row two with $-2R_1 + R_2$.

Replace row three with $-3R_1 + R_3$.

$$\begin{bmatrix} 1 & 2 & -3 & 9 \\ 0 & -5 & 8 & -26 \\ 0 & -7 & 5 & -24 \end{bmatrix}$$

Replace row two with $-\dfrac{1}{5}R_2$

$$\begin{bmatrix} 1 & 2 & -3 & 9 \\ 0 & 1 & -\dfrac{8}{5} & \dfrac{26}{5} \\ 0 & -7 & 5 & -24 \end{bmatrix}$$

Replace row three with $7R_2 + R_3$.

$$\begin{bmatrix} 1 & 2 & -3 & 9 \\ 0 & 1 & -\dfrac{8}{5} & \dfrac{26}{5} \\ 0 & 0 & -\dfrac{31}{5} & \dfrac{62}{5} \end{bmatrix}$$

Replace row 3 with $-\dfrac{5}{31}R_3$.

$$\begin{bmatrix} 1 & 2 & -3 & 9 \\ 0 & 1 & -\dfrac{8}{5} & \dfrac{26}{5} \\ 0 & 0 & 1 & -2 \end{bmatrix}$$

$z = -2$

$$y - \frac{8}{5}(-2) = \frac{26}{5}$$

$$y + \frac{16}{5} = \frac{26}{5}$$

$$y = 2$$

$$x + 2(2) - 3(-2) = 9$$

$$x + 4 + 6 = 9$$

$$x = -1$$

The solution set is $\{(-1, 2, -2)\}$.

**33.** $2x + 2y + 7z = -1$

$\quad 2x + y + 2z = 2$

$\quad 4x + 6y + z = 15$

$$\begin{bmatrix} 2 & 2 & 7 & -1 \\ 2 & 1 & 2 & 2 \\ 4 & 6 & 1 & 15 \end{bmatrix} \tfrac{1}{2}R_1$$

$$\begin{bmatrix} 1 & 1 & \tfrac{7}{2} & -\tfrac{1}{2} \\ 2 & 1 & 2 & 2 \\ 4 & 6 & 1 & 15 \end{bmatrix} -2R_1 + R_2$$

$$\begin{bmatrix} 1 & 1 & \tfrac{7}{2} & -\tfrac{1}{2} \\ 0 & -1 & -5 & 3 \\ 4 & 6 & 1 & 15 \end{bmatrix} -4R_1 + R_3$$

$$\begin{bmatrix} 1 & 1 & \tfrac{7}{2} & -\tfrac{1}{2} \\ 0 & -1 & -5 & 3 \\ 0 & 2 & -13 & 17 \end{bmatrix} -1R_2$$

$$\begin{bmatrix} 1 & 1 & \tfrac{7}{2} & -\tfrac{1}{2} \\ 0 & 1 & 5 & -3 \\ 0 & 2 & -13 & 17 \end{bmatrix} -2R_2 + R_3$$

$$\begin{bmatrix} 1 & 1 & \tfrac{7}{2} & -\tfrac{1}{2} \\ 0 & 1 & 5 & -3 \\ 0 & 0 & -23 & 23 \end{bmatrix} -\tfrac{1}{23}R_3$$

$$\begin{bmatrix} 1 & 1 & \tfrac{7}{2} & -\tfrac{1}{2} \\ 0 & 1 & 5 & -3 \\ 0 & 0 & 1 & -1 \end{bmatrix}$$

$z = -1$

$$y + 5z = -3$$
$$y + 5(-1) = -3$$
$$y - 5 = -3$$
$$y = 2$$

$$x + y + \frac{7}{2}z = -\frac{1}{2}$$
$$x + 2 + \frac{7}{2}(-1) = -\frac{1}{2}$$
$$x - \frac{3}{2} = -\frac{1}{2}$$
$$x = 1$$

The solution set is $\{(1, 2, -1)\}$.

**35.**
$$w + x + y + z = 4$$
$$2w + x - 2y - z = 0$$
$$w - 2x - y - 2z = -2$$
$$3w + 2x + y + 3z = 4$$

$$\begin{bmatrix} 1 & 1 & 1 & 1 & | & 4 \\ 2 & 1 & -2 & -1 & | & 0 \\ 1 & -2 & -1 & -2 & | & -2 \\ 3 & 2 & 1 & 3 & | & 4 \end{bmatrix} -2R_1 + R_2$$

$$\begin{bmatrix} 1 & 1 & 1 & 1 & | & 4 \\ 0 & -1 & -4 & -3 & | & -8 \\ 1 & -2 & -1 & -2 & | & -2 \\ 3 & 2 & 1 & 3 & | & 4 \end{bmatrix} -1R_1 + R_3$$

$$\begin{bmatrix} 1 & 1 & 1 & 1 & | & 4 \\ 0 & -1 & -4 & -3 & | & -8 \\ 0 & -3 & -2 & -3 & | & -6 \\ 3 & 2 & 1 & 3 & | & 4 \end{bmatrix} -3R_1 + R_4$$

$$\begin{bmatrix} 1 & 1 & 1 & 1 & | & 4 \\ 0 & -1 & -4 & -3 & | & -8 \\ 0 & -3 & -2 & -3 & | & -6 \\ 0 & -1 & -2 & 0 & | & -8 \end{bmatrix} -1R_2$$

$$\begin{bmatrix} 1 & 1 & 1 & 1 & | & 4 \\ 0 & 1 & 4 & 3 & | & 8 \\ 0 & -3 & -2 & -3 & | & -6 \\ 0 & -1 & -2 & 0 & | & -8 \end{bmatrix} 3R_2 + R_3$$

$$\begin{bmatrix} 1 & 1 & 1 & 1 & | & 4 \\ 0 & 1 & 4 & 3 & | & 8 \\ 0 & 0 & 10 & 6 & | & 18 \\ 0 & -1 & -2 & 0 & | & -8 \end{bmatrix} 1R_2 + R_4$$

$$\begin{bmatrix} 1 & 1 & 1 & 1 & | & 4 \\ 0 & 1 & 4 & 3 & | & 8 \\ 0 & 0 & 10 & 6 & | & 18 \\ 0 & 0 & 2 & 3 & | & 0 \end{bmatrix} \frac{1}{10}R_3$$

$$\begin{bmatrix} 1 & 1 & 1 & 1 & | & 4 \\ 0 & 1 & 4 & 3 & | & 8 \\ 0 & 0 & 1 & \frac{3}{5} & | & \frac{9}{5} \\ 0 & 0 & 2 & 3 & | & 0 \end{bmatrix} -2R_3 + R_4$$

$$\begin{bmatrix} 1 & 1 & 1 & 1 & | & 4 \\ 0 & 1 & 4 & 3 & | & 8 \\ 0 & 0 & 1 & \frac{3}{5} & | & \frac{9}{5} \\ 0 & 0 & 0 & \frac{9}{5} & | & -\frac{18}{5} \end{bmatrix} \frac{5}{9}R_4$$

$$\begin{bmatrix} 1 & 1 & 1 & 1 & | & 4 \\ 0 & 1 & 4 & 3 & | & 8 \\ 0 & 0 & 1 & \frac{3}{5} & | & \frac{9}{5} \\ 0 & 0 & 0 & 1 & | & -2 \end{bmatrix}$$

$$z = -2$$
$$y + \frac{3}{5}z = \frac{9}{5}$$
$$y + \frac{3}{5}(-2) = \frac{9}{5}$$
$$y - \frac{6}{5} = \frac{9}{5}$$
$$y = 3$$
$$x + 4y + 3z = 8$$
$$x + 4(3) + 3(-2) = 8$$
$$x + 6 = 8$$
$$x = 2$$
$$w + x + y + z = 4$$
$$w + 2 + 3 - 2 = 4$$
$$w + 3 = 4$$
$$w = 1$$

The solution set is $\{(1, 2, 3, -2)\}$.

**37.**
$$3w - 4x + y + z = 9$$
$$w + x - y - z = 0$$
$$2w + x + 4y - 2z = 3$$
$$-w + 2x + y - 3z = 3$$

$$\begin{bmatrix} 3 & -4 & 1 & 1 & | & 9 \\ 1 & 1 & -1 & -1 & | & 0 \\ 2 & 1 & 4 & -2 & | & 3 \\ -1 & 2 & 1 & -3 & | & 3 \end{bmatrix} R_1 \leftrightarrow R_2$$

$$\begin{bmatrix} 1 & 1 & -1 & -1 & | & 0 \\ 3 & -4 & 1 & 1 & | & 9 \\ 2 & 1 & 4 & -2 & | & 3 \\ -1 & 2 & 1 & -3 & | & 3 \end{bmatrix} -3R_1 + R_2$$

$$\begin{bmatrix} 1 & 1 & -1 & -1 & | & 0 \\ 0 & -7 & 4 & 4 & | & 9 \\ 2 & 1 & 4 & -2 & | & 3 \\ -1 & 2 & 1 & -3 & | & 3 \end{bmatrix} -2R_1 + R_3$$

$$\begin{bmatrix} 1 & 1 & -1 & -1 & | & 0 \\ 0 & -7 & 4 & 4 & | & 9 \\ 0 & -1 & 6 & 0 & | & 3 \\ -1 & 2 & 1 & -3 & | & 3 \end{bmatrix} 1R_1 + R_4$$

$$\begin{bmatrix} 1 & 1 & -1 & -1 & | & 0 \\ 0 & -7 & 4 & 4 & | & 9 \\ 0 & -1 & 6 & 0 & | & 3 \\ 0 & 3 & 0 & -4 & | & 3 \end{bmatrix} R_2 \leftrightarrow R_3$$

$$\begin{bmatrix} 1 & 1 & -1 & -1 & | & 0 \\ 0 & -1 & 6 & 0 & | & 3 \\ 0 & -7 & 4 & 4 & | & 9 \\ 0 & 3 & 0 & -4 & | & 3 \end{bmatrix} -R_2$$

$$\begin{bmatrix} 1 & 1 & -1 & -1 & | & 0 \\ 0 & 1 & -6 & 0 & | & -3 \\ 0 & -7 & 4 & 4 & | & 9 \\ 0 & 3 & 0 & -4 & | & 3 \end{bmatrix} 7R_2 + R_3$$

$$\begin{bmatrix} 1 & 1 & -1 & -1 & | & 0 \\ 0 & 1 & -6 & 0 & | & -3 \\ 0 & 0 & -38 & 4 & | & -12 \\ 0 & 3 & 0 & -4 & | & 3 \end{bmatrix} -3R_2 + R_4$$

$$\begin{bmatrix} 1 & 1 & -1 & -1 & | & 0 \\ 0 & 1 & -6 & 0 & | & -3 \\ 0 & 0 & -38 & 4 & | & -12 \\ 0 & 0 & 18 & -4 & | & 12 \end{bmatrix} -\tfrac{1}{38}R_3$$

$$\begin{bmatrix} 1 & 1 & -1 & -1 & | & 0 \\ 0 & 1 & -6 & 0 & | & -3 \\ 0 & 0 & 1 & -\tfrac{2}{19} & | & \tfrac{6}{19} \\ 0 & 0 & 18 & -4 & | & 12 \end{bmatrix} -18R_3 + R_4$$

$$\begin{bmatrix} 1 & 1 & -1 & -1 & | & 0 \\ 0 & 1 & -6 & 0 & | & -3 \\ 0 & 0 & 1 & -\tfrac{2}{19} & | & \tfrac{6}{19} \\ 0 & 0 & 0 & -\tfrac{40}{19} & | & \tfrac{120}{19} \end{bmatrix} -\tfrac{19}{40}R_4$$

$$\begin{bmatrix} 1 & 1 & -1 & -1 & | & 0 \\ 0 & 1 & -6 & 0 & | & -3 \\ 0 & 0 & 1 & -\tfrac{2}{19} & | & \tfrac{6}{19} \\ 0 & 0 & 0 & 1 & | & -3 \end{bmatrix}$$

$$z = -3$$
$$y - \frac{2}{19}z = \frac{6}{19}$$
$$y - \frac{2}{19}(-3) = \frac{6}{19}$$
$$y + \frac{6}{19} = \frac{6}{19}$$
$$y = 0$$
$$x - 6y = -3$$
$$x - 6(0) = -3$$
$$x = -3$$
$$w + x - y - z = 0$$
$$w - 3 + 0 + 3 = 0$$
$$w = 0$$

The solution set is $\{(0, -3, 0, -3)\}$.

**39.** $f(x) = ax^2 + bx + c$

Use the given function values to find three equations in terms of $a$, $b$, and $c$.

$$f(-2) = a(-2)^2 + b(-2) + c = -4$$
$$4a - 2b + c = -4$$
$$f(1) = a(1)^2 + b(1) + c = 2$$
$$a + b + c = 2$$
$$f(2) = a(2)^2 + b(2) + c = 0$$
$$4a + 2b + c = 0$$

System of equations:
$$4a - 2b + c = -4$$
$$a + b + c = 2$$
$$4a + 2b + c = 0$$

Matrix:

$$\begin{bmatrix} 4 & -2 & 1 & | & -4 \\ 1 & 1 & 1 & | & 2 \\ 4 & 2 & 1 & | & 0 \end{bmatrix}$$

This gives $a = -1$, $b = 1$, and $c = 2$.

Thus, $f(x) = -x^2 + x + 2$.

**41.** $f(x) = ax^3 + bx^2 + cx + d$

Use the given function values to find four equations in terms of $a$, $b$, $c$, and $d$.

$$f(-1) = a(-1)^3 + b(-1)^2 + c(-1) + d = 0$$
$$-a + b - c + d = 0$$

$$f(1) = a(1)^3 + b(1)^2 + c(1) + d = 2$$
$$a + b + c + d = 2$$

$$f(2) = a(2)^3 + b(2)^2 + c(2) + d = 3$$
$$8a + 4b + 2c + d = 3$$

$$f(3) = a(3)^3 + b(3)^2 + c(3) + d = 12$$
$$27a + 9b + 3c + d = 12$$

System of equations:
$$-a + b - c + d = 0$$
$$a + b + c + d = 2$$
$$8a + 4b + 2c + d = 3$$
$$27a + 9b + 3c + d = 12$$

Matrix:

$$\begin{bmatrix} -1 & 1 & -1 & 1 & | & 0 \\ 1 & 1 & 1 & 1 & | & 2 \\ 8 & 4 & 2 & 1 & | & 3 \\ 27 & 9 & 3 & 1 & | & 12 \end{bmatrix}$$

This gives $a = 1$, $b = -2$, $c = 0$, and $d = 3$.

Thus, $f(x) = x^3 - 2x^2 + 3$.

**43.** Let $A = \ln w$, $B = \ln x$, $C = \ln y$, and $D = \ln z$.

System of equations:
$$2A + B + 3C - 2D = -6$$
$$4A + 3B + C - D = -2$$
$$A + B + C + D = -5$$
$$A + B - C - D = 5$$

Matrix:

$$\begin{bmatrix} 2 & 1 & 3 & -2 & | & -6 \\ 4 & 3 & 1 & -1 & | & -2 \\ 1 & 1 & 1 & 1 & | & -5 \\ 1 & 1 & -1 & -1 & | & 5 \end{bmatrix}$$

This gives $A = -1$, $B = 1$, $C = -3$, and $D = -2$.

Substitute back to find $w$, $x$, $y$, and $z$.

$$A = -1 \qquad B = 1$$
$$\ln w = -1 \qquad \ln x = 1$$
$$w = e^{-1} \qquad x = e^1$$
$$w \approx 0.37 \qquad x \approx 2.72$$
$$C = -3 \qquad D = -2$$
$$\ln y = -3 \qquad \ln z = -2$$
$$y = e^{-3} \qquad z = e^{-2}$$
$$y \approx 0.05 \qquad z \approx 0.14$$

**45. a.** $s(t) = \dfrac{1}{2}at^2 + v_0 t + s_0$

Use the given function values to find three equations in terms of $a$, $v_0$, and $s_0$.

$$s(1) = \frac{1}{2}a(1)^2 + v_0(1) + s_0 = 40$$
$$\frac{1}{2}a + v_0 + s_0 = 40$$

$$s(2) = \frac{1}{2}a(2)^2 + v_0(2) + s_0 = 48$$
$$2a + 2v_0 + s_0 = 48$$

$$s(3) = \frac{1}{2}a(3)^2 + v_0(3) + s_0 = 24$$
$$\frac{9}{2}a + 3v_0 + s_0 = 24$$

System of equations:
$$\frac{1}{2}a + v_0 + s_0 = 40$$
$$2a + 2v_0 + s_0 = 48$$
$$\frac{9}{2}a + 3v_0 + s_0 = 24$$

Matrix:

$$\begin{bmatrix} \dfrac{1}{2} & 1 & 1 & | & 40 \\ 2 & 2 & 1 & | & 48 \\ \dfrac{9}{2} & 3 & 1 & | & 24 \end{bmatrix}$$

This gives $a = -32$, $v_0 = 56$, and $s_0 = 0$.

Thus, $s(t) = \dfrac{1}{2}(-32)t^2 + (56)t + (0)$

$$s(t) = -16t^2 + 56t$$

**b.** $s(t) = -16t^2 + 56t$

$$s(3.5) = -16(3.5)^2 + 56(3.5) = 0$$

This is the point $(3.5, 0)$.

The ball's height is 0 feet after 3.5 seconds.

This is the point $(3.5, 0)$.

**c.** The maximum occurs when $x = -\dfrac{b}{2a}$.

$$x = -\frac{b}{2a} = -\frac{v_0}{2a} = -\frac{56}{2(-16)} = 1.75$$

$$s(1.75) = -16(1.75)^2 + 56(1.75) = 49$$

At 1.75 seconds the ball will reach its maximum height of 49 feet.

**47.** Let $x$ = Food A
Let $y$ = Food B
Let $z$ = Food C

$$40x + 200y + 400z = 660$$
$$5x + 2y + 4z = 25$$
$$30x + 10y + 300z = 425$$
$$2x + 10y + 20z = 33$$
$$5x + 2y + 4z = 25$$
$$6x + 2y + 60z = 85$$

$$\begin{bmatrix} 2 & 10 & 20 & | & 33 \\ 5 & 2 & 4 & | & 25 \\ 6 & 2 & 60 & | & 85 \end{bmatrix} \tfrac{1}{2}R_1$$

$$\begin{bmatrix} 1 & 5 & 10 & | & \tfrac{33}{2} \\ 5 & 2 & 4 & | & 25 \\ 6 & 2 & 60 & | & 85 \end{bmatrix} -5R_1 + R_2$$

$$\begin{bmatrix} 1 & 5 & 10 & | & \tfrac{33}{2} \\ 0 & -23 & -46 & | & -\tfrac{115}{2} \\ 6 & 2 & 60 & | & 85 \end{bmatrix} -6R_1 + R_3$$

$$\begin{bmatrix} 1 & 5 & 10 & | & \tfrac{33}{2} \\ 0 & -23 & -46 & | & -\tfrac{115}{2} \\ 0 & -28 & 0 & | & -14 \end{bmatrix} -\tfrac{1}{23}R_2$$

$$\begin{bmatrix} 1 & 5 & 10 & | & \tfrac{33}{2} \\ 0 & 1 & 2 & | & \tfrac{5}{2} \\ 0 & -28 & 0 & | & -14 \end{bmatrix} 28R_2 + R_3$$

$$\begin{bmatrix} 1 & 5 & 10 & | & \tfrac{33}{2} \\ 0 & 1 & 2 & | & \tfrac{5}{2} \\ 0 & 0 & 56 & | & 56 \end{bmatrix} \tfrac{1}{56}R_3$$

$$\begin{bmatrix} 1 & 5 & 10 & | & \tfrac{33}{2} \\ 0 & 1 & 2 & | & \tfrac{5}{2} \\ 0 & 0 & 1 & | & 1 \end{bmatrix}$$

$$z = 1$$

$$y + 2z = \frac{5}{2}$$

$$y + 2 = \frac{5}{2}$$

$$2y + 4 = 5$$

$$2y = 1$$

$$y = \frac{1}{2}$$

$$x + 5y + 10z = \frac{33}{2}$$

$$x + \frac{5}{2} + 10 = \frac{33}{2}$$

$$2x + 5 + 20 = 33$$

$$2x + 25 = 33$$

$$2x = 8$$

$$x = 4$$

4 ounces of Food A

$\dfrac{1}{2}$ ounce of Food B

1 ounce of Food C

**49.** Let $w$ = number of Asians
Let $x$ = number of Africans
Let $y$ = number of Europeans
Let $z$ = number of Americans
Use the variables to model each sentence.

$$w + x + y + z = 183$$
$$w - x - y = 70$$
$$y - z = 15$$
$$2x - y - z = 23$$

$$\begin{bmatrix} 1 & 1 & 1 & 1 & | & 183 \\ 1 & -1 & -1 & 0 & | & 70 \\ 0 & 0 & 1 & -1 & | & 15 \\ 0 & 2 & -1 & -1 & | & 23 \end{bmatrix} -R_1 + R_2$$

$$\begin{bmatrix} 1 & 1 & 1 & 1 & | & 183 \\ 0 & -2 & -2 & -1 & | & -113 \\ 0 & 0 & 1 & -1 & | & 15 \\ 0 & 2 & -1 & -1 & | & 23 \end{bmatrix} R_4 + R_2$$

$$\begin{bmatrix} 1 & 1 & 1 & 1 & | & 183 \\ 0 & -2 & -2 & -1 & | & -113 \\ 0 & 0 & 1 & -1 & | & 15 \\ 0 & 0 & -3 & -2 & | & -90 \end{bmatrix} -\tfrac{1}{2}R_2$$

$$\begin{bmatrix} 1 & 1 & 1 & 1 & | & 183 \\ 0 & 1 & 1 & \tfrac{1}{2} & | & \tfrac{113}{2} \\ 0 & 0 & 1 & -1 & | & 15 \\ 0 & 0 & -3 & -2 & | & -90 \end{bmatrix} 3R_3 + R_4$$

$$\begin{bmatrix} 1 & 1 & 1 & 1 & | & 183 \\ 0 & 1 & 1 & \tfrac{1}{2} & | & \tfrac{113}{2} \\ 0 & 0 & 1 & -1 & | & 15 \\ 0 & 0 & 0 & -5 & | & -45 \end{bmatrix} -\tfrac{1}{5}R_4$$

$$\begin{bmatrix} 1 & 1 & 1 & 1 & | & 183 \\ 0 & 1 & 1 & \frac{1}{2} & | & \frac{113}{2} \\ 0 & 0 & 1 & -1 & | & 15 \\ 0 & 0 & 0 & 1 & | & 9 \end{bmatrix}$$

Back-substitute $z = 9$ to find $y$.

$y - z = 15$

$y - 9 = 15$

$y = 24$

Back-substitute to find $x$.

$x + y + \frac{1}{2}z = \frac{113}{2}$

$x + 24 + \frac{1}{2}(9) = \frac{113}{2}$

$x + 24 + \frac{1}{2}(9) = \frac{113}{2}$

$x + 28\frac{1}{2} = 56\frac{1}{2}$

$x = 28$

Back-substitute to find $w$.

$w + x + y + z = 183$

$w + 28 + 24 + 9 = 183$

$w + 28 + 24 + 9 = 183$

$w + 61 = 183$

$w = 122$

The number of Asians, Africans, Europeans, and Americans are 122, 28, 24, and 9, respectively.

**51. – 57.** Answers will vary.

**59.** makes sense

**61.** makes sense

**63.** false; Changes to make the statement true will vary. A sample change is: Multiplying a row by a negative fraction is permitted.

**65.** false; Changes to make the statement true will vary. A sample change is: When solving a system of three equations in three variables, we use row operations to obtain ones along the diagonal and zeros below the ones.

**67.** $y = ax^2 + bx + c$

$5900 = a(30)^2 + b(30) + c$

$5900 = 900a + 30b + c$

$7500 = a(50)^2 + b(50) + c$

$7500 = 2500a + 50b + c$

$4500 = a(100)^2 + b(100) + c$

$4500 = 10,000a + 100b + c$

$900a + 30b + c = 5900$

$2500a + 50b + c = 7500$

$10,000a + 100b + c = 4500$

$$\begin{bmatrix} 900 & 30 & 1 & | & 5900 \\ 2500 & 50 & 1 & | & 7500 \\ 10,000 & 100 & 1 & | & 4500 \end{bmatrix} R_1 \leftrightarrow R_2$$

$$\begin{bmatrix} 2500 & 50 & 1 & | & 7500 \\ 900 & 30 & 1 & | & 5900 \\ 10,000 & 100 & 1 & | & 4500 \end{bmatrix} \frac{1}{2500}R_1$$

$$\begin{bmatrix} 1 & \frac{1}{50} & \frac{1}{2500} & | & 3 \\ 900 & 30 & 1 & | & 5900 \\ 10,000 & 100 & 1 & | & 4500 \end{bmatrix} \begin{matrix} -900R_1 + R_2 \\ -10,000R_1 + R_3 \end{matrix}$$

$$\begin{bmatrix} 1 & \frac{1}{50} & \frac{1}{2500} & | & 3 \\ 0 & 12 & \frac{16}{25} & | & 3200 \\ 0 & -100 & -3 & | & -25,500 \end{bmatrix} R_2 \leftrightarrow R_3$$

$$\begin{bmatrix} 1 & \frac{1}{50} & \frac{1}{2500} & | & 3 \\ 0 & -100 & -3 & | & -25,500 \\ 0 & 12 & \frac{16}{25} & | & 3200 \end{bmatrix} \frac{-1}{100}R_2$$

$$\begin{bmatrix} 1 & \frac{1}{50} & \frac{1}{2500} & | & 3 \\ 0 & 1 & \frac{3}{100} & | & 255 \\ 0 & 12 & \frac{16}{25} & | & 3200 \end{bmatrix} \begin{matrix} -12R_2 + R_3 \\ -\frac{1}{50}R_2 + R_1 \end{matrix}$$

$$\begin{bmatrix} 1 & 0 & \frac{-1}{5000} & | & \frac{-21}{10} \\ 0 & 1 & \frac{3}{100} & | & 255 \\ 0 & 0 & \frac{7}{25} & | & 140 \end{bmatrix} \frac{25}{7}R_3$$

$$\begin{bmatrix} 1 & 0 & \frac{-1}{5000} & | & \frac{-21}{10} \\ 0 & 1 & \frac{3}{100} & | & 255 \\ 0 & 0 & 1 & | & 500 \end{bmatrix} \begin{matrix} \frac{-3}{100}R_3 + R_2 \\ \frac{1}{5000}R_3 + R_1 \end{matrix}$$

$$\begin{bmatrix} 1 & 0 & 0 & | & -2 \\ 0 & 1 & 0 & | & 240 \\ 0 & 0 & 1 & | & 500 \end{bmatrix}$$

$y = -2x^2 + 240x + 500$

$y = -2(x^2 - 120x) + 500$

$y = -2(x^2 - 120x + 3600) + 500 + 7200$

$y = -2(x - 60)^2 + 7700$

60 units produce $7700.

**68.** When $z = 0$, $(12z + 1, 10z - 1, z)$ is equivalent to $(12(0) + 1, 10(0) - 1, 0)$ or $(1, -1, 0)$.

Check $(1, -1, 0)$ in each equation.

$3x - 4y + 4z = 7$

$3(1) - 4(-1) + 4(0) = 7$

$7 = 7$, true

$x - y - 2z = 2$

$1 - (-1) - 2(0) = 2$

$2 = 2$, true

$2x - 3y - 2z = 5$

$2(1) - 3(-1) + 6(0) = 5$

$5 = 5$, true

$(1, -1, 0)$ satisfies each equation and, therefore, satisfies the system.

**69.** When $z = 1$, $(12z + 1, 10z - 1, z)$ is equivalent to $(12(1) + 1, 10(1) - 1, 1)$ or $(13, 9, 1)$.

Check $(13, 9, 1)$ in each equation.

$3x - 4y + 4z = 7$

$3(13) - 4(9) + 4(1) = 7$

$7 = 7$, true

$x - y - 2z = 2$

$13 - (9) - 2(1) = 2$

$2 = 2$, true

$2x - 3y - 2z = 5$

$2(13) - 3(9) + 6(1) = 5$

$5 = 5$, true

$(13, 9, 1)$ satisfies each equation and, therefore, satisfies the system.

**70.** **a.** Answers will vary. A sample answer is given selecting $z = 10$.

When $z = 10$, $(12z + 1, 10z - 1, z)$ is equivalent to $(12(10) + 1, 10(10) - 1, 10)$ or $(121, 99, 10)$.

Check $(121, 99, 10)$ in each equation.

$3x - 4y + 4z = 7$

$3(121) - 4(99) + 4(10) = 7$

$7 = 7$, true

$x - y - 2z = 2$

$121 - (99) - 2(10) = 2$

$2 = 2$, true

$2x - 3y - 2z = 5$

$2(121) - 3(99) + 6(10) = 5$

$5 = 5$, true

$(121, 99, 10)$ satisfies each equation and, therefore, satisfies the system.

**b.** This system has more than one solution.

**Section 6.2**

**Check Point Exercises**

**1.** $\begin{cases} x - 2y - z = 5 \\ 2x - 3y - z = 0 \\ 3x - 4y - z = 1 \end{cases} \rightarrow \begin{bmatrix} 1 & -2 & -1 & | & -5 \\ 2 & -3 & -1 & | & 0 \\ 3 & -4 & -1 & | & 1 \end{bmatrix}$

$\begin{bmatrix} 1 & -2 & -1 & | & -5 \\ 2 & -3 & -1 & | & 0 \\ 3 & -4 & -1 & | & 1 \end{bmatrix} \begin{matrix} \\ -2R_1 + R_2 \\ -3R_1 + R_3 \end{matrix}$

$\begin{bmatrix} 1 & -2 & -1 & | & -5 \\ 0 & 1 & 1 & | & 10 \\ 0 & 2 & 2 & | & 16 \end{bmatrix} -2R_2 + R_3$

$\begin{bmatrix} 1 & -2 & -1 & | & -5 \\ 0 & 1 & -1 & | & -10 \\ 0 & 0 & 0 & | & -4 \end{bmatrix}$

$0x + 0y + 0z = -4$ This equation can never be a true statement. Consequently, the system has no solution. The solution set is $\emptyset$, the empty set.

**2.** $\begin{cases} x - 2y - z = 5 \\ 2x - 5y + 3z = 16 \\ x - 3y + 4z = 1 \end{cases} \rightarrow \begin{bmatrix} 1 & -2 & -1 & | & 5 \\ 2 & -5 & 3 & | & 6 \\ 1 & -3 & 4 & | & 1 \end{bmatrix}$

$\begin{bmatrix} 1 & -2 & -1 & | & 5 \\ 2 & -5 & 3 & | & 6 \\ 1 & -3 & 4 & | & 1 \end{bmatrix} \begin{matrix} \\ -2R_1 + R_2 \\ -1R_1 + R_3 \end{matrix}$

$\begin{bmatrix} 1 & -2 & -1 & | & 5 \\ 0 & -1 & 5 & | & -4 \\ 0 & -1 & 5 & | & -4 \end{bmatrix} -1R_2$

$\begin{bmatrix} 1 & -2 & -1 & | & 5 \\ 0 & 1 & -5 & | & 4 \\ 0 & -1 & 5 & | & -4 \end{bmatrix} 1R_2 + R_3$

$\begin{bmatrix} 1 & -2 & -1 & | & 5 \\ 0 & 1 & -5 & | & 4 \\ 0 & 0 & 0 & | & 0 \end{bmatrix}$

$0x + 0y + 0z = 0$ or $0 = 0$

This equation, $0x + 0y + 0z = 0$ is *dependent* on the other two equations. Thus, it can be dropped from the system which can now be expressed in the form

$\begin{bmatrix} 1 & -2 & -1 & | & 5 \\ 0 & 1 & -5 & | & 4 \end{bmatrix}$

The original system is equivalent to the system

$$x - 2y - z = 5$$
$$y - 5z = 4$$

Solve for $x$ and $y$ in terms of $z$.

$$y = 5z + 4$$

Use back-substitution for $y$ in the previous equation.

$$x - 2(5z + 4) - z = 5$$
$$x - 10z - 8 - z = 5$$
$$x = 11z + 13$$

Finally, letting $z = t$ (or any letter of your choice), the solutions to the system are all of the form $x = 11t + 13$, $y = 5t + 4$, $z = t$, where $t$ is a real number. The solution set of the system with dependent equations can be written as $\{(11t + 13, 5t + 4, t)\}$.

**3.** $\begin{cases} x + 2y + 3z = 70 \\ x + y + z = 60 \end{cases} \rightarrow \begin{bmatrix} 1 & 2 & 3 & | & 70 \\ 1 & 1 & 1 & | & 60 \end{bmatrix}$

$\begin{bmatrix} 1 & 2 & 3 & | & 70 \\ 1 & 1 & 1 & | & 60 \end{bmatrix} \begin{matrix} \\ -1R_1 + R_2 \end{matrix}$

$\begin{bmatrix} 1 & 2 & 3 & | & 70 \\ 0 & -1 & -2 & | & -10 \end{bmatrix} \begin{matrix} \\ -1R_2 \end{matrix}$

$\begin{bmatrix} 1 & 2 & 3 & | & 70 \\ 0 & 1 & 2 & | & 10 \end{bmatrix} \rightarrow \begin{matrix} x + 2y + 3z = 70 \\ y + 2z = 10 \end{matrix}$

Express $x$ and $y$ in terms of $z$ using back-substitution.

$$y = -2z + 10$$
$$x + 2(-2z + 10) + 3z = 70$$
$$x - 4z + 20 + 3z = 70$$
$$x = z + 50$$

With $z = t$, the ordered solution $(x, y, z)$ enables us to express the system's solution set as $\{(t + 50, -2t + 10, t)\}$.

**4. a.** $I_1$: $10 + 5 = 15$ cars enter $I_1$, and $w + z$ cars leave $I_1$, then $w + z = 15$.

$I_2$: $20 + 10 = 30$ cars enter $I_2$ and $w + x$ cars leave $I_2$, then $w + x = 30$.

$I_3$: $15 + 30 = 45$ cars enter $I_3$ and $x + y$ cars leave $I_3$, then $x + y = 45$.

$I_4$: $10 + 20 = 30$ cars enter $I_4$ and $y + z$ cars leave $I_4$, then $y + z = 30$.

The system of equations that describes this situation is given by

$$w + z = 15$$
$$w + x = 30$$
$$x + y = 45$$
$$y + z = 30$$

**b.** $\begin{bmatrix} 1 & 0 & 0 & 1 & | & 15 \\ 1 & 1 & 0 & 0 & | & 30 \\ 0 & 1 & 1 & 0 & | & 45 \\ 0 & 0 & 1 & 1 & | & 30 \end{bmatrix} \begin{matrix} \\ -1R_1 + R_2 \\ \\ \end{matrix}$

$\begin{bmatrix} 1 & 0 & 0 & 1 & | & 15 \\ 0 & 1 & 0 & -1 & | & 15 \\ 0 & 1 & 1 & 0 & | & 45 \\ 0 & 0 & 1 & 1 & | & 30 \end{bmatrix} \begin{matrix} \\ -1R_2 + R_3 \\ \\ \end{matrix}$

$\begin{bmatrix} 1 & 0 & 0 & 1 & | & 15 \\ 0 & 1 & 0 & -1 & | & 15 \\ 0 & 0 & 1 & 1 & | & 30 \\ 0 & 0 & 1 & 1 & | & 30 \end{bmatrix} \begin{matrix} \\ -1R_3 + R_4 \\ \\ \end{matrix}$

$\begin{bmatrix} 1 & 0 & 0 & 1 & | & 15 \\ 0 & 1 & 0 & -1 & | & 15 \\ 0 & 0 & 1 & 1 & | & 30 \\ 0 & 0 & 0 & 0 & | & 0 \end{bmatrix}$

$$x + w = 15$$
$$y - w = 15$$
$$z + w = 30$$

The last row of the matrix shows that the system has dependent equations and infinitely many solutions.

Let $z$ be any real number.
Express $w$, $x$ and $y$ in terms of $z$:

$$w = 15 - z$$
$$x = 15 + z$$
$$y = 30 - z$$

With $w = t$, the ordered solution $(w, x, y, z)$ enables us to express the system's solution set as $\{(-t + 15, t + 15, -t + 30, t)\}$

**c.** If $z = 10$,

$$w = 15 - z$$
$$w = 15 - 10$$
$$= 5$$
$$x = 15 + z$$
$$x = 15 + 10$$
$$= 25$$
$$y = 30 - z$$
$$y = 30 - 10$$
$$= 20$$

If construction limits $z$ to 10 cars per minute, then the number of cars per minute that must pass between the other intersections to keep traffic flowing is $w = 5$, $x = 25$, and $y = 20$.

## Concept and Vocabulary Check 6.2

1. no solution

2. one solution

3. infinitely many solutions

4. true

5. $\begin{cases} x-y+z=5 \\ y-z=-3 \end{cases}$; $z-3$; $2$; $\{(2, z-3, z)\}$

## Exercise Set 6.2

1. $\begin{bmatrix} 5 & 12 & 1 & | & 10 \\ 2 & 5 & 2 & | & -1 \\ 1 & 2 & -3 & | & 5 \end{bmatrix} R_1 \leftrightarrow R_3$

$\begin{bmatrix} 1 & 2 & -3 & | & 5 \\ 2 & 5 & 2 & | & -1 \\ 5 & 12 & 1 & | & 10 \end{bmatrix} \begin{matrix} \\ -2R_1 + R_2 \\ -5R_1 + R_3 \end{matrix}$

$\begin{bmatrix} 1 & 2 & -3 & | & 5 \\ 0 & 1 & 8 & | & -11 \\ 0 & 2 & 16 & | & -15 \end{bmatrix} -2R_2 + R_3$

$\begin{bmatrix} 1 & 2 & 3 & | & 5 \\ 0 & 1 & 8 & | & -11 \\ 0 & 0 & 0 & | & 7 \end{bmatrix}$

From the last row, we see that the system has no solution. The solution set is $\varnothing$, the empty set.

3. $\begin{bmatrix} 5 & 8 & -6 & | & 14 \\ 3 & 4 & -2 & | & 8 \\ 1 & 2 & -2 & | & 3 \end{bmatrix} R_1 \leftrightarrow R_3$

$\begin{bmatrix} 1 & 2 & -2 & | & 3 \\ 3 & 4 & -2 & | & 8 \\ 5 & 8 & -6 & | & 14 \end{bmatrix} \begin{matrix} \\ -3R_1 + R_2 \\ -5R_1 + R_3 \end{matrix}$

$\begin{bmatrix} 1 & 2 & -2 & | & 3 \\ 0 & -2 & 4 & | & -1 \\ 0 & -2 & 4 & | & -1 \end{bmatrix} -1R_2 + R_3$

$\begin{bmatrix} 1 & 2 & -2 & | & 3 \\ 0 & -2 & 4 & | & -1 \\ 0 & 0 & 0 & | & 0 \end{bmatrix} -\frac{1}{2}R_2$

$\begin{bmatrix} 1 & 2 & -2 & | & 3 \\ 0 & 1 & -2 & | & \frac{1}{2} \\ 0 & 0 & 0 & | & 0 \end{bmatrix}$

The system $\begin{cases} x+2y-2z=3 \\ y-2z=\dfrac{1}{2} \end{cases}$ has no unique solution.

Express $x$ and $y$ in terms of $z$:

$y = 2z + \dfrac{1}{2}$

$x + 2\left(2z + \dfrac{1}{2}\right) - 2z = 3$

$x + 4z + 1 - 2z = 3$

$x + 2z + 1 = 3$

$x = -2z + 2$

With $z = t$, the complete solution to the system is

$\left\{ \left( -2t+2, \; 2t+\dfrac{1}{2}, \; t \right) \right\}$.

5. $\begin{bmatrix} 3 & 4 & 2 & | & 3 \\ 4 & -2 & -8 & | & -4 \\ 1 & 1 & -1 & | & 3 \end{bmatrix} R_1 \leftrightarrow R_3$

$\begin{bmatrix} 1 & 1 & -1 & | & 3 \\ 4 & -2 & -8 & | & -4 \\ 3 & 4 & 2 & | & 3 \end{bmatrix} \begin{matrix} \\ -4R_1 + R_2 \\ -3R_1 + R_3 \end{matrix}$

$\begin{bmatrix} 1 & 1 & -1 & | & 3 \\ 0 & -6 & -4 & | & -16 \\ 0 & 1 & 5 & | & -6 \end{bmatrix} R_2 \leftrightarrow R_3$

$\begin{bmatrix} 1 & 1 & -1 & | & 3 \\ 0 & 1 & 5 & | & -6 \\ 0 & -6 & -4 & | & -16 \end{bmatrix} 6R_2 + R_3$

$\begin{bmatrix} 1 & 1 & 1 & | & 3 \\ 0 & 1 & 5 & | & -6 \\ 0 & 0 & 26 & | & -52 \end{bmatrix} \frac{1}{26}R_3$

$\begin{bmatrix} 1 & 1 & -1 & | & 3 \\ 0 & 1 & 5 & | & -6 \\ 0 & 0 & 1 & | & -2 \end{bmatrix}$

This corresponds to the system
$$x + y - z = 3$$
$$y + 5z = -6$$
$$z = -2$$
Use back-substitution to find the values of $x$ and $y$:
$$y + 5(-2) = -6$$
$$y - 10 = -6$$
$$y = 4$$
$$x + 4 + 2 = 3$$
$$x + 6 = 3$$
$$x = -3$$
The solution to the system is $\{(-3, 4, -2)\}$.

**7.** $\begin{bmatrix} 8 & 5 & 11 & | & 30 \\ -1 & -4 & 2 & | & 3 \\ 2 & -1 & 5 & | & 12 \end{bmatrix} R_1 \leftrightarrow R_2$

$\begin{bmatrix} -1 & -4 & 2 & | & 3 \\ 8 & 5 & 11 & | & 30 \\ 2 & -1 & 5 & | & 12 \end{bmatrix} -1R_1$

$\begin{bmatrix} 1 & 4 & -2 & | & -3 \\ 8 & 5 & 11 & | & 30 \\ 2 & -1 & 5 & | & 12 \end{bmatrix} \begin{array}{l} -8R_1 + R_2 \\ -2R_1 + R_3 \end{array}$

$\begin{bmatrix} 1 & 4 & -2 & | & -3 \\ 0 & -27 & 27 & | & 54 \\ 0 & -9 & 9 & | & 18 \end{bmatrix} -\frac{1}{27}R_2$

$\begin{bmatrix} 1 & 4 & -2 & | & -3 \\ 0 & 1 & -1 & | & -2 \\ 0 & -9 & 9 & | & 18 \end{bmatrix} 9R_2 + R_3$

$\begin{bmatrix} 1 & 4 & -2 & | & -3 \\ 0 & 1 & -1 & | & -2 \\ 0 & 0 & 0 & | & 0 \end{bmatrix}$

The system $\begin{aligned} x + 4y - 2z &= -3 \\ y - z &= -2 \end{aligned}$ has no unique solution.

Express $x$ and $y$ in terms of $z$:
$y = -2 + z$
$x + 4(-2 + z) - 2z = -3$
$x - 8 + 4z - 2z = -3$
$x - 8 + 2z = -3$
$x = 5 - 2z$
With $z = t$, the complete solution to the system is
$\{(5 - 2t, -2 + t, t)\}$.

**9.** $\begin{bmatrix} 1 & -2 & -1 & -3 & | & -9 \\ 1 & 1 & -1 & 0 & | & 0 \\ 3 & 4 & 0 & 1 & | & 6 \\ 0 & 2 & -2 & 1 & | & 3 \end{bmatrix} \begin{array}{l} -1R_1 + R_2 \\ -3R_1 + R_3 \end{array}$

$\begin{bmatrix} 1 & -2 & -1 & -3 & | & -9 \\ 0 & 3 & 0 & 3 & | & 9 \\ 0 & 10 & 3 & 10 & | & 33 \\ 0 & 2 & -2 & 1 & | & 3 \end{bmatrix} \frac{1}{3}R_2$

$\begin{bmatrix} 1 & -2 & -1 & -3 & | & -9 \\ 0 & 1 & 0 & 1 & | & 3 \\ 0 & 10 & 3 & 10 & | & 33 \\ 0 & 2 & -2 & 1 & | & 3 \end{bmatrix} \begin{array}{l} -10R_2 + R_3 \\ -2R_2 + R_4 \end{array}$

$\begin{bmatrix} 1 & -2 & -1 & -3 & | & -9 \\ 0 & 1 & 0 & 1 & | & 3 \\ 0 & 0 & 3 & 0 & | & 3 \\ 0 & 0 & -2 & -1 & | & -3 \end{bmatrix} \frac{1}{3}R_3$

$\begin{bmatrix} 1 & -2 & -1 & -3 & | & -9 \\ 0 & 1 & 0 & 1 & | & 3 \\ 0 & 0 & 1 & 0 & | & 1 \\ 0 & 0 & -2 & -1 & | & -3 \end{bmatrix} 2R_3 + R_4$

$\begin{bmatrix} 1 & -2 & -1 & -3 & | & -9 \\ 0 & 1 & 0 & 1 & | & 3 \\ 0 & 0 & 1 & 0 & | & 1 \\ 0 & 0 & 0 & -1 & | & -1 \end{bmatrix} -1R_4$

$\begin{bmatrix} 1 & -2 & -1 & -3 & | & -9 \\ 0 & 1 & 0 & 1 & | & 3 \\ 0 & 0 & 1 & 0 & | & 1 \\ 0 & 0 & 0 & 1 & | & 1 \end{bmatrix}$

This corresponds to the system
$\begin{aligned} w - 2x - y - 3z &= -9 \\ x + z &= 3 \\ y &= 1 \\ z &= 1 \end{aligned}$

Use back-substitution to find the values of $w$ and $x$:
$x + 1 = 3$
$x = 2$
$w - 2(2) - 1 - 3(1) = -9$
$w - 4 - 1 - 3 = -9$
$w - 8 = -9$
$w = -1$
The solution to the system is $\{(-1, 2, 1, 1)\}$.

**11.** $\begin{bmatrix} 2 & 1 & -1 & 0 & | & 3 \\ 1 & -3 & 2 & 0 & | & -4 \\ 3 & 1 & -3 & 1 & | & 1 \\ 1 & 2 & -4 & -1 & | & -2 \end{bmatrix} R_1 \leftrightarrow R_2$

$\begin{bmatrix} 1 & -3 & 2 & 0 & | & -4 \\ 2 & 1 & -1 & 0 & | & 3 \\ 3 & 1 & -3 & 1 & | & 1 \\ 1 & 2 & -4 & -1 & | & -2 \end{bmatrix} \begin{array}{l} -2R_1 + R_2 \\ -3R_1 + R_3 \\ -1R_1 + R_4 \end{array}$

$\begin{bmatrix} 1 & -3 & 2 & 0 & | & -4 \\ 0 & 7 & -5 & 0 & | & 11 \\ 0 & 10 & -9 & 1 & | & 13 \\ 0 & 5 & -6 & -1 & | & 2 \end{bmatrix} \frac{1}{7}R_2$

$$\begin{bmatrix} 1 & -3 & 2 & 0 & | & -4 \\ 0 & 1 & -\frac{5}{7} & 0 & | & \frac{11}{7} \\ 0 & 10 & -9 & 1 & | & 13 \\ 0 & 5 & -6 & -1 & | & 2 \end{bmatrix} \begin{matrix} \\ \\ -10R_2 + R_3 \\ -5R_2 + R_4 \end{matrix}$$

$$\begin{bmatrix} 1 & -3 & 2 & 0 & | & -4 \\ 0 & 1 & -\frac{5}{7} & 0 & | & \frac{11}{7} \\ 0 & 0 & -\frac{13}{7} & 1 & | & -\frac{19}{7} \\ 0 & 0 & -\frac{17}{7} & -1 & | & -\frac{41}{7} \end{bmatrix} \begin{matrix} \\ \\ -\frac{7}{13}R_3 \\ \\ \end{matrix}$$

$$\begin{bmatrix} 1 & -3 & 2 & 0 & | & -4 \\ 0 & 1 & -\frac{5}{7} & 0 & | & \frac{11}{7} \\ 0 & 0 & 1 & -\frac{7}{13} & | & \frac{19}{13} \\ 0 & 0 & -\frac{17}{7} & -1 & | & -\frac{41}{7} \end{bmatrix} \begin{matrix} \\ \\ \frac{17}{7}R_3 + R_4 \\ \\ \end{matrix}$$

$$\begin{bmatrix} 1 & -3 & 2 & 0 & | & -4 \\ 0 & 1 & -\frac{5}{7} & 0 & | & \frac{11}{7} \\ 0 & 0 & 1 & -\frac{7}{13} & | & \frac{19}{13} \\ 0 & 0 & 0 & -\frac{30}{13} & | & -\frac{30}{13} \end{bmatrix} \begin{matrix} \\ \\ \\ -\frac{13}{30}R_4 \end{matrix}$$

$$\begin{bmatrix} 1 & -3 & 2 & 0 & | & -4 \\ 0 & 1 & -\frac{5}{7} & 0 & | & \frac{11}{7} \\ 0 & 0 & 1 & -\frac{7}{13} & | & \frac{19}{13} \\ 0 & 0 & 0 & 1 & | & 1 \end{bmatrix}$$

This corresponds to the system
$$w - 3x + 2y = -4$$
$$x - \frac{5}{7}y = \frac{11}{7}$$
$$y - \frac{7}{13}z = \frac{19}{13}$$
$$z = 1$$

Use back-substitution to find the values of $w$, $x$, and $y$:
$$y - \frac{7}{13}z = \frac{19}{13}$$
$$y - \frac{7}{13}(1) = \frac{19}{13}$$
$$y = 2$$
$$x - \frac{5}{7}(2) = \frac{11}{7}$$
$$x - \frac{10}{7} = \frac{11}{7}$$
$$x = 3$$
$$w - 3(3) + 2(2) = -4$$
$$w - 9 + 4 = -4$$
$$w - 5 = -4$$
$$w = 1$$
The solution to the system is $\{(1, 3, 2, 1)\}$.

**13.** $$\begin{bmatrix} 1 & -3 & 1 & -4 & | & 4 \\ -2 & 1 & 2 & 0 & | & -2 \\ 3 & -2 & 1 & -6 & | & 2 \\ -1 & 3 & 2 & -1 & | & -6 \end{bmatrix} \begin{matrix} 2R_1 + R_2 \\ -3R_1 + R_3 \\ R_1 + R_4 \\ \end{matrix}$$

$$\begin{bmatrix} 1 & -3 & 1 & -4 & | & 4 \\ 0 & -5 & 4 & -8 & | & 6 \\ 0 & 7 & -2 & 6 & | & -10 \\ 0 & 0 & 3 & -5 & | & -2 \end{bmatrix} \begin{matrix} \\ -\frac{1}{5}R_2 \\ \\ \end{matrix}$$

$$\begin{bmatrix} 1 & -3 & 1 & -4 & | & 4 \\ 0 & 1 & -\frac{4}{5} & \frac{8}{5} & | & -\frac{6}{5} \\ 0 & 7 & -2 & 6 & | & -10 \\ 0 & 0 & 3 & -5 & | & -2 \end{bmatrix} \begin{matrix} \\ \\ -7R_2 + R_3 \\ \end{matrix}$$

$$\begin{bmatrix} 1 & -3 & 1 & -4 & | & 4 \\ 0 & 1 & -\frac{4}{5} & \frac{8}{5} & | & -\frac{6}{5} \\ 0 & 0 & \frac{18}{5} & -\frac{26}{5} & | & -\frac{8}{5} \\ 0 & 0 & 3 & -5 & | & -2 \end{bmatrix} \begin{matrix} \\ \\ \frac{5}{18}R_3 \\ \end{matrix}$$

$$\begin{bmatrix} 1 & -3 & 1 & -4 & | & 4 \\ 0 & 1 & -\frac{4}{5} & \frac{8}{5} & | & -\frac{6}{5} \\ 0 & 0 & 1 & -\frac{13}{9} & | & -\frac{4}{9} \\ 0 & 0 & 3 & -5 & | & -2 \end{bmatrix} \begin{matrix} \\ \\ -3R_3 + R_4 \\ \end{matrix}$$

$$\begin{bmatrix} 1 & -3 & 1 & -4 & | & 4 \\ 0 & 1 & -\frac{4}{5} & \frac{8}{5} & | & -\frac{6}{5} \\ 0 & 0 & 1 & -\frac{13}{9} & | & -\frac{4}{9} \\ 0 & 0 & 0 & -\frac{2}{3} & | & -\frac{2}{3} \end{bmatrix} \begin{matrix} \\ \\ \\ -\frac{3}{2}R_4 \end{matrix}$$

$$\begin{bmatrix} 1 & -3 & 1 & -4 & | & 4 \\ 0 & 1 & -\frac{4}{5} & \frac{8}{5} & | & -\frac{6}{5} \\ 0 & 0 & 1 & -\frac{13}{9} & | & -\frac{4}{9} \\ 0 & 0 & 0 & 1 & | & 1 \end{bmatrix}$$

This corresponds to the system
$$w - 3x + y - 4z = 4$$
$$x - \frac{4}{5}y + \frac{8}{5}z = -\frac{6}{5}$$
$$y - \frac{13}{9}z = -\frac{4}{9}$$
$$z = 1$$

Use back-substitution to find the values of $w$, $z$, and $y$:
$$y - \frac{13}{9}(1) = -\frac{4}{9}$$
$$y = 1$$

$$x - \frac{4}{5}(1) + \frac{8}{5}(1) = -\frac{6}{5}$$
$$x + \frac{4}{5} = -\frac{6}{5}$$
$$x = -2$$
$$w - 3(-2) + 1 - 4 = 4$$
$$w + 6 - 3 = 4$$
$$w = 1$$

The solution to the system is $\{(1, -2, 1, 1)\}$.

**15.** $\begin{bmatrix} 2 & 1 & -1 & | & 2 \\ 3 & 3 & -2 & | & 3 \end{bmatrix} \frac{1}{2}R_1$ $\begin{bmatrix} 1 & \frac{1}{2} & -\frac{1}{2} & | & 1 \\ 3 & 3 & -2 & | & 3 \end{bmatrix} -3R_1 + R_2$

$\begin{bmatrix} 1 & \frac{1}{2} & -\frac{1}{2} & | & 1 \\ 0 & \frac{3}{2} & -\frac{1}{2} & | & 0 \end{bmatrix} \frac{2}{3}R_2$

$\begin{bmatrix} 1 & \frac{1}{2} & -\frac{1}{2} & | & 1 \\ 0 & 1 & -\frac{1}{3} & | & 0 \end{bmatrix}$

The system $x + \frac{1}{2}y - \frac{1}{2}z = 1$ has no unique solution.
$$y - \frac{1}{3}z = 0$$

Express $x$ and $y$ in terms of $z$:
$$y = \frac{1}{3}z$$
$$x + \frac{1}{2}\left(\frac{1}{3}z\right) - \frac{1}{2}z = 1$$
$$x + \frac{1}{6}z - \frac{1}{2}z = 1$$
$$x - \frac{1}{3}z = 1$$
$$x = 1 + \frac{1}{3}z$$

With $z = t$, the complete solution to the system is
$$\left\{\left(1 + \frac{1}{3}t, \frac{1}{3}t, t\right)\right\}.$$

**17.** The system $\begin{aligned} x + 2y + 3z &= 5 \\ y - 5z &= 0 \end{aligned}$ has no unique solution.

Express $x$ and $y$ in terms of $z$:
$$y = 5z$$
$$x + 2(5z) + 3z = 5$$
$$x + 10z + 3z = 5$$
$$x = -13z + 5$$
With $z = t$, the complete solution to the system is
$\{(-13t + 5, 5t, t)\}$.

**19.** $\begin{bmatrix} 1 & 1 & -2 & | & 2 \\ 3 & -1 & -6 & | & -7 \end{bmatrix} -3R_1 + R_2$

$\begin{bmatrix} 1 & 1 & -2 & | & 2 \\ 0 & -4 & 0 & | & -13 \end{bmatrix} -\frac{1}{4}R_2$

$\begin{bmatrix} 1 & 1 & -2 & | & 2 \\ 0 & 1 & 0 & | & \frac{13}{4} \end{bmatrix}$

$$x + y - 2z = 2$$
The system $y = \frac{13}{4}$ has no unique solution.

Express $x$ in terms of $z$:
$$x + \frac{13}{4} - 2z = 2$$
$$x = 2z - \frac{5}{4}$$
With $z = t$, the complete solution to the system is
$$\left\{\left(2t - \frac{5}{4}, \frac{13}{4}, t\right)\right\}.$$

**21.** $\begin{bmatrix} 1 & 1 & -1 & 1 & | & -2 \\ 2 & -1 & 2 & -1 & | & 7 \\ -1 & 2 & 1 & 2 & | & -1 \end{bmatrix} \begin{matrix} -2R_1 + R_2 \\ 1R_1 + R_3 \end{matrix}$

$\begin{bmatrix} 1 & 1 & -1 & 1 & | & -2 \\ 0 & -3 & 4 & -3 & | & 11 \\ 0 & 3 & 0 & 3 & | & -3 \end{bmatrix} R_2 \leftrightarrow R_3$

$\begin{bmatrix} 1 & 1 & -1 & 1 & | & -2 \\ 0 & 3 & 0 & 3 & | & -3 \\ 0 & -3 & 4 & -3 & | & 11 \end{bmatrix} \frac{1}{3}R_2$

$\begin{bmatrix} 1 & 1 & -1 & 1 & | & -2 \\ 0 & 1 & 0 & 1 & | & -1 \\ 0 & -3 & 4 & -3 & | & 11 \end{bmatrix} 3R_2 + R_3$

$\begin{bmatrix} 1 & 1 & -1 & 1 & | & -2 \\ 0 & 1 & 0 & 1 & | & -1 \\ 0 & 0 & 4 & 0 & | & 8 \end{bmatrix} \frac{1}{4}R_3$

$\begin{bmatrix} 1 & 1 & -1 & 1 & | & -2 \\ 0 & 1 & 0 & 1 & | & -1 \\ 0 & 0 & 1 & 0 & | & 2 \end{bmatrix}$

$$x + y - z + w = -2$$
The system $x + z = -1$ has no unique
$$y = 2$$

solution. Let $z = t$ and use back substitution to find remaining variables. The complete solution to the system is $\{(1, -t - 1, 2, t)\}$.

**23.** $\begin{bmatrix} 1 & 2 & 3 & -1 & | & 7 \\ 0 & 2 & -3 & 1 & | & 4 \\ 1 & -4 & 1 & 0 & | & 3 \end{bmatrix} -1R_1 + R_3$

$\begin{bmatrix} 1 & 2 & 3 & -1 & | & 7 \\ 0 & 2 & -3 & 1 & | & 4 \\ 0 & -6 & -2 & 1 & | & -4 \end{bmatrix} \frac{1}{2}R_2$

$\begin{bmatrix} 1 & 2 & 3 & -1 & | & 7 \\ 0 & 1 & -\frac{3}{2} & \frac{1}{2} & | & 2 \\ 0 & -6 & -2 & 1 & | & -4 \end{bmatrix} 6R_2 + R_3$

$\begin{bmatrix} 1 & 2 & 3 & -1 & | & 7 \\ 0 & 1 & -\frac{3}{2} & \frac{1}{2} & | & 2 \\ 0 & 0 & -11 & 4 & | & 8 \end{bmatrix} -\frac{1}{11}R_3$

$\begin{bmatrix} 1 & 2 & 3 & -1 & | & 7 \\ 0 & 1 & -\frac{3}{2} & \frac{1}{2} & | & 2 \\ 0 & 0 & 1 & -\frac{4}{11} & | & -\frac{8}{11} \end{bmatrix}$

The system has no unique solution. Let $z = t$ and use back substitution to find remaining variables. The complete solution to the system is

$$\left\{ \left( -\frac{2}{11}t + \frac{81}{11}, \frac{1}{22}t + \frac{10}{11}, \frac{4}{11}t - \frac{8}{11}, t \right) \right\}.$$

**25. a.** System of equations:
$4w - 2x + 2y - 3z = 0$
$7w - x - y - 3z = 0$
$w + x - y - z = 0$

**b.** Reduced system:
$w - 0.5z = 0$
$x = 0$
$y - 0.5z = 0$
Find $w$ and $y$ in terms of $z$.
$w - 0.5z = 0$
$w = 0.5z$
$y - 0.5z = 0$
$y = 0.5z$
The complete solution to the system is
$\{(0.5z, 0, 0.5z, z)\}$.

**27. a.** System of equations:
$w + 2x + 5y + 5z = 3$
$w + x + 3y + 4z = -1$
$w - x - y + 2z = 3$

**b.** Reduced system:
$w + y + 3z = 1$
$x + 2y + z = -2$

Find $w$ and $x$ in terms of $y$ and $z$.
$w + y + 3z = 1$
$w = 1 - y - 3z$
$x + 2y + z = -2$
$x = -2 - 2y - z$
The complete solution to the system is
$\{(1 - y - 3z, -2 - 2y - z, y, z)\}$.

**29.** $z + 12 = x + 6$

**31.** $x - y = 4$
$x - z = 6$
$y - z = 2$

$\begin{bmatrix} 1 & -1 & 0 & | & 4 \\ 1 & 0 & -1 & | & 6 \\ 0 & 1 & -1 & | & 2 \end{bmatrix} -1R_1 + R_2$

$\begin{bmatrix} 1 & -1 & 0 & 4 \\ 0 & -1 & 1 & -2 \\ 0 & 1 & -1 & 2 \end{bmatrix} -1R_2$

$\begin{bmatrix} 1 & -1 & 0 & 4 \\ 0 & 1 & -1 & 2 \\ 0 & 1 & -1 & 2 \end{bmatrix} \begin{matrix} -1R_2 + R_3 \\ 1R_2 + R_1 \end{matrix}$

$\begin{bmatrix} 1 & 0 & -1 & 6 \\ 0 & 1 & -1 & 2 \\ 0 & 0 & 0 & 0 \end{bmatrix}$

The system has no unique solution. Express $x$ and $y$ in terms of $z$:
$x - z = 6$
$y - z = 2$
$x = z + 6$
$y = z + 2$
With $z = t$, the complete solution to the system is
$\{(t + 6, t + 2, t)\}$.

**33. a.** From left to right along Palm Drive, then along Sunset Drive, we get the equations
$w + z = 200 + 180 = 380;$
$w + x = 400 + 200 = 600;$
$z + 70 = y + 20$ or $y - z = 50;$
$y + 200 = x + 30$ or $x - y = 170.$
The system is
$w + z = 380$
$w + x = 600$
$y - z = 50$
$x - y = 170$

**b.**
$$\begin{bmatrix} 1 & 0 & 0 & 1 & | & 380 \\ 0 & 1 & 0 & -1 & | & 220 \\ 0 & 0 & 1 & -1 & | & 50 \\ 0 & 1 & -1 & 0 & | & 170 \end{bmatrix} -1R_2 + R_4$$

$$\begin{bmatrix} 1 & 0 & 0 & 1 & | & 380 \\ 0 & 1 & 0 & -1 & | & 220 \\ 0 & 0 & 1 & -1 & | & 50 \\ 0 & 0 & -1 & 1 & | & -50 \end{bmatrix} 1R_3 + R_4$$

$$\begin{bmatrix} 1 & 0 & 0 & 1 & | & 380 \\ 0 & 1 & 0 & -1 & | & 220 \\ 0 & 0 & 1 & -1 & | & 50 \\ 0 & 0 & 0 & 0 & | & 0 \end{bmatrix}$$

The system has no unique solution. Express $x$ and $y$ and $z$ in terms of $z$:

$w = 380 - z$

$x = 220 + z$

$y = 50 + z$

With $z = t$, the complete solution to the system is $\{(380 - t, 220 + t, 50 + t, t)\}$.

**c.** Letting $z = 50$, the solution is

$w = 380 - 50 = 330$

$x = 220 + 50 = 270$

$y = 50 + 50 = 100$

**35.** Let $x$ = the number of ounces of Food 1,
$y$ = the number of ounces of Food 2, and
$z$ = the number of ounces of Food 3.
The amount of vitamin A is $20x + 30y + 10z$; the amount of iron is $20x + 10y + 10z$; the amount of calcium is $10x + 10y + 30z$.

**a.** Not having Food 1 means that all $x$ terms are left out. The vitamin A requirement can then be represented by $30y + 10z = 220$; the iron requirement is $10y + 10z = 180$; the calcium requirement is $10y + 30z = 340$.
The corresponding system is
$30y + 10z = 220$
$10y + 10z = 180$
$10y + 30z = 340$.
Dividing all of the numbers by 10, the matrix for this system is

$$\begin{bmatrix} 3 & 1 & | & 22 \\ 1 & 1 & | & 18 \\ 1 & 3 & | & 34 \end{bmatrix} R_1 \leftrightarrow R_2 \begin{bmatrix} 1 & 1 & | & 18 \\ 3 & 1 & | & 22 \\ 1 & 3 & | & 34 \end{bmatrix} \begin{matrix} -3R_1 + R_2 \\ -1R_1 + R_3 \end{matrix}$$

$$\begin{bmatrix} 1 & 1 & | & 18 \\ 0 & -2 & | & -32 \\ 0 & 2 & | & 16 \end{bmatrix} 1R_2 + R_3$$

$$\begin{bmatrix} 1 & 1 & | & 18 \\ 0 & -2 & | & -32 \\ 0 & 0 & | & -16 \end{bmatrix}.$$

From the last row, we see that the system has no solution, so there is no way to satisfy these dietary requirements with no Food 1 available.

**b.** With Food 1 available, and dropping the vitamin A requirement, the system is
$20x + 10y + 10z = 180$
$10x + 10y + 30z = 340$.
Dividing all of the numbers by 10, the matrix for this system is

$$\begin{bmatrix} 2 & 1 & 1 & | & 18 \\ 1 & 1 & 3 & | & 34 \end{bmatrix} R_1 \leftrightarrow R_2$$

$$\begin{bmatrix} 1 & 1 & 3 & | & 34 \\ 2 & 1 & 1 & | & 18 \end{bmatrix} -2R_1 + R_2$$

$$\begin{bmatrix} 1 & 1 & 3 & | & 34 \\ 0 & -1 & -5 & | & -50 \end{bmatrix} -1R_2$$

$$\begin{bmatrix} 1 & 1 & 3 & | & 34 \\ 0 & 1 & 5 & | & 50 \end{bmatrix}.$$

The system $\begin{array}{l} x + y + 3z = 34 \\ y + 5z = 50 \end{array}$ has no unique solution. Express $x$ and $y$ in terms of $z$:

$y = -5z + 50$

$x + (-5z + 50) + 3z = 34$

$\qquad x - 2z + 50 = 34$

$\qquad\qquad x = 2z - 16$

Now we can choose a value for $z$, i.e., an amount of Food 3, and find the corresponding values of $x$ and $y$. Note that negative amounts of food are not realistic, so $z \geq 0$, $y = -5z + 50 \geq 0$, and $x = 2z - 16 \geq 0$. These conditions are equivalent to $8 \leq z \leq 10$.

Using $z = 8$ and $z = 10$, two possibilities are 0 ounces of Food 1, 10 ounces of Food 2, and 8 ounces of Food 3 or 4 ounces of Food 1, 0 ounces of Food 2, and 10 ounces of Food 3. (Other answers are possible.)

**37. – 39.** Answers will vary.

**41.** does not make sense; Explanations will vary. Sample explanation: Row 3 indicates that this system has no solution. If eliminated, it would falsely appear that the system has an infinite number of solutions.

**43.** does not make sense; Explanations will vary. Sample explanation: In a nonsquare system, the number of equations differs from the number of variables.

**45.**
$$\begin{bmatrix} 1 & 3 & 1 & | & a^2 \\ 2 & 5 & 2a & | & 0 \\ 1 & 1 & a^2 & | & -9 \end{bmatrix} \begin{matrix} \\ -2R_1 + R_2 \\ -1R_1 + R_3 \end{matrix}$$

$$\begin{bmatrix} 1 & 3 & 1 & | & a^2 \\ 0 & -1 & 2a-2 & | & -2a^2 \\ 0 & -2 & a^2-1 & | & -9-a^2 \end{bmatrix} -1R_2$$

$$\begin{bmatrix} 1 & 3 & 1 & | & a^2 \\ 0 & 1 & 2-2a & | & 2a^2 \\ 0 & -2 & a^2-1 & | & -9-a^2 \end{bmatrix} 2R_2 + R_3$$

$$\begin{bmatrix} 1 & 3 & 1 & | & a^2 \\ 0 & 1 & 2-2a & | & 2a^2 \\ 0 & 0 & a^2-4a+3 & | & -9+3a^2 \end{bmatrix}$$

The system will be inconsistent when $a^2 - 4a + 3 = 0$ but $-9 + 3a^2 \neq 0$.
$a^2 - 4a + 3 = (a-1)(a-3) = 0$ when $a = 1$ or $a = 3$. $-9 + 3a^2 = 0$ when $a = \pm\sqrt{3}$.
Thus, the system is inconsistent when $a = 1$ or $a = 3$.

**47.** $-6 - (-5) = -6 + 5 = -1$

**48.** $1(-4) + 2(5) + 3(-6) = -4 + 10 - 18 = -12$

**49.** $\frac{1}{2}[8 - (-8)] = \frac{1}{2}[8+8] = \frac{1}{2}[16] = 8$

**Section 6.3**

**Check Point Exercises**

**1. a.** The matrix $A = \begin{bmatrix} 5 & -2 \\ -3 & \pi \\ 1 & 6 \end{bmatrix}$ has 3 rows and 2 columns, so it is of order $3 \times 2$.

**b.** The element $a_{12}$ is in the first row and second column. Thus, $a_{12} = -2$. The element $a_{31}$ is in the third row and first column. Thus, $a_{31} = 1$.

**2. a.** $\begin{bmatrix} -4 & 3 \\ 7 & -6 \end{bmatrix} + \begin{bmatrix} 6 & -3 \\ 2 & -4 \end{bmatrix}$

$= \begin{bmatrix} -4+6 & 3+(-3) \\ 7+2 & -6+(-4) \end{bmatrix} = \begin{bmatrix} 2 & 0 \\ 9 & -10 \end{bmatrix}$

**b.** $\begin{bmatrix} 5 & 4 \\ -3 & 7 \\ 0 & 1 \end{bmatrix} - \begin{bmatrix} -4 & 8 \\ 6 & 0 \\ -5 & 3 \end{bmatrix}$

$= \begin{bmatrix} 5-(-4) & 4-8 \\ -3-6 & 7-0 \\ 0-(-5) & 1-3 \end{bmatrix}$

$= \begin{bmatrix} 9 & -4 \\ -9 & 7 \\ 5 & -2 \end{bmatrix}$

**3. a.** $-6B = -6\begin{bmatrix} -1 & -2 \\ 8 & 5 \end{bmatrix}$

$= \begin{bmatrix} -6(-1) & -6(-2) \\ -6(8) & -6(5) \end{bmatrix}$

$= \begin{bmatrix} -6 & 12 \\ -48 & -30 \end{bmatrix}$

**b.** $3A + 2B = \begin{bmatrix} -4 & 1 \\ 3 & 0 \end{bmatrix} + 2 \begin{bmatrix} -1 & -2 \\ 8 & 5 \end{bmatrix}$

$= \begin{bmatrix} 3(-4) & 3(1) \\ 3(3) & 3(0) \end{bmatrix} + \begin{bmatrix} 2(-1) & 2(-2) \\ 2(8) & 2(5) \end{bmatrix}$

$= \begin{bmatrix} -12 & 3 \\ 9 & 0 \end{bmatrix} + \begin{bmatrix} -2 & -4 \\ 16 & 10 \end{bmatrix}$

$= \begin{bmatrix} -12 + (-2) & 3 + (-4) \\ 9 + 16 & 0 + 10 \end{bmatrix}$

$= \begin{bmatrix} -14 & -1 \\ 25 & 10 \end{bmatrix}$

**4.** $3X + A = B$

$3X = B - A$

$X = \dfrac{1}{3}(B - A)$

$X = \dfrac{1}{3}\left( \begin{bmatrix} -10 & 1 \\ -9 & 17 \end{bmatrix} - \begin{bmatrix} 2 & -8 \\ 0 & 4 \end{bmatrix} \right)$

$X = \dfrac{1}{3} \begin{bmatrix} -12 & 9 \\ -9 & 13 \end{bmatrix}$

$X = \begin{bmatrix} -4 & 3 \\ -3 & \dfrac{13}{3} \end{bmatrix}$

**5.** Given $A = \begin{bmatrix} 1 & 3 \\ 2 & 5 \end{bmatrix}$ and $B = \begin{bmatrix} 4 & 6 \\ 1 & 0 \end{bmatrix}$,

$AB = \begin{bmatrix} 1 & 3 \\ 2 & 5 \end{bmatrix} \cdot \begin{bmatrix} 4 & 6 \\ 1 & 0 \end{bmatrix} = \begin{bmatrix} 1(4)+3(1) & 1(6)+3(0) \\ 2(4)+5(1) & 2(6)+5(0) \end{bmatrix}$

$= \begin{bmatrix} 7 & 6 \\ 13 & 12 \end{bmatrix}$

**6.** If $A = \begin{bmatrix} 2 & 0 & 4 \end{bmatrix}$ and $B = \begin{bmatrix} 1 \\ 3 \\ 7 \end{bmatrix}$, then

$AB = \begin{bmatrix} 2 & 0 & 4 \end{bmatrix} \begin{bmatrix} 1 \\ 3 \\ 7 \end{bmatrix}$

$= \begin{bmatrix} 2(1) + 0(3) + 4(7) \end{bmatrix}$

$= \begin{bmatrix} 2 + 0 + 28 \end{bmatrix}$

$= \begin{bmatrix} 30 \end{bmatrix}$

and

$BA = \begin{bmatrix} 1 \\ 3 \\ 7 \end{bmatrix} \begin{bmatrix} 2 & 0 & 4 \end{bmatrix}$

$= \begin{bmatrix} 1(2) & 1(0) & 1(4) \\ 3(2) & 3(0) & 3(4) \\ 7(2) & 7(0) & 7(4) \end{bmatrix}$

$= \begin{bmatrix} 2 & 0 & 4 \\ 6 & 0 & 12 \\ 14 & 0 & 28 \end{bmatrix}.$

**7. a.** $\begin{bmatrix} 1 & 3 \\ 0 & 2 \end{bmatrix} \cdot \begin{bmatrix} 2 & 3 & -1 & 6 \\ 0 & 5 & 4 & 1 \end{bmatrix}$

$= \begin{bmatrix} 1(2)+3(0) & 1(3)+3(5) & 1(-1)+3(4) & 1(6)+3(1) \\ 0(2)+2(0) & 0(3)+2(5) & 0(-1)+2(4) & 0(6)+2(1) \end{bmatrix}$

$= \begin{bmatrix} 2 & 18 & 11 & 9 \\ 0 & 10 & 8 & 2 \end{bmatrix}$

**b.** $\begin{bmatrix} 2 & 3 & -1 & 6 \\ 0 & 5 & 4 & 1 \end{bmatrix} \begin{bmatrix} 1 & 3 \\ 0 & 2 \end{bmatrix}$

The number of columns in the first matrix does not equal the number of rows in the second matrix. Thus, the product of these two matrices is <u>undefined</u>.

**8.** Because the *L* is dark gray and the background is light gray, the digital photograph can be represented by the matrix

$\begin{bmatrix} 2 & 1 & 1 \\ 2 & 1 & 1 \\ 2 & 2 & 1 \end{bmatrix}$

We can make the *L* light gray by decreasing each 2 in the above matrix to 1. We can make the background black by increasing each 1 in the matrix to 3. This is accomplished using the following matrix addition.

$\begin{bmatrix} 2 & 1 & 1 \\ 2 & 1 & 1 \\ 2 & 2 & 1 \end{bmatrix} + \begin{bmatrix} -1 & 2 & 2 \\ -1 & 2 & 2 \\ -1 & -1 & 2 \end{bmatrix} = \begin{bmatrix} 1 & 3 & 3 \\ 1 & 3 & 3 \\ 1 & 1 & 3 \end{bmatrix}$

**9.** **a.**
$$\begin{bmatrix} 0 & 3 & 4 \\ 0 & 5 & 2 \end{bmatrix} + \begin{bmatrix} -3 & -3 & -3 \\ -1 & -1 & -1 \end{bmatrix} = \begin{bmatrix} -3 & 0 & 1 \\ -1 & 4 & 1 \end{bmatrix}$$

**b.**
$$2\begin{bmatrix} 0 & 3 & 4 \\ 0 & 5 & 2 \end{bmatrix} = \begin{bmatrix} 0 & 6 & 8 \\ 0 & 10 & 4 \end{bmatrix}$$

**c.**
$$\begin{bmatrix} 1 & 0 \\ 0 & -1 \end{bmatrix}\begin{bmatrix} 0 & 3 & 4 \\ 0 & 5 & 2 \end{bmatrix}$$
$$= \begin{bmatrix} (1)(0)+0(0) & (1)(3)+0(5) & (1)(4)+0(2) \\ 0(0)-1(0) & 0(3)-1(5) & 0(4)-1(2) \end{bmatrix}$$
$$= \begin{bmatrix} 0 & 3 & 4 \\ 0 & -5 & -2 \end{bmatrix}$$

Multiplication with $\begin{bmatrix} 1 & 0 \\ 0 & -1 \end{bmatrix}$ reflects the triangle over the *x*-axis.

## Concept and Vocabulary Check 6.3

**1.** third; fourth

**2.** $1 \times 3$

**3.** square

**4.** $-10$; 6

**5.** true

**6.** false

**7.** true

**8.** *m*; *p*; columns; rows

**9.** true

**10.** false

## Exercise Set 6.3

**1.** **a.** $2 \times 3$

**b.** $a_{32}$ does not exist (*A* only has 2 rows).
$a_{23} = -1$

**3.** **a.** $3 \times 4$

**b.** $a_{32} = \dfrac{1}{2}$; $a_{23} = -6$

**5.** $\begin{bmatrix} x \\ 4 \end{bmatrix} = \begin{bmatrix} 6 \\ y \end{bmatrix}$
$x = 6$
$y = 4$

**7.** $\begin{bmatrix} x & 2y \\ z & 9 \end{bmatrix} = \begin{bmatrix} 4 & 12 \\ 3 & 9 \end{bmatrix}$
$x = 4$
$2y = 12$
$y = 6$
$z = 3$

**9.** **a.** $A+B = \begin{bmatrix} 4+5 & 1+9 \\ 3+0 & 2+7 \end{bmatrix} = \begin{bmatrix} 9 & 10 \\ 3 & 9 \end{bmatrix}$

**b.** $A-B = \begin{bmatrix} 4-5 & 1-9 \\ 3-0 & 2-7 \end{bmatrix} = \begin{bmatrix} -1 & -8 \\ 3 & -5 \end{bmatrix}$

**c.** $-4A = \begin{bmatrix} -16 & -4 \\ -12 & -8 \end{bmatrix}$

**d.** $3A+2B = \begin{bmatrix} 12+10 & 3+18 \\ 9+0 & 6+14 \end{bmatrix} = \begin{bmatrix} 22 & 21 \\ 9 & 20 \end{bmatrix}$

**11.** **a.** $A+B = \begin{bmatrix} 1+2 & 3+(-1) \\ 3+3 & 4+(-2) \\ 5+0 & 6+1 \end{bmatrix} = \begin{bmatrix} 3 & 2 \\ 6 & 2 \\ 5 & 7 \end{bmatrix}$

**b.** $A-B = \begin{bmatrix} 1-2 & 3-(-1) \\ 3-3 & 4-(-2) \\ 5-0 & 6-1 \end{bmatrix} = \begin{bmatrix} -1 & 4 \\ 0 & 6 \\ 5 & 5 \end{bmatrix}$

**c.** $-4A = \begin{bmatrix} -4 & -12 \\ -12 & -16 \\ -20 & -24 \end{bmatrix}$

**d.** $3A+2B=\begin{bmatrix} 3+4 & 9-2 \\ 9+6 & 12-4 \\ 15+0 & 18+2 \end{bmatrix}=\begin{bmatrix} 7 & 7 \\ 15 & 8 \\ 15 & 20 \end{bmatrix}$

**13. a.** $A+B=\begin{bmatrix} 2+(-5) \\ -4+3 \\ 1+(-1) \end{bmatrix}=\begin{bmatrix} -3 \\ -1 \\ 0 \end{bmatrix}$

**b.** $A-B=\begin{bmatrix} 2-(-5) \\ -4-3 \\ 1-(-1) \end{bmatrix}=\begin{bmatrix} 7 \\ -7 \\ 2 \end{bmatrix}$

**c.** $-4A=\begin{bmatrix} -8 \\ 16 \\ -4 \end{bmatrix}$

**d.** $3A+2B=\begin{bmatrix} 6-10 \\ -12+6 \\ 3-2 \end{bmatrix}=\begin{bmatrix} -4 \\ -6 \\ 1 \end{bmatrix}$

**15. a.** $A+B=\begin{bmatrix} 2+6 & -10+10 & -2+(-2) \\ 14+0 & 12+(-12) & 10+(-4) \\ 4+(-5) & -2+2 & 2+(-2) \end{bmatrix}$

$=\begin{bmatrix} 8 & 0 & -4 \\ 14 & 0 & 6 \\ -1 & 0 & 0 \end{bmatrix}$

**b.** $A-B=\begin{bmatrix} 2-6 & -10-10 & -2-(-2) \\ 14-0 & 12-(-12) & 10-(-4) \\ 4-(-5) & -2-2 & 2-(-2) \end{bmatrix}$

$=\begin{bmatrix} -4 & -20 & 0 \\ 14 & 24 & 14 \\ 9 & -4 & 4 \end{bmatrix}$

**c.** $-4A=\begin{bmatrix} -8 & 40 & 8 \\ -56 & -48 & -40 \\ -16 & 8 & -8 \end{bmatrix}$

**d.** $3A+2B=\begin{bmatrix} 6+12 & -30+20 & -6-4 \\ 42+0 & 36-24 & 30-8 \\ 12-10 & -6+4 & 6-4 \end{bmatrix}$

$=\begin{bmatrix} 18 & -10 & -10 \\ 42 & 12 & 22 \\ 2 & -2 & 2 \end{bmatrix}$

**17.** $X - A = B$
$\qquad X = A + B$

$$X = \begin{bmatrix} -3 & -7 \\ 2 & -9 \\ 5 & 0 \end{bmatrix} + \begin{bmatrix} -5 & -1 \\ 0 & 0 \\ 3 & -4 \end{bmatrix} = \begin{bmatrix} -8 & -8 \\ 2 & -9 \\ 8 & -4 \end{bmatrix}$$

**19.** $2X + A = B$
$\qquad 2X = B - A$

$$X = \frac{1}{2}(B - A)$$

$$X = \frac{1}{2}\left( \begin{bmatrix} -5 & -1 \\ 0 & 0 \\ 3 & -4 \end{bmatrix} - \begin{bmatrix} -3 & -7 \\ 2 & -9 \\ 5 & 0 \end{bmatrix} \right) = \frac{1}{2}\begin{bmatrix} -2 & 6 \\ -2 & 9 \\ -2 & -4 \end{bmatrix} = \begin{bmatrix} -1 & 3 \\ -1 & \dfrac{9}{2} \\ -1 & -2 \end{bmatrix}$$

**21.** $3X + 2A = B$
$\qquad 3X = B - 2A$

$$X = \frac{1}{3}(B - 2A)$$

$$X = \frac{1}{3}\left( \begin{bmatrix} -5 & -1 \\ 0 & 0 \\ 3 & -4 \end{bmatrix} - 2\begin{bmatrix} -3 & -7 \\ 2 & -9 \\ 5 & 0 \end{bmatrix} \right) = \frac{1}{3}\begin{bmatrix} 1 & 13 \\ -4 & 18 \\ -7 & -4 \end{bmatrix} = \begin{bmatrix} \dfrac{1}{3} & \dfrac{13}{3} \\ -\dfrac{4}{3} & 6 \\ -\dfrac{7}{3} & -\dfrac{4}{3} \end{bmatrix}$$

**23.** $\quad B - X = 4A$
$\quad B - 4A = X$

$$X = \begin{bmatrix} -5 & -1 \\ 0 & 0 \\ 3 & -4 \end{bmatrix} - 4\begin{bmatrix} -3 & -7 \\ 2 & -9 \\ 5 & 0 \end{bmatrix} = \begin{bmatrix} -5 & -1 \\ 0 & 0 \\ 3 & -4 \end{bmatrix} + \begin{bmatrix} 12 & 28 \\ -8 & 36 \\ -20 & 0 \end{bmatrix} = \begin{bmatrix} 7 & 27 \\ -8 & 36 \\ -17 & -4 \end{bmatrix}$$

**25.** $\quad 4A + 3B = -2X$

$\quad -\dfrac{1}{2}(4A + 3B) = X$

$$X = -\frac{1}{2}\left( 4\begin{bmatrix} -3 & -7 \\ 2 & -9 \\ 5 & 0 \end{bmatrix} + 3\begin{bmatrix} -5 & -1 \\ 0 & 0 \\ 3 & -4 \end{bmatrix} \right) = -\frac{1}{2}\left( \begin{bmatrix} -12 & -28 \\ 8 & -36 \\ 20 & 0 \end{bmatrix} + \begin{bmatrix} -15 & -3 \\ 0 & 0 \\ 9 & -12 \end{bmatrix} \right) = -\frac{1}{2}\begin{bmatrix} -27 & -31 \\ 8 & -36 \\ 29 & -12 \end{bmatrix} = \begin{bmatrix} \dfrac{27}{2} & \dfrac{31}{2} \\ -4 & 18 \\ -\dfrac{29}{2} & 6 \end{bmatrix}$$

**27. a.** $AB = \begin{bmatrix} 1 & 3 \\ 5 & 3 \end{bmatrix}\begin{bmatrix} 3 & -2 \\ -1 & 6 \end{bmatrix} = \begin{bmatrix} (1)(3)+(3)(-1) & (1)(-2)+(3)(6) \\ (5)(3)+(3)(-1) & (5)(-2)+(3)(6) \end{bmatrix} = \begin{bmatrix} 3-3 & -2+18 \\ 15-3 & -10+18 \end{bmatrix} = \begin{bmatrix} 0 & 16 \\ 12 & 8 \end{bmatrix}$

**b.** $BA = \begin{bmatrix} 3 & -2 \\ -1 & 6 \end{bmatrix}\begin{bmatrix} 1 & 3 \\ 5 & 3 \end{bmatrix} = \begin{bmatrix} (3)(1)+(-2)(5) & (3)(3)+(-2)(3) \\ (-1)(1)+(6)(5) & (-1)(3)+(6)(3) \end{bmatrix} = \begin{bmatrix} 3-10 & 9-6 \\ -1+30 & -3+18 \end{bmatrix} = \begin{bmatrix} -7 & 3 \\ 29 & 15 \end{bmatrix}$

**29. a.** $AB = \begin{bmatrix} 1 & 2 & 3 & 4 \end{bmatrix} \begin{bmatrix} 1 \\ 2 \\ 3 \\ 4 \end{bmatrix} = [(1)(1) + (2)(2) + (3)(3) + (4)(4)] = [1 + 4 + 9 + 16] = [30]$

**b.** $BA = \begin{bmatrix} 1 \\ 2 \\ 3 \\ 4 \end{bmatrix} \begin{bmatrix} 1 & 2 & 3 & 4 \end{bmatrix} = \begin{bmatrix} (1)(1) & (1)(2) & (1)(3) & (1)(4) \\ (2)(1) & (2)(2) & (2)(3) & (2)(4) \\ (3)(1) & (3)(2) & (3)(3) & (3)(4) \\ (4)(1) & (4)(2) & (4)(3) & (4)(4) \end{bmatrix} = \begin{bmatrix} 1 & 2 & 3 & 4 \\ 2 & 4 & 6 & 8 \\ 3 & 6 & 9 & 12 \\ 4 & 8 & 12 & 16 \end{bmatrix}$

**31. a.** $AB = \begin{bmatrix} 1 & -1 & 4 \\ 4 & -1 & 3 \\ 2 & 0 & -2 \end{bmatrix} \begin{bmatrix} 1 & 1 & 0 \\ 1 & 2 & 4 \\ 1 & -1 & 3 \end{bmatrix}$

$= \begin{bmatrix} (1)(1)+(-1)(1)+(4)(1) & (1)(1)+(-1)(2)+(4)(-1) & (1)(0)+(-1)(4)+(4)(3) \\ (4)(1)+(-1)(1)+(3)(1) & (4)(1)+(-1)(2)+(3)(-1) & (4)(0)+(-1)(4)+(3)(3) \\ (2)(1)+(0)(1)+(-2)(1) & (2)(1)+(0)(2)+(-2)(-1) & (2)(0)+(0)(4)+(-2)(3) \end{bmatrix}$

$= \begin{bmatrix} 1-1+4 & 1-2-4 & 0-4+12 \\ 4-1+3 & 4-2-3 & 0-4+9 \\ 2+0-2 & 2+0+2 & 0+0-6 \end{bmatrix} = \begin{bmatrix} 4 & -5 & 8 \\ 6 & -1 & 5 \\ 0 & 4 & -6 \end{bmatrix}$

**b.** $BA = \begin{bmatrix} 1 & 1 & 0 \\ 1 & 2 & 4 \\ 1 & -1 & 3 \end{bmatrix} \begin{bmatrix} 1 & -1 & 4 \\ 4 & -1 & 3 \\ 2 & 0 & -2 \end{bmatrix}$

$= \begin{bmatrix} (1)(1)+(1)(4)+(0)(2) & (1)(-1)+(1)(-1)+(0)(0) & (1)(4)+(1)(3)+(0)(-2) \\ (1)(1)+(2)(4)+(4)(2) & (1)(-1)+(2)(-1)+(4)(0) & (1)(4)+(2)(3)+(4)(-2) \\ (1)(1)+(-1)(4)+(3)(2) & (1)(-1)+(-1)(-1)+(3)(0) & (1)(4)+(-1)(3)+(3)(-2) \end{bmatrix}$

$= \begin{bmatrix} 1+4+0 & -1-1+0 & 4+3+0 \\ 1+8+8 & -1-2+0 & 4+6-8 \\ 1-4+6 & -1+1+0 & 4-3-6 \end{bmatrix} = \begin{bmatrix} 5 & -2 & 7 \\ 17 & -3 & 2 \\ 3 & 0 & -5 \end{bmatrix}$

**33. a.** $AB = \begin{bmatrix} 4 & 2 \\ 6 & 1 \\ 3 & 5 \end{bmatrix} \begin{bmatrix} 2 & 3 & 4 \\ -1 & -2 & 0 \end{bmatrix} = \begin{bmatrix} (4)(2)+(2)(-1) & (4)(3)+(2)(-2) & (4)(4)+(2)(0) \\ (6)(2)+(1)(-1) & (6)(3)+(1)(-2) & (6)(4)+(1)(0) \\ (3)(2)+(5)(-1) & (3)(3)+(5)(-2) & (3)(4)+(5)(0) \end{bmatrix}$

$= \begin{bmatrix} 8-2 & 12-4 & 16+0 \\ 12-1 & 18-2 & 24+0 \\ 6-5 & 9-10 & 12+0 \end{bmatrix} = \begin{bmatrix} 6 & 8 & 16 \\ 11 & 16 & 24 \\ 1 & -1 & 12 \end{bmatrix}$

**b.** $BA = \begin{bmatrix} 2 & 3 & 4 \\ -1 & -2 & 0 \end{bmatrix} \begin{bmatrix} 4 & 2 \\ 6 & 1 \\ 3 & 5 \end{bmatrix} = \begin{bmatrix} (2)(4)+(3)(6)+(4)(3) & (2)(2)+(3)(1)+(4)(5) \\ (-1)(4)+(-2)(6)+(0)(3) & (-1)(2)+(-2)(1)+(0)(5) \end{bmatrix}$

$= \begin{bmatrix} 8+18+12 & 4+3+20 \\ -4-12+0 & -2-2+0 \end{bmatrix} = \begin{bmatrix} 38 & 27 \\ -16 & -4 \end{bmatrix}$

**35. a.** $AB = \begin{bmatrix} 2 & -3 & 1 & -1 \\ 1 & 1 & -2 & 1 \end{bmatrix} \begin{bmatrix} 1 & 2 \\ -1 & 1 \\ 5 & 4 \\ 10 & 5 \end{bmatrix} = \begin{bmatrix} (2)(1)+(-3)(-1)+(1)(5)+(-1)(10) & (2)(2)+(-3)(1)+(1)(4)+(-1)(5) \\ (1)(1)+(1)(-1)+(-2)(5)+(1)(10) & (1)(2)+(1)(1)+(-2)(4)+(1)(5) \end{bmatrix}$

$= \begin{bmatrix} 2+3+5-10 & 4-3+4-5 \\ 1-1-10+10 & 2+1-8+5 \end{bmatrix} = \begin{bmatrix} 0 & 0 \\ 0 & 0 \end{bmatrix}$

**37.** $4B - 3C = \begin{bmatrix} 20 & 4 \\ -8 & -8 \end{bmatrix} - \begin{bmatrix} 3 & -3 \\ -3 & 3 \end{bmatrix} = \begin{bmatrix} 20-3 & 4-(-3) \\ -8-(-3) & -8-3 \end{bmatrix} = \begin{bmatrix} 17 & 7 \\ -5 & -11 \end{bmatrix}$

**39.** $BC + CB = \begin{bmatrix} 5-1 & -5+1 \\ -2+2 & 2-2 \end{bmatrix} + \begin{bmatrix} 5+2 & 1+2 \\ -5-2 & -1-2 \end{bmatrix} = \begin{bmatrix} 4 & -4 \\ 0 & 0 \end{bmatrix} + \begin{bmatrix} 7 & 3 \\ -7 & -3 \end{bmatrix} = \begin{bmatrix} 11 & -1 \\ -7 & -3 \end{bmatrix}$

**41.** $A - C$ is not defined because $A$ is 3 x 2 and $C$ is 2 x 2.

**43.** $A(BC) = \begin{bmatrix} 4 & 0 \\ -3 & 5 \\ 0 & 1 \end{bmatrix} \begin{bmatrix} 5-1 & -5+1 \\ -2+2 & 2-2 \end{bmatrix} = \begin{bmatrix} 4 & 0 \\ -3 & 5 \\ 0 & 1 \end{bmatrix} \begin{bmatrix} 4 & -4 \\ 0 & 0 \end{bmatrix} = \begin{bmatrix} 16+0 & -16+0 \\ -12+0 & 12+0 \\ 0+0 & 0+0 \end{bmatrix} = \begin{bmatrix} 16 & -16 \\ -12 & 12 \\ 0 & 0 \end{bmatrix}$

**45.** $(A+B)(C-D) = \left( \begin{bmatrix} 1 & 0 \\ 0 & 1 \end{bmatrix} + \begin{bmatrix} 1 & 0 \\ 0 & -1 \end{bmatrix} \right) \left( \begin{bmatrix} -1 & 0 \\ 0 & 1 \end{bmatrix} - \begin{bmatrix} -1 & 0 \\ 0 & -1 \end{bmatrix} \right) = \begin{bmatrix} 2 & 0 \\ 0 & 0 \end{bmatrix} \begin{bmatrix} 0 & 0 \\ 0 & 2 \end{bmatrix} = \begin{bmatrix} 0 & 0 \\ 0 & 0 \end{bmatrix}$

**47.** Answers will vary.

**49.** $BZ = \begin{bmatrix} 1 & 0 \\ 0 & -1 \end{bmatrix} \begin{bmatrix} x \\ y \end{bmatrix} = \begin{bmatrix} x \\ -y \end{bmatrix}$ This reflects the graphic about the *x*-axis because all *y*-coordinates are negated.

**51. a.** $\begin{bmatrix} 1 & 3 & 1 \\ 3 & 3 & 3 \\ 1 & 3 & 1 \end{bmatrix}$

**b.** $\begin{bmatrix} 1 & 3 & 1 \\ 3 & 3 & 3 \\ 1 & 3 & 1 \end{bmatrix} + \begin{bmatrix} -1 & -1 & -1 \\ -1 & -1 & -1 \\ -1 & -1 & -1 \end{bmatrix} = \begin{bmatrix} 0 & 2 & 0 \\ 2 & 2 & 2 \\ 0 & 2 & 0 \end{bmatrix}$

**c.** $\begin{bmatrix} 1 & 3 & 1 \\ 3 & 3 & 3 \\ 1 & 3 & 1 \end{bmatrix} + \begin{bmatrix} 1 & -2 & 1 \\ -2 & -2 & -2 \\ 1 & -2 & 1 \end{bmatrix} = \begin{bmatrix} 2 & 1 & 2 \\ 1 & 1 & 1 \\ 2 & 1 & 2 \end{bmatrix}$

**53.** $\begin{bmatrix} 0 & 3 & 3 & 1 & 1 & 0 \\ 0 & 0 & 1 & 1 & 5 & 5 \end{bmatrix} + \begin{bmatrix} -2 & -2 & -2 & -2 & -2 & -2 \\ -3 & -3 & -3 & -3 & -3 & -3 \end{bmatrix} = \begin{bmatrix} -2 & 1 & 1 & -1 & -1 & -2 \\ -3 & -3 & -2 & -2 & 2 & 2 \end{bmatrix}$

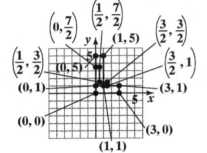

**55.** $0.5\begin{bmatrix} 0 & 3 & 3 & 1 & 1 & 0 \\ 0 & 0 & 1 & 1 & 5 & 5 \end{bmatrix} + \begin{bmatrix} 0 & 0 & 0 & 0 & 0 & 0 \\ 1 & 1 & 1 & 1 & 1 & 1 \end{bmatrix} = \begin{bmatrix} 0 & 1.5 & 1.5 & 0.5 & 0.5 & 0 \\ 0 & 0 & 0.5 & 0.5 & 2.5 & 2.5 \end{bmatrix} + \begin{bmatrix} 0 & 0 & 0 & 0 & 0 & 0 \\ 1 & 1 & 1 & 1 & 1 & 1 \end{bmatrix}$

$= \begin{bmatrix} 0 & 1.5 & 1.5 & 0.5 & 0.5 & 0 \\ 1 & 1 & 1.5 & 1.5 & 3.5 & 3.5 \end{bmatrix}$

**57. a.** $AB = \begin{bmatrix} 1 & 0 \\ 0 & -1 \end{bmatrix} \cdot \begin{bmatrix} 0 & 3 & 3 & 1 & 1 & 0 \\ 0 & 0 & 1 & 1 & 5 & 5 \end{bmatrix} = \begin{bmatrix} 0 & 3 & 3 & 1 & 1 & 0 \\ 0 & 0 & -1 & -1 & -5 & -5 \end{bmatrix}$

**b.**

Rotated L about the *x*-axis.

**59. a.** $AB = \begin{bmatrix} 0 & -1 \\ 1 & 0 \end{bmatrix} \cdot \begin{bmatrix} 0 & 3 & 3 & 1 & 1 & 0 \\ 0 & 0 & 1 & 1 & 5 & 5 \end{bmatrix} = \begin{bmatrix} 0 & 0 & -1 & -1 & -5 & -5 \\ 0 & 3 & 3 & 1 & 1 & 0 \end{bmatrix}$

**b.**

Rotated L 90° counterclockwise about the origin.

**61.** **a.** $A = \begin{bmatrix} 2 & 6 \\ 31 & 46 \end{bmatrix}$

**b.** $B = \begin{bmatrix} 9 & 29 \\ 65 & 77 \end{bmatrix}$

**c.** $B - A = \begin{bmatrix} 9 & 29 \\ 65 & 77 \end{bmatrix} - \begin{bmatrix} 2 & 6 \\ 31 & 46 \end{bmatrix} = \begin{bmatrix} 7 & 23 \\ 34 & 31 \end{bmatrix}$

This matrix represents the percentages of people completing the transition to adulthood in 1960 and 2000 by age and gender.

**63.** **a.** System 1: The midterm and final both count for 50% of the course grade.
System 2: The midterm counts for 30% of the course grade and the final counts for 70%.

**b.** $AB = \begin{bmatrix} 84 & 87.2 \\ 79 & 81 \\ 90 & 88.4 \\ 73 & 68.6 \\ 69 & 73.4 \end{bmatrix}$

System 1 grades are listed first (if different).
Student 1: B; Student 2: C or B; Student 3: A or B; Student 4: C or D; Student 5: D or C

**65. – 75.** Answers will vary.

**77.** makes sense

**79.** makes sense

**81.** Answers will vary.

**83.** $AB = \begin{bmatrix} 0 & -1 \\ 1 & 0 \end{bmatrix} \begin{bmatrix} 1 & 0 \\ 0 & -1 \end{bmatrix} = \begin{bmatrix} 0 & 1 \\ 1 & 0 \end{bmatrix}$

$-BA = \begin{bmatrix} 1 & 0 \\ 0 & -1 \end{bmatrix} \begin{bmatrix} 0 & -1 \\ 1 & 0 \end{bmatrix} = -\begin{bmatrix} 0 & -1 \\ -1 & 0 \end{bmatrix} = \begin{bmatrix} 0 & 1 \\ 1 & 0 \end{bmatrix}$

$AB = -BA$ so they are anticommutative.

**85.** $AB = \begin{bmatrix} a_{11} & a_{12} \\ a_{21} & a_{22} \end{bmatrix} \begin{bmatrix} 1 & 0 \\ 0 & 1 \end{bmatrix} = \begin{bmatrix} a_{11} & a_{12} \\ a_{21} & a_{22} \end{bmatrix}$

Nothing happens to the elements in the first matrix.

**86.** $\left[ \begin{array}{ccc|c} -1 & -1 & -1 & 1 \\ 4 & 5 & 0 & 0 \\ 0 & 1 & -3 & 0 \end{array} \right] 4R_1 + R_2$

$\left[ \begin{array}{ccc|c} -1 & -1 & -1 & 1 \\ 0 & 1 & -4 & 4 \\ 0 & 1 & -3 & 0 \end{array} \right] \begin{array}{c} R_2 + R_1 \\ \\ -R_2 + R_3 \end{array}$

$\left[ \begin{array}{ccc|c} -1 & 0 & -5 & 5 \\ 0 & 1 & -4 & 4 \\ 0 & 0 & 1 & -4 \end{array} \right] -R_1$

$\left[ \begin{array}{ccc|c} 1 & 0 & 5 & -5 \\ 0 & 1 & -4 & 4 \\ 0 & 0 & 1 & -4 \end{array} \right] \begin{array}{c} -5R_3 + R_1 \\ 4R_3 + R_2 \end{array}$

$\left[ \begin{array}{ccc|c} 1 & 0 & 0 & 15 \\ 0 & 1 & 0 & -12 \\ 0 & 0 & 1 & -4 \end{array} \right]$

The solution set is $\{(15, -12, -4)\}$.

**87.** $\begin{bmatrix} a_1 & b_1 & c_1 \\ a_2 & b_2 & c_2 \\ a_3 & b_3 & c_3 \end{bmatrix} \begin{bmatrix} x \\ y \\ x \end{bmatrix} = \begin{bmatrix} d_1 \\ d_2 \\ d_3 \end{bmatrix}$

$\begin{bmatrix} a_1 x + b_1 y + c_1 x \\ a_2 x + b_2 y + c_2 z \\ a_3 x + b_3 y + c_3 z \end{bmatrix} = \begin{bmatrix} d_1 \\ d_2 \\ d_3 \end{bmatrix}$

The linear system is written as follows.

$a_1 x + b_1 y + c_1 x = d_1$
$a_2 x + b_2 y + c_2 z = d_2$
$a_3 x + b_3 y + c_3 z = d_3$

**Mid-Chapter 6 Check Point**

**1.**
$$\begin{bmatrix} 1 & 2 & -3 & | & -7 \\ 3 & -1 & 2 & | & 8 \\ 2 & -1 & 1 & | & 5 \end{bmatrix} \begin{matrix} \\ -3R_1 + R_2 \\ -2R_1 + R_3 \end{matrix}$$

$$\begin{bmatrix} 1 & 2 & -3 & | & -7 \\ 0 & -7 & 11 & | & 29 \\ 0 & -5 & 7 & | & 19 \end{bmatrix} \begin{matrix} \\ -\frac{1}{7}R_2 \\ -\frac{1}{5}R_3 \end{matrix}$$

$$\begin{bmatrix} 1 & 2 & -3 & | & -7 \\ 0 & 1 & -\frac{11}{7} & | & -\frac{29}{7} \\ 0 & 1 & -\frac{7}{5} & | & -\frac{19}{5} \end{bmatrix} \begin{matrix} \\ \\ -R_2 + R_3 \end{matrix}$$

$$\begin{bmatrix} 1 & 2 & -3 & | & -7 \\ 0 & 1 & -\frac{11}{7} & | & -\frac{29}{7} \\ 0 & 0 & \frac{6}{35} & | & \frac{12}{35} \end{bmatrix} \begin{matrix} \\ \\ \frac{35}{6}R_3 \end{matrix}$$

$$\begin{bmatrix} 1 & 2 & -3 & | & -7 \\ 0 & 1 & -\frac{11}{7} & | & -\frac{29}{7} \\ 0 & 0 & 1 & | & 2 \end{bmatrix}$$

Back-substitute to find *y*.
$$y - \frac{11}{7}z = -\frac{29}{7}$$
$$y - \frac{11}{7}(2) = -\frac{29}{7}$$
$$y - \frac{22}{7} = -\frac{29}{7}$$
$$y = -\frac{7}{7}$$
$$y = -1$$

Back-substitute to find *x*.
$$x + 2y - 3z = -7$$
$$x + 2(-1) - 3(2) = -7$$
$$x - 2 - 6 = -7$$
$$x - 8 = -7$$
$$x = 1$$

The solution is $\{(1, -1, 2)\}$.

**2.**
$$\begin{bmatrix} 2 & 4 & 5 & | & 2 \\ 1 & 1 & 2 & | & 1 \\ 3 & 5 & 7 & | & 4 \end{bmatrix} R_1 \leftrightarrow R_3$$

$$\begin{bmatrix} 1 & 1 & 2 & | & 1 \\ 2 & 4 & 5 & | & 2 \\ 3 & 5 & 7 & | & 4 \end{bmatrix} \begin{matrix} \\ -2R_1 + R_2 \\ -3R_1 + R_3 \end{matrix}$$

$$\begin{bmatrix} 1 & 1 & 2 & | & 1 \\ 0 & 2 & 1 & | & 0 \\ 0 & 2 & 1 & | & 1 \end{bmatrix} \begin{matrix} \\ \\ -R_2 + R_3 \end{matrix}$$

$$\begin{bmatrix} 1 & 1 & 2 & | & 1 \\ 0 & 2 & 1 & | & 0 \\ 0 & 0 & 0 & | & 1 \end{bmatrix}$$

The third row of the matrix is equivalent to
$0x + 0y + 0z = 1$ which is false.

The solution is $\varnothing$.

**3.**
$$\begin{bmatrix} 1 & -2 & 2 & | & -2 \\ 2 & 3 & -1 & | & 1 \end{bmatrix} -2R_1 + R_2$$

$$\begin{bmatrix} 1 & -2 & 2 & | & -2 \\ 0 & 7 & -5 & | & 5 \end{bmatrix} \frac{1}{7}R_2$$

$$\begin{bmatrix} 1 & -2 & 2 & | & -2 \\ 0 & 1 & -\frac{5}{7} & | & \frac{5}{7} \end{bmatrix}$$

Back-substitute to find *y* in terms of *z*.
$$y - \frac{5}{7}z = \frac{5}{7}$$
$$y = \frac{5}{7}z + \frac{5}{7}$$

Back-substitute to find *x* in terms of *z*.
$$x - 2y + 2z = -2$$
$$x - 2\left(\frac{5}{7}z + \frac{5}{7}\right) + 2z = -2$$
$$x - \frac{10}{7}z - \frac{10}{7} + 2z = -2$$
$$x + \frac{4}{7}z - \frac{10}{7} = -2$$
$$x = -\frac{4}{7}z - \frac{4}{7}$$

The solution is $\left\{\left(-\frac{4}{7}z - \frac{4}{7}, \frac{5}{7}z + \frac{5}{7}, z\right)\right\}$.

**4.**
$$\begin{bmatrix} 1 & 1 & 1 & 1 & | & 6 \\ 1 & -1 & 3 & 1 & | & -14 \\ 1 & 2 & 0 & -3 & | & 12 \\ 2 & 3 & 6 & 1 & | & 1 \end{bmatrix} \begin{matrix} \\ -R_1 + R_2 \\ -R_1 + R_3 \\ -2R_1 + R_4 \end{matrix}$$

$$\begin{bmatrix} 1 & 1 & 1 & 1 & | & 6 \\ 0 & -2 & 2 & 0 & | & -20 \\ 0 & 1 & -1 & -4 & | & 6 \\ 0 & 1 & 4 & -1 & | & -11 \end{bmatrix} \begin{matrix} \\ R_2 \leftrightarrow R_3 \\ \\ \end{matrix}$$

$$\begin{bmatrix} 1 & 1 & 1 & 1 & | & 6 \\ 0 & 1 & -1 & -4 & | & 6 \\ 0 & -2 & 2 & 0 & | & -20 \\ 0 & 1 & 4 & -1 & | & -11 \end{bmatrix} \begin{matrix} \\ \\ 2R_2 + R_3 \\ -R_2 + R_4 \end{matrix}$$

$$\begin{bmatrix} 1 & 1 & 1 & 1 & | & 6 \\ 0 & 1 & -1 & -4 & | & 6 \\ 0 & 0 & 0 & -8 & | & -8 \\ 0 & 0 & 5 & 3 & | & -17 \end{bmatrix} \begin{matrix} \\ \\ R_3 \leftrightarrow R_4 \\ \end{matrix}$$

$$\begin{bmatrix} 1 & 1 & 1 & 1 & | & 6 \\ 0 & 1 & -1 & -4 & | & 6 \\ 0 & 0 & 5 & 3 & | & -17 \\ 0 & 0 & 0 & -8 & | & -8 \end{bmatrix} \begin{matrix} \\ \\ \frac{1}{5}R_3 \\ -\frac{1}{8}R_4 \end{matrix}$$

$$\begin{bmatrix} 1 & 1 & 1 & 1 & | & 6 \\ 0 & 1 & -1 & -4 & | & 6 \\ 0 & 0 & 1 & \frac{3}{5} & | & -\frac{17}{5} \\ 0 & 0 & 0 & 1 & | & 1 \end{bmatrix}$$

Back-substitute to find $y$ in terms of $z$.

$$y + \frac{3}{5}z = -\frac{17}{5}$$

$$y + \frac{3}{5}(1) = -\frac{17}{5}$$

$$y + \frac{3}{5} = -\frac{17}{5}$$

$$y = -\frac{20}{5}$$

$$y = -4$$

Back-substitute to find $x$ in terms of $z$.

$$x - y - 4z = 6$$

$$x - (-4) - 4(1) = 6$$

$$x + 4 - 4 = 6$$

$$x = 6$$

Back-substitute to find $w$ in terms of $z$.

$$w + x + y + z = 6$$

$$w + (6) + (-4) + (1) = 6$$

$$w + 3 = 6$$

$$w = 3$$

The solution is $\{(3, 6, -4, 1)\}$.

**5.**
$$\begin{bmatrix} 2 & -2 & 2 & | & 5 \\ 1 & -1 & 1 & | & 2 \\ 2 & 1 & -1 & | & 1 \end{bmatrix} \begin{matrix} -R_2 + R_1 \\ \\ -R_1 + R_3 \end{matrix}$$

$$\begin{bmatrix} 1 & -1 & 1 & | & 3 \\ 1 & -1 & 1 & | & 2 \\ 0 & 3 & -3 & | & -4 \end{bmatrix} \begin{matrix} \\ -R_1 + R_2 \\ \end{matrix}$$

$$\begin{bmatrix} 1 & -1 & 1 & | & 3 \\ 0 & 0 & 0 & | & -1 \\ 0 & 3 & -3 & | & -4 \end{bmatrix}$$

The second row of the matrix is equivalent to
$0x + 0y + 0z = -1$ which is false.

The solution is $\varnothing$.

**6.** $2C - \frac{1}{2}B$

$$= 2\begin{bmatrix} -1 & 0 \\ 0 & 1 \end{bmatrix} - \frac{1}{2}\begin{bmatrix} 4 & 1 \\ -6 & -2 \end{bmatrix}$$

$$= \begin{bmatrix} -2 & 0 \\ 0 & 2 \end{bmatrix} - \begin{bmatrix} 2 & \frac{1}{2} \\ -3 & -1 \end{bmatrix}$$

$$= \begin{bmatrix} -4 & -\frac{1}{2} \\ 3 & 3 \end{bmatrix}$$

**7.** $A(B + C)$

$$= \begin{bmatrix} 0 & 2 \\ -1 & 3 \\ 1 & 0 \end{bmatrix}\left(\begin{bmatrix} 4 & 1 \\ -6 & -2 \end{bmatrix} + \begin{bmatrix} -1 & 0 \\ 0 & 1 \end{bmatrix}\right)$$

$$= \begin{bmatrix} 0 & 2 \\ -1 & 3 \\ 1 & 0 \end{bmatrix}\left(\begin{bmatrix} 3 & 1 \\ -6 & -1 \end{bmatrix}\right)$$

$$= \begin{bmatrix} -12 & -2 \\ -21 & -4 \\ 3 & 1 \end{bmatrix}$$

**8.** $A(BC)$

$$= \begin{bmatrix} 0 & 2 \\ -1 & 3 \\ 1 & 0 \end{bmatrix}\left(\begin{bmatrix} 4 & 1 \\ -6 & -2 \end{bmatrix} \cdot \begin{bmatrix} -1 & 0 \\ 0 & 1 \end{bmatrix}\right)$$

$$= \begin{bmatrix} 0 & 2 \\ -1 & 3 \\ 1 & 0 \end{bmatrix}\left(\begin{bmatrix} -4 & 1 \\ 6 & -2 \end{bmatrix}\right)$$

$$= \begin{bmatrix} 12 & -4 \\ 22 & -7 \\ -4 & 1 \end{bmatrix}$$

**9.** The operation is not defined. Matrices must have the same dimensions in order to be added.

**10.** $2X - 3C = B$

$\qquad 2X = B + 3C$

$\qquad X = \tfrac{1}{2}(B + 3C)$

$\qquad = \tfrac{1}{2}\left(\begin{bmatrix} 4 & 1 \\ -6 & -2 \end{bmatrix} + 3\begin{bmatrix} -1 & 0 \\ 0 & 1 \end{bmatrix}\right)$

$\qquad = \tfrac{1}{2}\left(\begin{bmatrix} 4 & 1 \\ -6 & -2 \end{bmatrix} + \begin{bmatrix} -3 & 0 \\ 0 & 3 \end{bmatrix}\right)$

$\qquad = \tfrac{1}{2}\left(\begin{bmatrix} 1 & 1 \\ -6 & 1 \end{bmatrix}\right)$

$\qquad = \begin{bmatrix} \tfrac{1}{2} & \tfrac{1}{2} \\ -3 & \tfrac{1}{2} \end{bmatrix}$

## Section 6.4

## Check Point Exercises

**1.** We must show that: $AB = I_2 = \begin{bmatrix} 1 & 0 \\ 0 & 1 \end{bmatrix}$, and

$BA = I_2 = \begin{bmatrix} 1 & 0 \\ 0 & 1 \end{bmatrix}.$

$AB = \begin{bmatrix} 2 & 1 \\ 1 & 1 \end{bmatrix}\begin{bmatrix} 1 & -1 \\ -1 & 2 \end{bmatrix}$

$\quad = \begin{bmatrix} 2(1)+1(-1) & 2(-1)+1(2) \\ 1(1)+1(-1) & 1(-1)+1(2) \end{bmatrix}$

$\quad = \begin{bmatrix} 1 & 0 \\ 0 & 1 \end{bmatrix}$

$BA = \begin{bmatrix} 1 & -1 \\ -1 & 2 \end{bmatrix}\begin{bmatrix} 2 & 1 \\ 1 & 1 \end{bmatrix}$

$\quad = \begin{bmatrix} 1(2)+-1(1) & 1(1)+-1(1) \\ -1(2)+2(1) & -1(1)+2(1) \end{bmatrix}$

$\quad = \begin{bmatrix} 1 & 0 \\ 0 & 1 \end{bmatrix}$

Both products ($AB$ and $BA$) give the multiplicative identity matrix, $I_2$. Thus, $B$ is the multiplicative inverse of $A$.

**2.** Let us denote the multiplicative inverse of $A$ by $A^{-1} = \begin{bmatrix} w & x \\ y & z \end{bmatrix}$. Because $A$ is a $2 \times 2$ matrix, we use the equation $AA^{-1} = I_2$ to find values for $w, x, y$ and $z$.

$\begin{bmatrix} 5 & 7 \\ 2 & 3 \end{bmatrix}\begin{bmatrix} w & x \\ y & z \end{bmatrix} = \begin{bmatrix} 1 & 0 \\ 0 & 1 \end{bmatrix}$

$\begin{bmatrix} 5w+7y & 5x+7z \\ 2w+3y & 2x+3z \end{bmatrix} = \begin{bmatrix} 1 & 0 \\ 0 & 1 \end{bmatrix}$

$5w+7y = 1 \qquad 5x+7z = 0$

$2w+3y = 0 \qquad 2x+3z = 1$

Each of these systems can be solved using the addition method.

Multiply by –2: $5w+7y = 1 \rightarrow -10w-14y = -2$

Multiply by 5: $2w+3y = 0 \rightarrow \quad 10w+15y = 0$

Use back substitution: $w = 3, y = -2$

Multiply by –2: $5x+7z = 0 \rightarrow -10x-14z = 0$

Multiply by 5: $2x+3z = 1 \rightarrow 10x+15z = 5$

Use back substitution: $x = -7, z = 5$

Using these values, we have

$A^{-1} = \begin{bmatrix} w & x \\ y & z \end{bmatrix} = \begin{bmatrix} 3 & -7 \\ -2 & 5 \end{bmatrix}.$

**3.** $A^{-1} = \dfrac{1}{ad-bc}\begin{bmatrix} d & -b \\ -c & a \end{bmatrix}$

$\qquad = \dfrac{1}{3(1)-(-2)(-1)}\begin{bmatrix} 1 & -(-2) \\ -(-1) & 3 \end{bmatrix}$

$\qquad = \dfrac{1}{3-2}\begin{bmatrix} 1 & 2 \\ 1 & 3 \end{bmatrix}$

$\qquad = \dfrac{1}{1}\begin{bmatrix} 1 & 2 \\ 1 & 3 \end{bmatrix}$

$\qquad = \begin{bmatrix} 1 & 2 \\ 1 & 3 \end{bmatrix}$

**4.** The augmented matrix $[A \mid I_3]$ is

$\begin{bmatrix} 1 & 0 & 2 & 1 & 0 & 0 \\ -1 & 2 & 3 & 0 & 1 & 0 \\ 1 & -1 & 0 & 0 & 0 & 1 \end{bmatrix}.$

Perform row transformations on $[A \mid I_3]$ to obtain a matrix of the form $[I_3 \mid B]$.

$$\begin{bmatrix} 1 & 0 & 2 & | & 1 & 0 & 0 \\ -1 & 2 & 3 & | & 0 & 1 & 0 \\ 1 & -1 & 0 & | & 0 & 0 & 1 \end{bmatrix} 1R_1 / R_2$$

$$= \begin{bmatrix} 1 & 0 & 2 & | & 1 & 0 & 0 \\ 0 & 2 & 5 & | & 1 & 1 & 0 \\ 1 & -1 & 0 & | & 0 & 0 & 1 \end{bmatrix} -1R_3$$

$$= \begin{bmatrix} 1 & 0 & 2 & | & 1 & 0 & 0 \\ 0 & 2 & 5 & | & 1 & 1 & 0 \\ -1 & 1 & 0 & | & 0 & 0 & -1 \end{bmatrix} R_1 + R_3$$

$$= \begin{bmatrix} 1 & 0 & 2 & | & 1 & 0 & 0 \\ 0 & 2 & 5 & | & 1 & 1 & 0 \\ 0 & 1 & 2 & | & 1 & 0 & -1 \end{bmatrix} \tfrac{1}{2}R_2$$

$$= \begin{bmatrix} 1 & 0 & 2 & | & 1 & 0 & 0 \\ 0 & 1 & \tfrac{5}{2} & | & \tfrac{1}{2} & \tfrac{1}{2} & 0 \\ 0 & 1 & 2 & | & 1 & 0 & -1 \end{bmatrix} -1R_2 + R_3$$

$$= \begin{bmatrix} 1 & 0 & 2 & | & 1 & 0 & 0 \\ 0 & 1 & \tfrac{5}{2} & | & \tfrac{1}{2} & \tfrac{1}{2} & 0 \\ 0 & 0 & -\tfrac{1}{2} & | & \tfrac{1}{2} & -\tfrac{1}{2} & -1 \end{bmatrix} -2R_3$$

$$= \begin{bmatrix} 1 & 0 & 2 & | & 1 & 0 & 0 \\ 0 & 1 & \tfrac{5}{2} & | & \tfrac{1}{2} & \tfrac{1}{2} & 0 \\ 0 & 0 & 1 & | & -1 & 1 & 2 \end{bmatrix} \begin{matrix} -2R_3 + R_1 \\ -\tfrac{5}{2}R_3 + R_2 \end{matrix}$$

$$= \begin{bmatrix} 1 & 0 & 0 & | & 3 & -2 & -4 \\ 0 & 1 & 0 & | & 3 & -2 & -5 \\ 0 & 0 & 1 & | & 1 & 1 & 2 \end{bmatrix}$$

Thus, the multiplicative inverse of $A$ is

$$A^{-1} = \begin{bmatrix} 3 & -2 & -4 \\ 3 & -2 & -5 \\ -1 & 1 & 2 \end{bmatrix}.$$

**5.** The linear system can be written as $AX = B$.

$$\begin{bmatrix} 1 & 0 & 2 \\ -1 & 2 & 3 \\ 1 & -1 & 0 \end{bmatrix} \begin{bmatrix} x \\ y \\ z \end{bmatrix} = \begin{bmatrix} 6 \\ -5 \\ 6 \end{bmatrix}.$$

$$X = A^{-1}B = \begin{bmatrix} 3 & -2 & -4 \\ 3 & -2 & -5 \\ -1 & 1 & 2 \end{bmatrix} \begin{bmatrix} 6 \\ -5 \\ 6 \end{bmatrix}$$

$$= \begin{bmatrix} 3(6) + -2(-5) + -4(6) \\ 3(6) + -2(-5) + -5(6) \\ -1(6) + 1(-5) + 2(6) \end{bmatrix}$$

$$= \begin{bmatrix} 18 + 10 - 24 \\ 18 + 10 - 30 \\ -6 - 5 + 12 \end{bmatrix} = \begin{bmatrix} 4 \\ -2 \\ 1 \end{bmatrix}$$

Thus, $x = 4$, $y = -2$, and $z = 1$. The solution set is $\{(4, -2, 1)\}$.

**6.** The numerical representation of the word BASE is 2, 1, 19, 5. The $2 \times 2$ matrix formed is $\begin{bmatrix} 2 & 19 \\ 1 & 5 \end{bmatrix}$.

$$\begin{bmatrix} -2 & -3 \\ 3 & 4 \end{bmatrix} \begin{bmatrix} 2 & 19 \\ 1 & 5 \end{bmatrix}$$

$$= \begin{bmatrix} -2(2) + -3(1) & -2(19) + -3(5) \\ 3(2) + 4(1) & 3(19) + 4(5) \end{bmatrix}$$

$$= \begin{bmatrix} -4 - 3 & -38 - 15 \\ 6 + 4 & 57 + 20 \end{bmatrix} = \begin{bmatrix} -7 & -53 \\ 10 & 77 \end{bmatrix}$$

The encoded message is $-7$, 10, $-53$, 77.

**7.** Use the multiplicative inverse of the coding matrix. It is $\begin{bmatrix} 4 & 3 \\ -3 & -2 \end{bmatrix}$.

$$\begin{bmatrix} 4 & 3 \\ -3 & -2 \end{bmatrix} \begin{bmatrix} -7 & -53 \\ 10 & 77 \end{bmatrix}$$

$$= \begin{bmatrix} 4(-7) + 3(10) & 4(-53) + 3(77) \\ -3(-7) + -2(10) & -3(-53) + -2(77) \end{bmatrix}$$

$$= \begin{bmatrix} -28 + 30 & -212 + 231 \\ 21 - 20 & 159 - 154 \end{bmatrix} = \begin{bmatrix} 2 & 19 \\ 1 & 5 \end{bmatrix}$$

The numbers are 2, 1, 19, and 5. Using letters, the decoded message is BASE.

**Concept and Vocabulary Check 6.4**

1. $\begin{bmatrix} 1 & 0 \\ 0 & 1 \end{bmatrix}$

2. $\begin{bmatrix} 1 & 0 & 0 \\ 0 & 1 & 0 \\ 0 & 0 & 1 \end{bmatrix}$

3. multiplicative inverse

4. true

5. $ad - bc \neq 0$

6. singular

7. false

9. $A^{-1}$

10. $A^{-1}B$

**Exercise Set 6.4**

1. $A = \begin{bmatrix} 4 & -3 \\ -5 & 4 \end{bmatrix}$  $B = \begin{bmatrix} 4 & 3 \\ 5 & 4 \end{bmatrix}$

   $AB = \begin{bmatrix} 16-15 & 12-12 \\ -20+20 & -15+16 \end{bmatrix} = \begin{bmatrix} 1 & 0 \\ 0 & 1 \end{bmatrix}$

   $BA = \begin{bmatrix} 16-15 & -12+12 \\ 20-20 & -15+16 \end{bmatrix} = \begin{bmatrix} 1 & 0 \\ 0 & 1 \end{bmatrix}$

   Since $AB = I_2$ $BA = I_2$, $B = A^{-1}$.

3. $AB = \begin{bmatrix} 8+0 & -16+0 \\ -2+0 & 4+3 \end{bmatrix} = \begin{bmatrix} 8 & -16 \\ -2 & 7 \end{bmatrix}$

   $BA = \begin{bmatrix} 8+4 & 0+12 \\ 0+1 & 0+3 \end{bmatrix} = \begin{bmatrix} 12 & 12 \\ 1 & 3 \end{bmatrix}$

   If $B$ is the multiplicative inverse of $A$, both products ($AB$ and $BA$) will be the multiplicative identity matrix, $I_2$. Therefore, $B$ is not the multiplicative inverse of $A$. That is, $B \neq A^{-1}$.

5. $AB = \begin{bmatrix} -2+3 & -4+4 \\ \frac{3}{2}-\frac{3}{2} & 3-2 \end{bmatrix} = \begin{bmatrix} 1 & 0 \\ 0 & 1 \end{bmatrix}$

   $BA = \begin{bmatrix} -2+3 & 1-1 \\ -6+6 & 3-2 \end{bmatrix} = \begin{bmatrix} 1 & 0 \\ 0 & 1 \end{bmatrix}$

   Since $AB = I_2$ and $BA = I_2$, $B = A^{-1}$.

**7.** $A = \begin{bmatrix} 0 & 1 & 0 \\ 0 & 0 & 1 \\ 1 & 0 & 0 \end{bmatrix}$ $B = \begin{bmatrix} 0 & 0 & 1 \\ 1 & 0 & 0 \\ 0 & 1 & 0 \end{bmatrix}$

$AB = \begin{bmatrix} 0+1+0 & 0+0+0 & 0+0+0 \\ 0+0+0 & 0+0+1 & 0+0+0 \\ 0+0+0 & 0+0+0 & 1+0+0 \end{bmatrix} = \begin{bmatrix} 1 & 0 & 0 \\ 0 & 1 & 0 \\ 0 & 0 & 1 \end{bmatrix}$ $BA = \begin{bmatrix} 0+0+1 & 0+0+0 & 0+0+0 \\ 0+0+0 & 1+0+0 & 0+0+0 \\ 0+0+0 & 0+0+0 & 0+1+0 \end{bmatrix} = \begin{bmatrix} 1 & 0 & 0 \\ 0 & 1 & 0 \\ 0 & 0 & 1 \end{bmatrix}$

Since $AB = I_3$ and $BA = I_3$, $B = A^{-1}$.

**9.** $AB = \begin{bmatrix} \frac{7}{2}-1-\frac{3}{2} & -3+0+3 & \frac{1}{2}+1-\frac{3}{2} \\ \frac{7}{2}-\frac{3}{2}-2 & -3+0+4 & \frac{1}{2}+\frac{3}{2}-2 \\ \frac{7}{2}-2-\frac{3}{2} & -3+0+3 & \frac{1}{2}+2-\frac{3}{2} \end{bmatrix} = \begin{bmatrix} 1 & 0 & 0 \\ 0 & 1 & 0 \\ 0 & 0 & 1 \end{bmatrix}$

$BA = \begin{bmatrix} \frac{7}{2}-3+\frac{1}{2} & 7-9+2 & \frac{21}{2}-12+\frac{3}{2} \\ -\frac{1}{2}+0+\frac{1}{2} & -1+0+2 & -\frac{3}{2}+0+\frac{3}{2} \\ -\frac{1}{2}+1-\frac{1}{2} & -1+3-2 & -\frac{3}{2}+4-\frac{3}{2} \end{bmatrix} = \begin{bmatrix} 1 & 0 & 0 \\ 0 & 1 & 0 \\ 0 & 0 & 1 \end{bmatrix}$

Since $AB = I_3$ and $BA = I_3$, $B = A^{-1}$.

**11.** $AB = \begin{bmatrix} 0+0+0+1 & 0+0-2+2 & 0+0+0+0 & 0+0-2+2 \\ -1+0+0+1 & -2+0+1+2 & 0+0+0+0 & -3+0+1+2 \\ 0+0+0+0 & 0+1-1+0 & 0+1+0+0 & 0+1-1+0 \\ 1+0+0-1 & 2+0+0-2 & 0+0+0+0 & 3+0+0-2 \end{bmatrix} = \begin{bmatrix} 1 & 0 & 0 & 0 \\ 0 & 1 & 0 & 0 \\ 0 & 0 & 1 & 0 \\ 0 & 0 & 0 & 1 \end{bmatrix}$

$BA = \begin{bmatrix} 0-2+0+3 & 0+0+0+0 & -2+2+0+0 & 1+2+0-3 \\ 0-1+0+1 & 0+0+1+0 & 0+1-1+0 & 0+1+0-1 \\ 0-1+0+1 & 0+0+0+0 & 0+1+0+0 & 0+1+0-1 \\ 0-2+0+2 & 0+0+0+0 & -2+2+0+0 & 1+2+0-2 \end{bmatrix} = \begin{bmatrix} 1 & 0 & 0 & 0 \\ 0 & 1 & 0 & 0 \\ 0 & 0 & 1 & 0 \\ 0 & 0 & 0 & 1 \end{bmatrix}$

Since $AB = I_4$ and $BA = I_4$, $B = A^{-1}$.

**13.** $ad - bc = (2)(2) - (3)(-1) = 4 + 3 = 7$

$A^{-1} = \frac{1}{7}\begin{bmatrix} 2 & -3 \\ 1 & 2 \end{bmatrix} = \begin{bmatrix} \frac{2}{7} & -\frac{3}{7} \\ \frac{1}{7} & \frac{2}{7} \end{bmatrix}$

$AA^{-1} = \begin{bmatrix} \frac{4}{7}+\frac{3}{7} & -\frac{6}{7}+\frac{6}{7} \\ -\frac{2}{7}+\frac{2}{7} & \frac{3}{7}+\frac{4}{7} \end{bmatrix} = \begin{bmatrix} 1 & 0 \\ 0 & 1 \end{bmatrix}$ and $A^{-1}A = \begin{bmatrix} \frac{4}{7}+\frac{3}{7} & \frac{6}{7}-\frac{6}{7} \\ \frac{2}{7}-\frac{2}{7} & \frac{3}{7}+\frac{4}{7} \end{bmatrix} = \begin{bmatrix} 1 & 0 \\ 0 & 1 \end{bmatrix}$

**15.** $ad - bc = (3)(2) - (-1)(-4) = 6 - 4 = 2$

$A^{-1} = \frac{1}{2}\begin{bmatrix} 2 & 1 \\ 4 & 3 \end{bmatrix} = \begin{bmatrix} 1 & \frac{1}{2} \\ 2 & \frac{3}{2} \end{bmatrix}$

$AA^{-1} = \begin{bmatrix} 3-2 & \frac{3}{2}-\frac{3}{2} \\ -4+4 & -\frac{4}{2}+\frac{6}{2} \end{bmatrix} = \begin{bmatrix} 1 & 0 \\ 0 & 1 \end{bmatrix}$ and $A^{-1}A = \begin{bmatrix} 3-\frac{4}{2} & -1+\frac{2}{2} \\ 6-\frac{12}{2} & -2+\frac{6}{2} \end{bmatrix} = \begin{bmatrix} 1 & 0 \\ 0 & 1 \end{bmatrix}$

**17.** $ad - bc = (10)(1) - (-2)(-5) = 10 - 10 = 0$

Since division by zero is undefined, $A$ does not have an inverse.

For Problems 19–24, verification that $AA^{-1} = I$ and $A^{-1}A = I$ is left to the student.

For Problems 19–23, verification that $AA^{-1} = I$ and $A^{-1}A = I$ is left to the student.

**19.** $\begin{bmatrix} 2 & 0 & 0 & 1 & 0 & 0 \\ 0 & 4 & 0 & 0 & 1 & 0 \\ 0 & 0 & 6 & 0 & 0 & 1 \end{bmatrix}$

Divide row 1 by 2, divide row 2 by 4 and divide row 4 by 6.

$\begin{bmatrix} 1 & 0 & 0 & \frac{1}{2} & 0 & 0 \\ 0 & 1 & 0 & 0 & \frac{1}{4} & 0 \\ 0 & 0 & 1 & 0 & 0 & \frac{1}{6} \end{bmatrix}$

$A^{-1} = \begin{bmatrix} \frac{1}{2} & 0 & 0 \\ 0 & \frac{1}{4} & 0 \\ 0 & 0 & \frac{1}{6} \end{bmatrix}$

**21.** $\begin{bmatrix} 1 & 2 & -1 & 1 & 0 & 0 \\ -2 & 0 & 1 & 0 & 1 & 0 \\ 1 & -1 & 0 & 0 & 0 & 1 \end{bmatrix}$

Replace row 2 with $2R_1 + R_2$.
Replace row 3 with $R_1 - R_3$.

$\begin{bmatrix} 1 & 2 & -1 & 1 & 0 & 0 \\ 0 & 4 & -1 & 2 & 1 & 0 \\ 0 & 3 & -1 & 1 & 0 & -1 \end{bmatrix}$

Replace row 1 with $R_2 - 2R_1$.
Replace row 3 with $-3R_2 + 4R_3$.

$\begin{bmatrix} -2 & 0 & 1 & 0 & 1 & 0 \\ 0 & 4 & -1 & 2 & 1 & 0 \\ 0 & 0 & -1 & -2 & -3 & -4 \end{bmatrix}$

Replace row 1 with $R_3 + R_1$.
Replace row 2 with $R_2 - R_3$.
Replace row 3 with $-R_3$.

$\begin{bmatrix} -2 & 0 & 0 & -2 & -2 & -4 \\ 0 & 4 & 0 & 4 & 4 & 4 \\ 0 & 0 & 1 & 2 & 3 & 4 \end{bmatrix}$

Divide row 1 by $-2$ and divide row 2 by 4.

$\begin{bmatrix} 1 & 0 & 0 & 1 & 1 & 2 \\ 0 & 1 & 0 & 1 & 1 & 1 \\ 0 & 0 & 1 & 2 & 3 & 4 \end{bmatrix}$

$A^{-1} = \begin{bmatrix} 1 & 1 & 2 \\ 1 & 1 & 1 \\ 2 & 3 & 4 \end{bmatrix}$

**23.** $\begin{bmatrix} 2 & 2 & -1 & 1 & 0 & 0 \\ 0 & 3 & -1 & 0 & 1 & 0 \\ -1 & -2 & 1 & 0 & 0 & 1 \end{bmatrix} R_1 \leftrightarrow R_3$

$\begin{bmatrix} -1 & -2 & 1 & 0 & 0 & 1 \\ 0 & 3 & -1 & 0 & 1 & 0 \\ 2 & 2 & -1 & 1 & 0 & 0 \end{bmatrix} -1R_1$

$\begin{bmatrix} 1 & 2 & -1 & 0 & 0 & -1 \\ 0 & 3 & -1 & 0 & 1 & 0 \\ 2 & 2 & -1 & 1 & 0 & 0 \end{bmatrix} -2R_1 + R_3$

$\begin{bmatrix} 1 & 2 & -1 & 0 & 0 & -1 \\ 0 & 3 & -1 & 0 & 1 & 0 \\ 0 & -2 & 1 & 1 & 0 & 2 \end{bmatrix} \frac{1}{3}R_2$

$\begin{bmatrix} 1 & 2 & -1 & 0 & 0 & -1 \\ 0 & 1 & -\frac{1}{3} & 0 & \frac{1}{3} & 0 \\ 0 & -2 & 1 & 1 & 0 & 2 \end{bmatrix} \begin{matrix} -2R_2 + R_1 \\ 2R_2 + R_3 \end{matrix}$

$\begin{bmatrix} 1 & 0 & -\frac{1}{3} & 0 & -\frac{2}{3} & -1 \\ 0 & 1 & -\frac{1}{3} & 0 & \frac{1}{3} & 0 \\ 0 & 0 & \frac{1}{3} & 1 & \frac{2}{3} & 2 \end{bmatrix} \begin{matrix} 1R_3 + R_1 \\ 1R_2 + R_1 \end{matrix}$

$\begin{bmatrix} 1 & 0 & 0 & 1 & 0 & 1 \\ 0 & 1 & 0 & 1 & 1 & 2 \\ 0 & 0 & \frac{1}{3} & 1 & \frac{2}{3} & 2 \end{bmatrix} 3R_3$

$\begin{bmatrix} 1 & 0 & 0 & 1 & 0 & 1 \\ 0 & 1 & 0 & 1 & 1 & 2 \\ 0 & 0 & 1 & 3 & 2 & 6 \end{bmatrix}$

$A^{-1} = \begin{bmatrix} 1 & 0 & 1 \\ 1 & 1 & 2 \\ 3 & 2 & 6 \end{bmatrix}$

**25.** $\begin{bmatrix} 5 & 0 & 2 & 1 & 0 & 0 \\ 2 & 2 & 1 & 0 & 1 & 0 \\ -3 & 1 & -1 & 0 & 0 & 1 \end{bmatrix} \frac{1}{5}R_1$

$\begin{bmatrix} 1 & 0 & \frac{2}{5} & \frac{1}{5} & 0 & 0 \\ 2 & 2 & 1 & 0 & 1 & 0 \\ -3 & 1 & -1 & 0 & 0 & 1 \end{bmatrix} \begin{matrix} -2R_1 + R_2 \\ 3R_1 + R_3 \end{matrix}$

$\begin{bmatrix} 1 & 0 & \frac{2}{5} & \frac{1}{5} & 0 & 0 \\ 0 & 2 & \frac{1}{5} & -\frac{2}{5} & 1 & 0 \\ 0 & 1 & \frac{1}{5} & \frac{3}{5} & 0 & 1 \end{bmatrix} R_2 \leftrightarrow R_3$

$\begin{bmatrix} 1 & 0 & \frac{2}{5} & \frac{1}{5} & 0 & 0 \\ 0 & 1 & \frac{1}{5} & \frac{3}{5} & 0 & 1 \\ 0 & 2 & \frac{1}{5} & -\frac{2}{5} & 1 & 0 \end{bmatrix} -2R_2 + R_3$

$$\begin{bmatrix} 1 & 0 & \frac{2}{5} & \frac{1}{5} & 0 & 0 \\ 0 & 1 & \frac{1}{5} & \frac{3}{5} & 0 & 1 \\ 0 & 0 & -\frac{1}{5} & -\frac{8}{5} & 1 & -2 \end{bmatrix} \begin{matrix} 2R_3 + R_1 \\ \\ 1R_3 + R_2 \end{matrix}$$

$$\begin{bmatrix} 1 & 0 & 0 & -3 & 2 & -4 \\ 0 & 1 & 0 & -1 & 1 & -1 \\ 0 & 0 & -\frac{1}{5} & -\frac{8}{5} & 1 & -2 \end{bmatrix} -5R_3$$

$$\begin{bmatrix} 1 & 0 & 0 & -3 & 2 & -4 \\ 0 & 1 & 0 & -1 & 1 & -1 \\ 0 & 0 & 1 & 8 & -5 & 10 \end{bmatrix}$$

$$A^{-1} = \begin{bmatrix} -3 & 2 & -4 \\ -1 & 1 & -1 \\ 8 & -5 & 10 \end{bmatrix}$$

**27.**
$$\begin{bmatrix} 1 & 0 & 0 & 0 & 1 & 0 & 0 & 0 \\ 0 & -1 & 0 & 0 & 0 & 1 & 0 & 0 \\ 0 & 0 & 3 & 0 & 0 & 0 & 1 & 0 \\ 1 & 0 & 0 & 1 & 0 & 0 & 0 & 1 \end{bmatrix} -1R_1 + R_4$$

$$\begin{bmatrix} 1 & 0 & 0 & 0 & 1 & 0 & 0 & 0 \\ 0 & -1 & 0 & 0 & 0 & 1 & 0 & 0 \\ 0 & 0 & 3 & 0 & 0 & 0 & 1 & 0 \\ 0 & 0 & 0 & 1 & -1 & 0 & 0 & 1 \end{bmatrix} -1R_2$$

$$\begin{bmatrix} 1 & 0 & 0 & 0 & 1 & 0 & 0 & 0 \\ 0 & 1 & 0 & 0 & 0 & -1 & 0 & 0 \\ 0 & 0 & 3 & 0 & 0 & 0 & 1 & 0 \\ 0 & 0 & 0 & 1 & -1 & 0 & 0 & 1 \end{bmatrix} \frac{1}{3}R_3$$

$$\begin{bmatrix} 1 & 0 & 0 & 0 & 1 & 0 & 0 & 0 \\ 0 & 1 & 0 & 0 & 0 & -1 & 0 & 0 \\ 0 & 0 & 1 & 0 & 0 & 0 & \frac{1}{3} & 0 \\ 0 & 0 & 0 & 1 & -1 & 0 & 0 & 1 \end{bmatrix}$$

$$A^{-1} = \begin{bmatrix} 1 & 0 & 0 & 0 \\ 0 & -1 & 0 & 0 \\ 0 & 0 & \frac{1}{3} & 0 \\ -1 & 0 & 0 & 1 \end{bmatrix}$$

**29.** $\begin{bmatrix} 6 & 5 \\ 5 & 4 \end{bmatrix}\begin{bmatrix} x \\ y \end{bmatrix} = \begin{bmatrix} 13 \\ 10 \end{bmatrix}$

**31.** $\begin{bmatrix} 1 & 3 & 4 \\ 1 & 2 & 3 \\ 1 & 4 & 3 \end{bmatrix}\begin{bmatrix} x \\ y \\ z \end{bmatrix} = \begin{bmatrix} -3 \\ -2 \\ -6 \end{bmatrix}$

**33.** $4x - 7y = -3$
$2x - 3y = 1$

**35.** $2x - z = 6$
$3y = 9$
$x + y = 5$

**37. a.** $\begin{bmatrix} 2 & 6 & 6 \\ 2 & 7 & 6 \\ 2 & 7 & 7 \end{bmatrix}\begin{bmatrix} x \\ y \\ z \end{bmatrix} = \begin{bmatrix} 8 \\ 10 \\ 9 \end{bmatrix}$

$$\begin{bmatrix} \frac{7}{2} & 0 & -3 \\ -1 & 1 & 0 \\ 0 & -1 & 1 \end{bmatrix}\begin{bmatrix} 8 \\ 10 \\ 9 \end{bmatrix} = \begin{bmatrix} 28+0-27 \\ -8+10+0 \\ 0-10+9 \end{bmatrix} = \begin{bmatrix} 1 \\ 2 \\ -1 \end{bmatrix}$$

The solution to the system is $\{(1, 2, -1)\}$.

**39. a.** $\begin{bmatrix} 1 & -1 & 1 \\ 0 & 2 & -1 \\ 2 & 3 & 0 \end{bmatrix}\begin{bmatrix} x \\ y \\ z \end{bmatrix} = \begin{bmatrix} 8 \\ -7 \\ 1 \end{bmatrix}$

**b.** $\begin{bmatrix} 3 & 3 & -1 \\ -2 & -2 & 1 \\ -4 & -5 & 2 \end{bmatrix}\begin{bmatrix} 8 \\ -7 \\ 1 \end{bmatrix}$

$$= \begin{bmatrix} 24-21-1 \\ -16+14+1 \\ -32+35+2 \end{bmatrix} = \begin{bmatrix} 2 \\ -1 \\ 5 \end{bmatrix}$$

The solution to the system is $\{(2, -1, 5)\}$.

**41. a.**
$$\begin{bmatrix} 1 & -1 & 2 & 0 \\ 0 & 1 & -1 & 1 \\ -1 & 1 & -1 & 2 \\ 0 & -1 & 1 & -2 \end{bmatrix}\begin{bmatrix} w \\ x \\ y \\ z \end{bmatrix} = \begin{bmatrix} -3 \\ 4 \\ 2 \\ -4 \end{bmatrix}$$

**b.**
$$\begin{bmatrix} 0 & 0 & -1 & -1 \\ 1 & 4 & 1 & 3 \\ 1 & 2 & 1 & 2 \\ 0 & -1 & 0 & -1 \end{bmatrix}\begin{bmatrix} -3 \\ 4 \\ 2 \\ -4 \end{bmatrix}$$

$$= \begin{bmatrix} 0+0-2+4 \\ -3+16+2-12 \\ -3+8+2-8 \\ 0-4+0+4 \end{bmatrix} = \begin{bmatrix} 2 \\ 3 \\ -1 \\ 0 \end{bmatrix}$$

The solution to the system is
$\{(2, 3, -1, 0)\}$.

**43.** $A = \begin{bmatrix} e^x & e^{3x} \\ -e^{3x} & e^{5x} \end{bmatrix}$

$$A^{-1} = \frac{1}{ad-bc}\begin{bmatrix} d & -b \\ -c & a \end{bmatrix}$$

$$A^{-1} = \frac{1}{\left(e^x\right)\left(e^{5x}\right)-\left(e^{3x}\right)\left(-e^{3x}\right)}\begin{bmatrix} e^{5x} & -e^{3x} \\ -\left(-e^{3x}\right) & e^x \end{bmatrix}$$

$$A^{-1} = \frac{1}{e^{6x}+e^{6x}}\begin{bmatrix} e^{5x} & -e^{3x} \\ e^{3x} & e^x \end{bmatrix}$$

$$A^{-1} = \frac{1}{2e^{6x}}\begin{bmatrix} e^{5x} & -e^{3x} \\ e^{3x} & e^x \end{bmatrix}$$

$$A^{-1} = \begin{bmatrix} \frac{e^{5x}}{2e^{6x}} & \frac{-e^{3x}}{2e^{6x}} \\ \frac{e^{3x}}{2e^{6x}} & \frac{e^x}{2e^{6x}} \end{bmatrix}$$

$$A^{-1} = \begin{bmatrix} \frac{1}{2e^x} & -\frac{1}{2e^{3x}} \\ \frac{1}{2e^{3x}} & \frac{1}{2e^{5x}} \end{bmatrix} \text{ or } \begin{bmatrix} \frac{e^{-x}}{2} & -\frac{e^{-3x}}{2} \\ \frac{e^{-3x}}{2} & \frac{e^{-5x}}{2} \end{bmatrix}$$

Check:
$$\begin{bmatrix} e^x & e^{3x} \\ -e^{3x} & e^{5x} \end{bmatrix} \cdot \begin{bmatrix} \frac{1}{2}e^{-x} & -\frac{1}{2}e^{-3x} \\ \frac{1}{2}e^{-3x} & \frac{1}{2}e^{-5x} \end{bmatrix} = \begin{bmatrix} 1 & 0 \\ 0 & 1 \end{bmatrix}$$

**45.** $A = \begin{bmatrix} 8 & -5 \\ -3 & 2 \end{bmatrix}$

$$I - A = \begin{bmatrix} 1 & 0 \\ 0 & 1 \end{bmatrix} - \begin{bmatrix} 8 & -5 \\ -3 & 2 \end{bmatrix} = \begin{bmatrix} -7 & 5 \\ 3 & -1 \end{bmatrix}$$

$$(I-A)^{-1} = \frac{1}{(-7)(-1)-(5)(3)}\begin{bmatrix} -1 & -(5) \\ -(3) & -7 \end{bmatrix}$$

$$(I-A)^{-1} = \frac{1}{7-15}\begin{bmatrix} -1 & -5 \\ -3 & -7 \end{bmatrix}$$

$$(I-A)^{-1} = \frac{1}{-8}\begin{bmatrix} -1 & -5 \\ -3 & -7 \end{bmatrix}$$

$$(I-A)^{-1} = \begin{bmatrix} \frac{-1}{-8} & \frac{-5}{-8} \\ \frac{-3}{-8} & \frac{-7}{-8} \end{bmatrix}$$

$$(I-A)^{-1} = \begin{bmatrix} \frac{1}{8} & \frac{5}{8} \\ \frac{3}{8} & \frac{7}{8} \end{bmatrix}$$

**47.** $A = \begin{bmatrix} 2 & 1 \\ 3 & 1 \end{bmatrix}$  $B = \begin{bmatrix} 4 & 7 \\ 1 & 2 \end{bmatrix}$

$$A^{-1} = \begin{bmatrix} -1 & 1 \\ 3 & -2 \end{bmatrix} \quad B^{-1} = \begin{bmatrix} 2 & -7 \\ -1 & 4 \end{bmatrix}$$

$$AB = \begin{bmatrix} 2 & 1 \\ 3 & 1 \end{bmatrix}\begin{bmatrix} 4 & 7 \\ 1 & 2 \end{bmatrix} = \begin{bmatrix} 9 & 16 \\ 13 & 23 \end{bmatrix}$$

$$(AB)^{-1} = \left(\begin{bmatrix} 9 & 16 \\ 13 & 23 \end{bmatrix}\right)^{-1} = \begin{bmatrix} -23 & 16 \\ 13 & -9 \end{bmatrix}$$

$$A^{-1}B^{-1} = \begin{bmatrix} -1 & 1 \\ 3 & -2 \end{bmatrix}\begin{bmatrix} 2 & -7 \\ -1 & 4 \end{bmatrix} = \begin{bmatrix} -3 & 11 \\ 8 & -29 \end{bmatrix}$$

$$B^{-1}A^{-1} = \begin{bmatrix} 2 & -7 \\ -1 & 4 \end{bmatrix}\begin{bmatrix} -1 & 1 \\ 3 & -2 \end{bmatrix} = \begin{bmatrix} -23 & 16 \\ 13 & -9 \end{bmatrix}$$

Observe that $(AB)^{-1} = B^{-1}A^{-1}$.

**49.** 
$$\begin{bmatrix} a & 0 & 0 \\ 0 & b & 0 \\ 0 & 0 & c \end{bmatrix}\begin{bmatrix} \frac{1}{a} & 0 & 0 \\ 0 & \frac{1}{b} & 0 \\ 0 & 0 & \frac{1}{c} \end{bmatrix}$$

$$= \begin{bmatrix} (a)(\frac{1}{a})+(0)(0)+(0)(0) & (a)(0)+(0)(\frac{1}{b})+(0)(0) & (a)(0)+(0)(0)+(0)(\frac{1}{c}) \\ (0)(\frac{1}{a})+(b)(0)+(0)(0) & (0)(0)+(b)(\frac{1}{b})+(0)(0) & (0)(0)+(b)(0)+(0)(\frac{1}{c}) \\ (0)(\frac{1}{a})+(0)(0)+(c)(0) & (0)(0)+(0)(\frac{1}{b})+(c)(0) & (0)(0)+(0)(0)+(c)(\frac{1}{c}) \end{bmatrix}$$

$$= \begin{bmatrix} \frac{a}{a}+0+0 & 0+0+0 & 0+0+0 \\ 0+0+0 & 0+\frac{b}{b}+0 & 0+0+0 \\ 0+0+0 & 0+0+0 & 0+0+\frac{c}{c} \end{bmatrix} = \begin{bmatrix} 1 & 0 & 0 \\ 0 & 1 & 0 \\ 0 & 0 & 1 \end{bmatrix}$$

**51.** The numerical equivalent of HELP is
8, 5, 12, 16.
$$\begin{bmatrix} 4 & -1 \\ -3 & 1 \end{bmatrix}\begin{bmatrix} 8 \\ 5 \end{bmatrix} = \begin{bmatrix} 27 \\ -19 \end{bmatrix},$$
$$\begin{bmatrix} 4 & -1 \\ -3 & 1 \end{bmatrix}\begin{bmatrix} 12 \\ 16 \end{bmatrix} = \begin{bmatrix} 32 \\ -20 \end{bmatrix}$$
The encoded message is 27, –19, 32, –20.
$$\begin{bmatrix} 1 & 1 \\ 3 & 4 \end{bmatrix}\begin{bmatrix} 27 \\ -19 \end{bmatrix} = \begin{bmatrix} 8 \\ 5 \end{bmatrix}, \begin{bmatrix} 1 & 1 \\ 3 & 4 \end{bmatrix}\begin{bmatrix} 32 \\ -20 \end{bmatrix} = \begin{bmatrix} 12 \\ 16 \end{bmatrix}$$
The decoded message is 8, 5, 12, 16 or HELP.

**53.** 
$$\begin{bmatrix} 1 & -1 & 0 \\ 3 & 0 & 2 \\ -1 & 0 & -1 \end{bmatrix}\begin{bmatrix} 19 & 4 & 1 \\ 5 & 0 & 19 \\ 14 & 3 & 8 \end{bmatrix}$$

$$= \begin{bmatrix} 19-5+0 & 4+0+0 & 1-19+0 \\ 57+0+28 & 12+0+6 & 3+0+16 \\ -19+0-14 & -4+0-3 & -1+0-8 \end{bmatrix}$$

$$= \begin{bmatrix} 14 & 4 & -18 \\ 85 & 18 & 19 \\ -33 & -7 & -9 \end{bmatrix}$$
The encoded message is 14, 85, –33, 4, 18, –7, –18, 19, –9.
$$\begin{bmatrix} 0 & 1 & 2 \\ -1 & 1 & 2 \\ 0 & -1 & -3 \end{bmatrix}\begin{bmatrix} 14 & 4 & -18 \\ 85 & 18 & 19 \\ -33 & -7 & -9 \end{bmatrix}$$

$$= \begin{bmatrix} 0+85-66 & 0+18-14 & 0+19-18 \\ -14+85-66 & -4+18-14 & 18+19-18 \\ 0-85+99 & 0-18+21 & 0-19+27 \end{bmatrix}$$

$$= \begin{bmatrix} 19 & 4 & 1 \\ 5 & 0 & 19 \\ 14 & 3 & 8 \end{bmatrix}$$
The decoded message is 19, 5, 14, 4, 0, 3, 1, 19, 8 or SEND_CASH

**55. – 63.** Answers will vary.

**65.** Enter the matrix $\begin{bmatrix} 3 & -1 \\ -2 & 1 \end{bmatrix}$ as [A], then use $[A]^{-1}$.

$[A]^{-1} = \begin{bmatrix} 1 & 1 \\ 2 & 3 \end{bmatrix}$

Verify this result by showing that $[A][A]^{-1} = I_2$ and $[A]^{-1}[A] = I_2$.

**67.** Enter the matrix $\begin{bmatrix} -2 & 1 & -1 \\ -5 & 2 & -1 \\ 3 & -1 & 1 \end{bmatrix}$ as [A], then use $[A]^{-1}$.

$[A]^{-1} = \begin{bmatrix} 1 & 0 & 1 \\ 2 & 1 & 3 \\ -1 & 1 & 1 \end{bmatrix}$

Verify this result by showing that $[A][A]^{-1} = I_3$ and $[A]^{-1}[A] = I_3$.

**69.** Enter the matrix $\begin{bmatrix} 7 & -3 & 0 & 2 \\ -2 & 1 & 0 & -1 \\ 4 & 0 & 1 & -2 \\ -1 & 1 & 0 & -1 \end{bmatrix}$ as [A], then use $[A]^{-1}$. $[A]^{-1} = \begin{bmatrix} 0 & -1 & 0 & 1 \\ -1 & -5 & 0 & 3 \\ -2 & -4 & 1 & -2 \\ -1 & -4 & 0 & 1 \end{bmatrix}$

Verify this result by showing that $[A][A]^{-1} = I_4$ and $[A]^{-1}[A] = I_4$.

**71.** The system is $AX = B$ where

$A = \begin{bmatrix} 1 & -1 & 1 \\ 4 & 2 & 1 \\ 4 & -2 & 1 \end{bmatrix}$, $X = \begin{bmatrix} x \\ y \\ z \end{bmatrix}$, and $B = \begin{bmatrix} -6 \\ 9 \\ -3 \end{bmatrix}$. $X = \begin{bmatrix} 2 \\ 3 \\ -5 \end{bmatrix}$, so the solution to the system is $\{(2, 3, -5)\}$.

**73.** The system is $AX = B$ where $A = \begin{bmatrix} 3 & -2 & 1 \\ 4 & -5 & 3 \\ 2 & -1 & 5 \end{bmatrix}$, $X = \begin{bmatrix} x \\ y \\ z \end{bmatrix}$, and $B = \begin{bmatrix} -2 \\ -9 \\ -5 \end{bmatrix}$.

$X = \begin{bmatrix} 1 \\ 2 \\ -1 \end{bmatrix}$ so the solution to the system is $\{(1, 2, -1)\}$.

**75.** The system is $AX = B$ where $A = \begin{bmatrix} 1 & 0 & -3 & 0 & 1 \\ 0 & 1 & 0 & 1 & 0 \\ 0 & 0 & 1 & 0 & 1 \\ 1 & 1 & -1 & 4 & 0 \\ 1 & 1 & 1 & 1 & 1 \end{bmatrix}$, $X = \begin{bmatrix} v \\ w \\ x \\ y \\ z \end{bmatrix}$ and $B = \begin{bmatrix} -3 \\ -1 \\ 7 \\ -8 \\ 8 \end{bmatrix}$. $X = \begin{bmatrix} 2 \\ 1 \\ 3 \\ -2 \\ 4 \end{bmatrix}$, so the solution to the system

is $\{(2, 1, 3, -2, 4)\}$.

**77.** Answers will vary.

**79.** does not make sense; Explanations will vary. Sample explanation: Only square matrices have inverses.

**81.** makes sense

**83.** false; Changes to make the statement true will vary. A sample change is: Not all square matrices have inverses.

**85.** false; Changes to make the statement true will vary. A sample change is: You need to multiply the inverse of *A* and *B*.

**87.** false; Changes to make the statement true will vary. A sample change is: $(A+B)^{-1} \neq A^{-1} + B^{-1}$

**89.** Answers will vary.

**91.** Using the statement before problems 9–14, we want to find values for *a* such that
$(1)(4) - (a+1)(a-2) = 0$.
$(1)(4) - (a+1)(a-2) = 4 - (a^2 - a - 2)$
$\qquad = -a^2 + a + 6$
$0 = -a^2 + a + 6$
$0 = a^2 - a - 6$
$0 = (a-3)(a+2)$
$a = 3, -2$

**93.** $2(-5) - (-3)(4) = -10 + 12 = 2$

**94.** $\dfrac{2(-5) - 1(-4)}{5(-5) - 6(-4)} = \dfrac{-10 + 4}{-25 + 24} = \dfrac{-6}{-1} = 6$

**95.** $2(-30 - (-3)) - 3(6-9) + (-1)(1-15)$
$= 2(-27) - 3(-3) + (-1)(-14)$
$= -54 + 9 + 14$
$= -31$

## Section 6.5

### Check Point Exercises

**1. a.** $\begin{vmatrix} 10 & 9 \\ 6 & 5 \end{vmatrix} = 10 \cdot 5 - 6 \cdot 9 = 50 - 54 = -4$

**b.** $\begin{vmatrix} 4 & 3 \\ -5 & -8 \end{vmatrix} = 4 \cdot (-8) - (-5) \cdot (3)$
$\qquad\qquad = -32 + 15 = -17$

**2.** $5x + 4y = 12$
$3x - 6y = 24$

$D = \begin{vmatrix} 5 & 4 \\ 3 & -6 \end{vmatrix} = 5 \cdot (-6) - 3 \cdot 4$
$\qquad = -30 - 12 = -42$

$D_x = \begin{vmatrix} 12 & 4 \\ 24 & -6 \end{vmatrix} = 12(-6) - 24(4)$
$\qquad = -72 - 96 = -168$

$D_y = \begin{vmatrix} 5 & 12 \\ 3 & 24 \end{vmatrix} = 5(24) - 3(12)$
$\qquad = 120 - 36 = 84$

Thus, $x = \dfrac{D_x}{D} = \dfrac{-168}{-42} = 4$
$\qquad y = \dfrac{D_y}{D} = \dfrac{84}{-42} = -2$
The solution set is $\{(4, -2)\}$.

**3.** $\begin{bmatrix} 2 & 1 & 7 \\ -5 & 6 & 0 \\ -4 & 3 & 1 \end{bmatrix}$

The minor for 2 is $\begin{vmatrix} 6 & 0 \\ 3 & 1 \end{vmatrix}$.

The minor for –5 is $\begin{vmatrix} 1 & 7 \\ 3 & 1 \end{vmatrix}$.

The minor for –4 is $\begin{vmatrix} 1 & 7 \\ 6 & 0 \end{vmatrix}$.

$\begin{bmatrix} 2 & 1 & 7 \\ -5 & 6 & 0 \\ -4 & 3 & 1 \end{bmatrix} = 2 \begin{vmatrix} 6 & 0 \\ 3 & 1 \end{vmatrix} - (-5) \begin{vmatrix} 1 & 7 \\ 3 & 1 \end{vmatrix} - 4 \begin{vmatrix} 1 & 7 \\ 6 & 0 \end{vmatrix}$
$\qquad = 2(6 \cdot 1 - 3 \cdot 0) + 5(1 \cdot 1 - 3 \cdot 7) - 4(1 \cdot 0 - 6 \cdot 7)$
$\qquad = 2(6 - 0) + 5(1 - 21) - 4(0 - 42)$
$\qquad = 12 - 100 + 168$
$\qquad = 80$

**4.** $\begin{vmatrix} 6 & 4 & 0 \\ -3 & -5 & 3 \\ 1 & 2 & 0 \end{vmatrix} = 0 \begin{vmatrix} -3 & -5 \\ 1 & 2 \end{vmatrix} - 3 \begin{vmatrix} 6 & 4 \\ 1 & 2 \end{vmatrix} + 0 \begin{vmatrix} 6 & 4 \\ -3 & -5 \end{vmatrix}$
$\qquad = 0 - 3(6 \cdot 2 - 1 \cdot 4) + 0$
$\qquad = -3(12 - 4)$
$\qquad = -3(8)$
$\qquad = -24$

**5.** $3x - 2y + z = 16$

$2x + 3y - z = -9$

$x + 4y + 3z = 2$

$D = \begin{vmatrix} 3 & -2 & 1 \\ 2 & 3 & -1 \\ 1 & 4 & 3 \end{vmatrix}; \quad D_x = \begin{vmatrix} 16 & -2 & 1 \\ -9 & 3 & -1 \\ 2 & 4 & 3 \end{vmatrix}; \quad D_y = \begin{vmatrix} 3 & 16 & 1 \\ 2 & -9 & -1 \\ 1 & 2 & 3 \end{vmatrix}; \quad D_z = \begin{vmatrix} 3 & -2 & 16 \\ 2 & 3 & -9 \\ 1 & 4 & 2 \end{vmatrix}$

$D = \begin{vmatrix} 3 & -2 & 1 \\ 2 & 3 & -1 \\ 1 & 4 & 3 \end{vmatrix} = 3 \begin{vmatrix} 3 & -1 \\ 4 & 3 \end{vmatrix} - 2 \begin{vmatrix} -2 & 1 \\ 4 & 3 \end{vmatrix} + 1 \begin{vmatrix} -2 & 1 \\ 3 & -1 \end{vmatrix}$

$= 3[(3) \cdot 3 - 4 \cdot (-1)] - 2[(-2) \cdot 3 - 4 \cdot 1] + 1[(-2) \cdot (-1) - (3) \cdot 1]$

$= 3(9 + 4) - 2(-6 - 4) + 1(2 - 3)$

$= 39 + 20 - 1$

$= 58$

$D_x = \begin{vmatrix} 16 & -2 & 1 \\ -9 & 3 & -1 \\ 2 & 4 & 3 \end{vmatrix} = 1 \begin{vmatrix} -9 & 3 \\ 2 & 4 \end{vmatrix} - (-1) \begin{vmatrix} 16 & -2 \\ 2 & 4 \end{vmatrix} + 3 \begin{vmatrix} 16 & -2 \\ -9 & 3 \end{vmatrix}$

$= 1[(-9) \cdot 4 - 2 \cdot (3)] + 1[16 \cdot 4 - 2(-2)] + 3[16 \cdot (3) - (-9) \cdot (-2)]$

$= 1(-36 - 6) + 1(64 + 4) + 3(48 - 18)$

$= -42 + 68 + 90$

$= 116$

$D_y = \begin{vmatrix} 3 & 16 & 1 \\ 2 & -9 & -1 \\ 1 & 2 & 3 \end{vmatrix} = 3 \begin{vmatrix} -9 & -1 \\ 2 & 3 \end{vmatrix} - 2 \begin{vmatrix} 16 & 1 \\ 2 & 3 \end{vmatrix} + 1 \begin{vmatrix} 16 & 1 \\ -9 & -1 \end{vmatrix}$

$= 3[(-9) \cdot 3 - 2 \cdot (-1)] - 2[16 \cdot 3 - 2 \cdot 1] + 1[16(-1) - (-9) \cdot 1]$

$= 3(-27 + 2) - 2(48 - 2) + 1(-16 + 9)$

$= -75 - 92 - 7$

$= -174$

$D_z = \begin{vmatrix} 3 & -2 & 16 \\ 2 & 3 & -9 \\ 1 & 4 & 2 \end{vmatrix} = 3 \begin{vmatrix} 3 & -9 \\ 4 & 2 \end{vmatrix} - 2 \begin{vmatrix} -2 & 16 \\ 4 & 2 \end{vmatrix} + 1 \begin{vmatrix} -2 & 16 \\ 3 & -9 \end{vmatrix}$

$= 3[(3)2 - 4(-9)] - 2[(-2)2 - 4 \cdot 16] + 1[(-2)(-9) - (3) \cdot 16]$

$= 3(6 + 36) - 2(-4 - 64) + 1(18 - 48)$

$= 126 + 136 - 30$

$= 232$

$x = \dfrac{D_x}{D} = \dfrac{116}{58} = 2$

$y = \dfrac{D_y}{D} = \dfrac{-174}{58} = -3$

$z = \dfrac{D_z}{D} = \dfrac{232}{58} = 4$

The solution to the system is $\{(2, -3, 4)\}$.

**6.**  $|A| = \begin{vmatrix} 0 & 4 & 0 & -3 \\ -1 & 1 & 5 & 2 \\ 1 & -2 & 0 & 6 \\ 3 & 0 & 0 & 1 \end{vmatrix} = (-1)^{2+3}5\begin{vmatrix} 0 & 4 & -3 \\ 1 & -2 & 6 \\ 3 & 0 & 1 \end{vmatrix} = -5\begin{vmatrix} 0 & 4 & -3 \\ 1 & -2 & 6 \\ 3 & 0 & 1 \end{vmatrix}$

Evaluate the third-order determinant to get $|A| = -5(50) = -250$.

## Concept and Vocabulary Check 6.5

**1.**  5; 3; 2; 4; 15; 8; 7; determinant; 7

**2.**  $\dfrac{\begin{vmatrix} 8 & 1 \\ -2 & -1 \end{vmatrix}}{\begin{vmatrix} 1 & 1 \\ 1 & -1 \end{vmatrix}}$ ; $\dfrac{\begin{vmatrix} 1 & 8 \\ 1 & -2 \end{vmatrix}}{\begin{vmatrix} 1 & 1 \\ 1 & -1 \end{vmatrix}}$

**3.**  $\begin{vmatrix} 3 & 1 \\ 1 & 1 \end{vmatrix}$ ; $\begin{vmatrix} 2 & 1 \\ 1 & 1 \end{vmatrix}$ ; $\begin{vmatrix} 2 & 1 \\ 3 & 1 \end{vmatrix}$

**4.**  $\dfrac{\begin{vmatrix} 3 & -8 & 4 \\ 2 & 11 & -2 \\ 1 & 4 & -2 \end{vmatrix}}{\begin{vmatrix} 3 & 1 & 4 \\ 2 & 3 & -2 \\ 1 & -3 & -2 \end{vmatrix}}$

**5.**  column 3/the last column

## Exercise Set 6.5

**1.**  $\begin{vmatrix} 5 & 7 \\ 2 & 3 \end{vmatrix} = 5\cdot3 - 2\cdot7 = 15 - 14 = 1$

**3.**  $\begin{vmatrix} -4 & 1 \\ 5 & 6 \end{vmatrix} = (-4)6 - 5\cdot1 = -24 - 5 = -29$

**5.**  $\begin{vmatrix} -7 & 14 \\ 2 & -4 \end{vmatrix} = (-7)(-4) - 2(14) = 28 - 28 = 0$

**7.**  $\begin{vmatrix} -5 & -1 \\ -2 & -7 \end{vmatrix} = (-5)(-7) - (-2)(-1) = 35 - 2 = 33$

**9.**  $\begin{vmatrix} \frac{1}{2} & \frac{1}{2} \\ \frac{1}{8} & -\frac{3}{4} \end{vmatrix} = \frac{1}{2}\left(-\frac{3}{4}\right) - \frac{1}{8}\cdot\frac{1}{2} = -\frac{3}{8} - \frac{1}{16} = -\frac{7}{16}$

**11.** $D = \begin{vmatrix} 1 & 1 \\ 1 & -1 \end{vmatrix} = -1 - 1 = -2$

$D_x = \begin{vmatrix} 7 & 1 \\ 3 & -1 \end{vmatrix} = -7 - 3 = -10$

$D_y = \begin{vmatrix} 1 & 7 \\ 1 & 3 \end{vmatrix} = 3 - 7 = -4$

$x = \dfrac{D_x}{D} = \dfrac{-10}{-2} = 5$

$y = \dfrac{D_y}{D} = \dfrac{-4}{-2} = 2$

The solution set is $\{(5, 2)\}$.

**13.** $D = \begin{vmatrix} 12 & 3 \\ 2 & -3 \end{vmatrix} = -36 - 6 = -42$

$D_x = \begin{vmatrix} 15 & 3 \\ 13 & -3 \end{vmatrix} = -45 - 39 = -84$

$D_y = \begin{vmatrix} 12 & 15 \\ 2 & 13 \end{vmatrix} = 156 - 30 = 126$

$x = \dfrac{D_x}{D} = \dfrac{-84}{-42} = 2$

$y = \dfrac{D_y}{D} = \dfrac{126}{-42} = -3$

The solution set is $\{(2, -3)\}$.

**15.** $D = \begin{vmatrix} 4 & -5 \\ 2 & 3 \end{vmatrix} = 12 - (-10) = 22$

$D_x = \begin{vmatrix} 17 & -5 \\ 3 & 3 \end{vmatrix} = 51 - (-15) = 66$

$D_y = \begin{vmatrix} 4 & 17 \\ 2 & 3 \end{vmatrix} = 12 - 34 = -22$

$x = \dfrac{D_x}{D} = \dfrac{66}{22} = 3$

$y = \dfrac{D_y}{D} = \dfrac{-22}{22} = -1$

The solution set is $\{(3, -1)\}$.

**17.** $D = \begin{vmatrix} 1 & 2 \\ 5 & 10 \end{vmatrix} = 10 - 10 = 0$

$D_x = \begin{vmatrix} 3 & 2 \\ 15 & 10 \end{vmatrix} = 30 - 30 = 0$

$D_y = \begin{vmatrix} 1 & 3 \\ 5 & 15 \end{vmatrix} = 15 - 15 = 0$

Because all 3 determinants equal zero, the system is dependent.

**19.** $D = \begin{vmatrix} 3 & -4 \\ 2 & 2 \end{vmatrix} = 6 - (-8) = 14$

$D_x = \begin{vmatrix} 4 & -4 \\ 12 & 2 \end{vmatrix} = 8 - (-48) = 56$

$D_y = \begin{vmatrix} 3 & 4 \\ 2 & 12 \end{vmatrix} = 36 - 8 = 28$

$x = \dfrac{D_x}{D} = \dfrac{56}{14} = 4$

$y = \dfrac{D_y}{D} = \dfrac{28}{14} = 2$

The solution set is $\{(4, 2)\}$.

**21.** $D = \begin{vmatrix} 2 & -3 \\ 5 & 4 \end{vmatrix} = 8 - (-15) = 23$

$D_x = \begin{vmatrix} 2 & -3 \\ 51 & 4 \end{vmatrix} = 8 - (-153) = 161$

$D_y = \begin{vmatrix} 2 & 2 \\ 5 & 51 \end{vmatrix} = 102 - 10 = 92$

$x = \dfrac{D_x}{D} = \dfrac{161}{23} = 7$

$y = \dfrac{D_y}{D} = \dfrac{92}{23} = 4$

The solution set is $\{(7, 4)\}$.

**23.** $\begin{vmatrix} 3 & 0 & 0 \\ 2 & 1 & -5 \\ -2 & 5 & -1 \end{vmatrix} = 3\begin{vmatrix} 1 & -5 \\ 5 & -1 \end{vmatrix} - 0\begin{vmatrix} 2 & -5 \\ -2 & -1 \end{vmatrix} + 0\begin{vmatrix} 2 & 1 \\ -2 & 5 \end{vmatrix}$

$= 3[(1)(-1) - (5)(-5)]$
$= 3(-1 + 25) = 3(24)$
$= 72$

**25.** $\begin{vmatrix} 3 & 1 & 0 \\ -3 & 4 & 0 \\ -1 & 3 & -5 \end{vmatrix} = 0\begin{vmatrix} -3 & 4 \\ -1 & 3 \end{vmatrix} - 0\begin{vmatrix} 3 & 1 \\ -1 & 3 \end{vmatrix} + (-5)\begin{vmatrix} 3 & 1 \\ -3 & 4 \end{vmatrix}$

$= -5[3 \cdot 4 - (-3)(1)]$
$= -5(12 + 3) = -5(15)$
$= -75$

**27.** $\begin{vmatrix} 1 & 1 & 1 \\ 2 & 2 & 2 \\ -3 & 4 & -5 \end{vmatrix} -2R_1 + R_2$

$\begin{vmatrix} 1 & 1 & 1 \\ 0 & 0 & 0 \\ -3 & 4 & -5 \end{vmatrix} = 0$

**29.** $D = \begin{vmatrix} 1 & 1 & 1 \\ 2 & -1 & 1 \\ -1 & 3 & -1 \end{vmatrix}$

$= \begin{vmatrix} -1 & 1 \\ 3 & -1 \end{vmatrix} - \begin{vmatrix} 2 & 1 \\ -1 & -1 \end{vmatrix} + \begin{vmatrix} 2 & -1 \\ -1 & 3 \end{vmatrix}$

$= (1 - 3) - [-2 - (-1)] + (6 - 1)$

$= -2 - (-1) + 5 = -2 + 1 + 5 = 4$

$D_x = \begin{vmatrix} 0 & 1 & 1 \\ -1 & -1 & 1 \\ -8 & 3 & -1 \end{vmatrix} = (-1) \begin{vmatrix} -1 & 1 \\ -8 & -1 \end{vmatrix} + \begin{vmatrix} -1 & -1 \\ -8 & 3 \end{vmatrix}$

$= (-1)[1 - (-8)] + (-3 - 8) = (-1)(9) - 11$

$= -20$

$D_y = \begin{vmatrix} 1 & 0 & 1 \\ 2 & -1 & 1 \\ -1 & -8 & -1 \end{vmatrix} = \begin{vmatrix} -1 & 1 \\ -8 & -1 \end{vmatrix} + \begin{vmatrix} 2 & -1 \\ -1 & -8 \end{vmatrix}$

$= 1 - (-8) + (-16 - 1) = 1 + 8 - 17 = -8$

$D_z = \begin{vmatrix} 1 & 1 & 0 \\ 2 & -1 & -1 \\ -1 & 3 & -8 \end{vmatrix} = 1 \begin{vmatrix} -1 & -1 \\ 3 & -8 \end{vmatrix} - 1 \begin{vmatrix} 2 & -1 \\ -1 & -8 \end{vmatrix}$

$= 8 - (-3) - 1(-16 - 1) = 11 + 17 = 28$

$x = \dfrac{D_x}{D} = \dfrac{-20}{4} = -5$

$y = \dfrac{D_y}{D} = \dfrac{-8}{4} = -2$

$z = \dfrac{D_z}{D} = \dfrac{28}{4} = 7$

The solution to the system is $\{(-5, -2, 7)\}$.

**31.** $D = \begin{vmatrix} 4 & -5 & -6 \\ 1 & -2 & -5 \\ 2 & -1 & 0 \end{vmatrix} = 2 \begin{vmatrix} -5 & -6 \\ -2 & -5 \end{vmatrix} - (-1) \begin{vmatrix} 4 & -6 \\ 1 & -5 \end{vmatrix}$

$= 2(25 - 12) + [-20 - (-6)] = 2(13) + (-14)$

$= 26 - 14 = 12$

$D_x = \begin{vmatrix} -1 & -5 & -6 \\ -12 & -2 & -5 \\ 7 & -1 & 0 \end{vmatrix}$

$= 7 \begin{vmatrix} -5 & -6 \\ -2 & -5 \end{vmatrix} - (-1) \begin{vmatrix} -1 & -6 \\ -12 & -5 \end{vmatrix}$

$= 7(25 - 12) + (5 - 72) = 7(13) - 67$

$= 91 - 67 = 24$

$D_y = \begin{vmatrix} 4 & -1 & -6 \\ 1 & -12 & -5 \\ 2 & 7 & 0 \end{vmatrix} = 2 \begin{vmatrix} -1 & -6 \\ -12 & -5 \end{vmatrix} - 7 \begin{vmatrix} 4 & -6 \\ 1 & -5 \end{vmatrix}$

$= 2(5 - 72) - 7[-20 - (-6)]$

$= 2(-67) - 7(-14) = -134 + 98 = -36$

$D_z = \begin{vmatrix} 4 & -5 & -1 \\ 1 & -2 & -12 \\ 2 & -1 & 7 \end{vmatrix}$

$= 4 \begin{vmatrix} -2 & -12 \\ -1 & 7 \end{vmatrix} - (-5) \begin{vmatrix} 1 & -12 \\ 2 & 7 \end{vmatrix} + (-1) \begin{vmatrix} 1 & -2 \\ 2 & -1 \end{vmatrix}$

$= 4(-14 - 12) + 5[7 - (-24)] - [-1 - (-4)]$

$= 4(-26) + 5(31) - (3) = -104 + 155 - 3 = 48$

$x = \dfrac{D_x}{D} = \dfrac{24}{12} = 2, \quad y = \dfrac{D_y}{D} = \dfrac{-36}{12} = -3,$

$z = \dfrac{D_z}{D} = \dfrac{48}{12} = 4$

The solution set is $\{(2, -3, 4)\}$.

**33.** $D = \begin{vmatrix} 1 & 1 & 1 \\ 1 & -2 & 1 \\ 1 & 3 & 2 \end{vmatrix} = 1 \begin{vmatrix} -2 & 1 \\ 3 & 2 \end{vmatrix} - 1 \begin{vmatrix} 1 & 1 \\ 1 & 2 \end{vmatrix} + 1 \begin{vmatrix} 1 & -2 \\ 1 & 3 \end{vmatrix}$

$= -4 - 3 - (2 - 1) + [3 - (-2)]$

$= -7 - 1 + 5 = -3$

$D_x = \begin{vmatrix} 4 & 1 & 1 \\ 7 & -2 & 1 \\ 4 & 3 & 2 \end{vmatrix} = 4 \begin{vmatrix} -2 & 1 \\ 3 & 2 \end{vmatrix} - 1 \begin{vmatrix} 7 & 1 \\ 4 & 2 \end{vmatrix} + 1 \begin{vmatrix} 7 & -2 \\ 4 & 3 \end{vmatrix}$

$= 4(-4 - 3) - (14 - 4) + [21 - (-8)]$

$= 4(-7) - 10 + 29 = -28 + 19 = -9$

$D_y = \begin{vmatrix} 1 & 4 & 1 \\ 1 & 7 & 1 \\ 1 & 4 & 2 \end{vmatrix} = 1 \begin{vmatrix} 7 & 1 \\ 4 & 2 \end{vmatrix} - 1 \begin{vmatrix} 4 & 1 \\ 4 & 2 \end{vmatrix} + 1 \begin{vmatrix} 4 & 1 \\ 7 & 1 \end{vmatrix}$

$= 14 - 4 - (8 - 4) + (4 - 7) = 10 - 4 - 3 = 3$

$D_z = \begin{vmatrix} 1 & 1 & 4 \\ 1 & -2 & 7 \\ 1 & 3 & 4 \end{vmatrix} = 1 \begin{vmatrix} -2 & 7 \\ 3 & 4 \end{vmatrix} - 1 \begin{vmatrix} 1 & 4 \\ 3 & 4 \end{vmatrix} + 1 \begin{vmatrix} 1 & 4 \\ -2 & 7 \end{vmatrix}$

$= -8 - 21 - (4 - 12) + [7 - (-8)]$

$= -29 + 8 + 15 = -6$

$x = \dfrac{D_x}{D} = \dfrac{-9}{-3} = 3, \quad y = \dfrac{D_y}{D} = \dfrac{3}{-3} = -1,$

$z = \dfrac{D_z}{D} = \dfrac{-6}{-3} = 2$

The solution set is $\{3, -1, 2\}$.

**35.** $D = \begin{vmatrix} 1 & 0 & 2 \\ 0 & 2 & -1 \\ 2 & 3 & 0 \end{vmatrix} = \begin{vmatrix} 2 & -1 \\ 3 & 0 \end{vmatrix} + 2\begin{vmatrix} 0 & 2 \\ 2 & 3 \end{vmatrix}$

$= 0 - (-3) + 2(0 - 4) = 3 - 8 = -5$

$D_x = \begin{vmatrix} 4 & 0 & 2 \\ 5 & 2 & -1 \\ 13 & 3 & 0 \end{vmatrix} = 4\begin{vmatrix} 2 & -1 \\ 3 & 0 \end{vmatrix} + 2\begin{vmatrix} 5 & 2 \\ 13 & 3 \end{vmatrix}$

$= 4[0 - (-3)] + 2(15 - 26)$
$= 4(3) + 2(-11) = 12 - 22 = -10$

$D_y = \begin{vmatrix} 1 & 4 & 2 \\ 0 & 5 & -1 \\ 2 & 13 & 0 \end{vmatrix} = \begin{vmatrix} 5 & -1 \\ 13 & 0 \end{vmatrix} + 2\begin{vmatrix} 4 & 2 \\ 5 & -1 \end{vmatrix}$

$= 0 - (-13) + 2(-4 - 10)$
$= 13 + 2(-14) = 13 - 28 = -15$

$D_z = \begin{vmatrix} 1 & 0 & 4 \\ 0 & 2 & 5 \\ 2 & 3 & 13 \end{vmatrix} = \begin{vmatrix} 2 & 5 \\ 3 & 13 \end{vmatrix} + 4\begin{vmatrix} 0 & 2 \\ 2 & 3 \end{vmatrix}$

$= 26 - 15 + 4(0 - 4) = 11 + 4(-4)$
$= 11 - 16 = -5$

$x = \dfrac{D_x}{D} = \dfrac{-10}{-5} = 2,\ y = \dfrac{D_y}{D} = \dfrac{-15}{-5} = 3,$

$z = \dfrac{D_z}{D} = \dfrac{-5}{-5} = 1$

The solution set is $\{(2, 3, 1)\}$.

**37.** $\begin{vmatrix} 4 & 2 & 8 & -7 \\ -2 & 0 & 4 & 1 \\ 5 & 0 & 0 & 5 \\ 4 & 0 & 0 & -1 \end{vmatrix} = -2\begin{vmatrix} -2 & 4 & 1 \\ 5 & 0 & 5 \\ 4 & 0 & -1 \end{vmatrix} + 0\begin{vmatrix} 4 & 8 & -7 \\ 5 & 0 & 5 \\ 4 & 0 & -1 \end{vmatrix} - 0\begin{vmatrix} 4 & 8 & -7 \\ -2 & 4 & 1 \\ 4 & 0 & -1 \end{vmatrix} + 0\begin{vmatrix} 4 & 8 & -7 \\ -2 & 4 & 1 \\ 5 & 0 & 5 \end{vmatrix}$

$= (-2)\left[ (-4)\begin{vmatrix} 5 & 5 \\ 4 & -1 \end{vmatrix} + 0\begin{vmatrix} -2 & 1 \\ 4 & -1 \end{vmatrix} - 0\begin{vmatrix} -2 & 1 \\ 5 & 5 \end{vmatrix} \right] = (-2)(-4)[5(-1) - 4 \cdot 5] = 8(-5 - 20) = 8(-25) = -200$

**39.** $\begin{vmatrix} -2 & -3 & 3 & 5 \\ 1 & -4 & 0 & 0 \\ 1 & 2 & 2 & -3 \\ 2 & 0 & 1 & 1 \end{vmatrix} = -1\begin{vmatrix} -3 & 3 & 5 \\ 2 & 2 & -3 \\ 0 & 1 & 1 \end{vmatrix} + (-4)\begin{vmatrix} -2 & 3 & 5 \\ 1 & 2 & -3 \\ 2 & 1 & 1 \end{vmatrix} - 0\begin{vmatrix} -2 & -3 & 5 \\ 1 & 2 & -3 \\ 2 & 0 & 1 \end{vmatrix} + 0\begin{vmatrix} -2 & -3 & 3 \\ 1 & 2 & 2 \\ 2 & 0 & 1 \end{vmatrix}$

$= (-1)\left[ 0\begin{vmatrix} 3 & 5 \\ 2 & -3 \end{vmatrix} - 1\begin{vmatrix} -3 & 5 \\ 2 & -3 \end{vmatrix} + 1\begin{vmatrix} -3 & 3 \\ 2 & 2 \end{vmatrix} \right] - 4\left[ 2\begin{vmatrix} 3 & 5 \\ 2 & -3 \end{vmatrix} - 1\begin{vmatrix} -2 & 5 \\ 1 & -3 \end{vmatrix} + 1\begin{vmatrix} -2 & 3 \\ 1 & 2 \end{vmatrix} \right]$

$= (-1)\{(-1)[(-3)(-3) - 2 \cdot 5] + [(-3)(2) - 2 \cdot 3]\} - 4\{2[3(-3) - 2 \cdot 5] - [(-2)(-3) - 1 \cdot 5] + [(-2)(2) - 1 \cdot 3]\} = 195$

**41.** $\begin{vmatrix} \begin{vmatrix} 3 & 1 \\ -2 & 3 \end{vmatrix} & \begin{vmatrix} 7 & 0 \\ 1 & 5 \end{vmatrix} \\ \begin{vmatrix} 3 & 0 \\ 0 & 7 \end{vmatrix} & \begin{vmatrix} 9 & -6 \\ 3 & 5 \end{vmatrix} \end{vmatrix} = \begin{vmatrix} 3(3) - (-2)(1) & 7(5) - 1(0) \\ 3(7) - 0(0) & 9(5) - 3(-6) \end{vmatrix} = \begin{vmatrix} 9 + 2 & 35 - 0 \\ 21 - 0 & 45 + 18 \end{vmatrix} = \begin{vmatrix} 11 & 35 \\ 21 & 63 \end{vmatrix}$

$= 11(63) - 21(35) = 693 - 735 = -42$

**43.** From $D = \begin{vmatrix} 2 & -4 \\ 3 & 5 \end{vmatrix}$ we obtain the coefficients of the variables in our equations:

$2x - 4y = c_1$

$3x + 5y = c_2$

From $D_x = \begin{vmatrix} 8 & -4 \\ -10 & 5 \end{vmatrix}$ we obtain the constant coefficients: 8 and $-10$

$2x - 4y = 8$

$3x + 5y = -10$

**45.** $\begin{vmatrix} -2 & x \\ 4 & 6 \end{vmatrix} = 32$

$-2(6) - 4(x) = 32$

$-12 - 4x = 32$

$-4x = 44$

$x = -11$

The solution is $-11$.

**47.** $\begin{vmatrix} 1 & x & -2 \\ 3 & 1 & 1 \\ 0 & -2 & 2 \end{vmatrix} = -8$

$0 \begin{vmatrix} x & -2 \\ 1 & 1 \end{vmatrix} - (-2) \begin{vmatrix} 1 & -2 \\ 3 & 1 \end{vmatrix} + 2 \begin{vmatrix} 1 & x \\ 3 & 1 \end{vmatrix} = -8$

$2 \left[ 1(1) - 3(-2) \right] + 2 \left[ 1(1) - 3(x) \right] = -8$

$2(1 + 6) + 2(1 - 3x) = -8$

$2(7) + 2(1 - 3x) = -8$

$14 + 2 - 6x = -8$

$-6x = -24$

$x = 4$

The solution is 4.

**49.** Area $= \pm \dfrac{1}{2} \begin{vmatrix} 3 & -5 & 1 \\ 2 & 6 & 1 \\ -3 & 5 & 1 \end{vmatrix} = \pm \dfrac{1}{2} \begin{vmatrix} 3 & -5 & 1 \\ -1 & 11 & 0 \\ -6 & 10 & 0 \end{vmatrix} = \pm \dfrac{1}{2} \begin{vmatrix} -1 & 11 \\ -6 & 10 \end{vmatrix} = \pm \dfrac{1}{2} [-10 - (-66)] = \pm \dfrac{1}{2}(56) = 28$

The area is 28 square units.

The slope of the line through $(3, -5)$ and $(-3, 5)$ is $m = \dfrac{5 - (-5)}{-3 - 3} = \dfrac{10}{-6} = -\dfrac{5}{3}$.

The equation of the line is $y - (-5) = -\dfrac{5}{3}(x - 3)$ or $y = -\dfrac{5}{3}x$.

The line perpendicular to $y = -\dfrac{5}{3}x$ through $(2, 6)$ has equation $y - 6 = \dfrac{3}{5}(x - 2)$ or $y = \dfrac{3}{5}x + \dfrac{24}{5}$.

These lines intersect where $-\dfrac{5}{3}x = \dfrac{3}{5}x + \dfrac{24}{5}$.

$-\dfrac{36}{17} = x$ and $-\dfrac{24}{5} = \dfrac{34}{15}x$  $y = -\dfrac{5}{3}\left(-\dfrac{36}{17}\right) = \dfrac{60}{17}$

Using the side connecting (3, –5) and

(–3, 5) as the base, the height is the distance from (2, 6) to $\left(-\dfrac{36}{17}, \dfrac{60}{17}\right)$.

$b = \sqrt{[3-(-3)]^2 + (-5-5)^2}$

$\quad = \sqrt{36+100} = \sqrt{136} = 2\sqrt{34}$

$h = \sqrt{\left[2-\left(-\dfrac{36}{17}\right)\right]^2 + \left(6-\dfrac{60}{17}\right)^2}$

$\quad = \sqrt{\dfrac{4900}{289} + \dfrac{1764}{289}} = \dfrac{14\sqrt{34}}{17}$

$\dfrac{1}{2}bh = \dfrac{1}{2}\left(2\sqrt{34}\right)\left(\dfrac{14\sqrt{34}}{17}\right) = \dfrac{14(34)}{17}$

$\quad = 14(2) = 28$ square units

**51.** $\begin{vmatrix} 3 & -1 & 1 \\ 0 & -3 & 1 \\ 12 & 5 & 1 \end{vmatrix} = \begin{vmatrix} 3 & -1 & 1 \\ -3 & -2 & 0 \\ 9 & 6 & 0 \end{vmatrix} = \begin{vmatrix} -3 & -2 \\ 9 & 6 \end{vmatrix}$

$= -18 - (-18) = 0$

Yes, the points are collinear.

**53.** $\begin{vmatrix} x & y & 1 \\ 3 & -5 & 1 \\ -2 & 6 & 1 \end{vmatrix} = x\begin{vmatrix} -5 & 1 \\ 6 & 1 \end{vmatrix} - y\begin{vmatrix} 3 & 1 \\ -2 & 1 \end{vmatrix} + \begin{vmatrix} 3 & -5 \\ -2 & 6 \end{vmatrix} = x(-5-6) - y[3-(-2)] + (18-10)$

$= -11x - 5y + 8$

The equation of the line is $-11x - 5y + 8 = 0$. The equation of the line in slope-intercept form is $y = -\dfrac{11}{5}x + \dfrac{8}{5}$.

**55. – 63.** Answers will vary.

**65.** Input the matrix as [A], then use det[A] to find the determinant.

$\begin{vmatrix} 8 & 2 & 6 & -1 & 0 \\ 2 & 0 & -3 & 4 & 7 \\ 2 & 1 & -3 & 6 & -5 \\ -1 & 2 & 1 & 5 & -1 \\ 4 & 5 & -2 & 3 & -8 \end{vmatrix} = 13,200$

**67.** does not make sense; Explanations will vary. Sample explanation: Determinants must be square.

**69.** does not make sense; Explanations will vary. Sample explanation: The number of determinants needed is one greater than the number of variables.

**71.** In this exercise, expansions are all done about the first column of the matrix and the resulting products of 0 and a determinant are not shown.

**a.** $\begin{vmatrix} a & a \\ 0 & a \end{vmatrix} = a^2 - 0 = a^2$

**b.** $\begin{vmatrix} a & a & a \\ 0 & a & a \\ 0 & 0 & a \end{vmatrix} = a\begin{vmatrix} a & a \\ 0 & a \end{vmatrix} - 0 + 0$

$= a\left(a^2\right) = a^3$

**c.** $\begin{vmatrix} a & a & a & a \\ 0 & a & a & a \\ 0 & 0 & a & a \\ 0 & 0 & 0 & a \end{vmatrix} = a\begin{vmatrix} a & a & a \\ 0 & a & a \\ 0 & 0 & a \end{vmatrix} - 0 + 0 - 0$

$= a\left(a^3\right) = a^4$

**d.** Each determinant has zeros below the main diagonal and $a$'s everywhere else.

**e.** Each determinant equals $a$ raised to the power equal to the order of the determinant.

**73.** The sign of the value is changed when 2 columns are interchanged in a 2nd order determinant.

**75.** Evaluate the determinate and write the equation in slope intercept form.

$\begin{vmatrix} x & y & 1 \\ x_1 & y_1 & 1 \\ x_2 & y_2 & 1 \end{vmatrix} = 0$

$x\begin{vmatrix} y_1 & 1 \\ y_2 & 1 \end{vmatrix} - y\begin{vmatrix} x_1 & 1 \\ x_2 & 1 \end{vmatrix} + 1\begin{vmatrix} x_1 & y_1 \\ x_2 & y_2 \end{vmatrix} = 0$

$x\left(y_1 - y_2\right) - y\left(x_1 - x_2\right) + x_1 y_2 - x_2 y_1 = 0$

$-y\left(x_1 - x_2\right) = -x\left(y_1 - y_2\right) + x_2 y_1 - x_1 y_2$

$y\left(x_2 - x_1\right) = x\left(y_2 - y_1\right) + x_2 y_1 - x_1 y_2$

$y = \frac{y_2 - y_1}{x_2 - x_1}x + \frac{x_2 y_1 - x_1 y_2}{x_2 - x_1}$

$m = \frac{y_2 - y_1}{x_2 - x_1} \qquad b = \frac{x_2 y_1 - x_1 y_2}{x_2 - x_1}$

Write the slope-point equation of the line the in point slope form.

$y - y_1 = \frac{y_2 - y_1}{x_2 - x_1}(x - x_1)$

$y - y_1 = \frac{y_2 - y_1}{x_2 - x_1}x + \frac{-x_1 y_2 + x_1 y_1}{x_2 - x_1}$

$y = \frac{y_2 - y_1}{x_2 - x_1}x + \frac{-x_1 y_2 + x_1 y_1}{x_2 - x_1} + y_1$

$y = \frac{y_2 - y_1}{x_2 - x_1}x + \frac{-x_1 y_2 + x_1 y_1}{x_2 - x_1} + \frac{x_2 y_1 - x_1 y_1}{x_2 - x_1}$

$y = \frac{y_2 - y_1}{x_2 - x_1}x + \frac{x_2 y_1 - x_1 y_2}{x_2 - x_1}$

$m = \frac{y_2 - y_1}{x_2 - x_1} \qquad b = \frac{x_2 y_1 - x_1 y_2}{x_2 - x_1}$

Since both forms give the same slope and $y$-intercept, the determinant does give the equation of the line.

**77. a.** $\frac{x^2}{9} + \frac{y^2}{4} = 1$

$\frac{x^2}{9} + \frac{0^2}{4} = 1$

$\frac{x^2}{9} = 1$

$x^2 = 9$

$x = \pm 3$

**b.** $\frac{x^2}{9} + \frac{y^2}{4} = 1$

$\frac{0^2}{9} + \frac{y^2}{4} = 1$

$\frac{y^2}{4} = 1$

$y^2 = 4$

$y = \pm 2$

**78.** $25x^2 + 16y^2 = 400$

$\frac{25x^2}{400} + \frac{16y^2}{400} = \frac{400}{400}$

$\frac{x^2}{16} + \frac{y^2}{25} = 1$

**79.**
$$x^2 + y^2 - 2x + 4y = 4$$
$$x^2 - 2x + y^2 + 4y = 4$$
$$x^2 - 2x + 1 + y^2 + 4y + 4 = 4 + 1 + 4$$
$$(x-1)^2 + (y+2)^2 = 9$$
$$(x-1)^2 + (y+2)^2 = 3^2$$
Center: $(1, -2)$
Radius: 3

**Chapter 6 Review Exercises**

**1.**
$$\begin{bmatrix} 1 & 2 & 2 & | & 2 \\ 0 & 1 & -1 & | & 2 \\ 0 & 5 & 4 & | & 1 \end{bmatrix} -5R_2 + R_3$$
$$\begin{bmatrix} 1 & 2 & 2 & | & 2 \\ 0 & 1 & -1 & | & 2 \\ 0 & 0 & 9 & | & -9 \end{bmatrix}$$

**2.**
$$\begin{bmatrix} 2 & -2 & 1 & | & -1 \\ 1 & 2 & -1 & | & 2 \\ 6 & 4 & 3 & | & 5 \end{bmatrix} \frac{1}{2}R_1$$
$$\begin{bmatrix} 1 & -1 & \frac{1}{2} & | & -\frac{1}{2} \\ 1 & 2 & -1 & | & 2 \\ 6 & 4 & 3 & | & 5 \end{bmatrix}$$

**3.**
$$\begin{bmatrix} 1 & 2 & 3 & | & -5 \\ 2 & 1 & 1 & | & 1 \\ 1 & 1 & -1 & | & 8 \end{bmatrix} \begin{matrix} -2R_1 + R_2 \\ -1R_1 + R_3 \end{matrix}$$
$$\begin{bmatrix} 1 & 2 & 3 & | & -5 \\ 0 & -3 & -5 & | & 11 \\ 0 & -1 & -4 & | & 13 \end{bmatrix} R_2 \leftrightarrow R_3$$
$$\begin{bmatrix} 1 & 2 & 3 & | & -5 \\ 0 & -1 & -4 & | & 13 \\ 0 & -3 & -5 & | & 11 \end{bmatrix} -1R_2$$

$$\begin{bmatrix} 1 & 2 & 3 & | & -5 \\ 0 & 1 & 4 & | & -13 \\ 0 & -3 & -5 & | & 11 \end{bmatrix} 3R_2 + R_3$$
$$\begin{bmatrix} 1 & 2 & 3 & | & -5 \\ 0 & 1 & 4 & | & -13 \\ 0 & 0 & 7 & | & -28 \end{bmatrix} \frac{1}{7}R_3$$
$$\begin{bmatrix} 1 & 2 & 3 & | & -5 \\ 0 & 1 & 4 & | & -13 \\ 0 & 0 & 1 & | & -4 \end{bmatrix} -2R_2 + R_1$$
$$\begin{bmatrix} 1 & 0 & -5 & | & 21 \\ 0 & 1 & 4 & | & -13 \\ 0 & 0 & 1 & | & -4 \end{bmatrix} \begin{matrix} 5R_3 + R_1 \\ -4R_3 + R_2 \end{matrix}$$
$$\begin{bmatrix} 1 & 0 & 0 & | & 1 \\ 0 & 1 & 0 & | & 3 \\ 0 & 0 & 1 & | & -4 \end{bmatrix}$$
The solution set is $\{(1, 3, -4)\}$.

**4.**
$$\begin{bmatrix} 1 & -2 & 1 & | & 0 \\ 0 & 1 & -3 & | & -1 \\ 0 & 2 & 5 & | & -2 \end{bmatrix} -2R_2 + R_3$$
$$\begin{bmatrix} 1 & -2 & 1 & | & 0 \\ 0 & 1 & -3 & | & -1 \\ 0 & 0 & 11 & | & 0 \end{bmatrix} \frac{1}{11}R_3$$
$$\begin{bmatrix} 1 & -2 & 1 & | & 0 \\ 0 & 1 & -3 & | & -1 \\ 0 & 0 & 1 & | & 0 \end{bmatrix} 2R_2 + R_1$$
$$\begin{bmatrix} 1 & 0 & -5 & | & -2 \\ 0 & 1 & -3 & | & -1 \\ 0 & 0 & 1 & | & 0 \end{bmatrix} \begin{matrix} 3R_3 + R_2 \\ 5R_3 + R_1 \end{matrix}$$
$$\begin{bmatrix} 1 & 0 & 0 & | & -2 \\ 0 & 1 & 0 & | & -1 \\ 0 & 0 & 1 & | & 0 \end{bmatrix}$$
$x = -2; y = -1; z = 0$
The solution set is $\{(-2, -1, 0)\}$.

**5.**
$$\begin{bmatrix} 3 & 5 & -8 & 5 & \vert & -8 \\ 1 & 2 & -3 & 1 & \vert & -7 \\ 2 & 3 & -7 & 3 & \vert & -11 \\ 4 & 8 & -10 & 7 & \vert & -10 \end{bmatrix} R_1 \leftrightarrow R_2$$

$$\begin{bmatrix} 1 & 2 & -3 & 1 & \vert & -7 \\ 3 & 5 & -8 & 5 & \vert & -8 \\ 2 & 3 & -7 & 3 & \vert & -11 \\ 4 & 8 & -10 & 7 & \vert & -10 \end{bmatrix} \begin{matrix} -3R_1 + R_2 \\ -2R_1 + R_3 \\ -4R_1 + R_4 \end{matrix}$$

$$\begin{bmatrix} 1 & 2 & -3 & 1 & \vert & -7 \\ 0 & -1 & 1 & 2 & \vert & 13 \\ 0 & -1 & -1 & 1 & \vert & 3 \\ 0 & 0 & 2 & 3 & \vert & 18 \end{bmatrix} -1R_2$$

$$\begin{bmatrix} 1 & 2 & -3 & 1 & \vert & -7 \\ 0 & 1 & -1 & -2 & \vert & -13 \\ 0 & -1 & -1 & 1 & \vert & 3 \\ 0 & 0 & 2 & 3 & \vert & 18 \end{bmatrix} \begin{matrix} -2R_2 + R_1 \\ 1R_2 + R_3 \end{matrix}$$

$$\begin{bmatrix} 1 & 0 & -1 & 5 & \vert & 19 \\ 0 & 1 & -1 & -2 & \vert & -13 \\ 0 & 0 & -2 & -1 & \vert & -10 \\ 0 & 0 & 2 & 3 & \vert & 18 \end{bmatrix} -\frac{1}{2}R_3$$

$$\begin{bmatrix} 1 & 0 & -1 & 5 & \vert & 19 \\ 0 & 1 & -1 & -2 & \vert & -13 \\ 0 & 0 & 1 & \frac{1}{2} & \vert & 5 \\ 0 & 0 & 2 & 3 & \vert & 18 \end{bmatrix} \begin{matrix} 1R_3 + R_1 \\ 1R_3 + R_2 \\ -2R_3 + R_4 \end{matrix}$$

$$\begin{bmatrix} 1 & 0 & 0 & \frac{11}{2} & \vert & 24 \\ 0 & 1 & 0 & -\frac{3}{2} & \vert & -8 \\ 0 & 0 & 1 & \frac{1}{2} & \vert & 5 \\ 0 & 0 & 0 & 2 & \vert & 8 \end{bmatrix} \frac{1}{2}R_4$$

$$\begin{bmatrix} 1 & 0 & 0 & \frac{11}{2} & \vert & 24 \\ 0 & 1 & 0 & -\frac{3}{2} & \vert & -8 \\ 0 & 0 & 1 & \frac{1}{2} & \vert & 5 \\ 0 & 0 & 0 & 1 & \vert & 4 \end{bmatrix} \begin{matrix} -\frac{11}{2}R_4 + R_1 \\ \frac{3}{2}R_4 + R_2 \\ -\frac{1}{2}R_4 + R_3 \end{matrix}$$

$$\begin{bmatrix} 1 & 0 & 0 & 0 & \vert & 2 \\ 0 & 1 & 0 & 0 & \vert & -2 \\ 0 & 0 & 1 & 0 & \vert & 3 \\ 0 & 0 & 0 & 1 & \vert & 4 \end{bmatrix}$$

The solution set is $\{(2, -2, 3, 4)\}$.

**6. a.** The function must satisfy:
$$98 = 4a = 2b + c$$
$$138 = 16a + 4b + c$$
$$162 = 100a + 10b + c.$$

$$\begin{bmatrix} 4 & 2 & 1 & \vert & 98 \\ 16 & 4 & 1 & \vert & 138 \\ 100 & 10 & 1 & \vert & 162 \end{bmatrix} \frac{1}{4}R_1$$

$$\begin{bmatrix} 1 & \frac{1}{2} & \frac{1}{4} & \vert & \frac{49}{2} \\ 16 & 4 & 1 & \vert & 138 \\ 100 & 10 & 1 & \vert & 162 \end{bmatrix} \begin{matrix} -16R_1 + R_2 \\ -100R_1 + R_3 \end{matrix}$$

$$\begin{bmatrix} 1 & \frac{1}{2} & \frac{1}{4} & \vert & \frac{49}{2} \\ 0 & -4 & -3 & \vert & -254 \\ 0 & -40 & -24 & \vert & -2288 \end{bmatrix} -\frac{1}{4}R_2$$

$$\begin{bmatrix} 1 & \frac{1}{2} & \frac{1}{4} & \vert & \frac{49}{2} \\ 0 & 1 & \frac{3}{4} & \vert & \frac{127}{2} \\ 0 & -40 & -24 & \vert & -2288 \end{bmatrix} 40R_2 + R_3$$

$$\begin{bmatrix} 1 & \frac{1}{2} & \frac{1}{4} & \vert & \frac{49}{2} \\ 0 & 1 & \frac{3}{4} & \vert & \frac{127}{2} \\ 0 & 0 & 6 & \vert & 252 \end{bmatrix} \frac{1}{6}R_3$$

$$\begin{bmatrix} 1 & \frac{1}{2} & \frac{1}{4} & \vert & \frac{49}{2} \\ 0 & 1 & \frac{3}{4} & \vert & \frac{127}{2} \\ 0 & 0 & 1 & \vert & 42 \end{bmatrix} \begin{matrix} -\frac{1}{4}R_3 + R_1 \\ -\frac{3}{4}R_3 + R_2 \end{matrix}$$

$$\begin{bmatrix} 1 & \frac{1}{2} & 0 & \vert & 14 \\ 0 & 1 & 0 & \vert & 32 \\ 0 & 0 & 1 & \vert & 42 \end{bmatrix} -\frac{1}{2}R_3 + R_1$$

$$\begin{bmatrix} 1 & 0 & 0 & \vert & -2 \\ 0 & 1 & 0 & \vert & 32 \\ 0 & 0 & 1 & \vert & 42 \end{bmatrix}$$

The function is $y = -2x^2 + 32x + 42$ and $a = -2$, $b = 32$ and $c = 42$.

**b.** $y = -2x^2 + 32x + 42$ is a parabola.
The maximum occurs when
$$x = \frac{-32}{2(-2)} = \frac{-32}{-4} = 8.$$
The air pollution level is a maximum 8 hours after 6 A.M., which is 2 P.M.
When $x = 8$, $y = -2(64) + 32(8) + 42$
$$= -128 + 256 + 42.$$
$$= 170.$$
The maximum level is 170 parts per million at 2 P.M.

**7.** Write the equations.

$$w + x + y + z = 80$$
$$y - w - x = 18$$
$$y - z = 4$$
$$3x - w - y = 10$$

Rewrite the equations with terms in consistent order.

$$w + x + y + z = 80$$
$$-w - x + y = 18$$
$$y - z = 4$$
$$-w + 3x - y = 10$$

Write as a matrix and solve.

$$\begin{bmatrix} 1 & 1 & 1 & 1 & | & 80 \\ -1 & -1 & 1 & 0 & | & 18 \\ 0 & 0 & 1 & -1 & | & 4 \\ -1 & 3 & -1 & 0 & | & 10 \end{bmatrix} \begin{matrix} \\ R_1 + R_2 \\ \\ R_1 + R_4 \end{matrix}$$

$$\begin{bmatrix} 1 & 1 & 1 & 1 & | & 80 \\ 0 & 0 & 2 & 1 & | & 98 \\ 0 & 0 & 1 & -1 & | & 4 \\ 0 & 4 & 0 & 1 & | & 90 \end{bmatrix} \begin{matrix} \\ R_2 \leftrightarrow R_4 \\ \\ \end{matrix}$$

$$\begin{bmatrix} 1 & 1 & 1 & 1 & | & 80 \\ 0 & 4 & 0 & 1 & | & 90 \\ 0 & 0 & 1 & -1 & | & 4 \\ 0 & 0 & 2 & 1 & | & 98 \end{bmatrix} \begin{matrix} \\ \frac{1}{4}R_2 \\ \\ \end{matrix}$$

$$\begin{bmatrix} 1 & 1 & 1 & 1 & | & 80 \\ 0 & 1 & 0 & \frac{1}{4} & | & \frac{45}{2} \\ 0 & 0 & 1 & -1 & | & 4 \\ 0 & 0 & 2 & 1 & | & 98 \end{bmatrix} \begin{matrix} \\ \\ \\ -2R_3 + R_4 \end{matrix}$$

$$\begin{bmatrix} 1 & 1 & 1 & 1 & | & 80 \\ 0 & 1 & 0 & \frac{1}{4} & | & \frac{45}{2} \\ 0 & 0 & 1 & -1 & | & 4 \\ 0 & 0 & 0 & 3 & | & 90 \end{bmatrix} \begin{matrix} \\ \\ \\ \frac{1}{3}R_4 \end{matrix}$$

$$\begin{bmatrix} 1 & 1 & 1 & 1 & | & 80 \\ 0 & 1 & 0 & \frac{1}{4} & | & \frac{45}{2} \\ 0 & 0 & 1 & -1 & | & 4 \\ 0 & 0 & 0 & 1 & | & 30 \end{bmatrix}$$

Back-substitute $z = 30$ to find $y$.

$$y - z = 4$$
$$y - 30 = 4$$
$$y = 34$$

Back-substitute to find $x$.

$$x + \frac{1}{4}z = \frac{45}{2}$$
$$x + \frac{1}{4}(30) = \frac{45}{2}$$
$$x + \frac{15}{2} = \frac{45}{2}$$
$$x = 15$$

Back-substitute to find $x$.

$$w + x + y + z = 80$$
$$w + 15 + 34 + 30 = 80$$
$$w + 79 = 80$$
$$w = 1$$

Capitalist: 1%
Upper Middle: 15%
Lower Middle: 34%
Working: 30%

**8.**
$$\begin{bmatrix} 2 & -3 & 1 & | & 1 \\ 1 & -2 & 3 & | & 2 \\ 3 & -4 & -1 & | & 1 \end{bmatrix} R_1 \leftrightarrow R_2$$

$$\begin{bmatrix} 1 & -2 & 3 & | & 2 \\ 2 & -3 & 1 & | & 1 \\ 3 & -4 & -1 & | & 1 \end{bmatrix} \begin{matrix} \\ -2R_1 + R_2 \\ -3R_1 + R_3 \end{matrix}$$

$$\begin{bmatrix} 1 & -2 & 3 & | & 2 \\ 0 & 1 & -5 & | & -3 \\ 0 & 2 & -10 & | & -5 \end{bmatrix} \begin{matrix} \\ \\ -2R_2 + R_3 \end{matrix}$$

$$\begin{bmatrix} 1 & -2 & 3 & | & 2 \\ 0 & 1 & -5 & | & -3 \\ 0 & 0 & 0 & | & 1 \end{bmatrix}$$

From the last line, we see that the system has no solution. The solution set is $\varnothing$.

**9.**
$$\begin{bmatrix} 1 & -3 & 1 & | & 1 \\ -2 & 1 & 3 & | & -7 \\ 1 & -4 & 2 & | & 0 \end{bmatrix} \begin{matrix} \\ 2R_1 + R_2 \\ -1R_1 + R_3 \end{matrix}$$

$$\begin{bmatrix} 1 & -3 & 1 & | & 1 \\ 0 & -5 & 5 & | & -5 \\ 0 & -1 & 1 & | & -1 \end{bmatrix} \begin{matrix} \\ -\frac{1}{5}R_2 \\ \\ \end{matrix}$$

$$\begin{bmatrix} 1 & -3 & 1 & | & 1 \\ 0 & 1 & -1 & | & 1 \\ 0 & -1 & 1 & | & -1 \end{bmatrix} \begin{matrix} \\ \\ 1R_2 + R_3 \end{matrix}$$

$$\begin{bmatrix} 1 & -3 & 1 & | & 1 \\ 0 & 1 & -1 & | & 1 \\ 0 & 0 & 0 & | & 0 \end{bmatrix}$$

The system $\begin{array}{c} x-3y+z=1 \\ y-z=1 \end{array}$ has no unique solution.

Express $x$ and $y$ in terms of $z$:

$y = z + 1$

$x - 3(z + 1) + z = 1$

$\quad x - 3z - 3 + z = 1$

$\qquad\qquad x = 2z + 4$

With $z = t$, the complete solution to the system is $\{(2t + 4, t + 1, t)\}$.

**10.** $\begin{bmatrix} 1 & 4 & 3 & -6 & | & 5 \\ 1 & 3 & 1 & -4 & | & 3 \\ 2 & 8 & 7 & -5 & | & 11 \\ 2 & 5 & 0 & -6 & | & 4 \end{bmatrix} \begin{array}{l} -1R_1 + R_2 \\ -2R_1 + R_3 \\ -2R_1 + R_4 \end{array}$

$\begin{bmatrix} 1 & 4 & 3 & -6 & | & 5 \\ 0 & -1 & -2 & 2 & | & -2 \\ 0 & 0 & 1 & 7 & | & 1 \\ 0 & -3 & -6 & 6 & | & -6 \end{bmatrix} -1R_2$

$\begin{bmatrix} 1 & 4 & 3 & -6 & | & 5 \\ 0 & 1 & 2 & -2 & | & 2 \\ 0 & 0 & 1 & 7 & | & 1 \\ 0 & -3 & -6 & 6 & | & -6 \end{bmatrix} 3R_2 + R_4$

$\begin{bmatrix} 1 & 4 & 3 & -6 & | & 5 \\ 0 & 1 & 2 & -2 & | & 2 \\ 0 & 0 & 1 & 7 & | & 1 \\ 0 & 0 & 0 & 0 & | & 0 \end{bmatrix}$

The system $\begin{array}{c} x_1 + 4x_2 + 3x_3 - 6x_4 = 5 \\ x_2 + 2x_3 - 2x_4 = 2 \\ x_3 + 7x_4 = 1 \end{array}$

does not have a unique solution.

Express $x_1$, $x_2$, and $x_3$ in terms of $x_4$:

$x_3 = -7x_4 + 1$

$x_2 + 2(-7x_4 + 1) - 2x_4 = 2$

$\quad x_2 - 14x_4 + 2 - 2x_4 = 2$

$\qquad\qquad x_2 = 16x_4$

$x_1 + 4(16x_4) + 3(-7x_4 + 1) - 6x_4 = 5$

$\quad x_1 + 64x_4 - 21x_4 + 3 - 6x_4 = 5$

$\qquad\qquad x_1 = -37x_4 + 2$

With $x_4 = t$, the complete solution to the system is $\{(-37t + 2, 16t, -7t + 1, t)\}$.

**11.** $\begin{bmatrix} 2 & 3 & -5 & | & 15 \\ 1 & 2 & -1 & | & 4 \end{bmatrix} R_1 \leftrightarrow R_2$

$\begin{bmatrix} 1 & 2 & -1 & | & 4 \\ 2 & 3 & -5 & | & 15 \end{bmatrix} -2R_1 + R_2$

$\begin{bmatrix} 1 & 2 & -1 & | & 4 \\ 0 & -1 & -3 & | & 7 \end{bmatrix} -1R_2$

$\begin{bmatrix} 1 & 2 & -1 & | & 4 \\ 0 & 1 & 3 & | & -7 \end{bmatrix}$

The system $\begin{array}{c} x+2y-z=4 \\ y+3z=-7 \end{array}$ has no unique solution.

Express $x$ and $y$ in terms of $z$:

$y = -3z - 7$

$x + 2(-3z - 7) - z = 4$

$\quad x - 6z - 14 - z = 4$

$\qquad\qquad x = 7z + 18$

With $z = t$, the complete solution to the system is $\{(7t + 18, -3t - 7, t)\}$.

**12. a.** $350 + 400 = x + z$

$450 + z = y + 700$

$x + y = 300 + 200$

or

$x + z = 750$

$y - z = -250$

$x + y = 500$

**b.** $\begin{bmatrix} 1 & 0 & 1 & | & 750 \\ 0 & 1 & -1 & | & -250 \\ 1 & 1 & 0 & | & 500 \end{bmatrix} -1R_1 + R_3$

$\begin{bmatrix} 1 & 0 & 1 & | & 750 \\ 0 & 1 & -1 & | & -250 \\ 0 & 1 & -1 & | & -250 \end{bmatrix} -1R_2 + R_3$

$\begin{bmatrix} 1 & 0 & 1 & | & 750 \\ 0 & 1 & -1 & | & -250 \\ 0 & 0 & 0 & | & 0 \end{bmatrix}$

The system $\begin{array}{c} x+z=750 \\ y-z=-250 \end{array}$ has no unique solution.

Express $x$ and $y$ in terms of $z$:

$y = z - 250$

$x = -z + 750$

With $z = t$, the complete solution to the system is $\{(-t + 750, t - 250, t)\}$.

**c.** $x = -400 + 750 = 350$

$\quad y = 400 - 250 = 150$

**13.**  $2x = -10$
$x = -5$
$y + 7 = 13$
$y = 6$
$z = 6$
$x = -5; \; y = 6; \; z = 6$

**14.**  $A + D = \begin{bmatrix} 2-2 & -1+3 & 2+1 \\ 5+3 & 3-2 & -1+4 \end{bmatrix} = \begin{bmatrix} 0 & 2 & 3 \\ 8 & 1 & 3 \end{bmatrix}$

**15.**  $2B = \begin{bmatrix} 2(0) & 2(-2) \\ 2(3) & 2(2) \\ 2(1) & 2(-5) \end{bmatrix} = \begin{bmatrix} 0 & -4 \\ 6 & 4 \\ 2 & -10 \end{bmatrix}$

**16.**  $D - A = \begin{bmatrix} -2-2 & 3+1 & 1-2 \\ 3-5 & -2-3 & 4+1 \end{bmatrix}$
$= \begin{bmatrix} -4 & 4 & -1 \\ -2 & -5 & 5 \end{bmatrix}$

**17.**  Not possible since $B$ is $3 \times 2$ and $C$ is $3 \times 3$.

**18.**  $3A + 2D = \begin{bmatrix} 6 & -3 & 6 \\ 15 & 9 & -3 \end{bmatrix} + \begin{bmatrix} -4 & 6 & 2 \\ 6 & -4 & 8 \end{bmatrix}$
$= \begin{bmatrix} 2 & 3 & 8 \\ 21 & 5 & 5 \end{bmatrix}$

**19.**
$-2A + 4D = \begin{bmatrix} -4 & 2 & -4 \\ -10 & -6 & 2 \end{bmatrix} + \begin{bmatrix} -8 & 12 & 4 \\ 12 & -8 & 16 \end{bmatrix}$
$= \begin{bmatrix} -12 & 14 & 0 \\ 2 & -14 & 18 \end{bmatrix}$

**20.**  $-5(A + D) = -5\left( \begin{bmatrix} 0 & 2 & 3 \\ 8 & 1 & 3 \end{bmatrix} \right) = \begin{bmatrix} 0 & -10 & -15 \\ -40 & -5 & -15 \end{bmatrix}$

**21.**  $AB = \begin{bmatrix} 0-3+2 & -4-2-10 \\ 0+9-1 & -10+6+5 \end{bmatrix} = \begin{bmatrix} -1 & -16 \\ 8 & 1 \end{bmatrix}$

**22.**  $BA = \begin{bmatrix} 0-10 & 0-6 & 0+2 \\ 6+10 & -3+6 & 6-2 \\ 2-25 & -1-15 & 2+5 \end{bmatrix} = \begin{bmatrix} -10 & -6 & 2 \\ 16 & 3 & 4 \\ -23 & -16 & 7 \end{bmatrix}$

**23.**  $BD = \begin{bmatrix} 0-6 & 0+4 & 0-8 \\ -6+6 & 9-4 & 3+8 \\ -2-15 & 3+10 & 1-20 \end{bmatrix} = \begin{bmatrix} -6 & 4 & -8 \\ 0 & 5 & 11 \\ -17 & 13 & -19 \end{bmatrix}$

**24.**  $DB = \begin{bmatrix} 0+9+1 & 4+6-5 \\ 0-6+4 & -6-4-20 \end{bmatrix} = \begin{bmatrix} 10 & 5 \\ -2 & -30 \end{bmatrix}$

**25.**  Not possible since $AB$ is 2 x 2 and $BA$ is 3 x 3.

**26.**  $(A - D)C = \begin{bmatrix} 4 & -4 & 1 \\ 2 & 5 & -5 \end{bmatrix} \begin{bmatrix} 1 & 2 & 3 \\ -1 & 1 & 2 \\ -1 & 2 & 1 \end{bmatrix}$
$= \begin{bmatrix} 4+4-1 & 8-4+2 & 12-8+1 \\ 2-5+5 & 4+5-10 & 6+10-5 \end{bmatrix} = \begin{bmatrix} 7 & 6 & 5 \\ 2 & -1 & 11 \end{bmatrix}$

**27.**  $B(AC) = \begin{bmatrix} 0 & -2 \\ 3 & 2 \\ 1 & -5 \end{bmatrix} \begin{bmatrix} 2+1-2 & 4-1+4 & 6-2+2 \\ 5-3+1 & 10+3-2 & 15+6-1 \end{bmatrix}$
$= \begin{bmatrix} 0 & -2 \\ 3 & 2 \\ 1 & -5 \end{bmatrix} \begin{bmatrix} 1 & 7 & 6 \\ 3 & 11 & 20 \end{bmatrix}$
$= \begin{bmatrix} 0-6 & 0-22 & 0-40 \\ 3+6 & 21+22 & 18+40 \\ 1-15 & 7-55 & 6-100 \end{bmatrix}$
$= \begin{bmatrix} -6 & -22 & -40 \\ 9 & 43 & 58 \\ -14 & -48 & -94 \end{bmatrix}$

**28.**  $3X + A = B$
$3X = B - A$
$X = \dfrac{1}{3}(B - A)$
$X = \dfrac{1}{3}\left( \begin{bmatrix} -2 & -12 \\ 4 & 1 \end{bmatrix} - \begin{bmatrix} 4 & 6 \\ -5 & 0 \end{bmatrix} \right)$
$X = \dfrac{1}{3} \begin{bmatrix} -6 & -18 \\ 9 & 1 \end{bmatrix}$
$X = \begin{bmatrix} -2 & -6 \\ 3 & \dfrac{1}{3} \end{bmatrix}$

**29.**  $\begin{bmatrix} 2 & 2 & 2 \\ 1 & 2 & 1 \\ 1 & 2 & 1 \end{bmatrix}$

**30.**
$$\begin{bmatrix} 2 & 2 & 2 \\ 1 & 2 & 1 \\ 1 & 2 & 1 \end{bmatrix} + \begin{bmatrix} 1 & 1 & 1 \\ -1 & 1 & -1 \\ -1 & 1 & -1 \end{bmatrix} = \begin{bmatrix} 3 & 3 & 3 \\ 0 & 3 & 0 \\ 0 & 3 & 0 \end{bmatrix}$$

$$B = \begin{bmatrix} 3 & 3 & 3 \\ 0 & 3 & 0 \\ 0 & 3 & 0 \end{bmatrix}$$

**31.**
$$\begin{bmatrix} 0 & 2 & 2 \\ 0 & 0 & -4 \end{bmatrix} + \begin{bmatrix} -2 & -2 & -2 \\ 1 & 1 & 1 \end{bmatrix} = \begin{bmatrix} -2 & 0 & 0 \\ 1 & 1 & -3 \end{bmatrix}$$

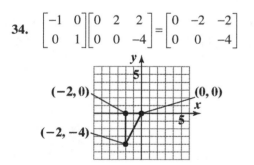

**32.**
$$\frac{1}{2}\begin{bmatrix} 0 & 2 & 2 \\ 0 & 0 & -4 \end{bmatrix} + \begin{bmatrix} 0 & 0 & 0 \\ -2 & -2 & -2 \end{bmatrix} = \begin{bmatrix} -2 & -1 & -1 \\ -2 & -2 & -4 \end{bmatrix}$$

**33.**
$$\begin{bmatrix} 1 & 0 \\ 0 & -1 \end{bmatrix}\begin{bmatrix} 0 & 2 & 2 \\ 0 & 0 & -4 \end{bmatrix} = \begin{bmatrix} 0 & 2 & 2 \\ 0 & 0 & 4 \end{bmatrix}$$

The triangle is reflected about the *x*-axis.

**34.**
$$\begin{bmatrix} -1 & 0 \\ 0 & 1 \end{bmatrix}\begin{bmatrix} 0 & 2 & 2 \\ 0 & 0 & -4 \end{bmatrix} = \begin{bmatrix} 0 & -2 & -2 \\ 0 & 0 & -4 \end{bmatrix}$$

The triangle is reflected about the *y*-axis.

**35.**
$$\begin{bmatrix} 0 & -1 \\ 1 & 0 \end{bmatrix}\begin{bmatrix} 0 & 2 & 2 \\ 0 & 0 & -4 \end{bmatrix} = \begin{bmatrix} 0 & 0 & 4 \\ 0 & 2 & 2 \end{bmatrix}$$

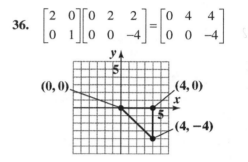

The triangle is rotated 90° counterclockwise about the origin.

**36.**
$$\begin{bmatrix} 2 & 0 \\ 0 & 1 \end{bmatrix}\begin{bmatrix} 0 & 2 & 2 \\ 0 & 0 & -4 \end{bmatrix} = \begin{bmatrix} 0 & 4 & 4 \\ 0 & 0 & -4 \end{bmatrix}$$

The triangle is stretched by a factor of 2 horizontally.

**37.**
$$AB = \begin{bmatrix} 8-7 & -14+21 \\ 4-4 & -7+12 \end{bmatrix} = \begin{bmatrix} 1 & 7 \\ 0 & 5 \end{bmatrix}$$

$$BA = \begin{bmatrix} 8-7 & 28-28 \\ -2+3 & -7+12 \end{bmatrix} = \begin{bmatrix} 1 & 0 \\ 1 & 5 \end{bmatrix}$$

If *B* is the multiplicative inverse of *A*, both products (*AB* and *BA*) will be the multiplicative identity matrix, $I_2$. Therefore, *B* is not the multiplicative inverse of *A*.

**38.**
$$AB = \begin{bmatrix} 1 & 0 & 0 \\ 0 & 2 & -7 \\ 0 & -1 & 4 \end{bmatrix}\begin{bmatrix} 1 & 0 & 0 \\ 0 & 4 & 7 \\ 0 & 1 & 2 \end{bmatrix} = \begin{bmatrix} 1 & 0 & 0 \\ 0 & 1 & 0 \\ 0 & 0 & 1 \end{bmatrix}$$

$$BA = \begin{bmatrix} 1 & 0 & 0 \\ 0 & 4 & 7 \\ 0 & 1 & 2 \end{bmatrix}\begin{bmatrix} 1 & 0 & 0 \\ 0 & 2 & -7 \\ 0 & -1 & 4 \end{bmatrix} = \begin{bmatrix} 1 & 0 & 0 \\ 0 & 1 & 0 \\ 0 & 0 & 1 \end{bmatrix}$$

If *B* is the multiplicative inverse of *A*, both products (*AB* and *BA*) will be the multiplicative identity matrix, $I_3$. Therefore, *B* is the multiplicative inverse of *A*.

**39.** $A^{-1} = \dfrac{1}{3-2}\begin{bmatrix} 3 & 1 \\ 2 & 1 \end{bmatrix} = \begin{bmatrix} 3 & 1 \\ 2 & 1 \end{bmatrix}$

$AA^{-1} = \begin{bmatrix} 1 & -1 \\ -2 & 3 \end{bmatrix}\begin{bmatrix} 3 & 1 \\ 2 & 1 \end{bmatrix}$

$\qquad = \begin{bmatrix} 3-2 & 1-1 \\ -6+6 & -2+3 \end{bmatrix}$

$\qquad = \begin{bmatrix} 1 & 0 \\ 0 & 1 \end{bmatrix}$

$A^{-1}A = \begin{bmatrix} 3 & 1 \\ 2 & 1 \end{bmatrix}\begin{bmatrix} 1 & -1 \\ -2 & 3 \end{bmatrix}$

$\qquad = \begin{bmatrix} 3-2 & 3+3 \\ 2-2 & -2+3 \end{bmatrix}$

$\qquad = \begin{bmatrix} 1 & 0 \\ 0 & 1 \end{bmatrix}$

**40.** $A^{-1} = \dfrac{1}{0-5}\begin{bmatrix} 3 & -1 \\ -5 & 0 \end{bmatrix}$

$\qquad = \dfrac{-1}{5}\begin{bmatrix} 3 & -1 \\ -5 & 0 \end{bmatrix}$

$\qquad = \begin{bmatrix} -\frac{3}{5} & \frac{1}{5} \\ 1 & 0 \end{bmatrix}$

$AA^{-1} = \begin{bmatrix} 0 & 1 \\ 5 & 3 \end{bmatrix}\begin{bmatrix} -\frac{3}{5} & \frac{1}{5} \\ 1 & 0 \end{bmatrix}$

$\qquad = \begin{bmatrix} 0+1 & 0+0 \\ -3+3 & 1+0 \end{bmatrix}$

$\qquad = \begin{bmatrix} 1 & 0 \\ 0 & 1 \end{bmatrix}$

$A^{-1}A = \begin{bmatrix} -\frac{3}{5} & \frac{1}{5} \\ 1 & 0 \end{bmatrix}\begin{bmatrix} 0 & 1 \\ 5 & 3 \end{bmatrix}$

$\qquad = \begin{bmatrix} 0+1 & -\frac{3}{5}+\frac{3}{5} \\ 0+0 & 1+0 \end{bmatrix}$

$\qquad = \begin{bmatrix} 1 & 0 \\ 0 & 1 \end{bmatrix}$

**41.** $\left[\begin{array}{ccc|ccc} 1 & 0 & -2 & 1 & 0 & 0 \\ 2 & 1 & 0 & 0 & 1 & 0 \\ 1 & 0 & -3 & 0 & 0 & 1 \end{array}\right]\begin{array}{l} -2R_1+R_2 \\ \\ -1R_1+R_3 \end{array}$

$\left[\begin{array}{ccc|ccc} 1 & 0 & -2 & 1 & 0 & 0 \\ 0 & 1 & 4 & -2 & 1 & 0 \\ 0 & 0 & -1 & -1 & 0 & 1 \end{array}\right]-1R_3$

$\left[\begin{array}{ccc|ccc} 1 & 0 & -2 & 1 & 0 & 0 \\ 0 & 1 & 4 & -2 & 1 & 0 \\ 0 & 0 & 1 & 1 & 0 & -1 \end{array}\right]\begin{array}{l} 2R_3+R_1 \\ \\ -4R_3+R_2 \end{array}$

$\left[\begin{array}{ccc|ccc} 1 & 0 & 0 & 3 & 0 & -2 \\ 0 & 1 & 0 & -6 & 1 & 4 \\ 0 & 0 & 1 & 1 & 0 & -1 \end{array}\right]$

$A^{-1} = \begin{bmatrix} 3 & 0 & -2 \\ -6 & 1 & 4 \\ 1 & 0 & -1 \end{bmatrix}$

$AA^{-1} = \begin{bmatrix} 1 & 0 & -2 \\ 2 & 1 & 0 \\ 1 & 0 & -3 \end{bmatrix}\begin{bmatrix} 3 & 0 & -2 \\ -6 & 1 & 4 \\ 1 & 0 & -1 \end{bmatrix}$

$\qquad = \begin{bmatrix} 3+0-2 & 0+0+0 & -2+0+2 \\ 6-6+0 & 0+1+0 & -4+4+0 \\ 3+0-3 & 0+0+0 & -2+0+3 \end{bmatrix}$

$\qquad = \begin{bmatrix} 1 & 0 & 0 \\ 0 & 1 & 0 \\ 0 & 0 & 1 \end{bmatrix}$

$A^{-1}A = \begin{bmatrix} 3 & 0 & -2 \\ -6 & 1 & 4 \\ 1 & 0 & -1 \end{bmatrix}\begin{bmatrix} 1 & 0 & -2 \\ 2 & 1 & 0 \\ 1 & 0 & -3 \end{bmatrix}$

$\qquad = \begin{bmatrix} 3+0-2 & 0+0+0 & -6+0+6 \\ -6+2+4 & 0+1+0 & 12+0-12 \\ 1+0-1 & 0+0+0 & -2+0+3 \end{bmatrix}$

$\qquad = \begin{bmatrix} 1 & 0 & 0 \\ 0 & 1 & 0 \\ 0 & 0 & 1 \end{bmatrix}$

**42.**
$$\left[\begin{array}{ccc|ccc} 1 & 3 & -2 & 1 & 0 & 0 \\ 4 & 13 & -7 & 0 & 1 & 0 \\ 5 & 16 & -8 & 0 & 0 & 1 \end{array}\right] \begin{array}{l} -4R_1 + R_2 \\ -5R_1 + R_3 \end{array}$$

$$\left[\begin{array}{ccc|ccc} 1 & 3 & -2 & 1 & 0 & 0 \\ 0 & 1 & 1 & -4 & 1 & 0 \\ 0 & 1 & 2 & -5 & 0 & 1 \end{array}\right] \begin{array}{l} -1R_2 + R_3 \\ -3R_2 + R_1 \end{array}$$

$$\left[\begin{array}{ccc|ccc} 1 & 0 & -5 & 13 & -3 & 0 \\ 0 & 1 & 1 & -4 & 1 & 0 \\ 0 & 0 & 1 & -1 & -1 & 1 \end{array}\right] \begin{array}{l} -1R_3 + R_2 \\ 5R_3 + R_1 \end{array}$$

$$\left[\begin{array}{ccc|ccc} 1 & 0 & 0 & 8 & -8 & 5 \\ 0 & 1 & 0 & -3 & 2 & -1 \\ 0 & 0 & 1 & -1 & -1 & 1 \end{array}\right]$$

$$A^{-1} = \begin{bmatrix} 8 & -8 & 5 \\ -3 & 2 & -1 \\ -1 & -1 & 1 \end{bmatrix}$$

$$AA^{-1} = \begin{bmatrix} 1 & 3 & -2 \\ 4 & 13 & -7 \\ 5 & 16 & -8 \end{bmatrix}\begin{bmatrix} 8 & -8 & 5 \\ -3 & 2 & -1 \\ -1 & -1 & 1 \end{bmatrix} = \begin{bmatrix} 8-9+2 & -8+6+2 & 5-3-2 \\ 32-39+7 & -32+26+7 & 20-13-7 \\ 40-48+8 & -40+32+8 & 25-16-8 \end{bmatrix} = \begin{bmatrix} 1 & 0 & 0 \\ 0 & 1 & 0 \\ 0 & 0 & 1 \end{bmatrix}$$

$$A^{-1}A = \begin{bmatrix} 8 & -8 & 5 \\ -3 & 2 & -1 \\ -1 & -1 & -1 \end{bmatrix}\begin{bmatrix} 1 & 3 & -2 \\ 4 & 13 & -7 \\ 5 & 16 & -8 \end{bmatrix} = \begin{bmatrix} 8-32+25 & 24-104+80 & -16+56-40 \\ -3+8-5 & -9+26-16 & 6-14+8 \\ -1-4+5 & -3-13+16 & 2+7-8 \end{bmatrix} = \begin{bmatrix} 1 & 0 & 0 \\ 0 & 1 & 0 \\ 0 & 0 & 1 \end{bmatrix}$$

**43. a.**
$$\begin{bmatrix} 1 & 1 & 2 \\ 0 & 1 & 3 \\ 3 & 0 & -2 \end{bmatrix}\begin{bmatrix} x \\ y \\ z \end{bmatrix} = \begin{bmatrix} 7 \\ -2 \\ 0 \end{bmatrix}.$$

**b.**
$$A^{-1}B = \begin{bmatrix} -2 & 2 & 1 \\ 9 & -8 & -3 \\ -3 & 3 & 1 \end{bmatrix}\begin{bmatrix} 7 \\ -2 \\ 0 \end{bmatrix} = \begin{bmatrix} -14-4+0 \\ 63+16+0 \\ -21-6+0 \end{bmatrix} = \begin{bmatrix} -18 \\ 79 \\ -27 \end{bmatrix}$$
The solution to the system is $\{(-18, 79, -27)\}$.

**44. a.**
$$\begin{bmatrix} 1 & -1 & 2 \\ 0 & 1 & -1 \\ 1 & 0 & 2 \end{bmatrix}\begin{bmatrix} x \\ y \\ z \end{bmatrix} = \begin{bmatrix} 12 \\ -5 \\ 10 \end{bmatrix}$$

**b.**
$$A^{-1}B = \begin{bmatrix} 2 & 2 & -1 \\ -1 & 0 & 1 \\ -1 & -1 & 1 \end{bmatrix}\begin{bmatrix} 12 \\ -5 \\ 10 \end{bmatrix} = \begin{bmatrix} 24-10-10 \\ -12+10 \\ -12+5+10 \end{bmatrix} = \begin{bmatrix} 4 \\ -2 \\ 3 \end{bmatrix}$$
The solution to the system is $\{(4, -2, 3)\}$.

**45.** R U L E has a numerical equivalent of 18, 21, 12, 5.

$$\begin{bmatrix} 3 & 2 \\ 4 & 3 \end{bmatrix}\begin{bmatrix} 18 & 12 \\ 21 & 5 \end{bmatrix} = \begin{bmatrix} 54+42 & 36+10 \\ 72+63 & 48+15 \end{bmatrix} = \begin{bmatrix} 96 & 46 \\ 135 & 63 \end{bmatrix}$$

The encoded message is 96, 135, 46, 63.

$$\begin{bmatrix} 3 & -2 \\ -4 & 3 \end{bmatrix}\begin{bmatrix} 96 & 46 \\ 135 & 63 \end{bmatrix} = \begin{bmatrix} 288-270 & 138-126 \\ -384+405 & -184+189 \end{bmatrix} = \begin{bmatrix} 18 & 12 \\ 21 & 5 \end{bmatrix}$$

The decoded message is 18, 21, 12, 5 or RULE.

**46.** $\begin{vmatrix} 3 & 2 \\ -1 & 5 \end{vmatrix} = 15 - (-2) = 17$

**47.** $\begin{vmatrix} -2 & -3 \\ -4 & -8 \end{vmatrix} = 16 - 12 = 4$

**48.** $\begin{vmatrix} 2 & 4 & -3 \\ 1 & -1 & 5 \\ -2 & 4 & 0 \end{vmatrix} = -2\begin{vmatrix} 4 & -3 \\ -1 & 5 \end{vmatrix} - 4\begin{vmatrix} 2 & -3 \\ 1 & 5 \end{vmatrix} + 0\begin{vmatrix} 2 & 4 \\ 1 & -1 \end{vmatrix}$

$$\begin{aligned} &= -2(20-3) - 4[10-(-3)] + 0 \\ &= -2(17) - 4(13) \\ &= -34 - 52 \\ &= -86 \end{aligned}$$

**49.** $\begin{vmatrix} 4 & 7 & 0 \\ -5 & 6 & 0 \\ 3 & 2 & -4 \end{vmatrix} = 4\begin{vmatrix} 6 & 0 \\ 2 & -4 \end{vmatrix} + 5\begin{vmatrix} 7 & 0 \\ 2 & -4 \end{vmatrix} + 3\begin{vmatrix} 7 & 0 \\ 6 & 0 \end{vmatrix}$

$$\begin{aligned} &= 4(-24-0) + 5(-28-0) + 3(0-0) \\ &= 4(-24) + 5(-28) + 0 \\ &= -236 \end{aligned}$$

**50.** $\begin{vmatrix} 1 & 1 & 0 & 2 \\ 0 & 3 & 2 & 1 \\ 0 & -2 & 4 & 0 \\ 0 & 3 & 0 & 1 \end{vmatrix} = \begin{vmatrix} 3 & 2 & 1 \\ -2 & 4 & 0 \\ 3 & 0 & 1 \end{vmatrix}$

$$\begin{aligned} &= 3\begin{vmatrix} 2 & 1 \\ 4 & 0 \end{vmatrix} + \begin{vmatrix} 3 & 2 \\ -2 & 4 \end{vmatrix} \\ &= 3(0-4) + [12-(-4)] \\ &= 3(-4) + 16 \\ &= -12 + 16 \\ &= 4 \end{aligned}$$

**51.**
$$\begin{vmatrix} 2 & 2 & 2 & 2 \\ 0 & 2 & 2 & 2 \\ 0 & 0 & 2 & 2 \\ 0 & 0 & 0 & 2 \end{vmatrix} = 2\begin{vmatrix} 2 & 2 & 2 \\ 0 & 2 & 2 \\ 0 & 0 & 2 \end{vmatrix}$$
$$= 2(2)\begin{vmatrix} 2 & 2 \\ 0 & 2 \end{vmatrix}$$
$$= 2(2)(4)$$
$$= 16$$

**52.** $D = \begin{vmatrix} 1 & -2 \\ 3 & 2 \end{vmatrix} = 2 - (-6) = 2 + 6 = 8$

$D_x = \begin{vmatrix} 8 & -2 \\ -1 & 2 \end{vmatrix} = 16 - 2 = 14$

$D_y = \begin{vmatrix} 1 & 8 \\ 3 & -1 \end{vmatrix} = -1 - 24 = -25$

$x = \dfrac{D_x}{D} = \dfrac{14}{8} = \dfrac{7}{4},\ y = \dfrac{D_y}{D} = \dfrac{-25}{8} = -\dfrac{25}{8}$

The solution to the system is $\left\{ \left( \dfrac{7}{4}, -\dfrac{25}{8} \right) \right\}$.

**53.** $D = \begin{vmatrix} 7 & 2 \\ 2 & 1 \end{vmatrix} = 7 - 4 = 3$

$D = \begin{vmatrix} 7 & 2 \\ 2 & 1 \end{vmatrix} = 7 - 4 = 3$

$D_x = \begin{vmatrix} 0 & 2 \\ -3 & 1 \end{vmatrix} = 0 - (-6) = 6$

$D_y = \begin{vmatrix} 7 & 0 \\ 2 & -3 \end{vmatrix} = -21 - 0 = -21$

$x = \dfrac{D_x}{D} = \dfrac{6}{3} = 2$

$y = \dfrac{D_y}{D} = \dfrac{-21}{3} = -7$

The solution to the system is $\{(2, -7)\}$.

**54.** $D = \begin{vmatrix} 1 & 2 & 2 \\ 2 & 4 & 7 \\ -2 & -5 & -2 \end{vmatrix} = \begin{vmatrix} 1 & 2 & 2 \\ 0 & 0 & 3 \\ 0 & -1 & 2 \end{vmatrix}$

$= \begin{vmatrix} 0 & 3 \\ -1 & 2 \end{vmatrix}$

$= 0 - (-3)$

$= 3$

$D_x = \begin{vmatrix} 5 & 2 & 2 \\ 19 & 4 & 7 \\ 8 & -5 & -2 \end{vmatrix}$

$= 5\begin{vmatrix} 4 & 7 \\ -5 & -2 \end{vmatrix} - 2\begin{vmatrix} 19 & 7 \\ 8 & -2 \end{vmatrix} + 2\begin{vmatrix} 19 & 4 \\ 8 & -5 \end{vmatrix}$

$= 5[-8 - (-35)] - 2(-38 - 56) + 2(-95 - 32)$

$= 5(27) - 2(-94) - 2(127)$

$= 135 + 188 - 254$

$= 69$

$D_y = \begin{vmatrix} 1 & 5 & 2 \\ 2 & 19 & 7 \\ -2 & 8 & -2 \end{vmatrix}$

$= \begin{vmatrix} 1 & 5 & 2 \\ 0 & 9 & 3 \\ 0 & 18 & 2 \end{vmatrix}$

$= \begin{vmatrix} 9 & 3 \\ 18 & 2 \end{vmatrix}$

$= 18 - 54$

$= -36$

$D_z = \begin{vmatrix} 1 & 2 & 5 \\ 2 & 4 & 19 \\ -2 & -5 & 8 \end{vmatrix}$

$= \begin{vmatrix} 1 & 2 & 5 \\ 0 & 0 & 9 \\ 0 & -1 & 18 \end{vmatrix}$

$= \begin{vmatrix} 0 & 9 \\ -1 & 18 \end{vmatrix}$

$= 0 - (-9)$

$= 9$

$x = \dfrac{D_x}{D} = \dfrac{69}{3} = 23,\ y = \dfrac{D_y}{D} = \dfrac{-36}{3} = -12,$

$z = \dfrac{D_z}{D} = \dfrac{9}{3} = 3$

The solution to the system is $\{(23, -12, 3)\}$.

**55.**

$$D = \begin{vmatrix} 2 & 1 & 0 \\ 0 & 1 & -2 \\ 3 & 0 & -2 \end{vmatrix}$$

$$= 2\begin{vmatrix} 1 & -2 \\ 0 & -2 \end{vmatrix} + 3\begin{vmatrix} 1 & 0 \\ 1 & -2 \end{vmatrix}$$

$$= 2(-2-0) + 3(-2-0)$$

$$= 2(-2) + 3(-2)$$

$$= -4 - 6$$

$$= -10$$

$$D_x = \begin{vmatrix} -4 & 1 & 0 \\ 0 & 1 & -2 \\ -11 & 0 & -2 \end{vmatrix}$$

$$= -1\begin{vmatrix} 0 & -2 \\ -11 & -2 \end{vmatrix} + 1\begin{vmatrix} -4 & 0 \\ -11 & -2 \end{vmatrix}$$

$$= -1(0-22) + 1(8-0)$$

$$= 22 + 8$$

$$= 30$$

$$D_y = \begin{vmatrix} 2 & -4 & 0 \\ 0 & 0 & -2 \\ 3 & -11 & -2 \end{vmatrix}$$

$$= 2\begin{vmatrix} 0 & -2 \\ -11 & -2 \end{vmatrix} + 3\begin{vmatrix} -4 & 0 \\ 0 & -2 \end{vmatrix}$$

$$= 2(0-22) + 3(8-0)$$

$$= 2(-22) + 3(8)$$

$$= -44 + 24$$

$$= -20$$

$$D_z = \begin{vmatrix} 2 & 1 & -4 \\ 0 & 1 & 0 \\ 3 & 0 & -11 \end{vmatrix}$$

$$= 2\begin{vmatrix} 1 & 0 \\ 0 & -11 \end{vmatrix} + 3\begin{vmatrix} 1 & -4 \\ 1 & 0 \end{vmatrix}$$

$$= 2(-11-0) + 3(0+4)$$

$$= 2(-11) + 3(+4) = -22 + 12$$

$$= -10$$

$$x = \frac{D_x}{D} = \frac{30}{-10} = -3$$

$$y = \frac{D_y}{D} = \frac{-20}{-10} = 2$$

$$z = \frac{D_z}{D} = \frac{-10}{-10} = 1$$

The solution to the system is $\{(-3, 2, 1)\}$.

**56.** The quadratic function must satisfy

$$f(20) = 400 = 400a + 20b + c$$
$$f(40) = 150 = 1600a + 40b + c$$
$$f(60) = 400 = 3600a + 60b + c$$

$$D = \begin{vmatrix} 400 & 20 & 1 \\ 1600 & 40 & 1 \\ 3600 & 60 & 1 \end{vmatrix}$$

$$= (400)(20)\begin{vmatrix} 1 & 1 & 1 \\ 4 & 2 & 1 \\ 9 & 3 & 1 \end{vmatrix}$$

$$= 8000\begin{vmatrix} 1 & 1 & 1 \\ 3 & 1 & 0 \\ 8 & 2 & 0 \end{vmatrix} = 8000\begin{vmatrix} 3 & 1 \\ 8 & 2 \end{vmatrix}$$

$$= 8000(6-8)$$

$$= 8000(-2)$$

$$= -16,000$$

$$D_a = \begin{vmatrix} 400 & 20 & 1 \\ 150 & 40 & 1 \\ 400 & 60 & 1 \end{vmatrix}$$

$$= (50)(20)\begin{vmatrix} 8 & 1 & 1 \\ 3 & 2 & 1 \\ 8 & 3 & 1 \end{vmatrix}$$

$$= 1000\begin{vmatrix} 8 & 1 & 1 \\ -5 & 1 & 0 \\ 0 & 2 & 0 \end{vmatrix}$$

$$= 1000\begin{vmatrix} -5 & 1 \\ 0 & 2 \end{vmatrix}$$

$$= 1000(-10-0)$$

$$= -10,000$$

$$D_b = \begin{vmatrix} 400 & 400 & 1 \\ 1600 & 150 & 1 \\ 3600 & 400 & 1 \end{vmatrix}$$

$$= (400)(50)\begin{vmatrix} 1 & 8 & 1 \\ 4 & 3 & 1 \\ 9 & 8 & 1 \end{vmatrix}$$

$$= 20,000\begin{vmatrix} 1 & 8 & 1 \\ 3 & -5 & 0 \\ 8 & 0 & 0 \end{vmatrix}$$

$$= 20,000\begin{vmatrix} 3 & -5 \\ 8 & 0 \end{vmatrix}$$

$$= 20,000[0 - (-40)]$$

$$= 20,000(40)$$

$$= 800,000$$

$$D_c = \begin{vmatrix} 400 & 20 & 400 \\ 1600 & 40 & 150 \\ 3600 & 60 & 400 \end{vmatrix}$$

$$= (400)(20)(50)\begin{vmatrix} 1 & 1 & 8 \\ 4 & 2 & 3 \\ 9 & 3 & 8 \end{vmatrix}$$

$$= 400,000\begin{vmatrix} 1 & 0 & 0 \\ 4 & -2 & -29 \\ 2 & -6 & -64 \end{vmatrix}$$

$$= 400,000\begin{vmatrix} -2 & -29 \\ -6 & -64 \end{vmatrix}$$

$$= 400,000(128 - 174)$$
$$= 400,000(-46)$$
$$= -18,400,000$$

$$a = \frac{D_a}{D} = \frac{-10,000}{-16,000} = \frac{5}{8},$$

$$b = \frac{D_b}{D} = \frac{800,000}{-16,000} = -50,$$

$$c = \frac{D_c}{D} = \frac{-18,400,000}{-16,000} = 1150$$

The model is $f(x) = \frac{5}{8}x^2 - 50x + 1150$.

$$f(30) = \frac{5}{8}(900) - 50(30) + 1150$$
$$= 562.5 - 1500 + 1150$$
$$= 212.5$$

$$f(50) = \frac{5}{8}(2500) - 50(50) + 1150$$
$$= 1562.8 - 2500 + 1150$$
$$= 212.5$$

30- and 50-year-olds are involved in an average of 212.5 automobile accidents per day.

## Chapter 6 Test

**1.** $\begin{bmatrix} 1 & 2 & -1 & | & -3 \\ 2 & -4 & 1 & | & -7 \\ -2 & 2 & -3 & | & 4 \end{bmatrix} \begin{matrix} -2R_1 + R_2 \\ 2R_1 + R_3 \end{matrix}$

$\begin{bmatrix} 1 & 2 & -1 & | & -3 \\ 0 & -8 & 3 & | & -1 \\ 0 & 6 & -5 & | & -2 \end{bmatrix} -\frac{1}{8}R_2$

$\begin{bmatrix} 1 & 2 & -1 & | & -3 \\ 0 & 1 & -\frac{3}{8} & | & \frac{1}{8} \\ 0 & 6 & -5 & | & -2 \end{bmatrix} -6R_2 + R_3$

$\begin{bmatrix} 1 & 2 & -1 & | & -3 \\ 0 & 1 & -\frac{3}{8} & | & \frac{1}{8} \\ 0 & 0 & -\frac{11}{4} & | & -\frac{11}{4} \end{bmatrix} -\frac{4}{11}R_3$

$\begin{bmatrix} 1 & 2 & -1 & | & -3 \\ 0 & 1 & -\frac{3}{8} & | & \frac{1}{8} \\ 0 & 0 & 1 & | & 1 \end{bmatrix}$

$$x + 2y - z = -3$$
$$y - \frac{3}{8}z = \frac{1}{8}$$
$$z = 1$$

Using back substitution,

$$y - \frac{3}{8}(1) = \frac{1}{8} \text{ and } x + 2\left(\frac{1}{2}\right) - 1 = -3.$$

$$y = \frac{1}{2} \qquad x + 1 - 1 = -3$$
$$x = -3$$

The solution to the system is $\left\{\left(-3, \frac{1}{2}, 1\right)\right\}$.

**2.** $\begin{bmatrix} 1 & -2 & 1 & | & 2 \\ 2 & -1 & -1 & | & 1 \end{bmatrix} -2R_1 + R_2$

$\begin{bmatrix} 1 & -2 & 1 & | & 2 \\ 0 & 3 & -3 & | & -3 \end{bmatrix} \frac{1}{3}R_2$

$\begin{bmatrix} 1 & -2 & 1 & | & 2 \\ 0 & 1 & -1 & | & -1 \end{bmatrix}$

The system $\begin{matrix} x - 2y + z = 2 \\ y - z = -1 \end{matrix}$ has no unique solution.

Express $x$ and $y$ in terms of $z$:
$$y = z - 1$$
$$x - 2(z - 1) + z = 2$$
$$x - 2z + 2 + z = 2$$
$$x = z$$

With $z = t$, the complete solution to the system is $\{(t, t - 1, t)\}$.

**3.** $2B + 3C = \begin{bmatrix} 2 & -2 \\ 4 & 2 \end{bmatrix} + \begin{bmatrix} 3 & 6 \\ -3 & 9 \end{bmatrix} = \begin{bmatrix} 5 & 4 \\ 1 & 11 \end{bmatrix}$

**4.** $AB = \begin{bmatrix} 3+2 & -3+1 \\ 1+0 & -1+0 \\ 2+2 & -2+1 \end{bmatrix} = \begin{bmatrix} 5 & -2 \\ 1 & -1 \\ 4 & -1 \end{bmatrix}$

**5.** $C^{-1} = \frac{1}{(1)(3) - (2)(-1)}\begin{bmatrix} 3 & -2 \\ 1 & 1 \end{bmatrix}$

$$= \frac{1}{3+2}\begin{bmatrix} 3 & -2 \\ 1 & 1 \end{bmatrix} = \begin{bmatrix} \frac{3}{5} & -\frac{2}{5} \\ \frac{1}{5} & \frac{1}{5} \end{bmatrix}$$

**6.** $BC = \begin{bmatrix} 1+1 & 2-3 \\ 2-1 & 4+3 \end{bmatrix} = \begin{bmatrix} 2 & -1 \\ 1 & 7 \end{bmatrix}$

$BC - 3B = \begin{bmatrix} 2 & -1 \\ 1 & 7 \end{bmatrix} - \begin{bmatrix} 3 & -3 \\ 6 & 3 \end{bmatrix} = \begin{bmatrix} -1 & 2 \\ -5 & 4 \end{bmatrix}$

**7.** $AB = \begin{bmatrix} -3+14-10 & 2-8+6 & 0+2-2 \\ -6+21-15 & 4-12+9 & 0+3-3 \\ -3-7+10 & 2+4-6 & 0-1+2 \end{bmatrix}$

$= \begin{bmatrix} 1 & 0 & 0 \\ 0 & 1 & 0 \\ 0 & 0 & 1 \end{bmatrix} = I_3$

$BA = \begin{bmatrix} -3+4+0 & -6+6+0 & -6+6+0 \\ 7-8+1 & 14-12-1 & 14-12-2 \\ -5+6-1 & -10+9+1 & -10+9+2 \end{bmatrix}$

$= \begin{bmatrix} 1 & 0 & 0 \\ 0 & 1 & 0 \\ 0 & 0 & 1 \end{bmatrix} = I_3$

**8. a.** $\begin{bmatrix} 3 & 5 \\ 2 & -3 \end{bmatrix}\begin{bmatrix} x \\ y \end{bmatrix} = \begin{bmatrix} 9 \\ -13 \end{bmatrix}$

**b.** $A^{-1} = \dfrac{1}{(3)(-3)-(5)(2)}\begin{bmatrix} -3 & -5 \\ -2 & 3 \end{bmatrix}$

$= \dfrac{1}{-19}\begin{bmatrix} -3 & -5 \\ -2 & 3 \end{bmatrix} = \begin{bmatrix} \frac{3}{19} & \frac{5}{19} \\ \frac{2}{19} & -\frac{3}{19} \end{bmatrix}$

**c.** $A^{-1}B = \begin{bmatrix} \frac{3}{19} & \frac{5}{19} \\ \frac{2}{19} & -\frac{3}{19} \end{bmatrix}\begin{bmatrix} 9 \\ -13 \end{bmatrix} = \begin{bmatrix} \frac{27}{19}-\frac{65}{19} \\ \frac{18}{19}+\frac{39}{19} \end{bmatrix} = \begin{bmatrix} -2 \\ 3 \end{bmatrix}$

The solution to the system is $\{(-2, 3)\}$.

**9.** $\begin{vmatrix} 4 & -1 & 3 \\ 0 & 5 & -1 \\ 5 & 2 & 4 \end{vmatrix} = 4\begin{vmatrix} 5 & -1 \\ 2 & 4 \end{vmatrix} + 5\begin{vmatrix} -1 & 3 \\ 5 & -1 \end{vmatrix}$

$= 4[20-(-2)]+5(1-15)$

$= 4(22)+5(-14)$

$= 88-70$

$= 18$

**10.** $D = \begin{vmatrix} 3 & 1 & -2 \\ 2 & 7 & 3 \\ 4 & -3 & -1 \end{vmatrix} = 3\begin{vmatrix} 7 & 3 \\ -3 & -1 \end{vmatrix} - 1\begin{vmatrix} 2 & 3 \\ 4 & -1 \end{vmatrix} - 2\begin{vmatrix} 2 & 7 \\ 4 & -3 \end{vmatrix}$

$= 3[-7-(-9)]-1(-2-12)-2(-6-28)$

$= 3(2)-1(-14)-2(-34)$

$= 6+14+68$

$= 88$

$D_x = \begin{vmatrix} -3 & 1 & -2 \\ 9 & 7 & 3 \\ 7 & -3 & -1 \end{vmatrix} = -3\begin{vmatrix} 7 & 3 \\ -3 & -1 \end{vmatrix} - 1\begin{vmatrix} 9 & 3 \\ 7 & -1 \end{vmatrix} - 2\begin{vmatrix} 9 & 7 \\ 7 & -3 \end{vmatrix}$

$= -3[-7-(-9)]-1(-9-21)-2(-27-49)$

$= -3(2)-1(-30)-2(-76)$

$= -6+30+152$

$= 176$

$x = \dfrac{D_x}{D} = \dfrac{176}{88} = 2$

## Cumulative Review Exercises (Chapters 1–6)

**1.** $2x^2 = 4 - x$

$2x^2 + x - 4 = 0$

$x = \dfrac{-1 \pm \sqrt{1^2 - 4(2)(-4)}}{2(2)}$

$x = \dfrac{-1 \pm \sqrt{1 - 32}}{4}$

$x = \dfrac{-1 \pm \sqrt{33}}{4}$

The solution set is $\left\{ \dfrac{-1+\sqrt{33}}{4}, \dfrac{-1-\sqrt{33}}{4} \right\}$.

**2.** $5x + 8 \le 7(1 + x)$

$5x + 8 \le 7 + 7x$

$-2x \le -1$

$x \ge \dfrac{1}{2}$

The solution set is $\left\{ x \mid x \ge \dfrac{1}{2} \right\}$ or $\left[ \dfrac{1}{2}, \infty \right)$.

**3.** $\sqrt{2x+4} - \sqrt{x+3} - 1 = 0$

$\sqrt{2x+4} = \sqrt{x+3} + 1$

$2x + 4 = x + 3 + 2\sqrt{x+3} + 1$

$x = 2\sqrt{x+3}$

$x^2 = 4(x+3)$

$x^2 = 4x + 12$

$x^2 - 4x - 12 = 0$

$(x-6)(x+2) = 0$

$x = 6$ or $x = -2$

$x = -2$ does not check. The solution set is $\{6\}$.

**4.** $3x^3 + 8x^2 - 15x + 4 = 0$

$p = \pm1, \pm2, \pm4$

$q = \pm1, \pm3$

$\dfrac{p}{q} = \pm1, \pm\dfrac{1}{3}, \pm2, \pm\dfrac{2}{3}, \ \pm4, \pm\dfrac{4}{3}$

$$
\begin{array}{r|rrrr}
-4 & 3 & 8 & -15 & 4 \\
   &   & -12 & 16 & -4 \\
\hline
   & 3 & -4 & 1 & 0
\end{array}
$$

$(x+4)(3x^2 - 4x + 1) = 0$

$(x+4)(3x-1)(x-1) = 0$

$x = -4, \ x = \dfrac{1}{3}, \ x = 1$

The solution set is $\left\{-4, \dfrac{1}{3}, 1\right\}$.

**5.** $e^{2x} - 14e^x + 45 = 0$   let $t = e^x$

$t^2 - 14t + 45 = 0$

$(t-5)(t-9) = 0$

$t = 5 \quad t = 9$

$e^x = 5 \quad e^x = 9$

$\ln e^x = \ln 5 \quad \ln e^x = \ln 9$

$x = e^x = \ln 5 \quad x = \ln 9$

The solution set is $\{\ln 5, \ln 9\}$.

**6.** $\log_3 x + \log_3(x+2) = 1$

$\log_3 x^2 + 2x = 1$

$3^1 = x^2 + 2x$

$x^2 + 2x - 3 = 0$

$(x-1)(x+3) = 0$

$x = 1, \ x = -3$

$x = -3$ does not check. The solution set is $\{1\}$.

**7.**
$$
\left[\begin{array}{rrr|r}
1 & -1 & 1 & 17 \\
2 & 3 & 1 & 8 \\
-4 & 1 & 5 & -2
\end{array}\right]
\begin{array}{l}
 \\ -2R_1 + R_2 \\ 4R_1 + R_3
\end{array}
$$

$$
\left[\begin{array}{rrr|r}
1 & -1 & 1 & 17 \\
0 & 5 & -1 & -26 \\
0 & -3 & 9 & 66
\end{array}\right]
\begin{array}{l}
 \\  \\ -\frac{1}{3}R_3
\end{array}
$$

$$
\left[\begin{array}{rrr|r}
1 & -1 & 1 & 17 \\
0 & 1 & -3 & -22 \\
0 & 5 & -1 & -26
\end{array}\right]
\begin{array}{l}
 \\ -5R_2 + R_3 \\ 1R_2 + R_1
\end{array}
$$

$$
\left[\begin{array}{rrr|r}
1 & 0 & -2 & -5 \\
0 & 1 & -3 & -22 \\
0 & 0 & 14 & 84
\end{array}\right]
\begin{array}{l}
 \\  \\ \frac{1}{14}R_3
\end{array}
$$

$$
\left[\begin{array}{rrr|r}
1 & 0 & -2 & -5 \\
0 & 1 & -3 & -22 \\
0 & 0 & 1 & 6
\end{array}\right]
\begin{array}{l}
 \\ 3R_3 + R_2 \\ 2R_3 + R_1
\end{array}
$$

$$
\left[\begin{array}{rrr|r}
1 & 0 & 0 & 7 \\
0 & 1 & 0 & -4 \\
0 & 0 & 1 & 6
\end{array}\right]
$$

$x = 7 \quad y = -4 \quad z = 6$

The solution set is $\{(7, -4, 6)\}$.

**8.** $D = \begin{vmatrix} 1 & -2 & 1 \\ 2 & 1 & -1 \\ 3 & 2 & -2 \end{vmatrix}$

$= 1\begin{vmatrix} 1 & -1 \\ 2 & -2 \end{vmatrix} - 2\begin{vmatrix} -2 & 1 \\ 2 & -2 \end{vmatrix} + 3\begin{vmatrix} -2 & 1 \\ 1 & 1 \end{vmatrix}$

$= 1(-2+2) - 2(4-2) + 3(2-1)$

$= 0 - 4 + 3$

$= -1$

$D_y = \begin{vmatrix} 1 & 7 & 1 \\ 2 & 0 & -1 \\ 3 & -2 & -2 \end{vmatrix} = 7\begin{vmatrix} 2 & -1 \\ 3 & -2 \end{vmatrix} - 2\begin{vmatrix} 1 & 1 \\ 2 & -1 \end{vmatrix}$

$= 7(-4+3) - 2(-1-2)$

$= -7 + 6 = 1$

$y = \dfrac{D_y}{D} = \dfrac{1}{-1} = -1$

$y = -1$

**9.** $y = \sqrt{4x-7}$

$x = \sqrt{4y-7}$

$x^2 = 4y - 7$

$x^2 + 7 = 4y$

$\dfrac{x^2 + 7}{4} = y$

$f^{-1}(x) = \dfrac{x^2 + 7}{4} \ (x \geq 0)$

**10.** $f(x) = \dfrac{x}{x^2 - 16}$

$f(0) = \dfrac{0}{-16} = 0$

*y*-intercept at 0

$0 = \dfrac{x}{x^2 - 16}$

$0 = x$

*x*-intercept at 0

$f(x) = \dfrac{x}{(x+4)(x-4)}$

vertical asymptotes at 4, –4
horizontal asymptote at 0

$f(x) = \dfrac{x}{x^2 - 16}$

**11.** $f(x) = 4x^4 - 4x^3 - 25x^2 + x + 6$

$$
\begin{array}{r|rrrrr}
-2 & 4 & -4 & -25 & 1 & 6 \\
   &   & -8 & 24 & 2 & -6 \\
\hline
3 & 4 & -12 & -1 & 3 & 0 \\
  &   & 12 & 0 & -3 & \\
\hline
  & 4 & 0 & -1 & 0 &
\end{array}
$$

$f(x) = (x+2)(x-3)(4x^2 - 1)$
$f(x) = (x+2)(x-3)(2x+1)(2x-1)$

**12.** $y = \log_2 x$

$2^y = x$

| x | y |
|---|---|
| 1 | 0 |
| 2 | 1 |
| $\frac{1}{2}$ | –1 |

$y = \log_2(x+1)$

Shift the graph of $y = \log_2{}^x$ left one unit.

**13. a.** $A = A_0 e^{kt}$

$450 = 900 e^{k(40)}$

$\dfrac{1}{2} = e^{40k}$

$\ln\left(\dfrac{1}{2}\right) = \ln\left(e^{40k}\right)$

$\ln\left(\dfrac{1}{2}\right) = 40k$

$k = \dfrac{\ln\left(\dfrac{1}{2}\right)}{40} \approx -0.017$

$A = 900 e^{-0.017t}$

**b.** $A = 900 e^{-0.017(10)}$

$A = 900 e^{-0.17}$

$A \approx 759.30$ grams

**14.** $\begin{bmatrix} 1 & -1 & 0 \\ 2 & 1 & 3 \end{bmatrix} \begin{bmatrix} 4 & -1 \\ 2 & 0 \\ 1 & 1 \end{bmatrix} = \begin{bmatrix} 4-2+0 & -1+0+0 \\ 8+2+3 & -2+0+3 \end{bmatrix}$

$= \begin{bmatrix} 2 & -1 \\ 13 & 1 \end{bmatrix}$

**15.** $\dfrac{3x^2 + 17x - 38}{(x-3)(x-2)(x+2)} = \dfrac{A}{x-3} + \dfrac{B}{x-2} + \dfrac{C}{x+2}$

$3x^2 + 17x - 38$
$\quad = A(x^2 - 4) + B(x^2 - x - 6) + C(x^2 - 5x + 6)$

$3x^2 + 17x - 38$
$\quad = Ax^2 - 4A + Bx^2 - Bx - 6B + Cx^2 - 5Cx + 6c$

$3x^2 + 17x - 38$
$\quad = (A+B+C)x^2 + (-B-5C)x - (4A+6B-6C)$

$\quad A + B + C = 3$
$\quad -B - 5C = 17$
$\quad 4A + 6B - 6C = 38$

$$\begin{bmatrix} 1 & 1 & 1 & | & 3 \\ 0 & -1 & -5 & | & 17 \\ 4 & 6 & -6 & | & 38 \end{bmatrix} -4R_1 + R_3$$

$$\begin{bmatrix} 1 & 1 & 1 & | & 3 \\ 0 & -1 & -5 & | & 17 \\ 0 & 2 & -10 & | & 26 \end{bmatrix} -1R_2$$

$$\begin{bmatrix} 1 & 1 & 1 & | & 3 \\ 0 & 1 & 5 & | & -17 \\ 0 & 2 & -10 & | & 26 \end{bmatrix} \begin{matrix} -2R_2 + R_3 \\ -1R_2 + R_1 \end{matrix}$$

$$\begin{bmatrix} 1 & 0 & -4 & | & 20 \\ 0 & 1 & 5 & | & -17 \\ 0 & 0 & -20 & | & 60 \end{bmatrix} -\tfrac{1}{20}R_3$$

$$\begin{bmatrix} 1 & 0 & -4 & | & 20 \\ 0 & 1 & 5 & | & -17 \\ 0 & 0 & 1 & | & -3 \end{bmatrix} \begin{matrix} -5R_3 + R_2 \\ 4R_3 + R_1 \end{matrix}$$

$$\begin{bmatrix} 1 & 0 & 0 & | & 8 \\ 0 & 1 & 0 & | & -2 \\ 0 & 0 & 1 & | & -3 \end{bmatrix}$$

$A = 8,\quad B = -2,\quad C = -3$

$$\frac{3x^2 + 17x - 38}{(x-3)(x-2)(x+2)} = \frac{8}{x-3} + \frac{-2}{x-2} + \frac{-3}{x+2}$$

**16.** $y = -\dfrac{2}{3}x - 1$

| x | y |
|---|---|
| 0 | -1 |
| 3 | -3 |
| -3 | 1 |

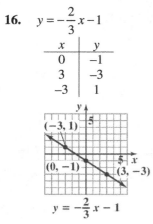

$$y = -\frac{2}{3}x - 1$$

**17.** $3x - 5y < 15$
$-5y < -3x + 15$
$y > \dfrac{3}{5}x - 3$

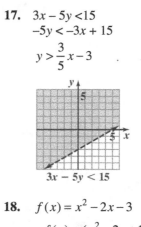

$3x - 5y < 15$

**18.** $f(x) = x^2 - 2x - 3$
$f(x) = (x^2 - 2x + 1) - 3 - 1$
$f(x) = (x-1)^2 - 4$

$f(x) = x^2 - 2x - 3$

**19.** $(x-1)^2 + (y+1)^2 = 9$
center $(1, -1)$
radius $= 3$

$(x-1)^2 + (y+1)^2 = 9$

**20.**

$$\begin{array}{r|rrrr} 2 & 1 & 0 & -6 & 4 \\ & & 2 & 4 & -4 \\ \hline & 1 & 2 & -2 & 0 \end{array}$$

$$\frac{x^3 - 6x + 4}{x - 2} = x^2 + 2x - 2$$

# Chapter 7
# Conic Sections

## Section 7.1

### Check Point Exercises

1. $\dfrac{x^2}{36}+\dfrac{y^2}{9}=1$

The center is $(0,\ 0)$.

$a^2=36,\ a=6$

$b^2=9,\ b=3$

$c^2=a^2-b^2$

$\quad=36-9$

$\quad=27$

$c=\sqrt{27}$

$\quad=3\sqrt{3}$

The foci are located at $(-3\sqrt{3},\ 0)$ and $(3\sqrt{3},\ 0)$.

The vertices are $(0\pm6,\ 0)$ which gives $(-6,\ 0)$ and $(6,\ 0)$.

The endpoints of the minor axis are $(0,\ 0\pm3)$ which gives $(0,-3)$ and $(0,3)$.

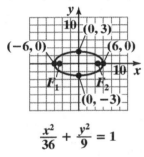

$$\dfrac{x^2}{36}+\dfrac{y^2}{9}=1$$

2. $16x^2+9y^2=144$

$\dfrac{16x^2}{144}+\dfrac{9y^2}{144}=\dfrac{144}{144}$

$\dfrac{x^2}{9}+\dfrac{y^2}{16}=1$

The center is $(0,\ 0)$.

$a^2=16,\ a=4$

$b^2=9,\ b=3$

$c^2=a^2-b^2$

$\quad=16-9$

$\quad=7$

$c=\sqrt{7}$

The foci are located at $(0,\ -\sqrt{7})$ and $(0,\ \sqrt{7})$.

The vertices are $(0,\ 0\pm4)$ which gives $(0,-4)$ and $(0,4)$.

The endpoints of the minor axis are $(0\pm3,\ 0)$ which gives $(-3,0)$ and $(3,0)$.

$$16x^2+9y^2=144$$

3. Because the foci are located at $(-2,0)$ and $(2,0)$, on the x-axis, the major axis is horizontal.
The center of the ellipse is midway between the foci, located at $(0,0)$.

Thus, the form of the equation is $\dfrac{x^2}{a^2}+\dfrac{y^2}{b^2}=1$.

We need to determine the values for $a^2$ and $b^2$.
The distance from the center, $(0,0)$, to either vertex is 3. Thus, $a=3$ and $a^2=9$.
The distance from the center, $(0,0)$, to either focus is 2. Thus, $c=2$ and $c^2=4$.

$b^2=a^2-c^2$

$\quad=9-4$

$\quad=5$

The equation is $\dfrac{x^2}{9}+\dfrac{y^2}{5}=1$.

4. $\dfrac{(x+1)^2}{9}+\dfrac{(y-2)^2}{4}=1$

The center is $(-1,\ 2)$.

$a^2=9,\ a=3$

$b^2=4,\ b=2$

$c^2=a^2-b^2$

$\quad=9-4$

$\quad=5$

$c=\sqrt{5}$

The foci are located at $(-1-\sqrt{5},\ 2)$ and $(-1+\sqrt{5},\ 2)$.

The vertices are $(-1\pm3,\ 2)$ which gives $(-4,\ 2)$ and $(2,\ 2)$.

The endpoints of the minor axis are $(-1, 2\pm2)$ which gives $(-1, 0)$ and $(-1, 4)$.

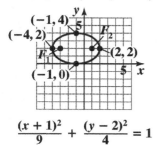

$$\frac{(x + 1)^2}{9} + \frac{(y - 2)^2}{4} = 1$$

5.  $\dfrac{x^2}{20^2}+\dfrac{y^2}{10^2}=1$

$\dfrac{x^2}{400}+\dfrac{y^2}{100}=1$

Since the truck is 12 feet wide, substitute $x = 6$ into the equation to find $y$.

$$\frac{6^2}{400}+\frac{y^2}{100}=1$$

$$400\left(\frac{36}{400}+\frac{y^2}{100}\right)=400(1)$$

$$36+4y^2=400$$

$$4y^2=364$$

$$y^2=91$$

$$y=\sqrt{91}$$

$$y\approx9.54$$

6 feet from the center, the height of the archway is 9.54 feet. Since the truck's height is 9 feet, it will fit under the archway.

### Concept and Vocabulary Check 7.1

1.  ellipse; foci; center

2.  25; −5; 5; (−5, 0); (5, 0); 9; −3; 3; (0,−3); (0, 3)

3.  25; −5; 5; (0, −5); (0, 5); 9; −3; 3; (−3, 0); (3, 0)

4.  5; $(0, -\sqrt{5})$; $(0, \sqrt{5})$

5.  (−1, 4)

6.  (−2,−2); (8,−2)

7.  (1, 9)

9.  4; 1; 16

### Exercise Set 7.1

1.  $\dfrac{x^2}{16}+\dfrac{y^2}{4}=1$

$a^2=16,\ a=4$

$b^2=4,\ b=2$

$c^2=a^2-b^2=16-4=12$

$c=\sqrt{12}=2\sqrt{3}$

The foci are located at $(-2\sqrt{3},\ 0)$ and $(2\sqrt{3},\ 0)$.

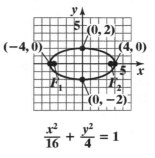

$$\frac{x^2}{16} + \frac{y^2}{4} = 1$$

3.  $a^2=36,\ a=6$

$b^2=9,\ b=3$

$c^2=a^2-b^2=36-9=27$

$c=\sqrt{27}=3\sqrt{3}$

The foci are located at $(0,\ -3\sqrt{3})$ and $(0,\ 3\sqrt{3})$.

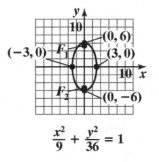

$$\frac{x^2}{9} + \frac{y^2}{36} = 1$$

**5.**  $a^2 = 64, a = 8$

$b^2 = 25, b = 5$

$c^2 = a^2 - b^2 = 64 - 25 = 39$

$c = \sqrt{39}$

The foci are located at $(0, -\sqrt{39})$ and $(0, \sqrt{39})$.

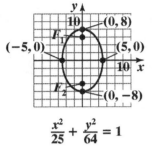

$$\frac{x^2}{25} + \frac{y^2}{64} = 1$$

**7.**  $a^2 = 81, a = 9$

$b^2 = 49, b = 7$

$c^2 = a^2 - b^2 = 81 - 49 = 32$

$c = \sqrt{32} = 4\sqrt{2}$

The foci are located at $(0, -4\sqrt{2})$ and $(0, 4\sqrt{2})$.

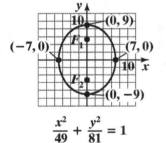

$$\frac{x^2}{49} + \frac{y^2}{81} = 1$$

**9.**  $\dfrac{x^2}{\frac{9}{4}} + \dfrac{y^2}{\frac{25}{4}} = 1$

$c^2 = \dfrac{25}{4} - \dfrac{9}{4}$

$c^2 = \dfrac{16}{4}$

$c^2 = 4$

$c = 2$

The foci are located at $(0, 2)$ and $(0, -2)$.

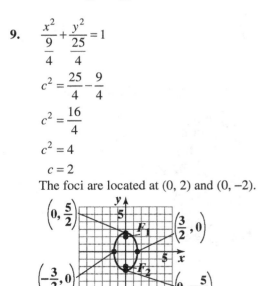

**11.**       $x^2 = 1 - 4y^2$

$x^2 + 4y^2 = 1$

$x^2 + \dfrac{y^2}{\frac{1}{4}} = 1$

$c^2 = 1 - \dfrac{1}{4}$

$c^2 = \dfrac{3}{4}$

$c = \pm\dfrac{\sqrt{3}}{2}$

$c \approx \pm 0.9$

The foci are located at $\left(\dfrac{\sqrt{3}}{2}, 0\right)$ and $\left(-\dfrac{\sqrt{3}}{2}, 0\right)$.

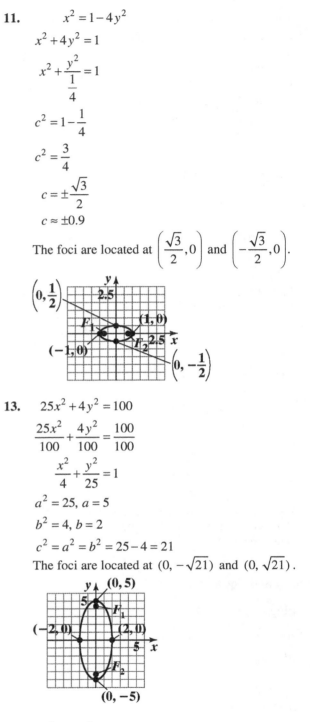

**13.**    $25x^2 + 4y^2 = 100$

$\dfrac{25x^2}{100} + \dfrac{4y^2}{100} = \dfrac{100}{100}$

$\dfrac{x^2}{4} + \dfrac{y^2}{25} = 1$

$a^2 = 25, a = 5$

$b^2 = 4, b = 2$

$c^2 = a^2 = b^2 = 25 - 4 = 21$

The foci are located at $(0, -\sqrt{21})$ and $(0, \sqrt{21})$.

**15.**   $4x^2 + 16y^2 = 64$

$$\frac{x^2}{16}+\frac{y^2}{4}=1$$

$a^2=16, a=4$

$b^2=4, b=2$

$c^2=16-4$

$c^2=12$

$c=\pm\sqrt{12}$

$c=\pm2\sqrt{3}$

$c\approx\pm3.5$

The foci are located at $(2\sqrt{3},0)$ and $(-2\sqrt{3},0)$.

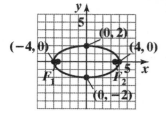

17.          $7x^2=35-5y^2$

$7x^2+5y^2=35$

$$\frac{x^2}{5}+\frac{y^2}{7}=1$$

$a^2=7, a=\sqrt{7}$

$b^2=5, b=\sqrt{5}$

$c^2=7-5$

$c^2=2$

$c=\pm\sqrt{2}$

$c\approx\pm1.4$

The foci are located at $(0,\sqrt{2})$ and $(0,-\sqrt{2})$.

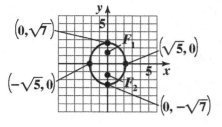

19.   $a^2=4, b^2=1$, center at $(0, 0)$

$$\frac{x^2}{4}+\frac{y^2}{1}=1$$

$c^2=a^2-b^2=4-1=3$

$c=\sqrt{3}$

The foci are at $(-\sqrt{3}, 0)$ and $(\sqrt{3}, 0)$.

21.   $a^2=4, b^2=1,$

center: $(0, 0)$

$$\frac{x^2}{1}+\frac{y^2}{4}=1$$

$c^2=a^2-b^2=4-1=3$

$c=\sqrt{3}$

The foci are at $(0, \sqrt{3})$ and $(0, -\sqrt{3})$.

23.   $\dfrac{(x+1)^2}{4}+\dfrac{(y-1)^2}{1}=1$

$a^2=4, \ b^2=1$

$c^2=4-1$

$c^2=3$

$c=\pm\sqrt{3}$

The foci are located at $(-1+\sqrt{3},1)$ and $(-1-\sqrt{3},1)$.

25.   $c^2=25, \ a^2=64$

$b^2=a^2-c^2=64-25=39$

$$\frac{x^2}{64}+\frac{y^2}{39}=1$$

27.   $c^2=16, \ a^2=49$

$b^2=a^2-c^2=49-16=33$

$$\frac{x^2}{33}+\frac{y^2}{49}=1$$

29.   $c^2=4, \ b^2=9$

$a^2=b^2+c^2=9+4=13$

$$\frac{x^2}{13}+\frac{y^2}{9}=1$$

31.   $2a=8, \ a=4, \ a^2=16$

$2b=4, \ b=2, \ b^2=4$

$$\frac{x^2}{16}+\frac{y^2}{4}=1$$

33.   $2a=10, \ a=5, \ a^2=25$

$2b=4, \ b=2, \ b^2=4$

$$\frac{(x+2)^2}{4}+\frac{(y-3)^2}{25}=1$$

**35.** length of the major axis = 9 – 3 = 6
$2a = 6$, $a = 3$ major axis is vertical
length of the minor axis = 9 – 5 = 4
$2b = 4$, $b = 2$
Center is at (7, 6).
$$\frac{(x-7)^2}{4} + \frac{(y-6)^2}{9} = 1$$

**37.** $a^2 = 9$, $a = 3$
$b^2 = 4$, $b = 2$
center: (2, 1)
$c^2 = a^2 - b^2 = 9 - 4 = 5$
$c = \sqrt{5}$
The foci are at $(2 - \sqrt{5}, 1)$ and $(2 + \sqrt{5}, 1)$.

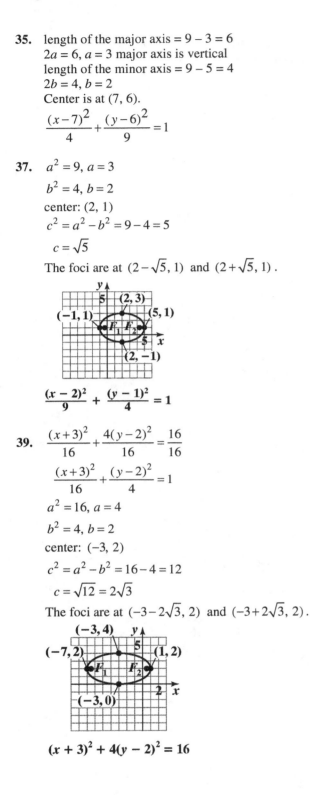

$$\frac{(x-2)^2}{9} + \frac{(y-1)^2}{4} = 1$$

**39.** $\dfrac{(x+3)^2}{16} + \dfrac{4(y-2)^2}{16} = \dfrac{16}{16}$

$\dfrac{(x+3)^2}{16} + \dfrac{(y-2)^2}{4} = 1$

$a^2 = 16$, $a = 4$
$b^2 = 4$, $b = 2$
center: (–3, 2)
$c^2 = a^2 - b^2 = 16 - 4 = 12$
$c = \sqrt{12} = 2\sqrt{3}$
The foci are at $(-3 - 2\sqrt{3}, 2)$ and $(-3 + 2\sqrt{3}, 2)$.

$$(x + 3)^2 + 4(y - 2)^2 = 16$$

**41.** $a^2 = 25$, $a = 5$
$b^2 = 9$, $b = 3$
center: (4, –2)
$c^2 = a^2 - b^2 = 25 - 9 = 16$
$c = 4$
The foci are at (4, 2) and (4, –6).

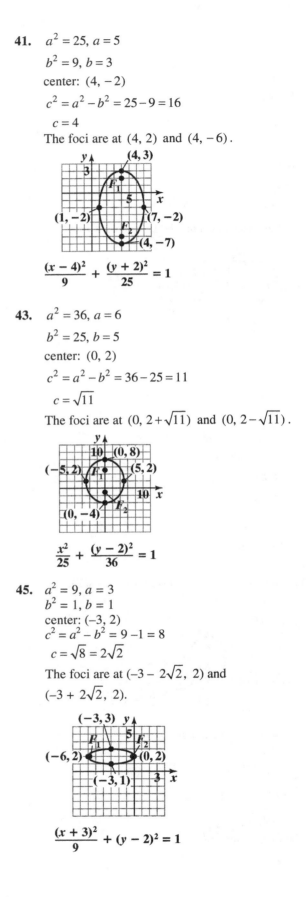

$$\frac{(x-4)^2}{9} + \frac{(y+2)^2}{25} = 1$$

**43.** $a^2 = 36$, $a = 6$
$b^2 = 25$, $b = 5$
center: (0, 2)
$c^2 = a^2 - b^2 = 36 - 25 = 11$
$c = \sqrt{11}$
The foci are at $(0, 2 + \sqrt{11})$ and $(0, 2 - \sqrt{11})$.

$$\frac{x^2}{25} + \frac{(y-2)^2}{36} = 1$$

**45.** $a^2 = 9$, $a = 3$
$b^2 = 1$, $b = 1$
center: (–3, 2)
$c^2 = a^2 - b^2 = 9 - 1 = 8$
$c = \sqrt{8} = 2\sqrt{2}$
The foci are at $(-3 - 2\sqrt{2}, 2)$ and
$(-3 + 2\sqrt{2}, 2)$.

$$\frac{(x+3)^2}{9} + (y - 2)^2 = 1$$

**47.** $c^2 = 5 - 2$

$c^2 = 3$

$c = \pm\sqrt{3}$

$c \approx \pm 1.7$

The foci are located at $(1, -3 + \sqrt{3})$ and $(1, -3 - \sqrt{3})$.

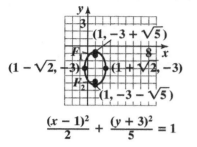

$$\frac{(x-1)^2}{2} + \frac{(y+3)^2}{5} = 1$$

**49.** $\dfrac{9(x-1)^2}{36} + \dfrac{4(y+3)^2}{36} = \dfrac{36}{36}$

$$\frac{(x-1)^2}{4} + \frac{(y+3)^2}{9} = 1$$

$a^2 = 9$, $a = 3$

$b^2 = 4$, $b = 2$

center: $(1, -3)$

$c^2 = a^2 - b^2 = 9 - 4 = 5$

$c = \sqrt{5}$

The foci are at $(1, -3 + \sqrt{5})$ and $(1, -3 - \sqrt{5})$.

$9(x-1)^2 + 4(y+3)^2 = 36$

**51.** $(9x^2 - 36x) + (25y^2 + 50y) = 164$

$9(x^2 - 4x) + 25(y^2 + 2y) = 164$

$9(x^2 - 4x + 4) + 25(y^2 + 2y + 1)$

$= 164 + 36 + 25$

$9(x-2)^2 + 25(y+1)^2 = 225$

$$\frac{9(x-2)^2}{225} + \frac{25(y+1)^2}{225} = \frac{225}{225}$$

$$\frac{(x-2)^2}{25} + \frac{(y+1)^2}{9} = 1$$

center: $(2, -1)$

$a^2 = 25$, $a = 5$

$b^2 = 9$, $b = 3$

$c^2 = a^2 - b^2 = 25 - 9 = 16$

$c = 4$

The foci are at $(-2, -1)$ and $(6, -1)$.

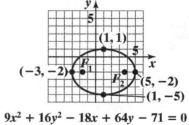

$9x^2 + 25y^2 - 36x + 50y - 164 = 0$

**53.** $(9x^2 - 18x) + (16y^2 + 64y) = 71$

$9(x^2 - 2x) + 16(y^2 + 4y) = 71$

$9(x^2 - 2x + 1) + 16(y^2 + 4y + 4)$

$= 71 + 9 + 64$

$9(x-1)^2 + 16(y+2)^2 = 144$

$$\frac{9(x-1)^2}{144} + \frac{16(y+2)^2}{144} = \frac{144}{144}$$

$$\frac{(x-1)^2}{16} + \frac{(y+2)^2}{9} = 1$$

center: $(1, -2)$

$a^2 = 16$, $a = 4$

$b^2 = 9$, $b = 3$

$c^2 = a^2 - b^2 = 16 - 9 = 7$

$c = \sqrt{7}$

The foci are at

$(1 - \sqrt{7}, -2)$ and $(1 + \sqrt{7}, -2)$.

$9x^2 + 16y^2 - 18x + 64y - 71 = 0$

**55.**
$$(4x^2 + 16x) + (y^2 - 6y) = 39$$
$$4(x^2 + 4x) + (y^2 - 6y) = 39$$
$$4(x^2 + 4x + 4) + (y^2 - 6y + 9) = 39 + 16 + 9$$
$$4(x+2)^2 + (y-3)^2 = 64$$
$$\frac{4(x+2)^2}{64} + \frac{(y-3)^2}{64} = \frac{64}{64}$$
$$\frac{(x+2)^2}{16} + \frac{(y-3)^2}{64} = 1$$
center: (–2, 3)
$a^2 = 64, a = 8$
$b^2 = 16, b = 4$
$c^2 = a^2 - b^2 = 64 - 16 = 48$
$c = \sqrt{48} = 4\sqrt{3}$
The foci are at $(-2, 3 + 4\sqrt{3})$ and $(-2, 3 - 4\sqrt{3})$.

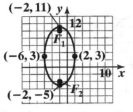

$4x^2 + y^2 + 16x - 6y - 39 = 0$

**57.** $x^2 + y^2 = 1$    $x^2 + 9y^2 = 9$
$$\frac{x^2}{9} + \frac{9y^2}{9} = \frac{9}{9}$$
$$\frac{x^2}{9} + \frac{y^2}{1} = 1$$

The first equation is that of a circle with center at the origin and $r = 1$. The second equation is that of an ellipse with center at the origin, horizontal major axis of length 6 units $(a = 3)$, and vertical minor axis of length 2 units $(b = 1)$.

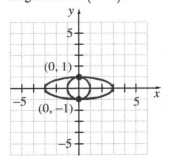

Check each intersection point.
The solution set is $\{(0,-1),(0,1)\}$.

**59.** $\frac{x^2}{25} + \frac{y^2}{9} = 1$    $y = 3$

The first equation is for an ellipse centered at the origin with horizontal major axis of length 10 units and vertical minor axis of length 6 units. The second equation is for a horizontal line with a $y$-intercept of 3.

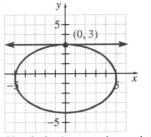

Check the intersection point.
The solution set is $\{(0,3)\}$.

**61.** $4x^2 + y^2 = 4$    $2x - y = 2$
$$\frac{4x^2}{4} + \frac{y^2}{4} = \frac{4}{4} \qquad -y = -2x + 2$$
$$\frac{x^2}{1} + \frac{y^2}{4} = 1 \qquad y = 2x - 2$$

The first equation is for an ellipse centered at the origin with vertical major axis of length 4 units $(b = 2)$ and horizontal minor axis of length 2 units $(a = 1)$. The second equation is for a line with slope 2 and $y$-intercept $-2$.

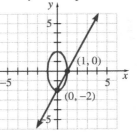

Check the intersection points.
The solution set is $\{(0,-2),(1,0)\}$.

**63.**
$$y^2 = \left(-\sqrt{16-4x^2}\right)^2$$
$$y^2 = 16 - 4x^2$$
$$4x^2 + y^2 = 16$$
$$\frac{x^2}{4} + \frac{y^2}{16} = 1$$

We want to graph the bottom half of an ellipse centered at the origin with a vertical major axis of length 8 units ($b = 4$) and horizontal minor axis of length 4 units $(a = 2)$.

$y = -\sqrt{16-4x^2}$

**65.** $a = 15, b = 10$
$$\frac{x^2}{225} + \frac{y^2}{100} = 1$$
Let $x = 4$
$$\frac{4^2}{225} + \frac{y^2}{100} = 1$$
$$900\left(\frac{16}{225} + \frac{y^2}{100}\right) = 900(1)$$
$$64 + 9y^2 = 900$$
$$9y^2 = 836$$
$$y = \sqrt{\frac{836}{9}} \approx 9.64$$

Yes, the truck only needs 7 feet so it will clear.

**67. a.** $a = 48, a^2 = 2304$
$b = 23, b^2 = 529$
$$\frac{x^2}{2304} + \frac{y^2}{529} = 1$$

**b.** $c^2 = a^2 - b^2 = 2304 - 529 = 1775$
$c = \sqrt{1775} \approx 42.13$
He situated his desk about 42 feet from the center of the ellipse, along the major axis.

**69. – 77.** Answers will vary.

**79.** does not make sense; Explanations will vary. Sample explanation: The foci are on the major axis.

**81.** does not make sense; Explanations will vary. Sample explanation: We must also know the other vertices.

**83.** $a = 6, a^2 = 36$
$$\frac{x^2}{b^2} + \frac{y^2}{36} = 1$$
When $x = 2$ and $y = -4$,
$$\frac{2^2}{b^2} + \frac{(-4)^2}{36} = 1$$
$$\frac{4}{b^2} + \frac{16}{36} = 1$$
$$\frac{4}{b^2} = \frac{5}{9}$$
$$36 = 5b^2$$
$$b^2 = \frac{36}{5}$$
$$\frac{x^2}{\frac{36}{5}} + \frac{y^2}{36} = 1$$

**85.** The large circle has radius 5 with center $(0, 0)$. Its equation is $x^2 + y^2 = 25$. The small circle has radius 3 with center $(0, 0)$. Its equation is $x^2 + y^2 = 9$.

**87.** $4x^2 - 9y^2 = 36$
$$\frac{4x^2}{36} - \frac{9y^2}{36} = \frac{36}{36}$$
$$\frac{x^2}{9} - \frac{y^2}{4} = 1$$

The terms are separated by subtraction rather than by addition.

**88.** $\dfrac{x^2}{16} - \dfrac{y^2}{9} = 1$

    **a.** Substitute 0 for $y$.

$$\frac{x^2}{16} - \frac{0^2}{9} = 1$$

$$\frac{x^2}{16} = 1$$

$$x^2 = 16$$

$$x = \pm 4$$

The $x$-intercepts are $-4$ and $4$.

    **b.** $\dfrac{0^2}{16} - \dfrac{y^2}{9} = 1$

$$-\frac{y^2}{9} = 1$$

$$y^2 = -9$$

The equation $y^2 = -9$ has no real solutions.

**89.** $\dfrac{y^2}{9} - \dfrac{x^2}{16} = 1$

    **a.** Substitute 0 for $x$.

$$\frac{y^2}{9} - \frac{0^2}{16} = 1$$

$$\frac{y^2}{9} = 1$$

$$y^2 = 9$$

$$y = \pm 3$$

The $y$-intercepts are $-3$ and $3$.

    **b.** $\dfrac{0^2}{9} - \dfrac{x^2}{16} = 1$

$$-\frac{x^2}{16} = 1$$

$$x^2 = -16$$

The equation $x^2 = -16$ has no real solutions.

## Section 7.2

### Check Point Exercises

**1.**   **a.** $\dfrac{x^2}{25} - \dfrac{y^2}{16} = 1$

$$a^2 = 25,\ a = 5$$

vertices: $(5, 0)$ and $(-5, 0)$

$$b^2 = 16$$

$$c^2 = a^2 + b^2$$

$$= 25 + 16$$

$$= 41$$

$$c = \sqrt{41}$$

The foci are at $(\sqrt{41}, 0)$ and $(-\sqrt{41}, 0)$.

    **b.** $\dfrac{y^2}{25} - \dfrac{x^2}{16} = 1$

$$a^2 = 25,\ a = 5$$

vertices: $(0, 5)$ and $(0, -5)$

$$b^2 = 16$$

$$c^2 = a^2 + b^2$$

$$= 25 + 16$$

$$= 41$$

$$c = \sqrt{41}$$

The foci are at $(0, \sqrt{41})$ and $(0, -\sqrt{41})$.

**2.** Because the foci are located at $(0, -5)$ and $(0, 5)$, on the $y$-axis, the transverse axis lies on the $y$-axis. The center of the hyperbola is midway between the foci, located at $(0, 0)$.

Thus, the form of the equation is $\dfrac{y^2}{a^2} - \dfrac{x^2}{b^2} = 1$.

We need to determine the values for $a^2$ and $b^2$. The distance from the center, $(0, 0)$, to either vertex is 3. Thus, $a = 3$ and $a^2 = 9$. The distance from the center, $(0, 0)$, to either focus is 5. Thus, $c = 5$ and $c^2 = 25$.

$$b^2 = c^2 - a^2$$

$$= 25 - 9$$

$$= 16$$

The equation is $\dfrac{y^2}{9} - \dfrac{x^2}{16} = 1$.

**3.** $\dfrac{x^2}{36} - \dfrac{y^2}{9} = 1$

$a^2 = 36,\ a = 6$

The vertices are (6, 0) and (–6, 0).

$b^2 = 9,\ b = 3$

asymptotes: $y = \pm\dfrac{b}{a}x = \pm\dfrac{3}{6}x = \pm\dfrac{1}{2}x$

$c^2 = a^2 + b^2$

$\quad = 36 + 9$

$\quad = 45$

$c = \sqrt{45}$

$\quad = 3\sqrt{5}$

The foci are at $(-3\sqrt{5},\ 0)$ and $(3\sqrt{5},\ 0)$.

**4.** $y^2 - 4x^2 = 4$

$\dfrac{y^2}{4} - \dfrac{4x^2}{4} = \dfrac{4}{4}$

$\dfrac{y^2}{4} - x^2 = 1$

$a^2 = 4,\ a = 2$

The vertices are (0, 2) and (0, –2).

$b^2 = 1,\ b = 1$

asymptotes: $y = \pm\dfrac{a}{b}x = \pm 2x$

$c^2 = a^2 + b^2$

$\quad = 4 + 1$

$\quad = 5$

$c = \sqrt{5}$

The foci are at $(0,\ \sqrt{5})$ and $(0,\ -\sqrt{5})$.

**5.** $\dfrac{(x-3)^2}{4} - \dfrac{(y-1)^2}{4} = 1$

center at (3, 1)

$a^2 = 4,\ a = 2$

$b^2 = 1,\ b = 1$

The vertices are (1, 1) and (5, 1).

asymptotes: $y - 1 = \pm\dfrac{1}{2}(x - 3)$

$c^2 = a^2 + b^2$

$\quad = 4 + 1$

$\quad = 5$

$c = \sqrt{5}$

The foci are at $(3 - \sqrt{5},\ 1)$ and $(3 + \sqrt{5},\ 1)$.

**6.** $4x^2 - 24x - 9y^2 - 90y - 153 = 0$

$4\left(x^2 - 6x \quad\right) - 9\left(y^2 + 10y \quad\right) = 153$

$4\left(x^2 - 6x + 9\right) - 9\left(y^2 + 10y + 25\right) = 153 + 36 + (-225)$

$4(x-3)^2 - 9(y+5)^2 = -36$

$\dfrac{4(x-3)^2}{-36} - \dfrac{9(y+5)^2}{-36} = \dfrac{-36}{-36}$

$-\dfrac{(x-3)^2}{9} + \dfrac{(y+5)^2}{4} = 1$

$\dfrac{(y+5)^2}{4} - \dfrac{(x-3)^2}{9} = 1$

center at (3, –5)

$a^2 = 4,\ a = 2$

$b^2 = 9,\ b = 3$

The vertices are (3, –3) and (3, –7).

asymptotes: $y + 5 = \pm\dfrac{2}{3}(x - 3)$

$c^2 = a^2 + b^2$

$\quad = 4 + 9$

$\quad = 13$

$c = \sqrt{13}$

The foci are at $(3, -5-\sqrt{13})$ and $(3, -5+\sqrt{13})$.

$4x^2 - 24x - 9y^2 - 90y - 153 = 0$

**7.**  $c = 5280$

$2a = 3300, a = 1650$

$b^2 = c^2 - a^2 = 5280^2 - 1650^2 = 25,155,900$

The explosion occurred somewhere at the right branch of the hyperbola given by

$$\frac{x^2}{2,722,500} - \frac{y^2}{25,155,900} = 1.$$

**Concept and Vocabulary Check 7.2**

**1.** hyperbola; foci; vertices; transverse

**2.** $(-5, 0)$; $(5, 0)$; $(-\sqrt{34}, 0)$; $(\sqrt{34}, 0)$

**3.** $(0, -5)$; $(0, 5)$; $(0, -\sqrt{34})$; $(0, \sqrt{34})$

**4.** asymptotes; center

**5.** dividing; 36

**6.** $y = -\dfrac{3}{2}x$;  $y = \dfrac{3}{2}x$

**7.** $y = -2x$;  $y = 2x$

**8.** $(-3, 3)$;  $(7, 3)$

**9.** $(7, -2)$

**10.** 16; 1; 128

**Exercise Set 7.2**

**1.**  $a^2 = 4, a = 2$
The vertices are $(2, 0)$ and $(-2, 0)$.
$b^2 = 1$
$c^2 = a^2 + b^2 = 4 + 1 = 5$
$c = \sqrt{5}$
The foci are located at $(\sqrt{5}, 0)$ and $(-\sqrt{5}, 0)$.
graph (b)

**3.**  $a^2 = 4, a = 2$
The vertices are $(0, 2)$ and $(0, -2)$.
$b^2 = 1$
$c^2 = a^2 + b^2 = 4 + 1 = 5$
$c = \sqrt{5}$
The foci are located at
$(0, \sqrt{5})$ and $(0, -\sqrt{5})$.
graph (a)

**5.**  $a = 1, c = 3$
$b^2 = c^2 - a^2 = 9 - 1 = 8$
$$y^2 - \frac{x^2}{8} = 1$$

**7.**  $a = 3, c = 4$
$b^2 = c^2 - a^2 = 16 - 9 = 7$
$$\frac{x^2}{9} - \frac{y^2}{7} = 1$$

**9.**  $2a = 6 - (-6)$
$2a = 12$
$a = 6$
$\dfrac{a}{b} = 2$
$\dfrac{6}{b} = 2$
$6 = 2b$
$3 = b$
Transverse axis is vertical.
$$\frac{y^2}{36} - \frac{x^2}{9} = 1$$

**11.**  $a = 2, c = 7 - 4 = 3$
$2^2 + b^2 = 3^2$
$4 + b^2 = 9$
$b^2 = 5$
Transverse axis is horizontal.
$$\frac{(x-4)^2}{4} - \frac{(y+2)^2}{5} = 1$$

**13.** $a^2 = 9, a = 3$
$b^2 = 25, b = 5$
vertices: $(3, 0)$ and $(-3, 0)$
asymptotes: $y = \pm\dfrac{b}{a}x = \pm\dfrac{5}{3}x$
$c^2 = a^2 + b^2 = 9 + 25 = 34$
$c = \sqrt{34}$ on $x$-axis
The foci are at $(\sqrt{34}, 0)$ and $(-\sqrt{34}, 0)$.

$$\frac{x^2}{9} - \frac{y^2}{25} = 1$$

**15.** $a^2 = 100, a = 10$
$b^2 = 64, b = 8$
vertices: $(10, 0)$ and $(-10, 0)$
asymptotes: $y = \pm\dfrac{b}{a}x = \pm\dfrac{8}{10}x$
or $y = \pm\dfrac{4}{5}x$
$c^2 = a^2 + b^2 = 100 + 64 = 164$
$c = \sqrt{164} = 2\sqrt{41}$ on $x$-axis
The foci are at $(2\sqrt{41}, 0)$ and $(-2\sqrt{41}, 0)$.

$$\frac{x^2}{100} - \frac{y^2}{64} = 1$$

**17.** $a^2 = 16, a = 4$
$b^2 = 36, b = 6$
vertices: $(0, 4)$ and $(0, -4)$
asymptotes: $y = \pm\dfrac{a}{b}x = \pm\dfrac{4}{6}x = \pm\dfrac{2}{3}x$
or $y = \pm\dfrac{2}{3}x$
$c^2 = a^2 + b^2 = 16 + 36 = 52$
$c = \sqrt{52} = 2\sqrt{13}$ on $y$-axis
The foci are at $(0, 2\sqrt{13})$ and $(0, -2\sqrt{13})$.

$$\frac{y^2}{16} - \frac{x^2}{36} = 1$$

**19.** $\dfrac{y^2}{\frac{1}{4}} - x^2 = 1$
$a^2 = \dfrac{1}{4}, a = \dfrac{1}{2}$
$b^2 = 1, b = 1$
$c^2 = a^2 + b^2$
$c^2 = \dfrac{1}{4} + 1$
$c^2 = \dfrac{5}{4}$
$c = \pm\dfrac{\sqrt{5}}{2}$
$c \approx \pm 1.1$

The foci are located at $\left(0, \dfrac{\sqrt{5}}{2}\right)$ and $\left(0, -\dfrac{\sqrt{5}}{2}\right)$.

asymptotes: $y = \pm\dfrac{\frac{1}{2}}{1}x$
$y = \pm\dfrac{1}{2}x$

$4y^2 - x^2 = 1$

**21.** $\dfrac{9x^2}{36} - \dfrac{4y^2}{36} = \dfrac{36}{36}$

$\dfrac{x^2}{4} - \dfrac{y^2}{9} = 1$

$a^2 = 4,\ a = 2$
$b^2 = 9,\ b = 3$
vertices: $(2, 0)$ and $(-2, 0)$

asymptotes: $y = \pm\dfrac{b}{a}x = \pm\dfrac{3}{2}x$

$c^2 = a^2 + b^2 = 4 + 9 = 13$

$c = \sqrt{13}$ on $x$-axis

The foci are at $(\sqrt{13}, 0)$ and $(-\sqrt{13}, 0)$.

**23.** $\dfrac{9y^2}{225} - \dfrac{25x^2}{225} = \dfrac{225}{225}$

$\dfrac{y^2}{25} - \dfrac{x^2}{9} = 1$

$a^2 = 25,\ a = 5$
$b^2 = 9,\ b = 3$
vertices: $(0, 5)$ and $(0, -5)$

asymptotes: $y = \pm\dfrac{a}{b}x = \pm\dfrac{5}{3}x$

$c^2 = a^2 + b^2 = 25 + 9 = 34$

$c = \sqrt{34}$ on $y$-axis

The foci are at $(0, \sqrt{34})$ and $(0, -\sqrt{34})$.

**25.** $y^2 = x^2 - 2$

$2 = x^2 - y^2$

$1 = \dfrac{x^2}{2} - \dfrac{y^2}{2}$

$a^2 = 2,\ a = \sqrt{2}$

$b^2 = 2,\ b = \sqrt{2}$

$c^2 = 2 + 2$

$c^2 = 4$

$c = 2$

The foci are located at $(2,0)$ and $(-2, 0)$.

asymptotes: $y = \pm\dfrac{\sqrt{2}}{\sqrt{2}}x$

$y = \pm x$

**27.** $a = 3, b = 5$

$\dfrac{x^2}{9} - \dfrac{y^2}{25} = 1$

**29.** $a = 2, b = 3$

$\dfrac{y^2}{4} - \dfrac{x^2}{9} = 1$

**31.** Center $(2, -3)$, $a = 2, b = 3$

$\dfrac{(x-2)^2}{4} - \dfrac{(y+3)^2}{9} = 1$

**33.** center: $(-4, -3)$
$a^2 = 9, a = 3$
$b^2 = 16, b = 4$
vertices: $(-7, -3)$ and $(-1, -3)$

asymptotes: $y + 3 = \pm\dfrac{4}{3}(x+4)$

$c^2 = a^2 + b^2 = 9 + 16 = 25$
$c = \pm 5$ parallel to $x$-axis
The foci are at $(-9, -3)$ and $(1, -3)$.

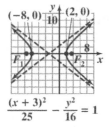

$$\dfrac{(x+4)^2}{9} - \dfrac{(y+3)^2}{16} = 1$$

**35.** center: $(-3, 0)$
$a^2 = 25, a = 5$
$b^2 = 16, b = 4$
vertices: $(2, 0)$ and $(-8, 0)$

asymptotes: $y = \pm\dfrac{4}{5}(x+3)$

$c^2 = a^2 + b^2 = 25 + 16 = 41$

$c = \sqrt{41}$

The foci are at $(-3+\sqrt{41}, 0)$ and $(-3-\sqrt{41}, 0)$.

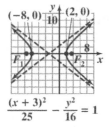

$$\dfrac{(x+3)^2}{25} - \dfrac{y^2}{16} = 1$$

**37.** center: $(1, -2)$
$a^2 = 4, a = 2$
$b^2 = 16, b = 4$
vertices: $(1, 0)$ and $(1, -4)$

asymptotes: $y + 2 = \pm\dfrac{1}{2}(x-1)$

$c^2 = a^2 + b^2 = 4 + 16 = 20$
$c = \sqrt{20} = 2\sqrt{5}$ parallel to $y$-axis
The foci are at $(1, -2+2\sqrt{5})$ and $(1, -2-2\sqrt{5})$.

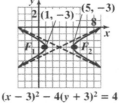

$$\dfrac{(y+2)^2}{4} - \dfrac{(x-1)^2}{16} = 1$$

**39.** $\dfrac{(x-3)^2}{4} - \dfrac{4(y+3)^2}{4} = \dfrac{4}{4}$

$\dfrac{(x-3)^2}{4} - (y+3)^2 = 1$

center: $(3, -3)$
$a^2 = 4, a = 2$
$b^2 = 1, b = 1$
vertices: $(1, -3)$ and $(5, -3)$

asymptotes: $y + 3 = \pm\dfrac{1}{2}(x-3)$

$c^2 = a^2 + b^2 = 4 + 1 = 5$

$c = \sqrt{5}$

The foci are at $(3+\sqrt{5}, -3)$ and $(3-\sqrt{5}, -3)$.

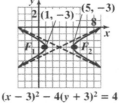

$$(x-3)^2 - 4(y+3)^2 = 4$$

**41.** $\dfrac{(x-1)^2}{3} - \dfrac{(y-2)^2}{3} = 1$

center: (1, 2)

$a^2 = 3, a = \sqrt{3}$

$b^2 = , b = \sqrt{3}$

vertices: (–1, 2) and (3, 2)

asymptotes: $y - 2 = \pm(x - 1)$

$c^2 = a^2 + b^2 = 3 + 3 = 6$

$c = \sqrt{6}$  parallel to *y*-axis

The foci are at $(1+\sqrt{6}, 2)$ and $(1-\sqrt{6}, 2)$.

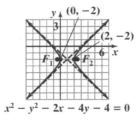

$(1-\sqrt{3}, 2)$ *y*  $(1+\sqrt{3}, 2)$

$(x-1)^2 - (y-2)^2 = 3$

**43.** $(x^2 - 2x) - (y^2 + 4y) = 4$

$(x^2 - 2x + 1) - (y^2 + 4y + 4) = 4 + 1 - 4$

$(x-1)^2 - (y+2)^2 = 1$

center: (1, –2)

$a^2 = 1, a = 1$

$b^2 = 1, b = 1$

$c^2 = a^2 + b^2 = 1 + 1 = 2$

$c = \sqrt{2}$

asymptotes:  $y + 2 = \pm(x - 1)$

The foci are at $(1+\sqrt{2}, -2)$ and $(1-\sqrt{2}, -2)$.

*y* (0, –2)

(2, –2)

$x^2 - y^2 - 2x - 4y - 4 = 0$

**45.** $(16x^2 + 64x) - (y^2 + 2y) = -67$

$16(x^2 + 4x + 4) - (y^2 + 2y + 1) = -67 + 64 - 1$

$16(x+2)^2 - (y+1)^2 = -4$

$\dfrac{16(x+2)^2}{-4} - \dfrac{(y+1)^2}{-4} = \dfrac{-4}{-4}$

$\dfrac{(y+1)^2}{4} - \dfrac{(x+2)^2}{\frac{1}{4}} = 1$

center: (–2, –1)

$a^2 = 4, a = 2$

$b^2 = \dfrac{1}{4}, b = \dfrac{1}{2}$

$c^2 = a^2 + b^2 = 4 + \dfrac{1}{4} = \dfrac{17}{4}$

$c = \sqrt{\frac{17}{4}} = \sqrt{4.25}$

asymptotes: $(y+1) = \pm\dfrac{2}{\frac{1}{2}}(x+2)$

$y + 1 = \pm 4(x+2)$

The foci are at $\left(-2, -1+\sqrt{4.25}\right)$ and $\left(-2, -1-\sqrt{4.25}\right)$.

*y* (–2, 1)

(–2, –3)

$16x^2 - y^2 + 64x - 2y + 67 = 0$

**47.** $(4x^2 - 16x) - (9y^2 - 54y) = 101$

$4(x^2 - 4x + 4) - 9(y^2 - 6y + 9) = 101 + 16 - 81$

$4(x-2)^2 - 9(y-3)^2 = 36$

$\dfrac{(x-2)^2}{9} - \dfrac{(y-3)^2}{4} = 1$

center: (2, 3)

$a^2 = 9, a = 3$

$b^2 = 4, b = 2$

$c^2 = a^2 + b^2 = 9 + 4 = 13$

$c = \sqrt{13}$

asymptotes:  $y - 3 = \pm\dfrac{2}{3}(x - 2)$

The foci are at $(2+\sqrt{13}, 3)$ and $(2-\sqrt{13}, 3)$.

*y* (–1, 3)

(5, 3)

$4x^2 - 9y^2 - 16x + 5y - 101 = 0$

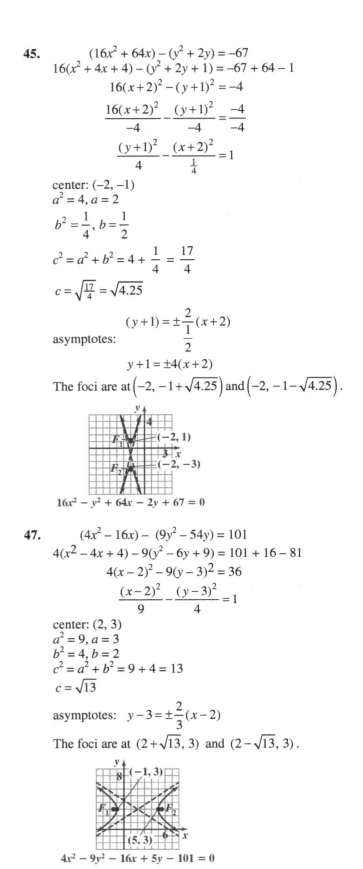

**49.**
$$(4x^2 - 32x) - 25y^2 = -164$$
$$4(x^2 - 8x + 16) - 25y^2 = -164 + 64$$
$$4(x - 4)^2 - 25y^2 = -100$$
$$\frac{4(x-4)^2}{-100} - \frac{25y^2}{-100} = \frac{-100}{-100}$$
$$\frac{y^2}{4} - \frac{(x-4)^2}{25} = 1$$

center: (4, 0)
$a^2 = 4$, $a = 2$
$b^2 = 25$, $b = 5$
$c^2 = a^2 + b^2 = 4 + 25 = 29$
$c = \sqrt{29}$

asymptotes: $y = \pm\dfrac{2}{5}(x - 4)$

The foci are at $(4, \sqrt{29})$ and $(4, -\sqrt{29})$.

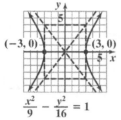

$4x^2 - 25y^2 - 32x + 164 = 0$

**51.**
$$\frac{x^2}{9} - \frac{y^2}{16} = 1$$

The equation is for a hyperbola in standard form with the transverse axis on the *x*-axis. We have $a^2 = 9$ and $b^2 = 16$, so $a = 3$ and $b = 4$.
Therefore, the vertices are at $(\pm a, 0)$ or $(\pm 3, 0)$.

Using a dashed line, we construct a rectangle using the ±3 on the *x*-axis and ±4 on the *y*-axis. Then use dashed lines to draw extended diagonals for the rectangle. These represent the asymptotes of the graph.

$\dfrac{x^2}{9} - \dfrac{y^2}{16} = 1$

From the graph we determine the following:
Domain: $\{x \mid x \le -3 \text{ or } x \ge 3\}$ or $(-\infty, -3] \cup [3, \infty)$
Range: $\{y \mid y \text{ is a real number}\}$ or $(-\infty, \infty)$

**53.**
$$\frac{x^2}{9} + \frac{y^2}{16} = 1$$
The equation is for an ellipse in standard form with major axis along the y-axis. We have $a^2 = 16$ and $b^2 = 9$, so $a = 4$ and $b = 3$. Therefore, the vertices are $(0, \pm a)$ or $(0, \pm 4)$. The endpoints of the minor axis are $(\pm b, 0)$ or $(\pm 3, 0)$.

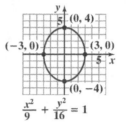

$\dfrac{x^2}{9} + \dfrac{y^2}{16} = 1$

From the graph we determine the following:
Domain: $\{x \mid -3 \le x \le 3\}$ or $[-3, 3]$
Range: $\{y \mid -4 \le y \le 4\}$ or $[-4, 4]$.

**55.**
$$\frac{y^2}{16} - \frac{x^2}{9} = 1$$
The equation is in standard form with the transverse axis on the *y*-axis. We have $a^2 = 16$ and $b^2 = 9$, so $a = 4$ and $b = 3$. Therefore, the vertices are at $(0, \pm a)$ or $(0, \pm 4)$. Using a dashed line, we construct a rectangle using the ±4 on the *y*-axis and ±3 on the *x*-axis. Then use dashed lines to draw extended diagonals for the rectangle. These represent the asymptotes of the graph.

$\dfrac{y^2}{16} - \dfrac{x^2}{9} = 1$

From the graph we determine the following:
Domain: $\{x \mid x \text{ is a real number}\}$ or $(-\infty, \infty)$
Range: $\{y \mid y \le -4 \text{ or } y \ge 4\}$ or $(-\infty, -4] \cup [4, \infty)$

**57.** $x^2 - y^2 = 4$

$x^2 + y^2 = 4$

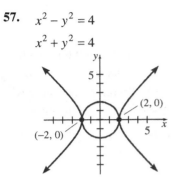

Check $(-2, 0)$:

$(-2)^2 - 0^2 = 4 \qquad (-2)^2 + 0^2 = 4$

$\qquad 4 - 0 = 4 \qquad\qquad 4 + 0 = 4$

$\qquad\qquad 4 = 4 \text{ true} \qquad\qquad 4 = 4 \text{ true}$

Check $(2, 0)$:

$(2)^2 - 0^2 = 4 \qquad (2)^2 + 0^2 = 4$

$\qquad 4 - 0 = 4 \qquad\qquad 4 + 0 = 4$

$\qquad\qquad 4 = 4 \text{ true} \qquad\qquad 4 = 4 \text{ true}$

The solution set is $\{(-2, 0), (2, 0)\}$.

**59.**

$9x^2 + y^2 = 9 \quad \text{or} \quad \dfrac{x^2}{1} + \dfrac{y^2}{9} = 1$

$y^2 - 9x^2 = 9 \qquad\qquad \dfrac{y^2}{9} - \dfrac{x^2}{1} = 1$

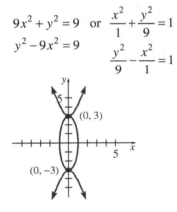

Check $(0, -3)$:

$9(0)^2 + (-3)^2 = 9 \qquad (-3)^2 - 9(0)^2 = 9$

$\qquad 0 + 9 = 9 \qquad\qquad\qquad 9 - 0 = 9$

$\qquad\qquad 9 = 9 \qquad\qquad\qquad\qquad 9 = 9$

$\qquad\qquad\quad \text{true} \qquad\qquad\qquad\qquad \text{true}$

Check $(0, 3)$:

$9(0)^2 + (3)^2 = 9 \qquad (3)^2 - 9(0)^2 = 9$

$\qquad 0 + 9 = 9 \qquad\qquad\qquad 9 - 0 = 9$

$\qquad\qquad 9 = 9 \qquad\qquad\qquad\qquad 9 = 9$

$\qquad\qquad\quad \text{true} \qquad\qquad\qquad\qquad \text{true}$

The solution set is $\{(0, -3), (0, 3)\}$.

**61.** $|d_2 - d_1| = 2a = (2 \text{ s})(1100 \text{ ft / s}) = 2200 \text{ ft}$

$a = 1100 \text{ ft}$

$2c = 5280 \text{ ft}, \ c = 2640 \text{ ft}$

$b^2 = c^2 - a^2 = (2640)^2 - (1100)^2$

$\qquad = 5{,}759{,}600$

$\dfrac{x^2}{(1100)^2} - \dfrac{y^2}{5{,}759{,}600} = 1$

$\dfrac{x^2}{1{,}210{,}000} - \dfrac{y^2}{5{,}759{,}600} = 1$

If $M_1$ is located 2640 feet to the right of the origin on the $x$-axis, the explosion is located on the right branch of the hyperbola given by the equation above.

**63.** $\qquad 625y^2 - 400x^2 = 250{,}000$

$\dfrac{625y^2}{250{,}000} - \dfrac{400x^2}{250{,}000} = \dfrac{250{,}000}{250{,}000}$

$\dfrac{y^2}{400} - \dfrac{x^2}{625} = 1$

$a^2 = 400, \ a = \sqrt{400} = 20$

$2a = 40$

The houses are 40 yards apart at their closest point.

**65.** **a.** ellipse

**b.** $x^2 + 4y^2 = 4$

**67. – 75.** Answers will vary.

**77.** $\dfrac{x^2}{4} - \dfrac{y^2}{9} = 0$

$y^2 = \dfrac{9}{4} x^2$

$y = \pm \dfrac{3}{2} x$

No; in general, the graph is two intersecting lines.

**79.** $4x^2 - 6xy + 2y^2 - 3x + 10y - 6 = 0$

$2y^2 + (10 - 6x)y + (4x^2 - 3x - 6) = 0$

$y = \dfrac{6x - 10 \pm \sqrt{(10 - 6x)^2 - 8(4x^2 - 3x - 6)}}{4}$

$y = \dfrac{6x - 10 \pm \sqrt{4(x^2 - 24x + 37)}}{4}$

$y = \dfrac{3x - 5 \pm \sqrt{x^2 - 24x + 37}}{2}$

The *xy*-term rotates the hyperbola. Separation of terms into ones containing only *x* or only *y* would not be possible.

**81.** does not make sense; Explanations will vary. Sample explanation: This would change the ellipse to a hyperbola.

**83.** makes sense

**85.** false; Changes to make the statement true will vary. A sample change is: If a hyperbola has a transverse axis along the *x*–axis and one of the branches is removed, the remaining branch does not define a function of *x*.

**87.** true

**89.** $\dfrac{c}{a}$ will be large when *a* is small. When this happens, the asymptotes will be nearly vertical.

**91.** If the asymptotes are perpendicular, then their slopes are negative reciprocals. For the hyperbola
$\dfrac{x^2}{a^2} - \dfrac{y^2}{b^2} = 1$, the asymptotes are $y = \pm \dfrac{b}{a}x$. The

slopes are negative reciprocals when $\dfrac{b}{a} = \dfrac{a}{b}$ (since one is already the negative of the other). This happens when $b^2 = a^2$,
so $a = b$. Any hyperbola where $a = b$, such as
$\dfrac{x^2}{4} - \dfrac{y^2}{4} = 1$, has perpendicular asymptotes.

**92.** $y = x^2 + 4x - 5$

Since $a = 1$ is positive, the parabola opens upward. The *x*-coordinate of the vertex is

$x = -\dfrac{b}{2a} = -\dfrac{4}{2(1)} = -2$. The *y*-coordinate of the

vertex is $y = (-2)^2 + 4(-2) - 5 = -9$.
Vertex: $(-2, -9)$.

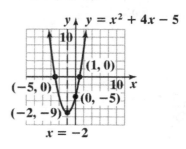

**93.** $y = -3(x - 1)^2 + 2$

Since $a = -3$ is negative, the parabola opens downward. The vertex of the parabola is
$(h, k) = (1, 2)$.
The *y*–intercept is $-1$.

**94.** $y^2 + 2y + 12x - 23 = 0$

$y^2 + 2y = -12x + 23$

$y^2 + 2y + 1 = -12x + 23 + 1$

$(y + 1)^2 = -12x + 24$

**Mid-Chapter 7 Check Point**

1.  Center: $(0,0)$

    Because the denominator of the $x^2$ – term is greater than the denominator of the $y^2$ – term, the major axis is horizontal. Since $a^2 = 25$, $a = 5$ and the vertices are $(-5,0)$ and $(5,0)$. Since $b^2 = 4$, $b = 2$ and endpoints of the minor axis are $(0,-2)$ and $(0,2)$.

    Foci: $\left(\pm\sqrt{21},0\right)$

    $$\frac{x^2}{25} + \frac{y^2}{4} = 1$$

2.  Divide both sides by 36 to get the standard form:

    $$\frac{x^2}{4} + \frac{y^2}{9} = 1$$

    Center: $(0,0)$

    Because the denominator of the $y^2$ – term is greater than the denominator of the $x^2$ – term, the major axis is vertical. Since $a^2 = 9$, $a = 3$ and the vertices are $(0,-3)$ and $(0,3)$. Since $b^2 = 4$, $b = 2$ and endpoints of the minor axis are $(-2,0)$ and $(2,0)$.

    Foci: $\left(0,\pm\sqrt{5}\right)$

    $$9x^2 + 4y^2 = 36$$

3.  Center: $(2,-1)$

    Because the denominator of the $y^2$ – term is greater than the denominator of the $x^2$ – term, the major axis is vertical. We have $a^2 = 25$ and $b^2 = 16$, so $a = 5$ and $b = 4$. The vertices lie 5 units above and below the center. The endpoints of the minor axis lie 4 units to the left and right of the center.

    Vertices: $(2,4)$ and $(2,-6)$

    Minor endpoints: $(-2,-1)$ and $(6,-1)$

    Foci: $(2,2)$, $(2,-4)$

    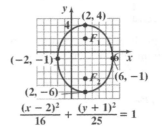

    $$\frac{(x-2)^2}{16} + \frac{(y+1)^2}{25} = 1$$

4.  Center: $(-2,1)$

    Because the denominator of the $x^2$ – term is greater than the denominator of the $y^2$ – term, the major axis is horizontal. We have $a^2 = 25$ and $b^2 = 16$, so $a = 5$ and $b = 4$. The vertices lie 5 units to the left and right of the center. The endpoints of the minor axis lie 4 units above and below the center.

    Vertices: $(-7,1)$ and $(3,1)$

    Minor endpoints: $(-2,5)$ and $(-2,-3)$

    Foci: $(-5,1)$, $(1,1)$

    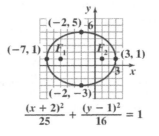

    $$\frac{(x+2)^2}{25} + \frac{(y-1)^2}{16} = 1$$

**5.**

$$x^2 - 4x\ + 9y^2 + 54y\ = -49$$

$$\left(x^2 - 4x\ \right) + 9\left(y^2 + 6y\ \right) = -49$$

$$\left(x^2 - 4x + 4\right) + 9\left(y^2 + 6y + 9\right) = -49 + 4 + 81$$

$$\left(x-2\right)^2 + 9\left(y+3\right)^2 = 36$$

$$\frac{\left(x-2\right)^2}{36} + \frac{9\left(y+3\right)^2}{36} = \frac{36}{36}$$

$$\frac{\left(x-2\right)^2}{36} + \frac{\left(y+3\right)^2}{4} = 1$$

Center: $(2,-3)$

Foci: $\left(2 \pm 4\sqrt{2}, -3\right)$

$$x^2 + 9y^2 - 4x + 54y + 49 = 0$$

**6.** The equation is for a hyperbola in standard form with the transverse axis on the $x$-axis. We have $a^2 = 9$ and $b^2 = 1$, so $a = 3$ and $b = 1$. Therefore, the vertices are at $(\pm a, 0)$ or $(\pm 3, 0)$. Using a dashed line, we construct a rectangle using the $\pm 3$ on the $x$-axis and $\pm 1$ on the $y$-axis. Then use dashed lines to draw extended diagonals for the rectangle. These represent the asymptotes of the graph.

Foci: $\left(\pm\sqrt{10}, 0\right)$

$$\frac{x^2}{9} - y^2 = 1$$

**7.** The equation is in the form $\dfrac{y^2}{a^2} - \dfrac{x^2}{b^2} = 1$ with $a^2 = 9$, and $b^2 = 1$. We know the transverse axis lies on the $y$-axis and the vertices are $(0,-3)$ and $(0,3)$. Because $a^2 = 9$ and $b^2 = 1$, $a = 3$ and $b = 1$. Construct a rectangle using $-1$ and 1 on the $x$-axis, and $-3$ and 3 on the $y$-axis. Draw extended diagonals to obtain the asymptotes.

Foci: $\left(0, \pm\sqrt{10}\right)$

$$\frac{y^2}{9} - x^2 = 1$$

**8.** $\dfrac{y^2}{16} - \dfrac{x^2}{4} = 1$

The equation is in the form $\dfrac{y^2}{a^2} - \dfrac{x^2}{b^2} = 1$ with $a^2 = 16$, and $b^2 = 4$. We know the transverse axis lies on the $y$-axis and the vertices are $(0,-4)$ and $(0,4)$. Because $a^2 = 16$ and $b^2 = 4$, $a = 4$ and $b = 2$. Construct a rectangle using $-2$ and 2 on the $x$-axis, and $-4$ and 4 on the $y$-axis. Draw extended diagonals to obtain the asymptotes.

Foci: $\left(0, \pm 2\sqrt{5}\right)$

$$y^2 - 4x^2 = 16$$

**9.** $\dfrac{x^2}{49} - \dfrac{y^2}{4} = 1$

The equation is for a hyperbola in standard form with the transverse axis on the $x$-axis. We have $a^2 = 49$ and $b^2 = 4$, so $a = 7$ and $b = 2$. Therefore, the vertices are at $(\pm a, 0)$ or $(\pm 7, 0)$.

Using a dashed line, we construct a rectangle using the $\pm 7$ on the $x$-axis and $\pm 2$ on the $y$-axis. Then use dashed lines to draw extended diagonals for the rectangle. These represent the asymptotes of the graph.

Foci: $\left(\pm\sqrt{53}, 0\right)$

$$4x^2 - 49y^2 = 196$$

**10.** The equation is for a hyperbola in standard form with center $(2, -2)$. We have $a^2 = 9$ and $b^2 = 16$, so $a = 3$ and $b = 4$.

Asymptotes: $y + 2 = \pm\dfrac{4}{3}(x - 2)$

Foci: $(-3, -2)$, $(7, -2)$

$$\dfrac{(x-2)^2}{9} - \dfrac{(y+2)^2}{16} = 1$$

**11.** Write the equation for the hyperbola in standard form:

$$4x^2 - y^2 + 8x + 6y + 11 = 0$$

$$4x^2 + 8x \quad - y^2 + 6y \quad = -11$$

$$4\left(x^2 + 2x \quad\right) - \left(y^2 - 6y \quad\right) = -11$$

$$4\left(x^2 + 2x + 1\right) - \left(y^2 - 6y + 9\right) = -11 + 4 - 9$$

$$4(x+1)^2 - (y-3)^2 = -16$$

$$\dfrac{4(x+1)^2}{-16} - \dfrac{(y-3)^2}{-16} = \dfrac{-16}{-16}$$

$$\dfrac{(y-3)^2}{16} - \dfrac{(x+1)^2}{4} = 1$$

Center $(-1, 3)$.

Asymptotes: $y - 3 = \pm 2(x + 1)$

Foci: $\left(-1, 3 \pm 2\sqrt{5}\right)$, $(7, -2)$

$$4x^2 - y^2 + 8x + 6y + 11 = 0$$

**12.** This is the equation of a circle centered at the origin with radius $r = \sqrt{4} = 2$.

We can plot points that are 2 units to the left, right, above, and below the origin and then graph the circle. The points are $(-2, 0)$, $(2, 0)$, $(0, 2)$, and $(0, -2)$.

$$x^2 + y^2 = 4$$

**13.** $x + y = 4$

$y = -x + 4$

This is the equation of a line with slope $m = -1$ and a $y$-intercept of 4. We can plot the point $(0, 4)$, use the slope to get an additional point, connect the points with a straight line and then extend the line to represent the graph of the equation.

$$x + y = 4$$

**14.** $x^2 - y^2 = 4$

$$\dfrac{x^2}{4} - \dfrac{y^2}{4} = 1$$

The equation is for a hyperbola in standard form with the transverse axis on the $x$-axis. We have $a^2 = 4$ and $b^2 = 4$, so $a = 2$ and $b = 2$.

Therefore, the vertices are at $(\pm a, 0)$ or $(\pm 2, 0)$.

Using a dashed line, we construct a rectangle using the $\pm 2$ on the $x$-axis and $\pm 2$ on the $y$-axis. Then use dashed lines to draw extended diagonals for the rectangle. These represent the asymptotes of the graph.

Graph the hyperbola.

$$x^2 - y^2 = 4$$

**15.** $x^2 + 4y^2 = 4$

$$\frac{x^2}{4} + \frac{y^2}{1} = 1$$

Center: $(0,0)$

Because the denominator of the $x^2$ – term is greater than the denominator of the $y^2$ – term, the major axis is horizontal. We have $a^2 = 4$ and $b^2 = 1$, so $a = 2$ and $b = 1$. The vertices lie 2 units to the left and right of the center. The endpoints of the minor axis lie 1 unit above and below the center.

Vertices: $(-2,0)$ and $(2,0)$

Minor endpoints: $(0,-1)$ and $(0,1)$

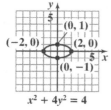

$x^2 + 4y^2 = 4$

**16.** Center: $(-1,1)$

Radius: $r = \sqrt{4} = 2$

We plot the points that are 2 units to the left, right, above and below the center.

These points are $(-3,1)$, $(1,1)$, $(-1,3)$, and $(-1.-1)$.

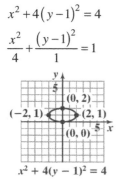

$(x + 1)^2 + (y - 1)^2 = 4$

**17.** $x^2 + 4(y-1)^2 = 4$

$$\frac{x^2}{4} + \frac{(y-1)^2}{1} = 1$$

$x^2 + 4(y - 1)^2 = 4$

**18.** $(x-1)^2 - (y-1)^2 = 4$

$$\frac{(x-1)^2}{4} - \frac{(y-1)^2}{4} = 1$$

The equation is for a hyperbola in standard form centered at (1, 1). We have $a^2 = 4$ and $b^2 = 4$, so $a = 2$ and $b = 2$.

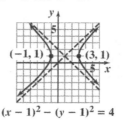

$(x - 1)^2 - (y - 1)^2 = 4$

**19.** The foci and vertices show that $c$ is 4 and $a$ is 5.

$$c^2 = a^2 - b^2$$
$$4^2 = 5^2 - b^2$$
$$b^2 = 25 - 16$$
$$b^2 = 9$$
$$\frac{x^2}{25} + \frac{y^2}{9} = 1$$

**20.** The endpoints show that the center is (1, 2).

Since $2a = 18$, $a = 9$ and $a^2 = 81$.

Since $2c = 10$, $c = 5$ and $c^2 = 25$.

$$c^2 = a^2 - b^2$$
$$25 = 81 - b^2$$
$$b^2 = 81 - 25$$
$$b^2 = 56$$
$$\frac{(x-1)^2}{81} + \frac{(y-2)^2}{56} = 1$$

**21.** The foci and vertices show that $c$ is 3 and $a$ is 2.

$$b^2 = c^2 - a^2$$
$$b^2 = 3^2 - 2^2$$
$$b^2 = 9 - 4$$
$$b^2 = 5$$
$$\frac{x^2}{4} - \frac{y^2}{5} = 1$$

**22.** The endpoints show that the center is $(-1, 5)$.

Since $2a = 4$, $a = 2$ and $a^2 = 4$.

Since $2c = 6$, $c = 3$ and $c^2 = 9$.

$$b^2 = c^2 - a^2$$
$$b^2 = 9 - 4$$
$$b^2 = 5$$
$$\frac{(x+1)^2}{4} + \frac{(y-5)^2}{5} = 1$$

**23.** $a = 15$, $b = 10$

$$\frac{x^2}{15^2} + \frac{y^2}{10^2} = 1$$
$$\frac{x^2}{225} + \frac{y^2}{100} = 1$$

Since the truck is 10 feet wide, substitute $x = 5$ into the equation to find $y$.

$$\frac{5^2}{225} + \frac{y^2}{100} = 1$$
$$\frac{25}{225} + \frac{y^2}{100} = 1$$
$$\frac{1}{9} + \frac{y^2}{100} = 1$$
$$900\left(\frac{1}{9} + \frac{y^2}{100}\right) = 900(1)$$
$$100 + 9y^2 = 900$$
$$9y^2 = 800$$
$$y^2 = 88.8889$$
$$y = \sqrt{88.8889}$$
$$y \approx 9.43$$

5 feet from the center, the height of the archway is 9.43 feet. Since the truck's height is 9.5 feet, it will not fit under the archway.

**24.** Find the distance between the foci.

Since $2a = 40$, $a = 20$ and $a^2 = 400$.

Since $2b = 20$, $b = 10$ and $b^2 = 100$.

$$c^2 = a^2 - b^2$$
$$c^2 = 400 - 100$$
$$c^2 = 300$$
$$c = \sqrt{300}$$
$$= 10\sqrt{3}$$
$$2c = 20\sqrt{3}$$
$$2c \approx 34.64$$

The kidney stone should be 34.64 cm from the electrode that sends the ultrasound waves.

**25. a.** Since $2c = 6$, $c = 3$ and $c^2 = 9$.

The ranger at the primary station heard the explosion 6 seconds before the other ranger. This means that the explosion occurred $6 \times 0.35 = 2.1$ miles closer to the primary station.

Since $2a = 2.1$, $a = 1.05$ and $a^2 = 1.1025$.

$$b^2 = c^2 - a^2$$
$$b^2 = 9 - 1.1025$$
$$b^2 = 7.8975$$
$$\frac{x^2}{1.1025} - \frac{y^2}{7.8975} = 1$$

**b.**

**Section 7.3**

**Check Point Exercises**

1. $y^2 = 8x$

   $4p = 8$

   $p = 2$

   focus: $(p,\ 0) = (2,\ 0)$

   directrix: $x = -p;\ x = -2$

2. $x^2 = -12y$

   $4p = -12$

   $p = -3$

   focus: $(0, p) = (0, -3)$

   directrix: $y = -p;\ y = 3$

3. The focus is (8, 0). Thus, the focus is on the $x$-axis. We use the standard form of the equation in which there is $x$-axis symmetry, namely, $y^2 = 4px$.

   The focus is 8 units to the right of the vertex, (0, 0). Thus, $p$ is positive and $p = 8$.

   $y^2 = 4px$

   $y^2 = 4 \cdot 8x$

   $y^2 = 32x$

4. $(x-2)^2 = 4(y+1)$

   From the equation we have $h = 2$ and $k = -1$.

   vertex: $(2, -1)$

   Find $p$: $4p = 4$

   $p = 1$

   focus: $(h, k+p) = (2, -1+1)$

   $= (2, 0)$

   directrix: $y = k - p$

   $y = -1 - 1$

   $y = -2$

5. $y^2 + 2y = -4x + 7$

   $y^2 + 2y + 1 = -4x + 7 + 1$

   $(y+1)^2 = -4(x-2)$

   From the equation we have $h = 2$ and $k = -1$.

   vertex: $(2, -1)$

   Find $p$: $4p = -4$

   $p = -1$

   focus: $(h+p, k) = (2-1, -1)$

   $= (1, -1)$

   directrix: $x = h - p$

   $x = 2 - (-1)$

   $x = 3$

**6.** $x^2 = 4py$

Let $x = 3$ and $y = 4$.

$3^2 = 4p \cdot 4$

$9 = 16p$

$p = \dfrac{9}{16}$

$x^2 = \dfrac{9}{4}y$

The light should be placed at $\left(0, \dfrac{9}{16}\right)$ or $\dfrac{9}{16}$ inch above the vertex.

**Concept and Vocabulary Check 7.3**

**1.** parabola; directrix; focus

**2.** a

**3.** $(-7, 0)$

**4.** $x = 7$

**5.** 28; $(-7, -14)$; $(-7, 14)$

**6.** d

**7.** $(-2, 0)$

**8.** $y = -2$

**9.** 4; $(-4, 0)$; $(0, 0)$

**Exercise Set 7.3**

**1.** $y^2 = 4x$
$4p = 4, p = 1$
vertex: $(0, 0)$
focus: $(1, 0)$
directrix: $x = -1$
graph (c)

**3.** $x^2 = -4y$
$4p = -4, p = -1$
vertex: $(0, 0)$
focus: $(0, -1)$
directrix: $y = 1$
graph (b)

**5.** $4p = 16, p = 4$
vertex: $(0, 0)$
focus: $(4, 0)$
directrix: $x = -4$

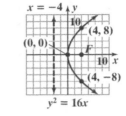

**7.** $4p = -8, p = -2$
vertex: $(0, 0)$
focus: $(-2, 0)$
directrix: $x = 2$

**9.** $4p = 12, p = 3$
vertex: $(0, 0)$
focus: $(0, 3)$
directrix: $y = -3$

**11.** $4p = -16, p = -4$
vertex: $(0, 0)$
focus: $(0, -4)$
directrix: $y = 4$

**13.**   $y^2 = 6x$

$4p = 6,\ p = \dfrac{6}{4} = \dfrac{3}{2}$

vertex: $(0, 0)$

focus: $\left(\dfrac{3}{2}, 0\right)$

directrix: $x = -\dfrac{3}{2}$

**15.**   $8x^2 = -4y$

$x^2 = -\dfrac{1}{2}y$

$4p = -\dfrac{1}{2}$

$p = -\dfrac{1}{8}$

focus: $\left(0, -\dfrac{1}{8}\right)$

directrix:   $y = \dfrac{1}{8}$

**17.**   $p = 7,\ 4p = 28$
$y^2 = 28x$

**19.**   $p = -5,\ 4p = -20$
$y^2 = -20x$

**21.**   $p = 15,\ 4p = 60$
$x^2 = 60y$

**23.**   $p = -25,\ 4p = -100$
$x^2 = -100y$

**25.**   $p = -5 - (-3) = -2$   Vertex, $(2, -3)$

$(x - 2)^2 = -8(y + 3)$

**27.**   vertex:  $(1, 2)$   $p = 2$
$(y - 2)^2 = 8(x - 1)$

**29.**   vertex:  $(-3, 3),\ p = 1$
$(x + 3)^2 = 4(y - 3)$

**31.**   $(y - 1)^2 = 4(x - 1)$
$4p = 4,\ p = 1$
vertex: $(1, 1)$
focus: $(2, 1)$
directrix: $x = 0$
graph (c)

**33.**   $(x + 1)^2 = -4(y + 1)$
$4p = -4,\ p = -1$
vertex: $(-1, -1)$
focus: $(-1, -2)$
directrix: $y = 0$
graph (d)

**35.**   $4p = 8,\ p = 2$
vertex: $(2, 1)$
focus: $(2, 3)$
directrix: $y = -1$

**37.**   $4p = -8,\ p = -2$
vertex: $(-1, -1)$
focus: $(-1, -3)$
directrix: $y = 1$

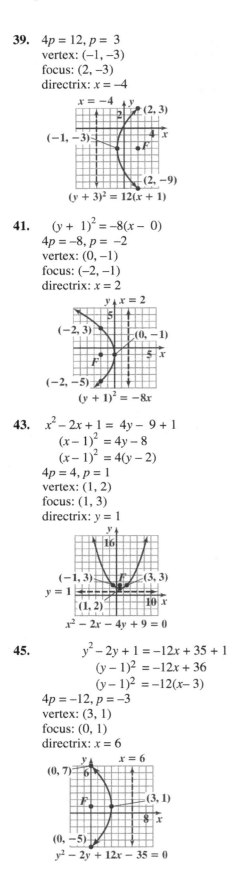

**39.** $4p = 12, p = 3$
vertex: $(-1, -3)$
focus: $(2, -3)$
directrix: $x = -4$

$(y + 3)^2 = 12(x + 1)$

**41.** $(y + 1)^2 = -8(x - 0)$
$4p = -8, p = -2$
vertex: $(0, -1)$
focus: $(-2, -1)$
directrix: $x = 2$

$(y + 1)^2 = -8x$

**43.** $x^2 - 2x + 1 = 4y - 9 + 1$
$(x - 1)^2 = 4y - 8$
$(x - 1)^2 = 4(y - 2)$
$4p = 4, p = 1$
vertex: $(1, 2)$
focus: $(1, 3)$
directrix: $y = 1$

$x^2 - 2x - 4y + 9 = 0$

**45.** $y^2 - 2y + 1 = -12x + 35 + 1$
$(y - 1)^2 = -12x + 36$
$(y - 1)^2 = -12(x - 3)$
$4p = -12, p = -3$
vertex: $(3, 1)$
focus: $(0, 1)$
directrix: $x = 6$

$y^2 - 2y + 12x - 35 = 0$

**47.** $x^2 + 6x = 4y - 1$
$x^2 + 6x + 9 = 4y - 1 + 9$
$(x + 3)^2 = 4(y + 2)$
$4p = 4, p = 1$
vertex: $(-3, -2)$
focus: $(-3, -1)$
directrix: $y = -3$

$x^2 + 6x - 4y + 1 = 0$

**49.** The y-coordinate of the vertex is
$$y = -\frac{b}{2a} = -\frac{6}{2(1)} = -3$$
The x-coordinate of the vertex is
$$x = (-3)^2 + 6(-3) + 5$$
$$= 9 - 18 + 5$$
$$= -4$$
The vertex is $(-4, -3)$.
Since the squared term is $y$ and $a > 0$, the graph opens to the right.
Domain: $\{x \mid x \geq -4\}$ or $[-4, \infty)$
Range: $\{y \mid y$ is a real number$\}$ or $(-\infty, \infty)$
The relation is not a function.

**51.** The x-coordinate of the vertex is
$$x = -\frac{b}{2a} = -\frac{(4)}{2(-1)} = 2$$
The y-coordinate of the vertex is
$$y = -(2)^2 + 4(2) - 3$$
$$= -4 + 8 - 3$$
$$= 1$$
The vertex is $(2, 1)$.
Since the squared term is $x$ and $a < 0$, the graph opens down.
Domain: $\{x \mid x$ is a real number$\}$ or $(-\infty, \infty)$
Range: $\{y \mid y \leq 1\}$ or $(-\infty, 1]$
The relation is a function.

**53.** The equation is in the form $x = a(y-k)^2 + h$

From the equation, we can see that the vertex is $(3,1)$.

Since the squared term is $y$ and $a < 0$, the graph opens to the left.

Domain: $\{x \mid x \le 3\}$ or $(-\infty, 3]$

Range: $\{y \mid y \text{ is a real number}\}$ or $(-\infty, \infty)$

The relation is not a function.

**55.**

Check $(-4, 2)$:

$-4 = (2-2)^2 - 4 \qquad 2 = -\dfrac{1}{2}(-4)$

$-4 = 0 - 4 \qquad\qquad\ 2 = 2$

$-4 = -4 \qquad\qquad\qquad\ \text{true}$

$\quad\ \text{true}$

Check $(0,0)$:

$0 = (0-2)^2 - 4 \qquad 0 = -\dfrac{1}{2}(0)$

$0 = 4 - 4 \qquad\qquad\ 0 = 0$

$0 = 0 \qquad\qquad\qquad\ \text{true}$

$\quad\ \text{true}$

The solution set is $\{(-4,2),(0,0)\}$.

**57.**

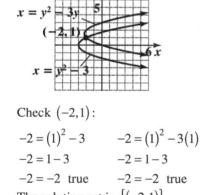

Check $(-2,1)$:

$-2 = (1)^2 - 3 \qquad -2 = (1)^2 - 3(1)$

$-2 = 1 - 3 \qquad\quad\ -2 = 1 - 3$

$-2 = -2 \ \text{true} \qquad -2 = -2 \ \text{true}$

The solution set is $\{(-2,1)\}$.

**59.**

The two graphs do not cross. Therefore, the solution set is the empty set, $\{\ \ \}$ or $\varnothing$.

**61.** $x^2 = 4py$

$2^2 = 4p(1)$

$4 = 4$

$p = 1$

The light bulb should be placed 1 inch above the vertex.

**63.** $x^2 = 4py$

$6^2 = 4p(2)$

$36 = 8p$

$p = \dfrac{36}{8} = \dfrac{9}{2} = 4.5$

The receiver should be located 4.5 feet from the base of the dish.

**65.** $x^2 = 4py$

$(640)^2 = 4p(160)$

$p = \dfrac{(640)^2}{640} = 640$

$x = 640 - 200 = 440$

$(440)^2 = 4(640)y$

$y = \dfrac{(440)^2}{4(640)} = 75.625$

The height is 76 meters.

**67.** $x^2 = 4py$

$\left(\dfrac{200}{2}\right)^2 = 4p(-50)$

$\dfrac{10,000}{-50} = 4p$

$4p = -200$

$x^2 = -200y$

$(30)^2 = -200y$

$y = \dfrac{900}{-200} = -4.5$

(height of bridge) $= 50 - 4.5 = 45.5$ feet.

Yes, the boat will clear the arch.

**69. – 75.** Answers will vary.

**77.** $y^2 + 2y - 6x + 13 = 0$

$y^2 + 2y + (-6x + 13) = 0$

$y = \dfrac{-2 \pm \sqrt{2^2 - 4(-6x + 13)}}{2}$

$y = \dfrac{-2 \pm \sqrt{24x - 48}}{2}$

$y = -1 \pm \sqrt{6x - 12}$

**79.** $16x^2 - 24xy + 9y^2 - 60x - 80y + 100 = 0$

$9y^2 - (24x + 80)y + (16x^2 - 60x + 100) = 0$

$y = \dfrac{24x + 80 \pm \sqrt{(24x + 80)^2 - 36(16x^2 - 60x + 100)}}{18}$

$y = \dfrac{24x + 80 \pm \sqrt{6000x + 2800}}{18}$

$y = \dfrac{24x + 80 \pm 20\sqrt{15x + 7}}{18}$

$y = \dfrac{12x + 40 \pm 10\sqrt{15x + 7}}{9}$

**81.** does not make sense; Explanations will vary. Sample explanation: Horizontal parabolas will rise without limit.

**83.** makes sense

**85.** false; Changes to make the statement true will vary. A sample change is: Because $a = -1$, the parabola will open to the left.

**87.** false; Changes to make the statement true will vary. A sample change is: If a parabola defines $y$ as a function of $x$, it will open up or down.

**89.** $Ax^2 + Ey = 0$

$Ax^2 = -Ey \qquad\qquad 4p = -\dfrac{E}{A}y$

$x^2 = -\dfrac{E}{A}y \qquad\qquad p = -\dfrac{E}{4A}y$

focus: $\left(0, -\dfrac{E}{4A}\right)$,

directrix: $y = \dfrac{E}{4A}$

**91.** Answers will vary.

**92.** For $n = 1$; $\dfrac{(-1)^n}{3^n - 1} = \dfrac{(-1)^1}{3^1 - 1} = \dfrac{-1}{3 - 1} = -\dfrac{1}{2}$

For $n = 2$; $\dfrac{(-1)^n}{3^n - 1} = \dfrac{(-1)^2}{3^2 - 1} = \dfrac{1}{9 - 1} = \dfrac{1}{8}$

For $n = 3$; $\dfrac{(-1)^n}{3^n - 1} = \dfrac{(-1)^3}{3^3 - 1} = \dfrac{-1}{27 - 1} = -\dfrac{1}{26}$

For $n = 4$; $\dfrac{(-1)^n}{3^n - 1} = \dfrac{(-1)^4}{3^4 - 1} = \dfrac{1}{81 - 1} = \dfrac{1}{80}$

**93.** $5 \cdot 4 \cdot 3 \cdot 2 \cdot 1 = 120$

**94.** For $i = 1$; $i^2 + 1 = 1^2 + 1 = 1 + 1 = 2$

For $i = 2$; $i^2 + 1 = 2^2 + 1 = 4 + 1 = 5$

For $i = 3$; $i^2 + 1 = 3^2 + 1 = 9 + 1 = 10$

For $i = 4$; $i^2 + 1 = 4^2 + 1 = 16 + 1 = 17$

For $i = 5$; $i^2 + 1 = 5^2 + 1 = 25 + 1 = 26$

For $i = 6$; $i^2 + 1 = 6^2 + 1 = 36 + 1 = 37$

$2 + 5 + 10 + 17 + 26 + 37 = 97$

**Chapter 7 Review Exercises**

**1.** $a^2 = 36, a = 6$
$b^2 = 25, b = 5$
$c^2 = a^2 - b^2 = 36 - 25 = 11$
$c = \sqrt{11}$
The foci are at $(\sqrt{11}, 0)$ and $(-\sqrt{11}, 0)$.

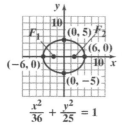

**2.** $a^2 = 25, a = 5$
$b^2 = 16, b = 4$
$c^2 = a^2 - b^2$
$c^2 = 25 - 16$
$c^2 = 9$
$c = 3$
The foci are $(0, 3)$ and $(0, -3)$.

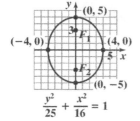

**3.** $\dfrac{4x^2}{16} + \dfrac{y^2}{16} = \dfrac{16}{16}$
$\dfrac{x^2}{4} + \dfrac{y^2}{16} = 1$
$b^2 = 4, b = 2$
$a^2 = 16, a = 4$
$c^2 = a^2 - b^2 = 16 - 4 = 12$
$c = \sqrt{12} = 2\sqrt{3}$
The foci are at $(0, 2\sqrt{3})$ and $(0, -2\sqrt{3})$.

**4.** $\dfrac{4x^2}{36} + \dfrac{9y^2}{36} = \dfrac{36}{36}$
$\dfrac{x^2}{9} + \dfrac{y^2}{4} = 1$
$a^2 = 9, a = 3$
$b^2 = 4, b = 2$
$c^2 = a^2 - b^2 = 9 - 4 = 5, c = \sqrt{5}$
The foci are at $(\sqrt{5}, 0)$ and $(-\sqrt{5}, 0)$.

**5.** $a^2 = 16\ a = 4$
$b^2 = 9\ b = 3$
$c^2 = 16 - 9 = 7, c = \sqrt{7}$
center: $(1, -2)$
The foci are at $(1+\sqrt{7}, -2)$ and $(1-\sqrt{7}, -2)$.

**6.** $a^2 = 16, a = 4$
$b^2 = 9, b = 3$
$c^2 = a^2 - b^2 = 16 - 9 = 7, c = \sqrt{7}$
center: $(-1, 2)$
The foci are at $(-1, 2+\sqrt{7})$ and $(-1, 2-\sqrt{7})$.

**7.** $4x^2 + 24x + 9y^2 - 36y = -36$

$4(x^2 + 6x + 9) + 9(y^2 - 4y + 4)$

$= -36 + 36 + 36$

$= 4(x + 3)^2 + 9(y - 2)^2 = 36$

$\dfrac{(x+3)^2}{9} + \dfrac{(y-2)^2}{4} = 1$

$c^2 = a^2 - b^2 = 5, \; c = \sqrt{5}$

center: $(-3, 2)$

The foci are at $(-3+\sqrt{5}, 2)$ and $(-3-\sqrt{5}, 2)$.

```
        y
       6
  (-3, 4)    F₂
F₁
(-6, 2)        (0, 2)
       (-3, 0)  2  x
```

$4x^2 + 9y^2 + 24x - 36y + 36 = 0$

**8.** $9x^2 - 18x + 4y^2 + 8y = 23$

$9(x^2 - 2x + 1) + 4(y^2 + 2y + 1) = 23 + 9 + 4$

$9(x - 1)^2 + 4(y + 1)^2 = 36$

$\dfrac{(x-1)^2}{4} + \dfrac{(y+1)^2}{9} = 1$

$c^2 = a^2 - b^2 = 9 - 4 = 5, \; c = \sqrt{5}$

center: $(1, -1)$

The foci are at $(1, -1 +\sqrt{5})$ and $(1, -1-\sqrt{5})$.

```
       y
      5
 F₁    (1, 2)
(-1, -1)    5  x
        (3, -1)
 F₂
      (1, -4)
```

$9x^2 + 4y^2 - 18x + 8y - 23 = 0$

**9.** $c = 4, \; c^2 = 16$

$a = 5, \; a^2 = 25$

$b^2 = a^2 - c^2 = 25 - 16 = 9$

$\dfrac{x^2}{25} + \dfrac{y^2}{9} = 1$

**10.** $c = 3, \; c^2 = 9$

$a = 6, \; a^2 = 36$

$b^2 = a^2 - c^2 = 36 - 9 = 27$

$\dfrac{x^2}{27} + \dfrac{y^2}{36} = 1$

**11.** $2a = 12, \; a = 6, \; a^2 = 36$

$2b = 4, \; b = 2, \; b^2 = 4$

$\dfrac{(x+3)^2}{36} + \dfrac{(y-5)^2}{4} = 1$

**12.** $2a = 20, \; a = 10, \; a^2 = 100$

$b = 6, \; b^2 = 36$

$\dfrac{x^2}{100} + \dfrac{y^2}{36} = 1$

**13.** $2a = 50, \; a = 25$

$b = 15$

$\dfrac{x^2}{625} + \dfrac{y^2}{225} = 1$

Let $x = 14$.

$\dfrac{(14)^2}{625} + \dfrac{y^2}{225} = 1$

$y^2 = 225\left(1 - \dfrac{196}{625}\right)$

$y \approx 15(0.8285) \approx 12.4 > 12$

Yes, the truck can drive under the archway.

**14.** The hit ball will collide with the other ball.

**15.** $c^2 = a^2 + b^2 = 16 + 1 = 17, \; c = \sqrt{17}$

The foci are at $(\sqrt{17}, 0)$ and $(-\sqrt{17}, 0)$.

Asymptotes: $y = \pm\dfrac{1}{4}x$

```
         y
        10
 (-4, 0)   (4, 0)
             10
              x
F₁        F₂
```

$\dfrac{x^2}{16} - y^2 = 1$

**16.** $c^2 = a^2 + b^2 = 16 + 1 = 17$

$c = \sqrt{17}$

The foci are at $(0, \sqrt{17})$ and $(0, -\sqrt{17})$.

Asymptotes: $y = \pm 4x$

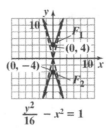

$\dfrac{y^2}{16} - x^2 = 1$

**17.** $\dfrac{x^2}{16} - \dfrac{y^2}{9} = 1$

$c^2 = a^2 + b^2 = 16 + 9 = 25, \ c = 5$

The foci are at $(5, 0)$ and $(-5, 0)$.

Asymptotes: $y = \pm \dfrac{3}{4} x$

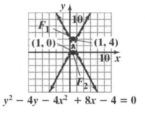

**18.** $\dfrac{y^2}{4} - \dfrac{x^2}{16} = 1$

$c^2 = a^2 + b^2 = 4 + 16 = 20$

$c = \sqrt{20} = 2\sqrt{5}$

The foci are at $(0, \ 2\sqrt{5})$ and $(0, -2\sqrt{5})$.

Asymptotes: $y = \pm \dfrac{1}{2} x$

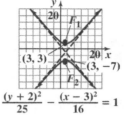

**19.** $c^2 = a^2 + b^2 = 25 + 16 = 41, c = \sqrt{41}$

center: $(2, -3)$

The foci are at $(2 + \sqrt{41}, \ -3)$ and $(2 - \sqrt{41}, \ -3)$.

Asymptotes: $y + 3 = \pm \dfrac{4}{5}(x - 2)$

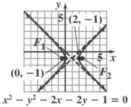

**20.** $c^2 = a^2 + b^2 = 25 + 16 = 41, \ c = \sqrt{41}$

center: $(3, -2)$

The foci are at $(3, \ -2 + \sqrt{41})$ and $(3, -2 - \sqrt{41})$.

Asymptotes: $y + 2 = \pm \dfrac{5}{4}(x - 3)$

**21.** 
$$y^2 - 4y - 4x^2 + 8x - 4 = 0$$
$$(y^2 - 4y + 4) - 4(x^2 - 2x + 1) = 4 + 4 - 4$$
$$(y - 2)^2 - 4(x - 1)^2 = 4$$
$$\dfrac{(y-2)^2}{4} - (x-1)^2 = 1$$

$c^2 = a^2 + b^2 = 4 + 1 = 5, \ c = \sqrt{5}$

center: $(1, 2)$

The foci are at $(1, \ 2 + \sqrt{5})$ and $(1, 2 - \sqrt{5})$.

Asymptotes: $y - 2 = \pm 2(x - 1)$

**22.** 
$$x^2 - 2x - y^2 - 2y = 1$$
$$(x^2 - 2x + 1) - (y^2 + 2y + 1) = 1 + 1 - 1$$
$$(x - 1)^2 - (y + 1)^2 = 1$$

$c^2 = a^2 + b^2 = 1 + 1 = 2, \ c = \sqrt{2}$

center: $(1, -1)$

The foci are at $(1 + \sqrt{2}, \ -1)$ and $(1 - \sqrt{2}, \ -1)$.

asymptotes: $y + 1 = \pm(x - 1)$

**23.** $c = 4, c^2 = 16$
$a = 2, a^2 = 4$
$b^2 = c^2 - a^2 = 16 - 4 = 12$

$$\frac{y^2}{4} - \frac{x^2}{12} = 1$$

**24.** $c = 8, c^2 = 64$
$a = 3, a^2 = 9$
$b^2 = c^2 - a^2 = 64 - 9 = 55$

$$\frac{x^2}{9} - \frac{y^2}{55} = 1$$

**25.** If the foci are at $(0, -2)$ and $(0, 2)$, then
$c = 2$. If the vertices are at $(0, -3)$ and $(0, 3)$ then $a = 3$. This is not possible since $c$ must be greater than $a$.

**26.** foci: $(\pm 100, 0), c = 100$

$$|d_1 - d_2| = \left(0.186 \frac{\text{mi}}{\mu s}\right)(500 \mu s) = 93 \text{ mi} = 2a$$

$$a = \frac{93}{2}$$

$$b^2 = c^2 - a^2 = (100)^2 - \left(\frac{93}{2}\right)^2 = 7837.75$$

$$\frac{x^2}{\left(\frac{93}{2}\right)^2} - \frac{y^2}{7837.75} = 1$$

$$\frac{x^2}{2162.25} - \frac{y^2}{7837.75} = 1$$

**27.** $4p = 8, p = 2$
vertex: $(0, 0)$
focus: $(2, 0)$
directrix: $x = -2$

**28.** $x^2 + 16y = 0$

$$x^2 = -16y$$
$4p = -16$
$p = -4$
vertex: $(0, 0)$
focus: $(0, -4)$
directrix: $y = 4$

**29.** $4p = -16$
$p = -4$
vertex: $(0, 2)$
focus: $(-4, 2)$
directrix: $x = 4$

**30.** $4p = 4, p = 1$
vertex: $(4, -1)$
focus: $(4, 0)$
directrix: $y = -2$

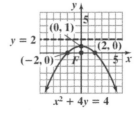

**31.** $x^2 = -4y + 4$
$x^2 = -4(y - 1)$
$4p = -4, p = -1$
vertex: $(0, 1)$
focus: $(0, 0)$
directrix: $y = 2$

**32.**
$$y^2 - 10y = 4x - 21$$
$$y^2 - 10y + 25 = 4x - 21 + 25$$
$$(y - 5)^2 = 4(x + 1)$$
$4p = 4, \ p = 1$
vertex: $(-1, 5)$
focus: $(0, 5)$
directrix: $x = -2$

**33.**
$$x^2 - 4x - 2y = 0$$
$$x^2 - 4x = 2y$$
$$(x^2 - 4x + 4) = 2y + 4$$
$$(x - 2)^2 = 2(y + 2)$$
$4p = 2, \ p = \dfrac{1}{2}$
vertex: $(2, -2)$
focus: $\left(2, -\dfrac{3}{2}\right)$
directrix: $y = -\dfrac{5}{2}$

**34.** $p = 12$
$y^2 = 48x$

**35.** $p = -11$
$x^2 = -44y$

**36.**
$$x^2 = 4py$$
$$(6)^2 = 4p(3)$$
$$p = 3$$
$$x^2 = 12y$$
Place the light 3 inches from the vertex at $(0, 3)$.

**37.**
$$x^2 = 4py$$
$$(1750)^2 = 4p(316)$$
$$4p \approx 9691$$
$$x^2 = 9691y$$
Let $x = 1750 - 1000 = 750$.
$$y = \frac{x^2}{9691} = \frac{(750)^2}{9691} \approx 58$$
The height is approximately 58 feet.

**38.** $x^2 = 4py$
$(150)^2 = 4p(44)$
$22{,}500 = 176p$
$p \approx 128$
The receiver should be placed approximately 128 feet from the base of the dish.

## Chapter 7 Test

**1.** $\dfrac{x^2}{4} - \dfrac{y^2}{9} = 1$
$c^2 = a^2 + b^2 = 4 + 9 = 13, \ c = \sqrt{13}$
hyperbola
The foci are at $\left(\sqrt{13}, 0\right)$ and $\left(-\sqrt{13}, 0\right)$.

Asymptotes: $y = \pm\dfrac{3}{2}x$

**2.** $4p = -8, \ p = -2$
parabola
vertex: $(0, 0)$
focus: $(0, -2)$
directrix: $y = 2$

**3.** The center is at $(-2, 5)$.
$c^2 = a^2 - b^2 = 25 - 9 = 16$, $c = 4$
ellipse
The foci are at $(-6, 5)$ and $(2, 5)$.

$$\frac{(x+2)^2}{25} + \frac{(y-5)^2}{9} = 1$$

**4.**
$$4x^2 - y^2 + 8x + 2y + 7 = 0$$
$$\left(4x^2 + 8x\right) - \left(y^2 - 2y\right) = -7$$
$$4\left(x^2 + 2x + 1\right) - \left(y^2 - 2y + 1\right) = -7 + 4 - 1$$
$$4(x+1)^2 - (y-1)^2 = -4$$
$$(y-1)^2 - 4(x+1)^2 = 4$$
$$\frac{(y-1)^2}{4} - (x+1)^2 = 1$$
$c^2 = a^2 + b^2 = 4 + 1 = 5$, $c = \sqrt{5}$
The center is at $(-1, 1)$.
Asymptotes: $y - 1 = \pm 2(x+1)$
hyperbola
The foci are at $\left(-1, 1 + \sqrt{5}\right)$ and $\left(-1, 1 - \sqrt{5}\right)$.

$4x^2 - y^2 + 8x + 2y + 7 = 0$

**5.** $4p = 8$, $p = 2$
parabola
vertex: $(-5, 1)$
focus: $(-5, 3)$
directrix: $y = -1$

$(x + 5)^2 = 8(y - 1)$

**6.** $c = 7$, $c^2 = 49$
$a = 10$, $a^2 = 100$
$b^2 = a^2 - c^2 = 100 - 49 = 51$
$$\frac{x^2}{100} + \frac{y^2}{51} = 1$$

**7.** $c = 10$, $c^2 = 100$
$a = 7$, $a^2 = 49$
$b^2 = c^2 - a^2 = 100 - 49 = 51$
$$\frac{y^2}{49} - \frac{x^2}{51} = 1$$

**8.** $p = 50$
$y^2 = 4px$
$y^2 = 200x$

**9.** $b = 24$, $b^2 = 576$
$2a = 80$, $a = 40$, $a^2 = 1600$
$c^2 = a^2 - b^2 = 1600 - 576 = 1024$
$c = \sqrt{1024} = 32$
The two people should each stand 32 feet from the center of the room, along the major axis.

**10. a.** $x^2 = 4py$
when $x = \pm 3$, $y = 3$
$9 = 4p(3)$
$3 = 4p$
$$\frac{3}{4} = p$$
$x^2 = 3y$

**b.** focus: $\left(0, \frac{3}{4}\right)$

The light is placed $\frac{3}{4}$ inch above the vertex.

**Cumulative Review Exercises (Chapters 1–7)**

**1.**   $2(x-3)+5x=8(x-1)$
$2x-6+5x=8x-8$
$7x-6=8x-8$
$-x=-2$
$x=2$
The solution set is $\{2\}$.

**2.**   $-3(2x-4)>2(6x-12)$
$-6x+12>12x-24$
$-18x>-36$
$x<2$
The solution set is $\{x\,|\,x<2\}$ or $(-\infty,2)$.

**3.**   $x-5=\sqrt{x+7}$
$(x-5)^2=x+7$
$x^2-10x+25=x+7$
$x^2-11x+18=0$
$(x-2)(x-9)=0$
$x=2$ or $x=9$
The solution $x=2$ is extraneous, so the only solution is $x=9$.
The solution set is $\{9\}$.

**4.**   $(x-2)^2=20$
$x-2=\pm\sqrt{20}$
$x-2=\pm2\sqrt5$
$x=2\pm2\sqrt5$
The solution set is $\{2\pm2\sqrt5\}$.

**5.**   $|2x-1|\ge7$
$2x-1\ge7$  or  $2x-1\le-7$
$2x\ge8$        $2x\le-6$
$x\ge4$  or   $x\le-3$
The solution set is $\{x\,|\,x\le-3\text{ or }x\ge4\}$ or $(-\infty,-3]\cup[4,\infty)$.

**6.**   $3x^3+4x^2-7x+2=0$
$p:\pm1,\pm2$
$q:\pm1,\pm3$
$\dfrac{p}{q}:\pm1,\pm2,\pm\dfrac13,\pm\dfrac23$
Let $f(x)=3x^3+4x^2-7x+2$.
Evaluate $f$ at the possible rational zeros to find
$f\left(\dfrac23\right)=0$.

$$\begin{array}{c|cccc}
\frac23 & 3 & 4 & -7 & 2\\
 & & 2 & 4 & -2\\
\hline
 & 3 & 6 & -3 & 0
\end{array}$$

$\left(x-\dfrac23\right)(3x^2+6x-3)=0$
$(3x-2)(x^2+2x-1)=0$
$x=\dfrac23$ or $x=\dfrac{-2\pm\sqrt{(2)^2-4(1)(-1)}}{2}$
$x=\dfrac{-2\pm\sqrt8}{2}$
$x=-1\pm\sqrt2$
The solution set is $\left\{\dfrac23,\,-1\pm\sqrt2\right\}$.

**7.**   $\log_2(x+1)+\log_2(x-1)=3$
$\log_2(x^2-1)=3$
$x^2-1=2^3$
$x^2=9$
$x=\pm3$
$x=-3$ is not a solution of the original equation. The solution set is $\{3\}$.

**8.**   $3x+4y=2$
$2x+5y=-1$
$6x+8y=4$
$-6x-15y=3$
$\overline{\qquad-7y=7}$
$y=-1$
$3x+4(-1)=2$
$3x=6$
$x=2$
The solution set is $\{(2,-1)\}$.

**9.**  $2x^2 - y^2 = -8$

$x - y = 6$

$x - y = 6$

$x = y + 6$

$x^2 = (y+6)^2 = y^2 + 12y + 36$

Substitute into first equation.

$2(y^2 + 12y + 36) - y^2 = -8$

$2y^2 + 24y + 72 - y^2 = -8$

$y^2 + 24y + 80 = 0$

$(y + 4)(y + 20) = 0$

$y = -4$ or $y = -20$

$x = 2 \qquad x = -14$

The solution set is $\{(2, -4), (-14, -20)\}$.

**10.** Set up the augmented matrix and use Gauss-Jordan reduction.

$$\begin{bmatrix} 1 & -1 & 1 & \bigm| & 17 \\ -4 & 1 & 5 & \bigm| & -2 \\ 2 & 3 & 1 & \bigm| & 8 \end{bmatrix}$$

$$\begin{bmatrix} 1 & -1 & 1 & \bigm| & 17 \\ 0 & -3 & 9 & \bigm| & 66 \\ 0 & 5 & -1 & \bigm| & -26 \end{bmatrix} \begin{matrix} \\ 4R_1 + R_2 \\ -2R_1 + R_3 \end{matrix}$$

$$\begin{bmatrix} 1 & -1 & 1 & \bigm| & 17 \\ 0 & 1 & -3 & \bigm| & -22 \\ 0 & 5 & -1 & \bigm| & -26 \end{bmatrix} -\frac{1}{3} R_2$$

$$\begin{bmatrix} 1 & 0 & -2 & \bigm| & -5 \\ 0 & 1 & -3 & \bigm| & -22 \\ 0 & 0 & 14 & \bigm| & 84 \end{bmatrix} \begin{matrix} R_2 + R_1 \\ \\ -5R_2 + R_3 \end{matrix}$$

$$\begin{bmatrix} 1 & 0 & -2 & \bigm| & -5 \\ 0 & 1 & -3 & \bigm| & -22 \\ 0 & 0 & 1 & \bigm| & 6 \end{bmatrix} \frac{1}{14} R_3$$

$$\begin{bmatrix} 1 & 0 & 0 & \bigm| & 7 \\ 0 & 1 & 0 & \bigm| & -4 \\ 0 & 0 & 1 & \bigm| & 6 \end{bmatrix} \begin{matrix} 2R_3 + R_1 \\ 3R_3 + R_2 \end{matrix}$$

$x = 7, y = -4, z = 6$

The solution set is $\{(7, -4, 6)\}$.

**11.** Parabola with vertex at $(1, -4)$.

$f(x) = (x - 1)^2 - 4$

**12.** Ellipse with center at $(0, 0)$ and vertices at $(3, 0)$ and $(-3, 0)$.

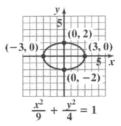

$$\frac{x^2}{9} + \frac{y^2}{4} = 1$$

**13.**  $5x + y \le 10 \qquad\qquad y \ge \frac{1}{4}x + 2$

$y \le -5x + 10$

Graph with solid line $y = -5x + 10$ and $y = \frac{1}{4}x + 2$. Shade the region that is below the line $y = -5x + 10$ and above the line $y = \frac{1}{4}x + 2$. Then dash the solid lines that do not contain the solution set.

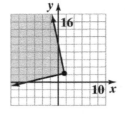

**14. a.**  $p: \pm 1, \pm 3$

$q: \pm 1, \pm 2, \pm 4, \pm 8, \pm 16, \pm 32$

$\dfrac{p}{q}: \pm 1, \pm 3, \pm \dfrac{1}{2}, \pm \dfrac{3}{2}, \pm \dfrac{1}{4}, \pm \dfrac{3}{4}, \pm \dfrac{1}{8},$

$\pm \dfrac{3}{8}, \pm \dfrac{1}{16}, \pm \dfrac{3}{16}, \pm \dfrac{1}{32}, \pm \dfrac{3}{32}$

**b.** $x = 1$ appears to be a root.

$$
\begin{array}{r|rrrr}
1 & 32 & -52 & 17 & 3 \\
  &    & 32 & -20 & -3 \\
\hline
  & 32 & -20 & -3 & 0
\end{array}
$$

$$32x^3 - 52x^2 + 17x + 3 = 0$$
$$(x-1)(32x^2 - 20x - 3) = 0$$
$$(x-1)(4x-3)(8x+1) = 0$$
$$x = 1 \text{ or } x = \frac{3}{4} \text{ or } x = -\frac{1}{8}$$

The solution set is $\left\{ -\dfrac{1}{8}, \dfrac{3}{4}, 1 \right\}$.

**15. a.** domain: $(-2, 2)$

range: $[-3, \infty)$

**b.** the relative minimum of $-3$ occurs at $x = 0$.

**c.** increasing: $(0, 2)$

**d.** $f(-1) - f(0) = 0 - (-3) = 3$

**e.** $(f \circ f)(1) = f\big(f(1)\big) = f(0) = -3$

**f.** $f(x) \to \infty$ as $x \to -2^{+}$ or as $x \to 2^{-}$

**g.**

$$g(x) = f(x-2) + 1$$

**h.**

$$h(x) = -f(2x)$$

**16.** $f(x) = x^2 - 4,\ g(x) = x + 2$

$$(g \circ f)(x) = g\big(x^2 - 4\big) = \big(x^2 - 4\big) + 2 = x^2 - 2$$

**17.** $\log_5 \dfrac{x^3 \sqrt{y}}{125} = \log_5 x^3 \sqrt{y} - \log_5 125$

$$= \log_5 x^3 + \log_5 \sqrt{y} - 3$$

$$= 3\log_5 x + \frac{1}{2}\log_5 y - 3$$

**18.** $m = \dfrac{y_2 - y_1}{x_2 - x_1} = \dfrac{8 - (-4)}{-5 - 1} = \dfrac{12}{-6} = -2$

$$y - y_1 = m(x - x_1)$$
$$y + 4 = -2(x - 1)$$
$$y + 4 = -2x + 2$$
$$y = -2x - 2$$

**19.** Let $R$ = the cost of a rental at Rent-a-Truck and let $A$ = the cost of a rental at Ace Truck Rentals.
$R = 39 + 0.16m$
$A = 25 + 0.24m$
where $m$ is the number of miles.
$$39 + 0.16m = 25 + 0.24m$$
$$14 = 0.08m$$
$$m = 175$$
$R = 39 + 0.16(175) = 67$
The cost will be the same when the number of miles driven is 175 miles. The cost will be \$67.

**20.** Let $x$ = cost of basic cable,
Let $y$ = cost of movie channel.
$$x + y = 35$$
$$x + 2y = 45$$
Multiply the first equation by $-1$ and then add the two equations.
$$-x - y = -35$$
$$\underline{x + 2y = 45}$$
$$y = 10$$
Use back-substitution to find $x$.
$$x + 10 = 35$$
$$x = 25$$
Basic cable costs \$25 and one movie channel costs \$10.

# Chapter 8
## Sequences, Induction, and Probability

**Check Point Exercises**

**1.** **a.** $a_n = 2n+5$

$a_1 = 2(1)+5 = 7$

$a_2 = 2(2)+5 = 9$

$a_3 = 2(3)+5 = 11$

$a_4 = 2(4)+5 = 13$

The first four terms are 7, 9, 11, and 13.

$a_n = \dfrac{(-1)^n}{2^n+1}$

$a_1 = \dfrac{(-1)^1}{2^1+1} = = \dfrac{-1}{3} - \dfrac{1}{3}$

$a_2 = \dfrac{(-1)^2}{2^2+1} = \dfrac{1}{5}$

$a_3 = \dfrac{(-1)^3}{2^3+1} = \dfrac{-1}{9} = -\dfrac{1}{9}$

**b.** $a_4 = \dfrac{(-1)^4}{2^4+1} = \dfrac{1}{17}$

The first four terms are $-\frac{1}{3}, \frac{1}{5}, -\frac{1}{9},$ and $\frac{1}{17}$.

**2.** $a_1 = 3$ and $a_n = 2a_{n-1}+5$ for $n \ge 2$

$a_2 = 2a_1+5$

$\quad = 2(3)+5 = 11$

$a_3 = 2a_2+5$

$\quad = 2(11)+5 = 27$

$a_4 = 2a_3+5$

$\quad = 2(27)+5 = 59$

The first four terms are 3, 11, 27, and 59.

**3.** $a_n = \dfrac{20}{(n+1)!}$

$a_1 = \dfrac{20}{(1+1)!} = \dfrac{20}{2!} = 10$

$a_2 = \dfrac{20}{(2+1)!} = \dfrac{20}{3!} = \dfrac{20}{6} = \dfrac{10}{3}$

$a_3 = \dfrac{20}{(3+1)!} = \dfrac{20}{4!} = \dfrac{20}{24} = \dfrac{5}{6}$

$a_4 = \dfrac{20}{(4+1)!} = \dfrac{20}{5!} = \dfrac{20}{120} = \dfrac{1}{6}$

The first four terms are 10, $\frac{10}{3}, \frac{5}{6},$ and $\frac{1}{6}$.

**4.** **a.** $\dfrac{14!}{2!\,12!} = \dfrac{14 \cdot 13 \cdot 12!}{2!\,12!} = \dfrac{14 \cdot 13}{2 \cdot 1} = 91$

**b.** $\dfrac{n!}{(n-1)!} = \dfrac{n \cdot (n-1)!}{(n-1)!} = n$

**5.** **a.** $\displaystyle\sum_{i=1}^{6} 2i^2$

$= 2(1)^2 + 2(2)^2 + 2(3)^2$

$\quad + 2(4)^2 + 2(5)^2 + 2(6)^2$

$= 2+8+18+32+50+72$

$= 182$

**b.** $\displaystyle\sum_{k=3}^{5}\left(2^k - 3\right)$

$= \left(2^3 - 3\right) + \left(2^4 - 3\right) + \left(2^5 - 3\right)$

$= (8-3) + (16-3) + (32-3)$

$= 5+13+29$

$= 47$

**c.** $\displaystyle\sum_{i=1}^{5} 4$

$= 4+4+4+4+4$

$= 20$

**6.** **a.** The sum has nine terms, each of the form $i^2$, starting at $i=1$ and ending at $i=9$.

$1^2 + 2^2 + 3^2 + \cdots + 9^2 = \displaystyle\sum_{i=1}^{9} i^2$

**b.** The sum has $n$ terms, each of the form $\frac{1}{2^{i-1}}$, starting at $i=1$ and ending at $i=n$.

$1 + \dfrac{1}{2} + \dfrac{1}{4} + \dfrac{1}{8} + \cdots + \dfrac{1}{2^{n-1}} = \displaystyle\sum_{i=1}^{n} \dfrac{1}{2^{i-1}}$

## Concept and Vocabulary Check 8.1

1. sequence; integers; terms

2. general

3. 4

4. $\dfrac{1}{15}$

5. 2

6. factorial; 5; 1; 1

7. $n+3$

8. $a_1$; $a_2$; $a_3$; $a_n$; index; upper limit; lower limit

## Exercise Set 8.1

1. $a_n = 3n+2$
   $a_1 = 3(1)+2 = 5$
   $a_2 = 3(2)+2 = 8$
   $a_3 = 3(3)+2 = 11$
   $a_4 = 3(4)+2 = 14$
   The first four terms are 5, 8, 11, and 14.

3. $a_n = 3^n$
   $a_1 = 3^1 = 3$
   $a_2 = 3^2 = 9$
   $a_3 = 3^3 = 27$
   $a_4 = 3^4 = 81$
   The first four terms are 3, 9, 27, and 81.

5. $a_n = (-3)^n$
   $a_1 = (-3)^1 = -3$
   $a_2 = (-3)^2 = 9$
   $a_3 = (-3)^3 = -27$
   $a_4 = (-3)^4 = 81$
   The first four terms are –3, 9, –27, and 81.

7. $a_n = (-1)^n (n+3)$
   $a_1 = (-1)^1 (1+3) = -4$
   $a_2 = (-1)^2 (2+3) = 5$
   $a_3 = (-1)^3 (3+3) = -6$
   $a_4 = (-1)^4 (4+3) = 7$
   The first four terms are –4, 5, –6, and 7.

9. $a_n = \dfrac{2n}{n+4}$
   $a_1 = \dfrac{2(1)}{1+4} = \dfrac{2}{5}$
   $a_2 = \dfrac{2(2)}{2+4} = \dfrac{4}{6} = \dfrac{2}{3}$
   $a_3 = \dfrac{2(3)}{3+4} = \dfrac{6}{7}$
   $a_4 = \dfrac{2(4)}{4+4} = \dfrac{8}{8} = 1$
   The first four terms are $\frac{2}{5}, \frac{2}{3}, \frac{6}{7},$ and $1$.

11. $a_n = \dfrac{(-1)^{n+1}}{2^n - 1}$
    $a_1 = \dfrac{(-1)^{1+1}}{2^1 - 1} = \dfrac{1}{1}$   $n = 1$
    $a_2 = \dfrac{(-1)^{2+1}}{2^2 - 1} = -\dfrac{1}{3}$
    $a_3 = \dfrac{(-1)^{3+1}}{2^3 - 1} = \dfrac{1}{7}$
    $a_4 = \dfrac{(-1)^{4+1}}{2^4 - 1} = -\dfrac{1}{15}$
    The first four terms are $1, -\frac{1}{3}, \frac{1}{7},$ and $-\frac{1}{15}$.

13. $a_1 = 7$ and $a_n = a_{n-1}+5$ for $n \geq 2$
    $a_2 = a_1 + 5 = 7+5 = 12$
    $a_3 = a_2 + 5 = 12+5 = 17$
    $a_4 = a_3 + 5 = 17+5 = 22$
    The first four terms are 7, 12, 17, and 22.

15. $a_1 = 3$ and $a_n = 4a_{n-1}$ for $n \geq 2$
    $a_2 = 4a_1 = 4(3) = 12$
    $a_3 = 4a_2 = 4(12) = 48$
    $a_4 = 4a_3 = 4(48) = 192$
    The first four terms are 3, 12, 48, and 192.

**17.** $a_1 = 4$ and $a_n = 2a_{n-1} + 3$

$a_2 = 2(4) + 3 = 11$

$a_3 = 2(11) + 3 = 25$

$a_4 = 2(25) + 3 = 53$

The first four terms are 4, 11, 25, and 53.

**19.** $a_n = \dfrac{n^2}{n!}$

$a_1 = \dfrac{1^2}{1!} = 1$

$a_2 = \dfrac{2^2}{2!} = 2$

$a_3 = \dfrac{3^2}{3!} = \dfrac{9}{6} = \dfrac{3}{2}$

$a_4 = \dfrac{4^2}{4!} = \dfrac{16}{24} = \dfrac{2}{3}$

The first four terms are 1, 2, $\frac{3}{2}$, and $\frac{2}{3}$.

**21.** $a_n = 2(n+1)!$

$a_1 = 2(1+1)! = 2(2) = 4$

$a_2 = 2(2+1)! = 2(6) = 12$

$a_3 = 2(3+1)! = 2(24) = 48$

$a_4 = 2(4+1)! = 2(120) = 240$

The first four terms are 4, 12, 48, and 240.

**23.** $\dfrac{17!}{15!} = \dfrac{17 \cdot 16 \cdot 15!}{15!} = 17 \cdot 16 = 272$

**25.** $\dfrac{16!}{2! \, 14!} = \dfrac{16 \cdot 15 \cdot 14!}{2! \, 14!} = \dfrac{16 \cdot 15}{2 \cdot 1} = \dfrac{8 \cdot 15}{1} = 120$

**27.** $\dfrac{(n+2)!}{n!} = \dfrac{(n+2)(n+1)n!}{n!} = (n+2)(n+1)$

**29.** $\displaystyle\sum_{i=1}^{6} 5i = 5\cdot1 + 5\cdot2 + 5\cdot3 + 5\cdot4 + 5\cdot5 + 5\cdot6$

$= 5 + 10 + 15 + 20 + 25 + 30$

$= 105$

**31.** $\displaystyle\sum_{i=1}^{4} 2i^2 = 2\cdot1^2 + 2\cdot2^2 + 2\cdot3^2 + 2\cdot4^2$

$= 2 + 8 + 18 + 32$

$= 60$

**33.** $\displaystyle\sum_{k=1}^{5} k(k+4) = 1(5) + 2(6) + 3(7) + 4(8) + 5(9)$

$= 5 + 12 + 21 + 32 + 45$

$= 115$

**35.** $\displaystyle\sum_{i=1}^{4}\left(\dfrac{-1}{2}\right)^i = \left(-\dfrac{1}{2}\right)^1 + \left(-\dfrac{1}{2}\right)^2 + \left(-\dfrac{1}{2}\right)^3 + \left(-\dfrac{1}{2}\right)^4$

$= -\dfrac{1}{2} + \dfrac{1}{4} - \dfrac{1}{8} + \dfrac{1}{16}$

$= -\dfrac{5}{16}$

**37.** $\displaystyle\sum_{i=5}^{9} 11 = 11 + 11 + 11 + 11 + 11 = 55$

**39.** $\displaystyle\sum_{i=0}^{4} \dfrac{(-1)^i}{i!}$

$= \dfrac{(-1)^0}{0!} + \dfrac{(-1)^1}{1!} + \dfrac{(-1)^2}{2!} + \dfrac{(-1)^3}{3!} + \dfrac{(-1)^4}{4!}$

$= 1 - 1 + \dfrac{1}{2} - \dfrac{1}{6} + \dfrac{1}{24}$

$= \dfrac{9}{24} = \dfrac{3}{8}$

**41.** $\displaystyle\sum_{i=1}^{5} \dfrac{i!}{(i-1)!} = \dfrac{1!}{0!} + \dfrac{2!}{1!} + \dfrac{3!}{2!} + \dfrac{4!}{3!} + \dfrac{5!}{4!}$

$= 1 + 2 + 3 + 4 + 5 = 15$

**43.** $1^2 + 2^2 + 3^2 + \cdots + 15^2 = \displaystyle\sum_{i=1}^{15} i^2$

**45.** $2 + 2^2 + 2^3 + 2^4 + \cdots + 2^{11} = \displaystyle\sum_{i=1}^{11} 2^i$

**47.** $1 + 2 + 3 + \cdots + 30 = \displaystyle\sum_{i=1}^{30} i$

**49.** $\dfrac{1}{2} + \dfrac{2}{3} + \dfrac{3}{4} + \cdots + \dfrac{14}{14+1} = \displaystyle\sum_{i=1}^{14} \dfrac{i}{i+1}$

**51.** $4 + \dfrac{4^2}{2} + \dfrac{4^3}{3} + \cdots + \dfrac{4^n}{n} = \displaystyle\sum_{i=1}^{n} \dfrac{4^i}{i}$

**53.** $1 + 3 + 5 + \cdots + (2n-1) = \displaystyle\sum_{i=1}^{n} (2i-1)$

**55.** $5 + 7 + 9 + \cdots + 31$

Possible answer: $\displaystyle\sum_{k=1}^{14} (2k+3)$

**57.** $a + ar + ar^2 + \cdots + ar^{12}$

Possible answer: $\displaystyle\sum_{k=0}^{12} ar^k$

**59.** $a + (a+d) + (a+2d) + \cdots + (a+nd)$

Possible answer: $\displaystyle\sum_{k=0}^{n}(a+kd)$

**61.** $\displaystyle\sum_{i=1}^{5}(a_i^2 + 1) = \left((-4)^2 + 1\right) + \left((-2)^2 + 1\right) + \left((0)^2 + 1\right) + \left((2)^2 + 1\right) + \left((4)^2 + 1\right)$

$$= 17 + 5 + 1 + 5 + 17$$
$$= 45$$

**63.** $\displaystyle\sum_{i=1}^{5}(2a_i + b_i) = \left(2(-4)+4\right) + \left(2(-2)+2\right) + \left(2(0)+0\right) + \left(2(2)+(-2)\right) + \left(2(4)+(-4)\right)$

$$= -4 + (-2) + 0 + 2 + 4 = 0$$

**65.** $\displaystyle\sum_{i=4}^{5}\left(\frac{a_i}{b_i}\right)^2 = \left(\frac{2}{-2}\right)^2 + \left(\frac{4}{-4}\right)^2 = (-1)^2 + (-1)^2 = 1 + 1 = 2$

**67.** $\displaystyle\sum_{i=1}^{5}a_i^2 + \sum_{i=1}^{5}b_i^2 = \left((-4)^2 + (-2)^2 + 0^2 + 2^2 + 4^2\right) + \left(4^2 + 2^2 + 0^2 + (-2)^2 + (-4)^2\right)$

$$= (16 + 4 + 0 + 4 + 16) + (16 + 4 + 0 + 4 + 16) = 80$$

**69.** **a.** $\displaystyle\sum_{i=1}^{8}a_i = 100 + 120 + 145 + 170 + 200 + 220 + 260 + 300 = 1515$

A total of 1515 thousand, or 1,515,000, autism cases were diagnosed in the United States from 2001 through 2008.

**b.** $\displaystyle\sum_{i=1}^{8}a_i = (28\cdot1+63) + (28\cdot2+63) + (28\cdot3+63) + (28\cdot4+63) + (28\cdot5+63) + (28\cdot6+63) + (28\cdot7+63) + (28\cdot8+63)$

$$= 91 + 119 + 147 + 175 + 203 + 231 + 259 + 287$$
$$= 1512$$

The model underestimates the actual sum by 3 thousand.

**71.** $a_n = 6000\left(1 + \dfrac{0.06}{4}\right)^n, n = 1, 2, 3, \cdots$

$a_{20} = 6000\left(1 + \dfrac{0.06}{4}\right)^{20} \approx 8081.13$

After five years, the balance is \$8081.13.

**73. – 79.** Answers will vary.

**81.** Most calculators give error message if the expression is entered directly.

However, $\dfrac{200!}{198!} = \dfrac{200\cdot199\cdot198!}{198!} = 200\cdot199 = 39{,}800$

**83.** $\dfrac{20!}{300} = 8{,}109{,}673{,}360{,}588{,}800$

However, most calculators give a rounded answer in scientific notation.

**85.** $\dfrac{54!}{(54-3)!3!} = 24,804$

**87.** Answers will vary.

**89.** $a_n = \dfrac{n}{n+1}$

As $n$ gets larger, $a_n$ approaches 1.

**91.** $a_n = \dfrac{2n^2 + 5n - 7}{n^3}$

As $n$ gets larger, $a_n$ approaches 0.

**93.** does not make sense; Explanations will vary. Sample explanation: There is nothing that implies that there is a negative number of sheep.

**95.** makes sense

**97.** false; Changes to make the statement true will vary. A sample change is: $\dfrac{n!}{(n-1)!} = \dfrac{n \cdot (n-1)!}{(n-1)!} = n$

**99.** false; Changes to make the statement true will vary. A sample change is:

$$\sum_{i=1}^{2} (-1)^i 2^i = (-1)^1 2^1 + (-1)^2 2^2 = -1(2) + 1(4) = -2 + 4 = 2$$

**101.** $a_n = \begin{cases} \dfrac{a_{n-1}}{2} & \text{if } a_{n-1} \text{ is even.} \\ 3a_n + 5 & \text{if } a_{n-1} \text{ is odd} \end{cases}$

      for $n \geq 2$.

$a_1 = 9$

Since 9 is odd, $a_2 = 3(9) + 5 = 32$.

Since 32 is even, $a_3 = \dfrac{32}{2} = 16$.

Similarly, $a_4 = \dfrac{16}{2} = 8$, $a_5 = \dfrac{8}{2} = 4$.

The first five terms of the sequence are 9, 32, 16, 8, and 4.

**103.** $a_2 - a_1 = 3 - 8 = -5$

$a_3 - a_2 = -2 - 3 = -5$

$a_4 - a_3 = -7 - (-2) = -5$

$a_5 - a_4 = -12 - (-7) = -5$

The difference between consecutive terms is always $-5$.

**104.** $a_2 - a_1 = (4(2) - 3) - (4(1) - 3) = 4$

$a_3 - a_2 = (4(3) - 3) - (4(2) - 3) = 4$

$a_4 - a_3 = (4(4) - 3) - (4(3) - 3) = 4$

$a_5 - a_4 = (4(5) - 3) - (4(4) - 3) = 4$

The difference between consecutive terms is always 4.

**105.** $a_n = 4 + (n - 1)(-7)$

$a_8 = 4 + (8 - 1)(-7) = 4 + (7)(-7) = 4 - 49 = -45$

## Section 8.2

### Check Point Exercises

**1.** $a_1 = 100$

$a_2 = a_1 - 30 = 100 - 30 = 70$

$a_3 = a_2 - 30 = 70 - 30 = 40$

$a_4 = a_3 - 30 = 40 - 30 = 10$

$a_5 = a_4 - 30 = 10 - 30 = -20$

$a_6 = a_5 - 30 = -20 - 30 = -50$

The first five terms are 100, 70, 40, 10, $-20$, $-50$.

**2.** $a_1 = 6, d = -5$

To find the ninth term, $a_9$, replace $n$ in the formula with 9, $a_1$ with 6, and $d$ with $-5$.

$a_n = a_1 + (n - 1)d$

$a_9 = 6 + (9 - 1)(-5)$

$= 6 + 8(-5)$

$= 6 + (-40)$

$= -34$

**3. a.** $a_n = a_1 + (n - 1)d$

$= 16 + (n - 1)0.35$

$= 0.35n + 15.65$

**b.** $a_n = 0.35n + 15.65$

$a_{21} = 0.35(21) + 15.65$

$= 23$

In 2030, it is projected that 23% of the U.S. population will be Latino.

**4.** 3, 6, 9, 12, ...

To find the sum of the first 15 terms, $S_{15}$, replace $n$ in the formula with 15.

$S_n = \frac{n}{2}(a_1 + a_n)$

$S_{15} = \frac{15}{2}(a_1 + a_{15})$

Use the formula for the general term of a sequence to find $a_{15}$. The common difference, $d$, is 3, and the first term, $a_1$, is 3.

$a_n = a_1 + (n - 1)d$

$a_{15} = 3 + (15 - 1)(3)$

$= 3 + 14(3)$

$= 3 + 42$

$= 45$

Thus, $S_{15} = \frac{15}{2}(3 + 45) = \frac{15}{2}(48) = 360$.

**5.** $\sum\limits_{i=1}^{30}(6i - 11) = (6 \cdot 1 - 11) + (6 \cdot 2 - 11)$

$+ (6 \cdot 3 - 11) + \ldots + (6 \cdot 30 - 11)$

$= -5 + 1 + 7 + \ldots + 169$

So the first term, $a_1$, is $-5$; the common difference, $d$, is $1 - (-5) = 6$; the last term, $a_{30}$, is 169.

Substitute $n = 30$, $a_1 = -5$, and $a_{30} = 169$ in the formula $S_n = \frac{n}{2}(a_1 + a_n)$.

$S_{30} = \frac{30}{2}(-5 + 169) = 15(164) = 2460$

Thus, $\sum\limits_{i=1}^{30}(6i - 11) = 2460$

**6.** $a_n = 1800n + 64,130$

$a_1 = 1800(1) + 64,130 = 65,930$

$a_{10} = 1800(10) + 64,130 = 82,130$

$S_n = \frac{n}{2}(a_1 + a_n)$

$S_{10} = \frac{10}{2}(a_1 + a_{10})$

$= 5(65,930 + 82,130)$

$= 5(148,060)$

$= \$740,300$

It would cost \$740,300 for the ten-year period beginning in 2014.

**Concept and Vocabulary Check 8.2**

1. arithmetic; common difference

2. $a_1 + (n-1)d$ ; first term; common difference

3. $\frac{n}{2}(a_1 + a_2)$ ; first term; $n$th term

4. 2; 116

5. 8; 13; 18; 5

**Exercise Set 8.2**

1. $a_1 = 200, \ d = 20$
   The first six terms are 200, 220, 240, 260, 280, and 300.

3. $a_1 = -7, \ d = 4$
   The first six terms are –7, –3, 1, 5, 9, and 13.

5. $a_1 = 300, \ d = -90$
   The first six terms are 300, 210, 120, 30, –60, and –150.

7. $a_1 = \dfrac{5}{2}, d = -\dfrac{1}{2}$

   The first six terms are $\dfrac{5}{2}, 2, \dfrac{3}{2}, 1, \dfrac{1}{2},$ and $0$.

9. $a_n = a_{n-1} + 6, \ a_1 = -9$
   The first six terms are –9, –3, 3, 9, 15, and 21.

11. $a_n = a_{n-1} - 10, \ a_1 = 30$
    The first six terms are 30, 20, 10, 0, –10, and –20.

13. $a_n = a_{n-1} - 0.4, \ a_1 = 1.6$
    The first six terms are 1.6, 1.2, 0.8, 0.4, 0, and –0.4.

15. $a_1 = 13, \ d = -4$
    $a_n = 13 + (n-1)4$
    $a_6 = 13 + 5(4) = 13 + 20 = 33$

17. $a_1 = 7, \ d = 5$
    $a_n = 7 + (n-1)2$
    $a_{50} = 7 + 49(5) = 252$

19. $a_1 = -40, \ d = 5$
    $a_n = -40 + (n-1)5$
    $a_{200} = -40 + (199)5 = 955$

21. $a_1 = 35, \ d = -3$
    $a_n = 35 - 3(n-1)$
    $a_{60} = 35 - 3(59) = -142$

23. 1, 5, 9, 13, ...
    $d = 5 - 1, \ = 4$
    $a_n = 1 + (n-1)4 = 1 + 4n - 4$
    $a_n = 4n - 3$
    $a_{20} = 4(20) - 3 = 77$

25. 7, 3, –1, –5, ...
    $d = 3 - 7 = -4$
    $a_n = 7 + (n-1)(-4) = 7 - 4n + 4$
    $a_n = 11 - 4n$
    $a_{20} = 11 - 4(20) = -69$

27. $a_1 = 9, \ d = 2$
    $a_n = 9 + (n-1)(2)$
    $a_n = 7 + 2n$
    $a_{20} = 7 + 2(20) = 47$

29. $a_1 = -20, \ d = -4$
    $a_n = -20 + (n-1)(-4)$
    $a_n = -20 - 4n + 4$
    $a_n = -16 - 4n$
    $a_{20} = -16 - 4(20) = -96$

31. $a_n = a_{n-1} + 3, \ a_1 = 4$
    $d = 3$
    $a_n = 4 + (n-1)(3)$
    $a_n = 1 + 3n$
    $a_{20} = 1 + 3(20) = 61$

33. $a_n = a_{n-1} - 10, \ a_1 = 30, \ d = -10$
    $a_n = 30 - 10(n-1) = 30 - 10n + 10$
    $a_n = 40 - 10n$
    $a_{20} = 40 - 10(20) = -160$

35. 4, 10, 16, 22, . . .
    $d = 10 - 4 = 6$
    $a_n = 4 + (n-1)(6)$
    $a_{20} = 4 + (19)(6) = 118$
    $S_{20} = \dfrac{20}{2}(4 + 118) = 1220$

**37.** $-10, -6, -2, 2, \ldots$

$$d = -6 - (-10) = -6 + 10 = 4$$

$$a_n = -10 + (n-1)4$$

$$a_{50} = -10 + (49)4 = 186$$

$$S_{50} = \frac{50}{2}(-10 + 186) = 4400$$

**39.** $1 + 2 + 3 + 4 + \cdots + 100$

$$S_{100} = \frac{100}{2}(1 + 100) = 5050$$

**41.** $2 + 4 + 6 + \cdots + 120$

$$S_{60} = \frac{60}{2}(2 + 120) = 3660$$

**43.** even integers between 21 and 45;
$22 + 24 + 26 + \cdots + 44$

$$S_{12} = \frac{12}{2}(22 + 44) = 396$$

**45.** $\sum\limits_{i=1}^{17}(5i + 3) = (5 + 3) + (10 + 3) + (15 + 3) + \cdots + (85 + 3) = 8 + 13 + 18 + \cdots + 88$

$$S_{17} = \frac{17}{2}(8 + 88) = 816$$

**47.** $\sum\limits_{i=1}^{30}(-3i + 5) = (-3 + 5) + (-6 + 5) + (-9 + 5) + \cdots + (-90 + 5) = 2 - 1 - 4 - \cdots - 85$

$$S_{30} = \frac{30}{2}(2 - 85) = -1245$$

**49.** $\sum\limits_{i=1}^{100} 4i = 4 + 8 + 12 + \cdots + 400$

$$S_{100} = \frac{100}{2}(4 + 400) = 20,200$$

**51.** First find $a_{14}$ and $b_{12}$ :

$$a_{14} = a_1 + (n-1)d$$
$$= 1 + (14 - 1)(-3 - 1) = -51$$
$$b_{12} = b_1 + (n-1)d$$
$$= 3 + (12 - 1)(8 - 3) = 58$$

So, $a_{14} + b_{12} = -51 + 58 = 7$ .

**53.** $a_n = a_1 + (n-1)d$

$-83 = 1 + (n-1)(-3-1)$

$-83 = 1 + -4(n-1)$

$-84 = -4n + 4$

$-88 = -4n$

$n = 22$

There are 22 terms.

**55.** $S_n = \dfrac{n}{2}(a_1 + a_n)$

For $\{a_n\}$: $S_{14} = \dfrac{14}{2}(a_1 + a_{14}) = 7(1 + (-51)) = -350$ For

$\{b_n\}$: $S_{14} = \dfrac{14}{2}(b_1 + b_{14}) = 7(3 + 68) = 497$

So $\displaystyle\sum_{n=1}^{14} b_n - \sum_{n=1}^{14} a_n = 497 - (-350) = 847$

**57.** Two points on the graph are $(1, 1)$ and $(2, -3)$.

Finding the slope of the line;

$m = \dfrac{y_2 - y_1}{x_2 - x_2} = \dfrac{-3-1}{2-1} = \dfrac{-4}{1} = -4$

Using the point-slope form of an equation of a line;

$y - y_2 = m(x - x_2)$

$y - 1 = -4(x - 1)$

$y - 1 = -4x + 4$

$y = -4x + 5$

Thus, $f(x) = -4x + 5$.

**59.** Using $a_n = a_1 + (n-1)d$ and $a_2 = 4$:

$a_2 = a_1 + (2-1)d$

$4 = a_1 + d$

And since $a_6 = 16$:

$a_6 = a_1 + (6-1)d$

$16 = a_1 + 5d$

The system of equations is

$4 = a_1 + d$

$16 = a_1 + 5d$

Solving the first equation for $a_1$:

$a_1 = 4 - d$

Substituting the value into the second equation and solving for $d$:

$16 = (4 - d) + 5d$

$16 = 4 + 4d$

$12 = 4d$

$3 = d$

Back-substitute:

$a_1 = 4 - d$

$a_1 = 4 - 3)$

$a_1 = 1$

Then $a_n = a_1 + (n-1)d$

$a_n = 1 + (n-1)3$

$a_n = 1 + 3n - 3$

$a_n = 3n - 2$

**61.** **a.** $a_n = a_1 + (n-1)d$

$a_n = 11.0 + (n-1)0.5$

$= 11.0 + 0.5n - 0.5$

$= 0.5n + 10.5$

**b.** $a_n = 0.5n + 10.5$

$= 0.5(50) + 10.5$

$= 35.5$

The percentage is projected to be 35.5% in 2019.

**63.** Company *A*

$a_n = 24000 + (n-1)1600$

$= 24000 + 1600n - 1600$

$= 1600n + 22400$

$a_{10} = 1600(10) + 22400$

$= 16000 + 22400 = 38400$

Company *B*

$a_n = 28000 + (n-1)1000$

$= 28000 + 1000n - 1000$

$= 1000n + 27000$

$a_{10} = 1000(10) + 27000$

$= 10000 + 27000 = 37000$

Company *A* will pay $1400 more in year 10.

**65.** **a.** Total cost:

$\$5836 + \$6185 + \$6585 + \$7020 = \$25,626$

**b.** $a_1 = 395(1) + 5419 = 5814$

$a_4 = 395(4) + 5419 = 6999$

$S_n = \dfrac{n}{2}(a_1 + a_n)$

$S_4 = \dfrac{4}{2}(5814 + 6999) = 2(12,813) = \$25,626$

The model gives actual sum of $25,626 obtained in part (a).

**67.** Answers will vary.

**69.** Company A:

$a_n = 19{,}000 + (n-1)2600$

$a_{10} = 19{,}000 + (9)2600 = \$42{,}200$

$S_{10} = \dfrac{10}{2}(19000 + 42400) = \$307{,}000$

Company B:

$a_n = 27{,}000 + (n-1)1200$

$a_{10} = 27{,}000 + (9)1200 = \$37{,}800$

$S_{10} = \dfrac{10}{2}(27{,}000 + 37{,}800) = \$324{,}000$

Company B pays the greater total amount.

**71.** $a_n = 20 + (n+1)3$

$a_{38} = 20 + (37)3 = 131$

$S_{38} = \dfrac{38}{2}(20 + 131) = 2869$

The theater has 2869 seats.

**73. – 77.** Answers will vary.

**79.** makes sense

**81.** makes sense

**83.** Degree days: 23, 25, 27, …

$a_1 = 23,\ d = 2$

$a_{10} = 23 + 9(2) = 41$

$S_{10} = \dfrac{10}{2}(a_1 + a_{10})$

$S_{10} = \dfrac{10}{2}(23 + 41) = 320$

There are 320 degree-days.

**85.** $\dfrac{a_2}{a_1} = \dfrac{-2}{1} = -2$

$\dfrac{a_3}{a_2} = \dfrac{4}{-2} = -2$

$\dfrac{a_4}{a_3} = \dfrac{-8}{4} = -2$

$\dfrac{a_5}{a_4} = \dfrac{16}{-8} = -2$

The ratio of a term to the term that directly precedes it is always $-2$.

**86.** $\dfrac{a_2}{a_1} = \dfrac{3\cdot 5^2}{3\cdot 5^1} = 5$

$\dfrac{a_3}{a_2} = \dfrac{3\cdot 5^3}{3\cdot 5^2} = 5$

$\dfrac{a_4}{a_3} = \dfrac{3\cdot 5^4}{3\cdot 5^3} = 5$

$\dfrac{a_5}{a_4} = \dfrac{3\cdot 5^5}{3\cdot 5^4} = 5$

The ratio of a term to the term that directly precedes it is always 5.

**87.** $a_n = a_1 3^{n-1}$

$a_7 = 11\cdot 3^{7-1} = 11\cdot 3^6 = 11\cdot 729 = 8019$

## Section 8.3

### Check Point Exercises

**1.** $a_1 = 12,\ r = \dfrac{1}{2}$

$a_2 = 12\left(\dfrac{1}{2}\right)^1 = 6$

$a_3 = 12\left(\dfrac{1}{2}\right)^2 = \dfrac{12}{4} = 3$

$a_4 = 12\left(\dfrac{1}{2}\right)^3 = \dfrac{12}{8} = \dfrac{3}{2}$

$a_5 = 12\left(\dfrac{1}{2}\right)^4 = \dfrac{12}{16} = \dfrac{3}{4}$

$a_6 = 12\left(\dfrac{1}{2}\right)^5 = \dfrac{12}{32} = \dfrac{3}{8}$

The first six terms are $12,\ 6,\ 3,\ \dfrac{3}{2},\ \dfrac{3}{4},$ and $\dfrac{3}{8}$.

**2.** $a_1 = 5,\ r = -3$

$a_n = 5r^{n-1}$

$a_7 = 5(-3)^{7-1} = 5(-3)^6 = 5(729) = 3645$

The seventh term is 3645.

**3.** 3, 6, 12, 24, 48, ...

$r = \dfrac{6}{3} = 2,\ a_1 = 3$

$a_n = 3(2)^{n-1}$

$a_8 = 3(2)^{8-1} = 3(2)^7 = 3(128) = 384$

The eighth term is 384.

**4.** $a_1 = 2,\ r = \dfrac{-6}{2} = -3$

$S_n = \dfrac{a_1(1 - r^r)}{1 - r}$

$S_9 = \dfrac{2\left(1 - (-3)^9\right)}{1 - (-3)} = \dfrac{2(19,684)}{4} = 9842$

The sum of the first nine terms is 9842.

**5.** $\displaystyle\sum_{i=1}^{8} 2 \cdot 3^i$

$a_1 = 2 \cdot (3)^1 = 6,\ r = 3$

$S_n = \dfrac{a_1(1 - r^n)}{1 - r}$

$S_8 = \dfrac{6\left(1 - 3^8\right)}{1 - 3} = \dfrac{6(-6560)}{-2} = 19,680$

Thus, $\displaystyle\sum_{i=1}^{8} 2 \cdot 3^i = 19,680$.

**6.** $a_1 = 30,000,\ r = 1.06$

$S_n = \dfrac{a_1(1 - r^n)}{1 - r}$

$S_{30} = \dfrac{30,000\left(1 - (1.06)^{30}\right)}{1 - 1.06} \approx 2,371,746$

The total lifetime salary is \$2,371,746.

**7.** **a.** $A = \dfrac{P\left[\left(1 + \frac{r}{n}\right)^{nt} - 1\right]}{\frac{r}{n}}$

$P = 100,\ r = 0.095,\ n = 12,\ t = 35$

$A = \dfrac{100\left[\left(1 + \dfrac{0.095}{12}\right)^{12 \cdot 35} - 1\right]}{\dfrac{0.095}{12}} \approx 333,946$

The value of the IRA will be \$333,946.

**b.** Interest = Value of IRA − Total deposits

$\approx \$333,946 - \$100 \cdot 12 \cdot 35$

$\approx \$333,946 - \$42,000$

$\approx \$291,946$

**8.** $3 + 2 + \dfrac{4}{3} + \dfrac{8}{9} + \cdots$

$a_1 = 3,\ r = \dfrac{2}{3}$

$S = \dfrac{a_1}{1 - r}$

$S = \dfrac{3}{1 - \frac{2}{3}} = \dfrac{3}{\frac{1}{3}} = 9$

The sum of this infinite geometric series is 9.

**9.** $0.\overline{9} = 0.9999\cdots = \dfrac{9}{10} + \dfrac{9}{100} + \dfrac{9}{1000} + \cdots$

$a_1 = \dfrac{9}{10},\ r = \dfrac{1}{10}$

$S = \dfrac{\frac{9}{10}}{1 - \frac{1}{10}} = \dfrac{\frac{9}{10}}{\frac{9}{10}} = 1$

An equivalent fraction for $0.\overline{9}$ is 1.

**10.** $a_1 = 1000(0.8) = 800,\ r = 0.8$

$S = \dfrac{800}{1 - 0.8} = 4000$

The total amount spent is \$4000.

**Concept and Vocabulary Check 8.3**

**1.** geometric; common ratio

**2.** $a_1 r^{n-1}$; first term; common ratio

**3.** $\dfrac{a_1(1 - r^n)}{1 - r}$; first term; common ratio

**4.** annuity; $P$; $r$; $n$

**5.** infinite geometric series; 1; $\dfrac{a_1}{1 - r}$; $|r| \geq 1$

**6.** 2; 4; 8; 16; 2

**7.** arithmetic

**8.** geometric

**9.** geometric

**10.** arithmetic

**Exercise Set 8.3**

**1.** $a_1 = 5,\ r = 3$
First five terms: 5, 15, 45, 135, 405.

**3.** $a_1 = 20,\ r = \dfrac{1}{2}$
First five terms: 20, 10, 5, $\frac{5}{2}$, $\frac{5}{4}$ $\frac{5}{4}$.

**5.** $a_n = -4a_{n-1},\ a_1 = 10$
First five terms: 10, −40, 160, −640, 2560.

**7.** $a_n = -5a_{n-1},\ a_1 = -6$
First five terms: −6, 30, −150, 750, −3750.

**9.** $a_1 = 6,\ r = 2$
$a_n = 6 \cdot 2^{n-1}$
$a_8 = 6 \cdot 2^7 = 768$

**11.** $a_1 = 5,\ r = -2$
$a_n = 5 \cdot (-2)^{n-1}$
$a_{12} = 5 \cdot (-2)^{11} = -10,240$

**13.** $a_1 = 1000,\ r = -\dfrac{1}{2}$
$a_n = 1000\left(-\dfrac{1}{2}\right)^{n-1}$
$a_{40} = 1000\left(-\dfrac{1}{2}\right)^{39}$
$\approx 0.000000002$

**15.** $a_1 = 1,000,000,\ r = 0.1$
$a_n = 1,000,000(0.1)^{n-1}$
$a_8 = 1,000,000(0.1)^7 = 0.1$

**17.** 3, 12, 48, 192, . . .
$r = \dfrac{12}{3} = 4$
$a_n = 3(4)^{n-1}$
$a_7 = 3(4)^6 = 12,288$

**19.** $19, 6, 2, \dfrac{2}{3}, \cdots$    $r = \dfrac{6}{18} = \dfrac{1}{3}$
$a_n = 18\left(\dfrac{1}{3}\right)^{n-1}$
$a_7 = 18\left(\dfrac{1}{3}\right)^6 = \dfrac{2}{81}$

**21.** 1.5, −3, 6, −12, . . .
$r = \dfrac{6}{-3} = -2$
$a_n = 1.5(-2)^{n-1}$
$a_7 = 1.5(-2)^6 = 96$

**23.** 0.0004, −0.004, 0.04, −0.4, . . .
$r = \dfrac{-0.004}{0.0004} = -10$
$a_n = 0.0004(-10)^{n-1}$
$a_7 = 0.0004(-10)^6 = 400$

**25.** 2, 6, 18, 54, . . .
$r = \dfrac{6}{2} = 3$
$S_{12} = \dfrac{2\left(1 - 3^{12}\right)}{1 - 3} = \dfrac{2(-531,440)}{-2} = 531,440$

**27.** 3, −6, 12, −24, . . .
$r = \dfrac{-6}{3} = -2$
$S_{11} = \dfrac{3\left[1 - (-2)^{11}\right]}{1 - (-2)} = \dfrac{3(2049)}{3} = 2049$

**29.** $-\dfrac{3}{2}, 3, -6, 12, \cdots$
$r = \dfrac{3}{\frac{-3}{2}} = -2$
$S_{14} = \dfrac{-\dfrac{3}{2}\left[1 - (-2)^{14}\right]}{1 - (-2)} = \dfrac{-\dfrac{3}{2}(-16,383)}{3} = \dfrac{16,383}{2}$

**31.** $\displaystyle\sum_{i=1}^{8} 3^i$
$r = 3,\ a_1 = 3$
$S_8 = \dfrac{3\left(1 - 3^8\right)}{1 - 3} = \dfrac{3(-6560)}{-2} = 9840$

**33.** $\displaystyle\sum_{i=1}^{10} 5 \cdot 2^i$

$r = 2, \quad a_1 = 10$

$S_{10} = \dfrac{10\left(1 - 2^{10}\right)}{1 - 2} = \dfrac{10(-1023)}{-1} = 10{,}230$

**35.** $\displaystyle\sum_{i=1}^{6} \left(\dfrac{1}{2}\right)^{i+1}$

$r = \dfrac{1}{2}, \quad a_1 = \dfrac{1}{4}$

$S_6 = \dfrac{\dfrac{1}{4}\left(1 - \left(\dfrac{1}{2}\right)^6\right)}{1 - \dfrac{1}{2}} = \dfrac{\dfrac{1}{4}\left(\dfrac{63}{64}\right)}{\dfrac{1}{2}} = \dfrac{63}{128}$

**37.** $r = \dfrac{1}{3}$

$S_\infty = \dfrac{1}{1 - \dfrac{1}{3}} = \dfrac{1}{\dfrac{2}{3}} = \dfrac{3}{2}$

**39.** $r = \dfrac{1}{4}$

$S_\infty = \dfrac{3}{1 - \dfrac{1}{4}} = \dfrac{3}{\dfrac{3}{4}} = 4$

**41.** $r = -\dfrac{1}{2}$

$S_\infty = \dfrac{1}{1 - \left(-\dfrac{1}{2}\right)} = \dfrac{1}{\dfrac{3}{2}} = \dfrac{2}{3}$

**43.** $r = -0.3$

$S_\infty = \dfrac{8}{1 - (-0.3)} = \dfrac{8}{1.3} \approx 6.15385$

**45.** $r = \dfrac{1}{10}$

$S_\infty = \dfrac{\dfrac{5}{10}}{1 - \dfrac{1}{10}} = \dfrac{\dfrac{5}{10}}{\dfrac{9}{10}} = \dfrac{5}{9}$

**47.** $r = \dfrac{1}{100}$

$S_\infty = \dfrac{\dfrac{47}{100}}{1 - \dfrac{1}{100}} = \dfrac{\dfrac{47}{100}}{\dfrac{99}{100}} = \dfrac{47}{99}$

**49.** $0.\overline{257} = \dfrac{257}{1000} + \dfrac{257}{10^6} + \dfrac{257}{10^9} + \cdots$

$r = \dfrac{1}{1000}$

$S_\infty = \dfrac{\dfrac{257}{1000}}{1 - \dfrac{1}{1000}} = \dfrac{\dfrac{257}{1000}}{\dfrac{999}{1000}} = \dfrac{257}{999}$

**51.** $a_n = n + 5$
arithmetic, $d = 1$

**53.** $a_n = 2^n$
geometric, $r = 2$

**55.** $a_n = n^2 + 5$
neither

**57.** First find $a_{10}$ and $b_{10}$:

$\begin{aligned} a_{10} &= a_1 r^{n-1} \\ &= (-5)\left(\dfrac{10}{-5}\right)^{10-1} = (-5)(-2)^9 \\ &= 2560 \\ b_{10} &= b_1 + (n-1)d \\ &= 10 + (10-1)(-5-10) \\ &= 10 + (9)(-15) = -125 \end{aligned}$

So, $a_{10} + b_{10} = 2560 + (-125) = 2435$.

**59.** For $\{a_n\}$, $r = \dfrac{10}{-5} = -2$ and

$\begin{aligned} S_{10} &= \dfrac{a_1(1 - r^n)}{1 - r} = \dfrac{(-5)\left(1 - (-2)^{10}\right)}{1 - (-2)} \\ &= \dfrac{(-5)(-1023)}{3} = 1705 \end{aligned}$

For $\{b_n\}$,

$$b_{10} = b_1 + (n-1)d$$
$$= 10 + (10-1)(-5-10)$$
$$= 10 + (9)(-15) = -125$$

$$S_n = \frac{n}{2}(b_1 + b_n)$$

$$S_{10} = \frac{n}{2}(b_1 + b_{10})$$

$$= \frac{10}{2}(10 + (-125))$$

$$= 5(-115) = -575$$

**61.** For $\{a_n\}$,

$$S_6 = \frac{a_1(1-r^n)}{1-r} = \frac{(-5)\left(1-(-2)^6\right)}{1-(-2)}$$

$$= \frac{(-5)(-63)}{3} = 105$$

For $\{c_n\}$,

$$S = \frac{a_1}{1-r} = \frac{-2}{1-\frac{1}{-2}} = \frac{-2}{\frac{3}{2}} = -\frac{4}{3}$$

So, $S_6 \cdot S = 105\left(-\frac{4}{3}\right) = -140$

**63.** It is given that $a_4 = 27$. Using the formula

$a_n = a_1 r^{n-1}$ when $n = 4$ we have

$$27 = 8r^{4-1}$$

$$\frac{27}{8} = r^3$$

$$r = \sqrt[3]{\frac{27}{8}} = \frac{3}{2}$$

Thus,

$$a_n = a_1 r^{n-1}$$

$$a_2 = 8\left(\frac{3}{2}\right)^{2-1} = 8\left(\frac{3}{2}\right) = 12$$

$$a_3 = 8\left(\frac{3}{2}\right)^{3-1} = 8\left(\frac{3}{2}\right)^2 = 8\left(\frac{9}{4}\right) = 18$$

**65.** 1, 2, 4, 8, . . .

$r = 2$

$$a_n = 2^{n-1}$$

$$a_{15} = 2^{14} = \$16,384$$

**67.** $a_1 = 3,000,000$

$r = 1.04$

$$a_n = 3,000,000(1.04)^{n-1}$$

$$a_7 = 3,000,000(1.04)^6 = \$3,795,957$$

**69.** **a.** $r_{2000 \text{ to } 2001} = \dfrac{34.21}{33.87} \approx 1.01$

$r_{2001 \text{ to } 2002} = \dfrac{34.55}{34.21} \approx 1.01$

$r_{2002 \text{ to } 2003} = \dfrac{34.90}{34.55} \approx 1.01$

$r_{2003 \text{ to } 2004} = \dfrac{35.25}{34.90} \approx 1.01$

$r_{2004 \text{ to } 2005} = \dfrac{35.60}{35.25} \approx 1.01$

$r_{2005 \text{ to } 2006} = \dfrac{36.00}{35.60} \approx 1.01$

$r_{2006 \text{ to } 2007} = \dfrac{36.36}{36.00} \approx 1.01$

$r_{2007 \text{ to } 2008} = \dfrac{36.72}{36.36} \approx 1.01$

$r_{2008 \text{ to } 2009} = \dfrac{37.09}{36.72} \approx 1.01$

$r_{2009 \text{ to } 2010} = \dfrac{37.25}{37.09} \approx 1.01$

$r$ is approximately 1.01 for all but one division.

**b.** $a_n = a_1 r^{n-1}$

$$a_n = 33.87(1.01)^{n-1}$$

**c.** Since year 2020 is the 21th term, find $a_{21}$.

$$a_n = 33.87(1.01)^{n-1}$$

$$a_{21} = 33.87(1.01)^{21-1} \approx 41.33$$

The population of California will be approximately 41.33 million in 2020.

**71.** 1, 2, 4, 8, . . .

$r = 2$

$$S_{15} = \frac{1(1-2^{15})}{1-2} = 32,767$$

The total savings is \$32,767.

**73.** $a_1 = 24,000$, $r = 1.05$

$$S_{20} = \frac{24,000\left[1-(1.05)^{20}\right]}{1-1.05} = 793,582.90$$

The total salary is \$793,583.

**75.** $r = 0.9$

$$S_{10} = \frac{20(1-0.9^{10})}{1-0.9} \approx 130.26$$

The total length is 130.26 inches.

**77.** **a.** $A = \dfrac{P\left[\left(1+\dfrac{r}{n}\right)^{nt}-1\right]}{\dfrac{r}{n}} = \dfrac{2000\left[\left(1+\dfrac{0.075}{1}\right)^{5}-1\right]}{\dfrac{0.075}{1}} \approx \$11,617$

    **b.** $\$11,617 - 5 \times \$2000 = \$1617$

**79.** **a.** $A = \dfrac{P\left[\left(1+\dfrac{r}{n}\right)^{nt}-1\right]}{\dfrac{r}{n}} = \dfrac{50\left[\left(1+\frac{0.055}{12}\right)^{12\times40}-1\right]}{\frac{0.055}{12}} \approx \$87,052$

    **b.** $\$87,052 - \$50 \cdot 12 \cdot 40 = \$63,052$

**81.** **a.** $A = \dfrac{P\left[\left(1+\frac{r}{n}\right)^{nt}-1\right]}{\frac{r}{n}} = \dfrac{10,000\left[\left(1+\frac{0.105}{4}\right)^{4\times10}-1\right]}{\frac{0.105}{4}} \approx \$693,031$

    **b.** $\$693,031 - \$10,000 \cdot 4 \cdot 10 = \$293,031$

**83.** Find the total value of the lump-sum investment.

$$A = P(1+r)^t = 30,000(1+0.05)^{20} \approx 79,599$$

Find the total value of the annuity.

$$A = \dfrac{P\left[\left(1+\dfrac{r}{n}\right)^{nt}-1\right]}{\dfrac{r}{n}} = \dfrac{1500\left[\left(1+\dfrac{0.05}{1}\right)^{20}-1\right]}{\dfrac{0.05}{1}} \approx 49,599$$

$\$79,599 - \$49,599 = \$30,000$

You will have $30,000 more from the lump-sum investment.

**85.** $r = 0.6$

$$S_\infty = \frac{6(0.6)}{1-0.6} = 9$$

The total economic impact is $9 million.

**87.** $r = \dfrac{1}{4}$

$$S_\infty = \frac{\frac{1}{4}}{1-\frac{1}{4}} = \frac{1}{4} \cdot \frac{4}{3} = \frac{1}{3}$$

**89. – 97.** Answers will vary.

**99.** $f(x) = \dfrac{2\left[1-\left(\dfrac{1}{3}\right)^{x}\right]}{1-\dfrac{1}{3}}$

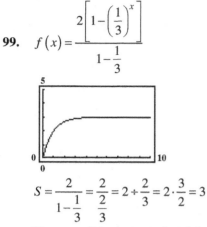

$$S = \dfrac{2}{1-\dfrac{1}{3}} = \dfrac{2}{\dfrac{2}{3}} = 2 \div \dfrac{2}{3} = 2 \cdot \dfrac{3}{2} = 3$$

The sum of the series is 3 and the asymptote of the function is $y = 3$.

**101.** makes sense

**103.** makes sense

**105.** false; Changes to make the statement true will vary. A sample change is: The sequence is not geometric. There is not a common ratio.

**107.** false; Changes to make the statement true will vary. A sample change is: The sum of the sequence is
$$\dfrac{10}{1-\left(-\dfrac{1}{2}\right)}.$$

**109.** Let $a_1$ equal the number of flies released each day.

On any day, the total number of flies is the number released that day, plus 90% of those released the day before, plus 90% of 90% of those released two days before, etc.:

$$S = \dfrac{a_1}{1-r}$$
$$20{,}000 = \dfrac{a_1}{1-.9}$$
$$20{,}000 = \dfrac{a_1}{.1}$$
$$2000 = a_1$$

2000 flies to be released each day.

**111.** Answers will vary.

**112.** $1+2+3 = \dfrac{3(3+1)}{2}$
$$6 = \dfrac{3(4)}{2}$$
$$6 = 6$$

**113.** $1+2+3+4+5 = \dfrac{5(5+1)}{2}$
$$15 = \dfrac{5(6)}{2}$$
$$15 = 15$$

**114.** $\dfrac{k(k+1)(2k+1)}{6} + (k+1)^2$
$$= \dfrac{k(k+1)(2k+1)}{6} + \dfrac{6(k+1)^2}{6}$$
$$= \dfrac{k(k+1)(2k+1) + 6(k+1)^2}{6}$$
$$= \dfrac{(k+1)[k(2k+1) + 6(k+1)]}{6}$$
$$= \dfrac{(k+1)[2k^2 + k + 6k + 6]}{6}$$
$$= \dfrac{(k+1)[2k^2 + 7k + 6]}{6}$$
$$= \dfrac{(k+1)(k+2)(2k+3)}{6}$$

**Mid-Chapter 8 Check Point**

**1.** $a_n = (-1)^{n+1}\dfrac{n}{(n-1)!}$

$a_1 = (-1)^{1+1}\dfrac{1}{(1-1)!} = (-1)^2\dfrac{1}{0!} = 1 \cdot 1 = 1$

$a_2 = (-1)^{2+1}\dfrac{2}{(2-1)!} = (-1)^3\dfrac{2}{1!} = (-1)(2) = -2$

$a_3 = (-1)^{3+1}\dfrac{3}{(3-1)!} = (-1)^4\dfrac{3}{2!} = 1 \cdot \dfrac{3}{2} = \dfrac{3}{2}$

$a_4 = (-1)^{4+1}\dfrac{4}{(4-1)!} = (-1)^5\dfrac{4}{3!} = (-1)\dfrac{4}{6} = -\dfrac{2}{3}$

$a_5 = (-1)^{5+1}\dfrac{5}{(5-1)!} = (-1)^6\dfrac{5}{4!} = 1 \cdot \dfrac{5}{24} = \dfrac{5}{24}$

**2.** Using $a_n = a_1 + (n-1)d$ ;

$a_1 = 5$

$a_2 = 5 + (2-1)(-3) = 5 + 1(-3) = 5 - 3 = 2$

$a_3 = 5 + (3-1)(-3) = 5 + 2(-3) = 5 - 6 = -1$

$a_4 = 5 + (4-1)(-3) = 5 + 3(-3) = 5 - 9 = -4$

$a_5 = 5 + (5-1)(-3) = 5 + 4(-3) = 5 - 12 = -7$

**3.** Using $a_n = a_1 r^{n-1}$ ;

$a_1 = 5$

$a_2 = 5(-3)^{2-1} = 5(-3)^1 = 5(-3) = -15$

$a_3 = 5(-3)^{3-1} = 5(-3)^2 = 5(9) = 45$

$a_4 = 5(-3)^{4-1} = 5(-3)^3 = 5(-27) = -135$

$a_5 = 5(-3)^{5-1} = 5(-3)^4 = 5(81) = 405$

**4.** $a_n = -a_{n-1} + 4$

$a_1 = 3$

$a_2 = -a_1 + 4 = -3 + 4 = 1$

$a_3 = -a_2 + 4 = -1 + 4 = 3$

$a_4 = -a_3 + 4 = -3 + 4 = 1$

$a_5 = -a_4 + 4 = -1 + 4 = 3$

**5.** $d = a_2 - a_1 = 6 - 2 = 4$

$\begin{aligned} a_n &= a_1 + (n-1)d \\ &= 2 + (n-1)4 \\ &= 2 + 4n - 4 \\ &= 4n - 2 \end{aligned}$

$a_{20} = 4(20) - 2 = 78$

**6.** $r = \dfrac{a_2}{a_1} = \dfrac{6}{3} = 2$

$\begin{aligned} a_n &= a_1 r^{n-1} \\ &= 3(2)^{n-1} \end{aligned}$

$\begin{aligned} a_{10} &= 3(2)^{10-1} \\ &= 3(2)^9 \\ &= 1536 \end{aligned}$

**7.** $d = a_2 - a_1 = 1 - \dfrac{3}{2} = -\dfrac{1}{2}$

$\quad a_n = a_1 + (n-1)d$

$\qquad = \dfrac{3}{2} + (n-1)\left(-\dfrac{1}{2}\right)$

$\qquad = \dfrac{3}{2} - \dfrac{1}{2}n + \dfrac{1}{2}$

$\qquad = -\dfrac{1}{2}n + 2$

$\quad a_{30} = -\dfrac{1}{2}(30) + 2$

$\qquad = -15 + 2$

$\qquad = -13$

**8.** $S_n = \dfrac{a_1(1 - r^n)}{1 - r}; \; r = \dfrac{a_2}{a_1} = \dfrac{10}{5} = 2$

$\quad S_{10} = \dfrac{5(1 - 2^{10})}{1 - 2} = \dfrac{5(-1023)}{-1} = 5115$

**9.** First find $a_{10}$;

$\quad d = a_2 - a_1 = 0 - (-2) = 2$

$\quad a_{50} = a_1 + (n-1)d = -2 + (50-1)(2) = -2 + 49(2) = 96$

$\quad S_{50} = \dfrac{n}{2}(a_1 + a_n) = \dfrac{50}{2}(-2 + 96) = 25(94) = 2350$

**10.** $r = \dfrac{a_2}{a_1} = \dfrac{40}{-20} = -2$

$\quad S_{10} = \dfrac{a_1(1 - r^n)}{1 - r} = \dfrac{-20(-1 - (-2)^{10})}{1 - (-2)} = \dfrac{-20(-1023)}{3} = \dfrac{20460}{3} = 6820$

**11.** First find $a_{100}$;

$\quad d = a_2 - a_1 = -2 - 4 = -6$

$\quad a_{100} = a_1 + (n-1)d = 4 + (100-1)(-6) = 4 + 99(-6) = -590$

$\quad S_{100} = \dfrac{n}{2}(a_1 + a_n) = \dfrac{100}{2}(4 - 590) = 50(-586) = -29{,}300$

**12.** $\displaystyle\sum_{i=1}^{4}(i+4)(i-1) = (1+4)(1-1) + (2+4)(2-1) + (3+4)(3-1) + (4+4)(4-1)$

$\qquad\qquad = 5(0) + 6(1) + 7(2) + 8(3) = 0 + 6 + 14 + 24 = 44$

**13.** $\displaystyle\sum_{i=1}^{50}(3i-2)=(3\cdot1-2)+(3\cdot2-2)+(3\cdot3-3)+...+(3\cdot50-2)$

$$=(3-2)+(6-2)+(9-3)+...+(150-2)$$

$$=1+4+6+...+148$$

The sum of this arithmetic sequence is given by $S_n=\dfrac{n}{2}(a_1+a_n)$ ;

$$S_{50}=\dfrac{50}{2}(1+148)=25(149)=3725$$

**14.** $\displaystyle\sum_{i=1}^{6}\left(\dfrac{3}{2}\right)^i=\left(\dfrac{3}{2}\right)^1+\left(\dfrac{3}{2}\right)^2+\left(\dfrac{3}{2}\right)^3+\left(\dfrac{3}{2}\right)^4+\left(\dfrac{3}{2}\right)^5+\left(\dfrac{3}{2}\right)^6=\dfrac{3}{2}+\dfrac{9}{4}+\dfrac{27}{8}+\dfrac{81}{16}+\dfrac{243}{32}+\dfrac{729}{64}=\dfrac{1995}{64}$

**15.** $\displaystyle\sum_{i=1}^{\infty}\left(-\dfrac{2}{5}\right)^{i-1}=\left(-\dfrac{2}{5}\right)^{1-1}+\left(-\dfrac{2}{5}\right)^{2-1}+\left(-\dfrac{2}{5}\right)^{3-1}+...=\left(-\dfrac{2}{5}\right)^0+\left(-\dfrac{2}{5}\right)^1+\left(-\dfrac{2}{5}\right)^2+...=1+\left(-\dfrac{2}{5}\right)+\dfrac{4}{25}+...$

This is an infinite geometric sequence with $r=\dfrac{a_2}{a_1}=\dfrac{-\frac{2}{5}}{1}=-\dfrac{2}{5}$ .

$$S=\dfrac{a_1}{1-r}=\dfrac{1}{1-\left(-\frac{2}{5}\right)}=\dfrac{1}{\frac{7}{5}}=\dfrac{5}{7}$$

**16.** $0.\overline{45}=\dfrac{a_1}{1-r}=\dfrac{\frac{45}{100}}{1-\frac{1}{100}}=\dfrac{\frac{45}{100}}{\frac{99}{100}}$

$$=\dfrac{45}{100}\div\dfrac{99}{100}=\dfrac{45}{100}\cdot\dfrac{100}{99}=\dfrac{45}{99}=\dfrac{5}{11}$$

**17.** Answers will vary. An example is $\displaystyle\sum_{i=1}^{18}\dfrac{i}{i+2}$ .

**18.** The arithmetic sequence is 16, 48, 80, 112, ....
First find $a_{15}$ where $d=a_2-a_1=48-16=32$ .
$a_{15}=a_1+(n-1)d=16+(15-1)(32)=16+14(32)=16+448=464$
The distance the skydiver falls during the 15$^{\text{th}}$ second is 464 feet.
$S_{15}=\dfrac{n}{2}(a_1+a_n)=\dfrac{15}{2}(16+464)=7.5(480)=3600$
The total distance the skydiver falls in 15 seconds is 3600 feet.

**19.** $r=0.10$

$A=P(1+r)^t$

$\quad=120000(1+0.10)^{10}$

$\quad\approx311249$

The value of the house after 10 years is \$311,249.

**Section 8.4**

**Check Point Exercises**

1.  **a.**  $S_1 : 2 = 1(1+1)$

    $S_k : 2+4+6+\cdots 2k = k(k+1)$

    $S_{k+1} : 2+4+6+\cdots+2(k+1) = (k+1)(k+2)$

    **b.**  $S_1 : 1^3 = \dfrac{1^2(1+1)^2}{4}$

    $S_k = 1^3 + 2^3 + 3^3 + \cdots + k^3 = \dfrac{k^2(k+1)^2}{4}$

    $S_{k+1} = 1^3 + 2^3 + 3^3 + \cdots + (k+1)^3 = \dfrac{(k+1)^2(k+2)^2}{4}$

2.  $S_1 : 2 = 1(1+1)$

    $2 = 2$ is true.

    $S_k : 2+4+6+\cdots+2k = k(k+1)$

    $S_{k+1} : 2+4+6+\cdots+2k+2(k+1) = (k+1)(k+2)$

    Add $2(k+1)$ to both sides of $S_k$:

    $2+4+6+\cdots+2k+2(k+1) = k(k+1)+2(k+1)$

    Simplify the right-hand side:

    $k(k+1)+2(k+1) = (k+1)(k+2)$

    If $S_k$ is true, then $S_{k+1}$ is true. The statement is true for all $n$.

3.  $S_1 : 1^3 = \dfrac{1^2(1+1)^2}{4}$

    $1 = \dfrac{4}{4}$

    $1 = 1$ is true.

    $S_k : 1^3 + 2^3 + 3^3 + \cdots + k^3 = \dfrac{k^2(k+1)^2}{4}$

    $S_{k+1} : 1^3 + 2^3 + 3^3 + \cdots + k^3 + (k+1)^3 = \dfrac{(k+1)^2(k+2)^2}{4}$

    Add $(k+1)^3$ to both sides of $S_k$:

    $1^3 + 2^3 + 3^3 + \cdots + k^3 + (k+1)^3 = \dfrac{k^2(k+1)^2}{4} + (k+1)^3$

    Simplify the right hand side:

    $\dfrac{k^2(k+1)^2}{4} + (k+1)^3 = \dfrac{k^2(k+1)^2 + 4(k+1)^3}{4} = \dfrac{(k+1)^2\left[k^2 + 4(k+1)\right]}{4} = \dfrac{(k+1)^2(k^2+4k+4)}{4}$

    $= \dfrac{(k+1)^2(k+2)^2}{4}$

    If $S_k$ is true, then $S_{k+1}$ is true. The statement is true for all $n$.

**4.**     $S_1$: 2 is a factor of $1^2 + 1 = 2$, since $2 = 2 \cdot 1$.

$S_k$: 2 is a factor of $k^2 + k$

$S_{k+1}$: 2 is a factor of $(k+1)^2 + (k+1)$
Simplify:
$$(k+1)^2 + (k+1) = k^2 + 2k + 1 + k + 1$$
$$= k^2 + 3k + 2$$
$$= k^2 + k + 2k + 2$$
$$= (k^2 + k) + 2(k+1)$$

Because we assume $S_k$ is true, we know 2 is a factor of $k^2 + k$. Since 2 is a factor of $2(k+1)$, we conclude 2 is a factor of the sum $(k^2 + k) + 2(k+1)$. If $S_k$ is true, then $S_{k+1}$ is true. The statement is true for all $n$.

## Concept and Vocabulary Check 8.4

**1.**     induction; 1; $k+1$

**2.**     $2 + 4 + 6 = 3(3+1)$;
$2 + 4 + 6 + \cdots + 2(k+1) = (k+1)(k+2)$

**3.**     $3 + 7 + 11 = 3(6+1)$;
$3 + 7 + 11 + \cdots + [4(k+1) - 1] = (k+1)[2(k+1) + 1]$

**4.**     4; 10; 18; $(k+1)^2 + 3(k+1)$; $k^2 + 5k + 4$

**5.**     $k+1$

## Exercise Set 8.4

**1.**     $S_n = 1 + 3 + 5 + \cdots + (n-1) = n^2$

$S_1 : 1 = 1^2$

$1 = 1$   true

$S_2 : 1 + 3 = 2^2$

$\qquad 4 = 4$ true

$S_3 : 1 + 3 + 5 = 3^2$

$\qquad 9 = 9$ true

**3.** $S_n$ : 2 is a factor of $n^2 - n$

$S_1$ : 2 is a factor of $1^2 - 1 = 0$

$0 = 0 \cdot 2$ so 2 is a factor of 0 is true.

$S_2$ : 2 is a factor of $2^2 - 2 = 2$

$2 = 1 \cdot 2$ so 2 is a factor of 2 is true.

$S_3$ : 2 is a factor of $3^2 - 3 = 6$

$6 = 3 \cdot 2$ so 2 is a factor of 6 is true.

**5.** $S_n : 4 + 8 + 12 + \cdots + 4n = 2n(n+1)$

$S_k : 4 + 8 + 12 + \cdots + 4k = 2k(k+1)$

$S_{k+1} : 4 + 8 + 12 + \cdots + 4(k+1) = 2(k+1)(k+1+1)$

$4 + 8 + 12 + \cdots + 4(k+1) = 2(k+1)(k+2)$

**7.** $S_n : 3 + 7 + 11 + \cdots + (4n-1) = n(2n+1)$

$S_k : 3 + 7 + 11 + \cdots + (4k-1) = k(2k+1)$

$S_{k+1} : 3 + 7 + 11 + \cdots + [4(k+1)-1] = (k+1)[2(k+1)+1]$

$3 + 7 + 11 + \cdots + (4k+3) = (k+1)(2k+3)$

**9.** $S_n$ : 2 is a factor of $n^2 - n + 2$

$S_k$ : 2 is a factor of $k^2 - k + 2$

$S_{k+1}$ : 2 is a factor of $(k+1)^2 - (k+1) + 2$

$k^2 + 2k + 1 - k - 1 + 2 = k^2 + k + 2$

$S_{k+1}$ : 2 is a factor of $k^2 + k + 2$.

**11.** $S_1 : 4 = 2(1)(1+1)$

$4 = 2(2)$

$4 = 4$ is true.

$S_k : 4 + 8 + 12 + \cdots + 4k = 2k(k+1)$

$S_{k+1} : 4 + 8 + 12 + \cdots 4(k+1) = 2(k+1)(k+1+1)$

Add $4(k+1)$ to both sides of $S_k$:

$4 + 8 + 12 + \cdots + 4(k+1) = 2k(k+1) + 4(k+1)$

Simplify the right-hand side:

$= 2k(k+1) + 4(k+1) = (2k+4)(k+1)$

$= 2(k+2)(k+1)$

$= 2(k+1)(k+1+1)$

If $S_k$ is true, then $S_{k+1}$ is true. The statement is true for all $n$.

**13.** $S_1 : 1 = 1^2$

$1 = 1$ is true.

$S_k : 1 + 3 + 5 + \cdots + (2k-1) = k^2$

$S_{k+1} : 1 + 3 + 5 + \cdots + (2k-1) + [2(k+1)-1] = (k+1)^2$

$1 + 3 + 5 + \cdots + (2k-1) + (2k+1) = (k+1)^2$

Add $(2k + 1)$ to both sides of $S_k$:

$$1+3+5+\cdots+(2k-1)+(2k+1) = k^2 +(2k+1)$$

Simplify the right-hand side:

$$= k^2 + (2k + 1)$$
$$= (k + 1)^2$$

If $S_k$ is true, then $S_{k+1}$ is true. The statement is true for all $n$.

**15.** $S_1 : 3 = 1[2(1)+1)]$

$\qquad 3 = 3$ is true.

$\qquad S_k : 3+7+11+\cdots+(4k-1) = k(2k+1)$

$S_{k+1} : 3+7+11+\cdots+(4k-1)+[4(k+1)-1] = (k+1)[2(k+1)+1]$

$\qquad\qquad 3+7+11+\cdots+(4k-1)+(4k+3) = (k+1)(2k+3)$

Add $(4k + 3)$ to both sides of $S_k$:

$3 + 7 + 11 + \ldots + (4k - 1) + (4k + 3) = k(2k + 1) + 4(k + 3)$

Simplify the right-hand side:

$$= k(2k+1)+(4k+3) = 2k^2 +k+4k+3$$

$$= 2k^2 +5k+3$$

$$= (k+1)(2k+3)$$

If $S_k$ is true, then $S_{k+1}$ is true. The statement is true for all $n$.

**17.** $S_1: 1 = 2^1 - 1$

$\qquad\qquad 1 = 1$ is true.

$\qquad S_k : 1+2+2^2 +\cdots+2^{k-1} = 2^k -1$

$S_{k+1} : 1+2+2^2 +\cdots+2^{k-1} +2^{k+1-1} = 2^{k+1} -1$

$\qquad\qquad 1+2+2^2 +\cdots+2^{k-1} +2^k = 2^{k+1} -1$

Add $2^k$ to both sides of $S_k$:

$$1+2+2^2 +\cdots+2^{k-1} +2^k = 2^k +2^k -1$$

Simplify the right-hand side:

$$= 2^k +2^k -1 = 2(2^k)-1$$

$$= 2^{k+1} -1$$

If $S_k$ is true, then $S_{k+1}$ is true. The statement is true for all $n$.

**19.** $S_1 : 2 = 2^{1+1} -2$

$\qquad 2 = 4-2$

$\qquad 2 = 2$ is true.

$\qquad S_k : 2+4+8+\cdots+2^k = 2^{k+1} -2$

$S_{k+1} : 2+4+8+\cdots+2^k +2^{k+1} = 2^{k+2} -2$

Add $2^{k+1}$ to both sides of $S_k$:

$$2+4+8+\cdots+2^k +2^{k+1} = 2^{k+1} +2^{k+1} -2$$

Simplify the right-hand side:

$$= 2^{k+1} +2^{k+1} -1 = 2(2^{k+1})-2$$

$$= 2^{k+2} -2$$

If $S_k$ is true, then $S_{k+1}$ is true. The statement is true for all $n$.

**21.** $S_1 : 1 \cdot 2 = \dfrac{1(1+1)(1+2)}{3}$

$2 = \dfrac{6}{3}$

$2 = 2$ is true.

$S_k : 1 \cdot 2 + 2 \cdot 3 + 3 \cdot 4 + \cdots + k(k+1) = \dfrac{k(k+1)(k+2)}{3}$

$S_{k+1} : 1 \cdot 2 + 2 \cdot 3 + 3 \cdot 4 + \cdots + k(k+1) + (k+1)(k+2) = \dfrac{(k+1)(k+2)(k+3)}{3}$

Add $(k+1)(k+2)$ to both sided of $S_k$:

$1 \cdot 2 + 2 \cdot 3 + 3 \cdot 4 + \cdots + k(k+1) + (k+1)(k+2) = \dfrac{k(k+1)(k+2)}{3} + (k+1)(k+2)$

Simplify the right-hand side:

$= \dfrac{k(k+1)(k+2)}{3} + (k+1)(k+2) = \dfrac{k(k+1)(k+2) + 3(k+1)(k+2)}{3}$

$= \dfrac{(k+1)(k+2)(k+3)}{3}$

If $S_k$ is true, then $S_{k+1}$ is true. The statement is true for all $n$.

**23.** $S_1 : \dfrac{1}{1 \cdot 2} = \dfrac{1}{1+1}$

$\dfrac{1}{2} = \dfrac{1}{2}$ is true.

$S_k : \dfrac{1}{1 \cdot 2} + \dfrac{1}{2 \cdot 3} + \dfrac{1}{3 \cdot 4} + \cdots + \dfrac{1}{k(k+1)} = \dfrac{k}{k+1}$

$S_{k+1} : \dfrac{1}{1 \cdot 2} + \dfrac{1}{2 \cdot 3} + \dfrac{1}{3 \cdot 4} + \cdots + \dfrac{1}{k(k+1)} + \dfrac{1}{(k+1)(k+2)} = \dfrac{k+1}{k+2}$

Add $\dfrac{1}{(k+1)(k+2)}$ to both sides of $S_k$:

$\dfrac{1}{1 \cdot 2} + \dfrac{1}{2 \cdot 3} + \dfrac{1}{3 \cdot 4} + \cdots + \dfrac{1}{k(k+1)} + \dfrac{1}{(k+1)(k+2)} = \dfrac{k}{k+1} + \dfrac{1}{(k+1)(k+2)}$

Simplify the right-hand side:

$\dfrac{k}{(k+1)} + \dfrac{1}{(k+1)(k+2)} = \dfrac{k(k+2)+1}{(k+1)(k+2)}$

$= \dfrac{k^2 + 2k + 1}{(k+1)(k+2)}$

$= \dfrac{(k+1)(k+1)}{(k+1)(k+2)}$

$= \dfrac{k+1}{k+2}$

If $S_k$ is true, then $S_{k+1}$ is true. The statement is true for all $n$.

**25.** $S_1 : 2$ is a factor of $1^2 - 1 = 0,$ since $0 = 2 \cdot 0.$

$\quad S_k : 2$ is a factor of $k^2 - k$

$\quad S_{k+1} : 2$ is a factor of $(k+1)^2 - (k+1)$

$\quad (k+1)^2 - (k-1) = k^2 + 2k + 1 - k - 1$

$$\qquad\qquad\qquad = k^2 + k$$

$$\qquad\qquad\qquad = k^2 - k + 2k$$

$$\qquad\qquad\qquad = (k^2 - k) + 2k$$

Because we assume $S_k$ is true, we know 2 as a factor of $k^2 - k$. Since 2 is a factor of 2k, we conclude 2 is factor of the sum $(k^2 + k) + 2k$. If $S_k$ is true, then $S_{k+1}$ is true. The statement is true for all *n*.

**27.** $S_1$: 6 is a factor of $1(1+1)(1+2) = 6$, since $6 = 6 \cdot 1.$

$\quad S_k$: 6 is a factor of $k(k + 1)(k + 2)$

$\quad S_{k+1}$: 6 is a factor of $(k + 1)(k + 2)(k + 3)$

$\quad (k+1)(k+2)(k+3) = k(k+1)(k+2) + 3(k+1)(k+2)$

Because we assume $S_k$ is true, we know 6 as a factor of $k(k+1)(k+2)$. Since either $k + 1$ or $k + 2$ must be even, the product $(k+1)(k+2)$ is even. Thus 2 is a factor of $(k+1)(k+2)$, and we can conclude that 6 is factor of $3(k+1)(k+2)$ If $S_k$ is true, then $S_{k+1}$ is true.

The statement is true for all *n*.

**29.** $\displaystyle\sum_{i=1}^{n} 5 \cdot 6^i = 6\left(6^n - 1\right)$

*Show that $S_1$ is true:* $\displaystyle\sum_{i=1}^{1} 5 \cdot 6^i = 6\left(6^1 - 1\right)$

$$5 \cdot 6^1 = 6(6-1)$$

$$5 \cdot 6 = 6 \cdot 5, \ \text{True}$$

*Show that if $S_k$ is true, then $S_{k+1}$ is true:*

Assume $S_k : \displaystyle\sum_{i=1}^{k} 5 \cdot 6^i = 6\left(6^k - 1\right)$ is true. Then,

$$\sum_{i=1}^{k} 5 \cdot 6^i + 5 \cdot 6^{k+1} = 6\left(6^k - 1\right) + 5 \cdot 6^{k+1}$$

$$\sum_{i=1}^{k+1} 5 \cdot 6^i = 6^{k+1} - 6 + 5 \cdot 6^{k+1}$$

$$\sum_{i=1}^{k+1} 5 \cdot 6^i = 6 \cdot 6^{k+1} - 6$$

$$\sum_{i=1}^{k+1} 5 \cdot 6^i = 6\left(6^{k+1} - 1\right)$$

The final statement is $S_{k+1}$. Thus, by mathematical induction, we have proven that $\displaystyle\sum_{i=1}^{n} 5 \cdot 6^i = 6\left(6^n - 1\right).$

**31.** $n+2>n$
Show that $S_1$ is true: $1+2>1$
$$3>1, \text{ True}$$
Show that if $S_k$ is true, then $S_{k+1}$ is true:
Assume $S_k : k+2>k$ is true. Then,
$$k+2+1>k+1$$
$$(k+1)+2>k+1$$
The final statement is $S_{k+1}$. Thus, by mathematical induction, we have proven that $n+2>n$.

**33.** $S_1 : (ab)^1 = a^1b^1$
$\quad ab = ab$ is true.
$S_k : (ab)^k = a^kb^k$
$S_{k+1} : (ab)^{k+1} = a^{k+1}b^{k+1}$
Multiply both sides of $S_k$ by $ab$:
$$(ab)^k (ab) = a^kb^k (ab)$$
$$(ab)^{k+1} = a^{k+1}b^{k+1}$$
If $S_k$ is true, then $S_{k+1}$ is true.
The statement is true for all $n$.

**35.** Answers will vary.

**37.** does not make sense; Explanations will vary. Sample explanation: We use mathematical induction to prove statements involving positive integers.

**39.** does not make sense; Explanations will vary. Sample explanation: It is necessary for all the dominoes to topple.

**41.** $n^2 > 2n + 1$ for $n \geq 3$
$\quad S_3: 3^2 > 2 \cdot 3 + 1$
$$\quad 9 > 7$$
$\quad S_k: k^2 > 2k + 1$ for $k \geq 3$
$S_{k+1}: (k+1)^2 > 2k + 3$.
Add $2k + 1$ to both sides of $S_k$.
$$k^2 + (2k+1) > 2k+1+(2k+1)$$
Write the left side of the inequalities as the square of a binomial and simplify the right side. $(k+1)^2 > 4k+2$
Since $4k + 2 > 2k + 3$ for $k \geq 3$, we can conclude that $(k+1)^2 > 4k+2 > 2k+3$.
By the transitive property,
$$(k+1)^2 > 2k+3$$
$$(k+1)^2 > 2(k+1)+1$$
If $S_k$ is true, then $S_{k+1}$ is true.
The statement is true for all $n$.

**43.** $S_1 = \dfrac{1}{4} = \dfrac{1}{4}$

$S_2 = \dfrac{1}{4} + \dfrac{1}{12} = \dfrac{1}{3}$

$S_3 = \dfrac{1}{4} + \dfrac{1}{12} + \dfrac{1}{24} = \dfrac{3}{8}$

$S_4 = \dfrac{1}{4} + \dfrac{1}{12} + \dfrac{1}{24} + \dfrac{1}{40} = \dfrac{2}{5}$

$S_5 = \dfrac{1}{4} + \dfrac{1}{12} + \dfrac{1}{24} + \dfrac{1}{40} + \dfrac{1}{60} = \dfrac{5}{12}$

$S_n = \dfrac{1}{4} + \dfrac{1}{12} + \dfrac{1}{24} + \cdots + \dfrac{1}{2n(n+1)} = \dfrac{n}{2n+2}$

$S_k = \dfrac{1}{4} + \dfrac{1}{12} + \dfrac{1}{24} + \cdots + \dfrac{1}{2k(k+1)} = \dfrac{k}{2k+2}$

$S_{k+1} = \dfrac{1}{4} + \dfrac{1}{12} + \dfrac{1}{24} + \cdots + \dfrac{1}{2k(k+1)} + \dfrac{1}{2(k+1)(k+2)} = \dfrac{k+1}{2k+4}$

Add $\dfrac{1}{2(k+1)(k+2)}$ to both sides of $S_k$:

$\dfrac{1}{4} + \dfrac{1}{12} + \dfrac{1}{24} + \cdots + \dfrac{1}{2k(k+1)} + \dfrac{1}{2(k+1)(k+2)} = \dfrac{k}{2k+2} + \dfrac{1}{2(k+1)(k+2)}$

Simplify the right-hand side:

$\dfrac{k}{2k+2} + \dfrac{1}{2(k+1)(k+2)} = \dfrac{k(k+2)+1}{2(k+1)(k+2)} = \dfrac{k^2 + 2k + 1}{2(k+1)(k+2)} = \dfrac{(k+1)^2}{2(k+1)(k+2)} = \dfrac{k+1}{2k+4}$

If $S_k$ is true, then $S_{k+1}$ is true. The conjecture is proven.

**45.** Answers will vary.

**46.** The exponents begin with the exponent on $a+b$ and decrease by 1 in each successive term.

**47.** The exponents begin with 0, increase by 1 in each successive term, and end with the exponent on $a+b$.

**48.** The sum of the exponents is the exponent on $a+b$.

## Section 8.5

### Check Point Exercises

**1.** **a.** $\dbinom{6}{3} = \dfrac{6!}{3!(6-3)!}$

$= \dfrac{6!}{3!3!}$

$= \dfrac{5 \cdot 4}{1}$

$= 20$

**b.** $\binom{6}{0} = \dfrac{6!}{0!(6-0)!}$

$= \dfrac{6!}{6!}$

$= 1$

**c.** $\binom{8}{2} = \dfrac{8!}{2!(8-2)!}$

$= \dfrac{8!}{2!6!}$

$= \dfrac{8 \cdot 7}{2}$

$= 28$

**d.** $\binom{3}{3} = \dfrac{3!}{3!(3-3)!}$

$= \dfrac{3!}{3!0!}$

$= \dfrac{3!}{3!}$

$= 1$

**2.** $(x+1)^4 = \binom{4}{0}x^4 + \binom{4}{1}x^3 + \binom{4}{2}x^2 + \binom{4}{1}x + \binom{4}{0} = x^4 + 4x^3 + 6x^2 + 4x + 1$

**3.** $(x-2y)^5 = \binom{5}{0}x^5(-2y)^0 + \binom{5}{1}x^4(-2y)^1 + \binom{5}{2}x^3(-2y)^2 + \binom{5}{3}x^2(-2y)^3 + \binom{5}{4}x(-2y)^4 + \binom{5}{5}x^0(-2y)^5$

$= x^5 - 5x^4(2y) + 10x^3(4y^2) - 10x^2(8y^3) + 5x(16y^4) - 32y^5$

$= x^5 - 10x^4y + 40x^3y^2 - 80x^2y^3 + 80xy^4 - 32y^5$

**4.** $(2x+y)^9$

fifth term $= \binom{9}{4}(2x)^5 y^4$

$= \dfrac{9!}{4!5!}(32x^5)y^4$

$= 4032x^5y^4$

**Concept and Vocabulary Check 8.5**

**1.** binomial

**2.** $\dfrac{8!}{2!6!}$

**3.** $\dfrac{n!}{r!(n-r)!}$

**4.** $\binom{5}{1}; \binom{5}{2}; \binom{5}{3}; \binom{5}{4}; \binom{5}{5}$

**5.** $\dbinom{n}{1}; \dbinom{n}{2}; \dbinom{n}{3}; \dbinom{n}{n}; n$

**6.** Binomial

**7.** $a^{n-r}b^r$

**Exercise Set 8.5**

**1.** $\dbinom{8}{3} = \dfrac{8!}{3!(8-3)!} = \dfrac{8 \cdot 7 \cdot 6}{3 \cdot 2 \cdot 1} = 56$

**3.** $\dbinom{12}{1} = \dfrac{12!}{1!11!} = 12$

**5.** $\dbinom{6}{6} = \dfrac{6!}{0!6!} = 1$

**7.** $\dbinom{100}{2} = \dfrac{100!}{2!98!} = \dfrac{100 \cdot 99}{2} = 4950$

**9.** $(x+2)^3 = \dbinom{3}{0}x^3 + \dbinom{3}{1}2x^2 + \dbinom{3}{2}4x + \dbinom{3}{3}8 = x^3 + 3x^2 \cdot 2 + 3x \cdot 4 + 8 = x^3 + 6x^2 + 12x + 8$

**11.** $(3x+y)^3 = \dbinom{3}{0}27x^3 + \dbinom{3}{1}9x^2y + \dbinom{3}{2}3xy^2 + \dbinom{3}{3}y^3 = 27x^3 + 27x^2y + 9xy^2 + y^3$

**13.** $(5x-1)^3 = \dbinom{3}{0}125x^3 - \dbinom{3}{1}25x^2 + \dbinom{3}{2}5x - \dbinom{3}{3} = 125x^3 - 75x^2 + 15x - 1$

**15.** $(2x+1)^4 = \dbinom{4}{0}16x^4 - \dbinom{4}{1}8x^3 + \dbinom{4}{2}4x^2 + \dbinom{4}{3}2x + \dbinom{4}{4} = 16x^4 + 32x^3 + 24x^2 + 8x + 1$

**17.** $(x^2+2y)^4 = \dbinom{4}{0}(x^2)^4 + \dbinom{4}{1}(x^2)^3(2y) + \dbinom{4}{2}(x^2)^2(2y)^2 + \dbinom{4}{3}(x^2)^1(2y)^3 + \dbinom{4}{4}(2y)^4$

$\qquad = 1(x^8) + 4(x^6)(2y) + 6(x^4)(4y^2) + 4x^2(8y^3) + 1(16y^4)$

$\qquad = x^8 + 8x^6y + 24x^4y^2 + 32x^2y^3 + 16y^4$

**19.** $(y-3)^4 = \dbinom{4}{0}y^4 + \dbinom{4}{1}y^3(-3) + \dbinom{4}{2}y^2(-3)^2 + \dbinom{4}{3}y(-3)^3 + \dbinom{4}{4}(-3)^4$

$\qquad = y^4 + 4(y^3)(-3) + 6(y^2)(9) + 4(y)(-27) + 81$

$\qquad = y^4 - 12y^3 + 54y^2 - 108y + 81$

**21.** $(2x^3-1)^4 = \dbinom{4}{0}(2x^3)^4 + \dbinom{4}{1}(2x^3)^3(-1) + \dbinom{4}{2}(2x^3)^2(-1)^2 + \dbinom{4}{3}(2x^3)(-1)^3 + \dbinom{4}{4}(-1)^4$

$\qquad = 16x^{12} - 4(8x^9) + 6(4x^6) - 4(2x^3) + 1$

$\qquad = 16x^{12} - 32x^9 + 24x^6 - 8x^3 + 1$

**23.** $(c+2)^5 = \binom{5}{0}c^5 + \binom{5}{1}c^4(2) + \binom{5}{2}c^3(2^2) + \binom{5}{3}c^2(2^3) + \binom{5}{4}c(2^4) + \binom{5}{5}(2^5)$

$= c^5 + 5c^4(2) + 10c^3(4) + 10c^2(8) + 5c(16) + 32$

$= c^5 + 10c^4 + 40c^3 + 80c^2 + 80c + 32$

**25.** $(x-1)^5 = \binom{5}{0}x^5 - \binom{5}{1}x^4 + \binom{5}{2}x^3 - \binom{5}{3}x^2 + \binom{5}{4}x - \binom{5}{5} = x^5 - 5x^4 + 10x^3 - 10x^2 + 5x - 1$

**27.** $(3x-y)^5 = \binom{5}{0}(3x)^5 - \binom{5}{1}(3x)^4 y + \binom{5}{2}(3x)^3 y^2 - \binom{5}{3}(3x)^2 y^3 + \binom{5}{4}3xy^4 - \binom{5}{5}y^5$

$= (1)243x^5 - 5(81x^4)y + 10(27x^3)y^2 - 10(9x^2)y^3 + 5(3x)y^4 - (1)y^5$

$= 243x^5 - 405x^4 y + 270x^3 y^2 - 90x^2 y^3 + 15xy^4 - y^5$

**29.** $(2a+b)^6 = \binom{6}{0}(2a)^6 + \binom{6}{1}(2a)^5 b + \binom{6}{2}(2a)^4 b^2 + \binom{6}{3}(2a)^3 b^3 + \binom{6}{4}(2a)^2 b^4 + \binom{6}{5}(2a)b^5 + \binom{6}{6}b^6$

$= 64a^6 + 6\left(32a^5\right)b + 15\left(16a^4\right)b^2 + 20\left(8a^3\right)b^3 + 15\left(4a^2\right)b^4 + 6(2a)b^5 + b^6$

$= 64a^6 + 192a^5 b + 240a^4 b^2 + 160a^3 b^3 + 60a^2 b^4 + 12ab^5 + b^6$

**31.** $(x+2)^8 = \binom{8}{0}x^8 + \binom{8}{1}x^7 2 + \binom{8}{3}x^6(2)^2 + \cdots$

$= x^8 + 16x^7 + 112x^6 + \cdots$

**33.** $(x-2y)^{10} = \binom{10}{0}x^{10} - \binom{10}{1}x^9(2y) + \binom{10}{2}x^8(2y)^2 - \cdots$

$= x^{10} - 20x^9 y + 180x^8 y^2 - \cdots$

**35.** $(x^2+1)^{16} = \binom{16}{0}(x^2)^{16} + \binom{16}{1}(x^2)^{15} + \binom{16}{2}(x^2)^{14} + \cdots$

$= x^{32} + 16x^{30} + 120x^{28} + \cdots$

**37.** $(y^3-1)^{20} = \binom{20}{0}\left(y^3\right)^{20} - \binom{20}{1}(y^3)^{19} + \binom{20}{2}(y^3)^{18} - \cdots$

$= y^{60} - 20y^{57} + 190y^{54} - \cdots$

**39.** $(2x+y)^6;$    third term $= \binom{6}{2}(2x)^4(y)^2 = 15\left(16x^4 y^2\right) = 240x^4 y^2$

**41.** $(x-1)^9;$    fifth term $= \binom{9}{4}x^5(-1)^4 = 126x^5$

**43.** $\left(x^2+y^3\right)^8;$    sixth term $= \binom{8}{5}\left(x^2\right)^3\left(y^3\right)^5 = 56x^6 y^{15}$

**45.** $\left(x-\tfrac{1}{2}\right)^9;$    fourth term $= \binom{9}{3}x^6\left(-\frac{1}{2}\right)^3 = 84x^6\left(-\frac{1}{8}\right) = -\frac{21}{2}x^6$

**47.** $\dbinom{22}{14}(x^2)^8 y^{14} = 319,770x^{16}y^{14}$

**49.**

$$\left(x^3 + x^{-2}\right)^4 = \dbinom{4}{0}(x^3)^4 + \dbinom{4}{1}(x^3)^3(x^{-2}) + \dbinom{4}{2}(x^3)^2(x^{-2})^2 + \dbinom{4}{3}(x^3)^1(x^{-2})^3 + \dbinom{4}{4}(x^{-2})^4$$

$$= \frac{4!}{0!(4-0)!}x^{12} + \frac{4!}{1!(4-1)!}x^9 x^{-2} + \frac{4!}{2!(4-2)!}x^6 x^{-4} + \frac{4!}{3!(4-3)!}x^3 x^{-6} + \frac{4!}{4!(4-4)!}x^{-8}$$

$$= \frac{4!}{0!\,4!}x^{12} + \frac{4\cdot 3!}{1!\,3!}x^7 + \frac{4\cdot 3\cdot 2!}{2\cdot 1\cdot 2!}x^2 + \frac{4\cdot 3!}{3!\cdot 1!}x^{-3} + \frac{4!}{4!\cdot 0!}x^{-8}$$

$$= x^{12} + 4x^7 + 6x^2 + \frac{4}{x^3} + \frac{1}{x^8}$$

**51.**

$$\left(x^{\frac{1}{3}} - x^{-\frac{1}{3}}\right)^3 = \left(x^{\frac{1}{3}} + \left(-x^{-\frac{1}{3}}\right)\right)^3 = \dbinom{3}{0}\left(x^{\frac{1}{3}}\right)^3 + \dbinom{3}{1}\left(x^{\frac{1}{3}}\right)^2\left(-x^{-\frac{1}{3}}\right) +$$

$$+ \dbinom{3}{2}\left(x^{\frac{1}{3}}\right)^1\left(-x^{-\frac{1}{3}}\right)^2 + \dbinom{3}{3}\left(-x^{-\frac{1}{3}}\right)^3$$

$$= \frac{3!}{0!(3-0)!}x^1 + \frac{3!}{1!(3-1)!}x^{\frac{2}{3}}\cdot -x^{-\frac{1}{3}} + \frac{3!}{2!(3-2)!}x^{\frac{1}{3}}x^{-\frac{2}{3}} + \frac{3!}{3!(3-3)!}\cdot -x^{-1}$$

$$= \frac{3!}{0!\,3!}x + \frac{3\cdot 2!}{1!\,2!}\cdot -x^{\frac{1}{3}} + \frac{\cdot 3\cdot 2!}{2!\cdot 1!}x^{-\frac{1}{3}} + \frac{3!}{3!\cdot 0!}\cdot -x^{-1}$$

$$= x - 3x^{\frac{1}{3}} + \frac{3}{x^{\frac{1}{3}}} - \frac{1}{x}$$

**53.** $f(x) = x^4 + 7$;

$$\frac{f(x+h)-f(x)}{h}$$

$$= \frac{(x+h)^4 + 7 - \left(x^4 + 7\right)}{h}$$

$$= \frac{\dbinom{4}{0}x^4 + \dbinom{4}{1}x^3 h + \dbinom{4}{2}x^2 h^2 + \dbinom{4}{3}xh^3 + \dbinom{4}{4}h^4 + 7 - x^4 - 7}{h}$$

$$= \frac{\frac{4!}{0!(4-0)!}x^4 + \frac{4!}{1!(4-1)!}x^3 h + \frac{4!}{2!(4-2)!}x^2 h^2 + \frac{4!}{3!(4-3)!}xh^3 + \frac{4!}{4!(4-4)!}h^4 - x^4}{h}$$

$$= \frac{\frac{4!}{0!\,4!}x^4 + \frac{4\cdot 3!}{1!\,3!}x^3 h + \frac{4\cdot 3\cdot 2!}{2!\cdot 2\cdot 1}x^2 h^2 + \frac{4\cdot 3!}{3!\,1!}xh^3 + \frac{4!}{4!\,0!}h^4 - x^4}{h}$$

$$= \frac{x^4 + 4x^3 h + 6x^2 h^2 + 4xh^3 + h^4 - x^4}{h}$$

$$= \frac{h(4x^3 + 6x^2 h + 4xh^2 + h^3)}{h}$$

$$= 4x^3 + 6x^2 h + 4xh^2 + h^3$$

**55.** Find the $(5+1)=6^{th}$ term.

$$\binom{n}{r}a^{n-r}b^r = \binom{10}{5}\left(\frac{3}{x}\right)^{10-5}\left(\frac{x}{3}\right)^5 = \frac{10!}{5!(10-5)!}\left(\frac{3}{x}\right)^5\left(\frac{x}{3}\right)^5$$

$$= \frac{10\cdot9\cdot8\cdot7\cdot6\cdot 5!}{5!\cdot 5\cdot4\cdot3\cdot 2\cdot1}\left(\frac{3}{x}\right)^5\left(\frac{x}{3}\right)^5 = 252\cdot\frac{3^5}{x^5}\cdot\frac{x^5}{3^5} = 252$$

**57.** $(0.28+0.72)^5$

Third Term $(r=2)$: $\binom{n}{r}a^{n-r}b^r = \binom{5}{2}0.28^{5-2}0.72^2 = \frac{5!}{2!(5-2)!}0.28^{5-2}0.72^2 = \frac{5!}{2!3!}0.28^3 0.72^2 \approx 0.1138$

**59. – 67.** Answers will vary.

**69.** $f_1(x) = (x+1)^4$

$f_2(x) = x^4$

$f_3(x) = x^4 + 4x^3$

$f_4(x) = x^4 + 4x^3 + 6x^2$

$f_5(x) = x^4 + 4x^3 + 6x^2 + 4x$

$f_6(x) = x^4 + 4x^3 + 6x^2 + 4x + 1$

$f_2$, $f_3$, $f_4$, and $f_5$ are approaching $f_1 = f_6$.

**71.** $f_1(x) = (x-2)^4$

$$= \binom{4}{0}x^4 + \binom{4}{1}x^3(-2) + \binom{4}{2}x^2(-2)^2 + \binom{4}{3}x(-2)^3 + \binom{4}{4}(-2)^4$$

$$= x^4 + 4x^3(-2) + 6x^2(4) + 4x(-8) + 16$$

$$= x^4 - 8x^3 + 24x^2 - 32x + 16$$

**73.** makes sense

**75.** does not make sense; Explanations will vary. Sample explanation: $\binom{n}{0}$ and $\binom{n}{1}$ are the coefficients of the first and second term.

**77.** false; Changes to make the statement true will vary. A sample change is: The binomial expansion for $(a+b)^n$ contains $n+1$ terms.

**79.** false; Changes to make the statement true will vary. A sample change is: The sum of the binomial coefficients in $(a+b)^n$ is $2^n$.

**81.** $\left(x^2+x+1\right)^3 = \left[x^2+(x+1)\right]^3$

$$= \binom{3}{0}\left(x^2\right)^3 + \binom{3}{1}\left(x^2\right)^2(x+1) + \binom{3}{2}x^2(x+1)^2 + \binom{3}{3}(x+1)^3$$

$$= x^6 + 3x^4(x+1) + 3x^2\left(x^2+2x+1\right) + x^3 + 3x^2 + 3x + 1$$

$$= x^6 + 3x^5 + 3x^4 + 3x^4 + 6x^3 + 3x^2 + x^3 + 3x^2 + 3x + 1$$

$$= x^6 + 3x^5 + 6x^4 + 7x^3 + 6x^2 + 3x + 1$$

**83.** $\displaystyle\binom{n}{r} = \frac{n!}{r!(n-r)!} = \frac{n!}{(n-r)!r!} = \frac{n!}{(n-r)![n-(n-r)]!} = \binom{n}{n-r}$

**85. a.** $S_1 : (a+b)^1 = \binom{1}{0}a^1 + \binom{1}{1}a^{1-1}b = a+b$

**b.** $S_k : (a+b)^k = \binom{k}{0}a^k + \binom{k}{1}a^{k-1}b + \binom{k}{2}a^{k-2}b^2 + \cdots + \binom{k}{k-1}ab^{k-1} + \binom{k}{k}b^k$

$S_{k+1} : (a+b)^{k+1} = \binom{k+1}{0}a^{k+1} + \binom{k+1}{1}a^k b + \binom{k+1}{2}a^{k-1}b^2 + \cdots + \binom{k+1}{k}ab^k + \binom{k+1}{k+1}b^{k+1}$

**c.** $(a+b)(a+b)^k$

$(a+b)^{k+1} = \binom{k}{0}a^{k+1} + \binom{k}{0}a^k b + \binom{k}{1}a^k b + \binom{k}{1}a^{k-1}b^2 + \binom{k}{2}a^{k-1}b^2 + \binom{k}{2}a^{k-2}b^3 + \cdots$

$\qquad = \binom{k}{k-1}a^2 b^{k-1} + \binom{k}{k-1}ab^k + \binom{k}{k}ab^k + \binom{k}{k}b^{k+1}$

**d.** $(a+b)^{k+1} = \binom{k}{0}a^{k+1} + \left[\binom{k}{0}+\binom{k}{1}\right]a^k b + \left[\binom{k}{1}+\binom{k}{2}\right]a^{k-1}b^2 + \left[\binom{k}{2}+\binom{k}{3}\right]a^{k-2}b^3 + \cdots$

$\qquad\qquad\qquad + \left[\binom{k}{k-1}+\binom{k}{k}\right]ab^k + \binom{k}{k}b^{k+1}$

**e.** $(a+b)^{k+1} = \binom{k}{0}a^{k+1} + \binom{k+1}{1}a^k b + \binom{k+1}{2}a^{k-1}b^2 + \binom{k+1}{3}a^{k-2}b^3 + \cdots + \binom{k+1}{k}ab^k + \binom{k}{k}b^{k+1}$

**f.** $\binom{k}{0} = \binom{k+1}{0}$ because both equal 1. $\binom{k}{k} = \binom{k+1}{k+1}$ also because both equal 1.

$S_{k+1} : (a+b)^{k+1} = \binom{k+1}{0}a^{k+1} + \binom{k+1}{1}a^k b + \binom{k+1}{2}a^{k-1}b^2 + \cdots + \binom{k+1}{k}ab^k + \binom{k+1}{k+1}b^{k+1}$

**86.** $\displaystyle\frac{n!}{(n-r)!} = \frac{20!}{(20-3)!} = \frac{20!}{17!} = \frac{20\cdot 19\cdot 18\cdot 17!}{17!} = 20\cdot 19\cdot 18 = 6840$

**87.** $\dfrac{n!}{(n-r)!\,r!} = \dfrac{8!}{(8-3)!3!} = \dfrac{8!}{5!3!} = \dfrac{8\cdot7\cdot6\cdot5!}{3\cdot2\cdot1\cdot5!} = \dfrac{8\cdot7\cdot6}{3\cdot2\cdot1} = 56$

**88.** true

## Section 8.6

### Check Point Exercises

**1.** We use the Fundamental Counting Principal to find the number of ways a one-topping pizza can be ordered.

Size : Crust : Topping:

   3  ×  4  ×   6  $= 72$

There are 72 different ways of ordering a one-topping pizza.

**2.** We use the Fundamental Counting Principal to find the number of ways we can answer the questions.

Question #1: Question #2: Question #3: Question #4: Question #5: Question #6:

  3   ×   3   ×   3   ×   3   ×   3   ×   3   $=\ 3^6 = 729$

There are 729 ways of answering the questions.

**3.** We use the Fundamental Counting Principal to find the number of different license plates that can be manufactured. Multiply the number of different letters, 26, for the first two places and the number of different digits, 10, for the next three places.

$26\cdot26\cdot10\cdot10\cdot10 = 26^2\cdot1000 = 676{,}000$ plates

There are 676,000 different license plates possible.

**4.** Your group is choosing $r = 4$ officers from a group of $n = 7$ people.

The order in which the officers are chosen matters because the four officers to be chosen have different responsibilities. Thus, we are looking for the number of permutations of 7 things taken 4 at a time.

We use the formula $_nP_r = \dfrac{n!}{(n-r)!}$ with $n = 7$ and $r = 4$. $_7P_4 = \dfrac{7!}{(7-4)!} = \dfrac{7!}{3!} = 840.$

Thus, there are 840 different ways of filling the four offices.

**5.** Because you are using all six of your books in every possible arrangement, you are arranging $r = 6$ books from a group of $n = 6$ books. Thus, we are looking for the number of permutations of 6 things taken 6 at a time. We use the formula

$_nP_r = \dfrac{n!}{(n-r)!}$ with $n = 6$ and $r = 6$.

$_6P_6 = \dfrac{6!}{(6-6)!} = \dfrac{6!}{0!} = 6! = 720.$

There are 720 different possible permutations.

Thus, you can arrange the books in 720 ways.

**6. a.** The order does not matter; this problem involves combinations.

    **b.** Since what place each runner finishes matters, this problem involves permutations.

**7.** The order in which the four people are selected does not matter. This is a problem of selecting $r = 4$ people from a group of $n = 10$ people. We are looking for the number of combinations of 10 things taken 4 at a time. We use the formula

$_nC_r = \dfrac{n!}{(n-r)!\,r!}$ with $n = 10$ and $r = 4$.

$_{10}C_4 = \dfrac{10!}{(10-4)!4!} = \dfrac{10!}{6!4!} = \dfrac{10\cdot9\cdot8\cdot7\cdot6!}{6!\,4\cdot3\cdot2\cdot1} = \dfrac{10\cdot9\cdot8\cdot7}{4\cdot3\cdot2\cdot1} = 210$

Thus, 210 committees of 4 people each can be found from 10 people at the conference on acupuncture.

8.  Because the order in which the 4 cards are dealt does not matter, this is a problem involving combinations. We are looking for the number of combinations of $n = 16$ cards drawn $r = 4$ at a time. We use the formula $_nC_r = \dfrac{n!}{(n-r)!\ r!}$ with $n = 16$ and $r = 4$.

$$_{16}C_4 = \frac{16!}{(16-4)!4!} = \frac{16!}{12!4!} = \frac{16\cdot15\cdot14\cdot13\cdot12!}{12!\,4\cdot3\cdot2\cdot1} = 1820$$

Thus, there are 1820 different 4-card hands possible.

**Concept and Vocabulary Check 8.6**

1.  $M \cdot N$

2.  multiplying; Fundamental Counting

3.  $\dfrac{n!}{(n-r)!}$

4.  $\dfrac{n!}{(n-r)!\ r!}$

5.  $r!$

**Exercise Set 8.6**

1.  $_9P_4 = \dfrac{9!}{5!} = 3024$

3.  $_8P_5 = \dfrac{8!}{3!} = 8\cdot7\cdot6\cdot5\cdot4 = 6720$

5.  $_6P_6 = \dfrac{6!}{0!} = 720$

7.  $_8P_0 = \dfrac{8!}{8!} = 1$

9.  $_9C_5 = \dfrac{9!}{4!5!} = \dfrac{9\cdot8\cdot7\cdot6}{4\cdot3\cdot2\cdot1} = \dfrac{3\cdot7\cdot6}{1} = 126$

11.  $_{11}C_4 = \dfrac{11!}{7!4!} = \dfrac{11\cdot10\cdot9\cdot8}{4\cdot3\cdot2\cdot1} = \dfrac{11\cdot10\cdot3}{1} = 330$

13.  $_7C_7 = \dfrac{7!}{0!7!} = 1$

15.  $_5C_0 = \dfrac{5!}{5!0!} = 1$

17.  combinations; The order in which the volunteers are chosen does not matter.

**19.** permutations; The order of the letters matters because ABCD is not the same as BADC.

**21.** $\dfrac{_7P_3}{3!} - {_7}C_3 = \dfrac{\dfrac{7!}{(7-3)!}}{3!} - \dfrac{7!}{(7-3)!3!} = \dfrac{\dfrac{7!}{4!}}{3!} - \dfrac{7!}{4!3!} = \dfrac{7!}{4!3!} - \dfrac{7!}{4!3!} = 0$

**23.** $1 - \dfrac{_3P_2}{_4P_3} = 1 - \dfrac{\dfrac{3!}{(3-2)!}}{\dfrac{4!}{(4-3)!}} = \dfrac{\dfrac{3!}{1!}}{\dfrac{4!}{1!}} = 1 - \dfrac{3!}{4!} = 1 - \dfrac{3!}{4 \cdot 3!} = 1 - \dfrac{1}{4} = \dfrac{3}{4}$

**25.** $\dfrac{_7C_3}{_5C_4} - \dfrac{98!}{96!} = \dfrac{\dfrac{7!}{(7-3)!3!}}{\dfrac{5!}{(5-4)!4!}} - \dfrac{98 \cdot 97 \cdot 96!}{96!} = \dfrac{\dfrac{7!}{4!3!}}{\dfrac{5!}{1!4!}} - 95067 = \dfrac{\dfrac{7 \cdot 6 \cdot 5 \cdot \cancel{4!}}{\cancel{4!}3 \cdot 2 \cdot 1}}{\dfrac{5 \cdot \cancel{4!}}{1! \cancel{4!}}} - 9506 = \dfrac{35}{5} - 9506 = 7 - 9506 = -9499$

**27.** $\dfrac{_4C_2 \cdot {_6}C_1}{_{18}C_3} = \dfrac{\dfrac{4!}{(4-2)!2!} \cdot \dfrac{6!}{(6-1)!1!}}{\dfrac{18!}{(18-3)!3!}} = \dfrac{\dfrac{4!}{2!2!} \cdot \dfrac{6!}{5!1!}}{\dfrac{18!}{15!3!}} = \dfrac{\dfrac{4 \cdot 3 \cdot \cancel{2!}}{\cancel{2!}2 \cdot 1} \cdot \dfrac{6 \cdot \cancel{5!}}{\cancel{5!}1!}}{\dfrac{18 \cdot 17 \cdot 16 \cdot \cancel{15!}}{\cancel{15!}3 \cdot 2 \cdot 1}} = \dfrac{36}{816} = \dfrac{3}{68}$

**29.** $9 \cdot 3 = 27$ ways

**31.** $2 \cdot 4 \cdot 5 = 40$ ways

**33.** $3^5 = 243$ ways

**35.** $8 \cdot 2 \cdot 9 = 144$ area codes

**37.** $5 \cdot 4 \cdot 3 \cdot 2 \cdot 1 \cdot 1 = 120$ ways

**39.** $1 \cdot 3 \cdot 2 \cdot 1 \cdot 1 = 6$ paragraphs

**41.** $_{10}P_3 = \dfrac{10!}{7!3!} = 10 \cdot 9 \cdot 8 = 720$ ways

**43.** $_{13}P_7 = \dfrac{13!}{6!} = 13 \cdot 12 \cdot 11 \cdot 10 \cdot 9 \cdot 8 \cdot 7$
$= 8,648,640$ ways

**45.** $_6P_3 = \dfrac{6!}{3!} = 6 \cdot 5 \cdot 4 = 120$ ways

**47.** $_9P_5 = \dfrac{9!}{4!} = 9 \cdot 8 \cdot 7 \cdot 6 \cdot 5 = 15,120$ lineups

**49.** $_6C_3 = \dfrac{6!}{3!3!} = \dfrac{6 \cdot 5 \cdot 4}{3 \cdot 2 \cdot 1} = 20$ ways

**51.** $_{12}C_4 = \dfrac{12!}{8!4!} = \dfrac{12 \cdot 11 \cdot 10 \cdot 9}{4 \cdot 3 \cdot 2 \cdot 1}$
$= 495$ collections

**53.** $_{17}C_8 = \dfrac{17!}{9!8!} = \dfrac{17 \cdot 16 \cdot 15 \cdot 14 \cdot 13 \cdot 12 \cdot 11 \cdot 10}{8 \cdot 7 \cdot 6 \cdot 5 \cdot 4 \cdot 3 \cdot 2 \cdot 1}$
$= 24{,}310$ groups

**55.** $_{53}C_6 = \dfrac{53!}{47!6!} = 22{,}957{,}480$ selections

**57.** $_6P_4 = \dfrac{6!}{2!} = 6 \cdot 5 \cdot 4 \cdot 3 = 360$ ways

**59.** $_{13}C_6 = \dfrac{13!}{7!6!} = \dfrac{13 \cdot 12 \cdot 11 \cdot 10 \cdot 9 \cdot 8}{6 \cdot 5 \cdot 4 \cdot 3 \cdot 2 \cdot 1}$
$= 1716$ ways

**61.** $_{20}C_3 = \dfrac{20!}{17!3!} = \dfrac{20 \cdot 19 \cdot 18}{3 \cdot 2 \cdot 1} = 1140$ ways

**63.** $_7P_4 = \dfrac{7!}{3!} = 840$ passwords

**65.** $_{15}P_3 = \dfrac{15!}{12!} = 15 \cdot 14 \cdot 13 = 2730$ cones

**67.** $_6P_6 = \dfrac{6!}{0!} = 6! = 720$ rankings

**69.** $_6C_3 = \dfrac{6!}{3!3!} = \dfrac{6 \cdot 5 \cdot 4}{3 \cdot 2 \cdot 1} = 20$ ways

**71.** $_4P_4 = \dfrac{4!}{0!} = 4! = 24$ ways

**73. – 81.** Answers will vary.

**83.** makes sense

**85.** does not make sense; Explanations will vary. Sample explanation: Since order matters use permutations.

**87.** false; Changes to make the statement true will vary. A sample change is: The number of ways to choose four questions out of ten questions is $_{10}C_4$.

**89.** true

**91.** $5 \cdot 5 \cdot 4 \cdot 4 \cdot 3 \cdot 3 \cdot 2 \cdot 2 \cdot 1 \cdot 1 = 14{,}400$ ways

**93.** $_{10}C_8 \cdot _5C_3 = \dfrac{10!}{(10-8)!8!} \cdot \dfrac{5!}{(5-3)!3!}$
$= 45 \cdot 10 = 450$
They can be chosen in 450 ways.

**95.** $\dfrac{4}{6}$ or $\dfrac{2}{3}$ are less than 5.

**96.** $\dfrac{2}{6}$ or $\dfrac{1}{3}$ are not less than 5.

**97.** 2, 4, 5, and 6 are even or greater than three. This is a fraction of $\dfrac{4}{6}$ or $\dfrac{2}{3}$.

**Section 8.7**

**Check Point Exercises**

1.  **a.**  $P(\text{positive test}) = \dfrac{\text{\# women w/positive test}}{\text{total number of women}}$

    $= \dfrac{7664}{100,000}$

    $= \dfrac{479}{6250}$

    $\approx 0.077$

    **b.**  $P(\text{positive test}) = \dfrac{\text{\# women w/breast cancer and positive test}}{\text{total number of women w/breast cancer}}$

    $= \dfrac{720}{800}$

    $= \dfrac{9}{10}$

    $= 0.9$

    **c.**  $P(\text{breast cancer}) = \dfrac{\text{\# women w/breast cancer and positive test}}{\text{total number of women w/positive test}}$

    $= \dfrac{720}{7664}$

    $= \dfrac{45}{479}$

    $= 0.094$

2.  The sample space of equally likely outcomes is $S = \{1, 2, 3, 4, 5, 6\}$. There are six outcomes in the sample space, so $n(S) = 6$. The event of getting a number greater than 4 can be represented by $E = \{5, 6\}$. There are two outcomes in this event, so $n(E) = 2$.

    The probability of rolling a number greater than 4 is $P(E) = \dfrac{n(E)}{n(S)} = \dfrac{2}{6} = \dfrac{1}{3}$.

3.  We have $n(S) = 36$. The phrase "getting a sum of 5" describes the event $E = \{(1,4), (2,3), (3,2), (4,1)\}$. This event has 4

    outcomes, so $n(E) = 4$. Thus, the probability of getting a sum of 5 is $P(E) = \dfrac{n(E)}{n(S)} = \dfrac{4}{36} = \dfrac{1}{9}$.

4.  Let $E$ be the event of being dealt a king. Because there are 4 kings in the deck, the event of being dealt a king can occur in 4 ways, i.e., $n(E) = 4$. With 52 cards in the deck, $n(S) = 52$.

    The probability of being dealt a king is $P(E) = \dfrac{n(E)}{n(S)} = \dfrac{4}{52} = \dfrac{1}{13}$.

5. Because the order of the six numbers does not matter, this is a situation involving combinations. With one lottery ticket, there is only one way of winning so $n(E) = 1$. Using the combinations formula $_nC_r = \dfrac{n!}{(n-r)!\,r!}$ to find the number of outcomes in the sample space, we are selecting $r = 6$ numbers from a collection of $n = 49$ numbers.

$_{49}C_6 = \dfrac{49!}{43!\,6!} = 13{,}983{,}816$ So $n(S) = 13{,}983{,}816$. If a person bought one lottery ticket, the probability of winning was

$P(E) = \dfrac{n(E)}{n(S)} = \dfrac{1}{13{,}983{,}816}$

The probability of winning the state lottery was 0.0000000715.

6. $P(\text{does not live in North America}) = 1 - P(\text{lives in North America})$

$$= 1 - \frac{550}{7000}$$
$$= \frac{7000}{7000} - \frac{550}{7000}$$
$$= \frac{6450}{7000}$$
$$= \frac{129}{140} \approx 0.921$$

7. We find the probability that either of these mutually exclusive events will occur by adding their individual probabilities.

$P(4 \text{ or } 5) = P(4) + P(5) = \dfrac{1}{6} + \dfrac{1}{6} = \dfrac{2}{6} = \dfrac{1}{3}$

The probability of selecting a 4 or a 5 is $\dfrac{1}{3}$.

8. It is possible for the pointer to land on a number that is odd and less than 5. Two of the numbers, 1 and 3, are odd and less than 5. These events are not mutually exclusive. The probability of landing on a number that is odd and less than 5 is

$P(\text{odd or less than } 5) = P(\text{odd}) + P(\text{less than } 5) - P(\text{odd and less than } 5)$

$$= \frac{4}{8} + \frac{4}{8} - \frac{2}{8}$$
$$= \frac{6}{8}$$
$$= \frac{3}{4}$$

The probability that the pointer will stop on an odd number or a number less than 5 is $\dfrac{3}{4}$.

9.   **a.**   These events are not mutually exclusive.
      $P(\text{married or female}) = P(\text{married}) + P(\text{female}) - P(\text{married and female})$

$$= \frac{130}{242} + \frac{124}{242} - \frac{65}{242}$$
$$= \frac{189}{242} \approx 0.78$$

   **b.**   These events are mutually exclusive.
      $P(\text{divorced or widowed}) = P(\text{divorced}) + P(\text{widowed})$

$$= \frac{24}{242} + \frac{14}{242}$$
$$= \frac{38}{242} \approx 0.16$$

**10.** The wheel has 38 equally likely outcomes and 2 are green.

Thus, the probability of a green occurring on a play is $\dfrac{2}{38}$, or $\dfrac{1}{19}$.

The result that occurs on each play is independent of all previous results.

Thus, $P$ (green and green) $= P$ (green) $\cdot$ $P$ (green) $= \dfrac{1}{19} \cdot \dfrac{1}{19} = \dfrac{1}{361} \approx 0.00277.$

The probability of green occurring on two consecutive plays is $\dfrac{1}{361}$ or 0.00277..

**11.** If two or more events are independent, we can find the probability of them all occurring by multiplying the probabilities.

The probability of a baby boy is $\dfrac{1}{2}$.

Thus, the probability of having four boys in a row is $P\left(4 \text{ boys in a row}\right) = \dfrac{1}{2} \cdot \dfrac{1}{2} \cdot \dfrac{1}{2} \cdot \dfrac{1}{2}$
$$= \dfrac{1}{16}.$$

**Concept and Vocabulary Check 8.7**

**1.** empirical

**2.** sample space

**3.** $P(E)$; number of outcomes in $E$; total number of possible outcomes

**4.** 52; Hearts; diamonds; clubs; spades

**5.** 1; combinations

**6.** $1 - P(E)$; $1 - P(\text{not } E)$

**7.** mutually exclusive; $P(A) + P(B)$

**8.** $P(A) + P(B) - P(A \text{ and } B)$

**9.** independent; $P(A) \cdot P(B)$

**Exercise Set 8.7**

**1.** $P(\text{divorced}) = \dfrac{\text{number of persons divorced}}{\text{total number of U.S. adults}} = \dfrac{24}{242} = \dfrac{12}{121} \approx 0.10$

**3.** $P(\text{female}) = \dfrac{\text{number of females}}{\text{total number of U.S. adults}} = \dfrac{124}{242} = \dfrac{62}{141} \approx 0.51$

**5.** $P(\text{widowed male}) = \dfrac{\text{number of widowed males}}{\text{total number of U.S. adults}} = \dfrac{3}{242} \approx 0.01$

**7.** $P(\text{selecting a woman from the divorced population}) = \dfrac{\text{number of divorced women}}{\text{number of persons divorced}} = \dfrac{14}{24} = \dfrac{7}{14} \approx 0.58$

**9.** $P(\text{selecting a married man from the adult male population}) = \dfrac{\text{number of married men}}{\text{number of males}} = \dfrac{65}{118} \approx 0.55$

**11.** $P(R) = \dfrac{n(E)}{n(S)} = \dfrac{1}{6}$

**13.** $P(E) = \dfrac{n(E)}{n(S)} = \dfrac{3}{6} = \dfrac{1}{2}$

**15.** $P(E) = \dfrac{n(E)}{n(S)} = \dfrac{2}{6} = \dfrac{1}{3}$

**17.** $P(E) = \dfrac{n(E)}{n(S)} = \dfrac{4}{52} = \dfrac{1}{13}$

**19.** $P(E) = \dfrac{n(E)}{n(S)} = \dfrac{12}{52} = \dfrac{3}{13}$

**21.** $P(E) = \dfrac{n(E)}{n(S)} = \dfrac{1}{4}$

**23.** $P(E) = \dfrac{n(E)}{n(S)} = \dfrac{7}{8}$

**25.** $P(E) = \dfrac{n(E)}{n(S)} = \dfrac{3}{36} = \dfrac{1}{12}$

**27.** Buying 1 ticket:
$P(E) = \dfrac{n(E)}{n(S)} = \dfrac{1}{{}_{51}C_6} = \dfrac{1}{18,009,460}$
Buying 100 tickets:
$P(E) = \dfrac{100}{18,009,460} = \dfrac{5}{900,473}$

**29. a.** ${}_{52}C_5 = \dfrac{52!}{47!5!}$
$= \dfrac{52\cdot 51\cdot 50\cdot 49\cdot 48}{5\cdot 4\cdot 3\cdot 2\cdot 1} = 2,598,960$

**b.** ${}_{13}C_5 = \dfrac{13!}{8!5!} = \dfrac{13\cdot 12\cdot 11\cdot 10\cdot 9}{5\cdot 4\cdot 3\cdot 2\cdot 1} = 1287$

**c.** $P(E) = \dfrac{n(E)}{n(S)} = \dfrac{1287}{2,598,960} \approx 0.0005$

**31.** $P(\text{not completed 4 years of college}) = 1 - P(\text{completed 4 years of college}) = 1 - \dfrac{45}{174} = \dfrac{43}{58}$

**33.** $P(\text{completed H.S. or less than 4 yrs college}) = P(\text{completed H.S}) + P(\text{less than 4 yrs college})$

$$= \frac{56}{174} + \frac{44}{174} = \frac{100}{174} = \frac{50}{87}$$

**35.** $P(\text{completed 4 yrs H.S. or man}) = P(\text{completed 4 yrs H.S.}) + P(\text{man}) - P(\text{man who completed 4 yrs H.S})$

$$= \frac{56}{174} + \frac{82}{174} - \frac{25}{174} = \frac{113}{174}$$

**37.** $P(\text{not king}) = 1 - P(\text{king}) = 1 - \frac{4}{52} = 1 - \frac{1}{13} = \frac{12}{13}$

**39.** $P(2 \text{ or } 3) = P(2) + P(3) = \frac{4}{52} + \frac{4}{52} = \frac{8}{52} = \frac{2}{13}$

**41.** $P(7 \text{ or red card}) = P(7) + P(\text{red card}) - P(7 \text{ and red}) = \frac{4}{52} + \frac{26}{52} - \frac{2}{52} = \frac{28}{52} = \frac{7}{13}$

**43.** $P(\text{odd or less than 6}) = P(\text{odd}) + P(\text{less than 6}) - P(\text{odd \# less than 6}) = \frac{4}{8} + \frac{5}{8} - \frac{3}{8} = \frac{6}{8} = \frac{3}{4}$

**45.** $P(\text{professor or male}) = P(\text{professor}) + P(\text{male}) - P(\text{male professor}) = \frac{19}{40} + \frac{22}{40} - \frac{8}{40} = \frac{33}{40}$

**47.** $P(2 \text{ and } 3) = P(2) \cdot P(3) = \frac{1}{6} \cdot \frac{1}{6} = \frac{1}{36}$

**49.** $P(\text{even and greater than 2}) = P(\text{even}) \cdot P(\text{greater than 2}) = \frac{3}{6} \cdot \frac{4}{6} = \frac{1}{3}$

**51.** $P(\text{all heads}) = \frac{1}{2} \cdot \frac{1}{2} \cdot \frac{1}{2} \cdot \frac{1}{2} \cdot \frac{1}{2} \cdot \frac{1}{2} = \left(\frac{1}{2}\right)^6 = \frac{1}{64}$

**53.**  **a.**  $P(\text{hit 2 yrs in a row}) = P(\text{hit in 1}^{\text{st}} \text{ year and 2}^{\text{nd}} \text{ year}) = P(\text{hit 1}^{\text{st}} \text{ year}) \cdot P(\text{hit 2}^{\text{nd}} \text{ year}) = \frac{1}{16} \cdot \frac{1}{16} = \frac{1}{256}$

**b.**  $P(\text{hit 3 yrs in a row}) = \frac{1}{16} \cdot \frac{1}{16} \cdot \frac{1}{16} = \frac{1}{4096}$

**c.**  First find the probability that South Florida will not be hit by a major hurricane in a single year:

$P(\text{not hit}) = 1 - \frac{1}{16} = \frac{15}{16}$

$P(\text{not hit in next 10 years}) = \left(\frac{15}{16}\right)^{10} \approx 0.524$

**d.**  $P(\text{hit at least once}) = 1 - P(\text{hit none}) = 1 - \left(\frac{15}{16}\right)^{10} \approx 0.476$

**55. – 63.**  Answers will vary.

**65.** does not make sense; Explanations will vary. Sample explanation: The probability of a Democrat winning and the probability of a Republican winning are not necessarily equal. For instance, it is possible that the probability of a Democrat winning is 0.6 and the probability of a Republican winning is 0.4.

**67.** makes sense

**69.** Answers will vary.

**71.** **a.** P(Democrat who is not a business major)

$$= \frac{\text{\# of students who are Democrats but not business majors}}{\text{\# of students}}$$

$$= \frac{29-5}{50} = \frac{24}{50} = \frac{12}{25}$$

**b.** P(neither Democrat nor business major)

$= 1 - P(\text{Democrat or business major})$

$= 1 - \left(P(\text{Democrat}) + P(\text{business major}) - P(\text{Democrat and business major})\right)$

$= 1 - \left(\frac{29}{50} + \frac{11}{50} - \frac{5}{50}\right) = 1 - \frac{35}{50} = \frac{15}{50} = \frac{3}{10}$

**73.** **a.** The first person can have any birthday in the year. The second person can have all but one birthday.

**b.** $\frac{365}{365} \cdot \frac{364}{365} \cdot \frac{363}{365} \approx 0.99$

**c.** $100\% - 99\% = 0.01$

**d.** $\frac{365}{365} \cdot \frac{364}{365} \cdot \frac{363}{365} \cdots \frac{346}{365} \approx 0.59$

$1 - 0.59 = 0.41$

**e.** With 23 people, the probability that at least two people have the same birthday is

$P(E) = 1 - \frac{365}{365} \cdot \frac{364}{365} \cdot \frac{363}{365} \cdots + \frac{342}{365}$

$\approx 1 - 0.4927 \approx 0.5073$

## Chapter 8 Review Exercises

**1.**  $a_n = 7n - 4$

$a_1 = 7 - 4 = 3$

$a_2 = 14 - 4 = 10$

$a_3 = 21 - 4 = 17$

$a_4 = 28 - 4 = 24$

First four terms: 3, 10, 17, 24.

**2.**  $a_n = (-1)^n \dfrac{n+2}{n+1}$

$a_1 = (-1)^1 \dfrac{1+2}{1+1} = -\dfrac{3}{2}$

$a_2 = (-1)^2 \dfrac{2+2}{2+1} = \dfrac{4}{3}$

$a_3 = (-1)^3 \dfrac{3+2}{3+1} = -\dfrac{5}{4}$

$a_4 = (-1)^4 \dfrac{4+2}{4+1} = \dfrac{6}{5}$

First four terms: $-\dfrac{3}{2}, \dfrac{4}{3}, -\dfrac{5}{4}, \dfrac{6}{5}$.

**3.**  $a_n = \dfrac{1}{(n-1)!}$

$a_1 = \dfrac{1}{0!} = 1$

$a_2 = \dfrac{1}{1!} = 1$

$a_3 = \dfrac{1}{2!} = \dfrac{1}{2}$

$a_4 = \dfrac{1}{3!} = \dfrac{1}{6}$

First four terms: $1, 1, \dfrac{1}{2}, \dfrac{1}{6}$.

**4.**  $a_n = \dfrac{(-1)^{n+1}}{2^n}$

$a_1 = \dfrac{(-1)^2}{2^1} = \dfrac{1}{2}$

$a_2 = \dfrac{(-1)^3}{2^2} = -\dfrac{1}{4}$

$a_3 = \dfrac{(-1)^4}{2^3} = \dfrac{1}{8}$

$a_4 = \dfrac{(-1)^5}{2^4} = -\dfrac{1}{16}$

First four terms: $\dfrac{1}{2}, -\dfrac{1}{4}, \dfrac{1}{8}, -\dfrac{1}{16}$.

**5.**  $a_1 = 9$ and $a_n = \dfrac{2}{3a_{n-1}}$

$a_1 = 9$

$a_2 = \dfrac{2}{3 \cdot 9} = \dfrac{2}{27}$

$a_3 = \dfrac{2}{3} \cdot \dfrac{27}{2} = \dfrac{54}{6} = 9$

$a_4 = \dfrac{2}{3 \cdot 9} = \dfrac{2}{27}$

First four terms: $9, \dfrac{2}{27}, 9, \dfrac{2}{27}$.

**6.**  $a_1 = 4$ and $a_n = 2a_{n-1} + 3$

$a_1 = 4$

$a_2 = 2 \cdot 4 + 3 = 8 + 3 = 11$

$a_3 = 2 \cdot 11 + 3 = 22 + 3 = 25$

$a_4 = 2 \cdot 25 + 3 = 50 + 3 = 53$

First four terms: 4, 11, 25, and 53.

**7.**  $\dfrac{40!}{4!\,38!} = \dfrac{40 \cdot 39 \cdot 38!}{4 \cdot 3 \cdot 2 \cdot 1 \cdot 38!} = 65$

**8.** $\displaystyle\sum_{i=1}^{5}\left(2i^2-3\right)=(2-3)+\left(2\cdot2^2-3\right)+\left(2\cdot3^2-3\right)+\left(2\cdot4^2-3\right)+\left(2\cdot5^2-3\right)$
$$=-1+5+15+29+47$$
$$=95$$

**9.** $\displaystyle\sum_{i=0}^{4}(-1)^{i+1}i!=(-1)^1 0!+(-1)^2 1!+(-1)^3 3!+(-1)^4 4!$
$$=-1+1-2+6-24$$
$$=-20$$

**10.** $\dfrac{1}{3}+\dfrac{2}{4}+\dfrac{3}{5}+\cdots+\dfrac{15}{17}=\displaystyle\sum_{i=1}^{15}\dfrac{i}{i+2}$

**11.** $4^3+5^3+6^3+\cdots+13^3=\displaystyle\sum_{i=1}^{10}(i+3)^3$

**12.** $a_1=7,\ d=4$
First six terms: 7, 11, 15, 19, 23, 27.

**13.** $a_1=-4,\ d=-5$
First six terms: $-4,-9,-14,-19,-24,-29$.

**14.** $a_1=\dfrac{3}{2},\ d=-\dfrac{1}{2}$
First six terms: $\dfrac{3}{2},1,\dfrac{1}{2},0,-\dfrac{1}{2},-1$.

**15.** $a_{n+1}=a_n+5,\ a_1=-2$
First six terms: $-2, 3, 8, 13, 18,\ 23$.

**16.** $a_1=5,\ d=3$
$a_n=5+(n-1)3$
$a_6=5+(5)3=20$

**17.** $a_1=-8,\ d=-2$
$a_n=-8+(n-1)(-2)$
$a_{12}=-8+11(-2)=-30$

**18.** $a_1=14,\ d=-4$
$a_n=14+(n-1)(-4)$
$a_{14}=14+(13)(-4)=-38$

**19.** $-7,-3,1,5,\ldots$
$d=-3-(-7)=4$
$a_n=-7+(n-1)(4)$
$a_n=4n-11$
$a_{20}=4(20)-11$
$a_{20}=69$

**20.**  $a_1 = 200, d = -20$

$a_n = 200 + (n-1)(-20)$

$a_n = 220 - 20n$

$a_{20} = 220 - 20(20)$

$a_{20} = -180$

**21.**  $a_n = a_{n-1} - 5, a_1 = 3$

$d = -5$

$a_n = 3 + (n-1)(-5) = 3 - 5n + 5$

$a_n = 8 - 5n$

$a_{20} = 8 - 5(20) = -92$

**22.**  5, 12, 19, 26, …

$d = 7$

$a_n = 5 + (n-1)(7)$

$a_{22} = 5 + 21(7) = 152$

$S_{22} = \dfrac{22}{2}(5 + 152) = 1727$

**23.**  −6, −3, 0, 3, …

$d = 3$

$a_n = -6 + (n-1)3$

$a_{15} = -6 + (14)3 = 36$

$S_{15} = \dfrac{15}{2}(-6 + 36) = 225$

**24.**  $3 + 6 + 9 + \ldots + 300$

$S_{100} = \dfrac{100}{2}(3 + 300) = 15{,}150$

**25.**  $\displaystyle\sum_{i=1}^{16}(3i + 2)$

$a_1 = 3 + 2 = 5$

$a_{16} = 3(16) + 2 = 50$

$S_{16} = \dfrac{16}{2}(5 + 50) = 440$

**26.**  $\displaystyle\sum_{i=1}^{25}(-2i + 6)$

$a_1 = -2 + 6 = 4$

$a_{25} = -2(25) + 6 = -44$

$S_{25} = \dfrac{25}{2}(4 - 44) = -500$

**27.**  $\displaystyle\sum_{i=1}^{30} -5i$

$a_1 = -5$

$a_{30} = -5(30) = -150$

$S_{30} = \dfrac{30}{2}(-5 - 150) = -2325$

**28.**  **a.**  $a_n = 34.5 + (n-1)(-0.3)$

$= 34.5 - 0.3n + 0.3$

$= 34.8 - 0.3n$

  **b.**  $a_n = 34.8 - 0.3n$

$a_{56} = 34.8 - 0.3(56)$

$= 18$

The model projects wives will devote 18 hours per week in 2020.

**29.**  $a_n = 31{,}500 + (n-1)2300$

$a_{10} = 31{,}500 + (9)2300 = 52{,}200$

$S_{10} = \dfrac{10}{2}(31{,}500 + 52{,}200) = 418{,}500$

The total salary is $418, 500.

**30.**  $a_n = 25 + (n-1)$

$a_{35} = 25 + 34 = 59$

$S_{35} = \dfrac{35}{2}(25 + 59) = 1470$

There are 1470 seats.

**31.**  $a_1 = 3, r = 2$

First five terms: 3, 6, 12, 24, 48.

**32.**  $a_1 = \dfrac{1}{2}, r = \dfrac{1}{2}$

First five terms: $\dfrac{1}{2}, \dfrac{1}{4}, \dfrac{1}{8}, \dfrac{1}{16}, \dfrac{1}{32}$.

**33.**  $a_1 = 16, r = -\dfrac{1}{2}$

First five terms: 16, −8, 4, −2, 1.

**34.**  $a_n = -5a_{n-1}, a_1 = -1$

First five terms: −1, 5, −25, 125, −625.

**35.**  $a_1 = 2, r = 3$

$a_n = 2 \cdot 3^{n-1}$

$a_7 = 2 \cdot 3^6 = 1458$

**36.** $a_1 = 16, r = \dfrac{1}{2}$

$a_n = 16\left(\dfrac{1}{2}\right)^{n-1}$

$a_6 = 16\left(\dfrac{1}{2}\right)^5 = \dfrac{16}{32} = \dfrac{1}{2}$

**37.** $a_1 = -3, r = 2$

$a_n = -3 \cdot 2^{n-1}$

$a_5 = -3 \cdot 2^4 = -48$

**38.** $1, 2, 4, 8, \ldots$

$a_1 = 1, r = \dfrac{2}{1} = 2$

$a_n = 2^{n-1}$

$a_8 = 2^7 = 128$

**39.** $100, 10, 1, \dfrac{1}{10}, \ldots$

$a_1 = 100, r = \dfrac{10}{100} = \dfrac{1}{10}$

$a_n = 100\left(\dfrac{1}{10}\right)^{n-1}$

$a_8 = 100\left(\dfrac{1}{10}\right)^7 = \dfrac{1}{100,000}$

**40.** $12, -4, \dfrac{4}{3}, -\dfrac{4}{9}\ldots$

$a_1 = 12, r = -\dfrac{4}{12} = -\dfrac{1}{3}$

$a_n = 12\left(-\dfrac{1}{3}\right)^{n-1}$

$a_8 = 12\left(-\dfrac{1}{3}\right)^7 = -\dfrac{4}{729}$

**41.** $5, -15, 45, -135, \ldots$

$r = \dfrac{-15}{5} = -3$

$S_{15} = \dfrac{5\left[1-(-3)^{15}\right]}{1-(-3)} = 17{,}936{,}135$

**42.** $r = \dfrac{1}{2}, a_1 = 8$

$S_{78} = \dfrac{8\left[1-\left(\dfrac{1}{2}\right)^7\right]}{1-\dfrac{1}{2}} = -16\left(1-\dfrac{1}{128}\right)$

$= -16\left(-\dfrac{127}{128}\right) = \dfrac{127}{8}$

**43.** $S_6 = \dfrac{5\left(1-5^6\right)}{1-5} = \dfrac{5(-15624)}{-4} = 19{,}530$

**44.** $\displaystyle\sum_{i=1}^{7} 3(-2)^i$

$a_1 = -6, r = -2$

$S_7 = \dfrac{-6\left[1-(-2)^7\right]}{1-(-2)} = \dfrac{-6(129)}{3} = -258$

**45.** $\displaystyle\sum_{i=1}^{5} 2\left(\tfrac{1}{4}\right)^{i-1}$

$a_1 = 2, r = \dfrac{1}{4}$

$S_5 = \dfrac{2\left[1-\left(\dfrac{1}{4}\right)^5\right]}{1-\dfrac{1}{4}} = \dfrac{2\left(\dfrac{1023}{1024}\right)}{\dfrac{3}{4}} = \dfrac{341}{128}$

**46.** $a_1 = 9, r = \dfrac{1}{3}$

$S_\infty = \dfrac{9}{1-\dfrac{1}{3}} = \dfrac{9}{\dfrac{2}{3}} = 9 \cdot \dfrac{3}{2} = \dfrac{27}{2}$

**47.** $a_1 = 2, r = -\dfrac{1}{2}$

$S_\infty = \dfrac{2}{1-\left(-\dfrac{1}{2}\right)} = \dfrac{2}{\dfrac{3}{2}} = \dfrac{4}{3}$

**48.** $a_1 = -6, r = -\dfrac{2}{3}$

$S_\infty = \dfrac{-6}{1-\left(-\dfrac{2}{3}\right)} = \dfrac{-6}{\dfrac{5}{3}} = -\dfrac{18}{5}$

**49.** $r = 0.8$

$S_\infty = \dfrac{4}{1-0.8} = 20$

**50.** $0.\overline{6} = 0.6 + 0.06 + 0.006 + \cdots$

$a_1 = \dfrac{6}{10}, r = \dfrac{1}{10}$

$S_\infty = \dfrac{\dfrac{6}{10}}{1 - \dfrac{1}{10}} = \dfrac{\dfrac{6}{10}}{\dfrac{9}{10}} = \dfrac{6}{9} = \dfrac{2}{3}$

**51.** $0.\overline{47} = 0.47 + 0.0047 + 0.000047 + \cdots$

$a_1 = \dfrac{47}{100}, r = \dfrac{1}{100}$

$S_\infty = \dfrac{\dfrac{47}{100}}{1 - \dfrac{1}{100}} = \dfrac{\dfrac{47}{100}}{\dfrac{99}{100}} = \dfrac{47}{99}$

**52. a.** Divide each value by the previous value:

$\dfrac{18.80}{18.44} \approx 1.02$

$\dfrac{18.44}{18.15} \approx 1.02$

$\dfrac{18.15}{17.86} \approx 1.02$

$\dfrac{17.86}{17.58} \approx 1.02$

$\dfrac{17.58}{17.30} \approx 1.02$

$\dfrac{17.30}{17.03} \approx 1.02$

$\dfrac{17.03}{16.76} \approx 1.02$

$\dfrac{16.76}{16.50} \approx 1.02$

$\dfrac{16.50}{16.24} \approx 1.02$

$\dfrac{16.24}{15.98} \approx 1.02$

The population is increasing geometrically with $r = 1.02$.

**b.** $a_n = 15.98(1.02)^n$

**c.** 2080 is 8 decades after 2000 so $n = 8$.

$a_n = 15.98(1.02)^{n-1}$

$a_{31} = 15.98(1.02)^{31-1}$

$\approx 28.95$

In 2030, the model predicts Florida's population will be 28.95 million.

**53.** $a_1 = 32,000,\ r = 1.06$

$a_6 = 32,000(1.06)^5 \approx \$42,823$

The sixth year salary is \$42, 823.

$$S_6 = \frac{32,000\left(1 - 1.06^6\right)}{1 - 1.06}$$

$$= \frac{32,000\left(1 - 1.06^6\right)}{-0.06}$$

$$\approx 223,210$$

The total salary paid is \$223, 210.

**54.** **a.** $A = \dfrac{P\left[\left(1 + \frac{r}{n}\right)^{nt} - 1\right]}{\frac{r}{n}}$

$P = \$520,\ r = 0.06,\ n = 1,\ t = 20$

$$A = \frac{\$520\left[\left(1 + \dfrac{0.06}{1}\right)^{1\cdot20} - 1\right]}{\dfrac{0.06}{1}} = \frac{\$520\left[(1.06)^{20} - 1\right]}{0.06} \approx \$19,129$$

The value of the annuity will be \$19,129.

**b.** Interest = Value of annuity − Total deposits

$\approx\ \$19,129\ -\ \$520 \cdot 20$

$\approx\ \$8729$

**55.** **a.** $A = \dfrac{P\left[\left(1 + \frac{r}{n}\right)^{nt} - 1\right]}{\frac{r}{n}}$

$P = 100,\ r = 0.055,\ n = 12,\ t = 30$

$$A = \frac{\$100\left[\left(1 + \dfrac{0.055}{12}\right)^{12\cdot30} - 1\right]}{\dfrac{0.055}{12}} \approx \$91,361$$

The value of the IRA will be \$91,361.

**b.** Interest = Value of IRA − Total deposits

$\approx\ \$91,361\ -\ \$100 \cdot 12 \cdot 30$

$\approx\ \$55,361$

**56.** $4(0.7) + 4(0.7)^2 + \cdots;\ r = 0.7$

$$S_\infty = \frac{4(0.7)}{1 - 0.7} = 9.\overline{3}$$

The total spending is $\$9\dfrac{1}{3}$ million.

**57.** $S_1 : 5 = \dfrac{5(1)(1+1)}{2}$

$\qquad 5 = \dfrac{5(2)}{2}$

$\qquad 5 = 5$ is true.

$S_k : 5 + 10 + 15 + \cdots + 5k = \dfrac{5k(k+1)}{2}$

$S_{k+1} : 5 + 10 + 15 + \cdots + 5k + 5(k+1)$

$\qquad = \dfrac{5(k+1)(k+2)}{2}$

Add $5(k+1)$ to both sides of $S_k$ :

$5 + 10 + 15 + \cdots + 5k + 5(k+1)$

$\qquad = \dfrac{5k(k+1)}{2} + 5k(k+1)$

Simplify the right-hand side:

$\qquad \dfrac{5k(k+1)}{2} + 5(k+1) = \dfrac{5k(k+1) + 10(k+1)}{2}$

$\qquad\qquad\qquad\qquad = \dfrac{(5k+10)(k+1)}{2}$

$\qquad\qquad\qquad\qquad = \dfrac{5(k+1)(k+2)}{2}$

If $S_k$ is true, then $S_{k+1}$ is true.
The statement is true for all $n$.

**58.** $S_1 : 1 = \dfrac{4^1 - 1}{3}$

$\qquad 1 = \dfrac{3}{3}$

$\qquad 1 = 1$ is true.

$S_k : 1 + 4 + 4^2 + \cdots + 4^{k-1} = \dfrac{4^k - 1}{3}$

$S_{k+1} : 1 + 4 + 4^2 + \cdots + 4^{k-1} + 4^k = \dfrac{4^{k+1} - 1}{3}$ Add $4^k$ to both sides of $S_k$ :

$S_k : 1 + 4 + 4^2 + \cdots + 4^{k-1} = \dfrac{4^k - 1}{3}$

$1 + 4 + 4^2 + \cdots + 4^{k-1} + 4^k = \dfrac{4^k - 1}{3} + 4^k$

Simplify the right-hand side:

$\qquad \dfrac{4^k - 1}{3} + 4^k = \dfrac{4^k - 1 + 3 \cdot 4^k}{3}$

$\qquad\qquad\qquad = \dfrac{4 \cdot 4^k - 1}{3}$

$\qquad\qquad\qquad = \dfrac{4^{k+1} - 1}{3}$

If $S_k$ is true, then $S_{k+1}$ is true.
The statement is true for all $n$.

**59.**    $S_1 : 2 = 2(1)^2$

$\qquad 2 = 2$ is true.

$\qquad S_k : 2+6+10+\cdots+(4k-2) = 2k^2$

$\qquad S_{k+1} : 2+6+10+\cdots+(4k-2)+(4k+2) = 2(k+1)^2$

Add $(4k+2)$ to both sides of $S_k$:

$2+6+10+\cdots+(4k-2)+(4k+2) = 2k^2+(4k+2)$

Simplify the right-hand side:

$\qquad 2k^2+4k+2 = 2(k^2+2k+1)$

$\qquad\qquad\qquad = 2(k+1)^2$

If $S_k$ is true, then $S_{k+1}$ is true. The statement is true for all $n$.

**60.**    $S_1 : 1 \cdot 3 = \dfrac{1(1+1)[2(1)+7]}{6}$

$\qquad 3 = \dfrac{2 \cdot 9}{6}$

$\qquad 3 = \dfrac{18}{6}$

$\qquad 3 = 3$ is true.

$\qquad S_k : 1\cdot 3+2\cdot 4+3\cdot 5+\cdots+k(k+2) = \dfrac{k(k+1)(2k+7)}{6}$

$\qquad S_{k+1} : 1\cdot 3+2\cdot 4+3\cdot 5+\cdots+k(k+2)+(k+1)(k+3) = \dfrac{(k+1)(k+2)(2k+9)}{6}$

Add $(k+1)(k+3)$ to both sides of $S_k$:

$1\cdot 3+2\cdot 4+3\cdot 5+\cdots+k(k+2)+(k+1)(k+3) = \dfrac{k(k+1)(2k+7)}{6}+(k+1)(k+3)$

Simplify the right-hand side:

$= \dfrac{k(k+1)(2k+7)}{6}+(k+1)(k+3)$

$= \dfrac{k(k+1)(2k+7)+6(k+1)(k+3)}{6}$

$= \dfrac{(k+1)[k(2k+7)+6(k+3)]}{6}$

$= \dfrac{(k+1)(2k^2+13k+18)}{6}$

$= \dfrac{(k+1)(k+2)(2k+9)}{6}$

If $S_k$ is true, then $S_{k+1}$ is true. The statement is true for all $n$.

**61.** $S_1$: 2 is a factor of $1^2 + 5(1) = 6$ since $6 = 2 \cdot 3$.

$S_k$: 2 is a factor of $k^2 + 5k$.

$S_{k+1}$: 2 is a factor of $(k+1)^2 + 5(k+1)$.

$$(k+1)^2 + 5(k+1) = k^2 + 2k + 1 + 5k + 5$$
$$= k^2 + 7k + 6$$
$$= k^2 + 5k + 2(k+3)$$
$$= (k^2 + 5k) + 2(k+3)$$

Because we assume $S_k$ is true, we know 2 is a factor of $k^2 + 5k$. Since 2 is a factor of $2(k+3)$, we conclude 2 is a factor of the sum $(k^2 + 5k) + 2(k+3)$. If $S_k$ is true, then $S_{k+1}$ is true. The statement is true for all $n$.

**62.** $\dbinom{11}{8} = \dfrac{11!}{3!8!} = \dfrac{11 \cdot 10 \cdot 9}{3 \cdot 2 \cdot 1} = 165$

**63.** $\dbinom{90}{2} = \dfrac{90!}{88!2!} = \dfrac{90 \cdot 89}{2 \cdot 1} = 4005$

**64.** $(2x+1)^3 = \dbinom{3}{0}(2x)^3 + \dbinom{3}{1}(2x)^2 \cdot 1 + \dbinom{3}{2}(2x)1^2 + \dbinom{3}{3}1^3$

$= 8x^3 + 3(4x^2) + 3(2x) + 1$

$= 8x^3 + 12x^2 + 6x + 1$

**65.** $(x^2-1)^4 = \dbinom{4}{0}(x^2)^4 + \dbinom{4}{1}(x^2)^3(-1) + \dbinom{4}{2}(x^2)^2(-1)^2 + \dbinom{4}{3}x^2(-1)^3 + \dbinom{4}{4}(-1)^4$

$= x^8 - 4x^6 + 6x^4 - 4x^2 + 1$

**66.** $(x+2y)^5 = \dbinom{5}{0}x^5 + \dbinom{5}{1}x^4(2y) + \dbinom{5}{2}x^3(2y)^2 + \dbinom{5}{3}x^2(2y)^3 + \dbinom{5}{4}x(2y)^4 + \dbinom{5}{5}(2y)^5$

$= x^5 + 5(2)x^4y + 10(4)x^3y^2 + 10(8)x^2y^3 + 5(16)xy^4 + 32y^5$

$= x^5 + 10x^4y + 40x^3y^2 + 80x^2y^3 + 80xy^4 + 32y^5$

**67.** $(x-2)^6 = \dbinom{6}{0}x^6 + \dbinom{6}{1}x^5(-2) + \dbinom{6}{2}x^4(-2)^2 + \dbinom{6}{3}x^3(-2)^3 + \dbinom{6}{4}x^2(-2)^4 + \dbinom{6}{5}x(-2)^5 \dbinom{6}{6}(-2)^6$

$= x^6 + 6x^5(-2) + 15x^4(4) + 20x^3(-8) + 15x^2(16) + 6x(-32) + 64$

$= x^6 - 12x^5 + 60x^4 - 160x^3 + 240x^2 - 192x + 64$

**68.** $(x^2+3)^8 = \dbinom{8}{0}(x^2)^8 + \dbinom{8}{1}(x^2)^7 3 + \dbinom{8}{2}(x^2)^6 3^2 + \cdots$

$= x^{16} + 8x^{14}3 + 28x^{12}9 + \cdots$

$= x^{16} + 24x^{14} + 252x^{12} + \cdots$

**69.** $(x-3)^9 = \binom{9}{0}x^9 + \binom{9}{1}x^8(-3) + \binom{9}{2}x^7(-3)^2 - \cdots$

$\qquad = x^9 + 9(-3)x^8 + 36(9)x^7 - \cdots$

$\qquad = x^9 - 27x^8 + 324x^7 - \cdots$

**70.** $(x+2)^5$

$\quad$ fourth term $= \binom{5}{3}x^2(2)^3$

$\qquad\qquad = 10(8)x^2 = 80x^2$

**71.** $(2x-3)^6$

$\quad$ fifth term $= \binom{6}{4}(2x)^2(-3)^4$

$\qquad\qquad = 15(4x^2)(81) = 4860x^2$

**72.** $_8P_3 = \dfrac{8!}{5!} = 8 \cdot 7 \cdot 6 = 336$

**73.** $_9P_5 = \dfrac{9!}{4!} = 9 \cdot 8 \cdot 7 \cdot 6 \cdot 5 = 15{,}120$

**74.** $_8C_3 = \dfrac{8!}{5!3!} = \dfrac{8 \cdot 7 \cdot 6}{3 \cdot 2 \cdot 1} = 56$

**75.** $_{13}C_{11} = \dfrac{13!}{2!11!} = \dfrac{13 \cdot 12}{2 \cdot 1} = 78$

**76.** $4 \cdot 5 = 20$ choices

**77.** $3^5 = 243$ possibilities

**78.** $_{15}P_4 = \dfrac{15!}{11!} = 15 \cdot 14 \cdot 13 \cdot 12 = 32{,}760$ ways

**79.** $_{20}C_4 = \dfrac{20!}{16!4!} = \dfrac{20 \cdot 19 \cdot 18 \cdot 17}{4 \cdot 3 \cdot 2 \cdot 1} = 4845$ ways

**80.** $_{20}C_3 = \dfrac{20!}{17!3!} = \dfrac{20 \cdot 19 \cdot 18}{3 \cdot 2 \cdot 1} = 1140$ sets

**81.** $_{20}P_4 = \dfrac{20!}{16!}$

$\qquad = 20 \cdot 19 \cdot 18 \cdot 17$

$\qquad = 116{,}280$ ways

**82.** $5! = 120$ ways

**83.**  $P(\text{public college}) = \dfrac{252}{350} = \dfrac{18}{25}$

**84.**  $P(\text{not from high-income family}) = 1 - P(\text{from high-income family}) = 1 - \dfrac{50}{350} = \dfrac{350}{350} - \dfrac{50}{350} = \dfrac{300}{350} = \dfrac{6}{7}$

**85.**  $P(\text{from middle-income family  or  high-income family}) = \dfrac{160 + 50}{350} = \dfrac{210}{350} = \dfrac{3}{5}$

**86.**  $P(\text{attended private college  or  is from a high income family})$

$= P(\text{private college}) + P(\text{high income family}) - P(\text{attended private college  and  is from a high income family})$

$= \dfrac{98}{350} + \dfrac{50}{350} - \dfrac{28}{350} = \dfrac{120}{350} = \dfrac{12}{35}$

**87.**  $P(\text{selecting a student from a low-income family from among those attending public college }) = \dfrac{120}{252} = \dfrac{10}{21}$

**88.**  $P(\text{selecting a student that attends a private college from among those in middle-income families}) = \dfrac{50}{160} = \dfrac{5}{16}$

**89.**  $P(\text{less than 5}) = P(\text{rolling a } 1, 2, 3, 4)$

$= \dfrac{4}{6} = \dfrac{2}{3}$

**90.**  $P(\text{less than 3 or greater than 4})$

$= P(\text{less than 3}) + P(\text{greater than 4})$

$= P(\text{rolling a } 1, 2) + P(\text{rolling a } 5, 6)$

$= \dfrac{2}{6} + \dfrac{2}{6} = \dfrac{4}{6} = \dfrac{2}{3}$

**91.**  $P(E) = \dfrac{4}{52} + \dfrac{4}{52} = \dfrac{8}{52} = \dfrac{2}{13}$

**92.**  $P(E) = \dfrac{4}{52} + \dfrac{26}{52} - \dfrac{2}{52} = \dfrac{28}{52} = \dfrac{7}{13}$

**93.**  $P(\text{not yellow}) = 1 - P(\text{yellow}) = 1 - \dfrac{1}{6} = \dfrac{5}{6}$

**94.**  $P\left(\text{red or greater than 3}\right) = \dfrac{3}{6} + \dfrac{3}{6} - \dfrac{1}{6} = \dfrac{5}{6}$

**95.**  $P\left(\text{green, then less than 4}\right) = \dfrac{2}{6} \cdot \dfrac{3}{6} = \dfrac{6}{36} = \dfrac{1}{6}$

**96.**  **a.**  $P(E) = \dfrac{n(E)}{n(S)} = \dfrac{1}{{}_{20}C_5} = \dfrac{1}{15,504}$

**b.**  $P(E) = \dfrac{100}{15,504} = \dfrac{25}{3876}$

**97.** $P(E) = \left(\dfrac{1}{2}\right)^5 = \dfrac{1}{32}$

**98. a.** $(0.2)^2 = 0.04$

   **b.** $(0.2)^3 = 0.008$

   **c.** $(1 - 0.2)^4 = (0.8)^4 = 0.4096$

## Chapter 8 Test

**1.** $a_n = \dfrac{(-1)^{n+1}}{n^2}$

$a_1 = \dfrac{(-1)^2}{1^2} = 1$

$a_2 = \dfrac{(-1)^3}{2^2} = -\dfrac{1}{4}$

$a_3 = \dfrac{(-1)^4}{3^2} = \dfrac{1}{9}$

$a_4 = \dfrac{(-1)^5}{4^2} = -\dfrac{1}{16}$

$a_5 = \dfrac{(-1)^6}{5^2} = \dfrac{1}{25}$

First five terms: $1, -\dfrac{1}{4}, \dfrac{1}{9}, -\dfrac{1}{16}, \dfrac{1}{25}$.

**2.** $\displaystyle\sum_{i=1}^{5}\left(i^2 + 10\right) = 11 + 14 + 19 + 26 + 35 = 105$

**3.** $\displaystyle\sum_{i=1}^{20}(3i - 4)$

$a_1 = 3 - 4 = -1$

$d = 3$

$a_n = -1 + (n - 1)3$

$a_{20} = -1 + (19)3 = 56$

$S_{20} = \dfrac{20}{2}(-1 + 56) = 550$

**4.** $\displaystyle\sum_{i=1}^{15}(-2)^i$

$a_1 = -2, \ r = -2$

$S_{15} = \dfrac{-2\left[1 - (-2)^{15}\right]}{1 - (-2)} = -21{,}846$

**5.** $\dbinom{9}{2} = \dfrac{9!}{7!2!} = \dfrac{9 \cdot 8}{2 \cdot 1} = 36$

**6.** $_{10}P_3 = \dfrac{10!}{7!} = 10 \cdot 9 \cdot 8 = 720$

**7.** $_{10}C_3 = \dfrac{10!}{7!3!} = \dfrac{10 \cdot 9 \cdot 8}{3 \cdot 2 \cdot 1} = 120$

**8.** $\dfrac{2}{3} + \dfrac{3}{4} + \dfrac{4}{5} + \cdots + \dfrac{21}{22} = \displaystyle\sum_{i=1}^{20} \dfrac{i+1}{i+2}$

**9.** $4, 9, 14, 19, \ldots$
$a_1 = 4, \ d = 5$
$a_n = 4 + (n-1) \cdot 5 = 4 + 5n - 1$
$a_n = 5n - 1$
$a_{12} = 5(12) - 1 = 59$

**10.** $16, 4, 1, \dfrac{1}{4}, \cdots$

$a_1 = 16, \ r = \dfrac{1}{4}$

$a_n = 16 \left( \dfrac{1}{4} \right)^{n-1}$

$a_{12} = 16 \left( \dfrac{1}{4} \right)^{11} = \dfrac{1}{262,144}$

**11.** $7, -14, 28, -56, \ldots$
$a_1 = 7, \ r = -2$

$S_{10} = \dfrac{7\left[1 - (-2)^{10}\right]}{1 - (-2)} = \dfrac{7(-1023)}{3} = -2387$

**12.** $-7, -14, -21, -28, \ldots$
$a_1 = -7, \ d = -7$
$a_n = -7 + (n-1)(-7)$
$a_{10} = -7 + 9(-7) = -70$
$S_{10} = \dfrac{10}{2}(-7 - 70) = -385$

**13.** $4 + \dfrac{4}{2} + \dfrac{4}{2^2} + \dfrac{4}{2^3} + \cdots$

$r = \dfrac{1}{2}$

$S_\infty = \dfrac{4}{1 - \dfrac{1}{2}} = 8$

**14.** $0.\overline{73} = 0.73 + 0.0073 + 0.000073 + \cdots$

$a_1 = \dfrac{73}{100}, r = \dfrac{1}{100}$

$S_\infty = \dfrac{\dfrac{73}{100}}{1 - \dfrac{1}{100}} = \dfrac{\dfrac{73}{100}}{\dfrac{99}{100}} = \dfrac{73}{99}$

**15.** $a_1 = 30{,}000, r = 1.04$

$S_8 = \dfrac{30{,}000\left[1 - (1.04)^8\right]}{1 - 1.04} \approx 276{,}426.79$

The total salary is \$276,427.

**16.** $S_1 : 1 = \dfrac{1[3(1) - 1]}{2}$

$1 = \dfrac{2}{2}$

$1 = 1$ is true.

$S_k : 1 + 4 + 7 + \cdots + (3k - 2) = \dfrac{k(3k - 1)}{2}$

$S_{k+1} : 1 + 4 + 7 + \cdots + (3k - 2) + (3k + 1) = \dfrac{(k+1)(3k+2)}{2}$

Add $(3k + 1)$ to both sides of $S_k$ :

$1 + 4 + 7 + \cdots + (3k - 2) + (3k + 1) = \dfrac{k(3k - 1)}{2} + (3k + 1)$

Simplify the right-hand side:

$\dfrac{k(3k - 1)}{2} + (3k + 1) = \dfrac{k(3k - 1) + 2(3k + 1)}{2}$

$= \dfrac{3k^2 + 5k + 2}{2}$

$= \dfrac{(k+1)(3k+2)}{2}$

If $S_k$ is true, then $S_{k+1}$ is true. The statement is true for all $n$.

**17.** $\left(x^2 - 1\right)^5 = \dbinom{5}{0}\left(x^2\right)^5 + \dbinom{5}{1}\left(x^2\right)^4(-1) + \dbinom{5}{2}\left(x^2\right)^3(-1)^2 + \dbinom{5}{3}\left(x^2\right)^2(-1)^3 + \dbinom{5}{4}x^2(-1)^4 + \dbinom{5}{5}(-1)^5$

$\qquad = x^{10} - 5x^8 + 10x^6 - 10x^4 + 5x^2 - 1$

**18.** $\left(x + y^2\right)^8$

First Term $\dbinom{n}{r-1}a^{n-r+1}b^{r-1} = \dbinom{8}{1-1}x^{8-1+1}\left(y^2\right)^{1-1} = \dbinom{8}{0}x^8\left(y^2\right)^0 = \dfrac{8!}{0!(8-0)!}x^8 \cdot 1 = \dfrac{\cancel{8!}}{0!\cancel{8!}}x^8$

$\qquad\qquad = x^8$

Second Term $\dbinom{n}{r-1}a^{n-r+1}b^{r-1} = \dbinom{8}{2-1}x^{8-2+1}\left(y^2\right)^{2-1} = \dbinom{8}{1}x^7\left(y^2\right)^1 = \dfrac{8!}{1!(8-1)!}x^7y^2 = \dfrac{8 \cdot \cancel{7!}}{1 \cdot \cancel{7!}}x^7y^2$

$\qquad\qquad = 8x^7y^2$

Third Term $\dbinom{n}{r-1}a^{n-r+1}b^{r-1} = \dbinom{8}{3-1}x^{8-3+1}\left(y^2\right)^{3-1} = \dbinom{8}{2}x^6\left(y^2\right)^2 = \dfrac{8!}{2!(8-2)!}x^6 y^4 = \dfrac{8\cdot 7\cdot \cancel{6!}}{2\cdot 1\cdot \cancel{6!}}x^6 y^4$

$$= 28x^6 y^4$$

$x^8 + 8x^7 y^2 + 28x^6 y^4 + \cdots$

**19.** $_{11}P_3 = \dfrac{11!}{8!} = 11\cdot 10\cdot 9 = 990$ ways

**20.** $_{10}C_4 = \dfrac{10!}{6!4!} = \dfrac{10\cdot 9\cdot 8\cdot 7}{4\cdot 3\cdot 2\cdot 1} = 210$ sets

**21.** Four digits are open: $10^4 = 10,000$

**22.** $P(\text{not brown eyes}) = 1 - P(\text{brown eyes})$

$$= 1 - \dfrac{40}{50}$$

$$= \dfrac{60}{100} = \dfrac{3}{5}$$

**23.** $P(\text{brown eyes or blue eyes})$

$= P(\text{brown eyes}) + P(\text{blue eyes})$

$= \dfrac{40}{100} + \dfrac{38}{100} = \dfrac{78}{100} = \dfrac{39}{50}$

**24.** $P(\text{female or green eyes})$

$= P(\text{female}) + P(\text{green eyes})$

$\qquad - P(\text{female with green eyes})$

$= \dfrac{50}{100} + \dfrac{22}{100} - \dfrac{12}{100} = \dfrac{60}{100} = \dfrac{3}{5}$

**25.** $P(\text{male, given blue eyes}) = \dfrac{18}{38} = \dfrac{9}{19}$

**26.** $_{15}C_6 = \dfrac{15!}{9!6!} = \dfrac{15\cdot 14\cdot 13\cdot 12\cdot 11\cdot 10}{6\cdot 5\cdot 4\cdot 3\cdot 2} = 5005$

$P(E) = \dfrac{50}{5005} = \dfrac{10}{1001}$

**27.** $P(E) = \dfrac{26}{52} + \dfrac{12}{52} - \dfrac{6}{52} = \dfrac{32}{52} = \dfrac{8}{13}$

**28.** $P(E) = \dfrac{25}{50} + \dfrac{20}{50} - \dfrac{15}{50} = \dfrac{30}{50} = \dfrac{3}{5}$

**29.** $P(E) = \left(\dfrac{1}{4}\right)^4 = \dfrac{1}{256}$

**30.** $P(E) = \dfrac{2}{8}\cdot\dfrac{2}{8} = \dfrac{1}{16}$

**Cumulative Review Exercises (Chapters 1–8)**

1.  domain: $[-4,1)$; range: $(-\infty, 2]$

2.  maximum of 2 at $x = -2$

3.  decreasing interval: $(-2,1)$

4.  neither

5.  $f(-3) = 1$ and $f(-1) = 1$

6.  $(f \circ f)(-4) = f(f(-4)) = f(0) = 0$

7.  $f(x) \to -\infty$ as $x \to 1^-$

8.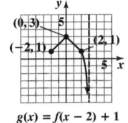

    $g(x) = f(x - 2) + 1$

9.

    $h(x) = -f(2x)$

10. $-2(x - 5) + 10 = 3(x + 2)$
    $-2x + 10 + 10 = 3x + 6$
    $\phantom{-2x + 10 + }14 = 5x$
    $\phantom{-2x + 10 + 1}x = \dfrac{14}{5}$

    The solution set is $\left\{ \dfrac{14}{5} \right\}$.

11. $3x^2 - 6x + 2 = 0$

    $x = \dfrac{6 \pm \sqrt{36 - 24}}{6}$

    $\phantom{x} = \dfrac{6 \pm \sqrt{12}}{6}$

    $\phantom{x} = \dfrac{6 \pm 2\sqrt{3}}{6}$

    $\phantom{x} = \dfrac{3 \pm \sqrt{3}}{3}$

    The solution set is $\left\{ \dfrac{3+\sqrt{3}}{3}, \dfrac{3-\sqrt{3}}{3} \right\}$.

12. $\log_2 x + \log_2 (2x - 3) = 1$
    $\phantom{\log_2 x}\log_2 x(2x - 3) = 1$
    $\phantom{\log_2 x \log_2}x(2x - 3) = 2$
    $\phantom{\log_2 x \log}2x^2 - 3x - 2 = 0$
    $\phantom{\log_2 x \log}(2x + 1)(x - 2) = 0$
    $2x + 1 = 0 \qquad$ or $\qquad x - 2 = 0$
    $\phantom{2x + 1 = 0 \quad or \quad}x = 2$
    $\phantom{2x + 1}x = -\dfrac{1}{2}$

    $x = -\dfrac{1}{2}$ does not check since $\log_2 \left( -\dfrac{1}{2} \right)$ does not exist.
    The solution set is $\{2\}$.

13. $x^{1/2} - 6x^{1/4} + 8 = 0$
    Let $t = x^{1/4}$.
    $\phantom{Let }t^2 - 6t + 8 = 0$
    $\phantom{Le}(t - 2)(t - 4) = 0$
    $\phantom{Le}t - 2 = 0 \quad$ or $\quad t - 4 = 0$
    $\phantom{Let xx}t = 2 \qquad\qquad t = 4$
    $\phantom{Let x}x^{1/4} = 2 \qquad\quad x^{1/4} = 4$
    $\phantom{Let xxx}x = 16 \qquad\quad x = 256$
    The solution set is $\{16, 256\}$.

**14.** $\sqrt{2x+4}-\sqrt{x+3}-1=0$

$$\left(\sqrt{2x+4}\right)^2=\left(\sqrt{x+3}+1\right)^2$$
$$2x+4=(x+3)+2\sqrt{x+3}+1$$
$$x=2\sqrt{x+3}$$
$$x^2=4(x+3)$$
$$x^2-4x-12=0$$
$$(x-6)(x+2)=0$$
$$x-6=0 \quad\text{or}\quad x+2=0$$
$$x=6 \qquad\qquad x=-2$$
$$x=-2 \text{ does not check.}$$

The solution set is $\{6\}$.

**15.** $|2x+1|\le 1$
$$-1\le 2x+1\le 1$$
$$-2\le 2x\le 0$$
$$-1\le x\le 0 \text{ or } [-1, 0]$$
The solution set is $\{x|-1\le x\le 0\}$ or $[-1, 0]$.

**16.** 
$$6x^2-6<5x$$
$$6x^2-5x-6<0$$
$$6x^2-5x-6=0$$
$$(3x+2)(2x-3)=0$$
$$3x+2=0 \text{ or } 2x-3=0$$
$$x=-\frac{2}{3} \qquad x=\frac{3}{2}$$

The test intervals are $\left(-\infty, -\frac{2}{3}\right), \left(-\frac{2}{3}, \frac{3}{2}\right),$ and $\left(\frac{3}{2}, \infty\right)$. Testing a point in each interval shows that the solution is $\left(-\frac{2}{3}, \frac{3}{2}\right)$.

**17.** $\dfrac{x-1}{x+3}\le 0$

The test intervals are $(-\infty, -3), (-3, 1)$ and $(1, \infty)$.
Testing a point in each interval shows that the solution is $(-3, 1]$.

**18.** $30e^{0.7x}=240$
$$e^{0.7x}=8$$
$$\ln e^{0.7x}=\ln 8$$
$$0.7x=\ln 8$$
$$x=\frac{\ln 8}{0.7}\approx 2.9706$$

The solution set is $\left\{\dfrac{\ln 8}{0.7}\right\}$ or $\{2.9706\}$.

**19.** $2x^3+3x^2-8x+3=0$
$p$: $\pm 1, \ \pm 3$
$q$: $\pm 1, \ \pm 2$
$\dfrac{p}{q}$: $\pm 1, \pm 3, \pm\dfrac{1}{2}, \pm\dfrac{3}{2}$

| 1 | 2 | 3 | −8 | 3 |
|---|---|---|----|---|
|   |   | 2 | 5 | −3 |
|   | 2 | 5 | −3 | 0 |

$$(x-1)(2x^2+5x-3)=0$$
$$(x-1)(2x-1)(x+3)=0$$
$$x=1 \text{ or } x=\frac{1}{2} \text{ or } x=-3$$

The solution set is $\left\{-3, \dfrac{1}{2}, 1\right\}$.

**20.** $4x^2+3y^2=48$
$3x^2+2y^2=35$
Multiply equation 1 by −2.
Multiply equation 2 by 3.
$$-8x^2-6y^2=-96$$
$$9x^2+6y^2=105$$
Add: $\qquad x^2=9$
$$x=\pm 3$$
Let $x=-3$:
$$4(-3)^2+3y^2=48$$
$$36+3y^2=48$$
$$3y^2=12$$
$$y^2=4$$
$$y=\pm 2$$
Let $x=3$:
$$4(3)^2+3y^2=48$$
$$36+3y^2=48$$
$$3y^2=12$$
$$y^2=4$$
$$y=\pm 2$$
The solution set is
$\{(3, 2), (3, -2), (-3, 2), (-3, -2)\}$.

**21.**
$$x - 2y + z = 16$$
$$2x - y - z = 14$$
$$3x + 5y - 4z = -10$$

$$\begin{bmatrix} 1 & -2 & 1 & | & 16 \\ 0 & -1 & -1 & | & 14 \\ 3 & 5 & -4 & | & -10 \end{bmatrix} \begin{matrix} \\ -2R_1 + R_2 \\ -3R_1 + R_3 \end{matrix}$$

$$\begin{bmatrix} 1 & -2 & 1 & | & 16 \\ 0 & 3 & -3 & | & -18 \\ 0 & 11 & -7 & | & -58 \end{bmatrix} \frac{1}{3}R_2$$

$$\begin{bmatrix} 1 & -2 & 1 & | & 16 \\ 0 & 1 & -1 & | & -6 \\ 0 & 11 & -7 & | & -58 \end{bmatrix} \begin{matrix} \\ 2R_2 + R_1 \\ -11R_2 + R_3 \end{matrix}$$

$$\begin{bmatrix} 1 & 0 & -1 & | & 4 \\ 0 & 1 & -1 & | & -6 \\ 0 & 0 & 4 & | & 8 \end{bmatrix} \frac{1}{4}R_3$$

$$\begin{bmatrix} 1 & 0 & -1 & | & 4 \\ 0 & 1 & -1 & | & -6 \\ 0 & 0 & 1 & | & 2 \end{bmatrix} \begin{matrix} R_3 + R_1 \\ R_2 + R_1 \\ \end{matrix}$$

$$\begin{bmatrix} 1 & 0 & 0 & | & 6 \\ 0 & 1 & 0 & | & -4 \\ 0 & 0 & 1 & | & 2 \end{bmatrix} \begin{matrix} R_3 + R_1 \\ R_2 + R_1 \\ \end{matrix}$$

The solution set is $\{(6, -4, 2)\}$.

**22.**
$$\begin{cases} x - y = 1 \\ x^2 - x - y = 1 \end{cases}$$

Solving $x - y = 1$ for y gives $y = x - 1$.
Substitute:
$$x^2 - x - y = 1$$
$$x^2 - x - \overbrace{(x-1)}^{y} = 1$$
$$x^2 - x - x + 1 = 1$$
$$x^2 - 2x + 1 = 1$$
$$x^2 - 2x = 0$$
$$x(x - 2) = 0$$
$x = 0$ or $x = 2$
If $x = 0$, $0 - y = 1$ so $y = -1$.
If $x = 2$, $2 - y = 1$ so $y = 1$.
The solution set is $\{(0, -1), (2, 1)\}$.

**23.** $100x^2 + y^2 = 25$

$$4x^2 + \frac{y^2}{25} = 1$$

$$\frac{x^2}{\left(\frac{1}{4}\right)} + \frac{y^2}{25} = 1$$

Ellipse
Foci on the $y$-axis

$a^2 = 25$ and $b^2 = \frac{1}{4}$, so $\frac{1}{4} = 25 - c^2$.

$$c^2 = \frac{99}{4}$$

$$c = \frac{3\sqrt{11}}{2}$$

Foci: $\left(0, -\frac{3\sqrt{11}}{2}\right), \left(0, \frac{3\sqrt{11}}{2}\right)$

**24.**
$$4x^2 - 9y^2 - 16x + 54y - 29 = 0$$
$$4(x^2 - 4x) - 9(y^2 - 6y) = 29$$
$$4(x^2 - 4x + 4) - 9(y^2 - 6y + 9) = 16 - 81 + 29$$
$$4(x - 2)^2 - 9(y - 3)^2 = -36$$
$$\frac{(y - 3)^2}{4} - \frac{(x - 2)^2}{9} = 1$$

Hyperbola with center at $(2, 3)$
Transverse axis vertical
$a^2 = 4$ and $b^2 = 9$, so $9 = c^2 - 4$.
$$c^2 = 13$$
$$c = \sqrt{13}$$
Foci: $\left(2, 3 - \sqrt{13}\right), \left(2, 3 + \sqrt{13}\right)$

$4x^2 - 9y^2 - 16x + 54y - 29 = 0$

**25.** Symmetry:

$$f(-x) = \frac{x^2 - 1}{-x - 2}$$

No symmetry since $f(-x) \neq f(x)$ and $f(-x) \neq -f(-x)$.

$x$-intercepts:

$$x^2 - 1 = 0$$
$$x = \pm 1$$

$y$-intercept:

$$f(0) = \frac{1}{2}$$
$$y = \frac{1}{2}$$

Vertical asymptote:

$$x - 2 = 0$$
$$x = 2$$

Horizontal asymptote:

$n > m$, so no horizontal asymptote.

Slant asymptote: $n = m + 1$

$$f(x) = x + 2 + \frac{3}{x - 2}$$
$$y = x + 2$$

$$f(x) = \frac{x^2 - 1}{x - 2}$$

**26.**

$2x - y \geq 4$
$x \leq 2$

**27.** $f(x) = x^2 - 4x - 5$

$$x = \frac{-b}{2a} = \frac{4}{2} = 2$$
$$f(2) = 2^2 - 8 - 5 = -9$$

vertex: $(2, -9)$

$x$-intercepts:

$$x^2 - 4x - 5 = 0$$
$$(x - 5)(x + 1) = 0$$
$$x = 5, -1$$

$y$-intercept: $f(0) = -5$

$f(x) = x^2 - 4x - 5$

**28.**

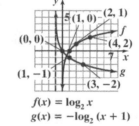

$$f(x) = \sqrt[3]{x + 4}$$
$$f^{-1}(x) = x^3 - 4$$

**29.**

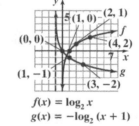

$$f(x) = \log_2 x$$
$$g(x) = -\log_2 (x + 1)$$

**30.** $f(x) = -x^2 - 2x + 1$, $g(x) = x - 1$

$$(f \circ g)(x) = f(g(x))$$
$$= -(x - 1)^2 - 2(x - 1) + 1$$
$$= -x^2 + 2x - 1 - 2x + 2 + 1$$
$$= -x^2 + 2$$
$$(g \circ f)(x) = g(f(x))$$
$$= \left(-x^2 - 2x + 1\right) - 1$$
$$= -x^2 - 2x$$

**31.** $f(x) = -x^2 - 2x + 1$

$$\frac{f(x+h) - f(x)}{h} = \frac{\left(-(x+h)^2 - 2(x+h) + 1\right) - \left(-x^2 - 2x + 1\right)}{h}$$

$$= \frac{-x^2 - 2xh - h^2 - 2x - 2h + 1 + x^2 + 2x - 1}{h}$$

$$= \frac{-2xh - h^2 - 2h}{h}$$

$$= \frac{h(-2x - h - 2)}{h}$$

$$= -2x - h - 2$$

**32.** $AB - 4A = \begin{bmatrix} 4 & 2 \\ 1 & -1 \\ 0 & 5 \end{bmatrix}\begin{bmatrix} 2 & 4 \\ 3 & 1 \end{bmatrix} - 4\begin{bmatrix} 4 & 2 \\ 1 & 1 \\ 0 & 5 \end{bmatrix}$

$$= \begin{bmatrix} 14 & 18 \\ -1 & 3 \\ 15 & 5 \end{bmatrix} - \begin{bmatrix} 16 & 8 \\ 4 & -4 \\ 0 & 20 \end{bmatrix} = \begin{bmatrix} -2 & 10 \\ -5 & 7 \\ 15 & -15 \end{bmatrix}$$

**33.** $\dfrac{2x^2 - 10x + 2}{(x-2)(x^2 + 2x + 2)} = \dfrac{A}{x-2} + \dfrac{Bx + C}{x^2 + 2x + 2}$

$2x^2 - 10x + 2 = A(x^2 + 2x + 2) + (Bx + C)(x - 2)$

$\qquad\qquad\quad = Ax^2 + 2Ax + 2A + Bx^2 - 2Bx + Cx - 2C$

$\qquad\qquad\quad = (A + B)x^2 + (2A - 2B + C)x + 2A - 2C$

Thus we have the following system of equations.

$\qquad A + B = 2$

$\quad 2A - 2B + C = -10$

$\qquad 2A - 2C = 2$

Add twice the first equation to the second equation.

$\qquad 2A + 2B = 4$

$\underline{\quad 2A - 2B + C = -10\quad}$

$\qquad\quad 4A + C = -6$

Add twice the resulting equation to the third equation.

$\quad 8A + 2C = -12$

$\underline{\quad 2A - 2C = 2\quad}$

$\qquad 10A = -10$

$\qquad\quad A = -1$

Back-substitute to find $B$ and $C$.

$2(-1) - 2C = 2$

$\quad -2 - 2C = 2$

$\qquad -2C = 4$

$\qquad\quad C = -2$

$-1 + B = 2$

$\qquad B = 3$

$\dfrac{-1}{x-2} + \dfrac{3x - 2}{x^2 + 2x + 2}$

**34.** $(x^3 + 2y)^5 = \binom{5}{0}(x^3)^5 + \binom{5}{1}(x^3)^4(2y) + \binom{5}{2}(x^3)^3(2y)^2 + \binom{5}{3}(x^3)^2(2y)^3 + \binom{5}{4}(x^3)(2y)^4 + \binom{5}{5}(2y)^5$

$= x^{15} + 5x^{12}(2y) + 10x^9(4y^2) + 10x^6(8y^3) + 5x^3(16y^4) + 32y^5$

$= x^{15} + 10x^{12}y + 40x^9y^2 + 80x^6y^3 + 80x^3y^4 + 32y^5$

**35.** $\displaystyle\sum_{i=1}^{50}(4i - 25)$

$a_1 = 4(1) - 25 = -21$

$a_{50} = 4(50) - 25 = 175$

$S_{50} = \dfrac{50}{2}(-21 + 175) = 3850$

**36.** Find slope:

$m = \dfrac{y_2 - y_1}{x_2 - x_1} = \dfrac{1 - 3}{-2 - 6} = \dfrac{-2}{-8} = \dfrac{1}{4}$

Find equation:

$y - y_1 = m(x - x_1)$

$y - 3 = \dfrac{1}{4}(x - 6)$

$y - 3 = \dfrac{1}{4}x - \dfrac{3}{2}$

$y = \dfrac{1}{4}x + \dfrac{3}{2}$

**37.** Find the slope:

$x - 5y - 20 = 0$

$-5y = -x + 20$

$\dfrac{-5y}{-5} = \dfrac{-x}{-5} + \dfrac{20}{-5}$

$y = \dfrac{1}{5}x - 4$

$m = -\dfrac{1}{\frac{1}{5}} = -5$

Find the equation:

$y - y_1 = m(x - x_1)$

$y - (-2) = -5(x - 0)$

$y + 2 = -5x$

$y = -5x - 2$

**38.** $200 + 0.05x = 0.15x$

$200 = 0.1x$

$\dfrac{200}{0.1} = \dfrac{0.1x}{0.1}$

$2000 = x$

At \$2000 in sales, the two earnings will be the same.

**39.** $2L + 2W = 300$
    $L = W + 50$
Rearrange the equations and add:
$L + W = 150$
$\underline{L - W = 50}$
    $2L = 200$
    $L = 100$
    $W = 50$
length: 100 yards, width 50 yards

**40.** $10x + 12y = 42$
    $5x + 10y = 29$
Multiply second equation by –2 and add:
    $10x + 12y = 42$
    $\underline{-10x - 20y = -58}$
        $-8y = -16$
            $y = 2$
Back substitute:
$5x + 10(2) = 29$
        $5x = 9$
        $x = 1.8$
pen: \$1.80,  pad: \$2

**41.** $s(t) = -16t^2 + 80t + 96$

a.    $-16t^2 + 80t + 96 = 0$
            $t^2 - 5t - 6 = 0$
        $(t + 1)(t - 6) = 0$
                $t = -1$ or $t = 6$
The ball will strike the ground after 6 seconds.

b.    $t = \dfrac{-b}{2a} = \dfrac{-80}{-32} = \dfrac{5}{2}$ or 2.5
$S(2.5) = -16(2.5)^2 + 80(2.5) + 96 = 196$
The ball reaches a maximum height of 196 feet, 2.5 seconds after it is thrown.

**42.** $I = \dfrac{k}{R}$
$5 = \dfrac{k}{22}$
$k = 110$
$I = \dfrac{110}{10} = 11$
11 amperes

**43.** Let $x$ represent the number of years after 2010.
The data from 2010 and 2011 are represented as $(0, 68.4)$ and $(1, 80.2)$
Find slope:  $m = \dfrac{80.2 - 68.4}{1 - 0} \approx 11.8$
Thus, $f(x) = 11.8x + 68.4$
Projections will vary.